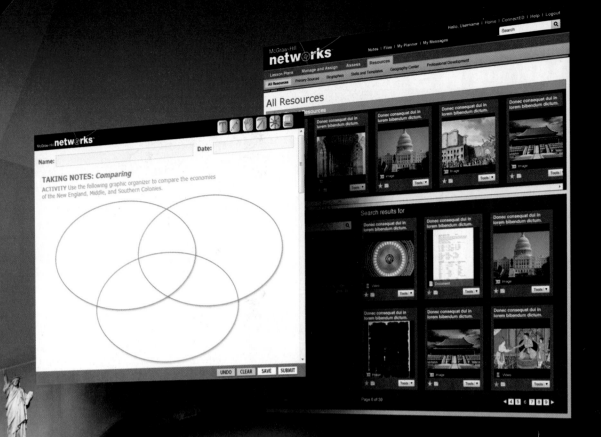

WHAT do you use?

Graphic Organizers • Primary Sources • Videos • Games • Photos

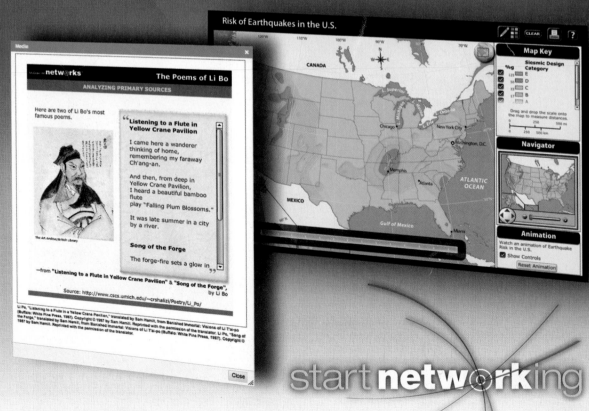

start **networ**ing

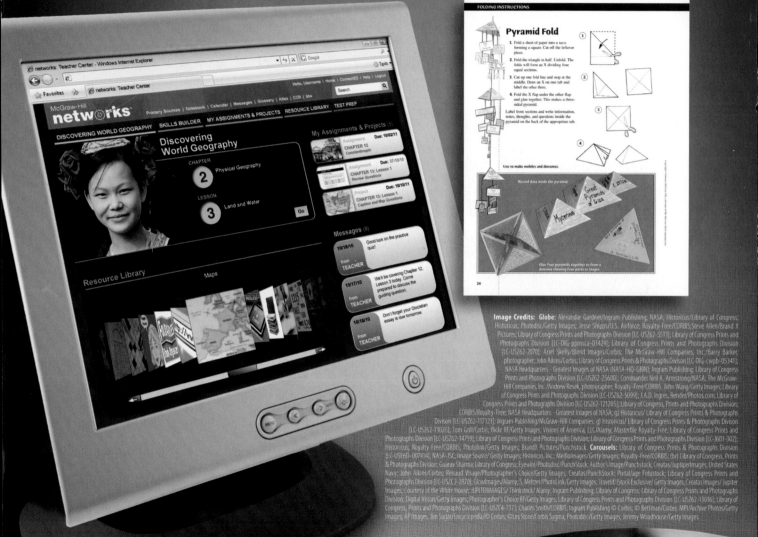

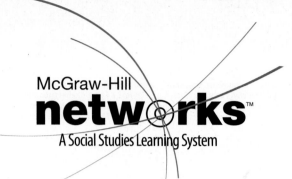

McGraw-Hill
netw♦rks™
A Social Studies Learning System

DISCOVERING
WORLD
GEOGRAPHY

Richard G. Boehm, Ph. D.

Mc
Graw
Hill
Education

Bothell, WA • Chicago, IL • Columbus, OH • New York, NY

www.mheonline.com/networks

Send all inquiries to:
McGraw-Hill Education
8787 Orion Place
Columbus, OH 43240

ISBN: 978-0-07-893619-7
MHID: 0-07-893619-5

Printed in the United States of America.

2 3 4 5 6 7 8 9 RJC 17 16 15 14 13

AUTHORS

SENIOR AUTHOR

Richard G. Boehm, Ph.D., was one of the original authors of *Geography for Life: National Geography Standards,* which outlined what students should know and be able to do in geography. He was also one of the authors of the *Guidelines for Geographic Education,* in which the Five Themes of Geography were first articulated. Dr. Boehm has received many honors, including "Distinguished Geography Educator" by the National Geographic Society (1990), the "George J. Miller Award" from the National Council for Geographic Education (NCGE) for distinguished service to geographic education (1991), "Gilbert Grosvenor Honors" in geographic education from the Association of American Geographers (2002), and the NCGE's "Distinguished Mentor Award" (2010). He served as president of the NCGE, has twice won the Journal of Geography award for best article, and also received the NCGE's "Distinguished Teaching Achievement." Presently, Dr. Boehm holds the Jesse H. Jones Distinguished Chair in Geographic Education at Texas State University in San Marcos, Texas, where he serves as director of The Gilbert M. Grosvenor Center for Geographic Education. His most current project includes the production of the video-based professional development series, *Geography: Teaching With the Stars.* Available programs may be viewed at www.geoteach.org.

CONTRIBUTING AUTHORS

Jay McTighe has published articles in a number of leading educational journals and has coauthored 10 books, including the best-selling *Understanding by Design* series with Grant Wiggins. McTighe also has an extensive background in professional development and is a featured speaker at national, state, and district conferences and workshops. He received his undergraduate degree from the College of William and Mary, earned a master's degree from the University of Maryland, and completed post-graduate studies at the Johns Hopkins University.

Dinah Zike, M.Ed., is an award-winning author, educator, and inventor recognized for designing three-dimensional, hands-on manipulatives and graphic organizers known as Foldables®. Foldables are used nationally and internationally by parents, teachers, and other professionals in the education field. Zike has developed more than 150 supplemental educational books and materials. Her two latest books, *Notebook Foldables®* and *Foldables®, Notebook Foldables®, & VKVs® for Spelling and Vocabulary 4th–12th* were each awarded *Learning Magazine's* Teachers' Choice Award for 2011. In 2004 Zike was honored with the CESI Science Advocacy Award. She received her M.Ed. from Texas A&M, College Station, Texas.

CONSULTANTS AND REVIEWERS

ACADEMIC CONSULTANTS

William H. Berentsen, Ph.D.
Professor of Geography and
European Studies
University of Connecticut
Storrs, Connecticut

David Berger, Ph.D.
Ruth and I. Lewis Gordon
Professor of Jewish History
Dean, Bernard Revel Graduate School
Yeshiva University
New York, New York

R. Denise Blanchard, Ph.D.
Professor of Geography
Texas State University–San Marcos
San Marcos, Texas

Brian W. Blouet, Ph.D.
Huby Professor of Geography and
International Education
The College of William and Mary
Williamsburg, Virginia

Olwyn M. Blouet, Ph.D.
Professor of History
Virginia State University
Petersburg, Virginia

Maria A. Caffrey, Ph.D.
Lecturer, Department of Geography
University of Tennessee
Knoxville, Tennessee

So-Min Cheong, Ph.D.
Associate Professor of Geography
University of Kansas
Lawrence, Kansas

Alasdair Drysdale, Ph.D.
Professor of Geography
University of New Hampshire
Durham, New Hampshire

Rosana Ferreira, Ph.D.
Assistant Professor of Geography
and Atmospheric Science
East Carolina University
Greenville, North Carolina

Eric J. Fournier, Ph.D.
Associate Professor of Geography
Samford University,
Birmingham, Alabama

Matthew Fry, Ph.D.
Assistant Professor of Geography
University of North Texas
Denton, Texas

Douglas W. Gamble, Ph.D.
Professor of Geography
University of North Carolina
Wilmington, North Carolina

Gregory Gaston, Ph.D.
Professor of Geography
University of North Alabama
Florence, Alabama

Jeffrey J. Gordon, Ph.D.
Associate Professor of Geography
Bowling Green State University
Bowling Green, Ohio

Alyson L. Greiner, Ph.D.
Associate Professor of Geography
Oklahoma State University
Stillwater, Oklahoma

William J. Gribb, Ph.D.
Associate Professor of Geography
University of Wyoming
Laramie, Wyoming

Joseph J. Hobbs, Ph.D.
Professor of Geography
University of Missouri
Columbia, Missouri

Ezekiel Kalipeni, Ph.D.
Professor of Geography and
Geography Information Science
University of Illinois
Urbana, Illinois

Pradyumna P. Karan, Ph.D.
Research Professor of Geography
University of Kentucky
Lexington, Kentucky

Christopher Laingen, Ph.D.
Assistant Professor of Geography
Eastern Illinois University
Charleston, Illinois

Jeffrey Lash, Ph.D.
Associate Professor of Geography
University of Houston–Clear Lake
Houston, Texas

Jerry T. Mitchell, Ph.D.
Research Professor of Geography
University of South Carolina
Columbia, South Carolina

Thomas R. Paradise, Ph.D.
Professor, Department of
Geosciences and the King Fahd
Center for Middle East Studies
University of Arkansas
Fayetteville, Arkansas

David Rutherford, Ph.D.
Assistant Professor of Public Policy
and Geography
Executive Director, Mississippi
Geographic Alliance
University of Mississippi
University, Mississippi

Dmitrii Sidorov, Ph.D.
Professor of Geography
California State University
Long Beach, California

Amanda G. Smith, Ph.D.
Professor of Education
University of North Alabama
Florence, Alabama

Jeffrey S. Ueland, Ph.D.
Associate Professor of Geography
Bemidji State University
Bemidji, Minnesota

Fahui Wang, Ph.D.
Professor of Geography
Louisiana State University
Baton Rouge, Louisiana

TEACHER REVIEWERS

Precious Steele Boyle, Ph.D.
Cypress Middle School
Memphis, TN

Jason E. Albrecht
Moscow Middle School
Moscow, ID

Jim Hauf
Berkeley Middle School
Berkeley, MO

Elaine M. Schuttinger
Trinity Catholic School
Columbus, OH

Mark Stahl
Longfellow Middle School
Norman, OK

Mollie Shanahan MacAdams
Southern Middle School
Lothian, MD

Sara Burkemper
Parkway West Middle Schools
Chesterfield, MO

Alicia Lewis
Mountain Brook Junior High School
Birmingham, AL

Steven E. Douglas
Northwest Jackson Middle School
Ridgeland, MS

LaShonda Grier
Richmond County Public Schools
Martinez, GA

Samuel Doughty
Spirit of Knowledge Charter
School
Worcester, MA

iv

CONTENTS

UNIT ONE

The World 1

The Geographer's World **15**

ESSENTIAL QUESTION

How does geography influence the way people live?

Physical Geography **39**

ESSENTIAL QUESTION

How does geography influence the way people live?

CCSS This icon indicates where reading skills and writing skills from the *Common Core State Standards for English Language Arts & Literacy in History/Social Studies, Science, and Technical Subjects* are practiced and reinforced.

CONTENTS

CONTENTS

(t) Christophe Boisvieux/age fotostock; (c) Dmitry Kostyukov/AFP/Getty Images; (b) Jon Arnold/Alamy

CONTENTS

CONTENTS

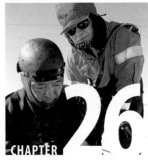

Think Again?

Thinking Like a Geographer

EXPLORE the CONTINENT

GLOBAL CONNECTIONS

What Do You **Think?**

MAPS

CHARTS, GRAPHS, DIAGRAMS, AND INFOGRAPHICS

networks ONLINE RESOURCES

Videos

Every lesson has a video to help you learn more about your world!

Infographics

Chapter 2
Lesson 3 Fresh and Salt Water in the World

Chapter 9
Lesson 2 How a Hacienda Works

Chapter 15
Lesson 2 The Spice Islands

Chapter 16
Lesson 2 Population of India

Interactive Charts/Graphs

Chapter 2
Lesson 1 Rain Shadow; Climate Zones

Chapter 3
Lesson 1 Understanding Population Pyramids
Lesson 3 Economic Questions

Chapter 14
Lesson 3 Chinese Lion Dance

Chapter 16
Lesson 1 Water Wells

Chapter 17
Lesson 3 Living in a Yurt

Chapter 18
Lesson 1 Oases
Lesson 2 Ziggurats

Chapter 20
Global Connections: Refugees in the Sudan

Chapter 24
Lesson 2 Australian Gold Rush

Animations

Chapter 1
Lesson 1 The Earth; Regions of Earth
Lesson 2 Elements of a Globe

Chapter 2
Lesson 1 Earth's Daily Rotation; Earth's Layers; Seasons on Earth
Lesson 3 How the Water Cycle Works

Chapter 3
Global Connections: Social Media

Chapter 5
Lesson 1 Chinooks

Chapter 6
Lesson 1 The St. Lawrence Seaway

Chapter 7
Lesson 1 Waterways as Political Boundaries
Global Connections: NAFTA

Chapter 8
Global Connections: Rain Forest Resources

Chapter 12
Lesson 1 How the Alps Formed
Global Connections: Aging of Europe's Population

Chapter 13
Lesson 2 Russian Historical Changes

Chapter 14
Lesson 3 Population Pyramid of East Asia
Global Connections: Tsunami!

Chapter 16
Lesson 1 Rivers in Southeast Asia

Chapter 18
Lesson 1 Why Much of the World's Oil Supply is in Southwest Asia

Chapter 19
Lesson 2 How the Pyramids Were Built

Chapter 20
Lesson 2 East African Independence
Global Connections: Refugee Camps

Chapter 24
Global Connections: Invasive Species

Jochen Schlenker/Photographer's Choice/Getty Images

Interactive Images

networks ONLINE RESOURCES

⌄ Games

REFERENCE ATLAS

ATLAS KEY

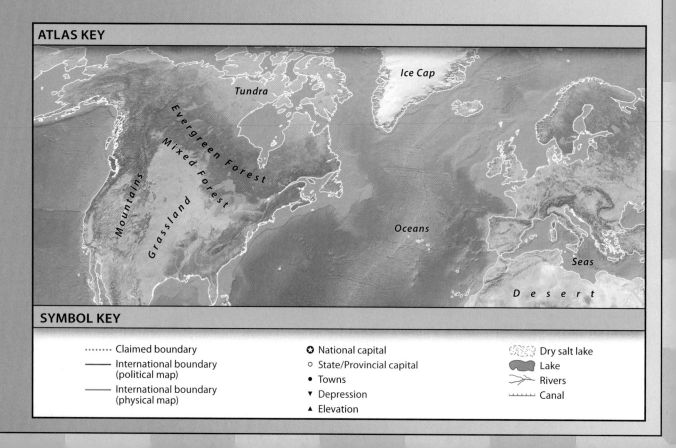

SYMBOL KEY

- ········ Claimed boundary
- —— International boundary (political map)
- —— International boundary (physical map)
- ✪ National capital
- ○ State/Provincial capital
- ● Towns
- ▼ Depression
- ▲ Elevation
- Dry salt lake
- Lake
- Rivers
- Canal

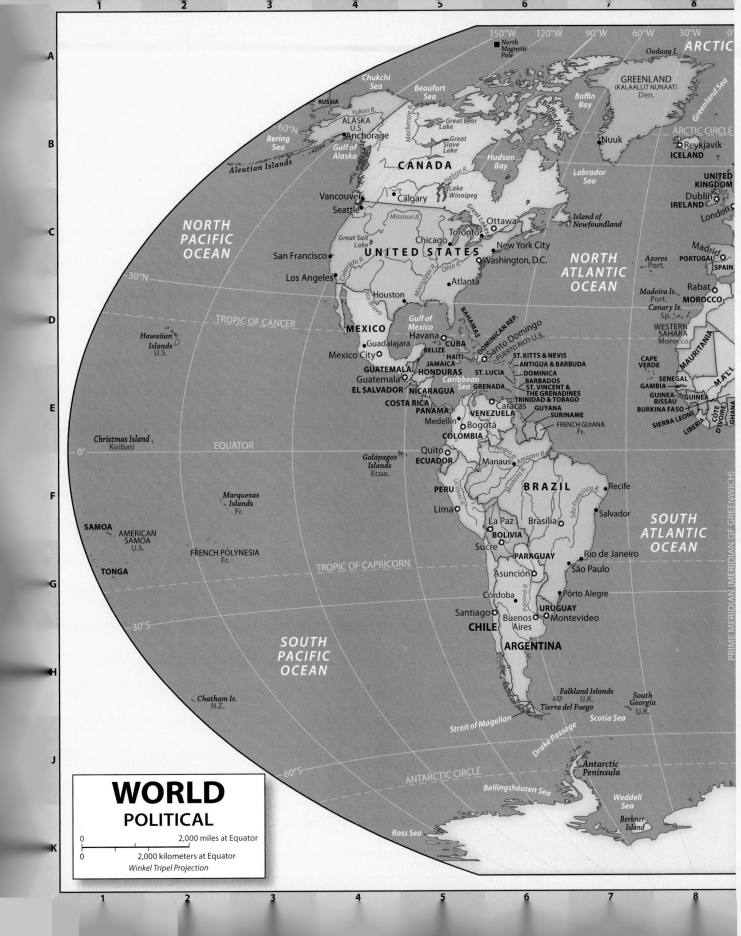

WORLD
POLITICAL

0   2,000 miles at Equator

0   2,000 kilometers at Equator

Winkel Tripel Projection

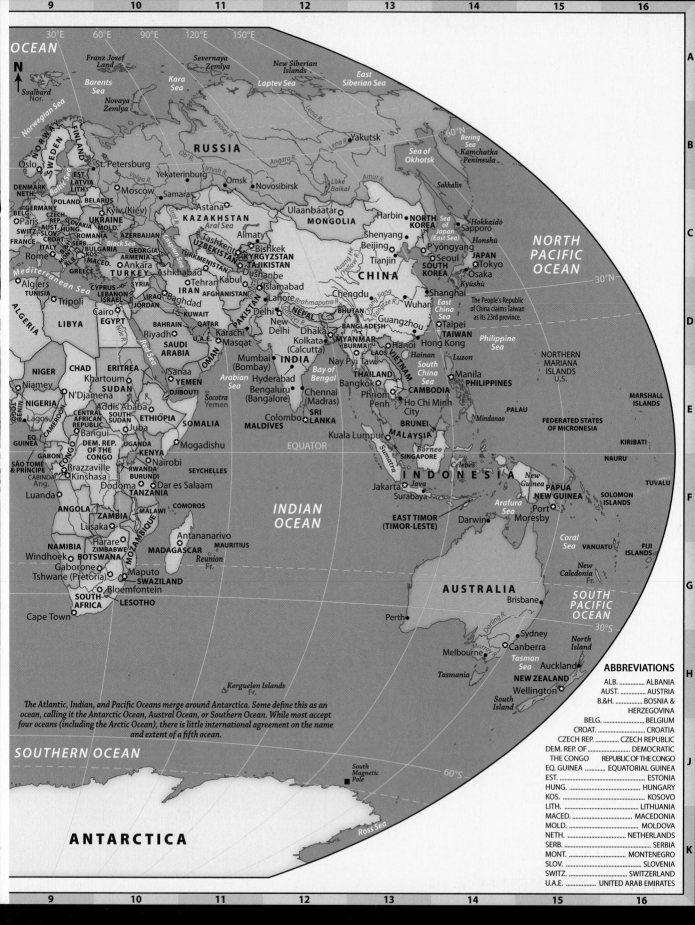

OCEAN

N

30°E   60°E   90°E   120°E   150°E

Svalbard
Nor.
Franz Josef Land
Novaya Zemlya
Severnaya Zemlya
New Siberian Islands
East Siberian Sea

Barents Sea
Kara Sea
Laptev Sea

Norwegian Sea

NORWAY   SWEDEN   FINLAND

RUSSIA

60°N

Bering Sea
Kamchatka Peninsula

Sea of Okhotsk

Oslo   St. Petersburg   Yekaterinburg   Omsk   Novosibirsk   Yakutsk

DENMARK   NETH.   EST. LATVIA LITH.   Moscow   Samara
GERMANY   POLAND   BELARUS
BELG.   CZECH REP.   SLOVAKIA   Kyiv (Kiev)   UKRAINE   MOLD.
Paris   AUST. HUNG.   ROMANIA   AZERBAIJAN
SWITZ. SLOV.   CROAT.   SERB.   BULGARIA   GEORGIA
FRANCE   ITALY   MONT. ALB. MACED.   GREECE   ARMENIA   Ankara   TURKEY

Sakhalin

Harbin   **NORTH KOREA**   *Hokkaidō* Sapporo

Sea of Japan (East Sea)

Honshū

Astana   **KAZAKHSTAN**   *Aral Sea*   Almaty
Tashkent   Bishkek   **KYRGYZSTAN**
UZBEKISTAN   TAJIKISTAN
TURKMENISTAN   Dushanbe

Ulaanbaatar   **MONGOLIA**

Shenyang   P'yŏngyang   **JAPAN**   Tokyo
Beijing   Seoul   Osaka
Tianjin   **SOUTH KOREA**   *Kyūshū*

30°N

Rome   *Mediterranean Sea*   CYPRUS LEBANON ISRAEL   SYRIA   IRAQ   IRAN
Algiers   TUNISIA   Tripoli   JORDAN   Baghdad   Tehran
ALGERIA   LIBYA   EGYPT   Cairo   KUWAIT   Kabul   Islamabad   Lahore

Ashkhabad   AFGHANISTAN   **CHINA**
Huang He (Yellow R.)   Chengdu   *Chang Jiang (Yangtze R.)*   Wuhan   Shanghai
Brahmaputra R.   NEPAL   BHUTAN   Guangzhou
Delhi   New Delhi   Dhaka   BANGLADESH

East China Sea
Taipei   **TAIWAN**

The People's Republic of China claims Taiwan as its 23rd province.

Philippine Sea

NORTHERN MARIANA ISLANDS U.S.

BAHRAIN   QATAR   U.A.E.   Masqat   OMAN
Riyadh   **SAUDI ARABIA**
Karachi   **PAKISTAN**
Mumbai (Bombay)   Hyderabad   **INDIA**

Kolkata (Calcutta)   MYANMAR (BURMA)   Hanoi   Hong Kong
Nay Pyi Taw   LAOS   VIETNAM   *Hainan*   *South China Sea*   *Luzon*

NIGER   CHAD   ERITREA   Khartoum   SUDAN   Sanaa   YEMEN   DJIBOUTI   *Arabian Sea*
Niamey   N'Djamena   *Red Sea*   *Nile R.*

Bay of Bengal   THAILAND   Bangkok   CAMBODIA   Manila   **PHILIPPINES**
Bengaluru (Bangalore)   Chennai (Madras)   Phnom Penh
Hyderabad   SRI LANKA   Ho Chi Minh City   BRUNEI

Socotra Yemen
Colombo   SRI LANKA   MALDIVES

Mindanao   PALAU

FEDERATED STATES OF MICRONESIA

MARSHALL ISLANDS

NIGERIA   Lagos   CENTRAL AFRICAN REPUBLIC   SOUTH SUDAN   ETHIOPIA   Addis Ababa
BENIN   CAMEROON   Bangui   Juba
EQ. GUINEA   GABON   CONGO   UGANDA   KENYA
SÃO TOMÉ & PRÍNCIPE   Brazzaville   Kinshasa   DEM. REP. OF THE CONGO   Nairobi   SEYCHELLES

Kuala Lumpur   MALAYSIA
SINGAPORE   *Borneo*
EQUATOR

KIRIBATI
NAURU

CABINDA Ang.   RWANDA   BURUNDI   Dodoma   Mogadishu   SOMALIA
Luanda   ANGOLA   Dar es Salaam   TANZANIA

Sumatra   **INDONESIA**   *New Guinea*   PAPUA NEW GUINEA
Jakarta   *Java*   Surabaya   *Celebes*   Port Moresby   SOLOMON ISLANDS

TUVALU

INDIAN OCEAN

Arafura Sea

ZAMBIA   MALAWI   COMOROS
Lusaka   NAMIBIA   ZIMBABWE   Harare   MOZAMBIQUE   Antananarivo   MADAGASCAR   MAURITIUS   *Reunion Fr.*
Windhoek   BOTSWANA

EAST TIMOR (TIMOR-LESTE)   Darwin

Coral Sea   VANUATU   FIJI ISLANDS
New Caledonia Fr.

Gaborone   Tshwane (Pretoria)   Maputo   SWAZILAND
SOUTH AFRICA   Bloemfontein   LESOTHO
Cape Town

AUSTRALIA   Brisbane

Perth   *Darling R.*   Sydney
Murray R.   Canberra   *North Island*

30°S

SOUTH PACIFIC OCEAN

Kerguelen Islands Fr.

Melbourne   *Tasman Sea*   Auckland
Tasmania   **NEW ZEALAND**   Wellington   *South Island*

The Atlantic, Indian, and Pacific Oceans merge around Antarctica. Some define this as an ocean, calling it the Antarctic Ocean, Austral Ocean, or Southern Ocean. While most accept four oceans (including the Arctic Ocean), there is little international agreement on the name and extent of a fifth ocean.

SOUTHERN OCEAN

South Magnetic Pole   60°S

ANTARCTICA   *Ross Sea*

ABBREVIATIONS	
ALB.	ALBANIA
AUST.	AUSTRIA
B.&H.	BOSNIA & HERZEGOVINA
BELG.	BELGIUM
CROAT.	CROATIA
CZECH REP.	CZECH REPUBLIC
DEM. REP. OF THE CONGO	DEMOCRATIC REPUBLIC OF THE CONGO
EQ. GUINEA	EQUATORIAL GUINEA
EST.	ESTONIA
HUNG.	HUNGARY
KOS.	KOSOVO
LITH.	LITHUANIA
MACED.	MACEDONIA
MOLD.	MOLDOVA
NETH.	NETHERLANDS
SERB.	SERBIA
MONT.	MONTENEGRO
SLOV.	SLOVENIA
SWITZ.	SWITZERLAND
U.A.E.	UNITED ARAB EMIRATES

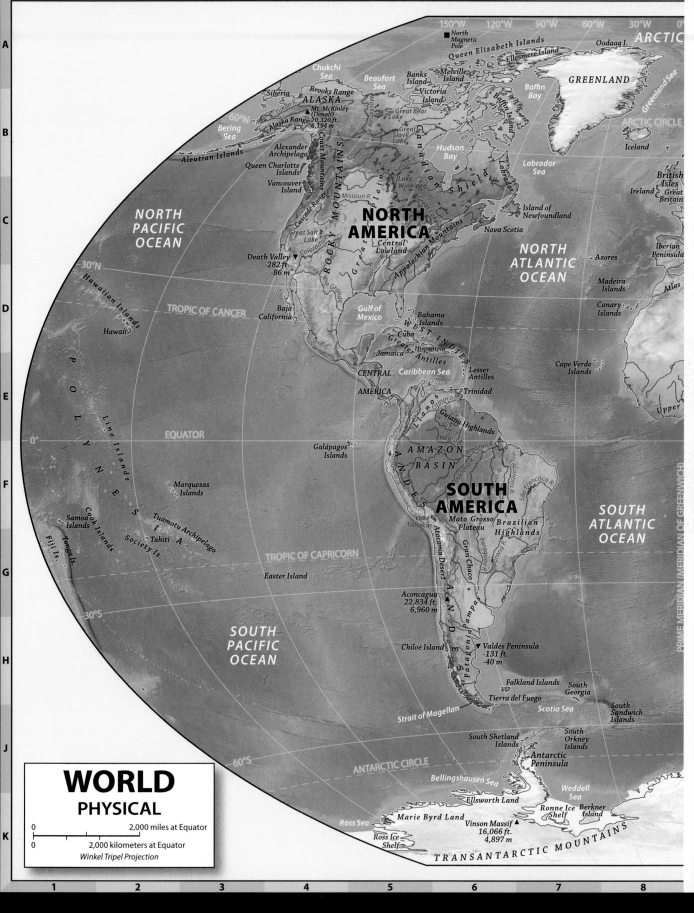

WORLD
PHYSICAL

0 — 2,000 miles at Equator
0 — 2,000 kilometers at Equator
Winkel Tripel Projection

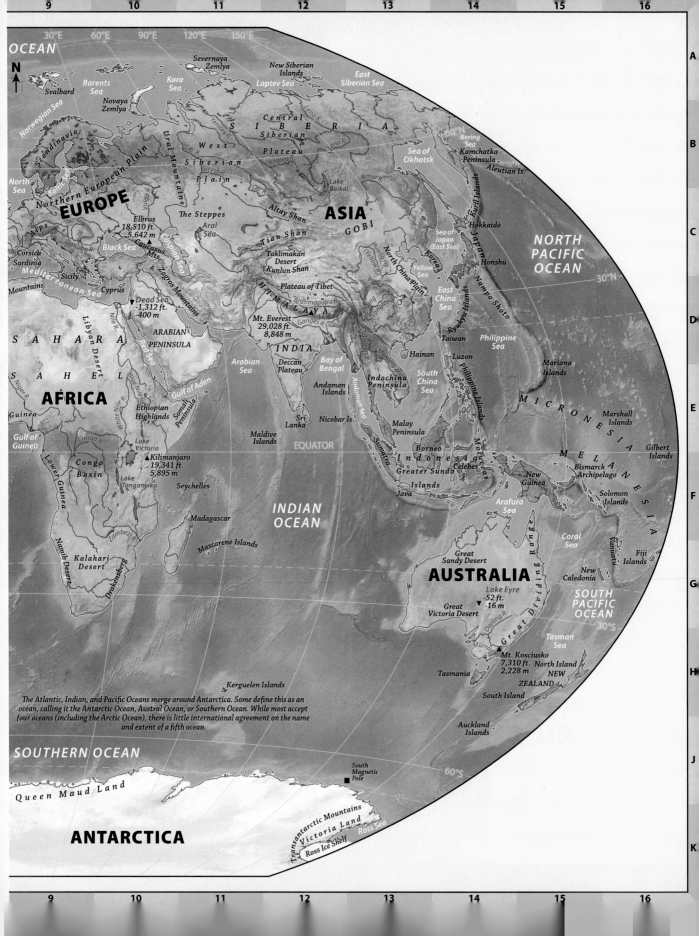

9   **10**   **11**   **12**   **13**   **14**   **15**   **16**

OCEAN

N

30°E   60°E   90°E   120°E   150°E

Svalbard

Norwegian Sea

Barents Sea

Novaya Zemlya

Kara Sea

Severnaya Zemlya

New Siberian Islands

Laptev Sea

East Siberian Sea

Scandinavia

North Sea

Baltic Sea

Northern European Plain

Ural Mountains

West Siberian Plain

Central Siberian Plateau

SIBERIA

Lena R.

Angara R.

Sea of Okhotsk

Bering Sea

Kamchatka Peninsula

Aleutian Is.

EUROPE

Volga R.

Yenisey R.

Ob R.

Irtysh R.

Ob R.

Lake Baikal

Amur R.

Kuril Islands

Alps

Danube R.

Black Sea

Caucasus Mts.

Elbrus 18,510 ft. 5,642 m

The Steppes

Aral Sea

Altay Shan

ASIA

GOBI

Sea of Japan (East Sea)

Hokkaidō

NORTH PACIFIC OCEAN

60°N

Corsica Sardinia

Mediterranean Sea

Sicily

Caspian Sea

Tian Shan

Taklimakan Desert

Kunlun Shan

North China Plain

Huang He (Yellow R.)

Korea

Honshū

Japan

30°N

Mountains

Cyprus

Zagros Mountains

Dead Sea -1,312 ft. -400 m

ARABIAN PENINSULA

Plateau of Tibet

HIMALAYA

Brahmaputra R.

Yellow Sea

East China Sea

Ryukyu Islands

Nampo Shoto

SAHARA

Libyan Desert

Red Sea

Nile R.

Mt. Everest 29,028 ft. 8,848 m

INDIA

Ganges R.

Mekong R.

Yangtze R.

Hainan

Taiwan

Luzon

Philippine Sea

SAHEL

Blue Nile R.

Gulf of Aden

Arabian Sea

Deccan Plateau

Bay of Bengal

Indochina Peninsula

South China Sea

Philippine Islands

Mariana Islands

AFRICA

White Nile R.

Ethiopian Highlands

Somali Peninsula

Sri Lanka

Andaman Islands

Andaman Sea

Nicobar Is.

Malay Peninsula

MICRONESIA

Guinea

N. Niger R.

Maldive Islands

EQUATOR

Sumatra

Marshall Islands

Gulf of Guinea

Congo R.

Congo Basin

Lake Victoria

Kilimanjaro 19,341 ft. 5,895 m

Indonesia

Greater Sunda Islands

Moluccas

Celebes

Borneo

Java

New Guinea

Bismarck Archipelago

MELANESIA

Gilbert Islands

Lower Guinea

Lake Tanganyika

Seychelles

Solomon Islands

Zambezi R.

Madagascar

INDIAN OCEAN

Arafura Sea

New Guinea

Vanuatu

Fiji Islands

Namib Desert

Mascarene Islands

Great Sandy Desert

Coral Sea

New Caledonia

Kalahari Desert

AUSTRALIA

SOUTH PACIFIC OCEAN

Drakensberg

Great Victoria Desert

Lake Eyre -52 ft. -16 m

Great Dividing Range

30°S

Murray R.

Darling R.

Mt. Kosciusko 7,310 ft. 2,228 m

North Island

Tasman Sea

Tasmania

South Island

NEW ZEALAND

The Atlantic, Indian, and Pacific Oceans merge around Antarctica. Some define this as an ocean, calling it the Antarctic Ocean, Austral Ocean, or Southern Ocean. While most accept four oceans (including the Arctic Ocean), there is little international agreement on the name and extent of a fifth ocean.

Kerguelen Islands

Auckland Islands

SOUTHERN OCEAN

South Magnetic Pole

60°S

Queen Maud Land

Transantarctic Mountains

Victoria Land

Ross Ice Shelf

Ross Sea

ANTARCTICA

9   **10**   **11**   **12**   **13**   **14**   **15**   **16**

A   B   C   D   E   F   G   H   J   K

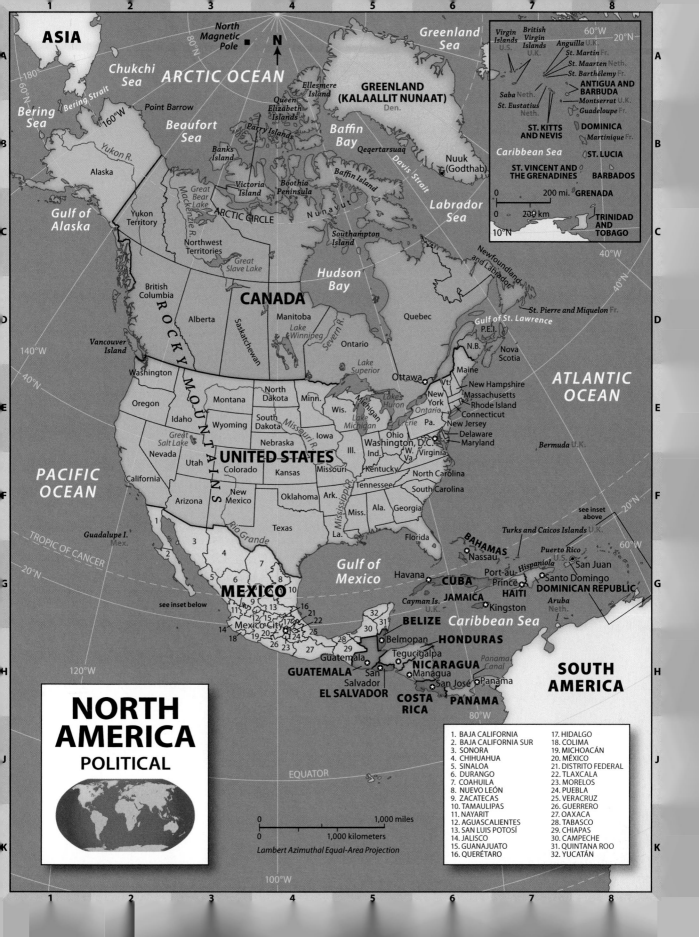

NORTH AMERICA
POLITICAL

1. BAJA CALIFORNIA
2. BAJA CALIFORNIA SUR
3. SONORA
4. CHIHUAHUA
5. SINALOA
6. DURANGO
7. COAHUILA
8. NUEVO LEÓN
9. ZACATECAS
10. TAMAULIPAS
11. NAYARIT
12. AGUASCALIENTES
13. SAN LUIS POTOSÍ
14. JALISCO
15. GUANAJUATO
16. QUERÉTARO
17. HIDALGO
18. COLIMA
19. MICHOACÁN
20. MÉXICO
21. DISTRITO FEDERAL
22. TLAXCALA
23. MORELOS
24. PUEBLA
25. VERACRUZ
26. GUERRERO
27. OAXACA
28. TABASCO
29. CHIAPAS
30. CAMPECHE
31. QUINTANA ROO
32. YUCATÁN

Lambert Azimuthal Equal-Area Projection

1,000 miles

1,000 kilometers

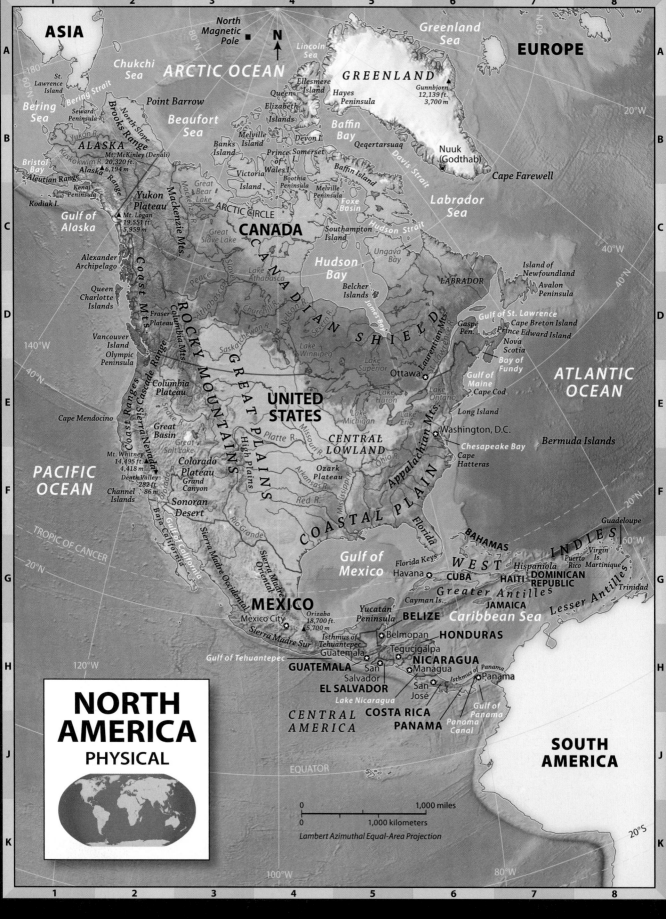

NORTH AMERICA
PHYSICAL

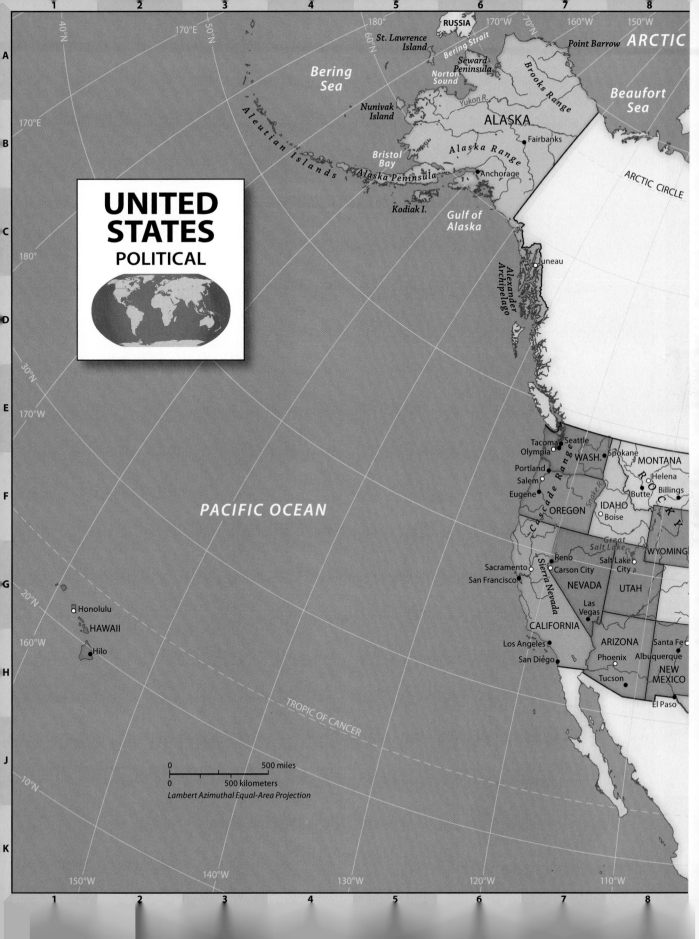

UNITED STATES
POLITICAL

Lambert Azimuthal Equal-Area Projection

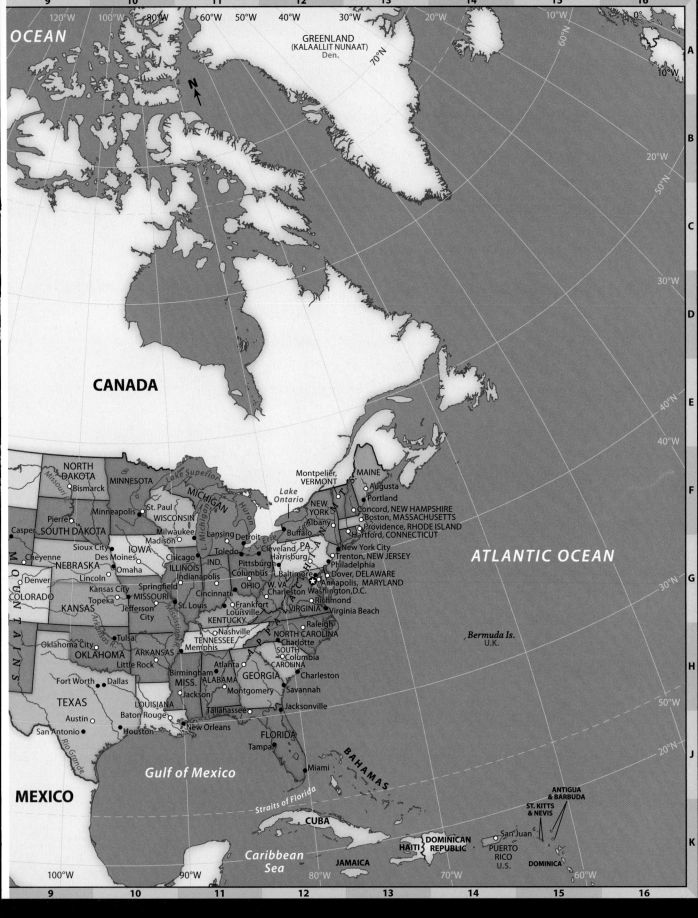

OCEAN

GREENLAND
(KALAALLIT NUNAAT)
Den.

70°N

10°W

20°W

50°N

30°W

CANADA

40°N

40°W

NORTH
DAKOTA
Bismarck

MINNESOTA
Lake Superior

MICHIGAN

Lake
Ontario

Huron

Montpelier,
VERMONT

MAINE

Augusta

Portland

ATLANTIC OCEAN

St. Paul
WISCONSIN

Minneapolis

Pierre
SOUTH DAKOTA

Milwaukee

Madison

Lansing

Detroit

Buffalo

NEW
YORK

Albany

Concord, NEW HAMPSHIRE

Boston, MASSACHUSETTS

Providence, RHODE ISLAND

Hartford, CONNECTICUT

Casper

Sioux City

IOWA

Chicago

Toledo

Cleveland

PA. A

New York City

Trenton, NEW JERSEY

Cheyenne

NEBRASKA

Des Moines

ILLINOIS

IND.

Pittsburgh

Harrisburg

Philadelphia

Denver

Lincoln

Omaha

Indianapolis

Columbus

OHIO

Baltimore

Dover, DELAWARE

Annapolis, MARYLAND

30°N

COLORADO

Kansas City

Topeka

Springfield

MISSOURI

Cincinnati

Frankfort

W. VA.

Charleston Washington,D.C.

KANSAS

Jefferson
City

St. Louis

Louisville

VIRGINIA

Richmond

Virginia Beach

KENTUCKY

Nashville

Raleigh

Bermuda Is.
U.K.

Oklahoma City

Tulsa

TENNESSEE

NORTH CAROLINA

Charlotte

OKLAHOMA

Little Rock

ARKANSAS

Memphis

SOUTH
CAROLINA

Columbia

Fort Worth

Dallas

MISS.

ALABAMA

Birmingham

GEORGIA

Charleston

Jackson

Atlanta

Montgomery

Savannah

TEXAS

LOUISIANA

Tallahassee

Jacksonville

50°N

Austin

Baton Rouge

New Orleans

20°N

San Antonio

Houston

FLORIDA

Tampa

Gulf of Mexico

Miami

BAHAMAS

MEXICO

Straits of Florida

ANTIGUA
& BARBUDA

ST. KITTS
& NEVIS

CUBA

Caribbean
Sea

JAMAICA

HAITI

DOMINICAN
REPUBLIC

San Juan

PUERTO
RICO
U.S.

DOMINICA

60°W

100°W

90°W

80°W

70°W

120°W

100°W

80°W

60°W

50°W

40°W

30°W

20°W

10°W

0°

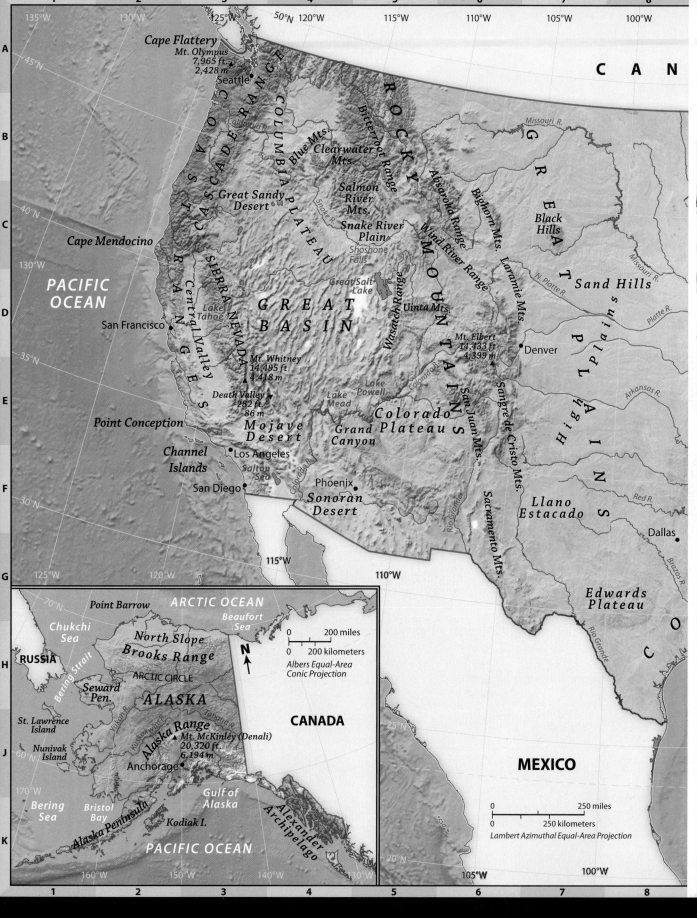

Cape Flattery
Mt. Olympus
7,965 ft.
2,428 m
Seattle

C A N

COLUMBIA RANGE
CASCADE RANGE
Blue Mts.
Clearwater Mts.
Bitterroot Range

R O C K Y

G R E A T

Great Sandy Desert
Salmon River Mts.
Snake River Plain
Shoshone Falls

Absaroka Range
Wind River Range
Bighorn Mts.
Laramie Mts.

Black Hills

COLUMBIA PLATEAU

Cape Mendocino

PACIFIC OCEAN

Great Salt Lake

Sand Hills

SIERRA NEVADA

Lake Tahoe

GREAT BASIN

Wasatch Range
Uinta Mts.

M O U N T A I N S

San Francisco

CENTRAL VALLEY

Mt. Whitney
14,495 ft.
4,418 m

Mt. Elbert
14,433 ft.
4,399 m

Denver

H i g h   P l a i n s

Death Valley
-282 ft.
-86 m

Lake Powell
Lake Mead

Colorado Plateau

Grand Canyon

Colorado R.

San Juan Mts.

Sangre de Cristo Mts.

Point Conception

Mojave Desert

Channel Islands

Los Angeles

Salton Sea

Phoenix

Sonoran Desert

Colorado R.

Sacramento Mts.

Llano Estacado

Red R.

Dallas

San Diego

Rio Grande

Brazos R.

Edwards Plateau

O

PACIFIC OCEAN

Point Barrow

ARCTIC OCEAN

Beaufort Sea

Chukchi Sea

North Slope

Brooks Range

RUSSIA

Bering Strait

Seward Pen.

ARCTIC CIRCLE

ALASKA

St. Lawrence Island

Nunivak Island

Yukon R.

Kuskokwim R.

Alaska Range

Mt. McKinley (Denali)
20,320 ft.
6,194 m

Anchorage

Tanana R.

CANADA

N

0   200 miles
0   200 kilometers
Albers Equal-Area Conic Projection

MEXICO

Bering Sea

Bristol Bay

Alaska Peninsula

Kodiak I.

Gulf of Alaska

Alexander Archipelago

PACIFIC OCEAN

C

0   250 miles
0   250 kilometers
Lambert Azimuthal Equal-Area Projection

RA10   Reference Atlas

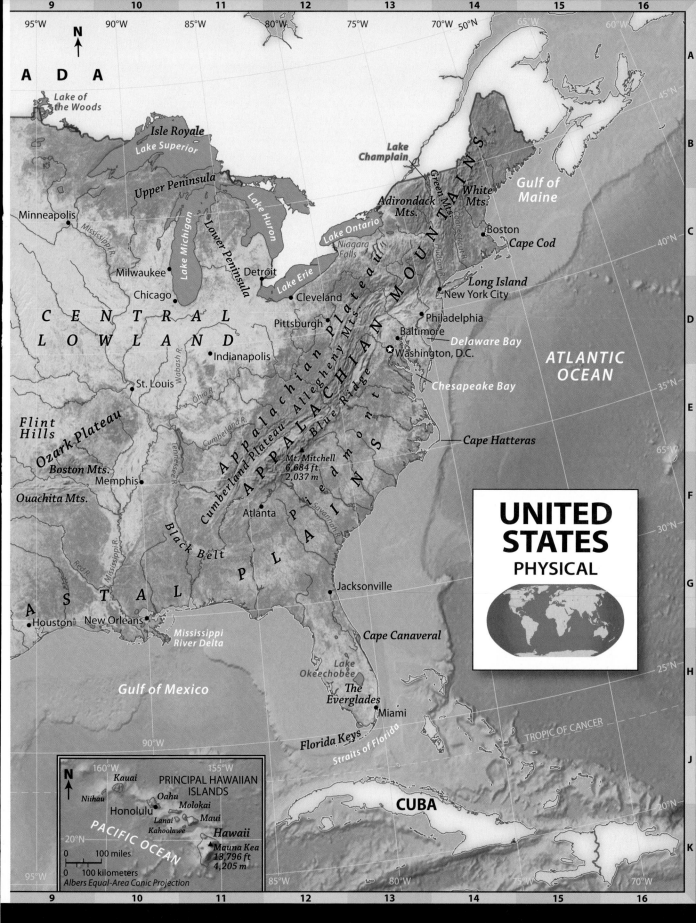

9 10 11 12 13 14 15 16

95°W   90°W   85°W   80°W   75°W   70°W   50°N   65°W   60°W

N

A D A

A

Lake of
the Woods

Isle Royale

Lake Superior

Upper Peninsula

Minneapolis

Mississippi R.

Lake Michigan

Lake Huron

Lower Peninsula

Milwaukee

Chicago

Detroit

Lake Erie

Cleveland

C E N T R A L

L O W L A N D

Wabash R.

Indianapolis

Ohio R.

St. Louis

Pittsburgh

Cumberland R.

Tennessee R.

Lake
Champlain

Green Mts.

*Adirondack
Mts.*

APPALACHIAN MOUNTAINS

*White
Mts.*

*Gulf of
Maine*

Boston

Cape Cod

Connecticut R.

Lake Ontario

*Niagara
Falls*

Hudson R.

Long Island

New York City

Appalachian Plateau

Allegheny Mts.

Philadelphia

Baltimore

Delaware Bay

Cumberland Plateau

Washington, D.C.

Chesapeake Bay

Blue Ridge

ATLANTIC
OCEAN

40°N

35°N

65°W

Flint
Hills

Ozark Plateau

Boston Mts.

Memphis

Ouachita Mts.

Black Belt

Atlanta

Mt. Mitchell
6,684 ft
2,037 m

Piedmont

Savannah R.

Cape Hatteras

APPALACHIAN

C O A S T A L

P L A I N

30°N

UNITED
STATES

PHYSICAL

Red R.

Mississippi R.

A S T A L

Houston

New Orleans

Jacksonville

Cape Canaveral

Mississippi
River Delta

Lake
Okeechobee

Gulf of Mexico

*The
Everglades*

Miami

25°N

Florida Keys

Straits of Florida

TROPIC OF CANCER

20°N

CUBA

N

160°W   155°W

Kauai

PRINCIPAL HAWAIIAN
ISLANDS

Niihau

Oahu

Molokai

Honolulu

Lanai

Maui

Kahoolawe

Hawaii

PACIFIC OCEAN

20°N

▲ Mauna Kea
13,796 ft
4,205 m

0   100 miles

0   100 kilometers
Albers Equal-Area Conic Projection

95°W   90°W   85°W   80°W   75°W   70°W

9 10 11 12 13 14 15 16

RUSSIA

ARCTIC OCEAN

N
Queen

Prince
Patrick I.

Elizabeth

ALASKA
U.S.

Islands

Melville
Island

Bathurst
Island

Beaufort
Sea

Banks
Island

Resolute

Somerset
Island
Prince of
Wales I.

Inuvik

Victoria
Island

Boothia
Pen.

Dawson

Fort
Good Hope

Cambridge Bay

Talòyoak

YUKON
TERRITORY

Kugluktuk

Mt. Logan
19,551 ft.
5,959 m

Yukon
Plateau

Great
Bear Lake

N   U   N

Whitehorse

NORTHWEST
TERRITORIES

Virginia
Falls

Great Slave
Lake

Yellowknife

Arviat

BRITISH
COLUMBIA

Fort Smith

Lake
Athabasca

Churchill

Prince Rupert

Fraser
Plateau

Fort St. John

Queen
Charlotte
Islands

Peace
River

Fort McMurray

Thompson

Prince
George

Edmonton

SASKATCHEWAN

MANITOBA

PACIFIC
OCEAN

ALBERTA

Lake
Winnipeg

Vancouver
Island

Calgary

Saskatoon

Vancouver

Lake
Winnipegosis

Regina

Winnipeg

Victoria

Brandon

Lake of
the Woods

0   500 miles

0   500 kilometers
Lambert Azimuthal Equal-Area Projection

UNITED STATES

CANADA
PHYSICAL / POLITICAL

ICELAND

**GREENLAND
(KALAALLIT NUNAAT)**
Den.

*Ellesmere
Island*

Devon Island

*Baffin
Bay*

Arctic Bay

Baffin Island

*Davis
Strait*

Igloolik

*Melville
Peninsula*

*Foxe
Basin*

N U N A V U T

Repulse Bay

Iqaluit

*Southampton
Island*

Hudson Strait

Chesterfield
Inlet

*Ungava
Bay*

*Labrador
Sea*

Kuujjuaq

Nain

*Hudson
Bay*

Cartwright

*Belcher
Islands*

Schefferville

Happy Valley-
Goose Bay

NEWFOUNDLAND AND LABRADOR

*Island of
Newfoundland*

Fort Severn

Kuujjuarapik

*Smallwood
Reservoir*

Churchill Falls

St. John's

Labrador City

*Avalon
Peninsula*

Q U E B E C

*Manicouagan
Reservoir*

Anticosti I.

St.-Pierre & Miquelon
Fr.

James Bay

Sept-Îles

*Gulf of
St. Lawrence*

S H I E L D

*Gaspé
Pen.*

**PRINCE
EDWARD
ISLAND**

Sydney

Cape Breton I.

O N T A R I O

*Lake
Nipigon*

Chicoutimi

**NEW
BRUNSWICK**

Charlottetown

**NOVA
SCOTIA**

Timmins

Rouyn-
Noranda

Quebec

Fredericton

**ATLANTIC
OCEAN**

Thunder
Bay

Saint John

Halifax

Lake Superior

Sudbury

North
Bay

Montreal

St. Lawrence R.

Bay of Fundy

Ottawa

*Lake
Huron*

Toronto

L. Ontario

Lake Michigan

London

Niagara Falls

Lake Erie

90°W

80°W

70°W

60°W

50°W

40°W

30°W

20°W

10°W

80°N

70°N

60°N

50°N

40°N

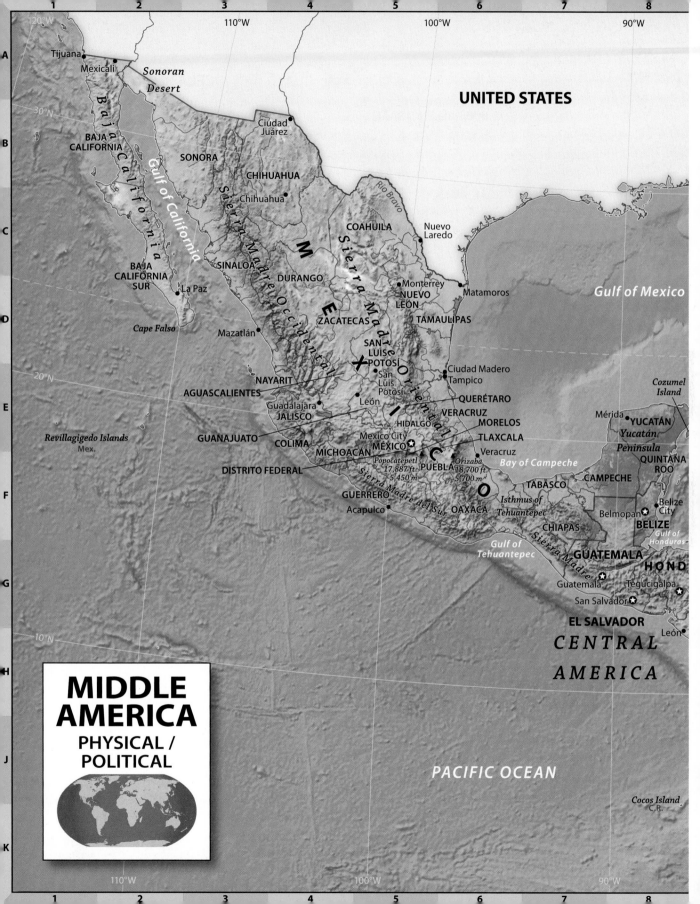

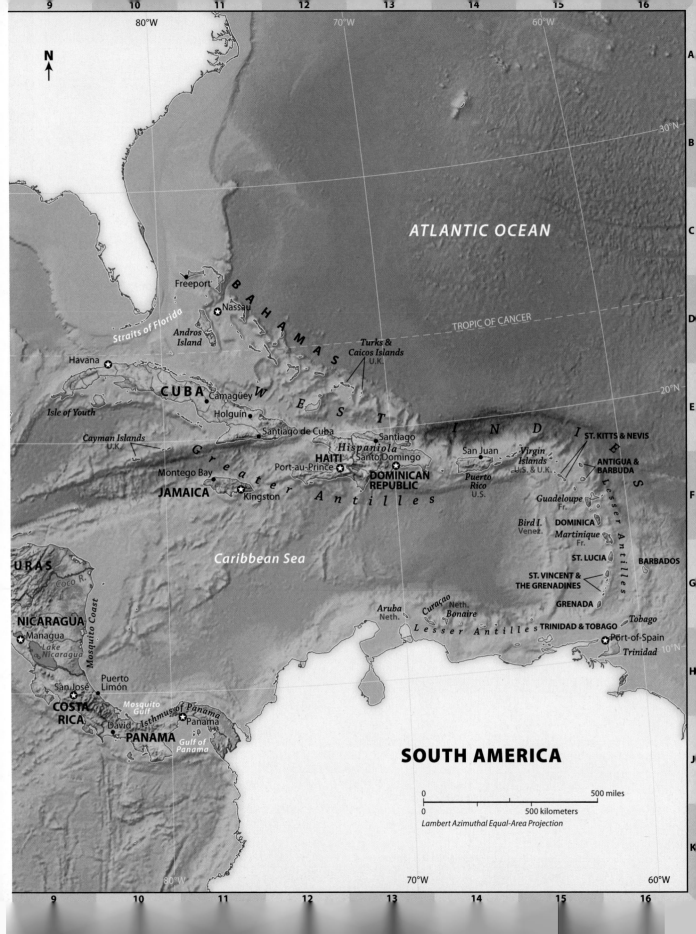

Caribbean Sea

N

0   1,000 miles
0   1,000 kilometers
Lambert Azimuthal Equal-Area Projection

Santa
Marta
Barranquilla   Maracaibo   Caracas
Cartagena   Valencia
     Lake   Ciudad Guayana   **GUYANA**
VENEZUELA   *Maracaibo*   **SURINAME**
Bucaramanga   San Cristóbal   Georgetown   Paramaribo
     Medellín   *Orinoco R.*   Cayenne
Malpelo I.   Bogotá   **FRENCH GUIANA**
/ Col.   Cali   Boa   *Fr.*
COLOMBIA   Vista
             Rio Negro   Boundary claimed   *Marajó*
Esmeraldas   by Suriname   *Island*
Quito   *A M A Z O N*   *Amazon R.*   EQUATOR
0°   **ECUADOR**   Belém   São Luís   0°
Guayaquil   Manaus   Santarém
     Iquitos   *B A S I N*   Fortaleza
     Marañón   Teresina   Natal
Amazon R.   Campina Grande
PERU   *Purus R.*   Recife
Río Branco   Pôrto Velho
Callao   Salvador
Lima   Machu   **BRAZIL**
Ayacucho   Picchu   Cuzco
     Trinidad   Brasília
     Lake   La Paz
Arequipa   *Titicaca*   **BOLIVIA**   Goiânia
     Oruro   Santa Cruz   Uberlândia
Arica   Sucre   Campo   Uberaba   Belo Horizonte
Iquique   Tarija   Grande
20°S   Campinas   Nova Iguaçu   20°S
Antofagasta   **PARAGUAY**   Londrina   Rio de Janeiro
TROPIC OF CAPRICORN   Salta   Asunción   São Paulo   Santos
CHILE   Curitiba
San Félix I. San Ambrosio I.   **ATLANTIC**
Chile   San Miguel   Uruguaiana   **OCEAN**
     de Tucumán   Santa   Pôrto Alegre
La Serena   Córdoba   Maria
Coquimbo   Rosario
Juan Fernández Is.   Valparaíso   Mendoza   **URUGUAY**
Chile   Santiago   Buenos Aires   Montevideo
     La Plata   *Río de la Plata*
Concepción   **ARGENTINA**   Mar del Plata
     Colorado R.   Bahía Blanca
Puerto Montt   *Negro R.*
40°S   40°S

PACIFIC
OCEAN   Comodoro Rivadavia

     Falkland Islands
     (Islas Malvinas)
Río Gallegos   Stanley   Administered by
            United Kingdom
Punta Arenas   Claimed by Arg.
Ushuaia
Strait of   *Cape Horn*   *South Georgia Island*
Magellan   *U.K.*

SOUTH
AMERICA
POLITICAL

Caribbean Sea

N

0 — 1,000 miles
0 — 1,000 kilometers
Lambert Azimuthal Equal-Area Projection

Caracas ✪
VENEZUELA
GUYANA
SURINAME
Lake Maracaibo
Orinoco R.
Georgetown ✪
Paramaribo ✪
Bogotá ✪
Angel Falls
Total drop
3,212 ft., 979 m
Cayenne ●
FRENCH GUIANA
L L A N O S
COLOMBIA
GUIANA HIGHLANDS
Quito ✪
Rio Negro
Marajó Island
ECUADOR
Boundary claimed by Suriname
EQUATOR
0°
0°
A M A Z O N
Amazon R.

Marañón R.
A N D E S
Amazon R.
B A S I N
Selvas
PERU
Ucayali R.
Purus R.
Madeira R.
Tapajós R.
Xingu R.
Araguaia R.
Tocantins R.
São Francisco R.
BRAZIL
Lima ✪
Machu Picchu ■
MATO GROSSO PLATEAU
B R A Z I L I A N
La Paz ✪
Lakes Titicaca
BOLIVIA
Brasília ✪
H I G H L A N D S
Altiplano
Sucre ✪
Salar de Uyuni
G R A N   C H A C O
Paraguay R.
Paraná R.
20°S
20°S
PARAGUAY
TROPIC OF CAPRICORN
San Ambrosio I.
Asunción ✪
Iguazú Falls
San Félix I.
Paraná R.
Uruguay R.
ATLANTIC OCEAN
A N D E S
P A M P A S
CHILE
Aconcagua ▲
22,834 ft.
6,960 m
G R A N   C H A C O
Juan Fernández Is.
Santiago ✪
Buenos Aires ✪
URUGUAY
Montevideo ✪
ARGENTINA
Río de la Plata

SOUTH AMERICA
PHYSICAL

Colorado R.
Negro R.
40°S
40°S
Chiloé Island
P A T A G O N I A
Valdés Peninsula
-131 ft.
-40 m
Gulf of San Jorge
PACIFIC OCEAN
Taitao Peninsula
Falkland Islands (Islas Malvinas)
Wellington I.
Stanley ●
Tierra del Fuego
South Georgia Island
Strait of Magellan
Cape Horn

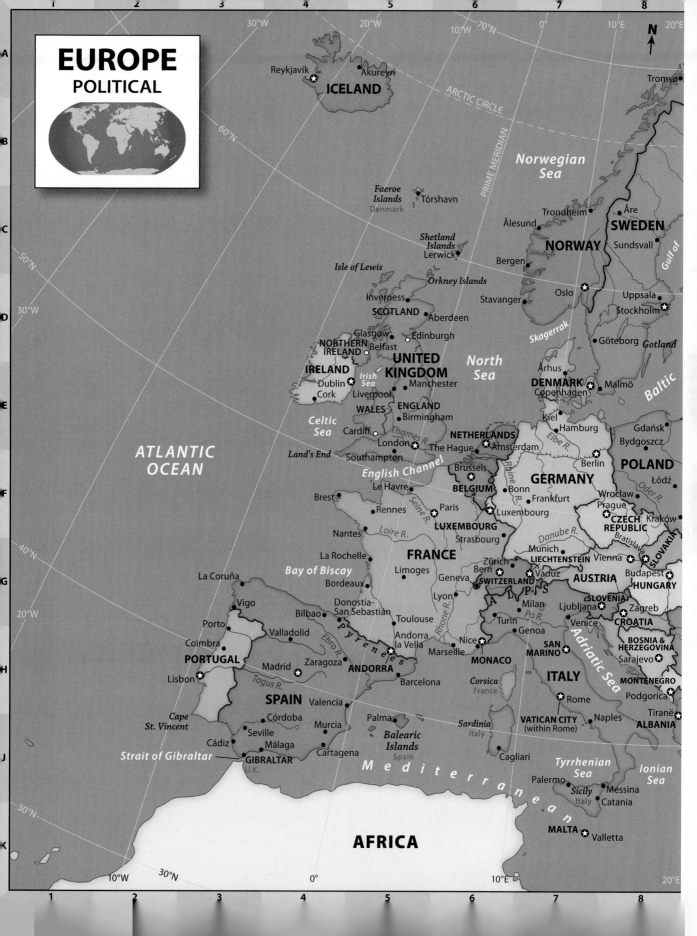

EUROPE
POLITICAL

N

ATLANTIC OCEAN

ARCTIC CIRCLE

PRIME MERIDIAN

Reykjavík • Akureyri
ICELAND

Tromsø

Norwegian Sea

Faeroe Islands
Denmark • Tórshavn

Trondheim • Åre
Ålesund •
SWEDEN
NORWAY • Sundsvall
Bergen •

Shetland Islands
Lerwick •

Isle of Lewis
Orkney Islands
Inverness • Stavanger • Oslo ✪ Uppsala •
SCOTLAND • Aberdeen Stockholm ✪

Glasgow • Edinburgh • Göteborg *Gotland*
NORTHERN IRELAND Belfast • *North Sea* Århus •
DENMARK ✪ Malmö *Baltic*
IRELAND Copenhagen
Dublin ✪ • Manchester Kiel •
Cork • Liverpool **ENGLAND** Hamburg • Gdańsk •
WALES Birmingham • Elbe R. Bydgoszcz •
Celtic Sea Cardiff • Berlin • **POLAND**
London ✪ The Hague • Amsterdam ✪ Łódź •
Land's End Southampton **NETHERLANDS** **GERMANY**
English Channel Brussels • Bonn • Wrocław •
Le Havre **BELGIUM** Frankfurt • Prague ✪ Kraków •
Brest • **LUXEMBOURG** Oder R. **CZECH REPUBLIC**
Rennes • Paris • Luxembourg ✪ Bratislava • **SLOVAKIA**
Strasbourg • *Danube R.* Vienna •
Nantes • Munich • **LIECHTENSTEIN** Budapest •
La Rochelle • **FRANCE** Zürich • Vaduz ✪ Vienna **HUNGARY**
Limoges • Bern ✪ **AUSTRIA**
La Coruña • *Bay of Biscay* Geneva • **SWITZERLAND** **SLOVENIA**
Vigo • Bordeaux • Lyon • Milan • Ljubljana ✪ Zagreb •
Porto • Bilbao • Donostia- Turin • Venice • **CROATIA**
Coimbra • Valladolid San Sebastián Genoa • **SAN** **BOSNIA & HERZEGOVINA**
PORTUGAL *Pyrenees* Toulouse • Nice ✪ **MARINO** Sarajevo •
Madrid • Zaragoza • **ANDORRA** Marseille • **MONACO** **MONTENEGRO**
Lisbon ✪ Ebro R. Andorra Podgorica •
SPAIN la Vella *Corsica* Tiranë •
Tagus R. Barcelona • *France* **ITALY** **ALBANIA**
Cape St. Vincent Valencia • Córdoba • Rome ✪ Naples •
Seville • Murcia • *Palma* **VATICAN CITY** *Adriatic Sea*
Cádiz • Málaga • *Balearic Islands* (within Rome)
Cartagena • *Spain* *Sardinia* Cagliari • Palermo •
Strait of Gibraltar *Italy* *Sicily* Messina •
GIBRALTAR *Tyrrhenian Sea* *Italy* Catania •
U.K. *Ionian Sea*
MALTA ✪ Valletta

M e d i t e r r a n e a n

AFRICA

A commonly accepted division between Asia and Europe—here marked by a gray line—is formed by the Ural Mountains, Ural River, Caspian Sea, Caucasus Mountains, and the Black Sea with its outlets, the Bosporus and the Dardanelles.

Europe/Asia boundary

ASIA

Barents Sea

Tobseda

Pechora

U R A L   M O U N T A I N S

Murmansk
Kirovsk
Ivalo
Kiruna
Kola Peninsula
Umba
White Sea

Kirsk
Kem'
Arkhangel'sk
Severodvinsk

Kemi
Luleå
Oulu
Umeå

Northern Dvina R.

Syktyvkar

FINLAND

Vaasa
Lake Onega
Perm'

Pori
Tampere
Lake Ladoga
Kirov
Ufa

Turku
Helsinki
St. Petersburg
RUSSIA
Kazan'
Orenburg

Sea
Tallinn
Novgorod
Yaroslavl'
Ural R.
ESTONIA
Nizhniy Novgorod
Samara

LATVIA
Tver'
Moscow
Oral

Riga
Daugavpils
Smolensk
Ryazan'
Penza
Saratov

LITHUANIA
Vitsyebsk
Vilnius
Bryansk
Don R.
Volga R.
KAZAKHSTAN
Kaunas
Minsk
Kursk
Kaliningrad
BELARUS
Homyel'
Chernihiv

Warsaw
Sumy
Kharkiv
Volgograd

Vistula R.
Kyiv (Kiev)
Poltava
Astrakhan

L'viv
Dnieper R.
Donets'k
Vinnytsya
UKRAINE
Dniester R.
Dnipropetrovs'k
Rostov

Carpathian Mts.
MOLDOVA
Caspian Sea

Chişinău
Odessa
Sea of Azov
Stavropol'

ROMANIA
Kerch
Grozny

Belgrade
Bucharest
Crimea
Simferopol'
Caucasus   Mountains
AZERBAIJAN
SERBIA
Sevastopol'
Yalta
GEORGIA
Baku

KOSOVO
Constanţa
Balkan Mts.
BULGARIA
Varna
Black Sea

Priština
Sofia

Skopje
İstanbul
MACEDONIA
T U R K E Y
Bosporus

Thessaloniki
Sea of Marmara

GREECE
Dardanelles

Aegean Sea
ASIA

Athens
Peloponnese

Rhodes
Nicosia

Iraklion
CYPRUS
S e a
Crete
Greece

0   400 miles
0   400 kilometers
Lambert Azimuthal Equal-Area Projection

EUROPE
PHYSICAL

N

Grid columns: 1, 2, 3, 4, 5, 6, 7, 8
Grid rows: A, B, C, D, E, F, G, H, J, K

30°W, 20°W, 10°W, 70°N, 0°, 10°E, 20°E

ARCTIC CIRCLE

PRIME MERIDIAN

60°N

50°N

40°N

30°N

Reykjavík
ICELAND

Faeroe Islands

Shetland Islands

Norwegian Sea

NORWAY

Oslo

SWEDEN

Stockholm

Gulf of

Gotland

Outer Hebrides

Orkney Islands

British Isles

Highlands

Edinburgh

Belfast

UNITED KINGDOM

IRELAND

Dublin

Irish Sea

North Sea

Skagerrak

Kattegat

Jutland

Zealand

DENMARK
Copenhagen

Baltic

Great Britain

Celtic Sea

Cardiff

London

Thames R.

Land's End

English Channel

NETHERLANDS
Amsterdam

Elbe R.

Berlin

N O R T H

POLAND

Brussels
BELGIUM

Rhine R.

GERMANY

Oder R.

Prague
CZECH REPUBLIC

ATLANTIC OCEAN

Brittany

Paris
Seine R.

Luxembourg
LUXEMBOURG

Loire R.

FRANCE

Danube R.

Bratislava
SLOVAKIA

LIECHTENSTEIN
Vaduz

Vienna

Budapest
HUNGARY

Bay of Biscay

Bern
SWITZERLAND

AUSTRIA

Drava R.

SLOVENIA
Ljubljana

Cantabrian Mountains

Mont Blanc
15,771 ft.
4,807 m

A L P S

Po R.

Zagreb
CROATIA

Douro R.

Ebro R.

Pyrenees

Andorra la Vella
ANDORRA

Rhône R.

Riviera

MONACO

SAN MARINO

Adriatic Sea

BOSNIA & HERZEGOVINA
Sarajevo

I B E R I A N

Madrid

MONTENEGRO
Podgorica

Lisbon
PORTUGAL

Tagus R.

SPAIN

P E N I N S U L A

Corsica

ITALY

Rome

Tiranë
ALBANIA

Cape St. Vincent

Baetic Mountains

Balearic Islands

VATICAN CITY
(within Rome)

Sardinia

A p e n n i n e s

Strait of Gibraltar

GIBRALTAR

M e d i t e r r a n e a n

Tyrrhenian Sea

Ionian Sea

Sicily

Etna
10,902 ft.
3,323 m

MALTA
Valletta

AFRICA

10°W, 30°N, 0°, 10°E, 20°E

North Cape

Barents Sea

LAPLAND
LAPLAND

Kola Peninsula

White Sea

Pechora R.

URAL MOUNTAINS

Europe/Asia boundary

ASIA

Bothnia

FINLAND

Lake Region

Northern Dvina R.

Lake Onega

RUSSIA

Lake Ladoga

Sea

Helsinki

Gulf of Finland

Tallinn

ESTONIA

LATVIA

Riga

Moscow

Ural R.

LITHUANIA

Vilnius

Minsk

CENTRAL

KAZAKHSTAN

RUSSIA

Volga R.

Caspian Depression

RUSSIA

BELARUS

UPLAND

Warsaw

Don R.

Vistula R.

(Kyiv) Kiev

Dnieper R.

UKRAINE

Dniester R.

Carpathian Mts.

MOLDOVA

Tisza R.

Chișinău

Sea of Azov

Crimea

Mt. Elbrus
18,510 ft.
5,642 m

Caucasus ▲ Mountains

Caspian Sea

Caspian

ROMANIA

Belgrade

Bucharest

Danube R.

AZERBAIJAN

GEORGIA

Baku

SERBIA

BALKAN

Black Sea

KOSOVO

Balkan Mts.

Priština Sofia

BULGARIA

PENINSULA

Bosporus

Skopje

MACEDONIA

TURKEY

GREECE

Dardanelles

Sea of Marmara

Aegean Sea

Athens

Peloponnese

ASIA

400 miles

0

0   400 kilometers

Lambert Azimuthal Equal-Area Projection

Rhodes

Nicosia

CYPRUS

Crete

Sea

30°E

40°E

50°E

60°E

70°E

80°E

60°N

50°N

70°E

60°E

40°N

30°N

50°E

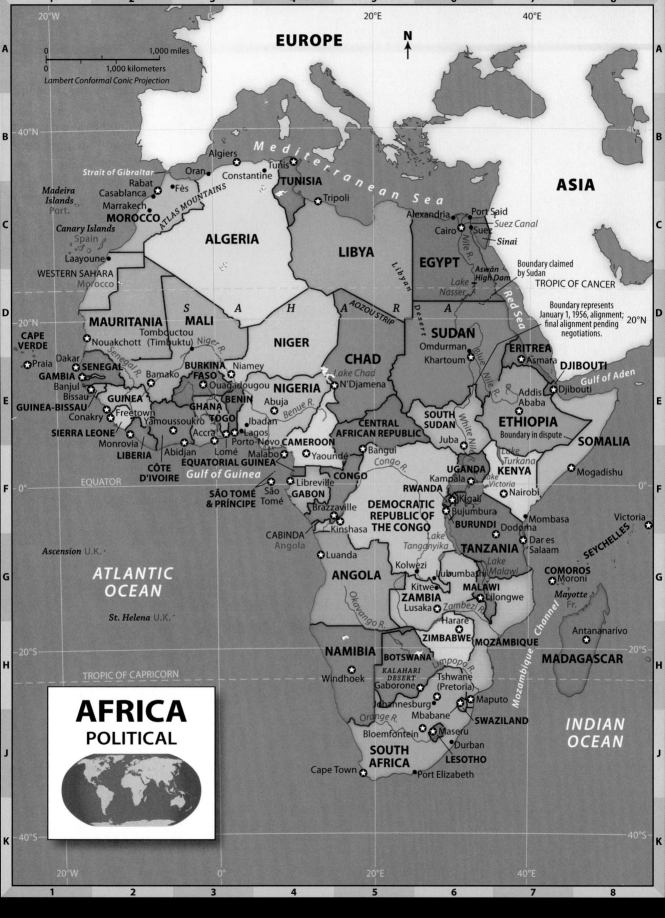

AFRICA
POLITICAL

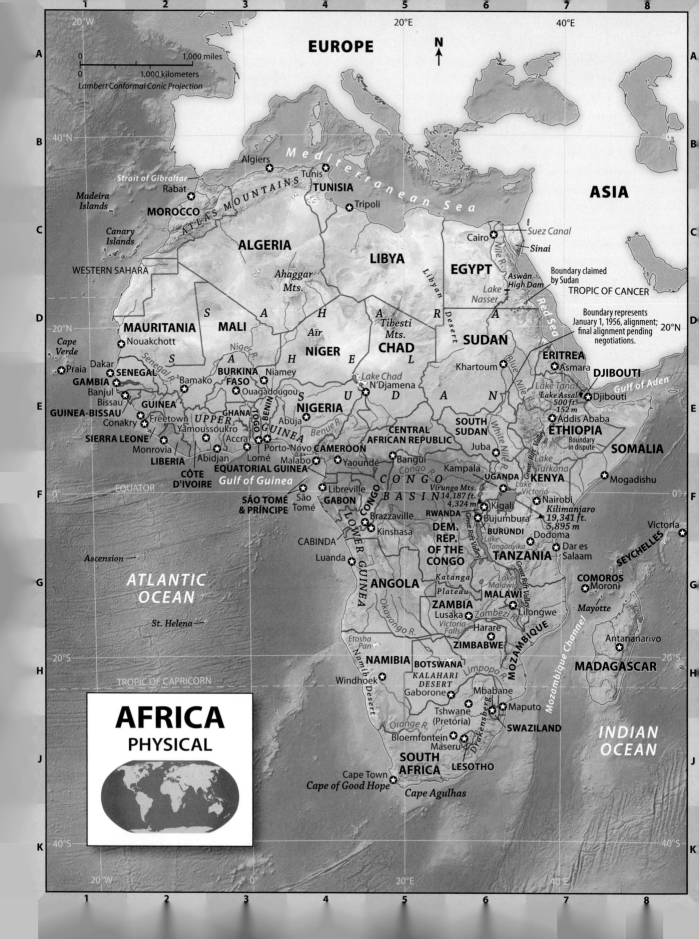

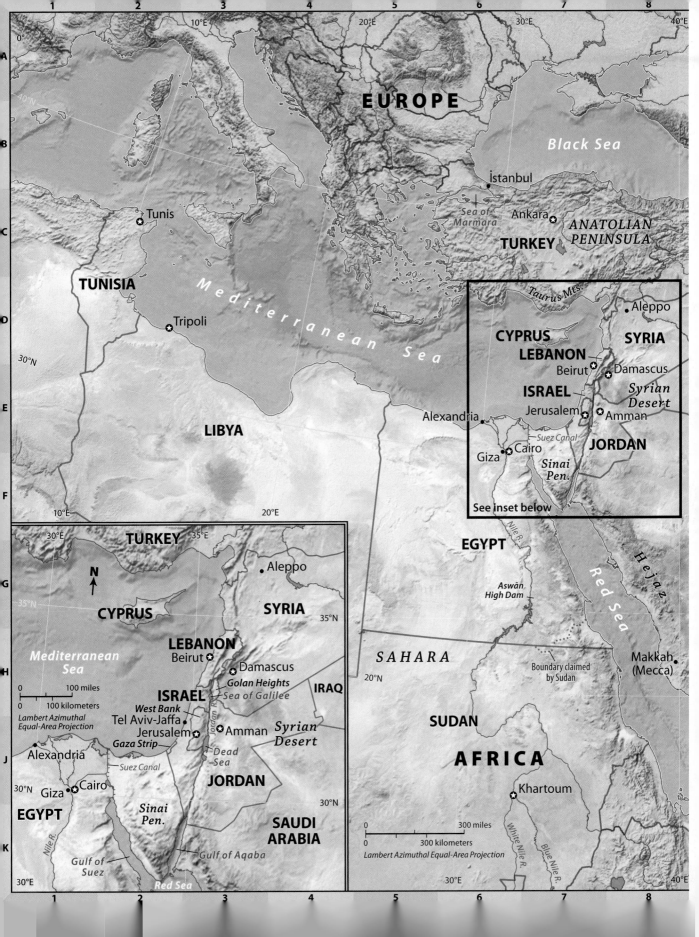

Main map labels

EUROPE

Black Sea

İstanbul

Sea of Marmara

Ankara

TURKEY

ANATOLIAN PENINSULA

Tunis

TUNISIA

Tripoli

Mediterranean Sea

Taurus Mts.

Aleppo

CYPRUS

SYRIA

LEBANON

Beirut

Damascus

Syrian Desert

ISRAEL

Jerusalem

Amman

Alexandria

Suez Canal

JORDAN

Giza Cairo

Sinai Pen.

See inset below

LIBYA

EGYPT

Nile R.

Aswān High Dam

Red Sea

Hejaz

SAHARA

Boundary claimed by Sudan

Makkah (Mecca)

SUDAN

AFRICA

Khartoum

White Nile R.

Blue Nile R.

0   300 miles

0   300 kilometers

Lambert Azimuthal Equal-Area Projection

Inset map labels

30°E   TURKEY   35°E

N

Aleppo

CYPRUS

SYRIA

35°N

LEBANON

Beirut

Mediterranean Sea

Damascus

Golan Heights

Sea of Galilee

IRAQ

ISRAEL

West Bank

Tel Aviv-Jaffa

Jerusalem

Amman

Jordan R.

Syrian Desert

Gaza Strip

Dead Sea

0   100 miles

0   100 kilometers

Lambert Azimuthal Equal-Area Projection

Alexandria

Suez Canal

30°N

JORDAN

Giza Cairo

EGYPT

Nile R.

Sinai Pen.

SAUDI ARABIA

Gulf of Suez

Gulf of Aqaba

Red Sea

30°E

Grid labels

Columns: 1 2 3 4 5 6 7 8

Rows: A B C D E F G H J K

0°   10°E   20°E   30°E   40°E

40°N   30°N

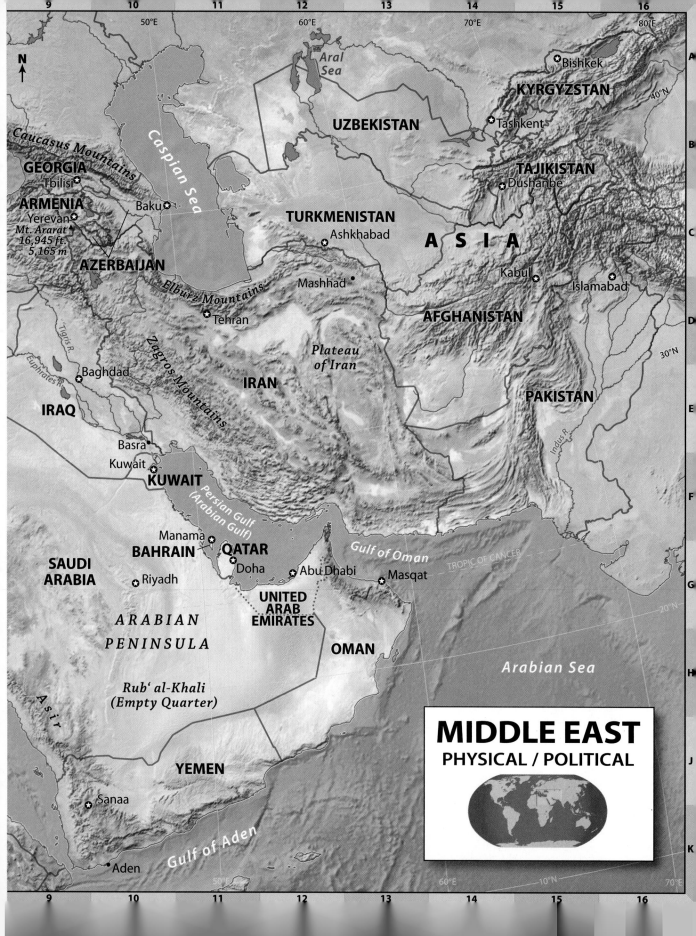

MIDDLE EAST
PHYSICAL / POLITICAL

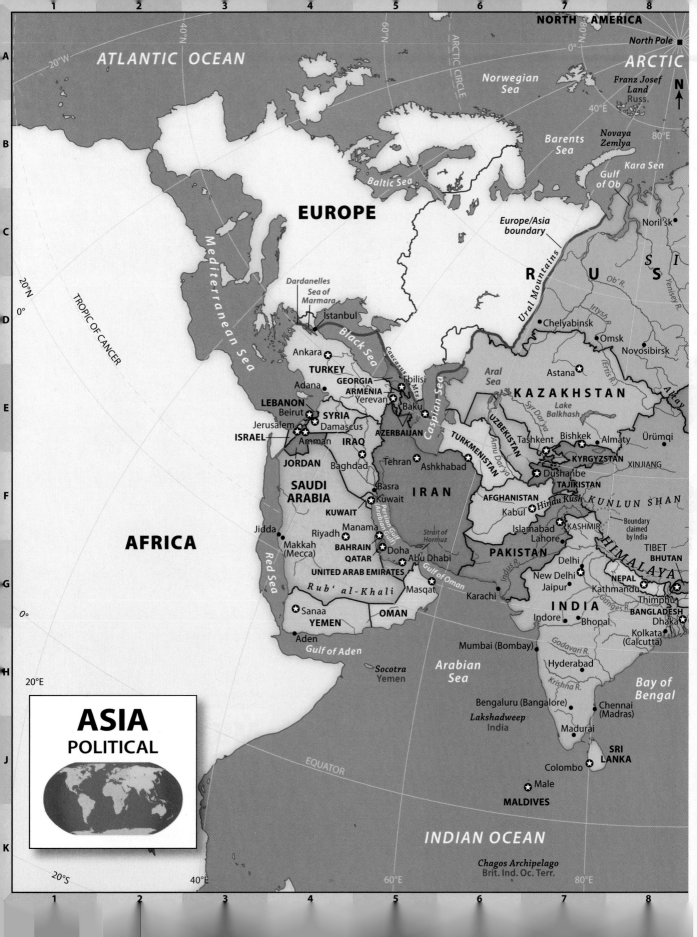

ASIA
POLITICAL

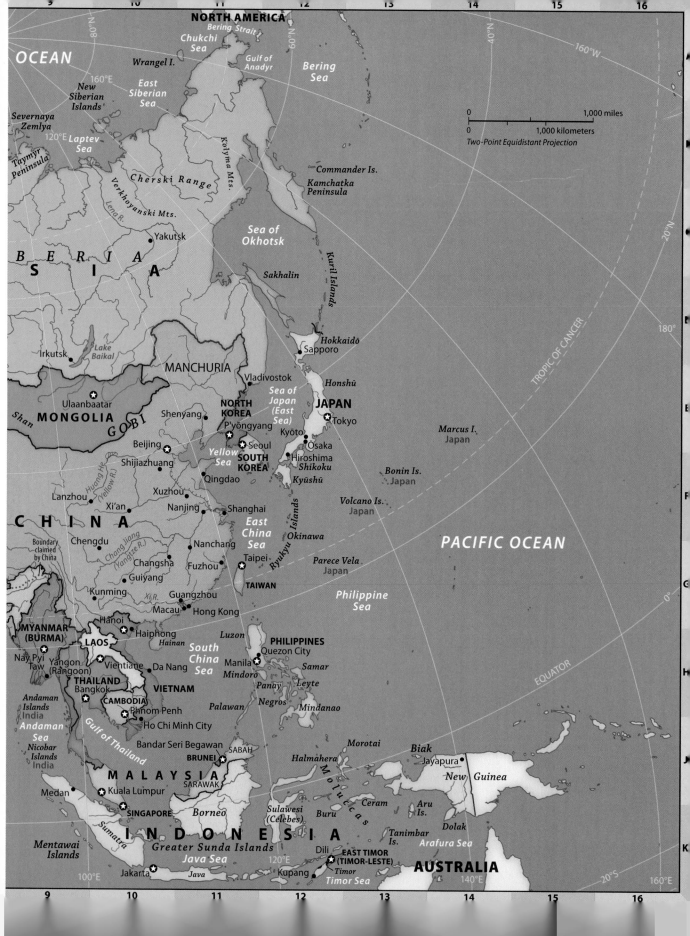

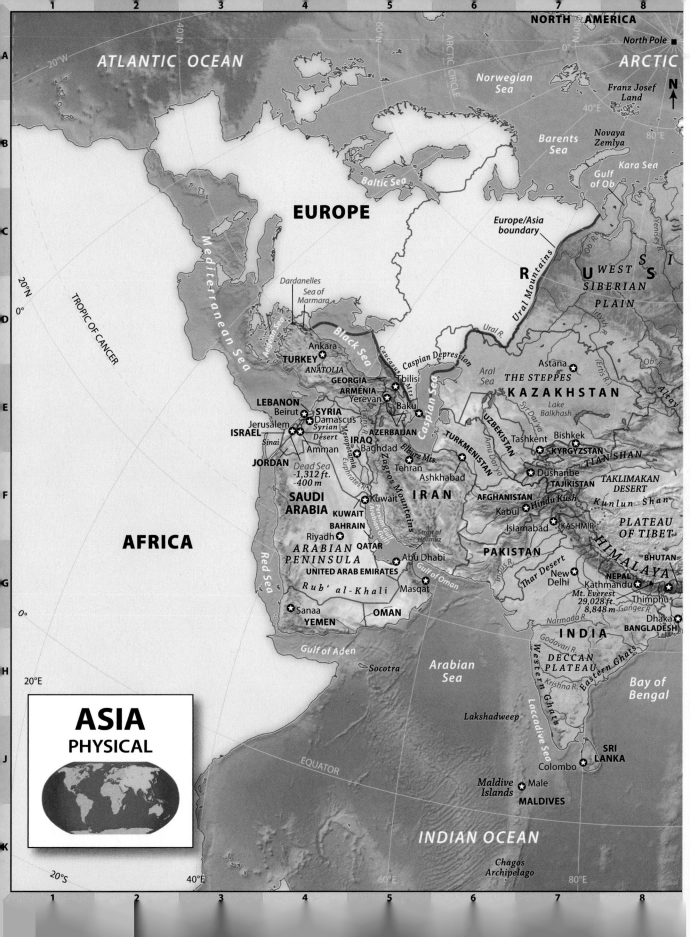

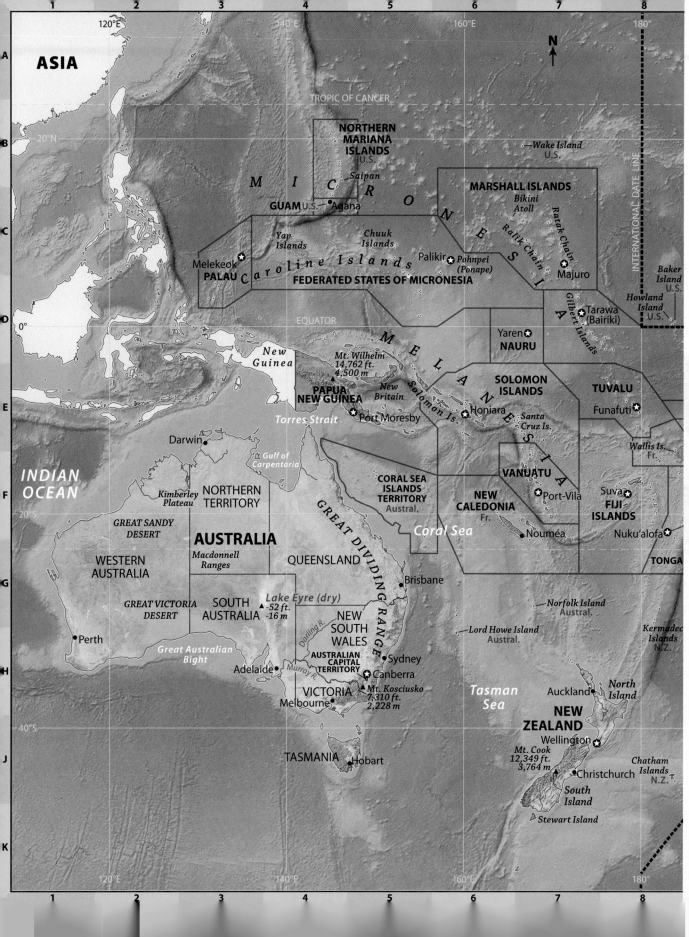

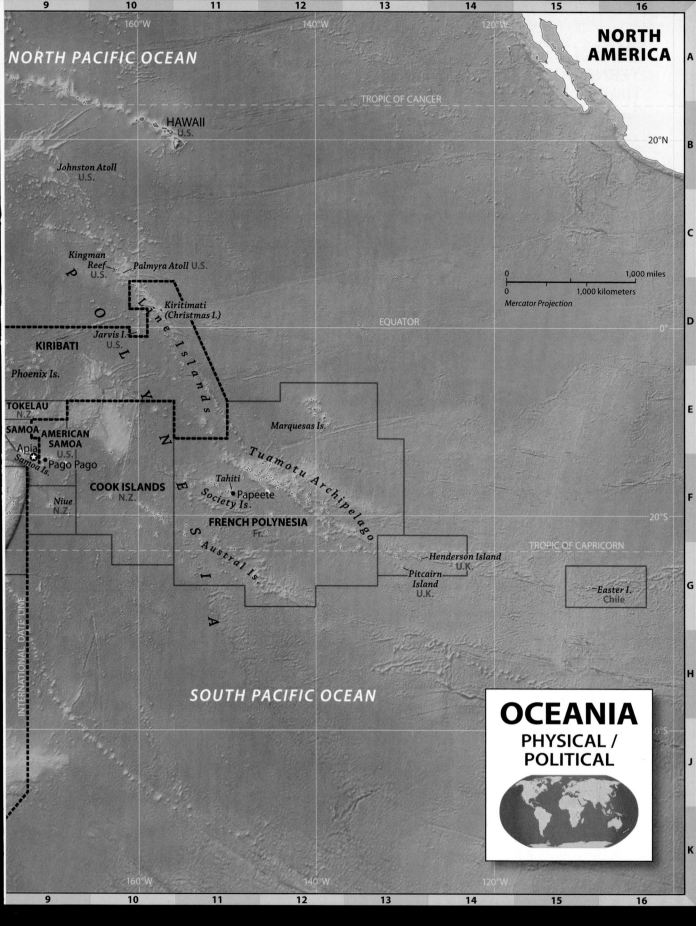

OCEANIA
PHYSICAL / POLITICAL

NORTH PACIFIC OCEAN

TROPIC OF CANCER

20°N

NORTH AMERICA

HAWAII
U.S.

Johnston Atoll
U.S.

Kingman
Reef
U.S.

Palmyra Atoll U.S.

P O L Y N E S I A

Kiritimati
(Christmas I.)

Line Islands

EQUATOR

0°

Jarvis I.
U.S.

KIRIBATI

Phoenix Is.

TOKELAU
N.Z.

SAMOA

AMERICAN
SAMOA
U.S.

Apia

Samoa Is.

Pago Pago

COOK ISLANDS
N.Z.

Niue
N.Z.

Marquesas Is.

Tuamotu Archipelago

Tahiti

Papeete

Society Is.

FRENCH POLYNESIA
Fr.

20°S

Austral Is.

TROPIC OF CAPRICORN

Henderson Island
U.K.

Pitcairn
Island
U.K.

Easter I.
Chile

INTERNATIONAL DATE LINE

SOUTH PACIFIC OCEAN

160°W

140°W

120°W

160°W

140°W

120°W

Mercator Projection

0   1,000 miles

0   1,000 kilometers

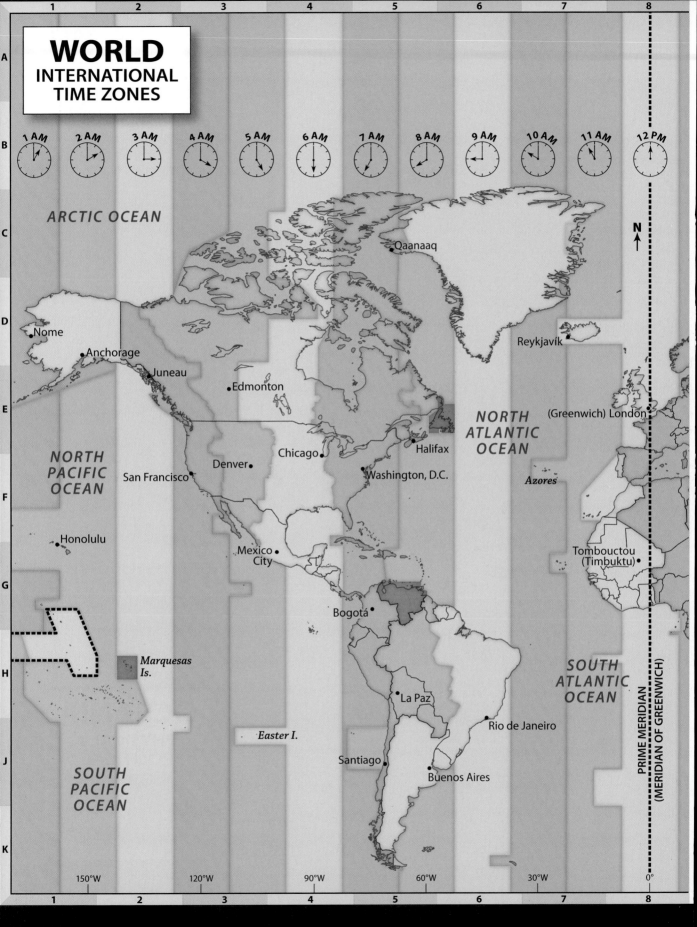

WORLD
INTERNATIONAL TIME ZONES

1 AM 2 AM 3 AM 4 AM 5 AM 6 AM 7 AM 8 AM 9 AM 10 AM 11 AM 12 PM

ARCTIC OCEAN

N

Qaanaaq

Nome

Anchorage

Juneau

Edmonton

Reykjavík

NORTH
ATLANTIC
OCEAN

(Greenwich) London

NORTH
PACIFIC
OCEAN

San Francisco

Denver

Chicago

Halifax

Washington, D.C.

Azores

Honolulu

Mexico
City

Tombouctou
(Timbuktu)

Marquesas
Is.

Bogotá

SOUTH
ATLANTIC
OCEAN

La Paz

Rio de Janeiro

Easter I.

Santiago

Buenos Aires

SOUTH
PACIFIC
OCEAN

PRIME MERIDIAN
(MERIDIAN OF GREENWICH)

150°W   120°W   90°W   60°W   30°W   0°

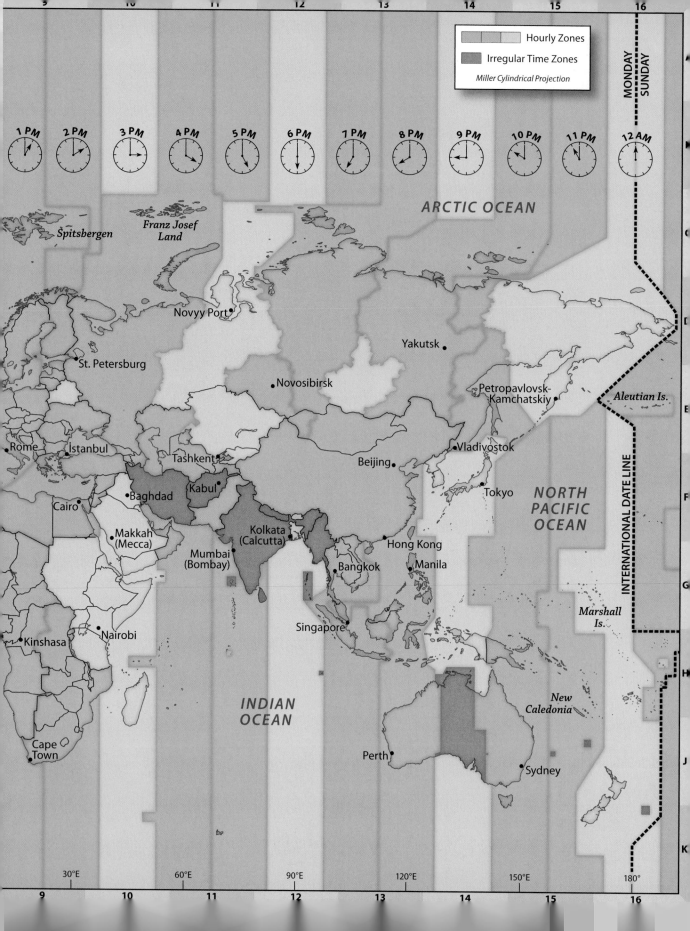

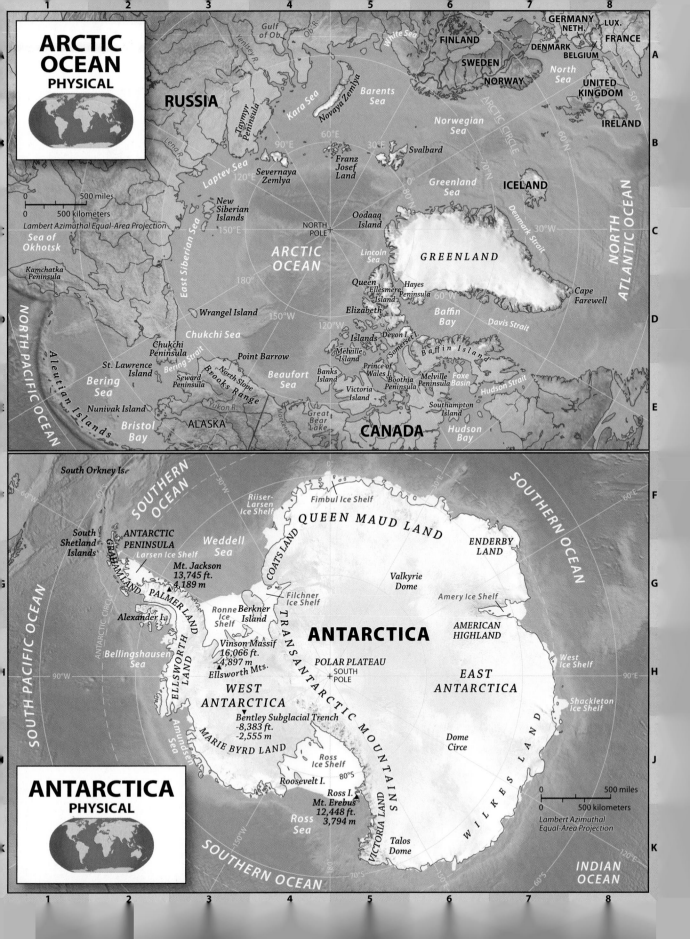

A WORLD OF EXTREMES

The largest continent
is Asia with an area of 17,139,445 sq. miles (44,391,162 sq. km).

The largest country
is Russia with an area of 6,592,812 sq. miles (17,075,383 sq. km).

The smallest country
is Vatican City with an area of 0.17 sq. mile (0.44 sq. km).

The deepest lake
is Lake Baikal with a maximum depth of 5,715 feet (1,742 m).

The highest waterfall
is Angel Falls with a height of 3,212 feet (979 m).

The largest desert
is the Sahara with an area of 3,500,000 sq. miles (9,065,000 sq. km).

The longest river
is the Nile River with a length of 4,160 miles (6,695 km).

The highest mountain
is Mount Everest with a height of 29,028 feet (8,848 m) above sea level.

The smallest continent
is Australia with an area of 2,967,909 sq. miles (7,686,884 sq. km).

Archipelago
Gulf
Reservoir
Volcano
Isthmus
Plateau
Canyon
Highlands
Cliff
Cape
Bay
Harbor
Reef
Island
Channel
Peninsula

rchipelago a group of islands

basin area of land drained by a given river and its
branches; area of land surrounded by lands of higher
elevations

bay part of a large body of water that extends into a
shoreline, generally smaller than a gulf

canyon deep and narrow valley with steep walls

cape point of land that extends into a river, lake, or ocean

channel wide strait or waterway between two
landmasses that lie close to each other; deep part of a
river or other waterway

cliff steep, high wall of rock, earth, or ice

continent one of the seven large landmasses on the
Earth

delta flat, low-lying land built up from soil carried
downstream by a river and deposited at its mouth

divide stretch of high land that separates river systems

downstream direction in which a river or stream flows
from its source to its mouth

scarpment steep cliff or slope between a higher and

glacier large, thick body of slowly moving ice

gulf part of a large body of water that extends into a
shoreline, generally larger and more deeply indented
than a bay

harbor a sheltered place along a shoreline where ships
can anchor safely

highland elevated land area such as a hill, mountain, or
plateau

hill elevated land with sloping sides and rounded
summit; generally smaller than a mountain

island land area, smaller than a continent, completely
surrounded by water

isthmus narrow stretch of land connecting two larger
land areas

lake a sizable inland body of water

lowland land, usually level, at a low elevation

mesa broad, flat-topped landform with steep sides;
smaller than a plateau

mountain land with steep sides that rises sharply (1,000
feet or more) from surrounding land; generally larger

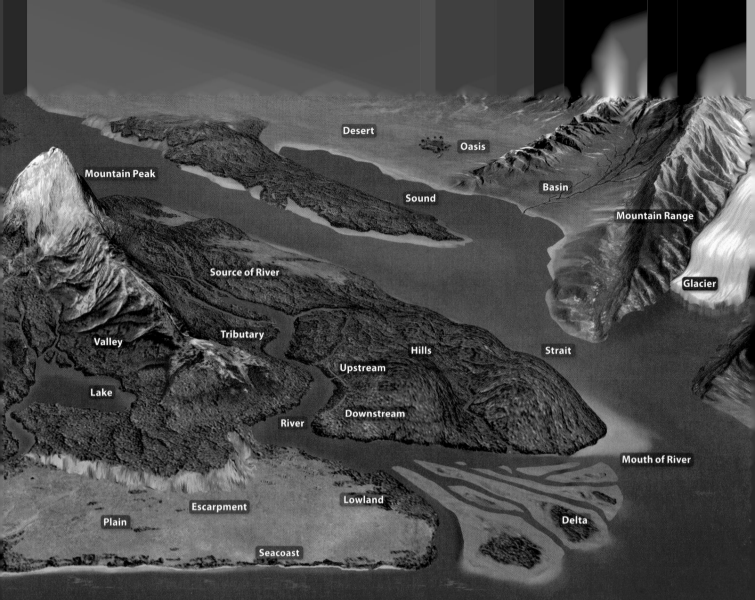

Desert

Oasis

Mountain Peak

Sound

Basin

Mountain Range

Source of River

Glacier

Tributary

Valley

Hills

Strait

Upstream

Lake

Downstream

River

Mouth of River

Escarpment

Lowland

Plain

Delta

Seacoast

mountain peak pointed top of a mountain

mountain range a series of connected mountains

mouth (of a river) place where a stream or river flows into a larger body of water

oasis small area in a desert where water and vegetation are found

ocean one of the four major bodies of salt water that surround the continents

ocean current stream of either cold or warm water that moves in a definite direction through an ocean

peninsula body of land jutting into a lake or ocean, surrounded on three sides by water

physical feature characteristic of a place occurring naturally, such as a landform, body of water, climate pattern, or resource

plain area of level land, usually at low elevation and often covered with grasses

plateau area of flat or rolling land at a high elevation, about 300 to 3,000 feet (90 to 900 m) high

river large natural stream of water that runs through the land

sea large body of water completely or partly surrounded by land

seacoast land lying next to a sea or an ocean

sound broad inland body of water, often between a coastline and one or more islands off the coast

source (of a river) place where a river or stream begins, often in highlands

strait narrow stretch of water joining two larger bodies of water

tributary small river or stream that flows into a large river or stream; a branch of the river

upstream direction opposite the flow of a river; toward the source of a river or stream

valley area of low land usually between hills or mountains

volcano mountain or hill created as liquid rock and ash erupt from inside the Earth

SCAVENGER HUNT

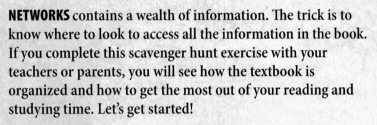

NETWORKS contains a wealth of information. The trick is to know where to look to access all the information in the book. If you complete this scavenger hunt exercise with your teachers or parents, you will see how the textbook is organized and how to get the most out of your reading and studying time. Let's get started!

1 How many lessons are in Chapter 2?

2 What does Unit 1 cover?

3 Where can you find the Essential Questions for each lesson?

4 In what three places can you find information on a Foldable?

5 How can you identify content vocabulary and academic vocabulary in the narrative?

6 Where do you find graphic organizers in your textbook?

7 You want to quickly find a map in the book about the world. Where do you look?

8 Where would you find the latitude and longitude for Dublin, Ireland?

9 If you needed to know the Spanish term for *earthquake*, where would you look?

10 Where can you find a list of all the charts in a unit?

THE WORLD

McGraw-Hill networks

UNIT 1

Chapter 1
The Geographer's World

Chapter 2
Physical Geography

Chapter 3
Human Geography

EXPLORE the WORLD

Geography is the study of Earth and all of its variety. When you study geography, you learn about the planet's land, water, plants, and animals. Some people call Earth "the water planet." Do you know why? Water—in the form of streams, rivers, lakes, seas, and oceans—covers nearly 70 percent of Earth's surface.

1 **BODIES OF WATER** Underseas explorers can still experience the thrill of investigating uncharted territory—one of Earth's last frontiers. Almost all of the Earth's water consists of a continuous body of water that circles the planet. This body of water makes up five oceans: the Pacific, the Atlantic, the Indian, the Southern, and the Arctic.

2 **LANDFORMS** Landforms are features of the land, such as mountains, valleys, and canyons. Landforms influence where people live and how they relate to their environment.

3 **NATURAL RESOURCES** Natural resources are products of Earth that people use to meet their needs. Solar energy is power produced by the heat of the sun. Sun and wind are renewable resources. These resources cannot be used up.

FAST
FACT

Earth's longest mountain range is underwater.

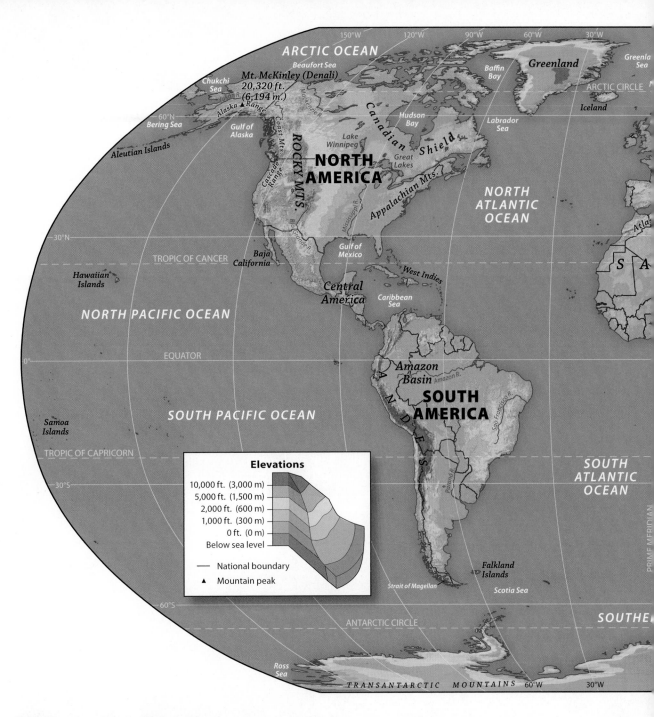

ARCTIC OCEAN

Beaufort Sea

Mt. McKinley (Denali)
20,320 ft.
(6,194 m.)

Chukchi
Sea

Alaska ▲ *Range*

60°N

Bering Sea

Aleutian Islands

Gulf of
Alaska

Coast Mts.

ROCKY MTS.

Cascade Range

Yukon R.

Mackenzie R.

Canadian *Shield*

Hudson
Bay

Baffin
Bay

Greenland

Greenla
Sea

ARCTIC CIRCLE

Iceland

Labrador
Sea

NORTH
AMERICA

Lake
Winnipeg

Great
Lakes

Appalachian Mts.

Mississippi R.

NORTH
ATLANTIC
OCEAN

Atla

30°N

TROPIC OF CANCER

Hawaiian
Islands

Baja
California

Rio Grande R.

Gulf of
Mexico

S A

Central
America

West Indies

NORTH PACIFIC OCEAN

Caribbean
Sea

EQUATOR

Amazon
Basin

Amazon R.

SOUTH PACIFIC OCEAN

SOUTH
AMERICA

São Francisco R.

SOUTH
ATLANTIC
OCEAN

Samoa
Islands

TROPIC OF CAPRICORN

A
N
D
E
S

Paraná R.

30°S

PRIME MERIDIAN

Elevations

10,000 ft. (3,000 m)
5,000 ft. (1,500 m)
2,000 ft. (600 m)
1,000 ft. (300 m)
0 ft. (0 m)
Below sea level

— National boundary
▲ Mountain peak

Falkland
Islands

Strait of Magellan

Scotia Sea

60°S

ANTARCTIC CIRCLE

SOUTHE

Ross
Sea

T R A N S A N T A R C T I C   M O U N T A I N S

60°W

30°W

150°W   120°W   90°W   60°W   30°W

THE WORLD

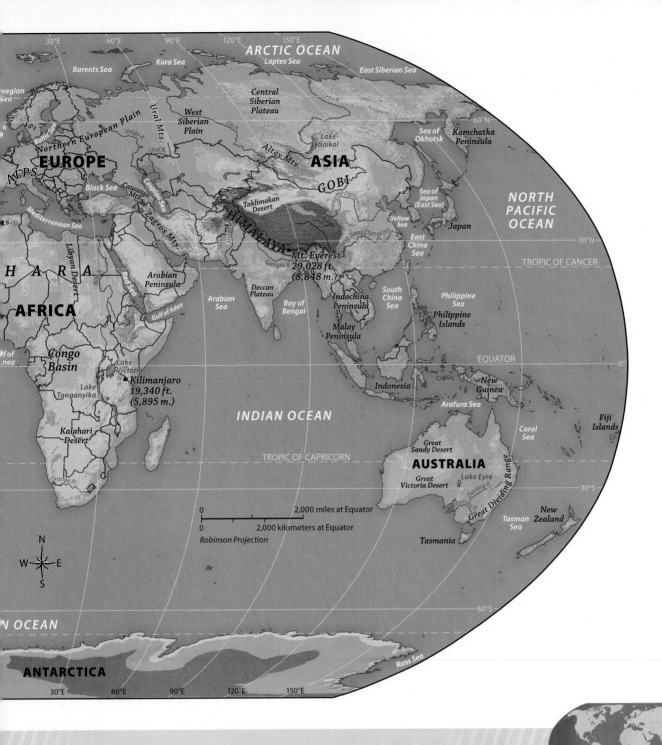

ARCTIC OCEAN

Barents Sea

Kara Sea

Laptev Sea

East Siberian Sea

30°E 60°E 90°E 120°E 150°E

Central Siberian Plateau

Lena R.

Ob R.

West Siberian Plain

Ural Mts.

Northern European Plain

Volga R.

Ural R.

Lake Baikal

Altay Mts.

60°N

Sea of Okhotsk

Kamchatka Peninsula

EUROPE

ASIA

GOBI

ALPS

Black Sea

Caucasus Mts.

Caspian Sea

Taklimakan Desert

Huang He (Yellow)

Sea of Japan (East Sea)

NORTH PACIFIC OCEAN

Mediterranean Sea

Zagros Mts.

HIMALAYA

Indus R.

Yellow Sea

Japan

30°N

H A R A

Libyan Desert

Red Sea

Arabian Peninsula

Ganges R.

▲Mt. Everest 29,028 ft. (8,848 m.)

Chang Jiang (Yangtze)

East China Sea

TROPIC OF CANCER

AFRICA

Nile R.

Gulf of Aden

Arabian Sea

Deccan Plateau

Bay of Bengal

Indochina Peninsula

South China Sea

Philippine Sea

Philippine Islands

f of nea

Congo Basin

Lake Victoria

▲Kilimanjaro 19,340 ft. (5,895 m.)

Malay Peninsula

Lake Tanganyika

INDIAN OCEAN

Indonesia

New Guinea

EQUATOR

0°

Arafura Sea

Fiji Islands

Kalahari Desert

Orange R.

TROPIC OF CAPRICORN

Great Sandy Desert

AUSTRALIA

Great Victoria Desert

Lake Eyre

Murray R.

Darling R.

Great Dividing Range

Coral Sea

30°S

Tasman Sea

New Zealand

0 2,000 miles at Equator
0 2,000 kilometers at Equator
Robinson Projection

N
W   E
S

Tasmania

ANTARCTICA

60°S

Ross Sea

N OCEAN

30°E 60°E 90°E 120°E 150°E

PHYSICAL

MAP SKILLS

1 **THE GEOGRAPHER'S WORLD** What part of South America has the highest elevation?

2 **THE GEOGRAPHER'S WORLD** What body of water is located west of Greenland?

3 **PLACES AND REGIONS** How would you describe southern Africa?

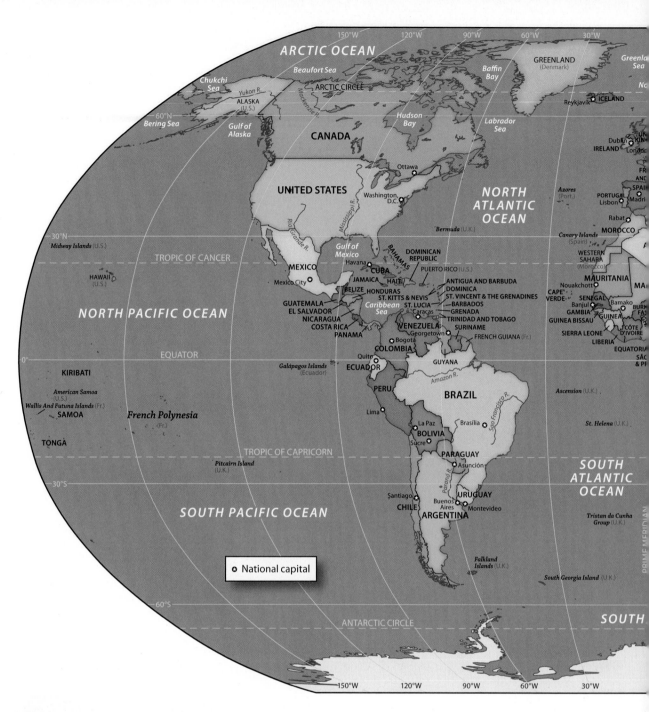

THE WORLD

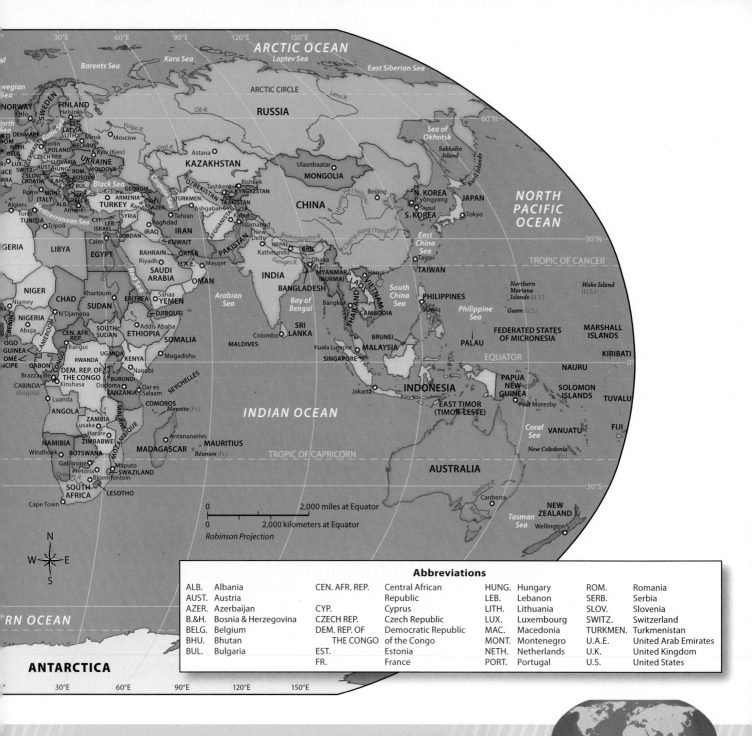

Abbreviations

ALB.	Albania	CEN. AFR. REP.	Central African	HUNG.	Hungary	ROM.	Romania
AUST.	Austria		Republic	LEB.	Lebanon	SERB.	Serbia
AZER.	Azerbaijan	CYP.	Cyprus	LITH.	Lithuania	SLOV.	Slovenia
B.&H.	Bosnia & Herzegovina	CZECH REP.	Czech Republic	LUX.	Luxembourg	SWITZ.	Switzerland
BELG.	Belgium	DEM. REP. OF	Democratic Republic	MAC.	Macedonia	TURKMEN.	Turkmenistan
BHU.	Bhutan	THE CONGO	of the Congo	MONT.	Montenegro	U.A.E.	United Arab Emirates
BUL.	Bulgaria	EST.	Estonia	NETH.	Netherlands	U.K.	United Kingdom
		FR.	France	PORT.	Portugal	U.S.	United States

POLITICAL

MAP SKILLS

1 **PLACES AND REGIONS** How would you describe the region north of Australia?

2 **THE GEOGRAPHER'S WORLD** Which country is located west of Egypt?

3 **THE GEOGRAPHER'S WORLD** What is the capital of Mongolia?

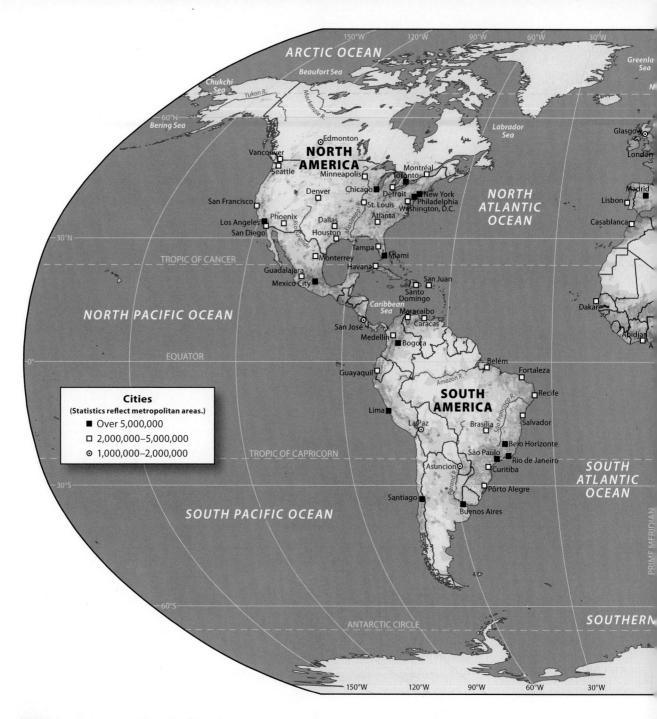

Cities
(Statistics reflect metropolitan areas.)
- ■ Over 5,000,000
- ☐ 2,000,000–5,000,000
- ◉ 1,000,000–2,000,000

THE WORLD

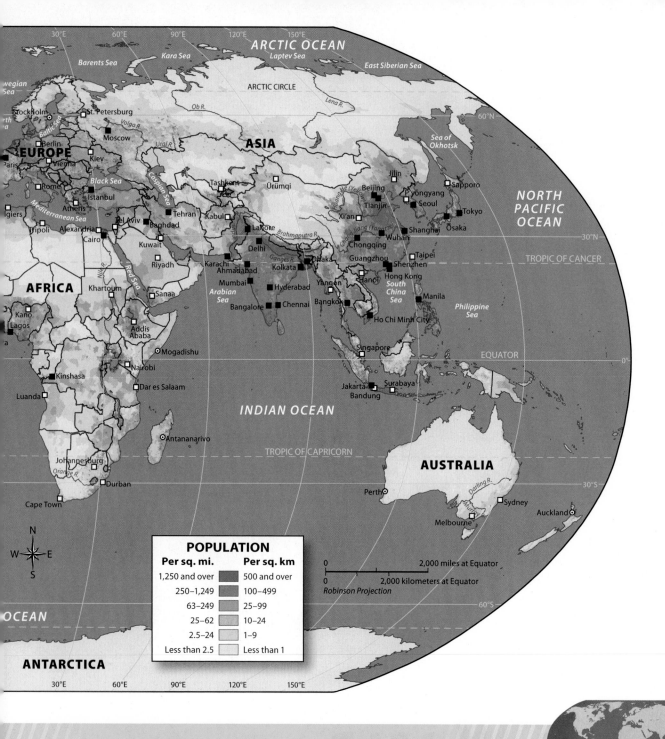

POPULATION

Per sq. mi.	Per sq. km
1,250 and over	500 and over
250–1,249	100–499
63–249	25–99
25–62	10–24
2.5–24	1–9
Less than 2.5	Less than 1

2,000 miles at Equator

2,000 kilometers at Equator

Robinson Projection

POPULATION DENSITY

MAP SKILLS

1 **PLACES AND REGIONS** What parts of South America are the most densely populated?

2 **PLACES AND REGIONS** Which part of Africa has the lowest population density?

3 **ENVIRONMENT AND SOCIETY** In general, what population pattern do you see in Europe?

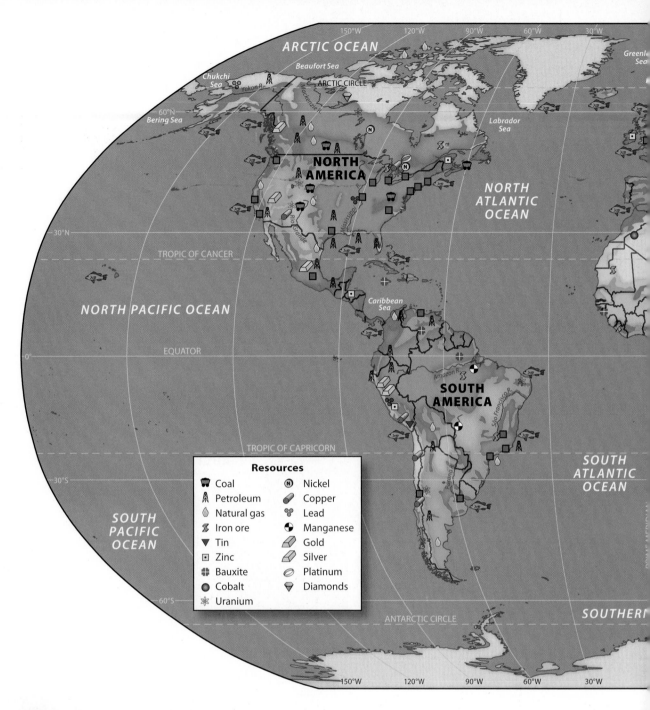

ARCTIC OCEAN

Beaufort Sea

Chukchi
Sea

ARCTIC CIRCLE

Yukon R.

Mackenzie R.

Greenle...
Sea

60°N
Bering Sea

Labrador
Sea

NORTH
AMERICA

Rio Grande

Mississippi

NORTH
ATLANTIC
OCEAN

30°N

TROPIC OF CANCER

NORTH PACIFIC OCEAN

Caribbean
Sea

EQUATOR

Amazon R.

SOUTH
AMERICA

São Francisco R.

SOUTH
ATLANTIC
OCEAN

TROPIC OF CAPRICORN

Resources

Coal		Nickel	
Petroleum		Copper	
Natural gas		Lead	
Iron ore		Manganese	
Tin		Gold	
Zinc		Silver	
Bauxite		Platinum	
Cobalt		Diamonds	
Uranium			

SOUTH
PACIFIC
OCEAN

30°S

60°S

ANTARCTIC CIRCLE

SOUTHERN

150°W 120°W 90°W 60°W 30°W

THE WORLD

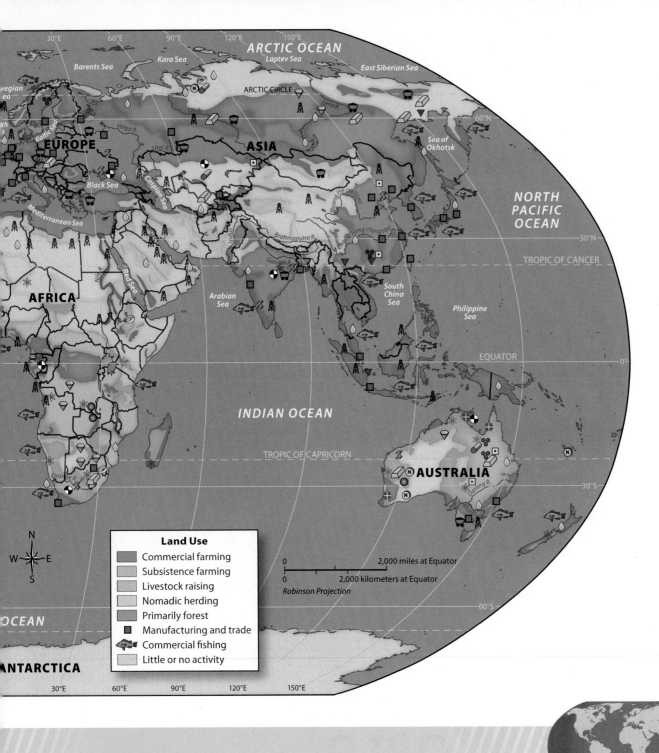

Land Use

- Commercial farming
- Subsistence farming
- Livestock raising
- Nomadic herding
- Primarily forest
- Manufacturing and trade
- Commercial fishing
- Little or no activity

0 | 2,000 miles at Equator
0 | 2,000 kilometers at Equator
Robinson Projection

ECONOMIC RESOURCES

MAP SKILLS

1 **ENVIRONMENT AND SOCIETY** What economic activity is found along most coastal regions?

2 **HUMAN GEOGRAPHY** Describe the general use of land in North Africa.

3 **PLACES AND REGIONS** Which area produces oil—North America or Australia?

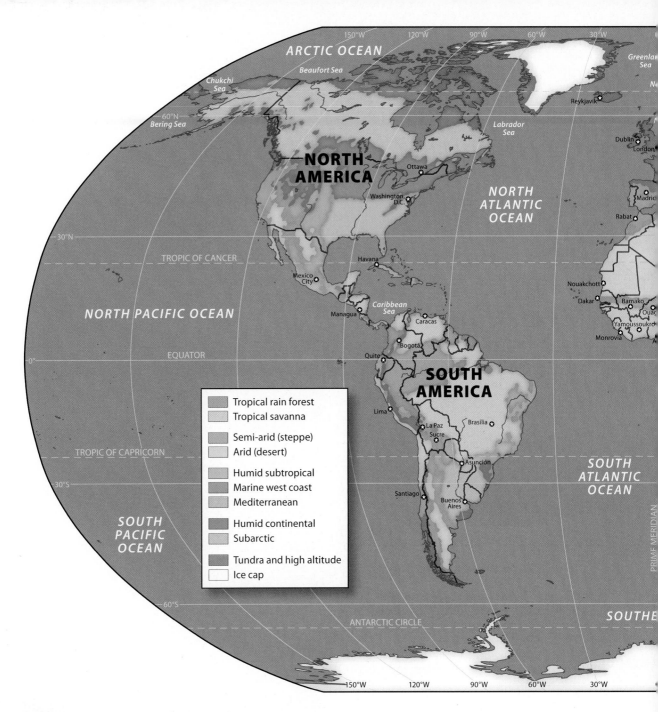

ARCTIC OCEAN

Beaufort Sea

Chukchi
Sea

Greenlan
Sea

Reykjavik

Bering Sea

60°N

Labrador
Sea

Dublin

London

NORTH
AMERICA

Ottawa

NORTH
ATLANTIC
OCEAN

Madric

Washington,
D.C.

30°N

Rabat

TROPIC OF CANCER

Havana

NORTH PACIFIC OCEAN

Mexico
City

Nouakchott

Dakar

Bamako

Managua

Caribbean
Sea

Ouac

EQUATOR

Caracas

Yamoussoukro

Bogotá

Monrovia

Quito

SOUTH
AMERICA

	Tropical rain forest
	Tropical savanna
	Semi-arid (steppe)
	Arid (desert)
	Humid subtropical
	Marine west coast
	Mediterranean
	Humid continental
	Subarctic
	Tundra and high altitude
	Ice cap

Lima

La Paz

Brasília

Sucre

TROPIC OF CAPRICORN

30°S

Asunción

SOUTH
ATLANTIC
OCEAN

SOUTH
PACIFIC
OCEAN

Santiago

Buenos
Aires

PRIME MERIDIAN

60°S

ANTARCTIC CIRCLE

SOUTHE

150°W   120°W   90°W   60°W   30°W

THE WORLD

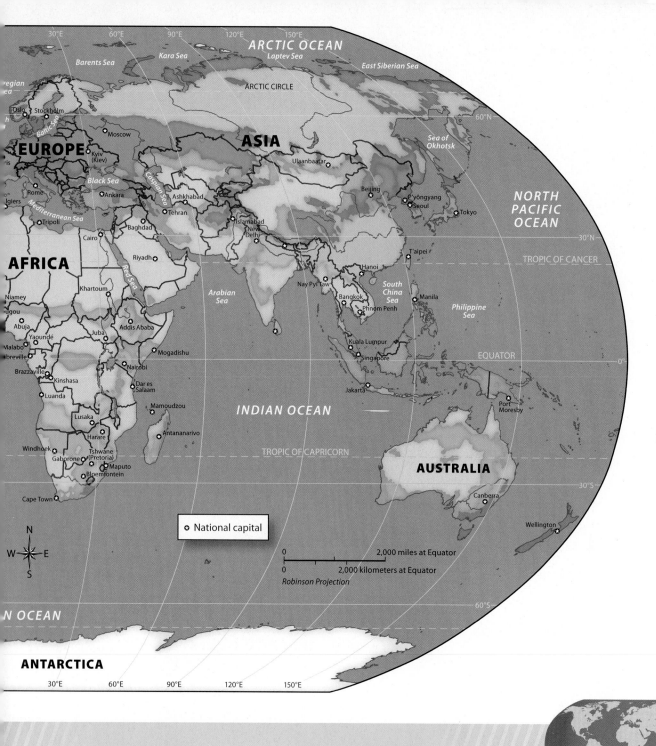

National capital

N W E S

0 ___ 2,000 miles at Equator
0 ___ 2,000 kilometers at Equator
Robinson Projection

CLIMATE

MAP SKILLS

1 **PHYSICAL GEOGRAPHY** Which climate zones appear in northern North America?

2 **THE GEOGRAPHER'S WORLD** Which continent receives more rain—Australia or South America? Why?

3 **PHYSICAL GEOGRAPHY** In general, how does the climate of southern Africa compare with the climate of northern Africa?

NASA/NOAA

THE GEOGRAPHER'S WORLD

ESSENTIAL QUESTION · *How does geography influence the way people live?*

The Story Matters...

Since ancient times, people have drawn maps to show their known world. As people explored, they came into contact with different places and people, which expanded their understanding of the world. Today, what we know about the world continues to grow as geographers study the world's environments with the latest technology. More importantly, by understanding the connections between humans and the environment, geographers can find solutions to significant problems.

Prisma/SuperStock

A geographer drills for an ice sample.

FOLDABLES
Study Organizer

Go to the Foldables® library in the back of your book to make a Foldable® that will help you take notes while reading this chapter.

○ Geographer's View
○ Geographer's Tools

THE GEOGRAPHER'S WORLD

Geography is the study of Earth in all of its variety. When you study geography, you learn about the physical features and the living things—humans, plants, and animals—that inhabit Earth.

Step Into the Place

MAP FOCUS Use the map to answer the following questions.

1 **THE GEOGRAPHER'S WORLD** What are the names of the large landmasses on the map?

2 **THE GEOGRAPHER'S WORLD** What are the names of the large bodies of water on the map?

3 **THE GEOGRAPHER'S WORLD** What do you think the blue lines are that appear within the landmasses?

4 **CRITICAL THINKING Analyzing** How is this world map similar to other maps you have seen?

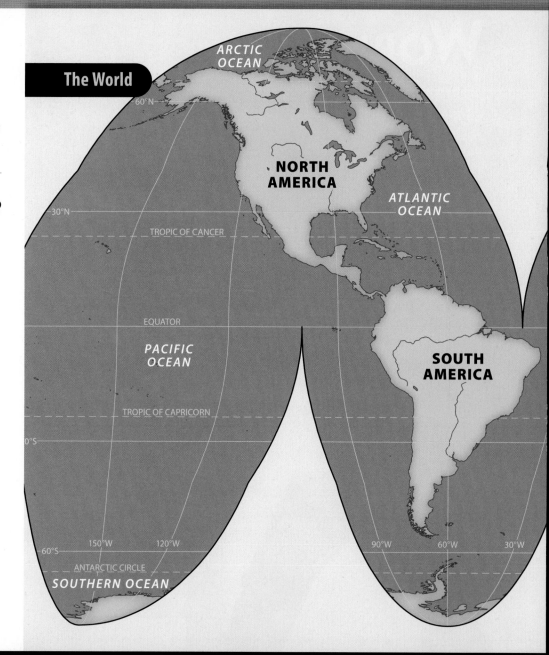

The World

ARCTIC OCEAN

60°N

NORTH AMERICA

ATLANTIC OCEAN

30°N

TROPIC OF CANCER

EQUATOR

PACIFIC OCEAN

TROPIC OF CAPRICORN

0°S

SOUTH AMERICA

60°S

150°W   120°W

90°W   60°W   30°W

ANTARCTIC CIRCLE

SOUTHERN OCEAN

Step Into the Time

DESCRIBING Choose an event from the time line and write a paragraph describing how it might have changed how people understood or viewed the world in which they lived.

150 Ptolemy creates atlas of known world

Universal Images Group/Getty Images

networks
There's More Online!

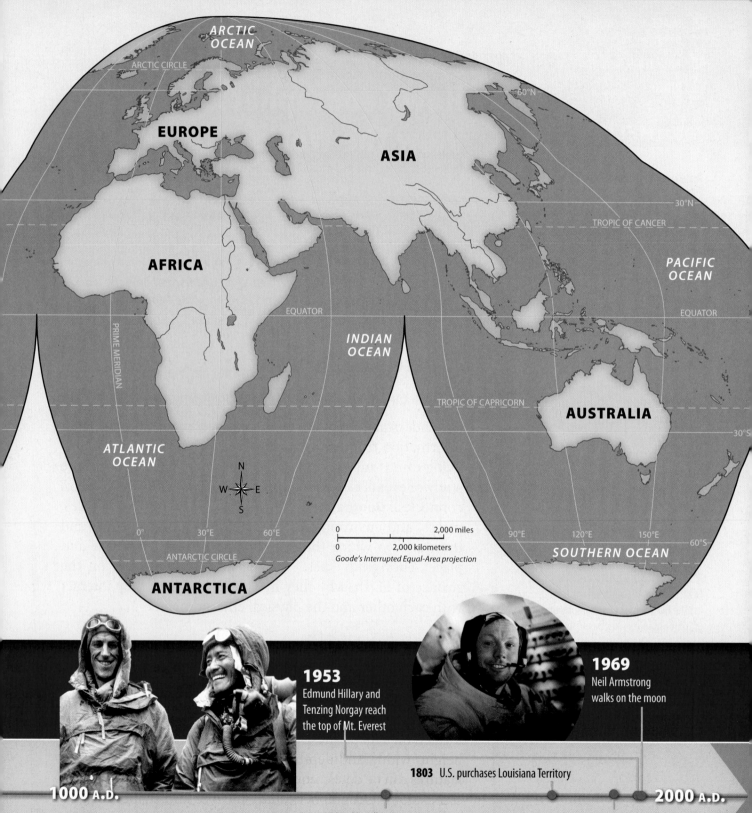

ARCTIC OCEAN

ARCTIC CIRCLE

60°N

EUROPE

ASIA

30°N

TROPIC OF CANCER

AFRICA

PACIFIC OCEAN

EQUATOR

EQUATOR

INDIAN OCEAN

PRIME MERIDIAN

TROPIC OF CAPRICORN

AUSTRALIA

ATLANTIC OCEAN

30°S

N
W E
S

0°   30°E   60°E

0   2,000 miles
0   2,000 kilometers
Goode's Interrupted Equal-Area projection

90°E   120°E   150°E

60°S

SOUTHERN OCEAN

ANTARCTIC CIRCLE

ANTARCTICA

1953
Edmund Hillary and
Tenzing Norgay reach
the top of Mt. Everest

1969
Neil Armstrong
walks on the moon

1803 U.S. purchases Louisiana Territory

1000 A.D.

2000 A.D.

1519 Magellan sets sail on
voyage around the world

1909 Robert Peary reaches
the North Pole

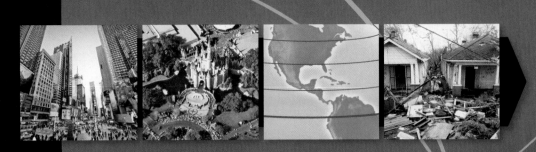

netw⭗rks

There's More Online!

☑ **CHART/GRAPH** Six Essential Elements of Geography

☑ **IMAGE** Places Change Over Time

☑ **ANIMATION** Regions of Earth

☑ **VIDEO**

Reading **HELP**DESK

Academic Vocabulary

- dynamic
- component

Content Vocabulary

- **geography**
- **spatial**
- **landscape**
- **relative location**
- **absolute location**
- **latitude**
- **Equator**
- **longitude**
- **Prime Meridian**
- **region**
- **environment**
- **landform**
- **climate**
- **resource**

TAKING NOTES: *Key Ideas and Details*

Identifying As you read the lesson, list the five themes of geography on a graphic organizer like the one below.

Themes

Lesson 1
How Geographers View the World

ESSENTIAL QUESTION · *How does geography influence the way people live?*

IT MATTERS BECAUSE
Thinking like a geographer helps you understand how the world works and appreciate the world's remarkable beauty and complexity.

Geographers Think Spatially

GUIDING QUESTION *What does it mean to think like a geographer?*

An understanding of the world is based on a combination of information from many sources. Biology is the study of how living things survive and relate to one another. History is the study of events that occur over time and how those events are connected. **Geography** is the study of Earth and its peoples, places, and environments. Geographers look at people and the world in which they live mainly in terms of space and place. They study such topics as where people live on the surface of Earth, why they live there, and how they interact with each other and the physical environment.

Thinking Spatially

Geography, then, emphasizes the spatial aspects of the world. **Spatial** refers to Earth's features in terms of their locations, their shapes, and their relationships to one another.

Physical features such as mountains and lakes can be located on a map. These features can be measured in terms of height, width, and depth. Distances and directions to other features can be determined. The human world also has spatial dimensions. Geographers study the size and shape of cities, states, and countries. They consider how close or far apart

18

these human features are to one another. Geographers also think about the relationships between human features and physical features.

But thinking spatially is more than just the study of the location or size of things. It means looking at the characteristics of Earth's features. Geographers ask what mountains in different locations are made of. They examine what kinds of fish live in different lakes. They study the layout of cities and think about how easy or difficult it is for people to move around in them.

The Perspective of Place

Locations on Earth are made up of different combinations of physical and human characteristics. Physical features such as climate, landforms, and vegetation combine with human features such as population, economic activity, and land use. These combinations create what geographers call places.

Places are locations on Earth that have distinctive characteristics that make them meaningful to people. The places where we live, work, and go to school are important to us. Our home is an important place. Even small places such as our bedroom or a classroom often have a unique and special meaning. In the same way, larger locations, such as our hometown, our country, or even Earth, are places that have meaning for people.

One way that geographers learn about places is by studying landscapes. **Landscapes** are portions of Earth's surface that can be viewed at one time and from one location. They can be as small as the view from the front porch of your home, or they can be as large as the view from a tall building that includes the city and surrounding countryside.

The geography theme of *place* describes all of the characteristics that give an area its own special quality.

▶ **CRITICAL THINKING**
Describing What are the characteristics that makes a place like Times Square in New York City special?

José Fuste Raga/age Fotostock

Whether we visit a landscape or we look at photographs of the landscape, it can tell us much about the people who live there. Geographers look at landscapes and try to explain their unique combinations of physical and human features. As you study geography, notice the great variety in the world's landscapes.

The Perspective of Experience

Geography is not something you learn about only in school or just from books. Geography is something you experience every day.

We all live in the world. We feel the change of the seasons. We hear the sounds of birds chirping and of car horns honking. We walk on sidewalks and in forests. We ride in cars along streets and highways. We shop in malls and grocery stores. We fly in airplanes to distant places. We surf the Internet or watch TV and learn about peoples and events in our neighborhood, our country, and the world.

This is all geography. By learning about geography in school, we can better appreciate and understand this world in which we live.

A Changing World

Earth is **dynamic**, or always changing. Rivers shift course. Volcanoes suddenly erupt, forming mountains or collapsing the peaks of mountains. The pounding surf removes sand from beaches.

The things that people make change, too. Farmers shift from growing one crop to another. Cities grow larger. Nations expand into new areas.

Geographers, then, study how places change over time. They try to understand what impact those changes have. What factors made a city grow? What effect did a growing city have on the people who live there? What effect did the city's growth have on nearby communities and on the land and water near it? Answering questions like these is part of the field of geography.

☑ **READING PROGRESS CHECK**

Describing How is geography related to history?

The Five Themes of Geography

GUIDING QUESTION *How can you make sense of a subject as large as Earth and its people?*

Geographers use five themes to organize information about the world. These themes help them view and understand Earth.

Location

Location is where something is found on Earth. There are two types of location. **Relative location** describes where a place is compared to another place. This approach often uses the cardinal directions—north, south, east, and west. A school might be on the east side of town. Relative location can also tell us about the characteristics of a

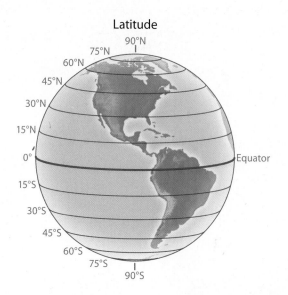

Latitude

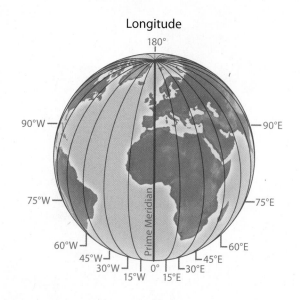

Longitude

place. For example, knowing that New Orleans is near the mouth of the Mississippi River helps us understand why the city became an important trading port.

Absolute location is the exact location of something. An address like 123 Main Street is an absolute location. Geographers identify the absolute location of places using a system of imaginary lines called latitude and longitude. Those lines form a grid for locating a place precisely.

Lines of **latitude** run east to west, but they measure distance on Earth in a north-to-south direction. One of these lines, the **Equator**, circles the middle of Earth. This line is equally distant from the North Pole and the South Pole. Other lines of latitude between the Equator and the North and South Poles are assigned a number from 1° to 90°. The higher the number, the farther the line is from the Equator. The Equator is 0° latitude. The North Pole is at 90° north latitude (90° N), and the South Pole is at 90° south latitude (90° S).

Lines of **longitude** run from north to south, but they measure distance on Earth in an east-to-west direction. They go from the North Pole to the South Pole. These lines are also called *meridians*. The **Prime Meridian** is the starting point for measuring longitude. It runs through Greenwich, England, and has the value of 0° longitude. There are 180 lines of longitude to the east of the Prime Meridian and 180 lines to the west. They meet at the meridian 180°, which is the International Date Line.

Geographers use latitude and longitude to locate anything on Earth. In stating absolute location, geographers always list latitude first. For example, the absolute location of Washington, D.C., is 38° N, 77° W.

Lines of latitude circle Earth parallel to the Equator and measure the distance north or south of the Equator in degrees. Lines of longitude circle the Earth from the North Pole to the South Pole. These lines measure distances east or west of the Prime Meridian.

▶ **CRITICAL THINKING**
The Geographer's World At what degree latitude is the Equator located?

Coordinates visible on image: 28°25'12"N, 81°35'00"W, 28°25'8"N, 81°34'52"W, 81°34'48"W, 28°25'16"N, 81°34'44"W

Disney World, located in Orlando, Florida, attracts millions of visitors every year.

▶ **CRITICAL THINKING**

Human Geography What effect do you think Disney World has on the surrounding communities?

Place

Another theme of geography is place. The features that help define a place can be physical or human.

Why is Denver called the "Mile High City"? Its location one mile above sea level gives it a special character. Why does New Orleans have the nickname "the Crescent City"? It is built on a crescent-shaped bend along the Mississippi River. That location has had a major impact on the city's growth and how its people live.

Region

Although places are unique, two or more places can share characteristics. Places that are close to one another and share some characteristics belong to the same **region**. For example, Los Angeles and San Diego are located in southern California. They have some features in common, such as nearness to the ocean. Both cities also have mostly warm temperatures throughout the year.

In the case of those two cities, the region is defined using physical characteristics. Regions can also be defined by human characteristics. For instance, the countries of North Africa are part of the same region. One reason is that most of the people living in these countries follow the same religion, Islam.

Geographers study region so they can identify the broad patterns of larger areas. They can compare and contrast the features in one region with those in another. They also examine the special features that make each place in a region distinct from the others.

Aerial Archives/Alamy; (inset) Ilene MacDonald/Alamy

Human-Environment Interaction

People and the environment interact. That is, they affect each other. The physical characteristics of a place affect how people live. Flat, rich, well-watered soil is good for farming. Mountains full of coal can be mined. The environment can present all kinds of hazards, such as floods, droughts, earthquakes, and volcanic eruptions.

People affect the environment, too. They blast tunnels through mountains to build roadways and drain swamps to make farmland. Although these actions can improve life for some people, they can also harm the environment. Exhaust from cars on the roadways can pollute the air, and turning swamps into farms destroys natural ecosystems and reduces biological diversity.

The **environment** is the natural surroundings of a place. It includes several key features. One is **landforms**, or the shape and nature of the land. Hills, mountains, and valleys are types of landforms. The environment also includes the presence or absence of a body of water. Cities located on coastlines, like New York City, have different characteristics than inland cities, like Dallas.

Weather and climate also play a role in how people interact with their environment. The average weather in a place over a long period of time is called its **climate**. Alaska's climate is marked by long, cold, wet winters and short, mild summers. Hawaii's climate is warm year-round. Alaskans interact with their environment differently in December than Hawaiians do.

Another **component**, or part, of the environment is **resources**. These are materials that can be used to produce crops or other products. Forests are a resource because the trees can be used to build homes and furniture. Oil is a resource because it can be used as a source of energy.

Movement

Geographers also look at how people, products, ideas, and information move from one place to another. People have many reasons for moving. Some move because they find a better job.

Academic Vocabulary

component part

In 2005, Hurricane Katrina devastated the Gulf Coast and the city of New Orleans (left). Years later, many houses remain abandoned (right).

▶ **CRITICAL THINKING**
Physical Geography What hazards does the environment present?

THE SIX ESSENTIAL ELEMENTS

Element	Definition
The World in Spatial Terms	Geography studies the location and spatial relationships among people, places, and environments. Maps reveal the complex spatial interactions.
Places and Regions	The identities of individuals and peoples are rooted in places and regions. Distinctive combinations of human and physical characteristics define places and regions.
Physical Systems	Physical processes, like wind and ocean currents, plate tectonics, and the water cycle, shape Earth's surface and change ecosystems.
Human Systems	Human systems are things like language, religion, and ways of life. They also include how groups of people govern themselves and how they make and trade products and ideas.
Environment and Society	Geography studies how the environment of a place helps shape people's lives. Geography also looks at how people affect the environment in positive and negative ways.
The Uses of Geography	Understanding geography and knowing how to use its tools and technologies helps people make good decisions about the world and prepares people for rewarding careers.

Being aware of the six essential elements will help you sort out what you are learning about geography.

▶ **CRITICAL THINKING**

Identifying The study of volcanoes, ocean currents, and climate is part of which essential element?

Sometimes, people are forced to move because of war, famine, or religious or racial prejudice. Movement by large numbers of people can have important effects. People may face shortages of housing and other services. If new arrivals to an area cannot find jobs, poverty levels can rise.

In our interconnected world, a vast number of products move from place to place. Apples from Washington State move to supermarkets in Texas. Clothes produced in Thailand end up in American shopping malls. Oil from Saudi Arabia powers cars and trucks across the United States. All this movement relies on transportation systems that use ships, railroads, airplanes, and trucks.

Ideas can move at an even faster pace than people and products. Communications systems, such as telephone, television, radio, and the Internet, carry ideas and information all around the Earth. Remote villagers on the island of Borneo watch American television shows and learn about life in the United States. Political protestors in Egypt use text messaging and social networking sites to coordinate their activities. The geography of movement affects us all.

The Six Essential Elements

The five themes are one way of thinking about geography. Geographers also divide the study of geography into six essential elements. Elements are the topics that make up a subject. Calling them *essential* means they are necessary to understanding geography.

☑ **READING PROGRESS CHECK**

Determining Central Ideas How is the theme of location related to the theme of place?

Skill Building

GUIDING QUESTION *How will studying geography help you develop skills for everyday life?*

Have you ever used a Web browser to find a route from your home to another place? If so, your search took you to a Web site that provides maps. If you followed that map to your destination, you were using a geography skill.

Interpreting Visuals

Maps are one tool geographers use to picture the world. They use other visual images, as well. These other visuals include graphs, charts, diagrams, and photographs.

Graphs are visual displays of numerical information. They can help you compare information. Charts display information in columns and rows. Diagrams are drawings that use pictures to represent something in the world or an abstract idea. A diagram might show the steps in a process or the parts that make up something.

Critical Thinking

Geographers ask analytical questions. For example, geographers might want to know why earthquakes are more likely in some places than in others. That question looks at causes. They might ask, How does climate affect the ways people live? Such questions examine effects.

Geographers might ask how the characteristics of a place have changed over time. That is a question of analysis. Or they could ask why people in different nations use their resources differently. That question calls on them to compare and contrast.

Learning how to ask—and answer—questions like these will help sharpen your mind. In addition to understanding geography better, you will also be able to use these skills in other subjects.

Include this lesson's information in your Foldable®.

 READING PROGRESS CHECK

Analyzing How do geographers use visuals?

LESSON 1 REVIEW (CCSS)

Reviewing Vocabulary
1. Why is it not possible to state the *absolute location* of a river?

Answering the Guiding Questions
2. ***Determining Central Ideas*** Why do geographers study more than a place's location and dimensions?

3. ***Analyzing*** Does the environment of a place involve physical or human characteristics?

4. ***Identifying*** What are two examples of a human system?

5. ***Analyzing*** Why do geographers need to use visuals other than maps?

6. ***Informative/Explanatory Writing*** Describe the physical and human characteristics of your community.

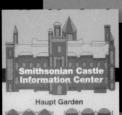

(l to r) Lana Sundman/Alamy; Antenna Audio, Inc./Getty Images; Kathy Collins/Getty Images; spacephotos.com/age fotostock; Chris Wallace/Alamy

networks

There's More Online!

☑ **SLIDE SHOW** History of Mapmaking

☑ **ANIMATION** Elements of a Globe

☑ **VIDEO**

Reading **HELP**DESK

Academic Vocabulary

- **sphere**
- **convert**
- **distort**

Content Vocabulary

- **hemisphere**
- **key**
- **scale bar**
- **compass rose**
- **map projection**
- **scale**
- **elevation**
- **relief**
- **thematic map**
- **technology**
- **remote sensing**

TAKING NOTES: *Key Ideas and Details*

Describing As you read the lesson, identify three parts of a map on a graphic organizer. Then, explain what each part shows.

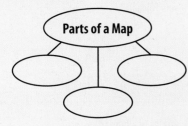

Parts of a Map

Lesson 2
The Geographer's Tools

ESSENTIAL QUESTION · *How does geography influence the way people live?*

IT MATTERS BECAUSE
The tools of geography help you understand the world.

Using Globes and Maps

GUIDING QUESTION *What is the difference between globes and maps?*

If you close your eyes, you can probably see your neighborhood in your mind. When you do, you are using a mental map. You are forming a picture of the buildings and other places and where each is located in relation to the others.

Making and using maps is a big part of geography. Of course, geographers make maps that have many parts. Their maps are more detailed than your mental map. Still, paper maps are essentially the same as your mental map. Both are a way to picture the world and show where things are located.

Globes

The most accurate way to show places on Earth is with a globe. Globes are the most accurate because globes, like Earth, are **spheres**; that is, they are shaped like a ball. As a result, globes represent the correct shapes of land and bodies of water. They show distances and directions between places more correctly than flat images of Earth.

The Equator and the Prime Meridian each divides Earth in half. Each half of Earth is called a **hemisphere**. The Equator divides Earth into sections called the Northern and Southern Hemispheres. The Prime Meridian, together with the International Date Line, splits Earth into the Eastern and Western Hemispheres.

Maps

Maps are not round like globes. Instead, maps are flat representations of the round Earth. They might be sketched on a piece of paper, printed in a book, or displayed on a computer screen. Wherever they appear, maps are always flat.

Maps **convert**, or change, a round space into a flat space. As a result, maps **distort** physical reality, or show it incorrectly. This is why maps are not as accurate as globes are, especially maps that show large areas or the whole world.

Despite this distortion problem, maps have several advantages over globes. Globes have to show the whole planet. Maps, though, can show only a part of it, such as one country, one city, or one mountain range. As a result, they can provide more detail than globes can. Think how large a globe would have to be to show the streets of a city. You could certainly never carry such a globe around with you. Maps make more sense if you want to study a small area. They can focus on just that area, and they are easy to store and carry.

Maps tend to show more kinds of information than globes. Globes generally show major physical and political features, such as landmasses, bodies of water, the countries of the world, and the largest cities. They cannot show much else without becoming too difficult to read or too large. However, some maps show these same features. But maps can also be specialized. One map might illustrate a large mountain range. Another might display the results of an election. Yet another could show the locations of all the schools in a city.

☑ **READING PROGRESS CHECK**

Analyzing What is the chief disadvantage of maps?

Academic Vocabulary

sphere a round shape like a ball

convert to change from one thing to another

distort to present in a manner that is misleading

A set of imaginary lines divides Earth into hemispheres.

▶ **CRITICAL THINKING**

The Geographer's World What line divides Earth into Eastern and Western Hemispheres?

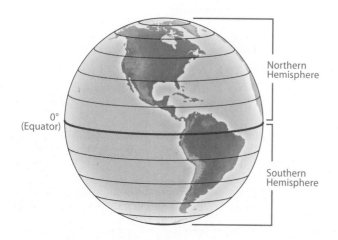

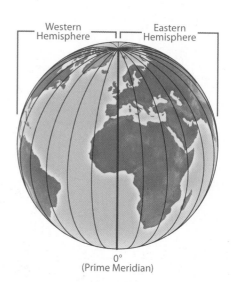

All About Maps

GUIDING QUESTION *How do maps work?*

You will find maps in many different places. You can see them in a subway station. Subway maps indicate the routes each train takes. In a textbook, a map might show new areas that were added to the United States at different times. At a company's Web site, a map can locate all its stores in a city. The map of a state park would tell visitors what activities they can enjoy in each area of the park. Each of these maps is different from the others, but they have some traits in common.

Parts of a Map

Maps have several important elements, or features. These features are the tools that convey information.

The map title tells what area the map will cover. It also identifies what kind of information the map presents about that area. The **key** unlocks the meaning of the map by explaining the symbols, colors, and lines. The **scale bar** is an important part of the map. It tells how a measured space on the map corresponds to actual distances on Earth. For example, by using the scale bar, you can determine how many miles in the real world each inch on the map represents. The **compass rose** shows direction. This map feature points out north, south, east, and west. Some maps include insets that show more detail for smaller areas, such as cities on a state map. Many maps show latitude and longitude lines to help you locate places.

Map Projections

To convert the round Earth to a flat map, geographers use **map projections**. A map projection distorts some aspects of Earth in order to represent other aspects as accurately as possible on a flat

Many different kinds of maps are available because maps are useful for showing a wide range of information.

▶ **CRITICAL THINKING**
Describing What is the difference between a large-scale and a small-scale map?

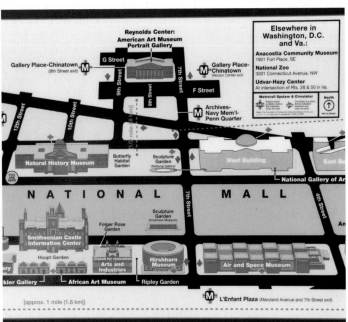

map. Some projections show the correct size of areas in relation to one another. Other map projections emphasize making the shapes of areas as accurate as possible.

Some projections break apart the world's oceans. By doing so, these maps show land areas more accurately. They clearly do not show the oceans accurately, though.

Mapmakers, known as cartographers, choose which projection to use based on the purpose of the map. Each projection distorts some parts of the globe more or less than other parts. Finally, mapmakers think about what part of Earth they are drawing and how large an area they want to cover.

Map Scale

Scale is another important feature of maps. As you learned, the scale bar relates distances on the map to actual distances on Earth. The scale bar is based on the scale at which the map is drawn. **Scale** is the relationship between distances on the map and on Earth.

Maps are either *large scale* or *small scale*. A large-scale map focuses on a smaller area. An inch on the map might correspond to 10 miles (16 km) on the ground. A small-scale map shows a relatively larger area. An inch on a small-scale map might be the same as 1,000 miles (1,609 km).

Each type of scale has benefits and drawbacks. Which scale to use depends on the map's purpose. Do you want to map your school and the streets and buildings near it? Then you need a large-scale map to show this small area in great detail. Do you want to show the entire United States? In that case, you need a small-scale map that shows the larger area but with less detail.

Types of Maps

The two types of maps are general purpose and thematic. The type depends on what kind of information is drawn on the map. General-purpose maps show a wide range of information about an area. They generally show either the human-made features of an area or its natural features, but not both.

Political maps are one common type of general-purpose map that shows human-made features. They show the boundaries of countries or the divisions within them, like the states of the United States. They also show the locations and names of cities.

Physical maps display natural features such as mountains and valleys, rivers, and lakes. They picture the location, size, and shape of these features. Many physical maps show **elevation**, or how much above or below sea level a feature is. Maps often use colors to present this information. A key on the map explains what height above or below sea level each color represents.

Physical maps usually show **relief**, or the difference between the elevation of one feature and the elevation of another feature near it.

Thinking Like a Geographer

Relief

Relief is the height of a landform compared to other nearby landforms. If a mountain 10,000 feet (3,048 m) high rises above a flat area at sea level, the relief of the mountain equals its elevation: 10,000 feet. If the 10,000-foot high mountain is in a highland region that is 4,000 feet (1,219 m) above sea level, its relief is *less than* its elevation— only 6,000 feet (1,829 m). The difference in height between it and the land around it is much less than its absolute height. *What would be the relief of a mountain 7,500 feet (2,286 m) high compared to its highest foothill, at 3,000 feet (914 m) high?*

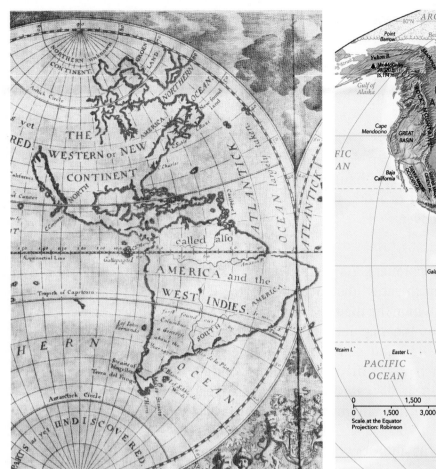

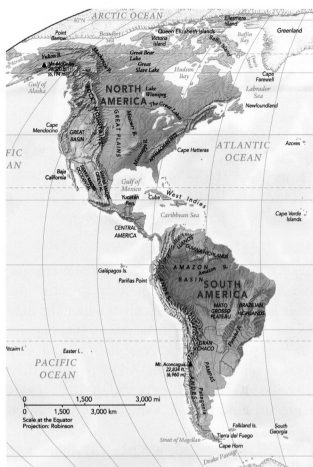

Cartography is the science of making maps. As knowledge of Earth grew, maps became increasingly accurate.

▶ CRITICAL THINKING

The Geographer's World Describe two ways in which the historical map differs from the present-day map.

Elevation is an absolute number, but relief is relative. It depends on other landforms that are nearby. The width of the colors on a physical map usually shows the relief. Colors that are narrow show steep places, and colors that are wide show gently sloping land.

Thematic maps show more specialized information. A thematic map might indicate the kinds of plants that grow in different areas. That kind of map is a vegetation map. Another could show where farming, ranching, or mining takes place. That kind of map is called a land-use map. Road maps show people how to travel from one place to another by car. Just about any physical or human feature can be displayed on a thematic map.

✔ **READING PROGRESS CHECK**

Describing How do the two main types of maps differ?

Geospatial Technologies

GUIDING QUESTION *How do geographers use geospatial technologies?*

Have you seen maps on cell phones and GPS devices in cars? These electronic maps are an example of geospatial technologies. **Technology** is any way that scientific discoveries are applied to practical use. Geospatial technologies can help us think spatially.

They provide practical information about the locations of physical and human features.

Global Positioning System

GPS devices work with a network called the Global Positioning System (GPS). This network was built by the U.S. government. Parts of it can be used only by the U.S. armed forces. Parts of it, though, can be used by ordinary people all over the world. The GPS has three elements.

The first element of this network is a set of more than 30 satellites that orbit Earth constantly. The U.S. government launched the satellites into space and maintains them. The satellites send out radio signals. Almost any spot on Earth can be reached by signals from at least four satellites at all times.

The second part of the network is the control system. Workers around the world track the satellites to make sure they are working properly and are on course. The workers reset the clocks on the satellites when needed.

The third part of the GPS system consists of GPS devices on Earth. These devices receive the signals sent by the satellites. By combining the signals from different satellites, a device calculates its location on Earth in terms of latitude and longitude. The more satellite signals the device receives at any time, the more accurately it can determine its location. Because satellites have accurate clocks, the GPS device also displays the correct time.

GPS is used in many ways. It is used to track the exact location and course of airplanes. That information helps ensure the safety of flights. Farmers use it to help them work their fields. Businesses use it to guide truck drivers. Cell phone companies use GPS to provide services. And of course, GPS in cars helps guide us to our destinations.

Geographic Information Systems

Another important geospatial technology is known as a geographic information system (GIS). These systems consist of computer hardware and software that gather, store, and analyze geographic information. The information is then shown on a computer screen. Sometimes it is displayed as maps. Sometimes the information is shown in other ways. Companies and governments around the world use this new tool.

A GIS is a powerful tool because it links data about all kinds of physical and human features with the locations of those features. Because computers can store and process so much data, the GIS can be accurate and detailed.

People select what features they want to study using the GIS. Then they can combine different features on the same map and analyze the patterns.

Think Again?

Geographers know the exact height of Mount Everest.

Not true. Mount Everest, on the border of Nepal and China, is the world's tallest mountain. It is said to be 29,028 feet (8,848 m) tall. Scientists disagree on this measurement, however, and the government of Nepal does not accept it. In 2011 Nepal launched a two-year effort to find the exact height using three GPS devices.

GPS satellites are used to measure, as well as determine, location on Earth. Some cell phones receive this satellite information, allowing people to locate places in a city (inset map).

▶ **CRITICAL THINKING**

Analyzing Why is it important for geographers to know exactly where places are located on Earth?

For instance, a farmer might want to compare the amount of moisture in the soil to the health of the plants. At the same time, he or she could add soil types around the farm to the comparison. The farmer could then use the results of the analysis to answer all kinds of questions. What plants should I plant in different locations? How much irrigation water should I use? How can I drive the tractor most efficiently?

Satellites and Sensors

Since the 1970s, satellites have gathered data about Earth's surface. They do so using remote sensing. **Remote sensing** simply means getting information from far away. Most early satellite sensors were used to gather information about the weather. Weather satellites helped save lives during disasters by providing warnings about approaching storms. Before satellites, tropical storms were often missed because they could not be tracked over open water.

Satellites gather information in different ways. They may use powerful cameras to take pictures of the land. They can also pick up other kinds of information, such as the amount of moisture in the soil, the amount of heat the soil holds, or the types of vegetation that

are present. In the early 2000s, scientists used satellites and GIS technology to help conserve the plants and animals that lived in the Amazon rain forest. Using the technology, scientists can compare data gathered from the ground to data taken from satellite pictures. Land use planners use this information to help local people make good decisions about how to use the land. These activities help prevent the rain forest from being destroyed.

Some satellites gather information regularly on every spot in the world. That way, scientists can compare the information from one year to another. They look for changes in the shape of the land or in its makeup, spot problems, and take steps to fix them.

Limits of Geospatial Technology

Geospatial technologies allow access to a wealth of information about the features and objects in the world and where those features and objects are located. This information can be helpful for identifying and navigating. By itself, however, the information does not answer questions about why features are located where they are. These questions lie at the heart of understanding our world. The answers are crucial for making decisions about this world in which we live.

It is important to go beyond the information provided by geospatial technologies. We must build understanding of peoples, places, and environments and the connections among them.

☑ **READING PROGRESS CHECK**

Analyzing How could remote sensing be used as part of a GIS?

Include this lesson's information in your Foldable®.

LESSON 2 REVIEW (CCSS)

Reviewing Vocabulary
1. Why do maps *distort* the way Earth's surface really looks?

Answering the Guiding Questions
2. *Identifying* Why are maps generally more useful than globes?

3. *Identifying* Suppose a map had the title "Russia: Land Use and Resources." What kinds of information will the map show and about what area?

4. *Identifying* Are road maps general-purpose maps or thematic maps? Explain your answer.

5. *Analyzing* How could GIS help businesses make better decisions?

6. *Informative/Explanatory Writing* Describe your neighborhood in a paragraph and then draw a map of it. Include the relevant features of a map and label them.

What Do You Think?

CCSS

Are Street-Mapping Technologies an Invasion of People's Privacy?

Suppose you are curious about a place you have never visited. Instead of going in person, you might be able to get a 360-degree view from your computer. Services like Google Street View and Bing Streetside display panoramic images of public roadways and buildings. The photos are taken by cameras attached to roving vehicles. They capture whatever is happening at the time, which means they sometimes capture random bystanders, too. Some people argue that street-level mapping programs violate the right to privacy. They point out that individuals' pictures can appear on the mapping Web sites without their knowledge or consent. Do tools like Street View intrude too much on people's privacy?

TEXT: Andrew Lavoie, "THE ONLINE ZOOM LENS: WHY INTERNET STREET-LEVEL MAPPING TECHNOLOGIES DEMAND RECONSIDERATION OF THE MODERN-DAY TORT NOTION OF PUBLIC PRIVACY," Georgia Law Review 00168300, Winter 2009, Vol. 43, Issue 2; PHOTO: (b) incamerastock/Alamy

Yes!

PRIMARY SOURCE

Users can view high-resolution imagery from Google Earth or Street View on their screens.

" Privacy encompasses the right to control information disseminated [spread] about oneself. . . . Personal behavior disclosures that occur as a result of Internet street-level mapping technologies almost certainly violate this personal right to choose which face to display to the world . . . A person may not mind that their friends and family know of their participation in certain socially stigmatizing [disapproved of] activities; an entirely new issue arises, however, should the entire public suddenly discover that the person is [doing something questionable.] . . . Internet street-level mapping scenes depart from being simply a record of what a member of the public could have seen on the street [because] on the Internet, images can be—and often are—saved onto users' hard drives for later dissemination. Thus, compromising [reputation-damaging] images, even if removed by Google after the fact, can be released to the public in an ever-widening wake [path]. "

—Andrew Lavoie, Georgia attorney

By Melissa Lafsky, June 5, 2007, in Freakonomics.com; PHOTO: (t) Silicon Valley Stock/ Alamy

A camera mounted on a car provides the technology for Google Street View.

No !

" At Google we take privacy very seriously. Street View only features imagery taken on public property and is not in real time. This imagery is no different from what any person can readily capture or see walking down the street. Imagery of this kind is available in a wide variety of formats for cities all around the world. While the Street View feature enables people to easily find, discover, and plan activities relevant to a location, we respect the fact that people may not want imagery they feel is objectionable featured on the service. We provide easily accessible tools for flagging inappropriate or sensitive imagery ... [U]sers can report objectionable images. Objectionable imagery includes nudity, certain types of locations (for example, domestic violence shelters) and clearly identifiable individuals ... We routinely review takedown requests and act quickly to remove objectionable imagery. "

—Stephen Chau, product manager for Google Maps

What Do You Think?   DBQ

1. **Identifying**   What types of images does Google consider inappropriate?

2. **Citing Text Evidence**   What points does Andrew Lavoie make to argue that Street View invades people's privacy?

Critical Thinking
3. **Identifying Point of View** Describe a situation when Street View could be useful and one when it could embarrass or endanger someone. Do you think the benefits outweigh the privacy concerns?

Directions: Write your answers on a separate piece of paper.

1 Use your **FOLDABLES** to Explore the Essential Question.

INFORMATIVE/EXPLANATORY WRITING Take a few moments to think about the physical geography and human geography where you live. Then write a short essay explaining how your area's geography impacts how your family lives, works, and plays.

2 **21st Century Skills**

INTEGRATING VISUAL INFORMATION What did your town or city look like 100 years ago? Work in groups of four or five and find historical photos or old maps of your community. Compare those to the way the community looks today. Write captions identifying the photos, and explain how the community has changed. Present your information as a slide show.

3 **Thinking Like a Geographer**

PLACES AND REGIONS Create a graphic organizer like the one shown here to describe the different types of maps geographers use.

4 **GEOGRAPHY ACTIVITY**

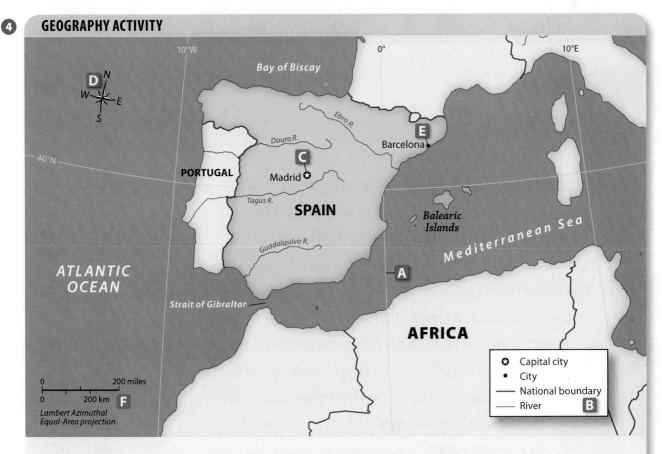

Reading Maps
Match the letters on the map with the numbered items listed below.

1. map key
2. compass rose
3. Prime Meridian
4. scale bar
5. city
6. capital

REVIEW THE GUIDING QUESTIONS

Directions: Choose the best answer for each question.

1 Lines of latitude measure distance on Earth
  A. in a north-to-south direction.
  B. in miles.
  C. in an east-to-west direction.
  D. in kilometers.

2 Which of these cities are located in the same region?
  F. Los Angeles and Dallas
  G. New York and Boston
  H. Chicago and Miami
  I. Atlanta and San Francisco

3 The average weather that occurs in a place over a long period of time is called its
  A. environment.
  B. relative location.
  C. climate.
  D. landscape.

4 What is the name of the line that divides Earth into sections called the Northern and Southern Hemispheres?
  F. the Prime Meridian
  G. the International Date Line
  H. the Tropic of Capricorn
  I. the Equator

5 Which part of a map shows the primary directions north, south, east, and west?
  A. key
  B. compass rose
  C. scale bar
  D. map projection

6 Which type of map would you use to locate your state's boundaries?
  F. political map
  G. land-use map
  H. road map
  I. physical map

DBQ ANALYZING DOCUMENTS

7 IDENTIFYING In the excerpt below, two geographers summarize the content of their geography book.

"*In this book we. . . investigate the world's great geographic realms [areas]. We will find that each of these realms possesses a special combination of cultural . . . and environmental properties [characteristics].*"

—from H.J. de Blij and Peter O. Muller, *Geography*

Which theme of geography is represented by these geographic realms?

A. human-environment interaction   C. movement

B. location   D. region

8 IDENTIFYING Which of these would be an example of the environmental characteristics of a realm?

F. plants and animals that live there

G. language and religion of the people who live there

H. nationalities of the people living there

I. type of government found in the realm

SHORT RESPONSE

"*Hurricanes, wildfires, floods, earthquakes, and other natural events affect the Nation's economy, . . . property, and lives. . . . The USGS gathers and disseminates [gives out] real-time hazard data to relief workers, conducts long-term monitoring and forecasting to help minimize the impacts of future events, and evaluates conditions in the aftermath of disasters.*"

—from United States Geological Service, *The National Map—Hazards and Disasters*

9 ANALYZING Why would the natural disasters named in this excerpt affect the nation's economy?

10 IDENTIFYING POINT OF VIEW How could relief workers benefit by having maps that show areas that were hit by a natural disaster?

EXTENDED RESPONSE

Write your answer on a separate piece of paper.

11 INFORMATIVE/EXPLANATORY WRITING Imagine that you were with Lewis and Clark when they explored and mapped the western United States. Choose any location along their route and write a journal entry describing the landforms, animals, plants, and people you would have seen.

Need Extra Help?

If You've Missed Question	1	2	3	4	5	6	7	8	9	10	11
Review Lesson	1	1	1	1	2	2	1	1	1	2	2

PHYSICAL GEOGRAPHY

ESSENTIAL QUESTION · *How does geography influence the way people live?*

networks

There's More Online about Earth's Physical Geography.

CHAPTER 2

Lesson 1
Planet Earth

Lesson 2
A Changing Earth

Lesson 3
Land and Water

Carsten Peter/National Geographic/Getty Images

A geologist prepares to enter the crater of Ambryn Island volcano.

The Story Matters...

Earth is part of a larger physical system called the solar system. Earth's position in the solar system makes life on our planet possible. The planet Earth has air, land, and water that make it suitable for plant, animal, and human life. Major natural forces inside and outside of our planet shape its surface. Some of these forces can move suddenly and violently, causing disasters that dramatically affect life on Earth.

FOLDABLES®
Study Organizer

Go to the Foldables® library in the back of your book to make a Foldable® that will help you take notes while reading this chapter.

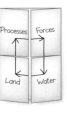

Continents sit on large bases called plates. As these plates move on top of Earth's fluid mantle, the continents move. Sometimes, the plates collide with each other or slide under each other, creating earthquakes or volcanoes.

Step Into the Place

MAP FOCUS Use the map to answer the following questions.

1 PHYSICAL GEOGRAPHY Where are most of the world's volcanoes located?

2 PHYSICAL GEOGRAPHY How many plates are underneath Australia?

3 CRITICAL THINKING **Integrating Visual Information** Why do you think the edge of the Pacific Ocean is often called the Ring of Fire?

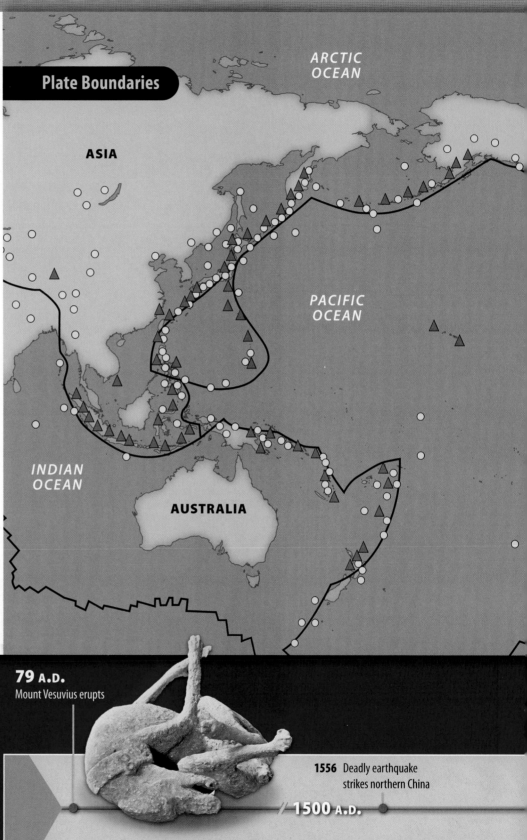

Plate Boundaries

ARCTIC OCEAN

ASIA

PACIFIC OCEAN

INDIAN OCEAN

AUSTRALIA

Step Into the Time

DRAWING EVIDENCE Choose one event from the time line and explain how the natural forces that shape the physical geography of a particular place can have a worldwide impact.

79 A.D. Mount Vesuvius erupts

1556 Deadly earthquake strikes northern China

/ **1500 A.D.**

ARCTIC
OCEAN

EUROPE

NORTH
AMERICA

ATLANTIC
OCEAN

AFRICA

INDIAN
OCEAN

PACIFIC
OCEAN

SOUTH
AMERICA

○ Earthquake
▲ Volcano
— Plate boundary

0		2,000 miles
0	2,000 kilometers	

Miller projection

2005
Hurricane Katrina
strikes southeastern
United States

2011
Earthquake, tsunami near
Japan triggers nuclear
accident

1931 Floods in China leave
80 million homeless

1700 A.D.　　　**1800** A.D.　　　**1900** A.D.　　　**2000**

1906 Earthquake, fire devastate
San Francisco

2010 Haiti earthquake kills more
than 220,000 people

netw🟊rks

There's More Online!

☑ **CHART/GRAPH** Climate Zones

☑ **ANIMATION** Earth's Rotation

☑ **IMAGE** Rain Shadow

☑ **SLIDE SHOW** Effects of Climate Change

☑ **VIDEO**

Lesson 1
Planet Earth

ESSENTIAL QUESTION • *How does geography influence the way people live?*

(l to r) NASA/NOAA/GSFC/Suomi NPP/VIIRS/Norman Kuring; Martin Pudd/Stone/Getty Images; Ariadne Van Zandbergen/Lonely Planet Images/Getty Images; Scott Warren/Aurora/Getty Images

Reading **HELP**DESK (CCSS)

Academic Vocabulary

- **accurate**

Content Vocabulary

- **orbit**
- **axis**
- **revolution**
- **atmosphere**
- **solstice**
- **equinox**
- **climate**
- **precipitation**
- **rain shadow**

TAKING NOTES: *Key Ideas and Details*

Summarize As you read, complete a graphic organizer about Earth's physical system.

Element	Description
Hydrosphere	
Lithosphere	
Atmosphere	
Biosphere	

IT MATTERS BECAUSE
We learn about the processes that change Earth.

Looking at Earth

GUIDING QUESTION *What is the structure of Earth?*

Earth is one planet among a group of planets that revolve around the sun. The sun is just one of hundreds of millions of stars in our galaxy. Because the sun is so large, its gravity causes the planets to constantly **orbit**, or move around, it. The sun is the center of the solar system in which we live. Earth is a member of the solar system—planets and the other bodies that revolve around our sun.

Earth and the Sun

Life on Earth could not exist without heat and light from the sun. Earth's orbit holds it close enough to the sun—about 93 million miles (150 million km)—to receive a constant supply of light and heat energy. The sun, in fact, is the source of all energy on Earth. Every plant and animal on the planet needs the sun's energy to survive. Without the sun, Earth would be a cold, dark, lifeless rock floating in space.

As Earth orbits the sun, it rotates, or spins, on its axis. The **axis** is an imaginary line that runs through Earth's center from the North Pole to the South Pole. Earth completes one rotation every 24 hours. As Earth rotates, different areas are in sunlight and in darkness. The part facing toward the sun experiences daylight, while the part facing away has night. Earth makes one **revolution**, or complete trip around the sun, in 365¼ days. This is what we define as one year. Every four

years, the extra fourths of a day are combined and added to the calendar as February 29th. A year that contains one of these extra days is called a leap year.

Inside Earth

Thousands of miles beneath your feet, Earth's heat has turned metal into liquid. You do not feel these forces, but what lies inside affects what lies on top. Mountains, deserts, and other landscapes were formed over time by forces acting below Earth's surface—and those forces are still changing the landscape.

If you cut an onion in half, you will see that it is made up of many layers. Earth is also made up of layers. An onion's layers are all made of onion, but Earth's layers are made up of many different materials.

Layers of Earth

The inside of Earth is made up of three layers: the core, the mantle, and the crust. The center of Earth—the core—is divided into a solid inner core and an outer core of melted, liquid metal. Surrounding the outer core is a thick layer of hot, dense rock called the mantle. Scientists calculate that the mantle is about 1,800 miles (2,897 km) thick. The mantel also has two parts. When volcanoes erupt, the glowing-hot lava that flows from the mouth of the volcano is magma from Earth's outer mantle. The inner mantle is solid, like the inner core. Magma is melted rock. The outer layer is the crust, a rocky shell forming the surface of Earth. The crust is thin, ranging from about 2 miles (3.2 km) thick under oceans to about 75 miles (121 km) thick under mountains.

DIAGRAM SKILLS >

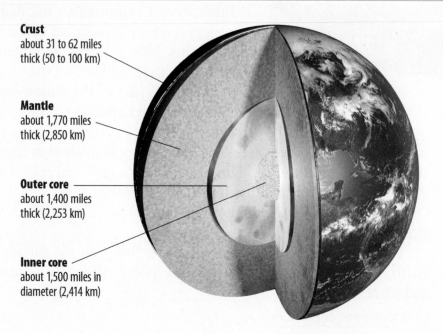

Crust
about 31 to 62 miles
thick (50 to 100 km)

Mantle
about 1,770 miles
thick (2,850 km)

Outer core
about 1,400 miles
thick (2,253 km)

Inner core
about 1,500 miles in
diameter (2,414 km)

EARTH'S LAYERS
Earth is comprised of several layers.

Identifying What is the innermost layer of Earth called?

Academic Vocabulary

accurate correct

The deepest hole ever drilled into Earth is about 8 miles (13 km) deep. That is still within Earth's crust. The farthest any human has traveled down into Earth's crust is about 2.5 miles (4 km) deep. Still, scientists have developed an **accurate** picture of the layers in Earth's structure. One important way that scientists do this is to study vibrations from deep within Earth. The vibrations are caused by earthquakes and explosions underground. From their observations, scientists have learned what materials are inside Earth and estimated the thickness and temperature of Earth's layers.

Earth's Physical Systems

Powerful processes operate below Earth's surface. Processes are also at work in the physical systems on the surface of Earth. Earth's physical systems consist of four major subsystems: the hydrosphere, the lithosphere, the atmosphere, and the biosphere.

About 71 percent of Earth's surface is water. The hydrosphere is the subsystem that consists of Earth's water. Water is found in oceans, seas, lakes, ponds, rivers, groundwater, and ice. Only 3 percent of the water on Earth is freshwater.

Only about 29 percent of Earth's surface is land. Land makes up the part of Earth called the lithosphere. Landforms are the shapes that occur on Earth's surface. Landforms include plains, hills, plateaus, mountains, and ocean basins, the land beneath the ocean.

The air we breathe is part of the **atmosphere**, the thin layer of gases that envelop Earth. The atmosphere is made up of about 78 percent nitrogen, 21 percent oxygen, and small amounts of other gases. The atmosphere is thickest at Earth's surface and gets thinner higher up. Ninety-eight percent of the atmosphere is found within 16 miles (26 km) of Earth's surface. Outer space begins at 100 miles (161 km) above Earth, where the atmosphere ends.

The biosphere is made up of all that is living on the surface of Earth, close to the surface, or in the atmosphere. All people, animals, and plants live in the biosphere.

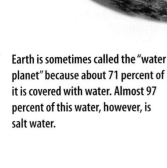

Earth is sometimes called the "water planet" because about 71 percent of it is covered with water. Almost 97 percent of this water, however, is salt water.

☑ **READING PROGRESS CHECK**

Identifying Which of Earth's layers is between the crust and the outer core?

NASA/NOAA/GSFC/Suomi NPP/VIIRS/Norman Kuring

Seasons

GUIDING QUESTION *How does Earth's orbit around the sun cause the seasons?*

Fruits such as strawberries, grapes, and bananas cannot grow in cold, icy weather. Yet grocery stores across America sell these ripe, colorful fruits all year, even in the middle of winter. Where in the world is it warm enough to grow fruit in January? To find the answer, we start with the tilt of Earth.

Earth is tilted 23.5 degrees on its axis. If you look at a globe that is attached to a stand, you will see what the tilt looks like. Because of the tilt, not all places on Earth receive the same amount of direct sunlight at the same time.

As Earth orbits the sun, it stays in its tilted position. This means that one-half of the planet is always tilted toward the sun, while the other half is tilted away. As a result, Earth's Northern and Southern Hemispheres experience seasons at different times.

On about June 21, the North Pole is tilted toward the sun. The Northern Hemisphere is receiving the direct rays of the sun. The sun appears directly overhead at the line of latitude called the Tropic of Cancer. This day is the summer **solstice**, or beginning of summer, in the Northern Hemisphere. It is the day of the year that has the most hours of sunlight during Earth's 24-hour rotation.

Six months later—about December 22—the North Pole is tilted away from the sun. The sun's direct rays strike the line of latitude known as the Tropic of Capricorn. This is the winter solstice—when winter occurs in the Northern Hemisphere and summer begins in the Southern Hemisphere. The days are short in the Northern Hemisphere but long in the Southern Hemisphere.

DIAGRAM SKILLS ›

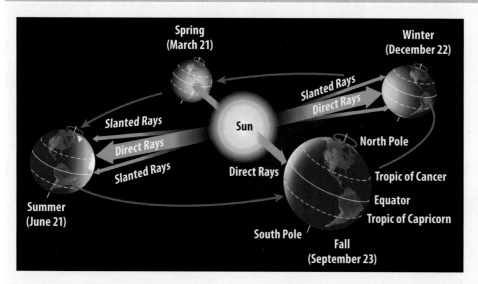

SEASONS
The tilt of Earth as it revolves around the sun causes the seasons to change.

▶ **CRITICAL THINKING**
Analyzing Why are the seasons reversed in the Northern and Southern Hemispheres?

Midway between the two solstices, about September 23 and March 21, the sun's rays are directly overhead at the Equator. These are **equinoxes**, when day and night in both hemispheres are of equal length—12 hours of daylight and 12 hours of nighttime everywhere on Earth.

☑ **READING PROGRESS CHECK**

Identifying When it is winter in the Southern Hemisphere, what season is it in the Northern Hemisphere?

Factors That Influence Climate

GUIDING QUESTION *How do elevation, wind and ocean currents, weather, and landforms influence climate?*

The sun's direct rays fall year-round at low latitudes near the Equator. This area, known as the Tropics, lies mainly between the Tropic of Cancer and the Tropic of Capricorn. The Tropics circle the globe like a belt. If you lived in the Tropics, you would experience hot, sunny weather most of the year because of the direct sunlight. Outside the Tropics, the sun is never directly overhead. Even when these high-latitude areas are tilted toward the sun, the sun's rays still hit Earth indirectly at a slant. This means that no sunlight at all shines on the high-latitude regions around the North and South Poles for as much as six months each year. Thus, climate in these regions is always cool or cold.

Water swirls down the street of a small town in India after heavy rains.

▶ **CRITICAL THINKING**

Analyzing How does elevation influence climate?

Martin Pudd/Stone/Getty Images

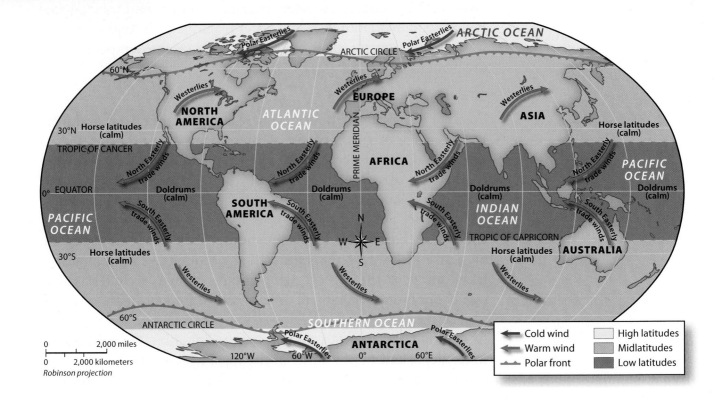

Elevation and Climate

At all latitudes, elevation influences climate. This is because Earth's atmosphere thins as altitude increases. Thinner air retains less heat. As elevation increases, temperatures decrease by about 3.5°F (1.9°C) for every 1,000 feet (305 m). For example, if the temperature averages 70°F (21.1°C) at sea level, the average temperature at 5,000 feet (1,524 m) is only 53°F (11.7°C). A high elevation will be colder than lower elevations at the same latitude.

Wind and Ocean Currents

In addition to latitude and elevation, the movement of air and water helps create Earth's climates. Moving air and water help circulate the sun's heat around the globe.

Movements of air are called winds. Winds are the result of changes in air pressure caused by uneven heating of Earth's surface. Winds follow prevailing, or typical, patterns. Warmer, low-pressure air rises higher into the atmosphere. Winds are created as air is drawn across the surface of Earth toward the low-pressure areas. The Equator is constantly warmed by the sun, so warm air masses tend to form near the Equator. This warm, low-pressure air rises, and then cooler, high-pressure air rushes in under the warm air, causing wind. This helps balance Earth's temperature.

MAP SKILLS

1 PHYSICAL GEOGRAPHY In what general direction does the wind blow over Africa?

2 PHYSICAL GEOGRAPHY What air currents flow over the midlatitudes?

Just as winds move in patterns, cold and warm streams of water, known as currents, circulate through the oceans. Warm water moves away from the Equator, transferring heat energy from the equatorial region to higher latitudes. Cold water from the polar regions moves toward the Equator, also helping to balance the temperature of the planet.

Weather and Climate

Weather is the state of the atmosphere at a given time, such as during a week, a day, or an afternoon. Weather refers to conditions such as hot or cold, wet or dry, calm or stormy, or cloudy or clear. Weather is what you can observe any time by going outside or looking out a window. **Climate** is the average weather conditions in a region or an area over a longer period. One useful measure for comparing climates is the average daily temperature. This is the average of the highest and lowest temperatures that occur in a 24-hour period. In addition to the average temperature, climate includes typical wind conditions and rainfall or snowfall that occur in an area year after year.

Rainfall and snowfall are types of precipitation. **Precipitation** is water deposited on the earth in the form of rain, snow, hail, sleet, or mist. Measuring the amount of precipitation in an area for one day provides data about the area's weather. Measuring the amount of precipitation for one full year provides data about the area's climate.

Landforms

It might seem strange to think that landforms such as mountains can affect weather and climate, but landforms and landmasses change the strength, speed, and direction of wind and ocean

RAIN SHADOW
A rain shadow affects the amount of rain a region receives.

1. *Determining Word Meanings*
   What is a rain shadow?

▶ **CRITICAL THINKING**
2. *Describing* How do landforms cause the formation of a rain shadow?

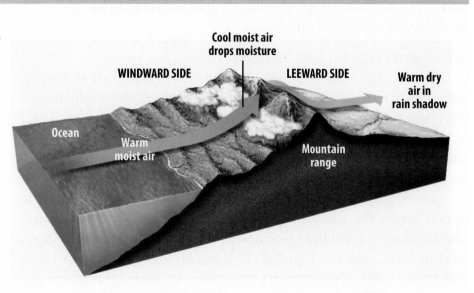

Cool moist air drops moisture

WINDWARD SIDE   LEEWARD SIDE   Warm dry air in rain shadow

Ocean   Warm moist air   Mountain range

currents. Wind and ocean currents carry heat and precipitation, which shape weather and climate. The sun warms the land and the surface of the world's oceans at different rates, causing differences in air pressure. As winds blow inland from the oceans, they carry moist air with them. As the land rises in elevation, the atmosphere cools. When masses of moist air approach mountains, the air rises and cools, causing rain to fall on the side of the mountain facing the ocean. The other side of the mountain receives little rain because of the rain shadow effect. A **rain shadow** is a region of reduced rainfall on one side of a high mountain; the rain shadow occurs on the side of the mountain facing away from the ocean.

☑ **READING PROGRESS CHECK**

Determining Word Meanings Do the terms *weather* and *climate* mean the same thing? Explain.

The climate in a zone affects how people live and work.
▶ **CRITICAL THINKING**
Identifying What are two useful measures for comparing climates in different areas?

Different Types of Climate Zones

GUIDING QUESTION *What are the characteristics of Earth's climate zones?*

Why do Florida and California have so many amusement parks? These places have cold or stormy weather at times, but their climates are generally warm, sunny, and mild, so parks can stay open all year.

The Zones

In the year 1900, German scientist Wladimir Köppen invented a system that divides Earth into five basic climate zones. Climate zones are regions of Earth classified by temperature, precipitation, and distance from the Equator. Köppen used names and capital letters to label the climate zones as follows: Tropical (A); Desert (B); Humid Temperate (C); Cold Temperate (D); and Polar (E). Years later, a sixth climate zone was added: High Mountain (F).

Each climate zone also can be divided into smaller subzones, but the areas within each zone have many similarities. Tropical areas are hot and rainy, oftentimes with dense forests. Desert areas are always dry, but they can be cold or hot, depending on their latitude. Humid temperate areas experience all types of weather with changing seasons. Cold temperate climates have a short summer season but are generally cold and windy. Polar climates are very cold, with ice and snow covering the ground most of the year. High mountain climates are found only at the tops of high mountain ranges such as the Rockies, the Alps, and the Himalaya. High mountain climates have variable conditions because the atmosphere cools with increasing elevation. Some of the highest mountaintops are cold and windy and stay white with snow all year.

Different types of plants grow best in different climates, so each climate zone has its own unique types of vegetation and animal life. These unique combinations form ecosystems of plants and animals that are adapted to environments within the climate zone. A biome is a type of large ecosystem with similar life-forms and climates. Earth's biomes include rain forest, desert, grassland, and tundra. All life is adapted to survive in its native climate zone and biome.

A bull moose stands alert on a tundra field in Alaska. Animals that live in that environment have unique adaptations that help them survive.

©Kennan Ward/Corbis

Changes to Climate

Many scientists say climates are changing around the world. If this is true, the world could experience new weather patterns. These changes might mean more extreme weather in some places and milder weather in others. Human activities can affect weather and climate. For example, people have cut down millions of square miles of rain forests in Central and South America. As a result, fewer trees are available to release moisture into the air. The result is a drier climate in the region.

Metal, asphalt, and concrete surfaces in cities absorb a huge amount of heat from the sun. An enormous mass of warmer air builds up in and around the city, affecting local weather.

In recent years, scientists have become aware of a problem called global warming. Global warming is an increase in the average temperature of Earth's atmosphere. Industries created by humans dump polluting chemicals into the atmosphere. Many scientists say that a buildup of this pollution is contributing to the increasing temperature of Earth's atmosphere.

If the temperature of the atmosphere continues to rise, all of Earth's climates could be affected. Changes in climate may result in altering many natural ecosystems. Another consequence is that the survival of some plant and animal species will be threatened. It is also likely to be expensive and difficult for humans to adapt to these changes.

☑ **READING PROGRESS CHECK**

Identifying In which one of the six major climate zones do you live?

| cold | cool | warm | hot |

A thermal image shows heat escaping from the roofs of buildings.
▶ **CRITICAL THINKING**
Analyzing Why do sites in large cities tend to get very warm?

 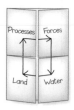
FOLDABLES Study Organizer
Include this lesson's information in your Foldable®.

LESSON 1 REVIEW CCSS

Reviewing Vocabulary

1. Write a sentence comparing the two terms below.
   a. revolution   **b.** solstice

Answering the Guiding Questions

2. *Identifying* What are the parts of our solar system?

3. *Describing* How does Earth's orbit around the sun cause the seasons?

4. *Determining Central Ideas* What factors determine the climate of an area?

5. *Describing* Identify Earth's six major climate zones. Describe the characteristics of the zones.

6. *Narrative Writing* Imagine that you live on a planet that is not tilted on its axis. What might this planet's seasons be like? How might life on this planet differ from life on Earth? Write a fictional narrative that addresses these questions.

Reading **HELP**DESK

Academic Vocabulary

- **intense**

Content Vocabulary

- **continent**
- **tectonic plates**
- **fault**
- **earthquake**
- **Ring of Fire**
- **tsunami**
- **weathering**
- **erosion**
- **glacier**

TAKING NOTES: *Key Ideas and Details*

Identify As you read, use a graphic organizer like this one to describe the external forces that have shaped Earth.

External Forces

Lesson 2
A Changing Earth

ESSENTIAL QUESTION • *How does geography influence the way people live?*

IT MATTERS BECAUSE
Internal and external forces change Earth, the setting for human life.

Forces of Change

GUIDING QUESTION *How was the surface of Earth formed?*

Since Earth was formed, the surface of the planet has been in constant motion. Landmasses have shifted and moved over time. Landforms have been created and destroyed. The way Earth looks from space has changed many times because of the movement of continents.

Earth's Surface

A **continent** is a large, continuous mass of land. Continents are part of Earth's crust. Earth has seven continents: Asia, Africa, North America, South America, Europe, Antarctica, and Australia. The region around the North Pole is not a continent because it is made of a huge mass of dense ice, not land. Greenland might seem as big as a continent, but it is classified as the world's largest island. Each of the seven continents has features that make it unique. Some of the most interesting features on the continents are landforms.

Even though you usually cannot feel it, the land beneath you is moving. This is because Earth's crust is not a solid sheet of rock. Earth's surface is like many massive puzzle pieces pushed close together and floating on a sea of boiling rock. The movement of these pieces is one of the major forces that create Earth's land features. Old mountains are worn down, while new mountains grow taller. Even the continents move.

Plate Movements

Earth's rigid crust is made up of 16 enormous pieces called **tectonic plates**. These plates vary in size and shape. They also vary in the amount they move over the more flexible layer of the mantle below them. Heat from deep within the planet causes plates to move. This movement happens so slowly that humans do not feel it. But some of Earth's plates move as much as a few inches each year. This might not seem like much, but over millions of years, it causes the plates to move thousands of miles.

Movement of surface plates changes Earth's surface features very slowly. It takes millions of years for plates to move enough to create landforms. Some land features form when plates are crushed together. At times, forces within Earth push the edge of one plate up over the edge of a plate beside it. This dramatic movement can create mountains, volcanoes, and deep trenches in the ocean floor.

At other times, plates are crushed together in a way that causes the edges of both plates to crumble and break. This event can form jagged mountain ranges. If plates on the ocean floor move apart, the space between them widens into a giant crack in Earth's crust. Magma rises through the crack and forms new crust as it hardens and cools. If enough cooled magma builds up that it reaches the surface of the ocean, an island will begin to form.

Powerful forces within Earth cause the Old Faithful geyser in Yellowstone National Park (left) to erupt. Those forces also cause lava to flow from Mount Etna volcano in Italy (right).

▶ **CRITICAL THINKING**

Describing What causes plates to move?

Sudden Changes

Change to Earth's surface also can happen quickly. Events such as earthquakes and volcanoes can destroy entire areas within minutes. Earthquakes and volcanoes are caused by plate movement. When two plates grind against each other, faults form. A **fault** results when the rocks on one side or both sides of a crack in Earth's crust have been moved by forces within Earth. **Earthquakes** are caused by plate movement along fault lines. Earthquakes also can be caused by the force of erupting volcanoes.

Various plates lie at the bottom of the Pacific Ocean. These include the huge Pacific Plate along with several smaller plates. Over time, the edges of these plates were forced under the edges of the plates surrounding the Pacific Ocean. This plate movement created a long, narrow band of volcanoes called the **Ring of Fire**. The Ring of Fire stretches for more than 24,000 miles (38,624 km) around the Pacific Ocean.

The **intense** vibrations caused by earthquakes and erupting volcanoes can transfer energy to Earth's surface. When this energy travels through ocean waters, it can cause enormous waves to form on the water's surface. A **tsunami** is a giant ocean wave caused by volcanic eruptions or movement of the earth under the ocean floor. Tsunamis have caused terrible flooding and damage to coastal areas. The forces of these mighty waves can level entire coastlines.

People attempt to cross a collapsed bridge after a powerful earthquake in the Philippines.
▶ **CRITICAL THINKING**
Describing What causes earthquakes?

Academic Vocabulary

intense great or strong

✓ **READING PROGRESS CHECK**

Determining Central Ideas Earth's surface plates are moving. Why don't we feel the ground moving under us?

Other Forces at Work

GUIDING QUESTION *How can wind, water, and human actions change Earth's surface?*

What happens when the tide comes in and washes over a sand castle on the beach? The water breaks down the sand castle. Similar changes take place on a larger scale across Earth's lithosphere. These changes happen much slower—over hundreds, thousands, or even millions of years.

Weathering

Some landforms are created when materials such as rocks and soil build up on Earth's surface. Other landforms take shape as rocks and soil break down and wear away over time. **Weathering** is a process by which Earth's surface is worn away by forces such as wind, rain, chemicals, and the movement of ice and flowing water. Even plants can cause weathering. Plant roots and small seeds can grow into tiny cracks in rock, gradually splitting the rock apart as the roots expand.

You may have seen the effects of weathering on an old building or statue. The edges become chipped and worn, and features such as raised lettering are smoothed down. Landforms such as mountains are affected by weathering, too. The Appalachian Mountains in the eastern United States have become rounded and crumbled after millions of years of weathering by natural forces.

Erosion

Erosion is a process that works with weathering to change surface features of Earth. **Erosion** is a process by which weathered bits of rock are moved elsewhere by water, wind, or ice. Rain and moving water can erode even the hardest stone over time. When material is broken down by weathering, it can easily be carried away by the action of erosion. For example, the Grand Canyon was formed by weathering and erosion caused by flowing water and blowing winds. Water flowed over the region for millions of years, weakening the surface of the rock. The moving water carried away tiny bits of rock. Over time, weathering and erosion carved a deep canyon into the rock. Erosion by wind and chemicals caused the Grand Canyon to widen until it became the amazing landform we see today.

Weathering and erosion cause different materials to break down at different speeds. Soft, porous rocks, such as sandstone and limestone, wear away faster than dense rocks like granite. The spectacular rock formations in Utah's Bryce Canyon were formed as different types of minerals within the rocks were worn away by erosion, some more quickly than others. The result is landforms with jagged, rough surfaces and unusual shapes.

Erosion created this rock formation, named Thor's Hammer, in Bryce Canyon National Park, Utah.
▶ **CRITICAL THINKING**
Determining Cause and Effect
How does weathering contribute to erosion?

Stockbyte/Getty Images

Pictured is one of the tunnel-boring machines used to dig the Channel Tunnel, or Chunnel. The Chunnel is an underseas rail tunnel connecting the United Kingdom and France.

▶ **CRITICAL THINKING**

Describing What are the dangers when humans change the natural course of land and waterways?

Buildup and Movement

The buildup of materials creates landforms such as beaches, islands, and plains. Ocean waves pound coastal rocks into smaller and smaller pieces until they are tiny grains of sand. Over time, waves and ocean currents deposit sand along coastlines, forming sandy beaches. Sand and other materials carried by ocean currents build up on mounds of volcanic rock in the ocean, forming islands. Rivers deposit soil where they empty into larger bodies of water, creating coastal plains and wetland ecosystems.

Entire valleys and plains can be formed by the incredible force and weight of large masses of ice and snow. These masses are often classified by size as glaciers, polar ice caps, or ice sheets. A **glacier**, the smallest of the ice masses, moves slowly over time, sometimes spreading outward on a land surface. Although glaciers are usually thought of as existing during the Ice Age, glaciers can still be found on Earth today.

Ice caps are high-altitude ice masses. Ice sheets, extending more than 20,000 square miles (51,800 sq. km), are the largest ice masses. Ice sheets cover most of Greenland and Antarctica.

Human Actions

Natural forces are awesome in their power to change the surface of Earth. Human actions, however, have also changed Earth in many ways. Activities such as coal mining have leveled entire mountains. Humans use explosives such as dynamite to blast tunnels through mountain ranges when building highways and railroads. Canals dug by humans change the natural course of waterways. Humans have cut down so many millions of acres of forests that deadly landslides and terrible erosion occur on the deforested lands.

Pollution caused by humans can change Earth, as well. When people burn gasoline and other fossil fuels, toxic chemicals are released into the air. These chemicals settle onto the surfaces of mountains, buildings, oceans, rivers, grasslands, and forests. The chemicals poison waterways, kill plants and animals, and cause erosion. The buildings in many cities show signs of being worn down by chemical erosion.

Studies show that humans have changed the environment of Earth faster and more broadly in the last 50 years than at any time in history. One major reason is demand for food and natural resources is greater than ever, and continues to grow.

Changes to Earth's surface caused by natural weathering and erosion happen slowly. They create different kinds of landforms that make our planet unique. Erosion and other changes caused by humans, however, can damage Earth's surface quickly. Their effects threaten our safety and survival. We need to protect our environment to ensure that our quality of life improves for future generations.

✔ **READING PROGRESS CHECK**

Describing Categorize each of the following events as a slow change or a sudden change: earthquake, glacier, tsunami, volcano eruption, wind erosion, water erosion, and plate movement.

 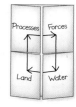

Include this lesson's information in your Foldable®.

LESSON 2 REVIEW

Reviewing Vocabulary

1. What causes the *erosion* of rocks on Earth's surface?

Answering the Guiding Questions

2. ***Identifying*** Identify Earth's seven continents.

3. ***Describing*** What evidence tells scientists that Earth's core is solid?

4. ***Distinguishing Fact From Opinion*** Is the following statement a fact or an opinion? Explain.
Earthquakes and tsunamis are caused by natural forces, so they cannot be prevented.

5. ***Analyzing*** Explain how you think the Ring of Fire got its name.

6. ***Identifying*** Which of the following is the best evidence that plants can cause weathering and erosion to rocks?
  a. A rock has vines growing tightly around it.
  b. A rock has a crack through its center and plant roots growing through the crack.
  c. A rock is covered in thick, green moss.

7. ***Citing Text Evidence*** Give one example of how human actions can change Earth.

8. ***Argument Writing*** Write an essay that argues whether a country should be allowed to develop Earth's resources without interference from other countries. Think about why resources are developed and the effects of using them. Be sure to address arguments in your essay that oppose your point of view.

Reading **HELP**DESK · CCSS

Academic Vocabulary

• transform

Content Vocabulary

• plateau
• plain
• isthmus
• continental shelf
• trench
• desalinization
• groundwater
• delta
• water cycle
• evaporation
• condensation
• acid rain

TAKING NOTES: *Key Ideas and Details*

Describing Using a chart like this one, describe two kinds of landforms and two bodies of water.

1.	1.
2.	2.

Lesson 3
Land and Water

ESSENTIAL QUESTION · *How does geography influence the way people live?*

IT MATTERS BECAUSE
Earth's landforms and bodies of water influence our ways of life.

Land Takes Different Forms

GUIDING QUESTION *What kinds of landforms cover Earth's surface?*

What is the land like where you live? Are unique landforms located in your area? Have you ever wondered how different kinds of landforms were created? The surface of Earth is covered with landforms and bodies of water. Our planet is filled with variety on land and under water.

Surface Features on Land

Earth has many different landforms. When scientists study landforms, they find it useful to group them by characteristics. One characteristic that is often used is elevation.

Elevation describes how far above sea level a landform or a location is. Low-lying areas, such as ocean coasts and deep valleys, may be just a few feet above sea level. Mountains and highland areas can be thousands of feet above sea level. Even flat areas of land can have high elevations, especially when they are located far inland from ocean shores.

Plateaus and plains are flat, but a **plateau** rises above the surrounding land. A steep cliff often forms at least one side of a plateau. **Plains** can be flat or have a gentle roll and can be found along coastlines or far inland. Some plains are home to grazing animals, such as horses and antelope. Farmers and ranchers use plains to raise crops and livestock. A valley is a lowland area between two higher sides. Some valleys are small, level places surrounded by hills or mountains. Other

Evape
from

valleys are huge expanses of land with highlands or mountain ranges on either side. Because they are often supplied with water runoff and topsoil from the higher lands around them, many valleys have rich soil and are used for farming and grazing livestock.

Another way to classify some landforms is to describe them in relation to bodies of water. Some types of landforms are surrounded by water. Continents are the largest of all landmasses. Most continents are bordered by land and water. Only Australia and Antarctica are completely surrounded by water. Islands are landmasses that are surrounded by water, but they are much smaller than continents.

A peninsula is a long, narrow area that extends into a river, a lake, or an ocean. Peninsulas at one end are connected to a larger landmass. An **isthmus** is a narrow strip of land connecting two larger land areas. One well-known isthmus is the Central American country of Panama. Panama connects two massive continents: North America and South America. Because it is the narrowest place in the Americas, the Isthmus of Panama is the location of the Panama Canal, a human-made canal connecting the Atlantic and Pacific Oceans.

The Ocean Floor

The ocean floor is also covered by different landforms. The ocean floor, like the ground we walk on, is part of Earth's crust. In many ways, the ocean floor and land are similar. If you could see an ocean without its water, you would see a huge expanse of plains, valleys, mountains, hills, and plateaus. Some of the landforms were shaped by the same forces that created the features we see on land.

This map of Central America and the Caribbean Sea reveals ridges that are underwater mountain chains.

A DROP IN THE OCEAN
SALT WATER VS FRESHWATER

Earth's surface is about 70 percent water. That seems like a lot of water, but how much can we humans actually use? Hint: probably less than you think.

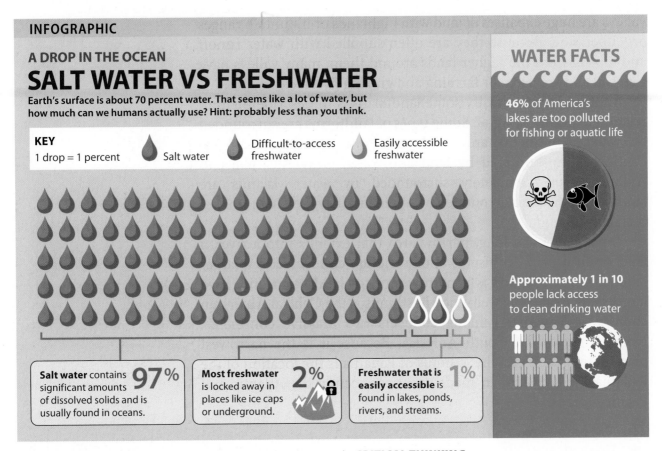

KEY
1 drop = 1 percent — Salt water — Difficult-to-access freshwater — Easily accessible freshwater

Salt water contains **97%** significant amounts of dissolved solids and is usually found in oceans.

Most freshwater **2%** is locked away in places like ice caps or underground.

Freshwater that is **1%** **easily accessible** is found in lakes, ponds, rivers, and streams.

WATER FACTS

46% of America's lakes are too polluted for fishing or aquatic life

Approximately 1 in 10 people lack access to clean drinking water

The surface of Earth is made up of water and land. Oceans, lakes, rivers, and other bodies of water make up a large part of Earth.

▶ **CRITICAL THINKING**

Analyzing Would you call freshwater a scarce or an abundant resource? Explain.

One type of ocean landform is the continental shelf. A **continental shelf** is an underwater plain that borders a continent. Continental shelves usually end at cliffs or downward slopes to the ocean floor.

When divers explore oceans, they sometimes find enormous underwater cliffs that drop off into total darkness. These cliffs extend downward for hundreds or even thousands of feet. The water below is so deep it is beyond the reach of the sun's light. The deepest location on Earth is the Mariana Trench in the Pacific Ocean. A **trench** is a long, narrow, steep-sided cut in the ground or on the ocean floor. At its deepest point, the Mariana Trench is more than 35,000 feet (10,668 m) below the ocean surface.

Other landforms on the ocean floor include volcanoes and mountains. When underwater volcanoes erupt, islands can form because layers of lava build up until they reach the ocean's surface. Mountains on the ocean floor can be as tall as Mount Everest. Undersea mountains can also form ranges. The Mid-Atlantic Ridge, the longest underwater mountain range, is longer than any mountain range on land.

☑ **READING PROGRESS CHECK**

Determining Word Meanings How is a valley similar to an ocean trench?

The Blue Planet

GUIDING QUESTION *What types of water are found on Earth's surface?*

Water exists around you in different forms. Water in each of the three states of matter—solid, liquid, and gas—can be found all over the world. Glaciers, polar ice caps, and ice sheets are large masses of water in solid form. Rivers, lakes, and oceans contain liquid water. The atmosphere contains water vapor, which is water in the form of a gas.

Two Kinds of Water

Water at Earth's surface can be freshwater or salt water. Salt water is water that contains a large percentage of salt and other dissolved minerals. About 97 percent of the planet's water is salt water. Salt water makes up the world's oceans and also a few lakes and seas, such as the Great Salt Lake and the Dead Sea.

Salt water supports a huge variety of plant and animal life, such as whales, fish, and other sea creatures. Because of its high concentration of minerals, humans and most animals cannot drink salt water. Humans have developed a way to remove minerals from salt water. **Desalinization** is a process that separates most of the dissolved chemical elements to produce water that is safe to drink. People who live in dry regions of the world use desalinization to process seawater into drinking water. But this process is expensive.

Freshwater makes up the remaining 3 percent of water on Earth. Most freshwater stays frozen in the ice caps of the Arctic and Antarctic. Only about 1 percent of all water on Earth is the liquid freshwater that humans and other living organisms use. Liquid freshwater is found in lakes, rivers, ponds, swamps, and marshes, and in the rocks and soil underground.

Water contained inside Earth's crust is called **groundwater**. Groundwater is an important source of drinking water, and it is used to irrigate crops. Groundwater often gathers in aquifers. These are underground layers of rock through which water flows. When humans dig wells down into rocks and soil, groundwater flows from the surrounding area and fills the well. Groundwater also flows naturally into rivers, lakes, and oceans.

Bodies of Water

You are probably familiar with some of the different kinds of bodies of water. Some bodies of water contain salt water, and others hold freshwater. The world's largest bodies of water are its five vast, salt water oceans.

A young girl from Gambia in West Africa pumps water from a well.
▶ CRITICAL THINKING
Describing What is groundwater? What is it used for?

Finnbarr Webster/Alamy

Humans use bodies of water for water sports and fishing.

▶ **CRITICAL THINKING**
Describing Describe three ways water affects the lives of people who live near it.

From largest to smallest, the oceans are the Pacific, Atlantic, Indian, Southern, and Arctic. The Pacific Ocean covers more area than all the Earth's land combined. The Southern Ocean surrounds the continent of Antarctica. Although it is convenient to name the different oceans, it is important to remember that these water bodies are actually connected and form one global ocean. Things that happen in one part of the ocean can affect the ocean all around the world.

When oceans meet landmasses, unique land features and bodies of water form. A coastal area where ocean waters are partially surrounded by land is called a bay. Bays are protected from rough ocean waves by the surrounding land, making them useful for docking ships, fishing, and boating. Larger areas of ocean waters partially surrounded by landmasses are called gulfs. The Gulf of Mexico is an example of ocean waters surrounded by continents and islands. Gulfs have many of the features of oceans but are smaller and are affected by the landmasses around them.

Bodies of water such as lakes, rivers, streams, and ponds usually hold freshwater. Freshwater contains some dissolved minerals, but only a small percentage. The fishes, plants, and other life-forms that live in freshwater cannot live in salty ocean water.

Freshwater rivers are found all over the world. Rivers begin at a source where water feeds into them. Some rivers begin where two other rivers meet; their waters flow together to form a larger river. Other rivers are fed by sources such as lakes, natural springs, and melting snow flowing down from higher ground.

A river's end point is called the mouth of the river. Rivers end where they empty into other bodies of water. A river can empty into a lake, another river, or an ocean. A **delta** is an area where sand, silt, clay, or gravel is deposited at the mouth of a river. Some deltas flow onto land, enriching the soil with the nutrients they deposit. River deltas can be huge areas with their own ecosystems.

Bodies of water of all kinds affect the lives of people who live near them. Water provides food, work, transportation, and recreation

to people in many parts of the world. People get food by fishing in rivers, lakes, and oceans. The ocean floor is mined for minerals and drilled for oil. All types of waters have been used for transportation for thousands of years. People also use water for sports and recreation, such as swimming, sailing, fishing, and scuba diving. Water is vital to human culture and survival.

☑ **READING PROGRESS CHECK**

Describing Describe three ways in which water affects your life.

The Water Cycle

GUIDING QUESTION *What is the water cycle?*

All living things need water. Humans and other mammals, birds, reptiles, insects, fishes, green plants, fungi, and bacteria must have water to survive. Water is essential for all life on Earth. To provide for the trillions of living organisms that use water every day, the planet needs a constant supply of fresh, clean water. Fortunately, water is recycled and renewed continually through Earth's natural systems of atmosphere, hydrosphere, lithosphere, and biosphere.

A Cycle of Balance

When it rains, puddles of water form on the ground. Have you noticed that after a day or two, puddles dry up and vanish? Where does the water go? It might seem as if water disappears and then new water is created, but this is not true. Water is not made or destroyed; it only changes form. When a puddle dries, the liquid water has turned into gas vapor that we cannot see. In time, the vapor will become liquid again, and perhaps it will fill another puddle someday.

Scientists believe the total amount of water on Earth has not changed since our planet formed billions of years ago. How can this be true? It is possible because the same water is being recycled. At all times, water is moving over, under, and above Earth's surface and changing form as it is recycled. Earth's water recycling system is called the **water cycle**. The water cycle keeps Earth's water supply in balance.

Water Changes Form

The sun's energy warms the surface of Earth, including the surface of oceans and lakes. Heat energy from the sun causes liquid water on Earth's surface to change into water vapor in a process called **evaporation**. Evaporation is happening all around us, at all times. Water in oceans, lakes, rivers, and swimming pools is constantly evaporating into the air. Even small amounts of water—in the soil, in the leaves of plants, and in the breath we exhale—evaporate to become part of the atmosphere.

Air that contains water vapor is less dense than dry air. This means that moist air tends to rise. As water evaporates, tiny droplets of water vapor rise into the atmosphere. Water vapor gathers into clouds of varying shapes and sizes. Sometimes clouds continue to build until they are saturated with water vapor and can hold no more. A process called **condensation** occurs, in which water vapor **transforms** into a denser liquid or a solid state.

Condensation causes water to fall back to Earth's surface as rain, hail, or snow. Hail and snow either build up and stay solid or melt into liquid water. Snow stays solid when it falls in cold climates or on frozen mountaintops. When snow melts, it flows into rivers and lakes or melts directly into the ground.

Liquid rainwater returns water to rivers, lakes, and oceans. Rainwater also soaks into the ground, supplying moisture to plants and refilling underground water supplies to wells and natural springs. Much of the rainwater that soaks into the ground filters through soil and rocks and trickles back into rivers, lakes, and oceans. In this way, water taken from Earth's surface during evaporation returns in the form of precipitation. This cycle repeats all over the world, recycling the water every living organism needs to survive.

Academic Vocabulary

transform to change

DIAGRAM SKILLS

Condensation

Clouds

Precipitation
(snow, sleet, hail, rain)

Surface
runoff

Evaporation
from lakes
and streams

Evaporation
from ocean

Groundwater to rivers and oceans

THE WATER CYCLE
Water is constantly moving—from the oceans to the air to the ground and finally back to the oceans.

▶ **CRITICAL THINKING**
Analyzing How does water enter the air during the water cycle?

Human actions have damaged the world's water supply. Waste from factories and runoff from toxic chemicals used on lawns and farm fields has polluted rivers, lakes, oceans, and groundwater. Chemicals such as pesticides and fertilizers seep into wells that hold drinking water, poisoning the water and causing deadly diseases.

Some of the fuels we burn release poisonous gases into the atmosphere. These gases combine with water vapor in the air to create toxic acids. These acids then fall to Earth as a deadly mixture called **acid rain**. Acid rain damages the environment in several ways. It pollutes the water humans and animals drink. The acids damage trees and other plants. As acid rain flows over the land and into waterways, it kills plant and animal life in bodies of water. This upsets the balance of the ecosystem.

 READING PROGRESS CHECK

Analyzing What are the causes and effects of acid rain?

Some human activities pollute our rivers and oceans.

▶ **CRITICAL THINKING**
Describing How does acid rain affect animal and plant life?

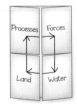

Include this lesson's information in your Foldable®.

(l to r) Comstock Images/Alamy; Carey Morishige/NOAA Marine Debris Program

---

LESSON 3 REVIEW (CCSS)

Reviewing Vocabulary

1. How are a *bridge* and an *isthmus* alike and different?

Answering the Guiding Questions

2. *Citing Text Evidence* Give examples of some of the landforms mentioned in the lesson by describing the landforms in your state or region.

3. *Analyzing* How is the ocean floor similar to the surface of dry land?

4. *Identifying* Identify these bodies of water as either freshwater or salt water: lake, river, ocean, pond, sea, bay, delta, gulf, groundwater.

5. *Determining Word Meanings* Explain how the processes of *evaporation* and *condensation* are both similar and different.

6. *Informative/Explanatory Writing* In your own words, explain the water cycle to someone you know. Use facts and details from the lesson in your writing.

Directions: Write your answers on a separate piece of paper.

① Use your **FOLDABLES** to explore the Essential Question.

INFORMATIVE/EXPLANATORY WRITING Identify physical features you see in your neighborhood or community. Choose one of the features and write a description of it. Include specific details such as sights and sounds associated with the feature.

② 21st Century Skills

INTEGRATING VISUAL INFORMATION Use a variety of resources to access information about one of the four seasons. Create a poster display, a slide show, or another multimedia presentation about the season. Explain how the movement of Earth creates seasons. Describe the typical seasonal weather where you live.

③ Thinking Like a Geographer

IDENTIFYING Create a two-column chart. Use it to list Earth's layers and their composition and depth, starting with the crust and working toward the center.

④ **GEOGRAPHY ACTIVITY**

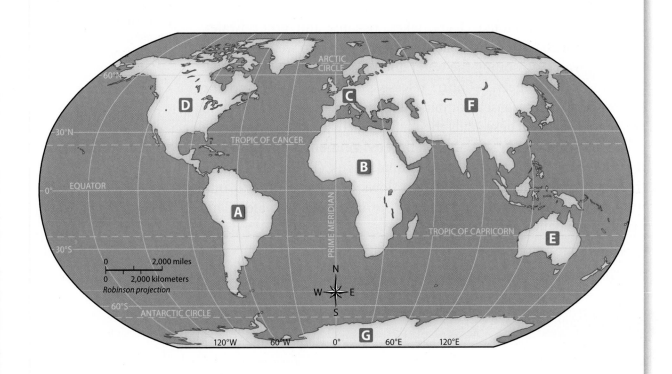

Locating Places
Match the letters on the map to the numbered continents below.

1. Africa   **3.** Europe   **5.** Asia   **7.** Antarctica

2. Australia   **4.** North America   **6.** South America

REVIEW THE GUIDING QUESTIONS

Directions: Choose the best answer for each question.

1 Earth's 24-hour cycle of day to night is caused by

   A. Earth's revolution around the sun.

   B. the pull of gravity from the moon.

   C. Earth's rotation on its axis.

   D. the sun's revolution around Earth.

2 Changes in air pressure caused by the uneven heating of Earth's surface produce

   F. ocean currents.

   G. wind.

   H. tropical storms.

   I. rain.

3 The world's largest island is

   A. Japan.

   B. Australia.

   C. Greenland.

   D. Indonesia.

4 Which forces created the Grand Canyon?

   F. wind and water

   G. snow and ice

   H. volcanoes and earthquakes

   I. glaciers and plate tectonics

5 What percentage of Earth's surface is covered in water?

   A. 40 percent

   B. 70 percent

   C. 50 percent

   D. 65 percent

6 Where is most of Earth's freshwater stored?

   F. in wells

   G. in lakes

   H. in the polar ice caps

   I. in rivers and streams

DBQ ANALYZING DOCUMENTS

7 **IDENTIFYING** In this paragraph, a science writer describes the two types of planets in our solar system.

"*We note that the planets can be divided quite easily into two categories—the inner planets, which are small and rocky, and the gas giants that circle through the outer reaches of the solar system. Within each class, the planets bear a striking resemblance to each other, but the two classes themselves are very different.*"

—from James S. Trefil, *Space, Time, Infinity: The Smithsonian Views the Universe*

Which of these is a rocky, inner planet?

A. Earth

B. Jupiter

C. Neptune

D. Saturn

8 **ANALYZING** Which characteristic(s) does the writer use to describe the two types of planets?

F. ability to support life

G. size and structure

H. presence of water

I. number of moons

SHORT RESPONSE

"*There was a great rattle and jar. . . . [Then] there came a really terrific shock; the ground seemed to roll under me in waves, interrupted by a violent joggling up and down, and there was a heavy grinding noise as of brick houses rubbing together.*"

—from Mark Twain, *Roughing It*

9 **IDENTIFYING** What natural event do you think Mark Twain is describing? Why?

10 **ANALYZING** What might cause the noises Twain describes hearing?

EXTENDED RESPONSE

11 **INFORMATIVE/EXPLANATORY WRITING** Explain how Earth's movements affect the length of the year. How do its movements bring about day and night?

Need Extra Help?

If You've Missed Question	**1**	**2**	**3**	**4**	**5**	**6**	**7**	**8**	**9**	**10**	**11**
Review Lesson	1	1	2	2	1	3	1	1	2	2	1

HUMAN GEOGRAPHY

ESSENTIAL QUESTIONS • *How do people adapt to their environment?*
• *What makes a culture unique?* • *Why do people make economic choices?*

Aurora Photos/Alamy

Young girl from the town of Dori in Burkina Faso

netw**rks**

There's More Online about Human Geography.

CHAPTER 3

Lesson 1
The World's People

Lesson 2
Culture

Lesson 3
The World's Economies

The Story Matters...

As part of our study of geography, we study culture, which is the way of life of people who share similar beliefs and customs. A particular culture can be understood by looking at the languages the people speak, what beliefs they hold, and what smaller groups form as parts of their society.

Study Organizer

Go to the Foldables® library in the back of your book to make a Foldable® that will help you take notes while reading this chapter.

Adaptations | Cultural Views | Basic Needs

HUMAN GEOGRAPHY ⓒⒸⓈⓈ

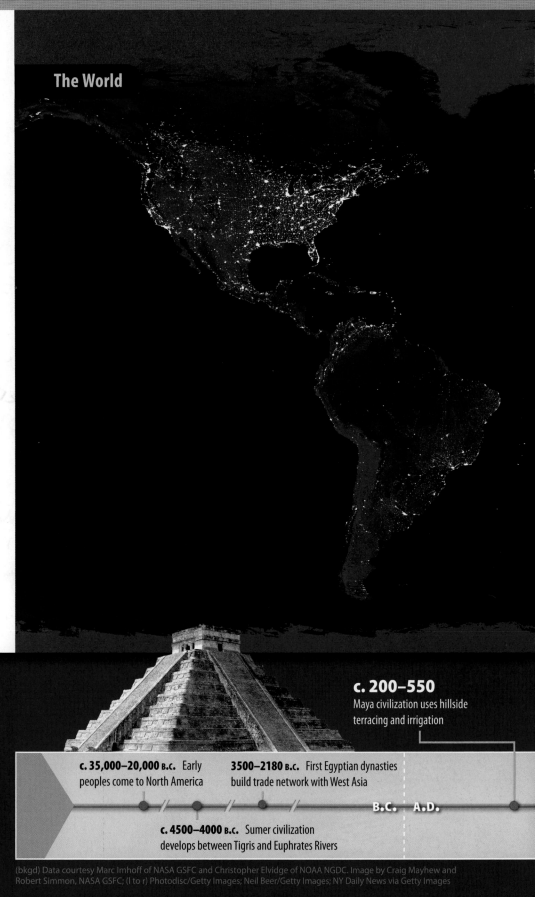

This image shows the world at night. Although the world's population is increasing, people still live on only a small part of Earth's surface. Some people live in highly urbanized areas, such as large cities. Others, however, live in areas where they may not have access to electricity or even running water.

The World

Step Into the Place

MAP FOCUS Use the image to answer the following questions.

1 HUMAN GEOGRAPHY
What do you think the brightly lit areas on the map represent?

2 HUMAN GEOGRAPHY
Why might some areas be brighter than others?

3 ENVIRONMENT AND SOCIETY Are the lights evenly distributed across the land? Where do more lights appear?

4 CRITICAL THINKING
Analyzing Do you think the darker areas have fewer people than brightly lit areas? Why or why not?

Step Into the Time

IDENTIFYING POINT OF VIEW Research one event from the time line. Write a journal entry describing the daily life of the time.

c. 200–550
Maya civilization uses hillside terracing and irrigation

c. 35,000–20,000 B.C. Early peoples come to North America

3500–2180 B.C. First Egyptian dynasties build trade network with West Asia

B.C. **A.D.**

c. 4500–4000 B.C. Sumer civilization develops between Tigris and Euphrates Rivers

(bkgd) Data courtesy Marc Imhoff of NASA GSFC and Christopher Elvidge of NOAA NGDC. Image by Craig Mayhew and Robert Simmon, NASA GSFC; (l to r) Photodisc/Getty Images; Neil Beer/Getty Images; NY Daily News via Getty Images

1492
Christopher Columbus
lands in Americas

1892–1924
Millions of immigrants
arrive at Ellis Island

1000

1200s West African kingdom
of Benin trade center

1845–1849 Ireland's Great
Famine leads to mass emigration

2011 South Sudan secedes,
forming new nation

2000

There's More Online!

☑ **CHART/GRAPH**
  Understanding Population
  Pyramids

☑ **MAP** Landforms, Waterways,
  and Population

☑ **SLIDE SHOW** Different Places in
  the World

☑ **VIDEO**

Reading **HELP**DESK ⓒⒸⓈⓈ

Academic Vocabulary

- **mature**

Content Vocabulary

- **death rate**
- **birthrate**
- **doubling time**
- **population distribution**
- **population density**
- **urban**
- **rural**
- **emigrate**
- **immigrate**
- **refugee**
- **urbanization**
- **megalopolis**

TAKING NOTES: *Key Ideas and Details*

Determining Cause and Effect
As you read, use a graphic organizer like this one to take notes about the causes of population growth and migration.

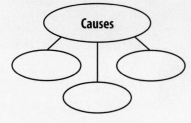

Lesson 1
The World's People

ESSENTIAL QUESTION • *How do people adapt to their environment?*

IT MATTERS BECAUSE
Billions of people share Earth. They have many different ways of life.

Earth's Growing Population

GUIDING QUESTION *What factors contribute to Earth's constantly rising population?*

How fast has the world's population grown? In 1800 about 860 million people lived in the world. During the next 100 years, the population doubled to nearly 1.7 billion. By 2012, the total passed 7 billion.

Causes of Population Growth

How has Earth's population become so large? What has caused our population to grow so quickly? Many factors cause populations to increase. One major cause of population growth is a falling death rate. The death rate is the number of deaths compared to the total number of individuals in a population at a given time. On average, about 154,080 people die each day worldwide. The **death rate** has decreased for many reasons. Better health care, more food, and cleaner water have helped more people—young and old—live longer, healthier lives.

Another major cause of population growth is the global birthrate. The **birthrate** is the number of babies born compared to the total number of individuals in a population at a given time. On average, about 215,120 babies are born each day worldwide. In time, the babies born today will **mature** and have children and grandchildren of their own.

This is how more and more people join the human population with each passing day.

However, during the past 60 years, the world's human birthrate has been decreasing slowly, although the global birthrate is still higher than the global death rate. This means that at any given time, such as a day or a year, more births than deaths occur. This results in population growth.

Growth Rates

In some countries, a high number of births has combined with low death rates to greatly increase population growth. As a result, **doubling time**, or the number of years it takes a population to double in size based on its current growth rate, is relatively short. In some parts of Asia and Africa, for example, the doubling time is relatively short—25 years or less. In contrast, the average doubling time of countries with slow growth rates, such as Canada, can be more than 75 years.

Despite the fact that the global population is growing, the rate of growth is gradually slowing. The United Nations Department of Economic and Social Affairs predicts that the world's population will peak at 9 billion by the year 2050. After that, the population will begin to decrease. This means that for the next few decades, Earth's population will continue to grow. In time, however, this growth trend might stop.

Academic Vocabulary

mature fully grown and developed as an adult; also refers to older adults

GRAPH SKILLS >

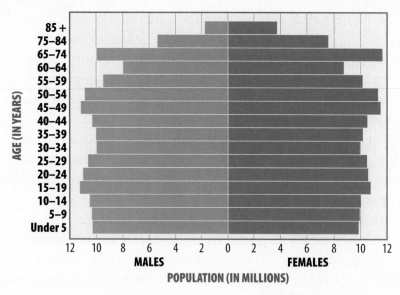

Source: U.S. Census Bureau, Statistical Abstract of the United States: 2012

POPULATION PYRAMID

Population pyramids use two bar graphs to show a country's population by age and gender. A growing population will be wider at the bottom. This represents a large number of young people. A declining population will be wider at the top. This means the country's population is mostly elderly. A stable population is represented by similar length bars over several age ranges. This graph shows the population of the U.S.

▶ **CRITICAL THINKING**

1. *Analyzing* Which age group has the most females?

2. *Analyzing* About how many males are in the 10-14 age group?

Farmers in many parts of the world clear land by cutting and burning forests.

▶ **CRITICAL THINKING**

Analyzing What is the purpose of slash-and-burn agriculture?

Population Challenges

When human populations grow, the places people inhabit can become crowded. In many parts of the world, cities, towns, and villages have grown and expanded beyond a comfortable capacity. Some cities are now so filled with people that they are becoming dangerously overcrowded.

When the population of an already crowded area continues to grow, serious problems can arise. For example, diseases spread quickly in crowded environments. Sometimes there is not enough work for everyone, and many households live in ongoing poverty. Where many people share tight living spaces, crime can be a serious problem and pollution can increase.

Effects on the Environment

On a global scale, rapid population growth can harm the environment. Each year, more people are sharing the same amount of space. People demand fuel for their cars and power for their homes. Miners drill and dig into the earth in a constant search for more energy resources. Forests are cut down to make farms for growing crops and raising livestock to feed hungry populations. Factory workers build cars, computers, and appliances. Some factories dump chemicals into waterways and vent poisonous smoke into the air.

Over time, and with many thousands of factories all over the world, chemical wastes have polluted Earth's atmosphere. Many groups and individuals are working to clean up the environment and restore once-polluted areas.

Humans have many methods of finding and using the resources we need for survival. Some of these methods are wasteful and destructive. However, humans are also creative in solving modern problems. People in all parts of the world have invented new ways to produce power and harvest resources. For example, some people install solar panels on the roofs of buildings. These panels collect energy from the sun, which can be used to produce heat and electric energy. Wind, solar, and geothermal energy are resources that do not pollute the environment. Humans are rising to the challenge of finding new methods of using these natural resources.

Population Growth Rates

Human populations grow at different rates in different areas of the world for a variety of reasons. Often, the number of children each family will have is influenced by the family's culture and religion. In some cultures, families are encouraged to have as many children as they can. Although birthrates have fallen greatly in many parts of Asia, Africa, and South America in recent decades, the rates are still higher than in industrialized nations.

In locations with the largest and fastest-growing populations, the need for resources, jobs, health care, and education is great. When millions of people living in a small area need food, water, and housing, there is sometimes not enough for everyone. People in many parts of the world go hungry or die of starvation. Water supplies in crowded cities are often polluted with wastes that can cause diseases. Some areas do not have enough land resources and materials for people to build safe, sturdy homes.

Children in areas affected by extreme poverty often do not receive an education. In areas with job shortages, people are forced to live on low incomes. People living in poverty often live in crowded neighborhoods called *slums*. Slums surround many of the world's cities. These places are often dirty and unsafe. Governments and organizations such as the United Nations are working to make these areas safer, healthier places to live. Because populations grow at different rates some areas experience more severe problems.

✓ **READING PROGRESS CHECK**

Determining Word Meanings What is the difference between the birthrate and death rate?

Solar panels collect energy from the sun.

Some cities, like Seoul, South Korea, have a high population density.

▶ **CRITICAL THINKING**

Analyzing How is population density calculated?

Population Patterns

GUIDING QUESTION *Why do more people live in some parts of the world than in others?*

Some families live in the same town or on the same land for generations. Other people move frequently from place to place.

Population Distribution

Population growth rates vary among Earth's regions. The **population distribution**, or the geographic pattern of where people live on Earth, is uneven as well. One reason people live in a certain place is work. During the industrial age, for example, people moved to places that had important resources such as coal or iron ore to make and operate machinery. People gather in other places because these places hold religious significance or because they are government or transportation centers.

Population Density

One way to look at population is by measuring **population density**—the average number of people living within a square mile or a square kilometer. To say an area is *densely populated* means the area has a large number of people living within it.

Keep in mind that a country's population density is the average for the entire country. Population is not distributed evenly throughout a country. As a result, some areas are more densely populated than their country's average indicates. In Egypt, for example, the population is concentrated along the Nile River; in China, along its eastern seaboard; and in Mexico, on the Central Plateau.

Much of the land surrounding this rural village in Austria is used for farming.

Where People Are Located

Urban areas are densely populated. **Rural** areas, in contrast, are sparsely populated. People inhabit only a small part of Earth. Remember that land covers about 30 percent of Earth's surface, and half of this land is not useful to humans. This means that only about 15 percent of Earth's surface is inhabitable. Large cities have dense populations, while deserts, oceans, and mountaintops are uninhabited.

The main reason people settle in some areas and not in others is the need for resources. People live where their basic needs can be met. People need shelter, food, water, and a way to earn a living. Some people live in cities, which have many places to live and work. Other people make their homes on open grasslands where they build their own shelters, grow their own food, and raise livestock.

☑ **READING PROGRESS CHECK**

Identifying Give one example of an urban area and one example of a rural area.

Population Movement

GUIDING QUESTION *What are the causes and effects of human migration?*

The populations of different areas change as people move from one area to another. When many people leave an area, that area's population decreases. When large numbers of people move into a city, a state, or a country, the population of that area increases. Moving from one place to another is called *migration*.

People in rural areas often obtain food in outdoor markets such as this Laotian market.

▶ **CRITICAL THINKING**

Analyzing What is the main reason people choose to settle in one area and not in another?

Migration is one of the main causes of population shifts in our world today. What causes people to leave their homelands and migrate to different parts of the world?

Causes of Migration

To **emigrate** means "to leave one's home to live in another place." Emigration can happen within the same nation, such as when people move from a village to a city inside the same country. Often, emigration happens when people move from one nation to another. For example, millions of people have emigrated from countries in Europe, Asia, and Africa to start new lives in the United States. The term *immigrate* is closely related to *emigrate*, but it does not mean the same thing. To **immigrate** means "to enter and live in a new country."

The reasons for leaving one area and going to another are called push-pull factors. *Push* factors drive people from an area. For example, when a war breaks out in a country or region, people emigrate from that place to escape danger. People who flee a country because of violence, war, or persecution are called **refugees**. Sometimes people emigrate from an area after a natural disaster such as a flood, an earthquake, or a tsunami has destroyed their homes and land. If the economy of a place becomes so weak that little or no work is available, people emigrate to seek new opportunities.

Lissa Harrison

Pull factors attract people to an area. Some people move to new places to be with friends or family members. Many young people move to cities or countries to attend universities or other schools. Some relocate in search of better jobs. Families sometimes move to places where their children will be able to attend better schools.

Effects of Migration

The movement of people to and from different parts of the world can affect the land, resources, culture, and economy of an area. Some of these effects are positive, but others can be harmful.

One positive effect of migration is cultural blending. As people from diverse cultures migrate to the same place and live close together, their cultures become mixed and blended. This blending creates new, unique cultures and ways of life. Artwork and music created in diverse urban areas is often an interesting mixture of styles and rhythms from around the world. Food, clothing styles, and languages spoken in urban areas change when people migrate into that area and bring new influences.

Some families and cultural groups work to preserve their original culture. These people want to keep their cultural traditions alive so they can be passed down to future generations. For example, the traditional Chinese New Year is an important celebration for many Chinese American families. Chinese Americans can be part of a blended American culture but still enjoy traditional Chinese foods, music, and arts, and celebrate Chinese holidays. It is possible to adapt to a local culture yet maintain strong ties to a home culture.

Some people migrate by choice. Others, such as the Libyan refugees shown here, are forced to flee to another country to live.
▶ **CRITICAL THINKING**
Identifying What are examples of "push" causes of migration?

Nearly half the world's people live in urban areas. Many live in very large cities such as Hong Kong.

▶ **CRITICAL THINKING**
Determining Word Meanings What is urbanization?

Causes and Effects of Urbanization

Another effect of migration is the growth of urban areas. **Urbanization** happens when cities grow larger and spread into surrounding areas. Migration is a primary reason that urbanization occurs.

People move to cities for many reasons. The most common reason is to find jobs. Transportation and trade centers draw people primarily by creating new opportunities for business. As the businesses grow and people move into an area, the need for services also grows. Workers fill positions in medical services, education, entertainment, housing, and food sectors.

As more people migrate to cities, urban areas become increasingly crowded. When populations within urban areas increase, cities grow and expand. Farmland is bought by developers to build homes, apartment buildings, factories, offices, schools, and stores to provide for the growing number of people. The loss of farmland means that food must be grown farther from cities, resulting in additional shipping and related pollution.

Urbanization is happening in cities all over the world. In some places, cities have grown so vast that they have reached the outer edges of other cities. The result is massive clusters of urban areas that continue for miles. A huge city or cluster of cities with an extremely large population is called a **megalopolis**. These huge cities are growing larger every day, and they face the challenges that come with population growth and urbanization.

Leung Cho Pan/Flickr/Getty Images

Delhi, one of India's largest cities, is a megalopolis. Its sprawling land area takes in a section called the Old City, dating from the mid-1600s. It also encompasses New Delhi, the modern capital city built by British colonial rulers in the early 1900s.

The largest megalopolis in the Americas is Mexico City. Because of its size and influence, Mexico City is a primate city, an urban area that dominates its country's economy and political affairs. Primate cities include Cairo, Egypt in Africa, Amman, Jordan in Asia, and Paris, France in Europe.

Examples of Urbanization

Urbanization takes place around the world, but for different reasons and at different rates.

Africa south of the Sahara, for example, is one of the least urbanized regions of the world. The region's urban areas, however, are growing so rapidly that Africa has a very high rate of urbanization. Many Africans leave rural villages in order to find better job opportunities, health care, and public services in urban areas. At the same time population growth has caused cities to spread out into the countryside.

Europe is highly urbanized. Beginning in the late 1700s, the Industrial Revolution transformed Europe from a rural, agricultural society to an urban, industrial society. The growth of industries and cities began first in western Europe. Later, after World War II, the process spread to eastern Europe.

☑ **READING PROGRESS CHECK**

Describing In your own words, briefly summarize the main reasons people emigrate from their homelands.

Include this lesson's information in your Foldable®.

LESSON 1 REVIEW

Reviewing Vocabulary

1. Why might a *refugee* move to a new area?

Answering the Guiding Questions

2. *Describing* Briefly describe three factors that have contributed to Earth's constantly rising population.

3. *Determining Central Ideas* Why do more people live in some parts of the world than in others?

4. *Identifying* List three causes and three effects of human migration.

5. *Narrative Writing* Imagine you have just emigrated from your homeland and are now living in a different part of the world. Write a letter to an imaginary friend back in your home country telling him or her your reasons for moving and describing some experiences you have had in your new location.

Reading **HELP**DESK (CCSS)

Academic Vocabulary

- **behalf**

Content Vocabulary

- **culture**
- **ethnic group**
- **dialect**
- **cultural region**
- **democracy**
- **representative democracy**
- **monarchy**
- **dictatorship**
- **human rights**
- **globalization**

TAKING NOTES: *Key Ideas and Details*

Organize On a graphic organizer like this one, take notes about the different forms of government.

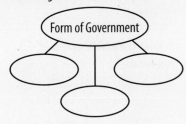

Form of Government

Lesson 2
Culture

ESSENTIAL QUESTION · *What makes a culture unique?*

IT MATTERS BECAUSE
Culture shapes the way people live and how they view the world.

Elements of Culture

GUIDING QUESTION *How is culture part of your life?*

What are some of your favorite foods? Do you like pizza, rice and beans, pasta, *samosas*, or corn on the cob? Have you ever thought about the people and cultures that invented the foods you enjoy eating? Millions of Americans eat foods created, grown, or developed by people of different cultures.

Culture is the set of beliefs, behaviors, and traits shared by a group of people. The term *culture* can also refer to the people of a certain culture. For example, saying "the Hindu culture" can mean the Hindu cultural traditions, the people who follow these traditions, or both.

You might be part of more than one culture. If your family has strong ties to a culture, such as that of a religion or a nation, you might follow this cultural tradition at home. You also might be part of a more mainstream American culture while at school and with friends.

If your family emigrated from Somalia to the United States, for example, you might speak the Somali language, wear traditional Somali clothing, and eat Somali foods. Your family might celebrate holidays observed in Somalia as well as American holidays, such as Thanksgiving and Independence Day. When you are with your friends, you might speak English, listen to American music, and watch American sports.

Ethnic Groups

We can look at members of a culture in terms of age, gender, or ethnic group. An **ethnic group** is a group of people with a common racial, national, tribal, religious, or cultural background. Members of the same Native American nation are an example of people of the same ethnic group. Other examples include the Maori of New Zealand and the Han Chinese. Large countries such as China, can be home to hundreds of different ethnic groups. Some ethnic groups in a country are minority groups—people whose race or ethnic origin is different from that of the majority group. The largest ethnic minority groups in the United States are Hispanic Americans and African Americans.

Members of a culture might have special roles or positions as part of their cultural traditions. In some cultures, women are expected to care for and educate children. Most cultures expect men to earn money to support their families or to provide in other ways, such as by hunting and farming. Many cultures respect the elderly and value their wisdom. The leaders of older, traditional cultures are often elderly men or women who have leadership experience. Most cultures have clearly defined roles for their members. From an early age, young people learn what their culture expects of them. It is possible, too, to be part of more than one culture.

Language

Language serves as a powerful form of communication. Through language, people communicate information and experience and pass on cultural beliefs and traditions. Thousands of different languages are spoken in the world. Some languages have become world languages, or languages that are commonly spoken in many parts of the world. Some languages are spoken differently in different regions or by different ethnic groups. A **dialect** is a regional variety of a language with unique features, such as vocabulary, grammar, or pronunciation. People who speak the same language can sometimes understand other dialects, but at times, the pronunciation, or accent, of a dialect can be nearly impossible for others to understand.

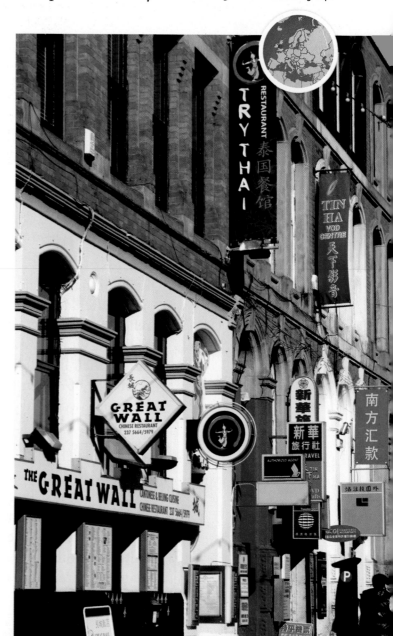

Cities often have communities within them that share a distinct and common culture, language, and customs.

▶ **CRITICAL THINKING**
Describing What is an ethnic group?

MAJOR WORLD RELIGIONS

Religion	Major Leader	Beliefs
Buddhism	Siddhārtha Gautama, the Buddha	Suffering comes from attachment to earthly things, which are not lasting. People become free by following the Eightfold Path, rules of right thought and conduct. People who follow the Path achieve nirvana—a state of endless peace and joy.
Christianity	Jesus Christ	The one God is Father, Son, and Holy Spirit. God the Son became human as Jesus Christ. Jesus died and rose again to bring God's forgiving love to sinful humanity. Those who trust in Jesus and follow his teachings of love for God and neighbor receive eternal life with God.
Hinduism	No one founder	One eternal spirit, Brahman, is represented as many deities. Every living thing has a soul that passes through many successive lives. Each soul's condition in a specific life is based on how the previous life was lived. When a soul reaches purity, it finally joins permanently with Brahman.
Islam	Muhammad	The one God sent a series of prophets, including the final prophet Muhammad, to teach humanity. Islam's laws are based on the Quran, the holy book, and the Sunnah, examples from Muhammad's life. Believers practice the five pillars—belief, prayer, charity, fasting, and pilgrimage—to go to an eternal paradise.
Judaism	Abraham	The one God made an agreement through Abraham and later Moses with the people of Israel. God would bless them, and they would follow God's laws, applying God's will in all parts of their lives. The main laws and practices of Judaism are stated in the Torah, the first five books of the Hebrew Bible.
Sikhism	Guru Nanak	The one God made truth known through 10 successive gurus, or teachers. God's will is that people should live honestly, work hard, and treat others fairly. The Sikh community, or Khalsa, bases its decisions on the principles of a sacred text, the Guru Granth Sahib.

Religion

Religion has a major influence on how people of a culture see the world. Religious beliefs are powerful. Some individuals see their religion as merely a tradition to follow during special occasions or holidays. Others view religion as the foundation and most important part of their life. Religious practices vary widely. Many cultures base their way of life on the spiritual teachings and laws of holy books. Religion is a central part of many of the world's cultures. Throughout history, religious stories and symbols have influenced painting, architecture, and music.

Customs

Customs are also an important outward display of culture. In many traditional cultures, a woman is not permitted to touch a man other than her husband, even for a handshake. In modern European cultures, polite greetings include kissing on the cheeks. People of many cultures bow to others as a sign of greeting, respect, and

goodwill. The world's many cultures have countless fascinating customs. Some are used only formally, and others are viewed as good manners and respectful, professional behavior.

History

History shapes how we view the world. We often celebrate holidays to honor the heroes and heroines who brought about successes. Stories about heroes reveal the personal characteristics that people think are important. Groups also remember the dark periods of history when they met with disaster or defeat. These experiences, too, influence how groups of people see themselves. Cultural holidays mark important events and enable people to celebrate their heritage.

The Arts and Sports

Dance, music, visual arts, and literature are important elements of culture. Nearly all cultures have unique art forms that celebrate their history and enrich people's lives. Some art forms, such as singing and dancing, are serious parts of religious ceremonies or other cultural events. Aboriginal peoples of the Pacific Islands have songs, dances, and chants that are vital parts of their cultural traditions. Art can be forms of personal expression or worship, entertainment, or even ways of retelling and preserving a culture's history.

In sports, as in many other aspects of culture, activities are adopted, modified, and shared. Many sports that we play today originated with different culture groups in the past. Athletes in ancient Japan, China, Greece, and Rome played a game similar to soccer. Scholars believe that the Maya of Mexico and Central America developed "ballgame," the first organized team sport. Playing on a 40- to 50-foot long (12.2 m to 15.2 m) recessed court, the athletes' goal was to kick a rubber ball through a goal.

Government

Government is another element of culture. Despite differences, governments around the world share certain features. They maintain order within an area and provide protection from outside dangers. Governments also provide services to citizens, such as education and transportation infrastructure. Different cultures have different ways of distributing power and making rules.

Soccer is one of the most popular international sports.

▶ CRITICAL THINKING

Identifying Which culture first played the game of soccer?

Economy

Economies control the use of natural resources and define how goods are produced and distributed to meet human needs. Some cultures have their own type of economy, but most follow the economy of the country or area where they live. This allows people of different cultures living in an area to trade and conduct other types of business with one another. For example, many people in Benin, West Africa, sell goods in open-air markets. Some people bring items to the markets to trade for the goods they need, but others pay for goods using paper money and coins.

Cultural Regions

A **cultural region** is a geographic area in which people have certain traits in common. People in a cultural region often live close to one another to share resources, for social reasons, and to keep their cultures and communities strong. Cultural regions can be large or relatively small. For example, one of the world's largest cultural areas stretches across northern Africa and Southwest Asia. This cultural region is home to millions of people of the Islamic, or Muslim, culture. A much smaller cultural region is Spanish Harlem in New York City. This cultural region is home to a large and growing Hispanic culture.

✓ READING PROGRESS CHECK

Identifying Point of View What cultural traditions do you practice? Make a list of the beliefs, behaviors, languages, foods, art, music, clothing, and other elements of culture that are part of your daily life.

On July 4, 1776, the Second Continental Congress approved the Declaration of Independence, establishing the United States as an independent country.

▶ CRITICAL THINKING

Describing What form of government does the United States have? How does that form of government work?

Forms of Government

GUIDING QUESTION *How does government affect way of life?*

All nations need some type of formal leadership. What differs among countries is how leaders are chosen, who makes the rules, how much freedom people have, and how much control governments have over people's lives. There are many different kinds of government systems that operate in the world today. Three of the most common are democracy, monarchy, and dictatorship.

Democracy

In a democracy, the people hold the power. Citizens of a nation make the decisions themselves. A **democracy** is a system of

©PoodlesRock/Corbis

government that is run by the people. In democratic systems of government, people are free to propose laws and policies. Citizens then vote to decide which laws and policies will be set in place. When people run the government, their rights and freedoms are protected.

In some democracies, the people elect leaders to make and carry out laws. A **representative democracy** is a form of democracy in which citizens elect government officials to represent the people; the government representatives make and carry out laws and policies on **behalf** of the people. The United States is an example of a representative democracy.

The queen is the symbolic head of the United Kingdom, but elected leaders hold the power to rule.

Monarchy

A **monarchy** is ruled by a king or a queen. In a monarchy, power and leadership are usually passed down from older to younger generations through heredity. The ruler of a monarchy, called a *monarch*, is usually a king, a queen, a prince, or a princess. In the past, monarchs had absolute power, or complete and unlimited power to rule the people. Today, most monarchs only represent, or stand for, a country's traditions and values, while elected officials run the government. The United Kingdom is a monarchy.

Dictatorship

A **dictatorship** is a form of government in which one person has absolute power to rule and control the government, the people, and the economy. People who live under a dictatorship often have few rights. With absolute power, a dictator can make laws with no concern for how just, fair, or practical the laws are. North Korea is an example of a dictatorship.

Some dictators abuse their power for personal gain. One negative consequence of abuse of power is lack of personal freedoms and human rights for the general public. **Human rights** are the rights that belong to all individuals. Those rights are the same for every human in every culture. Some basic human rights are the right to life, liberty, and fair treatment before the law.

✓ READING PROGRESS CHECK

Describing In your own words, describe the system of government used in the United States.

Academic Vocabulary

behalf in the interest of; in support of; in defense of

Ian Gavan/Getty Images Entertainment/Getty Images

Cargo containers are stockpiled and ready to be loaded onto ships in the port of Johor, Malaysia. International trade is the exchange of goods and services between countries. When people trade, they not only trade goods, they also trade customs and ideas.

Changing Cultures

GUIDING QUESTION *How do cultures change over time?*

Cultures change over time for many reasons. When people relocate, they bring their cultural traditions with them. The traditions often influence or blend with the cultures of the places where they settle. Over time, as people of many cultures move to a location, the culture of that location takes on elements of all the cultures within it. Cities, such as London and New York, are examples of areas that have richly diverse cultures.

Cultural Change

Change can also occur as a result of trade, travel, war, and exchange of ideas. Trade brings people to new areas to sell and barter goods. Whenever people travel, they bring their language, customs, and ideas with them. They also bring elements of foreign cultures back with them when they return home. Throughout history, traders and explorers have brought home new foods, clothing, jewelry, and other goods. Some of these, such as gold, chocolate, gunpowder, and silk, became popular all over the world. Trade in these items changed the course of history.

©Justin Guariglia/Corbis

Culture also can change as a result of technology. The telegraph, telephones, and e-mail have made communication increasingly faster and easier. Television and the Internet have given people in all parts of the world easy access to information and new ideas. Elements of culture such as language, clothing styles, customs, and behaviors spread quickly as people discover them by watching television and using the Internet.

Global Culture

Today's world is becoming more culturally blended every day. As cultures combine, new cultural elements and traditions are born. The spread of culture and ideas has caused our world to become globalized. **Globalization** is the process by which nations, cultures, and economies become integrated, or mixed. Globalization has had the positive effect of making people more understanding and accepting of other cultures. It also has helped spread ideas and innovations. Technology has made communication faster and easier. Travel also has become faster and easier, allowing more people to visit more places in less time. This is resulting in cultural blending on a wider scale than ever before.

The process of cultural blending through globalization is not always smooth and easy. Sometimes it produces tension and conflict as people from different cultures come into contact with one another. Some people do not want their cultures to change, or they want to control the amount of change. Sometimes the changes come too fast, and cultures can be damaged or destroyed.

Just as no one element defines a culture, no one culture can define the world. All cultures have value and add to the human experience. As the world becomes more globalized, people must continue to respect other ways of life. We have much to learn, and much to gain, from the many cultures that make our world a fascinating place.

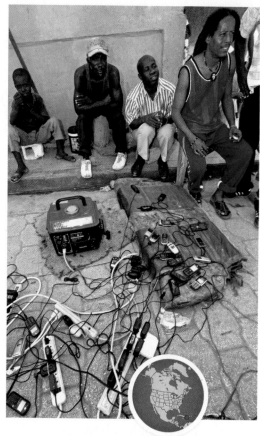

Widespread use of technology, such as cell phones, allows us to share information with a larger audience.

▶ **CRITICAL THINKING**
Analyzing How does technology help spread new ideas?

Include this lesson's information in your Foldable®.

☑ **READING PROGRESS CHECK**

Determining Word Meanings What is globalization?

LESSON 2 REVIEW (CCSS)

Reviewing Vocabulary

1. What type of leader could govern a *monarchy*?

Answering the Guiding Questions

2. *Analyzing* How do your clothing, behaviors, hairstyle, the language you speak, the foods you eat, and the music you listen to give clues about your culture?

3. *Determining Central Ideas* Explain some ways governments can affect people's ways of life.

4. *Describing* Give three examples, from your own experience or from the lesson, of how cultures change over time.

5. *Informative/Explanatory Writing* Write a short essay explaining your cultural traditions to your teacher. Include as many elements of culture as you can. Be sure to define, describe, and explain any words or concepts that your teacher might not be familiar with.

Social Media in a Changing World

News media, such as newspapers and radio, give us information. Social media, such as Facebook and Twitter, provide information, but they help us communicate, too.

Family and Friends The Internet helps us stay in touch with family and friends. It provides a good way for members of the military and their families and friends to share what is happening in their lives. Online connections also help service members in other ways. Where soldiers have access to the Internet, many have continued taking college courses, even in Iraq and Afghanistan.

In College Nearly one-third of higher-education students take at least one course online. For several years, online enrollment has grown faster than total higher-education enrollment.

> "It makes it easy to share information."

In an Emergency At times, social media provides the only way to communicate during an emergency like the tsunami that struck Japan in 2011. Through the use of sites such as Facebook and Twitter, people were able to contact family and friends. Social networks also posted information about shelters, medical help, and relief efforts.

Political Effects

The use of social media increases during times of political trouble or change, such as the "Arab Spring" protests. For example, the number of tweets from Egypt rose from 2,300 to 230,000 in the week leading to the resignation of Egyptian president Hosni Mubarak. The tweets spread the message of protest.

Safety Tips The Internet is a fun way to interact with friends. It makes it easy to share information. But some information should *not* be shared. Remember these safety tips:

- Keep your personal information to yourself. Don't post your full name, Social Security number, address, phone number, or bank account numbers.

- Don't accept as a friend anyone you do not know.

- Post only information that you are comfortable with others seeing—and knowing—about you.

networks

THERE'S MORE ONLINE

HEAR about China's Internet restrictions • **SEE** how people influence media • **WATCH** a video on piracy

These numbers and statistics can help you see how social media is changing the world.

840 Million

By 2012, Facebook had more than 840 million users. That's more than the population of every country in the world except China and India.

$1 Billion

In April 2012, Facebook, Inc. purchased the photo-sharing app called Instagram for $1 billion. Instagram allows users to apply a filter on a photograph and share it on any of a number of social networks. The company has 13 employees and has been in business since October 2010.

Fastest *to* 10 MILLION

Which social network was the quickest to reach 10 million users? That's Google+, taking only 16 days to reach that number. By contrast, Twitter reached 10 million users in 780 days. It took Facebook 852 days.

23 minutes

Social networks are taking up more and more of our time online. Users spend on average about 23 minutes of every hour of computer time on social networking sites.

62% Of online users worldwide, 62 percent use the Internet for social networking. Social media is most popular in Indonesia, where 83 percent take part. Although the use of social networking is growing, e-mail remains number one. About 85 percent of those surveyed say it remains their top online activity.

55 *and* older | Nielsen reported that users 55 and older are the fastest-growing group on social networks. However, people between ages 18 and 34 are the most active age group.

Singer from band Train takes a photo for Twitter during a concert.

190 Million That's the number of tweets sent on an average day.

April 23, 2005

On that day, the very first video was uploaded to YouTube. It was called "Me at the Zoo." A little more than a year later, more than 65,000 videos were being uploaded every day.

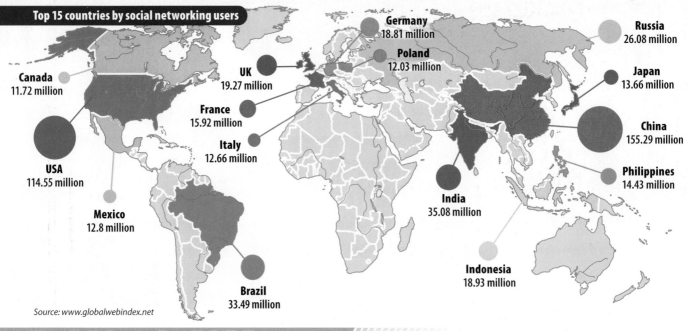

Top 15 countries by social networking users

Germany 18.81 million

Russia 26.08 million

Poland 12.03 million

Canada 11.72 million

UK 19.27 million

Japan 13.66 million

France 15.92 million

Italy 12.66 million

China 155.29 million

USA 114.55 million

Philippines 14.43 million

India 35.08 million

Mexico 12.8 million

Indonesia 18.93 million

Brazil 33.49 million

Source: www.globalwebindex.net

GLOBAL IMPACT

THE GEOGRAPHY OF SOCIAL MEDIA When social media and the Internet were developed, many people said that physical location would no longer matter. People are able to communicate as easily with someone half a world away as with the person next door. However, most communication on social media is with people who are close by. They could go to see the person they are addressing, but they prefer to send a message. Messages on Facebook can be sent to many people at the same time.

Facebook Users

The graph shows the number of Facebook users worldwide from 2004 to 2011.

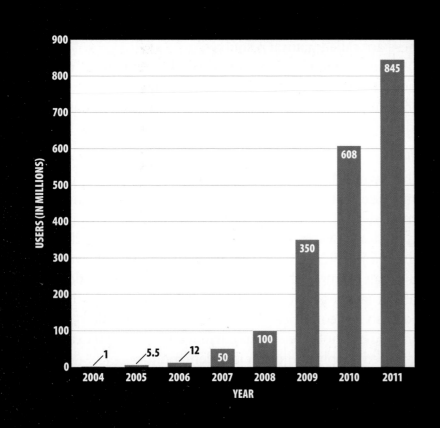

Thinking like a
Geographer

1. *Analyzing* Why is social media useful?

2. *Describing* How would you use social media to plan an event at your school?

3. *Integrating Visual Information* Research one country where "Arab Spring" protests took place in 2011. Prepare a PowerPoint® presentation that includes photographs, maps, and messages. Present your slide show to the class.

Reading **HELP**DESK ⓒⓒⓢⓢ

Academic Vocabulary

- **currency**

Content Vocabulary

- **renewable resource**
- **nonrenewable resource**
- **economic system**
- **traditional economy**
- **market economy**
- **command economy**
- **mixed economy**
- **gross domestic product**
- **standard of living**
- **productivity**
- **export**
- **import**
- **free trade**
- **sustainability**

TAKING NOTES: *Key Ideas and Details*

Organize As you read, summarize the key ideas about each economic system.

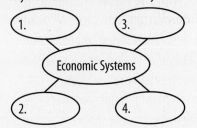

Lesson 3
The World's Economies

ESSENTIAL QUESTION · *Why do people make economic choices?*

IT MATTERS BECAUSE
People strive to meet their basic needs and their desires for a better life.

The Basic Economic Question

GUIDING QUESTION *How do people get the things they want and need?*

All human beings have wants and needs. How do you get the things you want and the things you need? To obtain these items, people use resources. Resources are the supplies that are used to meet our wants and needs. Some types of resources, such as water, soil, plants, and animals, come from the earth. These are called natural resources.

Other resources are supplied by humans. Human resources include the labor, skills, and talents people contribute. Countries also have wants and needs. Like individuals, nations must use resources to meet their needs.

Wants and Resources

What would happen if 14 students each wanted a glass of lemonade from a pitcher that contained only 12 glasses of lemonade? What if more students wanted a glass of lemonade? No matter how many people want lemonade, the pitcher still contains just 12 glasses. There is not enough for everyone. This is an example of a limited supply and unlimited demand. This situation is not uncommon. It happens to individuals and also to countries. You probably can think of many personal examples, as well as current and historical examples, of limited supply and unlimited demand.

One type of resource everyone needs is energy. Energy is the power to do work. Energy resources are the supplies that provide the power to do work. Many types of energy resources exist in our world. Energy resources can be renewable or nonrenewable. **Renewable resources** are resources that can be totally replaced or are always available naturally. They can be regenerated and replenished. Examples of renewable resources include water, trees, and energy from the wind and the sun.

In contrast are nonrenewable resources. **Nonrenewable resources** can not be totally replaced. Once nonrenewable resources are consumed, they are gone. Examples of nonrenewable resources include the fossil fuels oil, coal, and natural gas. These fuels received their name because they formed millions of years ago. Humans' increasing need for energy is taking its toll as supplies of nonrenewable resources shrink.

Making Choices

If the peoples of all nations have unlimited wants but face limited resources, what must happen? We must make choices. Do we continue to use nonrenewable resources? If so, at what rate should we be using them? Should we switch to renewable resources?

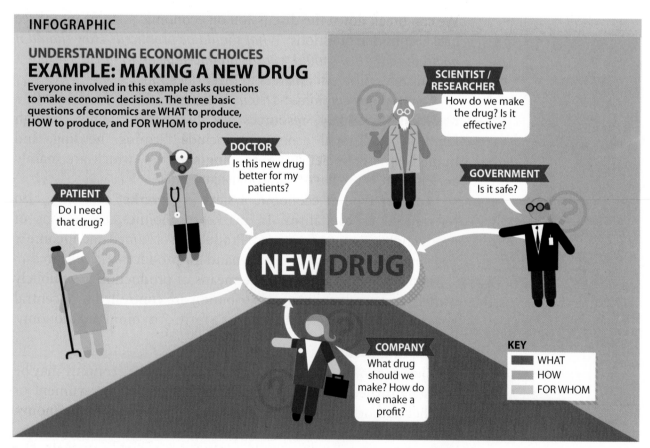

INFOGRAPHIC

UNDERSTANDING ECONOMIC CHOICES
EXAMPLE: MAKING A NEW DRUG
Everyone involved in this example asks questions to make economic decisions. The three basic questions of economics are WHAT to produce, HOW to produce, and FOR WHOM to produce.

SCIENTIST / RESEARCHER
How do we make the drug? Is it effective?

DOCTOR
Is this new drug better for my patients?

GOVERNMENT
Is it safe?

PATIENT
Do I need that drug?

NEW DRUG

COMPANY
What drug should we make? How do we make a profit?

KEY
WHAT
HOW
FOR WHOM

Every economic system must address three basic questions.

▶ **CRITICAL THINKING**
Determining Central Ideas What is scarcity? How are the three basic economic questions related to the problem of limited supply?

Think Again?

No countries today rely on a command economic system.

Not true. Some countries still have planned economies, including Cuba, Saudi Arabia, Iran, and North Korea. These nations have an economic system in which supply and prices are regulated by the government, not by the market.

We must weigh the opportunity cost, or the value of what we must give up to acquire something else, of using renewable resources versus nonrenewable resources. We must take into account these and many more considerations as we make choices now and in the future.

☑ **READING PROGRESS CHECK**

Describing How are renewable and nonrenewable resources alike, and how are they different?

Countries Build Economies

GUIDING QUESTION *What kinds of economic systems are used in our world today?*

Economic resources are another important resource. Economic resources include the goods and services a society provides and how they are produced, distributed, and used. How a society decides on the ownership and distribution of its economic resources is its **economic system**. Do you ever stop to think about the goods and services you use in a single day? How do these goods and services become available to you?

Different Economic Systems

We can break down the discussion on economic systems into three basic economic questions: *What should be produced? How should it be produced? How should what is produced be distributed?* Different nations have different answers to these questions.

One type of economic system is the traditional economy. In a **traditional economy**, resources are distributed mainly through families. Traditional economies include farming, herding, and hunter-gatherer societies. Developing societies, which are mainly agricultural, often have traditional economies.

Another type of economic system is a **market economy**, also referred to as capitalism. In market economies, the means of production are privately owned. Production is guided and income is distributed through sales and demand for products and resources.

In a **command economy**, the means of production are publicly owned, and production and distribution are controlled by a central governing authority. Communism is a form of a command economy.

What Is a Mixed Economy?

A **mixed economy** is just that—mixed. Parts of the economy may be privately owned, and parts may be owned by the government or another authority. The United States has a mixed market economy. Another economic system is socialism. In socialist societies, property and the distribution of goods and income are controlled by the community. How do individuals get the goods or services they need under each of these economic systems?

Parts of the Economy

We can break down the economy into parts. In economics, *land* is a factor of production that includes natural resources. Another factor of production is *labor*, which refers to all paid workers within a system. The other factor is *capital*, the human-made resources used to produce other goods. An *industry* is a branch of a business. For example, the *agriculture industry* grows crops and raises livestock. *Service industries* provide services rather than goods. Banking, retail, food service, transportation, and communications are examples of service industries.

Types of Economic Activities

Another way to view the parts of the economy is by the type of economic activity. Economists use the terms *primary sector*, *secondary sector*, and *tertiary sector* to group these activities. The primary sector includes activities that produce raw materials and basic goods. These activities include mining, fishing, agriculture, and logging.

The secondary sector makes finished goods. This sector includes home and building construction, food processing, and aerospace manufacturing. The tertiary sector of the economy is the service industry. Service sectors include sales, restaurants, banking, information technology, and health care.

Buyers and sellers come together in an outdoor market in Kunduz, Afghanistan.

▶ **CRITICAL THINKING**
Describing Is the seller involved in a primary, secondary, or tertiary economic activity?

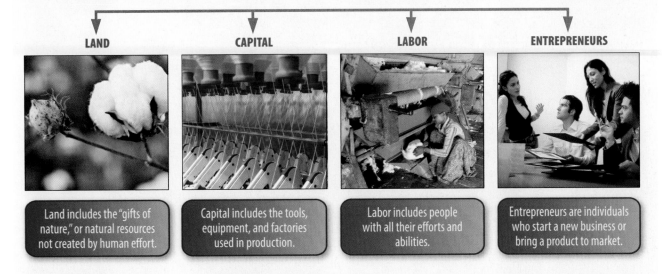

LAND	CAPITAL	LABOR	ENTREPRENEURS
Land includes the "gifts of nature," or natural resources not created by human effort.	Capital includes the tools, equipment, and factories used in production.	Labor includes people with all their efforts and abilities.	Entrepreneurs are individuals who start a new business or bring a product to market.

FACTORS OF PRODUCTION

The factors of production are broad categories of resources we need to produce the goods and services we want.

▶ **CRITICAL THINKING**

Identifying Machines, tools, and equipment are examples of what factor of production?

Economic Performance

Economic performance measures how well an economy meets the needs of society. Economic performance can be determined by several factors that measure economic success. The **gross domestic product (GDP)** is the total dollar value of all final goods and services produced in a country during a single year. The **standard of living** is the level at which a person, a group, or a nation lives as measured by the extent to which it meets its needs. These needs include food, shelter, clothing, education, and health care. Per capita income is the total national income divided by the number of people in the nation.

When referring to economics, **productivity** is a measurement of what is produced and what is required to produce it. Sustainable growth is the growth rate a business can maintain without having to borrow money. The employment rate is the percentage of the labor force that is employed. These factors help determine a nation's economic strength and performance.

Types of National Economies

National economies also can be classified by types. Developed countries are industrialized countries. Developing countries are less industrialized, agricultural countries that are working to become more advanced economically. Developing countries often have weak economies, and most of their population lives in poverty. Newly industrialized countries (NICs) are in the process of becoming developed and economically secure. Their economies are growing

and struggling to become fully developed, but they still face many economic and social challenges.

☑ **READING PROGRESS CHECK**

Describing How is standard of living a sign of economic performance?

A Global Economy

GUIDING QUESTION *How do the world's economies interact and affect one another?*

You have read about different economic systems and different types of economies. All the world's nations can be classified into the different economic categories. All nations must find ways to interact with one another. Look at the labels in your clothes or on other products you buy. How many different country names can you find? How and why do we get goods from far across the world?

Trade

Trade is the business of buying, selling, or bartering. When you buy something at a store, you are trading money for a product. On a much bigger scale, nations trade with each other. Countries have different resources. Resources can include raw materials, such as iron ore. Even labor may be cheaper in another country where workers earn lower wages. As a result, goods can be produced in some countries more easily or efficiently than in other countries.

Trade can benefit countries. One country can **export**, or send to another country, a product that it is able to produce. Another country **imports**, or buys that product from the exporting country.

GDP COMPARISON
Gross Domestic Product (GDP) represents the total dollar value of all goods and services produced over a specific period of time. The graph compares the GDP of eight countries over one year's time.

▶ **CRITICAL THINKING**

1. Integrating Visual Information What countries' GDP surpassed $6 trillion?

2. Integrating Visual Information Mexico's GDP has to grow by how many dollars to reach $2 trillion?

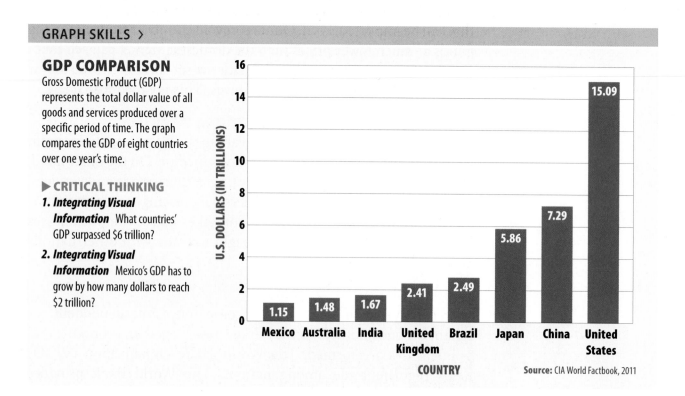

U.S. DOLLARS (IN TRILLIONS)

Country	GDP
Mexico	1.15
Australia	1.48
India	1.67
United Kingdom	2.41
Brazil	2.49
Japan	5.86
China	7.29
United States	15.09

COUNTRY

Source: CIA World Factbook, 2011

International trade involves preparing cargo for shipping (left) and transporting goods (right).

▶ **CRITICAL THINKING**

Describing What are the advantages of trade?

The country that imported the product can in turn export its products to another country. In global trade, extra fees are often added to the cost of importing products by a country's government. The extra money is a type of tax called a tariff. Governments often create tariffs to persuade their people to buy products made in their own country.

Sometimes a quota, a limit on the amount of one particular good that can be imported, is set. Quotas prevent countries from exporting goods at much lower prices than the domestic market can sell them for. A group of countries may decide to set little or no tariffs or quotas when trading among themselves. This is called **free trade**.

Advantages and Disadvantages of Trade

Trade has advantages and disadvantages. Trade can help build economic growth and increase a nation's income. On the other hand, jobs might be lost because of importing certain goods and services. With its benefits and its barriers, increasing trade leads to globalization. Economic globalization takes place when businesses move past national markets and begin to trade with other nations around the world.

Economic Organizations

In recent years, nations have become more interdependent, or reliant on one another. As they draw closer together, economic and political ties are formed. The World Trade Organization (WTO) helps regulate trade among nations. The World Bank provides

(l to r) Walter Hodges/Photodisc/Getty Images; Jens Kuhfs/Photographer's Choice/Getty Images

financing, advice, and research to developing nations to help them grow their economies. The International Monetary Fund (IMF) is a group which monitors economic development. The IMF also lends money to nations in need and provides training and technical help. One well-known policy and organization that promotes global trade is the North American Free Trade Agreement (NAFTA). NAFTA encourages free trade among the United States, Canada, and Mexico.

The European Union (EU) is a group of European countries that operate under one economic unit and one **currency,** or type of money—the euro. The Mercado Camon del Sur (formerly called MERCOSUR) is a group of South American countries that promote free trade, economic development, and globalization. The Mercado Camon del Sur helps countries make better use of their resources while preserving the environment.

The Association of Southeast Asian Nations (ASEAN) is a group of countries in Southeast Asia that promote economic, cultural, and political development. The Dominican Republic-Central America Free Trade Agreement (CAFTA-DR) is an agreement among the United States, five developing Central American countries, and the Dominican Republic. The agreement promotes free trade.

Whether a nation produces its own goods or trades, one basic principle exists: sustainability. The principle of **sustainability** is central to the discussion of resources. When a country focuses on sustainability, it works to create conditions where all the natural resources for meeting the needs of society are available.

What can countries do to ensure sustainability now and into the future? What can you do to plan for your future and the future of your community? Just as every nation is part of a global system, you are part of your community. The choices you make affect you and those around you. What can you do now to plan for a bright economic future?

Include this lesson's information in your Foldable®.

☑ **READING PROGRESS CHECK**

Analyzing How might globalization affect developing countries?

LESSON 3 REVIEW CCSS

Reviewing Vocabulary
1. Is coal a *renewable resource* or a *nonrenewable resource*?

Answering the Guiding Questions
2. *Describing* Why do people need natural resources?

3. *Identifying* What kinds of economic systems are used in our world today?

4. *Determining Central Ideas* How do the world's economies interact and affect one another?

5. *Argument Writing* Write a paragraph explaining the benefits of one economic system mentioned in this lesson. Make your opinions clear, and use facts from the lesson to present a strong argument as to why this system is better than other economic systems.

Directions: Write your answers on a separate piece of paper.

❶ Exploring the Essential Question

INFORMATIVE/EXPLANATORY Imagine that you are moving from a 12th-floor apartment in Chicago to a farm in Iowa. Write at least two paragraphs explaining what you might do to adapt.

❷ 21st Century Skills

ANALYZING Research the effects of human actions on an area where you live. Did the activities have a positive or negative impact on physical features and natural resources?

❸ Thinking Like a Geographer

INTEGRATING VISUAL INFORMATION Refer to the maps of the United States and China at the beginning of the chapter. Do you see any similarities or differences? What else do you know about the two countries from your textbook, newspapers, TV, or other media? Make a chart that lists ways in which the two countries are similar and different.

❹ GEOGRAPHY ACTIVITY

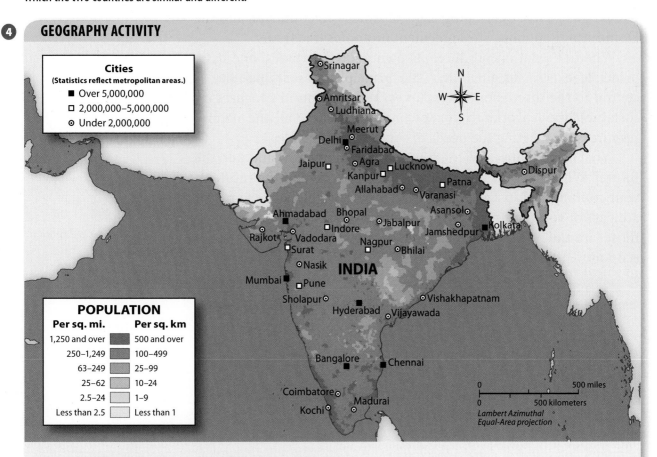

Study the map above and answer the following questions.

1. What is the population density around Mumbai, India?

2. About how many people live in Delhi?

3. What area in India has the highest population per square mile?

REVIEW THE GUIDING QUESTIONS

Directions: Choose the best answer for each question.

1 The world's rapidly increasing population is the result of
- A. global conflict.
- B. more births than deaths.
- C. urban growth.
- D. the need for more workers in some countries.

2 People who are forced to migrate from their homeland because of war are called
- F. refugees.
- G. migrants.
- H. rebels.
- I. patriots.

3 Which form of government gives one person absolute power to rule?
- A. constitutional monarchy
- B. dictatorship
- C. republic
- D. democracy

4 An example of a newly industrialized country is
- F. Greece.
- G. India.
- H. Australia.
- I. Russia.

5 A renewable resource is one that is
- A. inexpensive.
- B. a fossil fuel.
- C. always available naturally.
- D. easy to find.

DBQ ANALYZING DOCUMENTS

6 DETERMINING WORD MEANINGS A news story reports on India's population growth:

"*India . . . will surpass China to become the world's most populous country in less than two decades. The population growth will mean a nation full of working-age youth, which economists say could allow the already booming economy to maintain momentum.*"

—from Anjana Pasricha, "India Challenged to Provide Jobs, Education to Young Population," October 2011

What does "maintain momentum" refer to in this story?

A. continue population growth

B. produce more goods for export

C. keep up the economic boom

D. continue India's housing boom

7 IDENTIFYING Which of these issues do you think might be a challenge for India as a result of this population growth?

F. having enough workers for its companies

G. educating young people

H. facing the problems of an aging population

I. improving the economy

SHORT RESPONSE

"*A school that is . . . easy for students, teachers, [and] parents . . . to reach on foot or by bicycle helps reduce the air pollution from automobile use, protecting children's health. Building schools . . . in the neighborhoods they serve minimizes the amount of paved surface . . ., which can help protect water quality by reducing polluted runoff.*"

—from Environmental Protection Agency, "Smart Growth and Schools"

8 DETERMINING CENTRAL IDEAS What is the main idea of this paragraph?

9 ANALYZING How would building community schools affect energy use? Why?

EXTENDED RESPONSE

10 ARGUMENT WRITING Historically, one of the functions of government has been to invest in development for the well-being of future generations. Think about what might happen in the future if oil reserves run out. Of the alternative energy sources described in this chapter, choose one that you think the government should be developing. Research it and then write a letter to your senator or congressperson expressing your concern, sharing the results of your research, and asking your representative how he or she might address the problem of dwindling oil reserves.

Need Extra Help?

If You've Missed Question	❶	❷	❸	❹	❺	❻	❼	❽	❾	❿
Review Lesson	1	1	2	3	3	1	1	3	3	3

"India Challenged to Provide Jobs, Education to Young Population," by Anjana Pasricha, October 31, 2011. *Voice of America*, http://voanews.com; "Smart Growth and Schools," The United States Environmental Protection Agency official Web site. http://www.epa.gov/dced/schools.htm

NORTH AMERICA

UNIT **2**

EXPLORE the CONTINENT

NORTH AMERICA is made up of three large countries—Canada, the United States, Mexico—plus the Caribbean Islands and the countries of Central America. North America is a cultural kaleidoscope—a mixture of the many groups who have settled in the region.

1 **LANDFORMS** At 800 miles long (1,287 km), the Baja Peninsula of Mexico is one of the world's longest peninsulas. Running nearly its entire length is a chain of mountain ranges. Four deserts are also physical features of Baja.

2 **BODIES OF WATER** Boats line the harbor of a fishing village in Nova Scotia, along Canada's Atlantic coast. In addition to the waters of the Atlantic, the coastal waters of the Pacific Ocean and the Gulf of Mexico are important fisheries.

(bkgd) Ron and Patty Thomas/Photographer's Choice/Getty Images; (l to r) Rory T B Moore Images/Flickr/Getty Images; Blaine Harrington III/Corbis; Reto Stockli, NASA Earth Observatory

3

3 **NATURAL RESOURCES** Wheat is grown on the Great Plains of the United States, a region often called the Wheat Belt. The type of wheat grown depends on the climate. Farmers in the northern plains, with their short growing season, plant wheat in the spring and harvest it in the fall. Farther south, farmers plant winter wheat, which is harvested in early summer.

FAST
FACT

Canada is the world's second largest country in area.

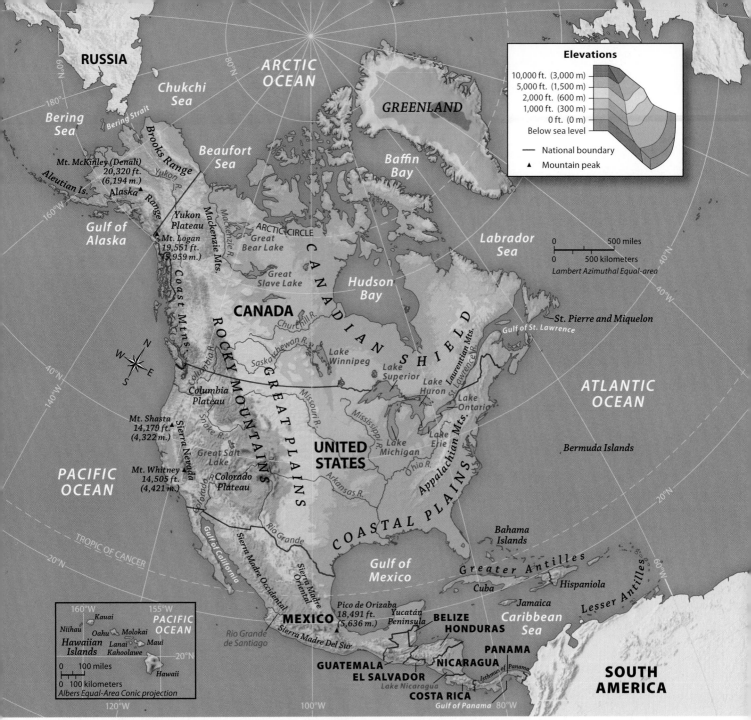

Elevations

| 10,000 ft. (3,000 m) |
| 5,000 ft. (1,500 m) |
| 2,000 ft. (600 m) |
| 1,000 ft. (300 m) |
| 0 ft. (0 m) |
| Below sea level |

— National boundary
▲ Mountain peak

RUSSIA

ARCTIC OCEAN

Chukchi Sea

Bering Sea

Bering Strait

GREENLAND

Baffin Bay

Beaufort Sea

Brooks Range

Mt. McKinley (Denali) 20,320 ft. (6,194 m.)

Aleutian Is.

Alaska Range

Gulf of Alaska

Yukon Plateau

Yukon R.

Mt. Logan 19,551 ft. (5,959 m.)

Mackenzie Mts.

ARCTIC CIRCLE

Great Bear Lake

Great Slave Lake

Hudson Bay

Labrador Sea

St. Pierre and Miquelon

Coast Mts.

CANADA

ROCKY MOUNTAINS

C A N A D I A N   S H I E L D

Churchill R.

Saskatchewan R.

Lake Winnipeg

Lake Superior

Lake Huron

Gulf of St. Lawrence

St. Lawrence R.

Laurentian Mts.

ATLANTIC OCEAN

Columbia Plateau

Columbia R.

G R E A T   P L A I N S

Missouri R.

Mississippi R.

Lake Michigan

Lake Ontario

Lake Erie

Bermuda Islands

Mt. Shasta 14,179 ft. (4,322 m.)

Sierra Nevada

Snake R.

Great Salt Lake

UNITED STATES

Appalachian Mts.

Mt. Whitney 14,505 ft. (4,421 m.)

Colorado Plateau

Colorado R.

Arkansas R.

Ohio R.

PACIFIC OCEAN

C O A S T A L   P L A I N S

TROPIC OF CANCER

Gulf of California

Rio Grande

Gulf of Mexico

Bahama Islands

Greater Antilles

Cuba

Hispaniola

Jamaica

Lesser Antilles

Sierra Madre Occidental

Sierra Madre Oriental

Pico de Orizaba 18,491 ft. (5,636 m.)

Yucatán Peninsula

MEXICO

BELIZE

HONDURAS

Caribbean Sea

Rio Grande de Santiago

Sierra Madre Del Sur

GUATEMALA

EL SALVADOR

NICARAGUA

Lake Nicaragua

COSTA RICA

PANAMA

Isthmus of Panama

Gulf of Panama

SOUTH AMERICA

0   500 miles
0   500 kilometers
Lambert Azimuthal Equal-area

160°W   155°W
PACIFIC OCEAN

Kauai
Niihau   Oahu   Molokai
Hawaiian   Lanai   Maui
Islands   Kahoolawe

20°N

Hawaii

0   100 miles
0   100 kilometers
Albers Equal-Area Conic projection

NORTH AMERICA

PHYSICAL

MAP SKILLS

1 **PHYSICAL GEOGRAPHY** What physical features probably acted as barriers to settlement in the United States and Canada?

2 **THE GEOGRAPHER'S WORLD** Which body of water is located east of Panama?

3 **PHYSICAL GEOGRAPHY** What mountain range lies in the eastern United States?

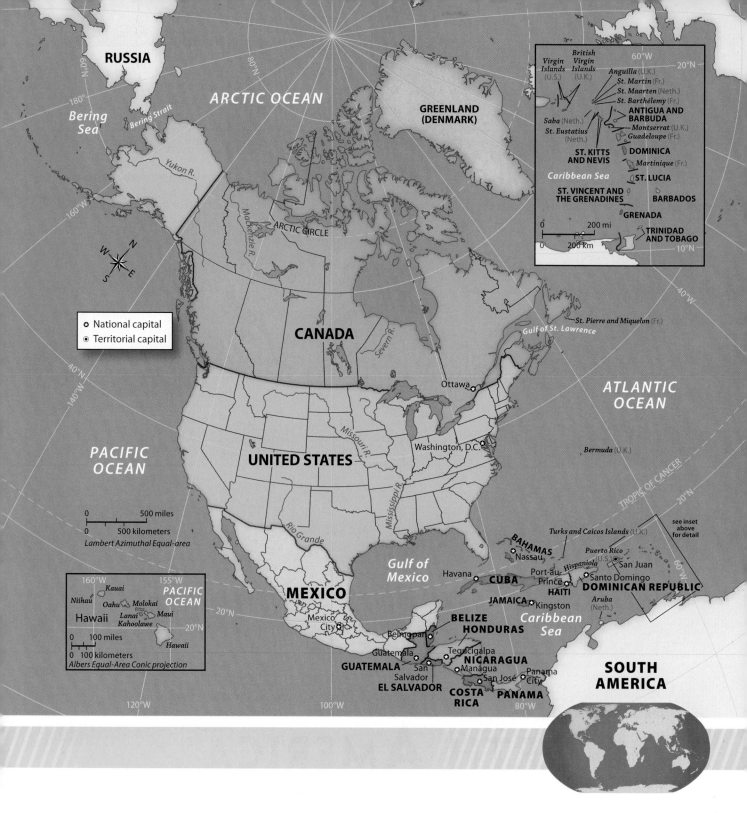

POLITICAL

MAP SKILLS

1 **PLACES AND REGIONS** What is the capital of Canada?

2 **THE GEOGRAPHER'S WORLD** What physical feature forms the natural boundary between the United States and Mexico?

3 **PLACES AND REGIONS** What is the capital of Cuba?

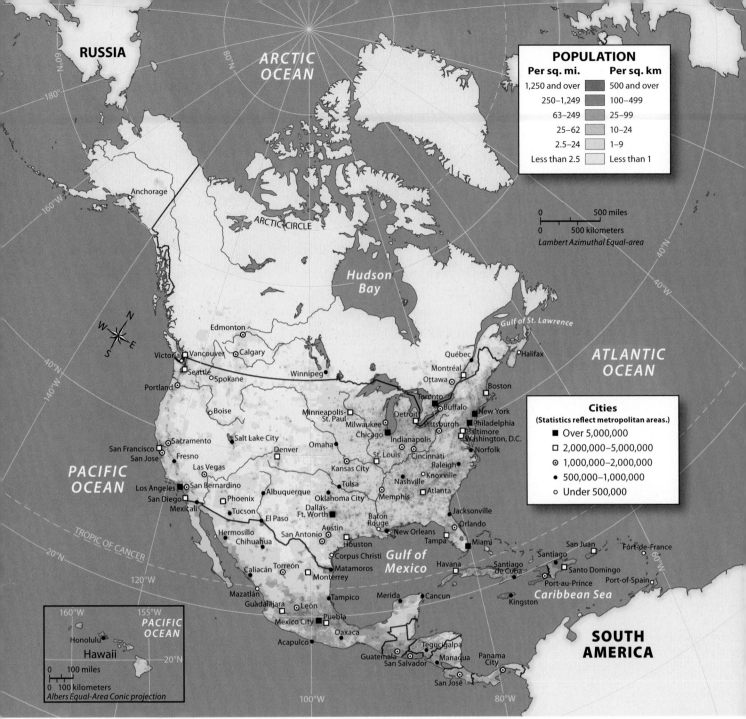

POPULATION

Per sq. mi.	Per sq. km
1,250 and over	500 and over
250–1,249	100–499
63–249	25–99
25–62	10–24
2.5–24	1–9
Less than 2.5	Less than 1

0   500 miles
0   500 kilometers
Lambert Azimuthal Equal-area

Cities
(Statistics reflect metropolitan areas.)
- ■ Over 5,000,000
- □ 2,000,000–5,000,000
- ⊙ 1,000,000–2,000,000
- • 500,000–1,000,000
- ○ Under 500,000

PACIFIC
OCEAN

160°W	155°W	
		PACIFIC OCEAN
Honolulu		
	Hawaii	20°N

0   100 miles
0   100 kilometers
Albers Equal-Area Conic projection

NORTH AMERICA

POPULATION DENSITY

MAP SKILLS

1 PLACES AND REGIONS Where is the greatest population density located in the United States?

2 PLACES AND REGIONS Contrast the population density of northern Mexico with southern Mexico.

3 ENVIRONMENT AND SOCIETY What generalizations can you make about the populations of the Caribbean islands?

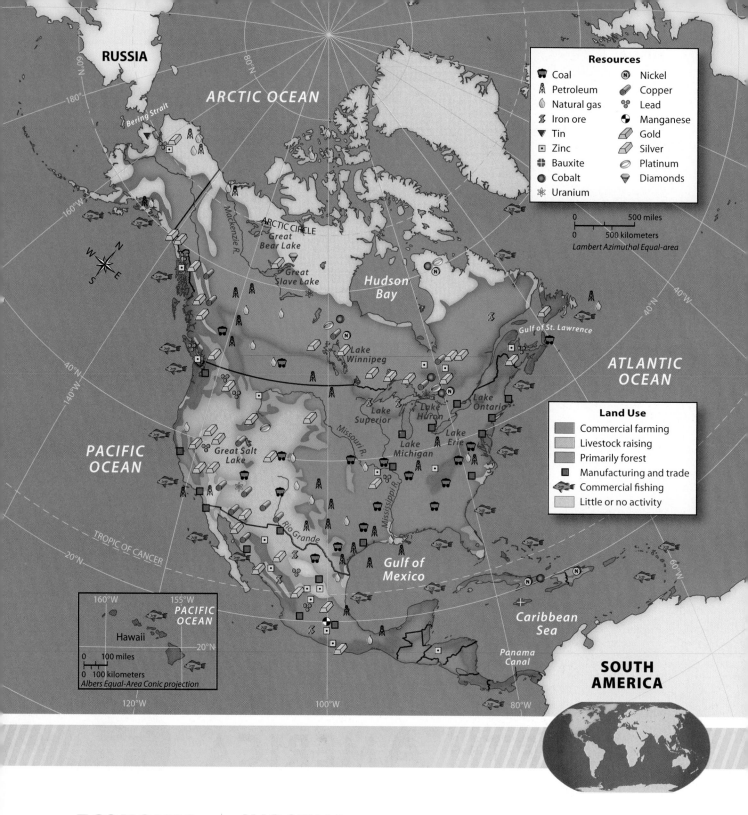

Resources

Coal		Nickel	
Petroleum		Copper	
Natural gas		Lead	
Iron ore		Manganese	
Tin		Gold	
Zinc		Silver	
Bauxite		Platinum	
Cobalt		Diamonds	
Uranium			

0   500 miles
0   500 kilometers
Lambert Azimuthal Equal-area

Land Use

- Commercial farming
- Livestock raising
- Primarily forest
- Manufacturing and trade
- Commercial fishing
- Little or no activity

RUSSIA

ARCTIC OCEAN

ATLANTIC OCEAN

PACIFIC OCEAN

ARCTIC CIRCLE
Great Bear Lake
Mackenzie R.
Great Slave Lake
Hudson Bay
Gulf of St. Lawrence
Lake Winnipeg
Lake Superior
Lake Huron
Lake Michigan
Lake Ontario
Lake Erie
Missouri R.
Great Salt Lake
Mississippi R.
Rio Grande
TROPIC OF CANCER
Gulf of Mexico
Caribbean Sea
Panama Canal
SOUTH AMERICA
Bering Strait

PACIFIC OCEAN
Hawaii
0   100 miles
0   100 kilometers
Albers Equal-Area Conic projection

ECONOMIC RESOURCES

MAP SKILLS

1 **HUMAN GEOGRAPHY** Describe how most land is used throughout the United States.

2 **HUMAN GEOGRAPHY** What is the main economic activity near Panama?

3 **HUMAN GEOGRAPHY** Contrast how land is used in northern Canada with Central America.

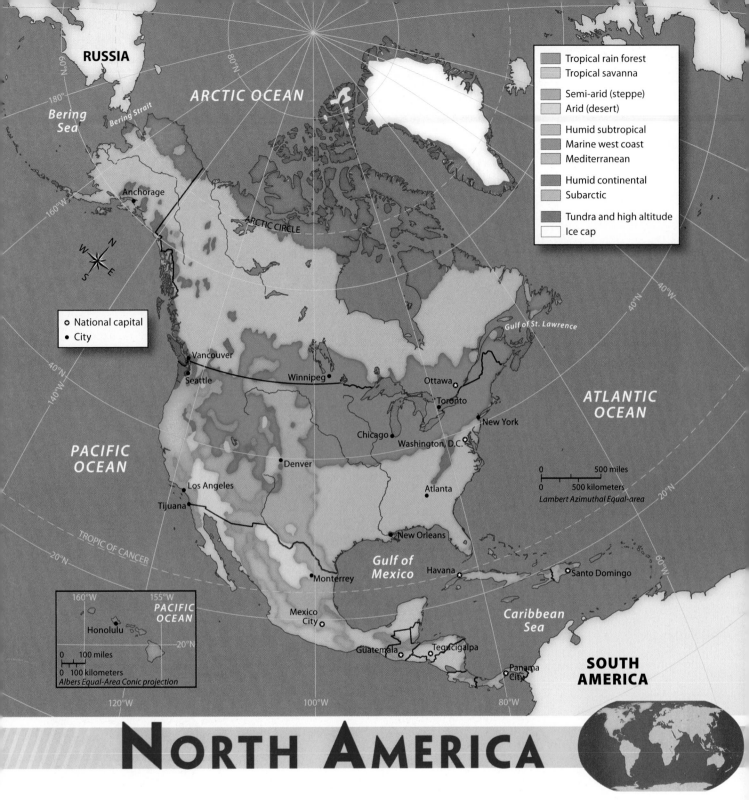

Legend

	Tropical rain forest
	Tropical savanna
	Semi-arid (steppe)
	Arid (desert)
	Humid subtropical
	Marine west coast
	Mediterranean
	Humid continental
	Subarctic
	Tundra and high altitude
	Ice cap

⊕ National capital
• City

RUSSIA

ARCTIC OCEAN

Bering Sea

Bering Strait

Anchorage

ARCTIC CIRCLE

PACIFIC OCEAN

Vancouver

Seattle

Winnipeg

Ottawa

Toronto

New York

Chicago

Washington, D.C.

Denver

Atlanta

Los Angeles

Tijuana

TROPIC OF CANCER

New Orleans

Monterrey

Gulf of Mexico

Havana

Santo Domingo

Caribbean Sea

Mexico City

Guatemala

Tegucigalpa

Panama City

SOUTH AMERICA

Gulf of St. Lawrence

ATLANTIC OCEAN

0   500 miles
0   500 kilometers
Lambert Azimuthal Equal-area

PACIFIC OCEAN

Honolulu

0   100 miles
0   100 kilometers
Albers Equal-Area Conic projection

NORTH AMERICA

CLIMATE

MAP SKILLS

1 PHYSICAL GEOGRAPHY What type of climate is most common throughout Canada?

2 PLACES AND REGIONS The Caribbean islands and central Mexico appear to have two things in common. What are these commonalities?

3 PHYSICAL GEOGRAPHY What is the overall climate of the southeastern United States?

THE UNITED STATES EAST OF THE MISSISSIPPI RIVER

network

There's More Online about The United States East of the Mississippi River.

CHAPTER 4

Lesson 1
Physical Features

Lesson 2
History of the Region

Lesson 3
Life in the Region

ESSENTIAL QUESTIONS · *How does geography influence the way people live?*
· *Why is history important?* · *What makes a culture unique?*

Christopher Morris/VII/Corbis

Worker at Michigan auto assembly plant

The Story Matters...

The eastern United States is a region of diverse physical features and many natural resources. Bordered by the Atlantic Ocean and the Mississippi River, this region is home to mighty rivers, the largest group of freshwater lakes in the world, and old-growth forests. The physical features and wealth of resources found in the eastern United States have played a key role in the region's history. They also influence where and how people live in this region today.

FOLDABLES®
Study Organizer

Go to the Foldables® library in the back of your book to make a Foldable® that will help you take notes while reading this chapter.

| Diversity |
| Geographic Barriers |
| East and West |
| The US East of the Mississippi |

THE UNITED STATES EAST OF THE MISSISSIPPI RIVER

The United States east of the Mississippi River is one of the two regions that make up the United States. As you study the map, identify the geographic features of the region.

Step Into the Place

MAP FOCUS Use the map to answer the following questions.

1 THE GEOGRAPHER'S WORLD Which body of water lies to the east of the United States?

2 THE GEOGRAPHER'S WORLD Which state has the longest coastline?

3 PLACES AND REGIONS Name the states that use the Mississippi River as all or part of their western border.

4 CRITICAL THINKING ANALYZING Why do you think rivers such as the Mississippi and the Ohio were used as state borders?

THE CAPITOL The U.S. Capitol in Washington, D.C., is the meeting place of Congress, the nation's legislature.

THE GREAT SMOKY MOUNTAINS Part of the Appalachian Mountain Range, the Smoky Mountains are named for the blue-gray mist that seems to hang above the peaks and valleys.

Step Into the Time

DESCRIBING Choose an event from the time line and write a paragraph describing a social, environmental, or economic effect that event had on the region, the country, or the world.

1620 Pilgrims' ship *Mayflower* arrives in Plymouth

1776 Declaration of Independence is signed

1800

1825 Erie Canal links New York City and the Great Lakes

1861 U.S. Civil War begins

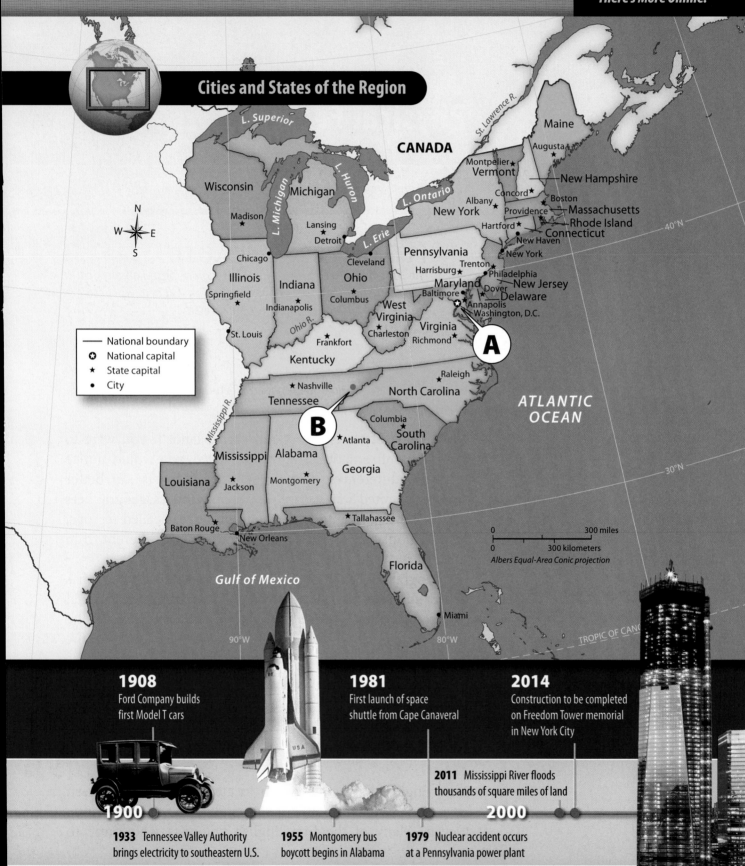

Cities and States of the Region

L. Superior

CANADA

St. Lawrence R.

Maine
Augusta ★

Montpelier ★
Vermont

New Hampshire

Wisconsin

Michigan

L. Michigan

L. Huron

L. Ontario

Concord ★

Boston

New York
Albany ★

Providence
Massachusetts

Rhode Island

Madison ★

Lansing ★
Detroit

L. Erie

Hartford ★
Connecticut

New Haven

Chicago ●

Cleveland ●

Pennsylvania

New York ●

Illinois

Indiana

Ohio

Harrisburg ★

Trenton ★

Philadelphia ●
New Jersey

Springfield ★

Indianapolis ★

Columbus ★

Maryland

Dover ★
Delaware

St. Louis ●

Ohio R.

West
Virginia

Baltimore ●

Annapolis ★
Washington, D.C.

N
W   E
S

Frankfort ★

Charleston ★

Virginia

Richmond ★

A

Kentucky

Mississippi R.

— National boundary
✪ National capital
★ State capital
● City

Raleigh ★

★ Nashville

North Carolina

Tennessee

B

Columbia ★

ATLANTIC
OCEAN

Mississippi

Alabama

★ Atlanta

South
Carolina

Georgia

Louisiana

Jackson ★

Montgomery ★

Baton Rouge ★

New Orleans ●

★ Tallahassee

0   300 miles
0   300 kilometers
Albers Equal-Area Conic projection

40°N

30°N

Florida

Gulf of Mexico

90°W

80°W

Miami ●

TROPIC OF CANCER

1908
Ford Company builds
first Model T cars

1981
First launch of space
shuttle from Cape Canaveral

2014
Construction to be completed
on Freedom Tower memorial
in New York City

2011 Mississippi River floods
thousands of square miles of land

1900

2000

1933 Tennessee Valley Authority
brings electricity to southeastern U.S.

1955 Montgomery bus
boycott begins in Alabama

1979 Nuclear accident occurs
at a Pennsylvania power plant

networks

There's More Online!

☑ **MAP** Atlantic Coast Fall Line

☑ **SLIDE SHOW** Agriculture East of the Mississippi River

☑ **ANIMATION** How the Great Lakes Formed

☑ **VIDEO**

Lesson 1
Physical Features

ESSENTIAL QUESTION • *How does geography influence the way people live?*

(l to r) Karl Weatherly/The Image Bank/Getty Images; Scenics of America/PhotoLink/Getty Images; J. R. Factor/Photo Researchers; Kennan Harvey/Getty Images

Reading HELPDESK (CCSS)

Academic Vocabulary

- **parallel**

Content Vocabulary

- **subregion**
- **lock**
- **tributary**
- **levee**
- **coastal plain**
- **fall line**
- **hurricane**

TAKING NOTES: *Key Ideas and Details*

Organize As you read about the region's physical landscape, take notes on a graphic organizer like this one.

U.S. East of the Mississippi
- Landscape
- Bodies of Water

IT MATTERS BECAUSE

The United States can be divided into regions based on physical characteristics. Learning about each region will help you better understand our nation's geographic diversity.

The Regions

GUIDING QUESTION *How do the physical features of the eastern United States make the region unique?*

The United States is a vast and varied land. If you were to view our entire nation from outer space, you would notice dramatic differences between its various parts. To better study the United States as well as other countries, geographers divide these parts into large geographic areas called regions. Each region's characteristics make it distinctly different from the others.

Geographers can further divide regions into smaller parts called **subregions**. Like a region, a subregion has special features that make it unique. The most basic way to divide the United States is into two regions: the United States east of the Mississippi and the United States west of the Mississippi. The Mississippi River is the dividing line between the two regions. In this lesson, you will learn about the United States east of the Mississippi and its four subregions: New England, the Mid-Atlantic, the Midwest, and the Southeast.

New England

New England is the subregion located in the northeastern corner of the United States, between Canada and the Atlantic Ocean. Many of the first English colonists who came to

America during the 1600s settled in this area. The settlers named the area New England in honor of their distant homeland. New England includes the states of Maine, New Hampshire, Vermont, Massachusetts, Rhode Island, and Connecticut.

The Mid-Atlantic

Located along the Atlantic coast, just south of New England, is the Mid-Atlantic subregion. The Mid-Atlantic includes the states of Delaware, Maryland, New Jersey, New York, and Pennsylvania. These states were part of America's original thirteen colonies. Our nation's capital, Washington, D.C., is also located in the Mid-Atlantic.

The Midwest

The states of Illinois, Indiana, Michigan, Ohio, and Wisconsin are part of the subregion called the Midwest. All five of these states share borders with one or more of the Great Lakes. The Midwest is nicknamed "the nation's breadbasket" because a large percentage of America's food crops are grown in its rich soil.

The Southeast

The Southeast is the largest subregion in the eastern United States. The Southeast is made up of 11 states: Alabama, Florida, Georgia, Kentucky, Louisiana, Mississippi, North Carolina, South Carolina, Tennessee, Virginia, and West Virginia. Some Southeastern states have long coastal borders where they meet the Atlantic Ocean or the Gulf of Mexico.

☑ **READING PROGRESS CHECK**

Determining Central Ideas Why do you think geographers divide the United States at the Mississippi River instead of dividing it through the middle into equal halves?

Bodies of Water

GUIDING QUESTION *Which of North America's major bodies of water are located east of the Mississippi?*

Oceans, lakes, and rivers have helped make this region prosperous. Oceans link the region to other countries for trade. An abundant supply of freshwater provides power for homes and industries.

MAP SKILLS

1 **PLACES AND REGIONS** Which subregion includes the state of Alabama?

2 **THE GEOGRAPHER'S WORLD** Which subregion extends the farthest north?

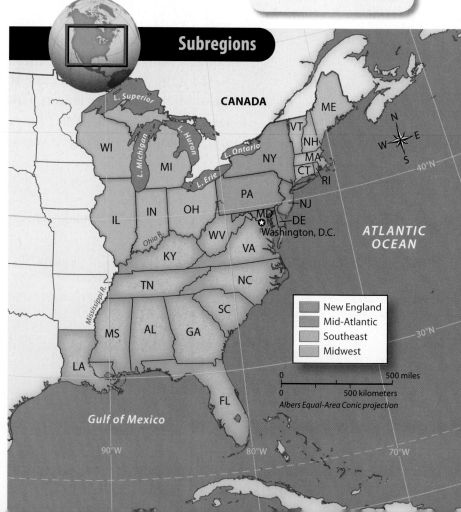

Subregions

CANADA

L. Superior

L. Michigan

L. Huron

L. Ontario

L. Erie

WI

MI

ME

VT

NH

NY

MA

CT

RI

IL

IN

OH

PA

NJ

MD

DE

Washington, D.C.

WV

VA

KY

Ohio R.

TN

NC

SC

Mississippi R.

MS

AL

GA

LA

FL

ATLANTIC OCEAN

Gulf of Mexico

40°N

30°N

90°W

80°W

70°W

New England
Mid-Atlantic
Southeast
Midwest

0   500 miles

0   500 kilometers

Albers Equal-Area Conic projection

The Atlantic Ocean and the Gulf Coast

The eastern United States is nearly surrounded by water. The largest body of water east of the Mississippi is the Atlantic Ocean. This enormous salt-water ocean borders the states along the East Coast. The East Coast is a shoreline that stretches for more than 2,000 miles (3,219 km), from Maine in the north to Florida in the south. It is jagged and rocky in New England but smooth and sandy in the Mid-Atlantic and Southeast. The Atlantic Ocean affects the region's land, weather, economy, and people in many ways.

The East Coast borders the Atlantic Ocean. The Gulf Coast borders a smaller body of water called the Gulf of Mexico. The Gulf of Mexico covers an area of about 600,000 square miles (1,550,000 sq. km) and is nearly surrounded by land. Several currents flow through the Gulf of Mexico like giant underwater rivers. One of the currents feeds into the Gulf Stream, a powerful current that flows through the Atlantic Ocean.

The Gulf Coast extends from Florida to Texas and Mexico. The land along the Gulf Coast varies from sandy beaches to marshes, bays, and lagoons. Waters in the Gulf of Mexico are warmer and generally calmer than those of the Atlantic.

The Great Lakes

The term *Great Lakes* refers to a cluster of five huge lakes located in the American Midwest and central Canada. These lakes were formed thousands of years ago when massive glaciers carved out the ground and melted over time. The Great Lakes form the largest group of freshwater lakes in the world. Together, the Great Lakes hold more liquid freshwater than any other location on Earth.

Moving from west to east, the Great Lakes are Lake Superior, Lake Michigan, Lake Huron, Lake Erie, and Lake Ontario. The five lakes are connected. Water flows west to east from one lake to the next, eventually making its way to the long St. Lawrence River.

Destin Beach is located on the Florida Panhandle along the Gulf of Mexico.

▶ **CRITICAL THINKING**

Describing How far does the Gulf Coast extend?

Karl Weatherly/The Image Bank/Getty Images

The St. Lawrence Seaway

The St. Lawrence River carries the water eastward for 750 miles (1,207 km), until it empties into the Atlantic Ocean. Because the Great Lakes border the United States and Canada, these nations work together to set up environmental programs for the region. Important goals include addressing population threats and protecting the health and safety of people living in the Great Lakes region.

During the 1950s, the United States and Canada worked together to build canals and gated passageways called **locks** between the Great Lakes and into the St. Lawrence River. The locks and canals made it possible for ships to travel the entire length of the Great Lakes and the St. Lawrence River. The final passageway, extending 2,340 miles (3,766 km) from Lake Superior to the Atlantic Ocean, is called the St. Lawrence Seaway. The St. Lawrence Seaway connects the Midwest to seaports all over the world. This has made it faster and easier for businesses in the Midwest to ship their products to buyers worldwide.

The Mississippi River

The "Mighty Mississippi" is one of the longest rivers in North America. Many people consider it the most important river in the United States. From its source in Minnesota, the Mississippi River winds its way southward for 2,350 miles (3,782 km). **Tributaries** such as the Missouri and Ohio rivers feed into the Mississippi, adding to its strength and volume. The Mississippi River ends at the point where it empties into the Gulf of Mexico.

Since early settlers arrived in America, the Mississippi River has affected the settlement patterns, the economy, and the lifestyles of countless Americans. People have used the river for transportation for hundreds of years.

The Welland Canal (left) begins on Lake Ontario. The St. Lawrence Seaway connects Chicago's harbors to the Atlantic Ocean. (right)

▶ CRITICAL THINKING

Describing How do canals aid ship travel?

The Mississippi River is the largest river system in North America.

▶ **CRITICAL THINKING**

Analyzing How do its many tributaries change the Mississippi River?

Ships and steamboats filled with passengers and cargo can follow the wide river and its tributaries for thousands of miles. This vast stretch makes the Mississippi one of the world's busiest commercial waterways.

In the past, the Mississippi would often flood its banks, dumping millions of tons of water and sediment onto the land. The sediment enriched the soil in farm fields, but the floods also destroyed homes and washed away entire fields of crops. The government built **levees**—embankments to control the flooding and reduce the damage to homes and crops. Unfortunately, levees also block the sediment that used to replenish farm fields.

The powerful Mississippi River has influenced the nation's history more than any other river. The river's importance, and the respect humans have for it, is shown in its name: *Mississippi* is a Choctaw word meaning "Great Water" or "Father of Waters."

Rivers as Boundaries

Rivers make natural boundaries. The Mississippi River forms much of the western border of the states of Wisconsin, Illinois, Kentucky, Tennessee, and Mississippi.

Rivers are boundaries for counties and cities, too. For example, the Tennessee River in northwest Alabama forms the border between Lauderdale County and Colbert County.

Rivers are examples of physical systems that form political boundaries. Mountains and lakes are other physical features that may be used as political boundaries. Structures that humans make, such as streets and roads, can also set boundaries.

Scenics of America/PhotoLink/Getty Images

The Ohio River

The Ohio River carries more water to the Mississippi River than any of the other tributaries. The Ohio River begins where the Allegheny and Monongahela rivers combine in western Pennsylvania. From Pennsylvania, the Ohio River flows westward for 981 miles (1,579 km), forming a wide, watery border that separates the states of Ohio, Indiana, and Illinois that lie along the north side of the river from West Virginia and Kentucky on its south side. Like the Mississippi, the Ohio River has long been an important shipping and transportation route. The Ohio River connects much of the Midwest to the Mississippi River. Both of these river systems have affected our nation's land, its people, and its history.

☑ **READING PROGRESS CHECK**

Determining Central Ideas How could a logging company in Kentucky send logs to a buyer on the Gulf of Mexico using an all-water route?

Physical Landscape

GUIDING QUESTION *What characteristics make the physical landscape east of the Mississippi unique?*

The Atlantic Coastal Plain

The East Coast of the United States sits at the edge of a huge continental platform. Most of this platform is underwater, forming a shelf around the Atlantic coastline. But over time, a large area of the platform rose above sea level. Ocean waves washed over the platform for millions of years, leaving behind layers of sandy sediment. As the sediment built up, a flat lowland called the **coastal plain** formed. The coastal plain stretches from the northeastern U.S. to Mexico. In places, the coastal plain was crushed under the weight of glaciers, pushing the land below sea level. These areas often become flooded by fierce storms and heavy rains.

The Appalachian Mountains

The Appalachian Mountain system is the oldest, longest chain of mountains in the United States east of the Mississippi River. It begins in Alabama and continues 1,500 miles (2,414 km) northeast to the Canadian border. Dense forests cover much of the Appalachian Mountains, which are known for their rugged beauty.

The mountains of the Appalachian system stand side by side in **parallel** ranges. Two of the most well-known Appalachian Mountain ranges are the Blue Ridge Mountains in Virginia and the Great Smoky Mountains in Tennessee.

Even though central New York is more than 200 miles (322 km) from the Atlantic Ocean, scientists have found fossils of marine organisms there.

▶ **CRITICAL THINKING**
Analyzing What is an explanation for finding the marine fossils so far from the ocean?

Academic Vocabulary

parallel extending side by side in the same direction, always the same distance apart

J. R. Factor/Photo Researchers

The Pisgah National Forest in western North Carolina is a land of heavy forests and many waterfalls.

▶ **CRITICAL THINKING**

Explaining What is a fall line?

These mountain ranges were formed from sedimentary rock by powerful upheavals within Earth's crust. The mountains have worn down over time because of natural erosion. Compared to younger mountain ranges such as the Rockies in the western United States, the Appalachian Mountains show their age in their worn, rounded appearance.

The Appalachian Mountains are home to many natural wonders. Old-growth forests are filled with diverse plant and animal life. Some of the most spectacular features of the Appalachians are the thousands of waterfalls that decorate the landscape. The many waterfalls are evidence that a fall line runs through the region. A **fall line** is an area where waterfalls flow from higher to lower ground. In this region, a fall line stretches for hundreds of miles between New Jersey and South Carolina. This fall line is a long, low cliff that runs parallel to the Atlantic coast. Throughout New England, the Mid-Atlantic, and the Southeast, waterfalls spill over this fall line. The fall line forms a boundary between higher, upland areas and the Atlantic coastal plain. Many cities originally located along the fall line because waterfalls provide water power, a renewable resource.

Climate in the Eastern United States

The climate of the eastern United States is as varied as the landscape. The changing seasons in most places east of the Mississippi are quite noticeable. New England and the Midwest see the most dramatic seasonal changes. These regions have cold winters and hot, humid summers. Autumn is cool and colorful as the leaves change color. Springtime brings rainy and snowy weather and strong storms.

Coastal areas tend to have mild climates. States along the East Coast still experience seasons, but temperatures are less extreme than they are inland. States located farther south experience milder changes in seasons.

Much of the Southeast has a humid subtropical climate. Summers are rainy and hot, and winters are cooler and drier. In general, climates of the eastern United States are more humid and rainy than climates of the West. In late summer and early autumn, **hurricanes**— ocean storms that span hundreds of miles with winds of at least 74 miles per hour (119 km per hour)—can pound the coastline. One of the most damaging hurricanes in history, Hurricane Katrina, struck the Gulf Coast in August 2005. More than 1,800 people died, and hundreds of thousands lost their homes.

Kennan Harvey/Getty Images

Minerals and Energy Resources

A wealth of resources is hidden below the surface of the region, and two of the most valuable materials are minerals and energy resources. Minerals are natural substances such as iron ore, gold, and zinc. These minerals can be processed into metals. Metals and other forms of minerals are used in manufacturing and construction.

Energy resources, such as coal, oil, and natural gas, are called fossil fuels. Burning coal can produce electricity. Oil is processed into fuel for cars and other vehicles. Natural gas is used to heat our homes and to generate electricity. The demand for mineral and energy resources is huge. Mining them is a major industry in the U.S. east of the Mississippi River. Minerals and energy resources are mined from inside mountains and from deep under the ground. Some mining methods harm the environment by damaging the land and polluting the water, soil, and air.

Farming and Industry

One of the most valuable resources east of the Mississippi River is farmland. The rich soil is excellent for growing crops such as grains, fruits, and vegetables. Sandy soils in the Southeast are good for growing cotton. Excellent growing conditions have helped the region become a major producer of meat, dairy foods, wood, cotton, sugar, corn, wheat, soybeans, and other food crops.

Industries such as logging, mining, and fishing are a way of life throughout the region. Products such as automobiles, electronics, and clothing are made in factories in these cities. Information technology (IT) and tourism are also important. Plentiful resources and hard-working people make the eastern United States one of the most productive regions in the world.

Include this lesson's information in your Foldable®.

☑ **READING PROGRESS CHECK**

Explaining Why is farmland considered a natural resource?

LESSON 1 REVIEW **CCSS**

Reviewing Vocabulary

1. How does a *tributary* affect the amount of water flowing through a river?

Answering the Guiding Questions

2. ***Determining Central Ideas*** Why do you think geographers divide the eastern U.S. into four subregions?

3. ***Identifying*** Which of the four subregions of the Eastern United States does not border the Atlantic Ocean?

4. ***Analyzing*** In what ways are the East Coast and the Eastern Gulf Coast alike? In what ways are they different?

5. ***Describing*** How do locks in the Great Lakes and the St. Lawrence Seaway affect transportation in the Midwest?

6. ***Analyzing*** Think about how mining minerals and energy resources can damage the environment. Brainstorm a creative solution to this problem. Describe your solution.

7. ***Identifying*** Crops of citrus fruits, such as oranges and lemons, will die in freezing temperatures. Which subregion of the Eastern U.S. has the best climate for growing citrus fruits?

8. ***Argument Writing*** Write a persuasive letter encouraging a friend or family member to visit one of the four subregions of the eastern U.S. Include details about the region's physical features, resources, and climate in your letter.

networkrks

There's More Online!

☑ **IMAGE** Early Settlements

☑ **MAP** Cultures and Communities
on the Mississippi River

☑ **VIDEO**

Reading HELPDESK CCSS

Academic Vocabulary

- **isolate**

Content Vocabulary

- **indigenous**
- **colonists**
- **agriculture**
- **industry**

TAKING NOTES: *Key Ideas and Details*

Organize Use the graphic organizer below to take notes about the effects of migration on the people and the land of the region.

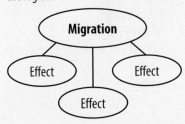

Lesson 2

History of the Region

ESSENTIAL QUESTION • *Why is history important?*

IT MATTERS BECAUSE
Learning about our nation's past helps us understand and appreciate its diversity and complexity.

Early America

GUIDING QUESTION *Who were the first peoples to live in the eastern United States?*

Looking back into history to learn about America's past is like examining a colorful quilt. A quilt is made up of bits and pieces of fabrics of many different colors and patterns. Each piece is made up of thousands of threads woven together. These threads were joined together into fabric pieces, and the pieces were joined together to make a quilt.

Like the quilt, our nation is a whole made up of many smaller parts. These parts are peoples and their cultures, interacting with places and environments. Over time, cultures, beliefs, and ways of living have become woven together like the threads of a quilt. The result is a unique and extraordinary nation.

Earlier, you read about the physical geography of the United States east of the Mississippi River and its many different kinds of landforms and bodies of water. This region has a variety of climates and resources. In this lesson, you will learn about the variety of peoples, cultures, and ways of life in the eastern United States, as well as how people have changed the land. You will discover how natural resources have influenced human settlement and how the land, the people, and the cultures of the eastern United States are connected.

Native Americans

Native American peoples were the first humans to settle in North America. Historians believe these peoples came to North America by crossing a land bridge from Asia around 14,000 to 19,000 years ago. Over time, these groups migrated in all directions. Hundreds of different groups settled in locations throughout North and South America. Each group developed a unique culture with its own language, religion, and lifestyle. Some of the groups that settled on lands east of the Mississippi River were the Cherokee, the Iroquois, the Miami, and the Shawnee. These groups are considered **indigenous** to North America. *Indigenous* means "living or occurring naturally in a particular place."

Native American peoples satisfied their needs by using the plants, animals, stones, water, and soil around them. Their way of life was shaped by their environment. For example, peoples who lived in northern woodland areas made their homes out of bark and wood. They burned wood to heat their homes during cold winters. They hunted woodland animals for food and used the animals' skins to make clothing.

Native peoples of the Americas built shelters suitable for the climates where they lived. People who lived in hot climates used grasses, vines, and reeds to build open-air homes. Other groups used stones, caves, and earth to build solid structures and mound cities. Their homes reflected their environments.

Some native groups were **isolated** from other groups, while others had contact with neighboring peoples. Sometimes groups interacted peacefully, such as when trading. Other times, wars over land and resources would develop.

For most of their history, Native Americans had little or no contact with people from other parts of the world. Native peoples lived off the land for thousands of years, resulting in only a minor impact on the natural environment. Then, suddenly, the land that they had relied on for everything was taken from them. When the first Europeans arrived in the Americas in the 1400s, native peoples' ways of life were changed forever.

European Colonization

Across the Atlantic Ocean, Europeans grew interested in the Americas. They heard tales about these wild, bountiful lands. Explorers who had been to the Americas told of endless forests, rivers overflowing with fish, and mountains filled with gold and silver. Kings and queens from England, France, Italy, and Spain wanted to claim land in North America. They wanted to control America's gold and natural resources.

Academic Vocabulary

isolated being alone or separated from others

Sequoyah spent many years developing 86 symbols to represent all the syllables of the Cherokee language.
▶ **CRITICAL THINKING**
Drawing Conclusions Why is having a written language essential for a culture?

In the 1500s, Spanish priests, soldiers, and settlers built military and religious outposts in the Americas. Included was St. Augustine in Florida. Originally founded as a settlement in 1565, settlers soon realized the need for protection after a series of pirate attacks. Settlers were also concerned by the arrival of English settlers. Construction on a stone fort began around 1672. This was the first permanent European settlement in what would become the United States. In the early 1600s, the English began to send **colonists** to the Americas. Colonists are people who are sent to live in a new place and claim land for their home country.

The first English colonists settled along the Atlantic coast of North America. They started early settlements in Jamestown, Virginia, in 1607 and Plymouth, Massachusetts, in 1620. Other settlements soon followed, built on lands that had been home to native peoples. The colonists turned native peoples' hunting grounds into farmland. They used many resources that were important to the Native Americans' survival.

Over the years, more and more Europeans journeyed across the Atlantic Ocean to America. Their colonies grew quickly. By 1650, about 52,000 colonists lived in America. Over the next 50 years, the

Visitors tour the Castillo de San Marcos National monument in St. Augustine, Florida. The fort is over 300 years old and took 23 years to complete.

▶ **CRITICAL THINKING**

Identifying Why do you think settlers chose a site along the water for the fort?

©Jose Fusta Raga/Corbis

European goods important to the survival of the colonists were received in Boston Harbor.

▶ **CRITICAL THINKING**
Determining Central Ideas Why were early colonists dependent on supplies from England?

number of colonists grew to about 250,000. By 1760, an estimated 1.7 million colonists lived in America. People built towns along the Atlantic coast. Large cities such as Boston and New York City started out as tiny settlements. By the 1750s, thirteen English colonies had been established in North America.

Life was hard for the first colonists. The food supplies they brought from England soon ran out. Many people got sick or died from starvation. In time, however, the colonists learned how to plant crops and hunt for food in this new land. They also adapted the natural resources they found there to make things they needed, such as candles, soap, pots, clothing, tools, and medicine.

The colonies built by English settlers were controlled by English rulers thousands of miles away. In 1707 England and Scotland united to form Great Britain. The colonists did not like the laws and taxes forced upon them by the British government. They made plans to break away and become free from British rule. In 1776, American colonists declared their independence from British rule, which led to the Revolutionary War. The war ended in 1781 when the British surrendered. The thirteen colonies became an independent nation called the United States of America, and the colonists called themselves *Americans*.

☑ **READING PROGRESS CHECK**

Identifying Point of View Why did European nations want to control land in North America?

Settling the Land

GUIDING QUESTION *How has the movement of people shaped the culture of the eastern United States?*

The new nation officially stretched from the Atlantic Ocean to the Mississippi River. However, the land west of the Appalachian Mountains was a mystery to most Americans. The few explorers and settlers who traveled west brought back stories of a wild and dangerous land. Many people were afraid to venture into the West. Others were willing to take their chances to seek new opportunities.

As more European settlers arrived, towns along the Atlantic coast became crowded. People began moving inland, away from the crowded coastal areas, to build new lives. They wanted to claim land for themselves. Many settlers packed everything they owned into wagons and headed west. Some looked for gold or silver. Some hunted animals and sold or traded the animals' skins. Most of these settlers stopped traveling, however, when they found land that looked suitable for farming.

These settlers quickly built homes and planted crops on the land. They hoped to grow enough food to feed their families. Many families of settlers were isolated from other people. They had no neighbors, and they were hundreds of miles from the closest town. These people had to make or grow everything they needed. Like the Native Americans, the early settlers lived off the land.

Daniel Boone helps lead settlers through the Cumberland Gap. This well-traveled path for settlers moving west later became known as the Wilderness Road.

▶ **CRITICAL THINKING**

Analyzing Primary Sources Why would the journey west have been difficult for early settlers?

©Bettmann/Corbis

New Territory

Between 1700 and 1800, thousands of settlers built homes along the Mississippi River. The Mississippi formed a natural boundary because it was wide, deep, and difficult to cross. When they arrived at the banks of the Mississippi, travelers either settled on the eastern side of the river or turned back. Those who did try to drive their horse-drawn wagons through the powerful river risked being swept away by its swift waters.

The young U.S. government wanted to claim as much of America's land as possible. The Land Ordinance of 1785 gave the United States legal claim to lands known as the Ohio Country. These lands were located north of the Ohio River and east of the Mississippi River. The Land Ordinance of 1785 allowed American settlers to buy sections of this land for one dollar per acre. The government divided up some land among settlers and also set aside land to be used for schools. Native Americans living on this land were forced to leave. The United States was slowly settling all territory east of the Mississippi River.

New Technology Changes Farming

Americans have been farming the land east of the Mississippi River for centuries. Much of our country's fruits, vegetables, grains, and cotton are raised on farms in this region. Growing crops and raising livestock to sell is called commercial **agriculture**. Good soil and frequent rains make the region one of the best places in the world for agriculture.

Until recently, planting and harvesting crops was hard work, performed mainly by hand and with the help of animals, such as horses and oxen. In the late 1700s, people designed and built new kinds of machines to do farmwork. Some machines planted large amounts of seeds quickly or harvested crops faster and more thoroughly than was possible by hand. For example, a Massachusetts man named Eli Whitney invented a machine called the cotton gin. The cotton gin made processing cotton faster and easier. This increased profits for cotton farmers.

Over the years, machines began to replace human workers on farms. With the help of machines, farmers planted and harvested more crops. This made farms more productive. As more and more farmwork was done with machines, however, fewer people were

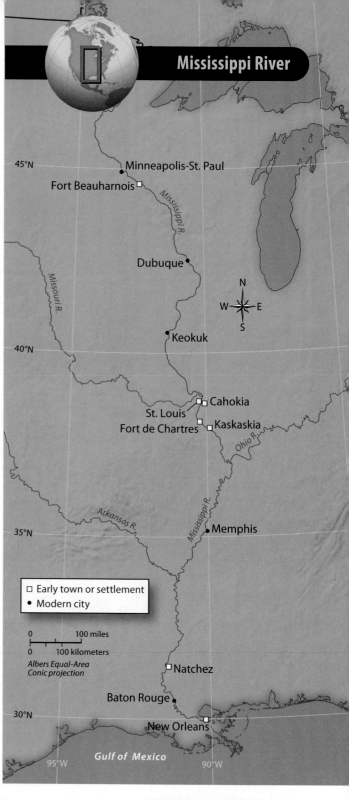

Mississippi River

Early town or settlement
Modern city

0 — 100 miles
0 — 100 kilometers
Albers Equal-Area
Conic projection

MAP SKILLS

Throughout history, many cities and towns were built and grew along the Mississippi River.

▶ CRITICAL THINKING

Describing How does a river contribute to the growth of a city or town?

needed to work on farms. This left thousands of people without work, and many moved to the cities. Today, most farmwork is done with the help of machines, such as tractors and combines.

Industrial Growth

New technology led to jobs in factories for millions of people. Manufacturing, making products to sell, is called **industry**.

Industry is an important part of the economy of the United States east of the Mississippi. This region has been one of the world's leading industrial regions for more than two centuries. Thousands of factories have been built all over the region during this time. Today, factories produce an endless variety of products including clothing, computer parts, shoes, baby formula, and medicines. Some factories process foods and bottled drinks. Cars are also built at automobile assembly plants. Products made east of the Mississippi River are shipped and sold all over the world.

Influence of Immigration

The history of the United States east of the Mississippi River is the history of many cultures. The cultures have become woven together over time, blending and changing into something new. Since the colonial days, people have been moving to the eastern United States from other countries. People immigrate to America because they want to be free. They also want jobs, education, and other opportunities. As a result, an amazing variety of languages, religions, cultures, and customs can be found east of the Mississippi. The cultural traditions brought by people from all over the world have made America a unique and diverse nation.

Forced Migration

In the 1830s, gold was discovered in the Southeast. Word quickly spread, and settlers poured into the southern states. Most did not find gold, but many stayed to start cotton farms in Alabama, Georgia, Mississippi, and the Carolinas. Some of these lands were home to the Cherokee, a large group of Native Americans. U.S. citizens wanted these lands—and their valuable resources.

Immigrants arrive in New York Harbor in the late 1800s.

▶ CRITICAL THINKING

Identifying State three reasons immigrants came to live in the United States.

Ingram Publishing

The government forced the Cherokee to leave their lands. In 1838 thousands of Cherokee men, women, and children were rounded up by U.S. soldiers. These people were forced to make the long and difficult journey to Indian Territory in Oklahoma, far west of the Mississippi River and 1,000 miles (1,609) from their homeland. The land they were given to live on was dry and difficult to farm, which was very different from the land they were forced to leave. Thousands of Cherokee died during and after the journey. This terrible event became known as the Trail of Tears.

The Trail of Tears was just one of many forced migrations in America's history. Native American groups all over the continent were forced to leave the lands that had shaped their ways of life for thousands of years. This movement destroyed some of their cultural traditions. As America gained territory, Native Americans lost their lands and their ways of life.

The Great Migration

The years after the Civil War were a time of change in the Southeast. Slavery became illegal in all states. Angry that slavery was outlawed, many states passed laws that took away the rights the freed Americans had recently gained. During the late 1800s, thousands of African Americans moved to states in the Mid-Atlantic, New England, and the Midwest. The relocation of people from the South to the North was called the Great Migration.

The Great Migration was part of the larger rural-to-urban migration occurring in the United States. This migration increased during the 1900s, when millions more people left rural areas and moved to cities to work in factories. This rural-to-urban movement of people is one of the largest migrations in America's history.

✓ **READING PROGRESS CHECK**

Analyzing How did the invention of farm machinery lead to unemployment in the eastern United States?

Include this lesson's information in your Foldable®.

LESSON 2 REVIEW **CCSS**

Reviewing Vocabulary

1. What makes the United States east of the Mississippi River a good region for *agriculture*?

Answering the Guiding Questions

2. ***Determining Central Ideas*** How did the early Native American peoples utilize resources from their environment?

3. ***Identifying Point of View*** What are reasons why English rulers sent colonists to America in the 1600s?

4. ***Analyzing*** What factors helped the U.S. economy change and grow?

5. ***Identifying Point of View*** How did laws ending slavery affect different populations of people in the Southeast after the Civil War?

6. ***Narrative Writing*** Imagine you are a young Native American living in the 1800s. Your family is being forced to leave its home as colonists take over the land. Write a narrative telling how you feel about what is happening to you and your people. Include details from the lesson in your narrative.

Reading **HELP**DESK (CCSS)

Academic Vocabulary

- **revenue**

Content Vocabulary

- **metropolitan area**
- **tourism**
- **civil rights**
- **Rust Belt**
- **service industry**

TAKING NOTES: *Key Ideas and Details*

Organize As you read about life in the eastern United States, write a one-sentence summary for each of the listed topics.

Topic	Summary Sentence
Metropolitan areas	
U.S. government	
Economy	

Lesson 3
Life in the Region

ESSENTIAL QUESTION · *What makes a culture unique?*

IT MATTERS BECAUSE

Learning about the human geography of the United States—its people, government, economy, and culture—can help you better understand and appreciate the nation as a whole.

Major Metropolitan Areas

GUIDING QUESTION *What is it like to live in a large metropolitan area in the eastern United States?*

People in the United States live in many different environments. Farmers and ranchers live in rural areas with open land for farming and livestock. Suburbs are popular places for people who want larger homes and their own pieces of property. People who enjoy urban environments often live in large cities called **metropolitan areas**. Metropolitan areas are centers of culture, education, business, and recreation. Large cities such as New York, Boston, Philadelphia, and Miami are home to millions. The populations of metropolitan areas in the eastern United States are large and diverse. The Boston-Washington corridor is a metropolitan area, home to about 50 million people. It is often referred to as a megalopolis.

Some metropolitan areas serve as hubs for international cooperation. The member countries of the United Nations (UN), located in New York City, work together to find and share solutions to problems related to education, science, and culture. For example, the UN's World Heritage program promotes and protects natural and cultural sites around the world. World Heritage sites in the eastern United States include the Everglades National Park in Florida and the Statue of Liberty in New York Harbor.

Tourism is one industry that provides jobs to people in and around metropolitan areas. The **tourism** business provides services to people who are traveling for enjoyment. Businesses such as restaurants, hotels, resorts, travel agencies, and tour companies are part of the tourism industry. The money tourists spend brings **revenue** to the state and local economies.

Some metropolitan areas on the East Coast began as port cities along the shores of the Atlantic Ocean. Port cities are large, busy towns where ships dock and depart. Trade between the United States and the rest of the world began in port cities such as Baltimore; Boston; Charleston, South Carolina; New Haven, Connecticut; and New York.

New York and Chicago

New York and Chicago are two of the largest metropolitan areas east of the Mississippi River. New York City, located on the Atlantic coast at the mouth of the Hudson River, is the most populous city in the United States. This famously diverse city is home to more than 9 million people, making it one of the most heavily populated cities in the world. New York began as a Dutch colonial port city called New Amsterdam. Although it is better known today as a hub of culture and commerce, New York is still home to one of the busiest ports in North America.

New York City is a dense cluster of urban areas, called boroughs, connected by streets, bridges, trains, and water passages. The five boroughs that make up the city are Manhattan, Brooklyn, Queens, the Bronx, and Staten Island. Though it is the smallest of the five boroughs, covering an area of only 22.6 square miles (58.5 sq. km), the island of Manhattan is the cultural, political, and economic center of New York City.

New York City is the most populous city in the United States.

▶ **CRITICAL THINKING**

Determining Central Ideas What are some of the advantages of living in a large city? What are some disadvantages?

Population Centers

Geographers study why certain cities became places where many people live. Many population centers in the United States lie in coastal areas where healthy economies support large populations. Some cities are important world trade centers because of their coastal or near-coastal locations. Some population centers are located inland, yet many are situated near rivers and lakes. Other inland cities such as Atlanta grew from agricultural and trading centers.

Manhattan is home to people of every racial, ethnic, and religious background in the world. It is also known worldwide as a center of finance, advertising, and entertainment. An endless variety of visual and performing arts—such as music, dance, drama, painting, sculpture, fashion, and architecture—bring the city to life.

More than 30 million tourists visit New York City each year. The city has an extensive public transportation system that includes subways and buses to help eliminate dependency on cars.

Chicago began as a small settlement between Lake Michigan and the Mississippi River in the early 1800s. The settlement developed into a thriving city after it was connected to the rest of the country by railroads and canals. By the mid-1800s, Chicago had become the center of all railroad travel in the United States. Chicago's importance as a transportation center increased when the St. Lawrence Seaway opened in the mid-1900s. Today, an elevated train system in the center of Chicago helps move its many tourists and residents.

Chicago remains one of the nation's most important centers of shipping, transportation, and industry. The city is one of America's leading producers of steel, machinery, and manufactured products. Several large printing and publishing companies are located in Chicago, as well as major financial institutions such as the Chicago Stock Exchange.

Atlanta and New Orleans

Atlanta and New Orleans are two of the most vibrant cities in the Southeast. Serving as Georgia's capital city, Atlanta is a historic city and a modern metropolis. Its location at the southern edge of the Appalachian Mountain range made it a popular passageway for settlers and other travelers. As railroads brought people and cargoes through the area, Atlanta grew into a thriving economic, cultural, and political center of the South.

During the Civil War, Atlanta served as a supply depot for the Confederate army. Most of the city's buildings were burned to the ground during a devastating invasion by the Union army in 1864. When Atlanta was rebuilt after the war, it became a symbol of strength and rebirth in the South.

Today, Atlanta is a strong center of transportation, industry, trade, education, and culture. It has been called the commercial center of the modern South. A wide array of industries, including publishing, telecommunications, banking, insurance, military supply, and manufacturing, have headquarters in Atlanta. The city's major factories produce electrical equipment, chemicals, packaged foods, paper products, and aircraft.

New Orleans began as a shipping town along the Mississippi River. During times of peace, New Orleans was an important center of transportation and trade. In times of war, such as during the Revolutionary War, the War of 1812, and the Civil War, New Orleans

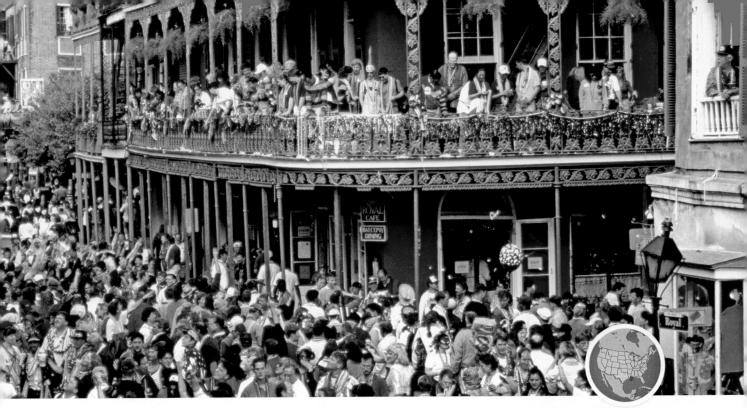

became a major stop along supply lines that served the military and civilians. Located only 110 miles (177 km) from the Gulf of Mexico, New Orleans eventually was connected to the Gulf by river channels, making it easier for ships to enter and leave the city.

In 2005, Hurricane Katrina struck the Gulf of Mexico's coast. The storm raged from Louisiana to Florida. In New Orleans, levees failed when the storm surged, flooding low-lying areas and trapping many people. Thousands of people were left homeless and many people died. Hurricane Katrina was one of the worst natural disasters in U.S. history. Some neighborhoods and areas of New Orleans have been rebuilt. Other areas have not recovered.

New Orleans remains an important commercial trade center in the eastern United States. Manufacturing and transportation still contribute to the city's economy. New Orleans is renowned for its rich cultural traditions including its spicy Cajun and Creole foods, original musical styles, and colorful celebrations such as Mardi Gras. New Orleans was one of the first centers of jazz music. The city and its people, art, music, language, and architecture are a bold and unique mixture of French, Spanish, Caribbean, and African cultures. Tourism has long been vital to the economy of New Orleans.

New York, Chicago, Atlanta, and New Orleans each have a unique character. These remarkable cities are just a few of the places that make the eastern United States such a fascinating region.

☑ **READING PROGRESS CHECK**

Citing Text Evidence How are New York and New Orleans alike? How are they different?

Millions visit the city of New Orleans every year to take part in the festivities of Mardi Gras.

▶ **CRITICAL THINKING**

Identifying Besides Mardi Gras, what attractions draw tourists to New Orleans?

Randy Wells/Stone/Getty Images

The Nation's Capital

The planned city of Washington, D.C., is the capital of the United States. Washington, D.C., is located along the banks of the Potomac River between the states of Virginia and Maryland. These two states gave land to the government to form the federal District of Columbia. Thus, the District was free of any single state's influence. *Why was the capital located in the eastern United States near the Atlantic Coast?*

The U.S. Government

GUIDING QUESTION *How have the government's actions affected the land and people of the United States?*

The United States declared its independence from Great Britain on July 4, 1776. On that day, Americans became citizens of a free and independent nation. Like all nations, the new country needed a system of government. It was important to Americans that their government protect the rights and freedoms of the people. The U.S. government was designed as a representative democracy, a system in which the people elect representatives to operate the government.

In 1787 representatives from each of the 13 states gathered to write a plan of government called a constitution. The U.S. Constitution is still the law of our country. The United States is a federal republic. The national government shares power with the states. Government leaders must promise to obey the Constitution. Amendments, or changes to the Constitution, have been made to meet the nation's changing needs. The first 10 amendments—the Bill of Rights—guarantee the basic rights of citizens.

The Three Branches of Government

One of the main functions of the U.S. Constitution is to make sure government power is shared. The men who wrote the Constitution did not want a single person or group to have all the power to make laws and decisions for the country. They divided the government into three separate but equal branches. Each branch has important functions, and each branch must work with the other two branches to govern the country. The three parts of the U.S. government are the executive, legislative, and judicial branches. The U.S. government's system of shared power was created to ensure that all parts of the government work together in a balance of power.

The legislative branch, called Congress, makes laws for the nation. Congress has two parts: the House of Representatives and the Senate. Members of Congress are elected by the people and come from all 50 states. To pass a law, the House of Representatives and the Senate must agree on what the law states.

The executive branch is the office of the president of the United States. The president's main duties are to carry out laws, to lead the military, to appoint judges to the Supreme Court, to plan the national budget, to meet with foreign leaders, and to appoint advisors to help make decisions for the nation.

The judicial branch is made up of state and federal courts. The role of the government's judicial branch is to decide if laws are fair and if they follow the Constitution. The Supreme Court, with nine judges called justices, is the most powerful court in America.

Government Actions Affect the Land

The three branches of the U.S. government make laws and decisions that affect our nation's land. Some laws and decisions protect the environment. For example, the president has the authority to set aside land for use as national parks. This protects plant and animal habitats and creates recreational areas. Local governments enforce laws that reduce water pollution and littering.

Another environmental program is Superfund. The purpose of the program is to clean up abandoned hazardous waste sites. Superfund was implemented after the discovery of toxic waste sites such as Love Canal and Times Beach in the 1970s. The federal government works in conjunction with the state government and the communities to implement cleanup plans.

Government Actions Affect People

Some actions of the United States government that affected Native American peoples had severe consequences. Other government actions—such as building roads and bridges, providing aid to people in need, and establishing national parks—help people and enrich our lives.

One example of how government actions have affected Americans positively is the civil rights movement. **Civil rights** are basic rights that belong to all citizens, such as the right to be treated equally under the law. To answer the demand for civil rights for African Americans, the federal government made laws ending segregation. These laws were meant to help people. Some local governments, however, did not agree with these laws and refused to enforce them.

National parks, such as Maine's Acadia National Park, are pieces of land that are protected by the federal government.

▶ **CRITICAL THINKING**

Describing What are the purposes of national parks?

Another example of positive action is how the government helps people during emergencies. Government agencies such as the Federal Emergency Management Agency (FEMA) provide food, water, medical care, and transportation to people affected by tornadoes, hurricanes, earthquakes, floods, terrorist attacks, and other disasters. FEMA has saved many lives, but some people claim the agency does too little to help and is slow to respond.

When thinking about how government actions affect people and the land, it is important to remember that no government is perfect. The U.S. government is made up of many different people doing many different jobs.

☑ **READING PROGRESS CHECK**

Determining Central Ideas Why is the U.S. government divided into three branches?

Everyday Life

GUIDING QUESTION *How has diversity shaped the culture of the United States?*

Religion and Ethnicity

In the United States, the religion with the largest number of followers is Christianity. The influence of the Christian faith in the United States has continued since colonial times. However, many Americans practice the Jewish, Muslim, Buddhist, or Hindu faith. Research shows that 16 percent of Americans do not participate in organized religion. Nearly half of all American adults say they have changed their faiths or beliefs at least once in their lives.

In gathering population statistics, the Census Bureau classifies the people of the United States into several different categories. The most recent data collected from the 2010 Census show that the four most populous ethnic groups are Caucasian, Latino, African American, and Asian American. In the last two censuses, people of Hispanic or Latin-American origin comprised the second largest ethnic group after Caucasians. In the Census of 1990, African Americans made up the second most-populous group.

Economy

America has one of the largest and strongest economies in the world. Advancements in technology help keep industries such as farming and manufacturing productive. The U.S. economy has slowed in the past decade, however. Many Americans remain unemployed. It could take years for the economy and employment rate to rebound.

Agriculture has long been an important industry. Today, the businesses of farming and raising animals for food are changing. Farms owned and operated by families are being replaced by corporate farms. They are managed by people who do not own or live on the land. Some Americans believe the growth of corporate farms is bad for the economy and the nation's people. They believe corporate farms agree to sell crops to large grocery-store chains at

---

GRAPH SKILLS ›

ETHNIC ORIGIN

The graph shows census statistics from 1980 to 2010.

▶ CRITICAL THINKING

1. Analyzing Which groups more than doubled in percentage of population from 1980 to 2010?

2. Analyzing Which group from 1980 to 2010 shows the biggest decrease in percentage of population?

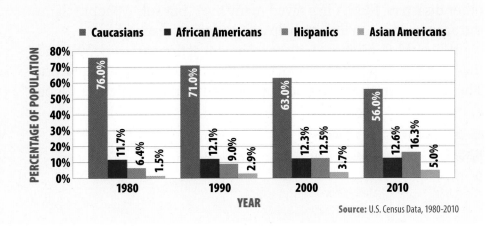

Source: U.S. Census Data, 1980-2010

such low prices that family farms can not compete. A growing number of Americans now support organic farming. Organic farms use only natural pesticides and fertilizers. Some people believe that organically-grown foods are safer and healthier than foods that have been treated with chemicals.

In the late 1800s and early 1900s, the U.S. shifted away from an agricultural-based economy to a more industrial economy. Factories producing automobiles, appliances, machinery, electronics, and other items have employed millions of Americans. During the 1980s, however, the manufacturing industry began to weaken. Businesses closed, and factories were abandoned. Workers lost their jobs. So many factories closed across the Midwest, the Mid-Atlantic, and New England that these areas earned the nickname "the **Rust Belt**."

In recent years, more Americans have found jobs in service industries. **Service industries** are businesses that provide services rather than products. Child care centers, restaurants, grocery stores, hair salons, electricians, moving companies, and auto repair shops are some examples of service industries. Other industries such as finance, insurance, and education still provide many jobs and revenue today. Many businesses, including banks and insurance companies, are located in cities east of the Mississippi River.

Different types of industries are important to the economy of the eastern U. S. Today, another shift is taking place as the nation moves from a service-based economy to an economy that is connected to the computer-information age.

Every March, Cuban Americans in Miami celebrate Calle Ocho, the single largest Latino celebration in the United States.

▶ **CRITICAL THINKING**
Analyzing Why are ethnic celebrations important for the community as well as for the people?

FOLDABLES
Study Organizer

Include this lesson's information in your Foldable®.

☑ **READING PROGRESS CHECK**

Citing Text Evidence What is the main difference between service industries and manufacturing industries?

LESSON 3 REVIEW **CCSS**

Reviewing Vocabulary

1. What are the names of the major *metropolitan* areas located in your state?

Answering the Guiding Questions

2. *Integrating Visual Information* What are some of the largest metropolitan areas located in the eastern United States?

3. *Determining Central Ideas* How have the government's actions affected the land and people of the United States?

4. *Determining Central Ideas* How have cultures from other parts of the world shaped the culture and character of the United States?

5. *Citing Text Evidence* What types of businesses and industries are important to the economy of the eastern United States today?

6. *Informative/Explanatory Writing* Write a paragraph explaining how tourism might affect the economy of a major metropolitan area.

What Do You Think?

CCSS

Is Fracking a Safe Method for Acquiring Energy Resources?

Fracking is a process for obtaining natural gas and oil through high-pressure blasting of underground rock. Fracking involves pumping a specially blended liquid—a mix of water, sand, and chemicals—into wells drilled deep below Earth's surface. The fluids pour in with such force that the rock formations fracture, or crack, releasing precious oil and natural gas. Oil and gas are key resources for heating our homes, creating electricity, and fueling our cars. But critics of fracking worry that the process is harming the environment and people's health. Is fracking a safe way to meet our energy needs?

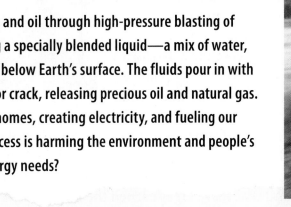

TEXT: "NEW HYDRAULIC FRACTURING REGULATIONS WON'T KEEP THE PUBLIC SAFE," by Sam Schabacker in *The Denver Post*, July 10, 2011 http://www.denverpost.com/opinion/ci_18436003?source=pkg. © Copyright Sam Schabacker 2012; PHOTO: Spencer Platt/Getty Images News/Getty Images

No!

PRIMARY SOURCE

" The form of natural gas drilling called fracking has caused livestock and crops to die from tainted water, people in small towns to black out and develop headaches from foul air, and flames to explode from kitchen taps. . . . [I]n recent years, we have learned that extracting gas through fracking poses unacceptable risks to the public. Fracking uses large quantities of water and a cocktail of toxic chemicals that have been shown to poison water resources. To date, thousands of cases of water contamination have been reported near drilling sites around the country. In many cases, residents can no longer drink from their taps, and in one instance, a home near a fracking site exploded after a gas well leaked methane into its tap water. . . . [S]tudies . . . found that 25 percent of fracking chemicals can cause cancer and 40 to 50 percent can affect the nervous, immune and cardiovascular [heart and blood vessel] systems. "

—Sam Schabacker, senior organizer for Food & Water Watch

Environmental activists protest fracking in New York.

Water is unloaded from trucks at a treatment plant. The plant will separate water, oil, and sediment that is mixed during the fracking process.

Yes !

PRIMARY SOURCE

❝ "Typically, steel pipe known as surface casing is cemented into place at the uppermost portion of a well for the explicit (specific) purpose of protecting the groundwater. . . . As the well is drilled deeper, additional casing (large pipe) is installed . . . which further protects groundwater. . . .

Casing and cementing are critical parts of the well construction that not only protect any water zones, but are also important to successful oil or natural gas production. . . . Industry well design practices protect sources of drinking water from . . . oil and natural gas well with multiple layers of impervious (hard to pass through) rock.

"While 99.5 percent of the fluids used consist of water and sand, some chemicals are added to improve the flow." ❞

—American Petroleum Institute

What Do You Think?   DBQ

1 *Identifying Point of View* According to Sam Schabacker, how does fracking put people's health at risk?

2 *Identifying Point of View* How does each side support its position?

Critical Thinking

3 *Analyzing* Some people believe that fracking should be halted until experts have studied the risks more thoroughly. Do you think a temporary ban on fracking is reasonable? What would be the advantages and disadvantages?

Directions: Write your answers on a separate piece of paper.

❶ Exploring the Essential Question

INFORMATIVE/EXPLANATORY WRITING Choose one of the subregions located in the United States east of the Mississippi. Write an essay explaining how the physical geography of the subregion influenced its settlement and economic development.

❷ 21st Century Skills

INTEGRATING VISUAL INFORMATION Use a map to trace the Ohio River from its birthplace at the mouths of the Allegheny and Monongahela Rivers in Pennsylvania and the Mississippi River from its birthplace in Minnesota to the point where they meet. Then follow the path of the Mississippi to the Gulf of Mexico. List the states that border these two rivers and five cities along the rivers' banks.

❸ Thinking Like a Geographer

INTEGRATING VISUAL INFORMATION Create a graph showing how quickly the population of the colonies grew between 1650 and 1700 and from 1700 to 1750.

❹ GEOGRAPHY ACTIVITY

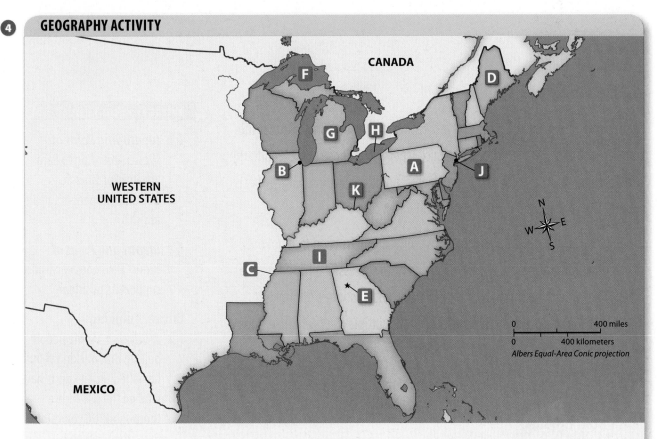

Locating Places

Match the letters on the map with the numbered places below.

1. New York City	**4.** Maine	**7.** Tennessee	**10.** Pennsylvania
2. Chicago	**5.** Mississippi River	**8.** Lake Erie	**11.** Lake Superior
3. Atlanta	**6.** Ohio River	**9.** Michigan	

REVIEW THE GUIDING QUESTIONS

Directions: Choose the best answer for each question.

1 In which subregion of the United States is the nation's capital located?

A. Southeast

B. Midwest

C. Mid-Atlantic

D. New England

2 What is the name of the river system that connects the Great Lakes to the Atlantic Ocean?

F. the Hudson River

G. the St. Lawrence Seaway

H. the Ohio and Mississippi rivers

I. the Monongahela and Allegheny rivers

3 Why did Europe's kings and queens claim lands in North America?

A. They wanted to spread Catholicism to the Western Hemisphere.

B. They wanted control over America's gold and natural resources.

C. They wanted to reduce the population in overcrowded cities.

D. They wanted to protect the native population from exploitation.

4 Why were the Cherokee people forced to leave their land in the Southeast and walk 1,000 miles (1,609 km) to Oklahoma?

F. A hurricane destroyed their homes.

G. Their land was needed for forts to protect America from the Spanish.

H. White settlers wanted their land and the resources on it.

I. Locusts destroyed their crops.

5 New York City is made up of five unique urban areas called

A. counties.

B. boroughs.

C. townships.

D. districts.

6 Whose job is it to carry out the laws passed by Congress, appoint federal judges, lead the military, meet with foreign leaders, and plan the budget?

F. the Senate

G. the Speaker of the House of Representatives

H. the president of the United States

I. the secretary of state

DBQ ANALYZING DOCUMENTS

7 DETERMINING CENTRAL IDEAS The government reports on the future need for workers who create software:

"*Employment of software developers is projected to grow . . . much faster than the average for all occupations. . . . The main reason . . . is a large increase in the demand for computer software. Mobile technology requires new applications. Also, the healthcare industry is greatly increasing its use of computer systems.*"

—from the *Occupational Outlook Handbook*

Which generalization can you make from the information in this quote?

A. Computer use is expected to go down in the future.

B. More software developers will be needed than most other jobs.

C. Software developers are not likely to work for health care companies.

D. The government will be the biggest employer of software developers.

8 ANALYZING What will happen to the number of software developer jobs if smartphone sales go down in the future?

F. increase at the expected rate

G. increase more rapidly than expected

H. stay the same instead of increase

I. decrease or increase at a slower rate

SHORT RESPONSE

"*We the People of the United States, in Order to form a more perfect Union, establish Justice, insure domestic Tranquility [calm], provide for the common defence, promote the general Welfare, and secure the Blessings of Liberty . . ., do . . . establish this Constitution.*"

—from the United States Constitution

9 IDENTIFYING What is the purpose of this section of the United States Constitution?

10 DETERMINING WORD MEANINGS What do you think the writers of the Constitution meant by "the Blessings of Liberty"?

EXTENDED RESPONSE

11 INFORMATIVE/EXPLANATORY WRITING If you could choose to live in any state east of the Mississippi River, which one would it be and why? Have you ever lived in or visited that state? Do you have friends or family who live there? Write a report explaining your choice. Be sure to give details about the state's features that appeal to you. Consider things like climate, job opportunities, education, and recreational opportunities when making your choice.

Need Extra Help?

If You've Missed Question	1	2	3	4	5	6	7	8	9	10	11
Review Lesson	1	1	2	2	3	3	3	3	3	3	1

THE UNITED STATES WEST OF THE MISSISSIPPI RIVER

networks

There's More Online about The United States West of the Mississippi River.

CHAPTER 5

ESSENTIAL QUESTIONS • *How does geography influence the way people live?* • *How do people make economic choices?* • *How does technology change the way people live?*

Modern-day cowhands still ride in the cattle country of the West.

Just One Film/The Image Bank/Getty Images

Lesson 1
Physical Features

Lesson 2
History of the Region

Lesson 3
Life in the United States West of the Mississippi

The Story Matters...

Within the region are several mountain ranges, including the Rocky Mountains, the longest mountain range in North America. Its many mountains, plateaus, basins, and valleys mean that this region contains a range of elevations. The region is rich in land, mineral, and energy resources, all of which contributed to westward expansion to the Pacific Ocean in the 1800s, and continues to influence the way of life of its residents.

FOLDABLES
Study Organizer

Go to the Foldables® library in the back of your book to make a Foldable® that will help you take notes while reading this chapter.

| Geography | History | Economy |
| Know \| Learned | Know \| Learned | Know \| Learned |

THE UNITED STATES WEST OF THE MISSISSIPPI RIVER

The United States west of the Mississippi River is one of the two regions that make up the United States. As you study the map, identify the states and cities of the region.

Step Into the Place

MAP FOCUS Use the map to answer the following questions.

1 THE GEOGRAPHER'S WORLD What is the name of the state just north of Missouri?

2 THE GEOGRAPHER'S WORLD Which four states meet at one point?

3 PLACES AND REGIONS What natural feature separates Texas from Mexico?

4 CRITICAL THINKING Identifying The contiguous United States consists of the states between Canada and Mexico. Which two states are not contiguous?

MONUMENT VALLEY Buttes are a common landform in Arizona's Monument Valley Navajo Tribal Park.

GOLDEN GATE BRIDGE Named after the Golden Gate Strait, the bridge stands where water from the Pacific Ocean enters San Franciso Bay.

Step Into the Time

DESCRIBING Select one event on the time line and write a paragraph describing how social, political, ecological, and/or economic factors of the time period led to the occurrence of that event.

1598 Spain settles Santa Fe

1846 The Mexican-American War begins

1800

1803 The U.S. purchases the Louisiana Territory from France

Cities and States of the Region

CANADA

Legend:
- National boundary
- ★ State capital
- • City

Washington — Seattle, Tacoma, Olympia★, Portland•, Salem★

Oregon

Montana — Helena★

Idaho — Boise★

Wyoming

North Dakota — Bismark★

South Dakota — Pierre★

Minnesota — St. Paul★

Nevada — Carson City★

Utah — Salt Lake City★, Provo•

Colorado — Denver★, Cheyenne★

Nebraska — Omaha•, Lincoln★

Iowa — Des Moines★

Kansas — Topeka★, Kansas City•

Missouri — Jefferson City★

California — Sacramento★, San Francisco•, Las Vegas•, Los Angeles•, Long Beach•

Arizona — Phoenix★

New Mexico — Santa Fe★

Oklahoma — Oklahoma City★

Arkansas — Little Rock★

Texas — Ft. Worth•, Dallas•, Austin★, Houston•

Louisiana — Baton Rouge★

PACIFIC OCEAN

Great Salt Lake

Missouri R.
Mississippi R.
Arkansas R.
Colorado R.
Rio Grande

40°N
30°N
120°W

0 — 400 miles
0 — 400 kilometers
Albers Equal-Area Conic projection

MEXICO

Gulf of Mexico

TROPIC OF CANCER

Alaska — Valdez, Juneau, Arctic Circle, 70° N, 60° N
0 — 300 miles
0 — 300 kilometers
160° W, 150° W, 140° W

Hawaii — Honolulu
110°W, 100°W, 90°W, 20° N, 160° W
0 — 200 miles
0 — 200 kilometers

1869
Transcontinental Railroad completed

1959
Alaska becomes a state

1900

2000

1930s Dust storms destroy farmland in Great Plains

1989 *Exxon Valdez* oil spill damages environment

networks
There's More Online!

Reading **HELP**DESK ⓒCSS

Academic Vocabulary

- significant
- create

Content Vocabulary

- cordillera
- timberline
- contiguous
- Continental Divide
- irrigation
- chinook
- ethanol
- national park

TAKING NOTES: *Key Ideas and Details*

Organize As you read, take notes on different characteristics of the Great Plains.

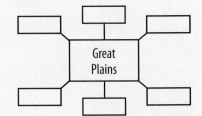

Lesson 1
Physical Features

ESSENTIAL QUESTION • *How does geography influence the way people live?*

IT MATTERS BECAUSE
The region includes many resources, but its rapidly growing population is causing overuse of some of them.

Physical Landscape

GUIDING QUESTION *How do the physical features of the western United States make the region unique?*

The Mississippi River divides the United States into two parts. These two regions are not equal in size. The area west of the Mississippi River is larger than the area to the east.

The states to the west differ from each other in some ways. At the same time, they have more in common with each other than with the states east of the Mississippi River. The western states are typically larger than the eastern states. Their human populations are generally more spread out across these vast distances. Their landforms are steeper and rockier than in the eastern states, and their climates overall are much drier. Many of the western states are rich in natural resources.

The Great Plains

Just west of the Mississippi River lie the Great Plains. In many places, the Plains appear flat. In other places, the Plains are gently rolling land. In spite of their flat appearance, however, the Plains are tilted downward toward the east. In eastern Nebraska, for example, the elevation of the land is less than 1,500 feet (457 m) above sea level. But on western edge of Nebraska, the land rises to about 6,000 feet (1,829 m).

The Great Plains were once covered by wild grasses. Vast herds of bison and pronghorn—an American antelope—grazed there. Today, the Great Plains are covered by farms and ranches.

Mountains and Hills

Toward the north, the Great Plains are interrupted by the Black Hills. These hills were once mountains, but over time they eroded. Evergreen trees appear to darken the hills, giving them their name.

West of the Great Plains tower the Rocky Mountains. The Rockies are not a single mountain chain, but a cordillera. A **cordillera** is a region of parallel mountain chains. The Rockies include dozens of different mountain systems. They extend from the Canadian border to the Mexican border. Peaks soar up to 14,000 feet (4,267 m), and valleys plunge thousands of feet below. Many of the mountains are snow capped. Trees cover the slopes, but not above the **timberline**. At that elevation, the climate is too cold for trees to grow.

Several different mountain ranges tower over the Pacific coast. Many of these mountains formed because plate tectonics exert pressure on Earth's lithosphere. This causes the lithosphere to crack and the broken land to rise into steep, rugged mountains. Among them are the Olympic Mountains of Washington. Heavy rainfall and cold temperatures form glaciers on these mountains.

About 150 miles (241 km) east of the Pacific coast are two higher ranges. The Cascades run from north to south through western Oregon and Washington State. The Cascades are volcanic, and some of the volcanoes are still active. The Sierra Nevada range runs along the California-Nevada border. The name *Sierra Nevada* comes from the Spanish for "snowy mountains." These mountains include Mount Whitney, which at about 14,500 feet (4,419 m) is the highest point in the 48 contiguous states. **Contiguous** means "connected to." The contiguous states are those that stretch from the Atlantic to the Pacific oceans. Alaska and Hawaii are not among them.

Think Again?

The bison and buffalo are the same animal.

Not true. The bison and buffalo are animals that belong to the *Bovidae* biological family, but they differ in their physical appearance and habitat. The buffalo is native to Asia and Africa. The bison is native to North and South America. *Bison* is the correct scientific name for the American animal, but the term *buffalo* is widely used.

Today, bison live mainly in parks and reserves.

Identifying In what area do most bison live?

©Corbis Premium RF/Alamy

Hikers stroll toward the peaks called the Maroon Bells. The peaks are part of the Elk Mountains near the ski town of Aspen, Colorado.

Identifying What three sections of landforms are located between the Rocky Mountains and the mountain ranges along the Pacific Coast?

Between the coastal mountains and the line of the Sierras and the Cascades are long, low valleys. This lowland area is called Central Valley in California. In Oregon, it is called the Willamette Valley.

Basins and Plateaus

Between the Rockies and the Sierra Nevada and Cascade Range is a mix of landforms. They can be grouped into three sections.

To the south and east is the Colorado Plateau. A plateau is a large area of generally flat land. This highland area is marked by smaller, flat-topped features called mesas that are sometimes separated by canyons. In addition, many canyons cut deep into the Colorado Plateau. Among these canyons is the Grand Canyon. Winding along the canyon floor, more than a mile (1.6 km) below the rim, is the Colorado River. Rising up to the plateau are rocks of many colors and shapes. The spectacular sight attracts more than 4 million visitors every year.

West of the Colorado Plateau and extending to the north is the Basin and Range region. This name refers to a pattern on the land in which clusters of steep, high mountains are separated by low-lying basins.

To the north is the Columbia Basin. This large area was formed mainly by vast amounts of lava that flowed from volcanoes and then cooled and hardened. Much of the area is flat, but rivers cut deep valleys and canyons.

Landforms of Alaska

Alaska—the largest U.S. state in land area—lies to the west of Canada. Mountains run along its southern and northern edges. The Alaska Range, also in the south, is the home of the highest point in the United States, Mount McKinley. Also called Denali, the mountain soars 20,320 feet (6,194 m) high. Lowland plains cover the area between the Alaska Range and the northern mountains.

Landforms of Hawaii

Nearly 2,400 miles (3,862 km) southwest of California is Hawaii. An archipelago, Hawaii includes more than 130 islands. The eight largest ones are in the eastern part of this chain of islands. Volcanoes formed these islands. Two volcanoes—Mauna Loa and Kilauea—are still active. Wind and the sea have eroded some mountains to make steep cliffs. Along the shore, some islands have sandy beaches that draw many tourists.

☑ **READING PROGRESS CHECK**

Citing Text Evidence What are two ways in which the states west of the Mississippi River are similar to each other?

Bodies of Water

GUIDING QUESTION *How do the bodies of water in the region affect people's lives?*

The United States west of the Mississippi River is much drier than the eastern part of the country. As a result, it has fewer rivers and lakes. Those that do exist play **significant** roles in the area's economy.

Ocean and Gulf

The chief body of water for this region is the Pacific Ocean. This vast ocean meets the western shores of the continental United States and of Alaska. Hawaii sits in its midst. Inlets from the ocean **create** many excellent harbors along the Pacific coast. As a result, the coast has many major ports. They include San Diego, Long Beach, and Los Angeles, California; Portland, Oregon; and Seattle and Tacoma, Washington. Valdez, Alaska, and Honolulu are also important Pacific ports.

Louisiana and Texas border the Gulf of Mexico. They both have several major ports, including those of New Orleans and Houston. The coastlines of the Pacific Ocean and the Gulf of Mexico are very different. Plate tectonics are active along the Pacific coast. This not only causes earthquakes but also makes for a steep coastline with rocky cliffs. The Gulf coast is much more stable and very flat. It has many swamps and shallow water.

Lakes of the Region

Because the western United States has a fairly dry climate, lakes are not as common in this region as east of the Mississippi. But thousands of years ago, the climate was wetter. A huge, freshwater lake covered many of the basins that are located in what are now Utah, Idaho, and Nevada. Today, only a few isolated lakes remain.

©Doug Sherman/Geofile

Tall cliffs surround volcanic Crater Lake in Oregon. A volcanic hill rises out of the middle of the lake.

▶ CRITICAL THINKING

Analyzing Are lakes more plentiful in the western United States than in the eastern region? Explain.

Hoover Dam is located on the Colorado River along the border between Arizona and Nevada.

▶ **CRITICAL THINKING**

Describing How do dams benefit the people of the region?

One of them is Utah's Great Salt Lake. It is the largest salt lake in the Americas. The lake changes in size gradually because a varying amount of water evaporates from it. The lake is salty because it does not have an outlet. Tributary rivers bring salt to the lake. Evaporation takes the water away but leaves the salt behind.

Two other important lakes in the region are Lake Tahoe and Lake Mead. Lake Tahoe sits high in the Sierra Nevada. Lake Mead is a human-made lake formed when Hoover Dam was built on the Colorado River. Both lakes are used for boating and other water recreation.

Rivers

Rivers are important in the western United States. The Colorado is one of the major rivers of the region. It begins along the western slope of the Rocky Mountains and twists its way south and west to the Gulf of California. At many places along its course, the Colorado has been dammed. So much of the river's water is used for farming and in cities that no water at all reaches the Gulf of California. To the north, the Columbia River flows from the Rocky Mountains to the Pacific Ocean. It has been dammed in several places to provide hydroelectric power. The Snake and Willamette rivers feed into this river.

All these rivers flow west. But some of the rivers in the region flow east toward the Gulf of Mexico. The **Continental Divide**, an imaginary line through the Rocky Mountains, separates these two sets of rivers. The eastward-flowing rivers include the Missouri, the Platte, the Kansas, the Arkansas, and the Rio Grande. The first four of these rivers feed into the Mississippi.

The dams on all the rivers bring much benefit to the people in the region. They control floods, generate hydroelectric power, and provide water for urban and rural areas. The dams also greatly affect the hydrosphere and biosphere. Without dams, the rivers flow fast and cold during springtime when snows are melting. Then in summer, they flow slowly and are warmer.

✓ **READING PROGRESS CHECK**

Integrating Visual Information What landform is crossed by the rivers that flow from the Rocky Mountains to the Mississippi River?

©ThinkStock/SuperStock

Climates of the Region

GUIDING QUESTION *What factors influence the climates of the region?*

The United States west of the Mississippi River has many different climates. Tropical rain forests cover parts of Washington and Oregon because of the many storms that come from the North Pacific Ocean. Dry, hot deserts cover large parts of the Southwest.

Coastal and Highland Climates

High mountains play a role in forming climates in the region. The western mountains cause what is called a rain shadow. West-facing slopes of the Pacific Coast Ranges, the Cascades, and the Sierra Nevada receive plentiful rain and snow from Pacific Ocean storms. Heavy rains give northern California, Oregon, and Washington a marine west coast climate. Vast forests grow there.

The valleys to the east of the coastal ranges lie in the rain shadow, so they are dry. Although California's Central Valley lies in the rain shadow, it is a major farming region. It is hot and dry, but it is located near the western slopes of the Sierra Nevada, which receive abundant rain. Mountain rainwater is used for irrigation in the valley. **Irrigation** is the process by which water is supplied to dry land.

Climates in the Interior

The high mountains of the Cascades and the Sierra Nevada produce a rain shadow effect that keeps the interior of the region dry. This dry climate is what causes so much evaporation from Great Salt Lake. Some areas are covered by large deserts.

Advances in technology have produced new methods of watering farmland.

Determining Word Meaning What is irrigation?

Don Farrall/Stockbyte/Getty Images

The mountain regions in Colorado can have large variations in climate. Snow-covered mountains have cold nighttime temperatures in winter. Bright sunshine can make summer days comfortably warm.

Identifying What area of the region has a humid subtropical climate?

The western Great Plains have a semiarid climate. A semiarid area receives more rain than a desert but not enough for trees to grow. Instead, semiarid areas have bushes and grasslands. Temperatures in the Great Plains get hot in the summer and cold in the winter. This is what is known as a humid continental climate. Sometimes in the winter, a dry wind called the **chinook** blows over the region. It originates in the mountains where it is cold. But the air heats up as it blows down the eastern slopes of the mountain.

The eastern half of the Plains has two climate types. The northern part has a humid continental climate. It is influenced by cold air masses moving down from the Arctic. The climate of the southern part is humid subtropical. It is shaped by warm, moist air from the Gulf of Mexico. It has higher temperatures than the northern Plains. When cold, dry air from the north collides with warm, moist air from the south, thunderstorms and even tornadoes can result.

Climates of Alaska and Hawaii

Climates in Alaska are generally moderate but cool to cold. More moderate temperatures occur toward the south and colder ones to the north. Winters are cold in the far north. Snow can be heavy in the south and southeast. Valdez can receive as much as 200 inches (508 cm) of snow per year.

Hawaii has a tropical rain forest climate with high temperatures and high levels of rainfall. Rain tends to be heaviest in the winter. More rain falls on the northeastern side of mountains because the moist wind comes from that direction. Steady ocean breezes keep the air comfortable even when temperatures are high.

☑ **READING PROGRESS CHECK**

Citing Text Evidence Why does so much water evaporate from Great Salt Lake?

Resources of the Region

GUIDING QUESTION *What resources does the region have?*

The United States west of the Mississippi River has a great variety of natural resources. In addition to land that supports raising livestock, rich reserves of petroleum, minerals, and a variety of energy sources are found here.

Energy Resources

The United States west of the Mississippi River has large reserves of energy resources. Petroleum is found in the Gulf of Mexico, near Louisiana and Texas; in the southern Great Plains; in California; and in Alaska. Natural gas is found in the same areas. Coal is abundant in Wyoming.

A growing source of energy coming from the region is ethanol. **Ethanol** is a liquid fuel made from plants. In the United States, ethanol is made from corn and blended with gasoline. The United States is one of the world's leading producers of ethanol.

Hydroelectric power is an important source of energy in this region. Dams along the Columbia and Colorado rivers supply this power. Wind power is a growing source of energy here. South Dakota gets nearly a quarter of its electricity from wind power—more than any other state. Solar power is also becoming more important.

Minerals and Other Resources

The Rocky Mountains are important sources of gold, silver, copper, zinc, and lead. Timber is an important resource, too. Fertile soil makes the Plains, California's Central Valley, and parts of Oregon and Washington major farming regions.

Another important resource in the region is its natural beauty. Large areas of great natural beauty have been set aside in **national parks**. These parks attract millions of visitors every year.

☑ READING PROGRESS CHECK

Analyzing Does this region of the United States rely too much on one energy resource? Explain?

Think Again?

Nothing grows in a desert.

Not true. Deserts can support plant life—and some desert plants can reach large sizes. A fully grown saguaro cactus can be as much as 50 feet (15 m) high. Desert plants have to be well-suited to the dry conditions, though. They need large root systems and thick leaves that trap and retain moisture.

FOLDABLES® Study Organizer

Include this lesson's information in your Foldable®.

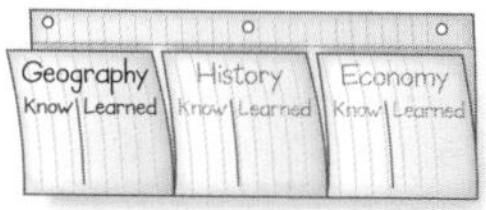

LESSON 1 REVIEW CCSS

Reviewing Vocabulary

1. Why is *irrigation* needed in California's Central Valley?

Answering the Guiding Questions

2. ***Determining Central Ideas*** How would the landforms and climate of the region affect where people live?
3. ***Identifying*** What are two characteristics that Washington and Hawaii share?
4. ***Describing*** What can result in the Great Plains when cold, dry air collides with warm, moist air?
5. ***Analyzing*** Are you more likely to find hydroelectric power in Washington and Oregon or on the Great Plains? Why?
6. ***Informative/Explanatory Writing*** In a paragraph, describe the scenery you would see on a drive from the Mississippi River to the Pacific coast.

There's More Online!

☑ **IMAGE** Chicago: Before and After Industrialization

☑ **TIME LINE** Notable Events: West of the Mississippi

☑ **VIDEO**

Reading HELPDESK (CCSS)

Academic Vocabulary

- **establish**
- **data**

Content Vocabulary

- **nomadic**
- **pueblo**
- **mission**
- **frontier**
- **Manifest Destiny**
- **annex**
- **extinct**
- **reservation**

TAKING NOTES: *Key Ideas and Details*

Organize As you read about westward expansion, use a graphic organizer like the one below to identify two land acquisitions and how the land was acquired.

Land Acquired	How Acquired

Lesson 2

History of the Region

ESSENTIAL QUESTION · *How do people make economic choices?*

IT MATTERS BECAUSE
Westward expansion is an important story in U.S. history.

Early Settlements

GUIDING QUESTION *How did life in the region change for Native Americans?*

The first people to live in the western states of what is now the United States were Native Americans. Native Americans belonged to dozens of different groups. Each group had its own language and culture and followed a lifestyle well-suited to the area where they lived.

Native American Ways of Life

The tribes of the Great Plains adopted different ways to live on these grasslands. Some farmed and hunted. They settled along rivers, where they tended fields that grew corn, squash, and other foods.

Other Native Americans of the Plains hunted the herds of bison. Along with obtaining meat, the people used other parts of the animals for clothing and homes. These peoples were **nomadic**, always on the move. Few trees grew there, so the nomadic peoples of the Plains built homes called teepees, using animal hides stretched over long poles. Teepees could be folded up and moved fairly easily, allowing Plains peoples to take their homes with them as they traveled.

The Pueblo people of the Southwest lived in villages that the Spanish called **pueblos** ("towns" or "villages" in Spanish). These villages' multistoried homes were made of dried mud. The Pueblo practiced dry farming, conserving scarce water to grow corn, beans, and squash.

(l to r) ©George H.H. Huey/Corbis; ©Bettmann/Corbis; ©Corbis; ©Corbis

In the Northwest, Native Americans fished for salmon and hunted sea mammals. On land, they hunted small game. Taking advantage of the thick forests that grew in the region's climate, they built large homes of wood.

When Europeans came to North America, the lives of Native Americans changed dramatically. For example, Europeans introduced new animals such as horses and sheep. Horses made hunting bison easier for the Plains peoples. The Navajo of the Southwest began herding sheep. But Europeans also brought diseases that killed large numbers of Native Americans.

Colonial Times

The Spanish were the first Europeans to come to this region. In 1598 they **established** the first European settlement in the region near what is now El Paso, Texas. By the early 1600s, they had founded Santa Fe, New Mexico. Soon they spread out along the upper reaches of the Rio Grande.

In the 1700s, the Spanish settled parts of California and Texas. Central to some settlements were **missions**. These church-based communities led by Catholic priests were meant to house native peoples. The priests hoped the Native Americans would adopt Christianity and the Spanish way of life. They relied on the work of the native peoples to grow food.

The land and climate across much of the region were similar to what they had left in Spain. The Spanish settlers introduced numerous types of crops that grew in Spain but did not exist in the Western Hemisphere. These included oranges, grapes, apples, peaches, pears, and olives. The Spanish also adopted crops that grew in the Western Hemisphere but not in Spain. These crops included corn, tomatoes, and avocados.

(l to r) ©George H.H. Huey/Corbis; Peter Pearson/Getty Images

Academic Vocabulary

establish to start

This ancient pueblo (left) is near Taos, New Mexico. Located near Tucson, Arizona, San Xavier del Bac (right) was founded as a Catholic mission by Father Eusebio Kino.

▶ **CRITICAL THINKING**

Describing In what ways was life for Native Americans in pueblos similar to life in the mission? In what ways was it different?

Thinking Like a
Geographer

Into the Unknown

What lay in the vast Louisiana Territory? How far was it to the Pacific Ocean? While exploring the Louisiana Territory, William Clark recorded his observations in a journal. The explorers carefully mapped the entire trip, which covered about 8,000 miles (12,875 km). The pages of Clark's journal also include drawings of the animals the expedition encountered.

Academic Vocabulary

data information

In the 1680s, France claimed the land drained by the Mississippi River, which included much of the land east of the Rocky Mountains. It called this vast area Louisiana. Over the years, the French placed a few settlements along the Mississippi River. The most important was New Orleans. It was founded in the early 1700s as a port for shipping goods from the river's valley.

☑ **READING PROGRESS CHECK**

Determining Central Ideas What was the Native American lifestyle like before Europeans came to the region?

Westward Expansion

GUIDING QUESTION *Why and how did Americans move into this region?*

When the American Revolution ended in 1783, the territory of the United States was entirely east of the Mississippi River. Much of this land was still unsettled by white Americans. But after the Revolution, they quickly began moving to the frontier in large numbers. A **frontier** is a region just beyond or at the edge of a settled area. Soon, Americans turned their eyes west of the river, eager for more land.

Exploring the West

In 1803 President Thomas Jefferson purchased the vast Louisiana Territory from France. The Louisiana Purchase gave the United States most of the land between the Mississippi River and the Rocky Mountains.

Soon after, Meriwether Lewis and William Clark led nearly 50 men to explore parts of the area. The Lewis and Clark expedition lasted more than two years, as they traveled from St. Louis to what is now the coast of Oregon and back. They traveled along the Missouri River as far as they could, and then they proceeded overland by horseback. They mapped the land and rivers they saw. They recorded **data** about the plants and animals living there. They also made peaceful contact with Native American peoples. By reaching the Pacific Ocean, they helped set an American claim to Oregon and Washington.

Over the next decades, other explorers helped open new areas. Meanwhile, hardy adventurers called mountain men began to trap beavers in the Rocky Mountains. Their travels added more knowledge about the geography of the American West. They also discovered ways through the mountains that settlers would use later.

Settling the West

By the 1830s, some Americans had come to believe in the idea of **Manifest Destiny**. According to this concept, the United States had a right to extend its boundaries to the Pacific Ocean. This belief helped promote the nation's westward movement.

Settlers began moving to the rich farmlands in what is now Oregon. Traveling in wagons, they took a long route called the Oregon Trail. It carried them across the Great Plains and through passes in the Rocky Mountains. Over the years, thousands of people moved to Oregon. What is now Oregon and Washington, however, were claimed by both the United States and Great Britain. In 1846 the two countries reached an agreement. Under the deal, the United States gained control of those two future states. Britain kept control of lands to the north, which became part of Canada.

Meanwhile, some American settlers had moved to what is now Texas, which belonged to Mexico at the time. In the next decade, they declared independence. They set up an independent country, though many Texans wanted to join the United States. In 1845 the United States **annexed**, or took control of, Texas.

Some Americans hoped to gain California and other lands that were part of Mexico. This desire led to a war with Mexico, which the United States won. In the Treaty of Guadalupe Hidalgo (1848), Mexico gave the United States a vast area that later formed all of California, Utah, and Nevada and parts of Colorado, Arizona, Wyoming, and New Mexico. This territory added a sizable Spanish-speaking population to the United States.

MAP SKILLS

1 **THE GEOGRAPHER'S WORLD** Where did the Spanish found most of their western settlements?

2 **ENVIRONMENT AND SOCIETY** How were settlers able to reach Salt Lake City from the central part of the United States?

Westward Expansion

PURCHASED FROM GREAT BRITAIN, 1818

1851 - Seattle

1811 - Astoria, Oregon

OREGON TERRITORY, 1846

Oregon Trail

LOUISIANA PURCHASE, 1803

40°N

Mormon Trail

1847 - Salt Lake City

1839 - Sacramento

1776 - San Francisco (founded by Spanish)

1777 - San Jose (founded by Spanish)

MEXICAN CESSION, 1848

1858 - Denver

PACIFIC OCEAN

1781 - Los Angeles (founded by Spanish)

Santa Fe Trail

30°N

1868 - Phoenix

1769 - Mission San Diego (founded by Spanish)

TEXAS ANNEXATION, 1845

1841 - Dallas

130°W

• Settlement/City
1845 Year acquired

GADSDEN PURCHASE, 1853

1718 - San Antonio (founded by Spanish)

0   400 miles
0   400 kilometers
Albers Equal-Area Conic projection

120°W

MEXICO

Gulf of Mexico

110°W

100°W

90°W

When gold was discovered in California, thousands of people streamed there in hopes of making their fortunes. This mass migration is called the California Gold Rush. San Francisco and other cities grew rapidly as a result.

Western Lands in the Late 1800s

Later in the 1800s came more discoveries of mineral wealth in other areas in the region. Each new discovery brought more people to the region in hopes of becoming wealthy. A huge reserve of silver lured them to Nevada in the 1870s. Also during that decade, gold attracted people to South Dakota's Black Hills and to Colorado.

At the same time, the nation was building railroads to join the eastern and western areas. The first line from the Mississippi River to the Pacific Ocean was completed in 1869. Others followed. Trains carried settlers to the western states. Some started farms or ranches. Others settled in towns that sprang up along rail lines or near mines.

As a result of these changes, the population of the West grew rapidly. In 1900 more than twice as many people lived in the West as in 1880, just 20 years earlier.

The Great Plains changed dramatically during this time. The vast grasslands were turned into farms and ranches. Settlers hunted the huge herds of bison and other animals. Some of the animals became **extinct**, or disappeared from Earth. These were huge changes to the biosphere of the Plains.

Passengers board stagecoaches in Virginia City, Nevada. The discovery of silver drew settlers to the town and made the area wealthy almost overnight.

Identifying What other western areas drew settlers as a result of the discovery of minerals?

©Bettmann/Corbis

The spread of white settlements came at the expense of Native Americans. Native Americans suffered from the changes to the environment and the growth of the population. Farms, ranches, railroads, and mines took away land that Native Americans had farmed or hunted on. Some Native American groups resisted the changes, but they were outnumbered. Finally, Native Americans were forced to live on **reservations**. These are lands that were set aside for them. Reservations were often located in areas with poor soil that made farming difficult.

Gaining New Lands

During the late 1800s, the United States made its last land acquisitions. The first new territorial gain was Alaska. In 1867, the United States purchased Alaska from Russia for just over $7 million. The future state was so large that the cost was only about two cents per acre. The purchase of Alaska was not entirely popular. Some newspaper editors criticized Secretary of State William Seward for agreeing to the sale. They called the area "Seward's Icebox" or "Seward's Folly." In 1898, however, gold and copper were found in Alaska. These discoveries awakened new interest in the land.

Americans also took an interest in Hawaii in the late 1800s. Businesspeople began to grow sugar there. By the late 1880s, American sugar planters feared Hawaii's royal family would take away the power and land they had acquired. Instead, they seized the government and requested that the United States annex Hawaii. In 1900 the government agreed to do so.

☑ **READING PROGRESS CHECK**

Analyzing How were the acquisitions of Texas and Hawaii similar?

Shows as He Goes, a Native American chief (left), fought U.S. pioneers and soldiers on the Great Plains. Native American boys in uniform (right) attended a white-run school opened in Pennsylvania during the late 1800s.

▶ **CRITICAL THINKING**
Determining Central Ideas How did Native Americans in the West live before the arrival of white settlers? How did they live after whites settled the area?

(l to r) ©Corbis; ©Bettmann/Corbis

Agriculture and Industry

GUIDING QUESTION *How did people in the states west of the Mississippi live?*

As people moved into the states west of the Mississippi, they developed various ways of earning a living. For many decades, their choices depended on the resources of the area where they settled. Most made their living in primary economic activities that extract resources directly from the earth. These include farming, ranching, mining, lumbering, and fishing.

Farming and Ranching

In the 1800s, many Americans were farmers. Many dreamed of starting farms in the West. In 1862 Congress made that easier by passing the Homestead Act. This law made public land in the western states free to anyone who claimed the land, built a farm, and stayed on it for five years. Hundreds of thousands of people settled on the Great Plains to start farms.

Life on these farms was not easy. The lack of trees made it difficult to find wood to build homes. People covered homes with sod—chunks of soil held together by the roots of grasses.

Another important activity in the western states was raising cattle. Cowboys in places such as Texas herded the cattle and drove them north to towns in Colorado and Kansas that had railroad

Cowboys on horseback round up cattle near Colorado's Cimarron River.

▶ **CRITICAL THINKING**

Describing How did railroads help the cattle industry grow?

©Corbis

stations. There, the animals were shipped east to cities like Chicago. Meatpacking companies butchered the animals into meat that could be sold in growing eastern cities. Cities like Denver; Kansas City, Missouri; and Omaha, Nebraska, became major centers for processing crops and meat.

Industry

The first industries in the western states were also primary economic activities. The industries concentrated on using the resources of the region. Companies set up silver and copper mines. Others cut trees in the Northwest to provide lumber for building homes and ships. Fish canneries were important along the Pacific coast and the Gulf coast.

In the 1900s, new primary industries developed. The growing popularity of cars created rising demand for oil. The oil industry boomed in Texas, Oklahoma, and California. Oil in northern Alaska became usable in the 1970s with completion of a major construction project: the Trans-Alaska Pipeline System (TAPS). That project built a pipeline from the northern oil fields to the port of Valdez in the south so that oil could be shipped to the continental states.

Although primary economic activities remained dominant as the region developed, more secondary economic activities began to emerge. These are industries that turn raw materials into manufactured goods. Many cities across the region became major manufacturing centers. Los Angeles specialized in machine tools and automobiles. San Francisco became a major shipbuilding center.

Recreation and Entertainment

Other new industries focused on recreation and entertainment. Southern California became home to the movie industry. Las Vegas turned into a major resort city known for its casino-hotels, shops, and restaurants. Areas in the Rockies with great natural beauty—or excellent slopes for skiing—became favored vacation spots.

Include this lesson's information in your Foldable®.

☑ **READING PROGRESS CHECK**

Determining Central Ideas What resources attracted Americans to the western region in the 1800s?

LESSON 2 REVIEW (CCSS)

Reviewing Vocabulary

1. How did *pueblos* and *missions* differ?

Answering the Guiding Questions

2. *Describing* What was one change in the way of life for Native Americans after the Spanish and French came to this region?

3. *Determining Central Ideas* Why was the Lewis and Clark expedition important?

4. *Identifying* Why did the movement of white settlements into the West cause problems for Native Americans?

5. *Analyzing* In what way was the recreation industry in the region similar to farming, ranching, and mining in earlier times?

6. *Informative/Explanatory Writing* In a paragraph, explain how the industries that developed in the region late in the 1900s were different from those of earlier times.

(l to r) Somos/Veer/Jupiterimages; Kevork Djansezian/Getty Images News/Getty Images; Roy Delgado/www.Cartoonstock.com; Stephen Brashear/Getty Images News/Getty Images

Reading **HELP**DESK (CCSS)

Academic Vocabulary

- **annual**
- **decline**

Content Vocabulary

- **Mormon**
- **Dust Bowl**
- **topsoil**
- **agribusiness**
- **aerospace**

TAKING NOTES: *Key Ideas and Details*

Organize As you study the lesson, take notes on the topics shown below.

Urban and Rural Life
Challenges
The Economy

Lesson 3
Life in the United States West of the Mississippi

ESSENTIAL QUESTION · *How does technology change the way people live?*

IT MATTERS BECAUSE
The states west of the Mississippi are a source of technological change.

The Region's Cities and Rural Areas

GUIDING QUESTION *Where do the people of the region live?*

Modern cities are in many ways similar. Glass, steel, and concrete skyscrapers rise into the sky. Networks of highways carry heavy traffic. Cities in this region have distinct characters, though. The French flavor of New Orleans differs from the Spanish style of Santa Fe, New Mexico. Denver, near towering mountains, is unlike Omaha, Nebraska, on the relatively flat Plains. What could be more different than tropical Honolulu and cold Anchorage?

One characteristic common to almost all cities in the western United States is dependence on the automobile for transportation. Most western cities have limited or no subway or light-rail systems. In addition, most large cities are spread out, and the distances between cities are often great.

Major Port Cities

The region has many major ports. Los Angeles and Long Beach in California are essentially one port. They handle more than one-half the value of all imports into the United States that come through Pacific ports. San Diego and San Francisco in California, and Seattle and Tacoma in Washington State, are also vital to U.S. trade with Asia.

Although only a small part of the region is located along the Gulf of Mexico, Gulf ports are important. Three of the nation's top 10 ports in terms of the **annual**, or yearly, value of goods they handle are the Texas port cities of Houston, Beaumont, and Corpus Christi.

Many of these cities are diverse. Los Angeles—the nation's second-largest city—is home to people who collectively speak about 90 languages other than English at home. Los Angeles County has more Latinos and Native Americans than any other county in the United States. The city of Los Angeles has more people from South Korea and Nicaragua than any other city outside those nations.

Interior Cities

Denver has an unusual location for a major city. It is not a seaport or on the navigable part of a river. Denver owes its vibrance to the mountains nearby. It originally grew as a mining town. In the late 1900s, the city attracted people who wanted to enjoy the mountains. Its economy is based on software, finance, and communications.

Some of the nation's most rapidly growing cities are in the interior of the United States west of the Mississippi. They include Austin, Texas; Boise, Idaho; Las Vegas; Phoenix; Provo, Utah; and Riverside, California. In Texas, the location of San Antonio, Dallas, and Fort Worth near Mexico makes them important to trade with that country.

Somos/Veer/Jupiterimages

Many urban schools today reflect the growing ethnic diversity of America's cities.

Life in Rural Areas

Small towns and villages remain home to millions of people in the states west of the Mississippi. Many rural Americans rely on the rich resources of the land and sea. They might be farmers of the Plains or fishers in rural Alaska. They may drill for oil in Texas or run a ranch in Montana. Although these occupations have existed for centuries, modern technology often makes these jobs easier. For example, farmers and ranchers use GPS devices to map regions, manage the land, and track cattle.

☑ **READING PROGRESS CHECK**

Citing Text Evidence In what ways are Las Vegas and Phoenix different from Los Angeles and Seattle?

Challenges Facing the Region

GUIDING QUESTION *What issues will face the region in the coming years?*

Americans celebrate their ethnic and religious diversity. Diversity has long been a strength of the nation. While diversity enriches American life, other population changes pose challenges to the region's future.

A man from East Asia takes the oath to become a U.S. citizen. He was part of a group of 7,000 candidates who became U.S. citizens in a ceremony held in Los Angeles, California.

▶ **CRITICAL THINKING**

Identifyng What are some of the ethnic groups that make up the diverse population of the western United States? Why do you think many people from East Asia have settled in this region?

Kevork Djansezian/Getty Images News/Getty Images

Population Changes

In recent decades, much of the growth in U.S. population has taken place in the states west of the Mississippi River. This region attracts new residents because of the mild climate and growing businesses. But more people means a strain on natural resources such as water, which is already scarce in much of the region.

Ethnic and racial diversity are common in the states in this region. In Hawaii, Asian Americans and other distinct ethnic or racial groups form the majority of the population. In California, New Mexico, and Texas, Latinos constitute a large part of the population. In these four states, non-Hispanic whites are in the minority.

The United States west of the Mississippi is also marked by great religious diversity. A few Christian groups are particularly important in this region. Lutherans are numerous in the northern Great Plains states. Catholics are prominent in the states from New Mexico to California. Utah and some neighboring states have many Mormons. **Mormons** are members of the Church of Jesus Christ of Latter Day Saints. Large numbers of Mormons settled in Utah in the mid-1800s.

Another population trend raises important economic challenges. The share of the population over age 65 has been growing. Many older people require extensive health care. That increases the costs for Medicare, a government-run program that pays the health care costs of the elderly. Most older people collect monthly retirement checks from the Social Security system. Meeting the costs of Social Security and Medicare will be a challenge in the coming years.

Relations With Neighbors

Relations between the United States and neighboring Canada and Mexico are strong. In the early 1990s, the three nations signed a trade agreement called the North American Free Trade Agreement (NAFTA). In that treaty, they pledged to remove all barriers to trade among themselves. This created the world's largest free trade area.

Today, Canada and Mexico are the largest markets for exports from the United States. They are also the second- and third-largest sources of U.S. imports, behind China. States west of the Mississippi form a vital part of this trade. Their food products and manufactured goods form a share of the exports to these nations. In addition, much of the trade that takes place flows into and out of the United States through ports in this region.

Open borders help trade. They also make it possible for people to enter the country illegally. In the late 1900s, the problem of illegal immigrants drew a great deal of attention. The U.S. government has taken steps to reduce the flow of illegal immigrants. These efforts have had some success. Illegal immigration dropped from about 550,000 people in 2005 to around 300,000 in 2008. Changes in Mexico also help explain this reduction. A better economy, smaller families, and better education have meant better chances of landing good jobs in Mexico.

The Water Problem

Most of the United States west of the Mississippi usually receives little rain. Years of low rainfall can easily lead to drought. Between 1930 and 1940, a severe drought dried the southern Great Plains so thoroughly that crops died. Strong winds carried dry soil away, covering other areas with dust. The area came to be called the **Dust Bowl**. Many farmers lost their homes and left the area looking for work.

Thinking Like a
Geographer

Making the Desert Bloom

Why did Mormon pioneers settle in Utah? Utah in the mid-1800s was a harsh desert land. The Mormons, however, needed a safe, isolated place where they could practice their religion free of persecution. In 1847 the first Mormon settlers reached Utah after traveling 1,000 difficult miles (1,609 km) from the Midwest. The land was dry and wild. Nevertheless, the Mormons stayed in Utah. They built irrigation canals to support farms and towns. Life at first was difficult, but the Mormons made their Utah communities prosper because of their hard work and determination to succeed. By 1860, many other Mormons had arrived, and numerous Mormon settlements dotted the Utah region.

" Bad news . . we've run out of unlimited resources. "

An editorial cartoon makes a point about a political issue or event.

▶ **CRITICAL THINKING**

Analyzing What issue do you think this cartoon is about? What do you think the cartoonist's opinion on this issue is?

The Dust Bowl resulted in new practices in land use and farming. These practices were intended to reduce the chance of another dust bowl. Nevertheless, unusually low rainfall for several years could cause similar difficulties in the future.

The lack of water is more of a challenge because of population growth. As more and more people settle in the area, more water is needed. When farmers expand their operations, they also need more water. This strains the limited amount of water present in the region, which could limit economic growth.

Limited water resources have prompted scientists to develop ways to remove salt from seawater to make it usable for drinking and farming. This process, called desalination, is expensive but might provide a long-term solution to the problem in the future.

Human Actions and the Environment

Oil—another precious resource—is also a potential cause of environmental damage. In 1989 the oil tanker *Exxon Valdez* ran aground in Alaska's Prince William Sound. Its shattered hull released more than 250,000 barrels of oil into the sea. The oil killed plants and animals and severely hurt the local economy. Even worse was the Deepwater Horizon disaster of 2010. An explosion destroyed an oil drilling platform deep in the Gulf of Mexico. By the time the underwater leak was stopped, 5 million barrels of oil had gushed into the sea. It devastated the coastal economy.

Erosion and the Environment

Erosion is another environmental problem in the region. Harvesting trees for lumber has left some mountain slopes bare. With no tree roots to hold the soil in place, soil runs off with the rain. This runoff affects the surface of the mountain. Pieces of the mountain become smaller pieces and go down the side. Erosion of rich **topsoil**, the fertile soil that crops depend on to grow, is a problem in some farming areas.

The region west of the Mississippi River also experiences a variety of natural disasters. Washington, Alaska, and Hawaii have active volcanoes that can cause damage if they erupt. Wildfires and

Roy Delgado/www.Cartoonstock.com

HAL GARB/AFP/Getty Images

mud slides can also strike the region. The wildfires occur when wooded areas become too dry. Mud slides can result from heavy rains or severe shaking from an earthquake. During a mud slide, the soil moves like a liquid, flowing downhill in a sea of mud that dislodges trees and buries houses in its path.

Some parts of the region are prone to earthquakes. A major fault system called the San Andreas Fault, cuts through western California. Movement along this fault has caused several major earthquakes over the years. Scientists think more are likely to occur. As a result, severe damage is possible—as well as the loss of many lives—in major cities such as Los Angeles and San Francisco.

✔ **READING PROGRESS CHECK**

Determining Central Ideas Why is the high likelihood of a major earthquake along the San Andreas Fault so worrisome?

In 1994 the deadly Northridge earthquake struck near Los Angeles. It resulted in 57 deaths and was one of the costliest natural disasters in U.S. history.
Citing Text Evidence Why is California so prone to earthquakes?

The Economy

GUIDING QUESTION *How do the people of the region make their living?*

Land and resources have attracted many Americans to the western states. Some people of the region still rely on these advantages to earn their living. Most people, however, work in modern settings that are far removed from the land.

Modern Agriculture

Agriculture remains an important part of life west of the Mississippi River. The Great Plains are the center of the nation's wheat industry with eight of the nation's top wheat-producing states. Spring wheat—planted in spring and harvested in autumn—is grown on the

northern Plains. Winter wheat—planted in the fall and harvested in the spring—grows in the southern Plains. Wheat plants would wither in this southern region if they faced the hot summer temperatures. Other important crops in the Plains include cotton, corn, hay, and sorghum, another grain. Cattle and sheep ranching are major activities from Montana to New Mexico.

Other areas in the region are known for other crops. Washington and Oregon produce dairy products and fruit. Idaho is famous for potatoes. California, with its warm, year-round Mediterranean climate, provides fruits and vegetables throughout the year.

Today, the number of small family farms in the region is **declining**. They are being replaced by large **agribusinesses**, firms that rely on machines, advanced technology, and mass-production methods to farm large areas.

The Mining Industry

Mining the region's vast resources remains a vital part of its economy. Oil and gas fields in the southern Plains, Wyoming, and southern California provide energy. Mines in Montana furnish copper, and those in the Colorado Plateau provide uranium. The Northwest is still a major producer of timber.

The Aerospace Industry

In the early 1900s, the airplane industry was born. Decades later, the states of Washington, California, and Texas became important to the aerospace industry. The **aerospace** industry makes vehicles that

Stephen Brashear/Getty Images News/Getty Images

Workers install an engine cover on a passenger plane at an aerospace factory in Washington state. *Identifying* What other states became important centers of the aerospace industry?

travel in the air and in outer space. This industry brought many engineers and other highly skilled workers to the United States west of the Mississippi.

The aerospace industry remains important in southern California and around Seattle. Employment in this industry has declined in recent years, however. Taking its place has been the computer industry. The first area to experience rapid growth from this field was California's Silicon Valley. Today, centers of computer research and manufacturing are also found in Texas, Washington, and New Mexico.

Service Industries in the Region

Advanced technologies, such as robotics and computerized automation, have transformed manufacturing in the region. As with farming, the region's factories produce greater quantities of goods with fewer workers than in the past.

The growth of computer technology has spawned other industries. Software and information science companies are based in the same areas that are home to computer manufacturing. Utah and Colorado have also become important in these fields. Los Angeles, San Francisco, Denver, Dallas, and Seattle have become major financial centers. Telecommunications—telephone and related services—is an important industry in Denver and Dallas.

Over time, new industries have focused on tertiary economic activities such as retail sales, entertainment, and tourism. Visitors flock to the area to see the soaring mountains, stunning rock formations, and dense forests. They enjoy unusual features like Alaska's glaciers and Hawaii's tropical beaches. The western region includes some of the most-visited national parks in the country.

☑ **READING PROGRESS CHECK**

Citing Text Evidence Why did tourism and the tourist industry develop in the United States west of the Mississippi?

FOLDABLES®
Study Organizer

Include this lesson's information in your Foldable®.

LESSON 3 REVIEW (CCSS)

Reviewing Vocabulary
1. Why would the loss of *topsoil* threaten *agribusiness*?

Answering the Guiding Questions
2. ***Determining Central Ideas*** Why is it unusual that Denver developed into a major city?

3. ***Identifying*** What are the advantages and disadvantages of open borders?

4. ***Analyzing*** What steps could be taken to prevent the damage caused by mud slides? Explain your answer.

5. ***Describing*** Would you describe the economy of these states as diversified? Explain.

6. ***Informative/Explanatory Writing*** Write a short essay explaining why water conservation is important in the region.

Directions: Write your answers on a separate piece of paper.

1 Use your **FOLDABLES** to explore the Essential Question.
INFORMATIVE/EXPLANATORY WRITING Write an essay explaining why the physical geography of the area west of the Mississippi River made settlement of the region difficult.

2 21st Century Skills
ANALYZING Working in small groups, research and prepare a presentation explaining how an alternative resource can be used to reduce America's dependence on foreign oil.

3 Thinking Like a Geographer
CITING TEXT EVIDENCE Create a time line like the one shown and place these six events in the correct order on it.

- Lewis and Clark expedition
- The American Revolution
- Louisiana Purchase
- Spanish establish settlements in the West
- Alaska is purchased from Russia
- The Pueblo flourish in the Southwest

4 **GEOGRAPHY ACTIVITY**

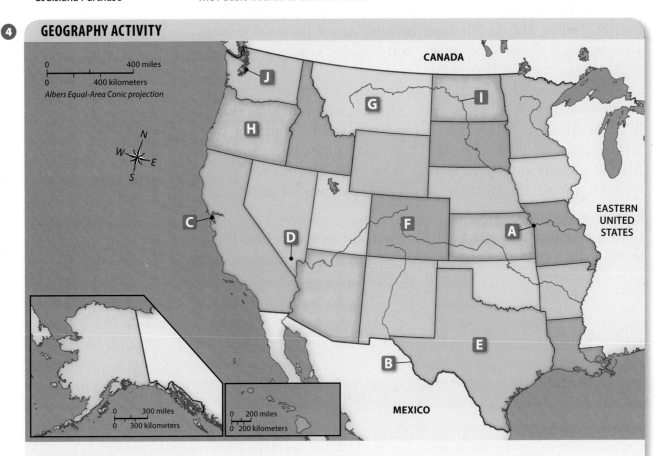

Locating Places
Match the letters on the map with the numbered places listed below.

1. Oregon
2. Missouri River
3. San Francisco
4. Texas
5. Las Vegas
6. Colorado
7. Montana
8. Rio Grande
9. Kansas City
10. Seattle

REVIEW THE GUIDING QUESTIONS

Directions: Choose the best answer for each question.

1 The Rocky Mountains are
   A. the shortest mountain range in the United States.
   B. a cordillera that extends from Canada to Mexico.
   C. an archipelago.
   D. worn down from erosion.

2 Which of the following is a contiguous state?
   F. Puerto Rico
   G. Alaska
   H. Colorado
   I. Hawaii

3 How did western settlement affect Native Americans?
   A. They were able to sell their lands for a great deal of money.
   B. Those who did not die of disease were forced onto reservations.
   C. They moved east to work in coal mines and factories.
   D   Many became cowboys or joined the army.

4 During the Civil War, Congress passed the Homestead Act, which
   F. guaranteed low-interest loans to home buyers.
   G. established the border with Canada.
   H. set aside land to be used for schools.
   I. gave away western lands to people who were willing to move there and build farms.

5 Population in the states west of the Mississippi River is
   A. growing slowly.
   B. declining.
   C. growing rapidly.
   D. showing little change.

6 Which resource is the most critical factor to further development in the western states?
   F. oil
   G. natural gas
   H. solar power
   I. freshwater

{Chapter 5 ASSESSMENT (continued)}

DBQ **ANALYZING DOCUMENTS**

7 **DETERMINING WORD MEANING** Read the following quotation about the American population.

" *With an overall 20 percent growth rate, the [population of the] West grew more rapidly than any other region. The South was the second fastest growing region, increasing 17 percent. The Midwest and the Northeast grew almost 8 percent and 6 percent, respectively.* "

—from *National Atlas of the United States*

What does the term *growth rate* in the passage refer to?

A. the number of states in one year compared to another

B. the amount of land in one year compared to another

C. the number of people in one year compared to another

D. the economic output in one year compared to another

8 **CITING TEXT EVIDENCE** How did population growth in the West compare to that in the Northeast?

F. much lower

G. more than three times higher

H. about the same

I. nearly twice as high

SHORT RESPONSE

" *In the Western United States, the availability of water has become a serious concern. . . . The climate . . . in the West . . . is best known for its low precipitation, aridity [dryness], and drought. . . .The potential for departures from average climatic conditions threatens to disrupt society and local to regional economies.* "

—from Mark T. Anderson and Lloyd H. Woosley, Jr.,
Water Availability in the Western United States

9 **IDENTIFYING** Why might a change in climate threaten to disrupt societies and economies?

10 **ANALYZING** What do you think the people of the western states should do in light of this problem? Why would it work?

EXTENDED RESPONSE

11 **INFORMATIVE/EXPLANATORY WRITING** Imagine that you are the governor of any one of the states west of the Mississippi River and you want to bring more tourism dollars into your state. Use what you have learned from the chapter and do additional Internet research on that state. Write a press release to newspaper travel editors that promotes all the reasons people should vacation in your state.

Need Extra Help?

If You've Missed Question	**1**	**2**	**3**	**4**	**5**	**6**	**7**	**8**	**9**	**10**	**11**
Review Lesson	1	1	2	2	3	3	3	3	3	3	1

From "Water Availability for the Western United States—Key Scientific Challenges" by Mark T. Anderson and Lloyd H. Woosley, Jr., U.S. Geological Survey Circular 1261. Department of the Interior/USGS. The USGS home page is http://www.usgs.gov.

CANADA

ESSENTIAL QUESTIONS · *How do people adapt to their environment?*
· *What makes a culture unique?*

Michelle Gilders Canada West/Alamy

Blackfoot girl takes part in First Nations Pow Wow held in Alberta, Canada.

networks

There's More Online about Canada.

The Story Matters...

Nearly 4 million square miles (10.4 million sq. km) in total area, Canada is the second-largest country in the world. The landscape of this immense country is known for its beauty and bountiful natural resources. Even though Canada is immense in size, it is sparsely populated. Canada's population of First Nations and Inuit people, French, English, and immigrants from around the world reflects its history and diversity.

Study Organizer

Go to the Foldables® library in the back of your book to make a Foldable® that will help you take notes while reading this chapter.

North | South

Past and Present | World Relations

In land area, Canada is the second-largest country in the world. It occupies most of the northern part of North America and is bordered by three major oceans. As you study the map, look for the geographic features that make Canada unique.

Step Into the Place

MAP FOCUS Use the map to answer the following questions.

1 **THE GEOGRAPHER'S WORLD** Which Canadian province borders the Great Lakes?

2 **THE GEOGRAPHER'S WORLD** What body of water divides Canada nearly in half?

3 **PLACES AND REGIONS** What is the national capital of Canada?

4 **CRITICAL THINKING Analyzing** What direction would you travel when flying from Winnipeg to Yellowknife?

A

CANADIAN ARCTIC Canada's polar bears live in icy Arctic terrain surrounded by open water.

B

VICTORIA, BRITISH COLUMBIA British Columbia's Legislative Assembly assembles in the Parliament Buildings to pass laws for the province.

Step Into the Time

DESCRIBING Select one event from the time line and write a paragraph describing how that event changed Canada.

1000 Vikings arrive in Canada

1534 Cartier explores Gulf of St. Lawrence

1608 Quebec founded

1670 Hudson's Bay Company formed

1700

Canada

ARCTIC
OCEAN

Greenland
(Denmark)

Alaska
(U.S.)

Yukon
★ Whitehorse

Great Bear
Lake

Mackenzie R.

Northwest
Territories
★ Yellowknife

Great
Slave Lake

Nunavut

Iqaluit ★

ATLANTIC
OCEAN

Newfoundland and Labrador

A

Hudson
Bay

British
Columbia

Alberta

Lake
Athabasca

Saskatchewan

Manitoba

Edmonton ★

Calgary •

Regina ★

Lake
Winnipeg

Ontario

Québec

St. John's ★

Gulf of
St. Lawrence

Prince
Edward
Island

Fredericton ★ Charlottetown ★

Nova Scotia

Halifax ★

New Brunswick

Victoria ★ • Vancouver

B

PACIFIC
OCEAN

★ Winnipeg

Severn R.

Lake
Superior

Lake
Huron

Lake
Michigan

Montréal •

Ottawa ✪

Toronto ★

Lake
Ontario

Lake Erie

St. Lawrence R.

Québec ★

UNITED STATES

National boundary
✪ National capital
★ Provincial/territorial capital
• City

0   500 miles
0   500 kilometers
Lambert Azimuthal Equal-Area projection

120°W   110°W   100°W   90°W   80°W   60°W

70°N

ARCTIC CIRCLE

60°N

50°N

40°N

1896
Gold discovered
in Yukon Territory

2010
Vancouver hosts
Winter Olympics

1763 France cedes
Canada to Britain

1800

1900

2000

1774 Quebec Act becomes law

1993 Canada ratifies NAFTA

1999 Inuit win rights to their own territory

The Physical Geography of Canada

Academic Vocabulary

- **comprise**
- **access**

Content Vocabulary

- **province**
- **territory**
- **shield**
- **coniferous**
- **deciduous**
- **archipelago**
- **tundra**
- **fishery**

TAKING NOTES: *Key Ideas and Details*

Summarizing As you read about Canada's regions, take notes about them using the graphic organizer below.

Region	Provinces or Territories	Key Facts
Atlantic Provinces		

ESSENTIAL QUESTION · *How do people adapt to their environment?*

IT MATTERS BECAUSE
Canada shares many physical features with the United States.

Canada's Physical Landscape

GUIDING QUESTION *How is Canada's physical geography similar to and different from that of the United States?*

For millions of U.S. residents, another nation, Canada, is a short drive away. Canada is a vast and sprawling land of some 3.86 million square miles (10 million sq. km). That size makes it six times larger than the state of Alaska. Canada is larger than every nation in the world except Russia.

Canada and the United States share the longest undefended border in the world. Several of the crossings along this 5,523-mile (8,888-km) border are busy. About 400,000 people and $1.4 billion worth of goods cross it every day.

Overview of Canada

Although huge, Canada is home to relatively few people. Its population ranks thirty-sixth in the world. The great majority of those people live in the southern part of the country. Almost 90 percent of them live within 150 miles (241 km) of the Canada-U.S. border.

Canada is divided into smaller units. The main divisions are Canada's 10 provinces. **Provinces** are administrative units similar to states. Each province has its own government. Canada also has three territories. The **territories** are lands administered by the national government. Canada's provinces and territories can be grouped into five regions.

(l to r) Design Pics/Bilderbuch; Design Pics/Richard Wear; ©Dave Broscha/First Light/Corbis; Anita Erdmann/Flickr/Getty Images

The Atlantic Provinces

Nova Scotia, New Brunswick, Prince Edward Island, and Newfoundland and Labrador are called the Atlantic Provinces. These four relatively small provinces border the Atlantic Ocean. Except for Newfoundland and Labrador, they are distinguished mainly by lowlands and plateaus. Highlands cover western Newfoundland and Labrador.

Much of the Atlantic Provinces has a humid continental climate. Winters are cold, and summers are warm. Toward the north, winters can last as long as six months. To the south, winters last only four months. More rain falls here than in Canada's interior.

Quebec and Ontario

The provinces of Quebec and Ontario reach from the St. Lawrence River and Great Lakes to northern Canada. They are the heart of Canada and home to more than 6 out of every 10 Canadians.

The land along the St. Lawrence River and the Great Lakes tends to be lowland plains with fertile soil. A massive plateau called the Canadian Shield covers the northern area of these two provinces. A **shield** is a large area of relatively flat land **comprised** of ancient, hard rock. This plateau extends south, east, and west of Hudson Bay. The Shield holds many valuable minerals, such as iron ore, uranium, gold, and copper.

Like the Atlantic Provinces, the southern parts of Ontario and Quebec have a humid continental climate. Temperatures are colder to the north than to the south, and winters are longer. In the northern part of these two provinces, a subarctic climate prevails. These areas have long, cold winters and mild summers. Much of the northern part of these provinces is covered by forests that include coniferous and deciduous trees. **Coniferous** evergreen trees produce cones that hold seeds, and they have needles instead of leaves. **Deciduous** trees shed their leaves in the autumn.

Rainfall and relatively mild temperatures give the southern edge of these two provinces a longer growing season than areas in northern Canada. That, combined with fertile soil, makes it a productive farming area.

MAP SKILLS

1 PLACES AND REGIONS
What subregion lies directly north of the Great Lakes?

2 THE GEOGRAPHER'S WORLD What subregions border the Pacific Ocean?

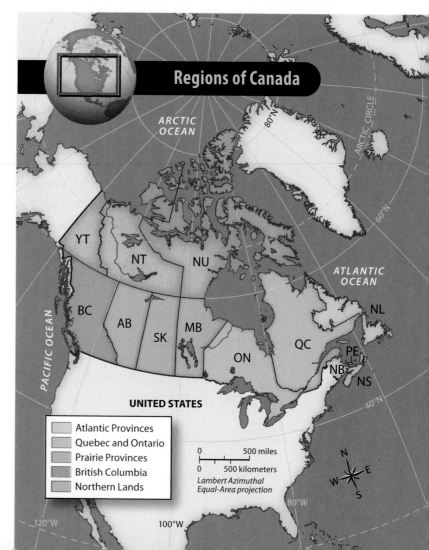

Regions of Canada

Legend:
- Atlantic Provinces
- Quebec and Ontario
- Prairie Provinces
- British Columbia
- Northern Lands

0   500 miles
0   500 kilometers

Lambert Azimuthal Equal-Area projection

Cape Breton Highlands National Park (left) is located in Nova Scotia. Alberta's Jasper National Park (right) is one of Canada's oldest and largest national parks.

▶ **CRITICAL THINKING**

Describing What is the climate of Alberta and the other Prairie Provinces?

The Prairie Provinces

The Prairie Provinces comprise Manitoba, Saskatchewan, and Alberta. They are covered chiefly by plains extending north from the United States. As in the United States, the plains are tilted from west to east. Their elevation is higher in Alberta than in Manitoba.

Highland areas rim the plains to the northeast and the west. The northeastern highlands, in Manitoba, are part of the Canadian Shield. The western highlands, in Alberta, are the eastern edge of the Canadian Rockies.

The flat land of the Prairie Provinces receives cold blasts of Arctic air from the north. The average daily temperature in January for Regina, Saskatchewan, is –1°F (–18°C). Summers are much warmer, with the average temperature in July being 67°F (19°C). The prairies have reached the highest temperature ever recorded in Canada. On one hot day in 1937, two communities in Saskatchewan had temperatures of 113°F (45°C).

Precipitation in the Prairie Provinces is light. The area receives only about 15 inches (38 cm) of rain or snow each year.

British Columbia

The landscape of British Columbia is like that of the northwestern United States. The Rocky Mountains tower over the eastern edge of the province. A deep valley and plateaus separate the mountains from another high range farther west, the Coast Mountains. The Canadian Rockies include 30 mountain peaks that reach more than 10,000 feet (3,048 m). They are also home to five of Canada's national parks. The Coast Mountains rise even higher—several of them soar more than 15,000 feet (4,572 m).

(l to r) Design Pics/Bilderbuch; Design Pics/Richard Wear

Inlets of the Pacific Ocean cut into the mountains along British Columbia's coast. Steep cliffs rise directly from the water to heights of more than 7,000 feet (2,134 m).

British Columbia has a marine west coast climate. Temperatures are mild because warm air blows west from the ocean. Rainfall is heavy in some areas—as much as 100 inches (254 cm) per year.

Northern Lands

Canada's three territories—Nunavut, Northwest Territories, and Yukon Territory—lie to the north. The Arctic Ocean laps their northern shores. Within that ocean lies an archipelago. An **archipelago** is a group of islands. Some of the roughly 1,000 islands are tiny. Baffin Island, though, is nearly 200,000 square miles (518,000 sq. km)—larger than the state of California.

The vast center of these territories is covered by lowland plains. To the east, the rim of the Canadian Shield rises. The Shield continues onto the islands north of Hudson Bay.

The western part of the far north has high mountains. Among them is Mount Logan, Canada's highest peak, at 19,524 feet (5,951 m).

Much of the land of the territories, called the far north, is covered by a subarctic climate zone marked by cold winters and mild summers. The areas farthest north, however, have a tundra climate. The name of this climate zone comes from the landscape. A **tundra** is a flat, treeless plain with permanently frozen ground.

☑ **READING PROGRESS CHECK**

Determining Central Ideas Why do most of Canada's people live in southern Canada?

Cold and ice do not stop Yellowknife residents in the Northwest Territories from enjoying outdoor activities (left). Canada's wettest area, in terms of rain and snow, is the Pacific coast.

▶ **CRITICAL THINKING**
Describing What is the climate of the far north?

Think Again?

Northern Canada receives a huge amount of snow.

Not true. The cold Arctic air that controls the climate of northern Canada holds little moisture. The dry air brings little snowfall to northern Canada. The Rocky Mountains and the Gulf of St. Lawrence region receive more snow than the far north.

Bodies of Water

GUIDING QUESTION *What bodies of water are important to Canada?*

Some bodies of water shape Canada's climate. Others have economic importance.

Oceans, Bays, and Gulfs

Three oceans border Canada—the Atlantic, the Pacific, and the Arctic. The Pacific Ocean brings rain and mild temperatures to western Canada. Cold air blows over the Arctic Ocean to chill northern and central Canada. The Atlantic Ocean moderates the temperatures of eastern Canada.

The Atlantic and Pacific oceans are also important economically. Ships cross the oceans to bring goods to and from Canada. They are important for another reason, too. East of the Atlantic Provinces is an area called the Grand Banks. This part of the Atlantic Ocean is one of the world's great fisheries. A **fishery** is an area where fish come to feed in huge numbers. The Grand Banks is visited by fishing fleets from all over the world. However, overfishing has severely hurt the populations of some kinds of fish in recent years.

Tugboats tow logs along the Fraser River to sawmills. Forest industries have always been an important part of the country's economy.

North Light Images/age fotostock

The Pacific coast is another important fishery. Many transitions are occurring in the Arctic Ocean because of climate change. Less ice is forming, which allows more ship traffic through the region. It is also altering how native peoples live.

The Gulf of St. Lawrence is another important body of water. It is an extension of the Atlantic Ocean and serves as the mouth of the St. Lawrence River. It provides **access**, or a way in, to the interior of Canada.

An inland sea called Hudson Bay covers much of east central Canada. Native peoples live around its shores. They catch fish and hunt sea mammals for food.

Lakes and Waterways

The Gulf of St. Lawrence connects to the St. Lawrence River, which flows into the Great Lakes. For centuries, rapids and steep drops in elevation blocked ships from moving along the river west of Montreal. In the 1950s, though, the United States and Canada worked together to build the St. Lawrence Seaway. They made canals and a system of locks that can raise and lower ships from water at two different levels. As a result, oceangoing ships can reach as deep into the interior as western Lake Superior.

Canada shares four of the five Great Lakes with the United States. The one it does not share is Lake Michigan. Three other major lakes are found in lowland areas west of the Canadian Shield. They are Great Slave Lake and Great Bear Lake in the Northwest Territories, and Lake Winnipeg in Manitoba.

The Mackenzie River and its tributaries dominate much of the lowlands of the far north. Beginning at Great Slave Lake, the Mackenzie flows north and west to empty into the Arctic Ocean. It is a wide river—from 1 mile to 4 miles (1.6 km to 6.4 km) across—and flows for more than 1,000 miles (1,609 km). The Fraser River flows through the mountains of British Columbia. The river basin is important for many activities such as farming, fishing, and mining.

☑ READING PROGRESS CHECK

Analyzing Why is the St. Lawrence River economically more important to Canada than the Mackenzie River?

Academic Vocabulary

access a way to reach a distant area

Include this lesson's information in your Foldable®.

LESSON 1 REVIEW (CCSS)

Reviewing Vocabulary

1. What is the difference between a *province* and a *territory* in Canada?

Answering the Guiding Questions

2. *Identifying* What landforms in Canada are similar to those in the United States?

3. *Analyzing* Why do most of Canada's people live in southern Ontario and Quebec?

4. *Identifying Point of View* Which body of water do you think is most important to Canada? Why?

5. *Analyzing* Which of Canada's regions do you think has benefited most from the fisheries of the Grand Banks? Why?

6. *Informative/Explanatory Writing* Write a paragraph describing the physical geography of one of Canada's regions and the impact of its landforms and climate on the people in that region.

Lesson 2
The History of Canada

ESSENTIAL QUESTION · *What makes a culture unique?*

IT MATTERS BECAUSE
Canada has one of the world's largest economies.

The First Nations of Canada

GUIDING QUESTION *How did native peoples of Canada live before Europeans came to the area?*

In the United States, the **aboriginal**, or native, peoples who lived in North America before Europeans are called Native Americans. Canadians call them the First Nations.

Coming to Canada

The first people to arrive in Canada came from Asia. They came during a long period of intense cold called the Ice Age. The first groups to arrive moved south because Canada was covered with ice. Over thousands of years, though, Earth's climate warmed. The ice sheets over most of Canada melted. As Canada warmed, people **occupied** the land there. Their way of life depended on the resources where they lived.

Different Ways of Life

The peoples of the eastern woodlands lived by farming, hunting, and fishing. They built villages where they lived for most of the year. They traded with one another. Two important eastern groups were the Huron and the Iroquois. The two nations were rivals and often fought.

The peoples of the Pacific coast lived in a region of plenty. Rivers and the waters of the Pacific Ocean provided fish and sea mammals. They hunted game animals in the forests. They used wood from the region's trees to make houses. They also made oceangoing canoes, which they used to hunt and fish.

Life was more difficult in the far north. There, peoples like the Inuit had to find food in a land where few plants grow. Shelters had to be built without using trees. They had to protect themselves from fierce cold. They hunted caribou, a large animal related to deer, on land and seals and whales on the water. Caribou skins were used to make clothes and shoes to keep them warm in winter.

☑ READING PROGRESS CHECK

Determining Cause and Effect How did the presence and absence of ice affect the early settlement of Canada?

Exploration and Settlement

GUIDING QUESTION *How did migration and settlement change Canada?*

The first Europeans to reach what is now Canada were Vikings. They began their travels around A.D. 1000 and settled in southern Newfoundland. They soon abandoned their settlements and left, however. More than 500 years later, other Europeans came to the Americas.

Europeans in Canada

The next Europeans to explore Canada were the French. An explorer named Jacques Cartier sailed up the St. Lawrence River in the 1530s. He claimed the St. Lawrence and the lands around it for France. The whole area of French control **eventually** was called New France.

In the 1600s, the French made the first serious effort to settle the region. Explorer Samuel de Champlain founded the first French settlement, Quebec, in 1608.

Over time, more French settlers **migrated** to Canada. Some became fur traders. These traders exchanged European goods with the Huron, a First Nations people. The French received beaver furs that could be shipped back to Europe. Some settlers were priests. They came to Canada to minister to the French settlers, who were Roman Catholic. They also hoped to convert native peoples to Christianity. Some settlers farmed. Their crops fed the other settlers.

©Bettmann/Corbis

Academic Vocabulary

occupy to settle in a place
eventually at some later time
migrate to move to an area to settle

Inuit and other native North Americans lived in Canada long before European settlers arrived.
▶ **CRITICAL THINKING**
Describing How did the lives of the First Nations people of the far north differ from the lives of the Pacific coast people?

British Canada

France was a powerful nation in the 1600s and 1700s. It had a rival for power, however. Britain competed with France in the Americas. In the late 1600s, some British merchants formed a company called the Hudson's Bay Company. They set up trading posts around the bay in the hope of gaining some of the profitable fur trade.

The two nations fought wars in the 1700s. As a result of a British victory in 1763, France was forced to give up much of its land in North America. However, the Quebec Act, passed by the British in 1774, gave French settlers in Canada the right to keep their language, religion, and system of laws.

The next big change in Canada came in the 1770s and 1780s. When the American Revolution broke out, thousands of Americans remained loyal to Britain. During and after the war, many of them moved to Canada. Some settled in what are now the Atlantic Provinces and others in modern Ontario.

During the early 1800s, English and French communities disagreed over colonial government policies. Fears of a U.S. takeover, however, forced them to work together. In 1867 the British colonies of Quebec, Ontario, Nova Scotia, and New Brunswick united as provinces of the Dominion of Canada. This new nation was partly self-governing within the British Empire. Other territories would join Canada over the next 100 years.

Known as outstanding commanders, the British General Wolfe (below) and the French General Montcalm lost their lives in the Battle of Quebec. The war ended in 1763.

▶ **CRITICAL THINKING**

Analyzing What was the result of the war?

World History Archive/Alamy

Canada Expands Westward

Canada's leaders began looking westward. They hoped to expand the nation all the way to the Pacific. In 1869 Canada gained the vast territory held by the Hudson's Bay Company. Many Métis lived on some of this land. **Métis** are the children of French and native peoples. They wanted more say in governing themselves. The province of Manitoba was created to give them that chance.

By this time, settlers had already arrived in British Columbia on the Pacific coast. There, they traded furs and searched for gold. The people of British Columbia agreed to join Canada in 1871. In the 1880s, Canadians built a **transcontinental**, or continent-crossing, railroad that united the eastern and western parts of their nation.

New Provinces

Canada's leaders made agreements with some native peoples of the west. According to the government, the native peoples agreed to give up their lands in exchange for aid. Native people believed they gave settlers only the right to use the land for farming. Canada's westward expansion came at a price. New settlers pushed native peoples off their lands. In 1905 Saskatchewan and Alberta entered Canada as provinces.

Meanwhile, gold had been discovered in the Yukon Territory. This discovery led to a gold rush. Helping create order in the west was a police force that formed in 1873. Called the North West Mounted Police, the group is known today as the Royal Canadian Mounted Police.

✔ **READING PROGRESS CHECK**

Determining Central Ideas How did European rivalries affect the development of Canada?

Canada honored two African Canadian heroes in 2012. John Ware (left) was important in starting the ranching industry in western Canada. Viola Desmond (right) worked to repeal unjust laws.

▶ **CRITICAL THINKING**
Describing Who were the Métis? What right did they obtain?

Thinking Like a Geographer

Why was a mounted police force needed in western Canada?

The vast size and sparse population of the region explain it. Police needed to be able to ride horses to travel quickly from one place to another. Because settlements were far-flung, they had to travel often. The Mounties filled an important need in a time before car travel.

Canada Grows and Unites

GUIDING QUESTION *How did Canada change in the 1900s?*

During the 1900s, Canada grew into a prosperous, independent country. Its population grew and became more diverse. Canada also reached the territorial size it is today. In 1949 Newfoundland and Labrador became the last territory to join Canada as a province.

Economic Growth and Immigration

Early in the 1900s, however, Canada had many economic problems. The country's economy was based on agriculture and mining. When prices for those resources fell, Canada suffered. In the late 1900s, Canada became an industrial nation. Canadians built factories and took advantage of their mineral resources. They developed hydroelectric projects and transportation systems. Agriculture also grew as farmers in the west increased production. Granaries stored wheat to feed the growing population. A **granary** is a building used to store harvested grain.

To help increase the nation's industrial power, Canada needed more workers. Canada's leaders made it easier for people to enter the country. Canada's population began to grow, particularly after World War II, which helped Canada meet its need for more workers. Canada's population jumped from 12 million people in 1945 to nearly 35 million in 2012.

The Royal Canadian Mounted Police take part in Canada Day celebrations in Ottawa. The annual celebration commemorates when Canada became a self-governing dominion within the British Empire.

▶ **CRITICAL THINKING**

Identifying What major change in Canada occurred in 1931?

©Zou Zheng/Xinhua Press/Corbis

In recent decades, Canada has welcomed immigrants from all over the world. Today it is a multicultural society. People descended from settlers who came from the British Isles or France account for only about half the population. Another 15 percent trace their background to some other European country. About 6 percent of the people have an African or an Asian background. About 2 percent of Canada's people are from the First Nations. More than a quarter of Canadians have mixed backgrounds.

Independence

A major change came to Canada in 1931. That year, Britain granted almost complete independence to Canada. Canadians were now able to make their own laws without interference from Britain. The British government, however, maintained the right to approve changes to Canada's constitution. This link to Britain finally ended in 1982.

During the later 1900s, Canada became less connected to Britain. Instead, Canada developed closer ties to the United States. It also became active in international bodies, such as the United Nations.

The Growth of Industry

Canada's industries boomed after World War II. Part of the boom was the result of increased demand for Canada's mineral resources. Part was the result of industrial growth. Part, too, came from population growth and movement. More people meant more demand for more goods. Also, people were moving from rural areas to cities and suburbs. That created a need for construction of homes, stores, offices, and roads.

Construction of the St. Lawrence Seaway helped the economy grow, too. Canada could more easily—and cheaply—ship its products around the world. It also could bring in needed goods at lower costs. At the same time, agriculture expanded. Farmers in the Prairie Provinces boosted production. Granaries were built. By 2013, Canada's economy was strong in all sectors: agriculture, industry, services, and information technology (IT).

☑ **READING PROGRESS CHECK**

Identifying What are two ways that Canada changed in the 1900s?

Include this lesson's information in your Foldable®.

LESSON 2 REVIEW

Reviewing Vocabulary
1. Why do the Prairie Provinces have *granaries*?

Answering the Guiding Questions
2. *Describing* Why did people of the First Nations have different ways of life in different parts of Canada?

3. *Analyzing* Why was the province of Manitoba created?

4. *Analyzing* Why do you think people settled what is now British Columbia before settling the Prairie Provinces?

5. *Determining Central Ideas* Why has Canada developed closer ties to the United States?

6. *Informative/Explanatory* Write a summary that highlights the key points of Canada's development as a nation.

Reading **HELP**DESK ⓒⓒⓢⓢ

Academic Vocabulary

- **via**
- **vary**

Content Vocabulary

- **metropolitan area**
- **bilingual**
- **peacekeeping**
- **separatist**
- **autonomy**
- **acid rain**

TAKING NOTES: *Key Ideas and Details*

Summarizing As you read about life in Canada, take notes using the graphic organizer below.

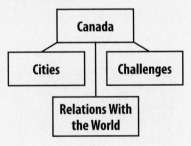

Lesson 3
Life in Canada

ESSENTIAL QUESTION · *What makes a culture unique?*

IT MATTERS BECAUSE
Canada plays a major role in world affairs.

City and Country Life

GUIDING QUESTION *Where and how do Canadians live?*

Where do you think most Canadians live? Do you think of a fishing village or a farm? Do you picture a remote settlement in the far north? Actually, four out of every five Canadians live in cities or suburbs. Canada is a modern urban society.

Ottawa and Toronto

Ottawa is located in the province of Ontario. It is Canada's capital city and home to the national government. Canada's government is similar to those of the United States and the United Kingdom. Like the United States, Canada has national and regional governments. In Canada, those regions are the provinces. Like the United Kingdom, Canada has a parliamentary system. Voters elect members of the legislature, or Parliament. The party with the most members chooses the prime minister, who carries out the laws.

Another Ontario city, Toronto, is Canada's largest metropolitan area. A **metropolitan area** is a city and its surrounding suburbs. Metropolitan Toronto has more than 5 million people. Toronto has access to the Atlantic Ocean **via** the St. Lawrence Seaway. As a result, it is a major port. Rail lines and highways make it possible to ship imported goods from the seaway across Canada. About half of Canada's manufactured goods are made in Ontario. Toronto is an ideal location for shipping Canadian products around the world.

Each summer, Toronto is home to the Canadian National Exhibition, a combination fair and business meeting. The city also has major sports teams and a wide array of cultural opportunities. Its restaurants offer food from many different ethnic groups.

Montreal and Quebec

Montreal and Quebec are the major cities of Quebec Province. Canada's French heritage remains strong throughout this province. Canada is a **bilingual** nation, meaning it has two official languages—English and French. Most people in Quebec speak French.

Montreal is Canada's second-largest city. It is the economic hub, or center of activity, of Quebec Province. Like Toronto, it is a major port because of the St. Lawrence Seaway. Montreal is an important center of manufacturing, banking, and insurance. It offers a special combination of historic European charm and modern life.

Quebec, the capital of Quebec Province, attracts tourists who are eager to see buildings that reflect its 400-year history. Costumed performers reenact life in the early days of New France to make that history come alive.

Western Cities

Canada's third-largest metropolitan area is Vancouver, in British Columbia. It is Canada's busiest port. Vancouver's port ships food products from the nearby Prairie Provinces. Because of its Pacific Ocean location, it is a vital center of trade with Asia. Many people have moved to the city from Asian nations in recent years.

Canada's other major western cities are Calgary and Edmonton in Alberta. Both benefit from Alberta's large oil and natural gas reserves, which have fueled an economic boom in the province. The cities have grown rapidly in recent years. Both are centers for processing the grain and meat produced in the province.

Life in Rural Areas

Life in rural Canada **varies** from place to place. People of the First Nations who live in the far north live in a harsh landscape. Many follow traditional ways, although modern aspects of life can be found as well. For example, they may travel on snowmobiles instead of dogsleds. Fishing villages in the Atlantic Provinces have suffered in recent years. Overfishing has reduced fish stocks—and thus income from fishing.

✔ **READING PROGRESS CHECK**

Analyzing Why are Toronto and Vancouver more important to trade than Calgary and Edmonton?

Academic Vocabulary

via through
vary to differ

Canada's National Tower in Toronto is a symbol of the country, and it is the tallest, free-standing structure in the Western Hemisphere.
▶ **CRITICAL THINKING**
Identifying What are Canada's two most populous cities?

Outdoor winter sports are a popular pastime in both Canada and the northern United States.

▶ **CRITICAL THINKING**

Analyzing Why is Canada's government making a strong effort to promote the nation's culture?

Economic and Political Relationships

GUIDING QUESTION *What is Canada's relationship with other nations?*

Canada has one of the world's largest economies. It also plays a leading role in many world issues.

Canada and the United States

Canada has close economic, defense, and cultural ties to the United States. In the early 1990s, Canada joined the United States and Mexico in signing the North American Free Trade Agreement (NAFTA). This agreement eliminated trade barriers in North America. In 2010 three-quarters of all of Canada's exports went to the United States. The same year, three-quarters of its imports came from the United States.

Canada and the United States also cooperate in defense. They work together to defend the air space against possible attack from planes or missiles. They also work together to combat terrorism.

Canada and the United States share many cultural features. Canadians watch American movies and television shows. Many Canadian singers and actors enjoy success in the United States. Canadians worry, though, about cultural dominance from the United States. Canada's government promotes the production of Canadian movies and television shows.

Canada and Britain

Canada's government is modeled on Britain's. So are its laws, except in Quebec Province. Canadian culture draws on British culture. Many Canadians trace their ancestry to Britain. For these reasons, Canada maintains close ties with that nation. The British king or queen is officially Canada's king or queen, as well. A Canadian official called the governor-general acts in his or her place, though.

Canada and the World

Canada is active in many world organizations. It is a member of the United Nations and the North Atlantic Treaty Organization (NATO). NATO links Canada, the United States, and many nations of Europe in defense matters. Canada has worked in recent years to connect more closely with Asian nations.

Canada plays a major role in efforts to aid poorer nations. It has often taken part in peacekeeping efforts. **Peacekeeping** is sending trained members of the military to crisis spots to maintain peace and order.

☑ READING PROGRESS CHECK

Analyzing Why is Canada more similar to the United States and the United Kingdom than to other nations?

An efficient and comfortable train system is an important part of Canada's economic development.

▶ **CRITICAL THINKING**
Identifying Point of View The nation's transcontinental train makes only eight planned stops en route from Toronto to Vancouver. How does this help travelers?

Canada's Challenges

GUIDING QUESTION *What challenges do Canadians face?*

Canada's biggest challenge might be staying together as a nation. Some people in Quebec want to create their own nation.

Unity and Diversity

Canada's constitution guarantees the rights of French-speaking people in Quebec and elsewhere. Still, tension is evident. English speakers have dominated the nation. They controlled the economy of Quebec for many years. As a result, some French speakers felt they were treated as second-class citizens.

In the late 1900s, some Quebec leaders launched a separatist movement. **Separatists** are those who want to break away from control by a dominant group. Voters in Quebec have twice defeated attempts to make it independent. Still, the issue remains unsettled.

The government did succeed in giving more power to people of the First Nations. In 1999 the government created the new territory called Nunavut. Most of its people are from the First Nations. The government gave them greater **autonomy**, or self-government, than they had in the past.

A Montreal rally calls for a "No" vote on independence for Quebec.

▶ **CRITICAL THINKING**
Identifying Point of View Why do some Canadians support independence for Quebec?

Environmental Challenges

Some scientists worry about the effects of climate change. Milder weather threatens plants and animals that are adapted to the cold of the far north. Experts also believe fisheries will suffer further. In addition, they fear water shortages and more extreme weather. Extreme weather includes long periods of drought and sudden damaging storms.

Canada's government is taking some steps to reduce its use of fossil fuels such as oil, coal, and natural gas. Burning these fuels is thought to contribute to climate change. The government is also encouraging research into clean energy.

Nevertheless, Canada depends greatly on fossil fuels to power its industries, transportation systems, and homes. Primary economic activities such as fossil fuel production are a large part of Canada's economy. Fossil fuel extraction has become an environmental concern in Athabasca Tar Sands in northwestern Alberta. This area has sand located near Earth's surface that contains a form of crude oil that is as thick as tar. The tar can be refined into oil, but the process is difficult and requires a great deal of energy and water.

Acid Rain

Another environmental problem for Canada is acid rain. **Acid rain** is produced when chemicals from air pollution combine with precipitation. When the rain falls to Earth, the chemicals may kill fish, land animals, and trees. Even when they do not kill living things, the chemicals weaken them. That makes them more vulnerable to damage from pests, disease, or severe weather.

Damage from acid rain has been particularly bad in eastern Canada. The government has made efforts to reduce acid rain. But many of the chemicals that cause acid rain in Canada enter the air in the United States. Canada cannot solve its problem without U.S. help. Canada's government wants to reach an agreement that includes tough steps to prevent acid rain.

☑ **READING PROGRESS CHECK**

Determining Central Ideas Why can Canada not meet its environmental challenges by itself?

Include this lesson's information in your Foldable®.

LESSON 3 REVIEW (CCSS)

Reviewing Vocabulary
1. What does it mean to say that Canada is a *bilingual* nation?

Answering the Guiding Questions
2. *Determining Central Ideas* How do the cultures of Quebec and Montreal reflect Canada's history?

3. *Describing* How does life in Calgary and Edmonton relate to Alberta's resources?

4. *Describing* In what ways has Canada tried to build a better world?

5. *Identifying Point of View* Why is separatism in Quebec an ongoing problem for Canada?

6. *Argument Writing* Take the role of Canada's prime minister. Write a letter to the president of the United States urging stronger American action to reduce air pollution that causes acid rain.

Directions: Write your answers on a separate piece of paper.

1 **Use your** FOLDABLES **to explore the Essential Questions.**
ANALYZING Look at the population map of Canada at the beginning of the unit. In a short essay, explain why Canada's major cities and population centers developed where they did.

2 **21st Century Skills**
ANALYZING In a small group, research to find a historical painting about an important event in Canada's history. Find a secondary source about the same event. Compare the two sources and develop an answer to the question: Did one of the sources help you better understand the event? Why?

Benefits	Disadvantages

3 **Thinking Like a Geographer**
INTEGRATING VISUAL INFORMATION Reseach the benefits and disadvantages of separatism for Quebec and list them in a chart like the one shown.

4 **GEOGRAPHY ACTIVITY**

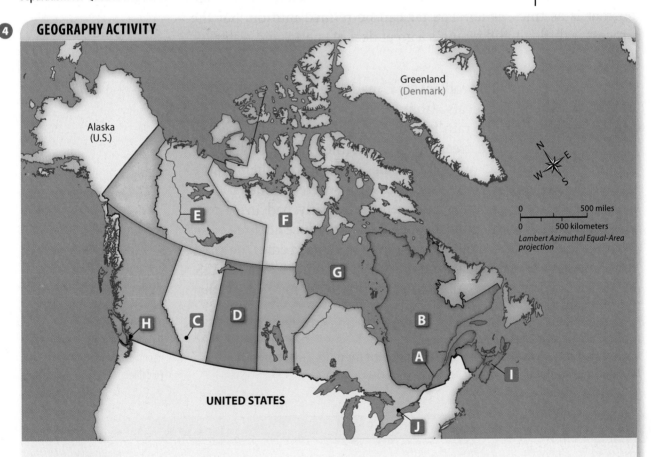

Locating Places
Match the letters on the map with the provinces, territories, or locations listed below.

1. Vancouver
2. Quebec
3. Calgary
4. Toronto

5. Mackenzie River
6. Nunavut
7. Hudson Bay
8. St. Lawrence River

9. Nova Scotia
10. Saskatchewan

REVIEW THE GUIDING QUESTIONS

Directions: Choose the best answer for each question.

1 Manitoba, Saskatchewan, and Alberta are called the

   A. archipelago.

   B. Northwest Territories.

   C. Prairie Provinces.

   D. Atlantic Provinces.

2 Canada's highest mountain is

   F. Mount McKinley.

   G. Mount Logan.

   H. Hudson Mountain.

   I. Pikes Peak.

3 The man who founded Quebec and led the French effort to settle Canada was

   A. Henry Hudson.

   B. Father Junipero Serra.

   C. Samuel de Champlain.

   D. the Métis.

4 Which of Canada's provinces is predominately French in language and culture?

   F. Nova Scotia

   G. Prince Edward Island

   H. Quebec

   I. Yukon Territory

5 Canada's largest metropolitan area, which includes the city and its surrounding suburbs, is

   A. Ottawa.

   B. Vancouver.

   C. Montreal.

   D. Toronto.

6 The territory created in 1999 to give the First Nations greater self-government is

   F. Nunavut.

   G. Saskatchewan.

   H. Northwest Territory.

   I. Yukon Territory.

DBQ DOCUMENT-BASED QUESTIONS

7 **DETERMINING CENTRAL IDEAS** Read the following statement about Canada's people.

"*Immigration . . . was responsible for two-thirds of [Canada's] population growth in the period 2001 to 2006. . . . The effect of immigration is mostly felt in Canada's largest urban [centers] and their surrounding municipalities.*"

—from *The Atlas of Canada*

Which generalization can be made from this quotation?

A. Without immigration, Canada's population would have declined.

B. The immigration rate is higher in Canada today than ever before.

C. The immigration rate to Canada is higher than to the United States.

D. Immigrants were the major reason for Canada's population growth.

8 **DETERMINING CENTRAL IDEAS** What does the second sentence in the quotation tell you about Canada's immigrants?

F. their occupations

G. their origin in urban or rural areas

H. their settlement patterns

I. their economic status

SHORT RESPONSE

"*[Samuel de Champlain] thought that the conquering spirit in which the Spanish usually approached the new Indian groups was itself mistaken. . . . A new colony should seek friends and allies amidst the indigenous peoples.*"

—from Arthur Quinn, *A New World*

9 **ANALYZING** How did Champlain benefit from France not being the first European nation to place colonies in the Americas?

10 **IDENTIFYING** In what ways could French colonists benefit from peaceful relations with Native Americans?

EXTENDED RESPONSE

11 **INFORMATIVE/EXPLANATORY WRITING** Choose the Canadian province or territory where you think you would be most comfortable living. Write an essay describing the province or territory, and explain why you chose it over the others. Your essay should show your understanding of the similarities and parallels between life in Canada and the United States.

Need Extra Help?

If You've Missed Question	**1**	**2**	**3**	**4**	**5**	**6**	**7**	**8**	**9**	**10**	**11**
Review Lesson	1	1	2	3	3	3	2	2	2	2	1

FROM *The Atlas of Canada* Web site, http://atlas.nrcan.gc.ca/auth/english/maps/peopleandsociety/immigration, Natural Resources Canada.

MEXICO, CENTRAL AMERICA, AND THE CARIBBEAN ISLANDS

ESSENTIAL QUESTIONS · *How does geography influence the way people live?* · *Why does conflict develop?* · *Why do people trade?*

Girl from the highlands of Guatemala

networks

There's More Online about Mexico, Central America, and the Caribbean Islands.

CHAPTER 7

Lesson 1
Physical Geography

Lesson 2
History of the Region

Lesson 3
Life in the Region

The Story Matters...

Early advanced civilizations developed in this region of the Americas. Their people developed economies based on farming and trade. They built planned cities and developed highly organized societies and governments. The arrival of the Spanish and other Europeans had a dramatic impact on the region and its indigenous peoples. The influence of European colonial rule and the struggles for independence can still be seen in the economies, politics, and cultures of the region today.

FOLDABLES
Study Organizer

Go to the Foldables® library in the back of your book to make a Foldable® that will help you take notes while reading this chapter.

MEXICO, CENTRAL AMERICA, AND THE CARIBBEAN ISLANDS

Mexico, Central America, and the Caribbean islands sit between North America and South America. The region is surrounded by oceans and seas and is located close to the Equator. As you study the map, look for the geographic features that make this area unique.

Step Into the Place

MAP FOCUS Use the map to answer the following questions.

1 **THE GEOGRAPHER'S WORLD** What is the largest country in this region?

2 **ENVIRONMENT AND SOCIETY** Why was the Panama Canal built where it is?

3 **THE GEOGRAPHER'S WORLD** Which of the Caribbean islands is part of the United States?

4 **CRITICAL THINKING**
Analyzing Given their location, what might be a key economic industry of the Caribbean islands?

HISTORIC CITY Willemstad, capital of the Caribbean island of Curacao, was founded in 1634 by Dutch settlers.

MAYAN RUINS Early Americans known as the Maya built cities in the rain forests of southern Mexico.

Step Into the Time

DESCRIBING Select one location on the time line and describe the impact of European colonization on the lives and environment of the people who lived there.

1325 Aztec found Tenochtitlán

1492 Christopher Columbus arrives in Americas

1300

1500

Map of the Region

networks
There's More Online!

UNITED STATES

Rio Grande

Gulf of California

MEXICO

Mexico City

PACIFIC OCEAN

Gulf of Mexico

ATLANTIC OCEAN

Bermuda (U.K.)

TROPIC OF CANCER

Nassau
BAHAMAS

Havana

CUBA

DOMINICAN REPUBLIC

Cayman Islands (U.K.)

Port-au-Prince
Kingston
JAMAICA
HAITI
Santo Domingo

Puerto Rico (U.S.)

see inset to left for detail

B

BELIZE
Belmopan

GUATEMALA
Guatemala
San Salvador
EL SALVADOR

HONDURAS
Tegucigalpa
NICARAGUA
Managua

Caribbean Sea

A

San José

Panama City

COSTA RICA

PANAMA
Panama Canal

SOUTH AMERICA

40°N
30°N
20°N
10°N

110°W · 100°W · 90°W · 80°W · 60°W

EQUATOR · 0°

Inset (Lesser Antilles):

20°N

Anguilla (U.K.)
St. Martin (Fr.)
St. Maarten (Neth.)
St. Barthélemy (Fr.)

British Virgin Islands (U.K.)
San Juan
Puerto Rico (U.S.)
Virgin Islands (U.S.)

ANTIGUA AND BARBUDA
Montserrat (U.K.)
Guadeloupe (Fr.)

ST. KITTS AND NEVIS
DOMINICA
Martinique (Fr.)

Caribbean Sea

ST. LUCIA
ST. VINCENT AND THE GRENADINES
BARBADOS
GRENADA
TRINIDAD AND TOBAGO

0   200 mi
0   200 km

10°N
60°W

—— National boundary
✪ National capital
○ Territorial capital

0   500 miles
0   500 kilometers
Lambert Azimuthal Equal-Area projection

Timeline:

1804 Led by Toussaint-Louverture, Haiti achieves independence from France

1810 Father Hidalgo leads Mexico rebellion

1914 Panama Canal links Atlantic and Pacific Oceans

2010 Deadly earthquake strikes Haiti

1600 · 1700 · 1800 · 1900 · 2000

1848 Mexico cedes large areas of territory to U.S.

1959 Fidel Castro takes power in Cuba

There's More Online!

☑ **GRAPHIC ORGANIZER**
Landforms and Waterways

☑ **IMAGE** 360° View: Mexico City

☑ **ANIMATION** Waterways as
Political Boundaries

☑ **VIDEO**

Reading **HELP**DESK

Academic Vocabulary

- **similar**
- **benefit**

Content Vocabulary

- **isthmus**
- *tierra caliente*
- *tierra templada*
- *tierra fría*
- **bauxite**
- **extinct**
- **dormant**

TAKING NOTES: *Key Ideas and Details*

Organize As you read about the region, take notes on the physical geography using a graphic organizer like the one below.

Area	Landforms
Mexico	
Central America	
Caribbean islands	

Lesson 1
Physical Geography

ESSENTIAL QUESTION · *How does geography influence the way people live?*

IT MATTERS BECAUSE
Mexico and Central America are southern neighbors to the United States.

Physical Geography of Mexico and Central America

GUIDING QUESTION *What landforms and waterways do Mexico and Central America have?*

Mexico and the seven nations of Central America act like a bridge between two worlds. Geographically, they form an isthmus that connects North and South America. An **isthmus** is a narrow piece of land that connects two larger landmasses. Culturally, they join with South America and some Caribbean islands to make up Latin America. Latin America is a region of the Americas where the Spanish and Portuguese languages, based on the Latin language of ancient Rome, are spoken. Economically, the nations have close ties to the United States. They also trade with their Latin American neighbors.

Shaped like a funnel, the region is wider in the north than in the south. To the north, Mexico has a 1,951-mile (3,140-km) border with the United States. At the southern end, the Central American country of Panama is only about 40 miles (64 km) wide.

Land Features

Mexico is the largest nation of the region, occupying about two-thirds of the land. Imagine a backwards *y* along the western and eastern coasts of Mexico, with the tail to the south. That backwards *y* neatly traces the mountain systems

on Mexico's two coasts and south central region. The coastal ranges are called the Sierra Madre Occidental (Spanish for "western") and the Sierra Madre Oriental (Spanish for "eastern"). They join in the southern highlands. Coastal plains flank the western and eastern mountains. The eastern plain is wider.

Between the two arms of the *y* is a vast highland region called the Central Plateau. It is the heartland of Mexico. This plateau is home to Mexico City, which is the capital, and a large share of the nation's people.

Mexico has two peninsulas. The Yucatán Peninsula bulges northeast into the Gulf of Mexico. Baja California (*baja* means "lower" in Spanish) extends to the south in western Mexico.

Central America has landforms **similar** to those of south central Mexico. Mountains run down the center of these countries. Narrow coastal lowlands flank them on the east and west.

Mexico and Central America lie along the Ring of Fire that rims the Pacific Ocean. Earthquakes and volcanoes are common in the Ring of Fire. The Sierra Madre Occidental are made of volcanic rocks, but they have no active volcanoes. The mountains in the southern part of the central plateau and in Central America, however, do have numerous active volcanoes. These volcanoes bring a **benefit**. Volcanic materials weather into fertile, productive soils.

Academic Vocabulary

similar much like
benefit advantage

Mountains appear on the hazy horizon of Mexico City. Mexico's capital lies about 7,800 feet (2,377 m) above sea level.

Popocatépetl volcano stands in the background as a farmer plows the land. Popocatépetl, also known as "smoking mountain," has experienced eruptions since ancient times.

▶ **CRITICAL THINKING**
Explaining Why are earthquakes common in some parts of Mexico and Central America?

Earthquakes are common in the area, too. A magnitude 8.0 earthquake that hit Mexico City in 1985 killed thousands of people. One that struck El Salvador in 2001 produced another kind of disaster. A hill weakened by the earth's movement collapsed onto the town of Las Colinas. It crushed homes and killed hundreds of people.

Bodies of Water

Mexico and Central America are bordered by the Pacific Ocean to the west. The Gulf of California, an inlet of that ocean, separates Baja California from the rest of Mexico. To the east, the region is surrounded by the waters of two arms of the Atlantic Ocean. They are the Gulf of Mexico and the Caribbean Sea.

The region has few major rivers. In the northern half of Mexico, the climate is dry. This means that few rivers flow across the rocky landscape. Southern Mexico and Central America receive more rain, but the landscape is steep and mountainous, and the rivers are short. An important river is the Río Bravo. In the United States, this river is called the Rio Grande. The largest lake in the region is Lake Nicaragua, in Nicaragua.

©Francisco Guasco/epa/Corbis

An important waterway in the region is not a river, but a feature built by people. It is the Panama Canal, built in the early 1900s. The Panama Canal makes it possible for ships to pass between the Atlantic and Pacific oceans without journeying around South America. It saves thousands of miles of travel, which in turn saves time and money. It is one of the world's most important waterways.

Climates

Most of Mexico and Central America lie in the Tropics. Because of their location near the Equator, it might seem that the climate would be hot. Although the coastal lowlands are hot, areas with higher elevation are not. The highlands are much cooler.

Nearly the entire region can be divided into three vertical climate zones. Soil, crops, animals, and climate change from zone to zone. The *tierra caliente*, or "hot land," is the warmest zone. It reaches from sea level to about 2,500 feet (762 m) above sea level. Major crops grown here are bananas, sugarcane, and rice.

Next highest is the *tierra templada*, or "temperate land." This climate zone has cooler temperatures. Here farmers grow such crops as coffee, corn, and wheat. Most of the region's people live in this climate zone.

Higher in elevation is the *tierra fría*, or "cold land." This region has chilly nights. It can be used only for dairy farming and to grow hearty crops such as potatoes, barley, and wheat.

Jeremy Woodhouse/age fotostock

Think Again?

Because the Panama Canal connects the Atlantic and Pacific oceans, it must go east to west.

Not really! Central America twists to the east where Panama is located. As a result, the Panama Canal is cut from the north to the south.

The Rio Bravo, or Rio Grande, carves its way through rugged countryside. It forms part of the border between Mexico and the United States.

▶ CRITICAL THINKING

Describing Why are there so few rivers in the northern part of Mexico?

CLIMATE ZONES

Although the region is located in the Tropics, many inland areas of Mexico and Central America have relatively cool climates.

▶ **CRITICAL THINKING**

1. *Identifying* What products are grown in *tierra caliente*?

2. *Analyzing* Why are many inland areas of Mexico and Central America relatively cool?

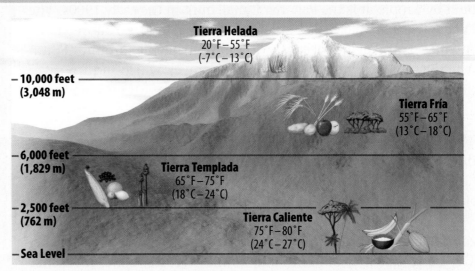

Tierra Helada
20°F–55°F
(−7°C–13°C)

Tierra Fría
55°F–65°F
(13°C–18°C)

—10,000 feet
(3,048 m)

—6,000 feet
(1,829 m)

Tierra Templada
65°F–75°F
(18°C–24°C)

—2,500 feet
(762 m)

Tierra Caliente
75°F–80°F
(24°C–27°C)

—Sea Level

Geographers also designate other vertical climate zones. Few human activities take place on the *tierra helada,* or "frozen land". This vertical climate zone is more common in other regions of the Americas.

Tropical Wet/Dry Climate

Much of Mexico and Central America have a tropical wet/dry climate. The climate is characterized by two distinct seasons. The wet season, during the summer months, is when most of the precipitation falls. The dry season occurs during the winter months. The dry season is longer in the areas farther from the Equator and closer to the polar regions.

The region's tropical location exposes it to another natural hazard. Ferocious hurricanes can strike in the summer and early autumn months. These storms do great damage. For example, a 1998 hurricane killed more than 9,000 people in Honduras and destroyed 150,000 homes.

Natural Resources

Oil and natural gas are Mexico's most important resources. They are found along the coast of the Gulf of Mexico and in the gulf waters. Mexico is an important oil-producing country. It has enough oil and gas to meet its own needs and still export a large amount. The exports help fuel the nation's economy. However, Mexico's oil production has declined since 2004. Many oil fields are old and are starting to run out of oil.

When Spanish explorers first came to Mexico, they were attracted to the area's gold and silver. Mexico still produces silver, which is mined in the central and north central parts of the country. Gold also is still mined in Mexico. Other minerals include copper, iron ore, and **bauxite**. Bauxite is used to make aluminum.

The seven smaller nations of Central America have few mineral resources. Nicaragua is an exception, with gold, silver, iron ore, lead, zinc, and copper. The nation is so poor, however, that it has not been able to take advantage of these deposits. Guatemala also has some oil, and its mountains produce nickel.

☑ **READING PROGRESS CHECK**

Analyzing Why are different climate zones found in this region, even though most of the region is in the Tropics?

Physical Geography of the Caribbean Islands

GUIDING QUESTION *How are the Caribbean islands alike and different from one another?*

Hundreds of islands dot the Caribbean Sea. The islands are home to more than 30 countries or territories belonging to other countries. Some are large, with millions of people living on them. Others are tiny and home to only thousands.

Major Islands

The Caribbean islands can be segmented into three different groups. The first group is the Greater Antilles. The four islands, the largest Caribbean islands, include Cuba, Jamaica, Hispaniola, and Puerto Rico. Cuba and Jamaica are independent countries. Hispaniola is home to two countries: Haiti in the west and the Dominican Republic

This scenic bay in the Caribbean island of Antigua provides an ideal harbor for yachts and other sailing ships. Antigua is part of the Lesser Antilles.

▶ **CRITICAL THINKING**

Describing What islands make up the Greater Antilles?

Pixtal/age fotostock

Why are the Caribbean islands hit so often by earthquakes and volcanoes?

The islands of the Caribbean are not along the edge of the Pacific Ocean. So why do they experience these disasters? The Ring of Fire is not the only place where tectonic plates meet. Many Caribbean islands are located at the boundaries of different plates where volcanoes form and earthquakes occur. Hence, the islands are vulnerable to these disasters. A 2010 earthquake near Haiti's capital of Port-au-Prince might have killed as many as 200,000 people and forced 1 million others out of their homes.

in the east. Puerto Rico is a commonwealth of the United States. Although it is a possession of the United States, it has its own government. The people of Puerto Rico are American citizens. They can travel freely between their island and the United States.

The second group of islands is the Lesser Antilles. Dozens of smaller islands make up this group. They form an arc moving east and south from Puerto Rico to northern South America. Most of the islands are now independent countries. At one time, they were colonies of France, Britain, Spain, or the Netherlands. Each has a culture reflecting its colonial period.

The third island group is the independent nation of the Bahamas. The islands lie north of the Greater Antilles and east of Florida. The Bahamas include more than 3,000 islands, although people live on only about 30 of them.

The Greater Antilles are a mountain chain, much of which is under water. On a map, you can see that this chain extends eastward from Mexico's Yucatán Peninsula. These islands include some mountains, such as the Sierra Maestra in the eastern part of Cuba and the Blue Mountains of Jamaica. The highest point in the Caribbean is Duarte Peak, in the Dominican Republic. The Lesser Antilles are formed by volcanic mountains. Many of the volcanoes are **extinct**, or no longer able to erupt. Some islands have **dormant** volcanoes, or ones that can still erupt but show no signs of activity.

The Caribbean Sea

The Caribbean Sea is a western arm of the Atlantic Ocean. In the past, sailing ships traveling west from Europe followed trade winds blowing east to west to reach the sea. Christopher Columbus used the winds to reach the Bahamas in 1492. There, he first sighted land in the Americas. Columbus explored the Caribbean, too. These voyages sparked European settlement of the Americas.

The warm waters of the Caribbean help feed the Gulf Stream. This current carries warm water up the eastern coast of the United States.

The Climate of the Caribbean Islands

The Caribbean islands have a tropical wet/dry climate. Temperatures are high year-round, though ocean breezes make life comfortable. Humidity is generally high, but rainfall is seasonal and varies significantly. Islands like Bonaire receive only about 10 inches (25 cm) of rain per year. Dominica, on the other hand, receives about 350 inches (899 cm) of rain each year. That is an average of almost an inch of rain every day.

Like Central America and Mexico, the Caribbean islands are prone to hurricanes. These storms are more likely to occur in the northern areas, toward the Gulf of Mexico, than to the south. On average, seven hurricanes strike the Caribbean islands each year.

Natural Resources

The waters of the Caribbean are rich in fish. Some are fished for food and others for sport. The islands have few timber resources today. People have cut down most of the trees already to use for fuel or to make farmland.

Mineral resources are generally lacking too, although some Caribbean islands have important resources. Trinidad and Tobago has reserves of oil and natural gas. The Dominican Republic exports nickel, gold, and silver. Cuba is a major producer of nickel. Jamaica has large amounts of bauxite.

Perhaps the most important resources of the Caribbean are its climate and people. Warm temperatures and gracious hosts attract millions of tourists to the region each year. Some enjoy the white sandy beaches and clear blue water. Some scuba dive to see the colorful fish darting through coral reefs.

 READING PROGRESS CHECK

Citing Text Evidence How did the islands of the Caribbean form?

Tourists on a boat near the French-ruled island of Guadeloupe learn to dive in Caribbean waters.

Identifying What natural resource of the Caribbean Sea benefits the region's island nations?

 FOLDABLES
Study Organizer

Include this lesson's information in your Foldable®.

Ingolf Pompe/LOOK-foto/Glow Images

LESSON 1 REVIEW **CCSS**

Reviewing Vocabulary

1. What is the *tierra templada*? Why do most people in Mexico and Central America live in this vertical climate zone?

Answering the Guiding Questions

2. ***Describing*** How are the physical geography of Mexico and Central America similar?

3. ***Analyzing*** What impact does the Panama Canal have on the cost of shipping goods? Why?

4. ***Determining Central Ideas*** How do the locations of Mexico and Central America increase the possibility of natural hazards striking the region?

5. ***Determining Word Meanings*** Why are some islands in the Caribbean called the *Greater* Antilles and others called the *Lesser* Antilles?

6. ***Narrative Writing*** Imagine you are taking a cruise that stops at a Caribbean island, a port in Central America, and a port in Mexico. Write three diary entries describing what you would see in each place.

Lesson 2
History of the Regions

Reading **HELP**DESK **CCSS**

Academic Vocabulary

- feature
- transform

Content Vocabulary

- **staple**
- **surplus**
- **conquistador**
- **colonialism**
- **revolution**
- **plantation**
- **cash crop**
- **caudillo**
- **Columbian Exchange**

TAKING NOTES: *Key Ideas and Details*

Summarize As you read about the history of Mexico, take notes using the graphic organizer below.

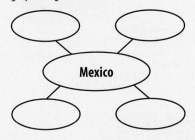

ESSENTIAL QUESTION · *Why does conflict develop?*

IT MATTERS BECAUSE
The region was home to highly developed Native American civilizations.

Mexico's History

GUIDING QUESTION *How did economic and governmental relationships between Spanish and Native Americans in Mexico change over time?*

Mexico was first inhabited by Native American groups. Later, Spanish soldiers conquered the groups and ruled them. Since the early 1800s, Mexico has been independent. Its history is long and rich, and its accomplishments are many.

Early Civilizations

Native peoples first grew corn in Mexico about 7,000 years ago. They also grew other foods that have become **staples**, or foods that are eaten regularly, such as corn, squash, chilies, and avocados. Farming allowed people to produce food **surpluses**, or more than they needed to survive. Surpluses helped people specialize in jobs other than getting food.

About 3,000 years ago, the Maya formed the major civilization in the region. They lived mainly in the lowland plains of Mexico's Yucatán Peninsula and in what is now Guatemala and Belize. One **feature** of their culture was great cities. The Maya erected pyramids with stepped sides and temples on top. They invented a complex system of writing. By studying astronomy, they were able to make accurate calendars. The height of Maya civilization was from about A.D. 300 to A.D. 900. Then their power suddenly collapsed. Archaeologists do not know exactly why.

The Aztec ruled the region next. They settled in central Mexico in about 1300. Their impressive capital city was Tenochtitlán. Mexico City occupies the site where it once stood. Tenochtitlán was built on an island in the middle of a lake. Causeways connected it to the mainland.

The Aztec had a complex social and religious system. They conquered many of their neighbors and made slaves of captured soldiers. Priests performed rituals to win the favor of their gods. The Aztec were also skilled farmers. They built up land in the lake to form small islands called *chinampas,* which they used to grow crops.

The Spanish Arrive

In the early 1500s, a rival power appeared. Around 1520, Hernán Cortés led a small force of Spanish **conquistadors**, or conquerors, to Mexico. Within two years, these explorers and soldiers had defeated the Aztec and taken control of their empire.

How could the Spanish conquer the Aztec with only a few hundred men? Spanish guns and armor were better weapons than Aztec spears. Another major factor was European diseases. The diseases did not exist in the Americas until Europeans unknowingly brought them. Native Americans had no resistance to them, so the diseases killed many thousands. Cortés also took advantage of the anger of other native peoples who resented Aztec rule. Several groups joined him as allies.

Winning the Aztec Empire brought Spain riches in gold and silver mines. The conquest completely **transformed** life in Mexico. Roman Catholic priests converted native peoples to Catholicism. Conquistadors forced native peoples to work on farms or in mines. Spanish rule in Mexico was an example of colonialism.

DEA/G. DAGLI ORTI/DEA PICTURE LIBRARY/Getty Images

Academic Vocabulary

feature a characteristic
transform to change

The Aztec city of Tenochtitlán was linked by canals, bridges, and raised streets built across the water.
▶ CRITICAL THINKING
Describing How were the Aztec able to build a city and farms in an area covered by a lake?

Under **colonialism**, one nation takes control of an area and dominates its government, economy, and society. The colonial power uses the colony's resources to make itself wealthier. In colonial Mexico, settlers from Spain had the most wealth.

Independence and Conflict

After almost 300 years of Spanish rule, a priest named Miguel Hidalgo led a rebellion in Mexico in 1810. The goal of the rebellion was to win independence from Spain. Some people hoped it would also create a more nearly equal society. The Spanish captured and executed Hidalgo, but by 1821 Mexico had gained its independence. Spanish rulers, though, were replaced by wealthy Mexican landowners. Native peoples remained poor.

Through much of the 1800s, Mexico was troubled by political conflict. Rival groups fought one another for power. Most of Mexico's people remained poor.

Revolution and Stability

By the early 1900s, dissatisfaction was widespread. A revolution erupted in Mexico. A **revolution** is a period of violent social and political change. One change was the land reform plan, which divided large estates into parcels of land that were then given to poor people to farm. National public schools were established, and a new constitution was written detailing the responsibilities of the government toward the people. Only one political party, however, held power until the 1990s.

☑ **READING PROGRESS CHECK**

Determining Central Ideas How were the Spanish able to conquer the Aztec?

A History of Central America

GUIDING QUESTION *How did the nations of Central America develop?*

The nations of Central America developed in similar ways to Mexico. But there were differences, as well.

Early Civilizations and Conquest

The Maya had flourished in Guatemala and Belize, as well as in southern Mexico. Even after their great cities were abandoned, the Maya continued to live in the region. After conquering Mexico, the Spanish moved south. By the 1560s, Spain had seized control of most of Central America. During the early 1800s, Britain claimed the area that is now Belize.

Independence

Central America gained its independence soon after Mexico. In 1823 the territories of Central America united to form one government. By 1840, they had separated into five independent

countries: Guatemala, Honduras, El Salvador, Nicaragua, and Costa Rica. The area that is now Belize was still a British colony. Panama was part of Colombia.

Central American countries were subjected to economic colonialism. This means that foreign interests dominate a people economically. These foreign interests were large companies from other countries. They set up **plantations**, or large farms, where poorly paid workers produced cash crops. **Cash crops** are crops sold for profit. The most important were bananas, coffee, and sugarcane.

Heading the governments for much of this time were military strongmen called **caudillos**. The caudillos helped ensure the foreigners' success. In turn, the foreigners made sure that the caudillos remained in power.

Conflict in Modern Times

Around 1900, Panama gained its independence from Colombia. It was helped by the United States, which wanted to build a canal there. The United States controlled the canal until 2000. Then, by agreement, Panama took control of the canal.

The late 1900s was a time of conflict. New wealth came to the upper classes, but most people remained poor. Various groups demanded reforms. Several countries were ravaged by civil wars. Only Costa Rica and Belize remained peaceful. One of Costa Rica's presidents, Óscar Arias Sánchez, helped bring peace to the region.

☑ **READING PROGRESS CHECK**

Analyzing How did Central America and Mexico's history differ?

Built by the United States in the early 1900s, the Panama Canal is now owned and operated by the Republic of Panama.

Identifying Which country held Panama shortly before the United States built the Panama Canal?

Russell Kord/Alamy

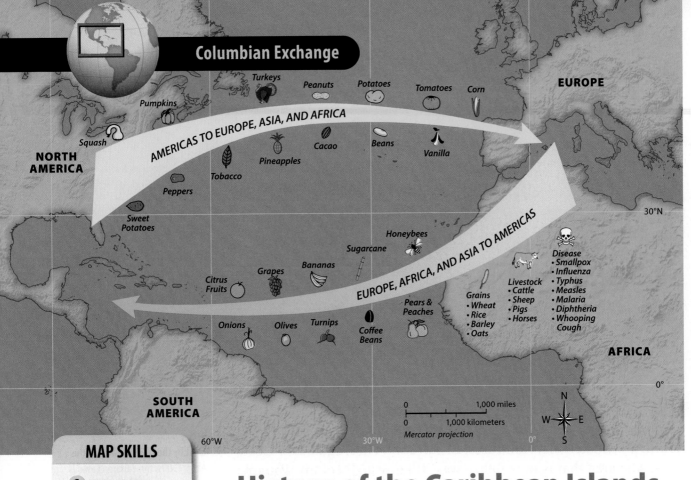

EUROPE

NORTH AMERICA

Turkeys
Pumpkins
Peanuts
Potatoes
Tomatoes
Corn
Squash
Tobacco
Pineapples
Cacao
Beans
Vanilla
Peppers
Sweet Potatoes

AMERICAS TO EUROPE, ASIA, AND AFRICA

30°N

Honeybees
Sugarcane
Grapes
Bananas
Citrus Fruits
Onions
Olives
Turnips
Coffee Beans
Pears & Peaches

EUROPE, AFRICA, AND ASIA TO AMERICAS

Grains
• Wheat
• Rice
• Barley
• Oats

Livestock
• Cattle
• Sheep
• Pigs
• Horses

Disease
• Smallpox
• Influenza
• Typhus
• Measles
• Malaria
• Diphtheria
• Whooping Cough

AFRICA

0°

SOUTH AMERICA

60°W   30°W   0°

0   1,000 miles
0   1,000 kilometers
Mercator projection

N W E S

MAP SKILLS

1 THE USE OF GEOGRAPHY What happened as a result of the Columbian Exchange?

2 HUMAN GEOGRAPHY How did the Columbian Exchange affect African peoples?

History of the Caribbean Islands

GUIDING QUESTION *How did the Caribbean islands develop?*

The history of the Caribbean islands is similar to that of Mexico and Central America. The islands have greater diversity, though, because several European countries ruled them as colonies.

Indigenous Peoples and European Settlers

Europeans changed the way the native peoples of the Caribbean lived. Like the Native Americans of the mainland, they suffered from diseases carried by the Europeans. This is why their numbers declined sharply soon after the arrival of the Europeans. Overwork and starvation also reduced their numbers. The Spanish set up colonies in what are now Cuba, the Dominican Republic, and Puerto Rico. Later, the French settled in what is now Haiti and on other smaller islands. The British and Dutch had some colonies, too.

Colonialism

During the 1600s, the Caribbean colonies became the center of the growing sugar industry. European landowners hoped to make money by selling the sugar in Europe. Because so many Native American workers had died, Europeans brought in hundreds of thousands of enslaved Africans to work the plantations.

The transport of enslaved Africans was part of the **Columbian Exchange**. This term refers to the transfer of plants, animals, and

people between Europe, Asia, and Africa on one side and the Americas on the other. Foods such as wheat, rice, grapes, and apples were introduced to the Americas as were cattle, sheep, pigs, and horses. At the same time, products from the Americas were introduced into Europe, Africa, and Asia. They included corn, chocolate, and the potato. The Columbian Exchange also resulted in the introduction of new diseases into different parts of the world.

Independence

The first area in the Caribbean to gain independence was Haiti, then called Saint Domingue. Led by Toussaint-Louverture, Haiti gained its independence from France in 1804. The Dominican Republic won its independence in 1844. Cuba and Puerto Rico remained Spanish until 1898. When Spain lost the Spanish-American War, it gave independence to Cuba. Puerto Rico passed into American hands. Other islands of the Caribbean did not win the right to self-government until the middle 1900s.

Turmoil in the Twentieth Century

Independence did not mean freedom or prosperity. Rule by caudillos and widespread poverty have remained a problem in Haiti and the Dominican Republic.

Cuba, too, was often subject to dictatorial rule following its independence. Then in 1959, revolutionaries led by Fidel Castro took over. Castro soon cut all ties with the United States. He said his government would follow the ideas of communism. Communism involves government control of all areas of the economy and society. His rule did not bring economic success to Cuba.

The other islands of the Caribbean have had their own difficulties. Some countries in the region are trying to improve conditions and bring economic benefits to all their citizens. Small and with few resources, they have been unable to develop strong economies. Many of the islands depend on aid from the governments that used to run them as colonies.

Include this lesson's information in your Foldable®.

 READING PROGRESS CHECK

Analyzing What caused the population of the Caribbean islands to grow in colonial times?

LESSON 2 REVIEW (CCSS)

Reviewing Vocabulary

1. What is the difference between a *conquistador* and a *caudillo*?

Answering the Guiding Questions

2. *Describing* How did the Maya and the Aztec differ?

3. *Analyzing* How were relations between people with European and Native American heritage similar in Mexico during colonial times and the 1800s?

4. *Determining Central Ideas* How did economic colonialism affect the nations of Central America?

5. *Analyzing* How was the development of Cuba and of Haiti similar and different?

6. *Informative/Explanatory Writing* Write a summary of the history of Mexico, Central America, or the Caribbean islands after independence.

(l to r) ©Patrick Frilet/Hemis/Corbis; Christopher Pillitz/Photonica World/Getty Images; Steve Bly/Photographer's Choice/Getty Images; James P. Blair/National Geographic/Getty Images

Reading **HELP**DESK (CCSS)

Academic Vocabulary

- circumstance
- initiate

Content Vocabulary

- **maquiladora**
- **mural**
- **dependence**
- **free-trade zone**
- **remittance**
- **reggae**

TAKING NOTES: *Key Ideas and Details*

Summarize As you read about Mexico, use the graphic organizer below to take notes about its economy and culture.

Mexico

Economy   Culture

Lesson 3
Life in the Region

ESSENTIAL QUESTION • *Why do people trade?*

IT MATTERS BECAUSE
Mexico and other countries in the region have close ties to the United States.

Modern Mexico

GUIDING QUESTION *What is life like in Mexico today?*

When you think of Mexico, you might think of Mexican food like tacos. You might think of mariachi musicians playing lively music and wearing large sombreros. But Mexico has a rich and complex culture and is a rising economic power.

The Economy

Mexico has close economic ties to the United States and Canada. These ties are a result of joining with them in the North American Free Trade Agreement (NAFTA). About 80 percent of Mexico's exports go to NAFTA partners. More than 60 percent of Mexico's imports come from members of NAFTA. Most of this trade is with the United States.

In recent decades, Mexico has developed its manufacturing industry. Factories account for about a third of Mexico's output. Some of them are **maquiladoras**. These are factories where parts made elsewhere are assembled into products. Many of the factories are located in northern Mexico. The goods are then exported. Food processing is another major industry in Mexico. The textile and clothing industries are important, too. Mexico also has heavy manufacturing, producing iron, steel, and automobiles.

Farming remains important. Cotton and wheat are grown in the dry north using irrigation. Along the southeastern coast, farms produce coffee, sugarcane, and fruit. On the

central plateau, farmers grow corn, wheat, and fruits and vegetables. In the poor south, many farmers engage in subsistence farming—growing just enough food to feed themselves and their families.

Service industries are important in Mexico. Banking helps finance economic growth. A major service industry is tourism. Visitors from around the world come to visit ancient Maya sites or to see the architecture of Spanish colonial cities. Tourists also come to relax in resorts along the warm and scenic tropical coasts.

Culture

Mexicans are proud of their blend of Spanish and native cultures. They have long celebrated the folk arts that reflect native traditions. In the early 1900s, several Mexican painters drew on these traditions to paint impressive murals celebrating Mexico's history and people. **Murals** are large paintings made on walls. The Ballet Folklorico performs Mexican dances.

Sports reflect Mexico's ties to Spain and the United States. Soccer is popular there, as it is in Spain. So is baseball.

Challenges

With nearly 9 million people, Mexico City is one of the largest cities in the world. Including the city's suburbs, it has more than 21 million people—nearly 20 percent of Mexico's population. Overcrowding is a major problem.

Pollution is another problem, particularly air pollution. Because Mexico City is at a high elevation, the air has less oxygen than at sea level. This makes breathing difficult for some people in normal **circumstances**, but conditions in Mexico City are not normal. A great deal of exhaust from cars and factories is released into the air. The polluted air is held in place by the mountains around the city.

Think Again?

All Mexican food is spicy.

Not true! Mexican food varies from region to region. Most dishes are based on traditional native foods —corn, beans, and squash—along with rice, introduced from Spain. Not all Mexican food is spicy, although chilies are used in many dishes.

Academic Vocabulary

circumstance a condition

Farmers sell produce at a village market in Mexico.
▶ **CRITICAL THINKING**
Describing What is farming like in southern Mexico?

Nicaraguan workers make clothing in a factory.

▶ **CRITICAL THINKING**

Identifying What types of products have Central American manufacturers recently begun to make?

Sometimes a layer of cold air high in the atmosphere keeps the pollution from rising. The result can be a serious threat to health.

Another challenge facing Mexico is the power of criminals who sell illegal drugs. Drug lords use violence to fight police and to intimidate people. Mexico has mounted a major effort to battle this problem with some success.

Poverty is yet another major challenge facing Mexico. Anywhere from one-fifth to nearly half of Mexico's people are poor. Continued economic growth would help, and seems to be working. Some economists are predicting that Mexico will overtake Brazil in the 2010s as the leading economy in Latin America.

☑ **READING PROGRESS CHECK**

Analyzing How have close ties with the United States helped Mexico's economy?

Modern Central America

GUIDING QUESTION *What is life like in Central America?*

The nations of Central America have fewer resources than Mexico. The region must also deal with political problems.

Central America's Economies

The countries of Central America long showed **dependence**, or too much reliance, on cash crops. In recent years, some have begun to escape this trap. A good sign is the growth of manufacturing. This consists mostly of food processing and production of clothing and textiles. Tourism has grown as well. Tourists come to Belize and Guatemala to see ancient Maya sites. They travel to Costa Rica to see the varied plants and animals in its rain forests.

Panama benefits economically from the Panama Canal. Working for additional benefit, Panama **initiated** a major building program to expand the canal so it can accept larger cargo ships.

Academic Vocabulary

initiate to begin

High rates of population growth create an economic challenge. The countries need to grow their economies fast to provide enough jobs. One hope for promoting growth is trade agreements between the countries of the region and other countries.

In the 2000s, the United States and the Dominican Republic signed a series of agreements with five Central American countries (Costa Rica, El Salvador, Guatemala, Honduras, and Nicaragua). The agreement, called the Central America Free Trade Agreement (CAFTA-DR), was the first agreement among the United States and smaller developing economies. CAFTA-DR creates a **free-trade zone** that lowers trade barriers between the countries. Often, however, such trade agreements help the United States more than the other countries.

Challenges Facing the Region

Another challenge to the area is natural disasters. Earthquakes and hurricanes can have a devastating effect on the region's fragile economies. Nicaragua was making some economic progress in the 1990s when Hurricane Mitch hit. The destruction set the nation's economy back significantly.

The need to solve long-standing political problems also holds the region back. The civil wars of the 1980s and 1990s are over, but some of the issues that caused them remain unsolved. If these issues again become more severe, conflict may resume.

Culture

The culture of Central America is strongly influenced by European and native traditions. Spanish is the chief language in all countries except Belize, where English is the official language. English is spoken in many cities in the region as well. In rural Guatemala, native languages are common.

The population is mainly of mixed European and native heritage. Some people of African and Asian descent live there as well. Most people of the region are Roman Catholics. In recent years, however, Protestant faiths have gained followers.

☑ READING PROGRESS CHECK

Identifying What are the causes of poverty in Central America?

A tourist descends from a tree in a rain forest in Costa Rica.

▶ CRITICAL THINKING
Identifying Point of View Why do visitors travel to Costa Rica's rain forests?

Haiti's brown, barren landscape contrasts sharply with the richly forested terrain of the neighboring Dominican Republic.

▶ **CRITICAL THINKING**

Analyzing Why has economic development been held back in Haiti?

The Caribbean Islands

GUIDING QUESTION *What is life like on the Caribbean islands?*

The Caribbean islands are mostly small countries with small populations and few resources. Although they have a rich and vibrant culture, they face many challenges.

Island Economies

The biggest challenge for the islands is to develop economically. Many people on the islands are poor. Even in Puerto Rico, a large share of the population lives in poverty. One reason for the poverty is high unemployment.

Cuba's economy is in poor condition after decades of communism. The government has been unable to promote economic development. It relied on aid first from the Soviet Union and more recently from Venezuela. Conditions are worse now than in the 1980s. Cubans also have little political freedom. Those who criticize the government are often arrested.

In Haiti, a history of poor political leadership has held back economic development. Haiti ranks among the world's poorest nations. Poverty is not the country's only problem. Widespread disease is another threat. In addition, as many as one in eight Haitians have left the country. Many of those who emigrated were among Haiti's most educated people. This loss hurts efforts to improve the economy. Finally, the country has not yet recovered from a deadly 2010 earthquake. Despite these problems, Haiti's people are determined to succeed.

Trinidad and Tobago has one of the more successful economies in the region. Sales of its oil and natural gas have funded economic development. Its location near Venezuela and Brazil has helped

James P. Blair/National Geographic/Getty Images

make its ports busy. The smaller Caribbean islands have had more political success than the larger ones. Governments are democratic and stable, but the economies are plagued by few resources and poverty.

Another important economic factor in the region is remittances. A **remittance** is money sent back to the homeland by people who migrated someplace else to find work. Many Dominicans came to the United States for work and send money home to support their families.

Tourism is a major part of the economy of several islands. Resorts in the Bahamas, Jamaica, and other islands invite tourists to come and relax in pleasant surroundings. The resorts often separate tourists from the lifestyle of the islanders, but they provide jobs for island citizens.

Island Cultures

The cultures of the Caribbean islands show a mix of mainly European and African influences. Large numbers of Asians also came to some of the islands in the 1800s and 1900s. Those from China went mainly to Cuba. South Asians settled in Jamaica, Guadalupe, and Trinidad and Tobago.

The languages spoken on the islands reflect their colonial heritage. English is the language of former British colonies such as the Bahamas and Jamaica. Spanish is spoken in Cuba, the Dominican Republic, and Puerto Rico. English is also taught in Puerto Rico's schools. French and Creole, a blend of French and African languages, are spoken in Haiti.

The Caribbean islands have strongly influenced world music. Much of the music blends African and European influences. Cuba is famous for its salsa, and Jamaica for reggae. Both forms of music rely on complex drum rhythms. **Reggae** has become popular around the world not only for its musical qualities but also for lyrics that protest poverty and lack of equal rights.

Include this lesson's information in your Foldable®.

☑ **READING PROGRESS CHECK**

Citing Text Evidence How do economic conditions in Jamaica relate to the development of reggae?

LESSON 3 REVIEW

Reviewing Vocabulary

1. What is a *free-trade zone*, and why do the nations of the region want to be in one?

Answering the Guiding Questions

2. *Determining Central Ideas* Do you think Mexico has a strong economy? Why or why not?

3. *Identifying* What challenges does Mexico face?

4. *Describing* How have the economies of the Central American countries changed in recent years?

5. *Analyzing* How do the languages of the Caribbean islands reflect their colonial history?

6. *Argument Writing* Take the role of a government official in one of these countries. Write a brief report to the nation's president explaining whether you think promoting tourism is good or bad for the nation's economy. Give reasons.

NAFTA and Its Effects

The North American Free Trade Agreement (NAFTA) was created to grow trade among the United States, Canada, and Mexico to help these countries become more competitive in global markets.

Why Do Nations Trade?

No country produces all the goods and services it needs. Because most countries have more than they need of some things but not enough of others, trade is important.

Trade Barriers

Sometimes countries try to protect their industries from competition by setting up trade barriers such as tariffs and quotas. A tariff is a tax on imports. Tariffs raise the prices of imported goods so a country's own industries can produce and sell those goods at competitive prices. A quota restricts the amount of certain goods that can be imported from other countries.

Free Trade

The United States, Mexico, and Canada agreed to free trade, or getting rid of trade barriers, in 1994. The United States also has free trade agreements with 17 other nations, including Australia, Israel, and Peru.

> **"No country produces all the goods and services it needs."**

Disadvantages

Critics of NAFTA say that the agreement has cost U.S. jobs. Workers in Mexico are paid less. As a result, many U.S. industries moved all or part of their production to Mexico. Critics also argue that NAFTA hurt Mexican farmers. Mexico imported more corn and other grains when tariffs on those items were removed. Small farmers in Mexico could not compete with technologically advanced U.S. farms.

Advantages

Supporters of NAFTA say that it creates the largest free trade area in the world. With tariffs removed, the NAFTA countries can trade with one another at lower cost. It allows the 463 million people in the three countries greater choice in the marketplace. The three countries produced an estimated $18 trillion worth of goods and services in 2011.

A worker assembles parts at a U.S. automobile plant. Critics argued that NAFTA resulted in U.S. job losses, especially in the manufacturing industry. ▶

(tr) Joe Raedle/Getty Images News/Getty Images; (bl) David McNew/Getty Images News/Getty Images; (r) Bloomberg/Getty Images

THERE'S MORE ONLINE

SEE the political boundaries of the U.S., Canada, and Mexico • **WATCH** changes in migration patterns

These numbers and statistics can help you learn about the effects of NAFTA.

Growth Triples

In 1993, the year before NAFTA went into effect, U.S. trade with Mexico and Canada totaled $276.1 billion. In 2010 U.S. exports and imports of goods with its NAFTA partners amounted to $918 billion.

$1 million a minute

Almost 400,000 people—truckers, businesspeople, commuters, and tourists—cross the U.S.-Canada border daily. U.S.-Canada two-way trade amounts to $1.4 billion a day. That's almost a million dollars every minute.

700,000 Jobs

Critics of NAFTA say the agreement has cost the jobs of U.S. workers. According to the Economic Policy Institute, the transfer of production to Canada, Mexico, and other countries has resulted in the loss of about 700,000 jobs in the United States since NAFTA began.

One-fifth

Canada is the world's largest supplier of energy to the United States. Canada provides 20 percent of U.S. oil imports and 18 percent of U.S. natural gas imports.

U.S. Surplus in Services

A trade deficit occurs when a nation imports (buys) more goods and services than it exports (sells). In 2010 the U.S. experienced a trade deficit of $94.6 billion in *goods* with its NAFTA partners. A trade surplus occurs when a nation exports (sells) more than it buys. In 2009 the United States had a $28.3 billion trade surplus with its NAFTA partners in the value of *services*. The main services exported are financial services and insurance.

21,444

the number of U.S. Border Patrol agents in 2011. This is double the number of agents in 2003.

FIFTEEN THOUSAND

This is the number of workers at the new Volkswagen plant in Puebla, Mexico, making it one of the country's largest employers.

TOP 10 COUNTRIES IN EXPORTS
Countries with exports in excess of $300 billion

CHINA

U.S.

GERMANY

JAPAN

NETHERLANDS

FRANCE

ITALY

U.K.

SOUTH KOREA

RUSSIA

KEY:

■ $1 trillion or more ■ $500 billion–$999 billion ■ $300 billion–$499 billion

U.S. TRADE WITH OTHER NATIONS
Trade in billions of dollars
(imports and exports combined, 2011)

CANADA
$597*

CHINA
$503

MEXICO
$461*

JAPAN
$195

SOUTH KOREA
$100

NETHERLANDS
$66

GERMANY
$148

U.K.
$107

BRAZIL
$74

SAUDI ARABIA
$61

*NAFTA countries

GLOBAL IMPACT

EXPORTS AND IMPORTS Based on 2011 statistics, the exports of three countries—China, the United States, and Germany—exceeded $1 trillion in value. China's major exports to the U.S. include electrical machinery and toys, guns, and sports equipment. Top U.S. exports to China include oil seeds, fruits, vehicles, and aircraft.

The U.S. did more trade, if exports and imports are combined, with Canada in 2011 than with any other nation.

NAFTA Signing 1992

Mexican President Carlos Salinas, U.S. President George H.W. Bush, and Canadian Prime Minister Brian Mulroney look on as the chief trade representatives sign the NAFTA agreement in 1992. NAFTA was ratified by the three countries in 1993.

©Bettmann/Corbis

Thinking Like a Geographer

1. **Human Geography** What is the purpose of NAFTA?

2. **The Uses of Geography** Find a product in a store or at home that has a label in another language in addition to English. Is that language used in one of the NAFTA countries? Why would a product be labeled in more than one language?

3. **Human Geography** Hold a debate in your class on this statement: NAFTA has been good for U.S. workers and consumers.

Directions: Write your answers on a separate piece of paper.

1 Use your FOLDABLES to explore the Essential Question.
INFORMATIVE/EXPLANATORY WRITING Write a couple of paragraphs explaining how geographical features led the Aztec and then much later the founders of Mexico City to build their cities on the same site.

2 **21st Century Skills**
INTEGRATING VISUAL INFORMATION Choose a country or one of the islands mentioned in this chapter. Find out more about it by researching it on the Internet. Use the information to create a travel poster or slide show highlighting the country's or island's best features and include any places you would like to see.

3 **Thinking Like a Geographer**
INTEGRATING VISUAL INFORMATION Draw a graphic organizer like the one shown here and use it to record information about the islands of the Caribbean.

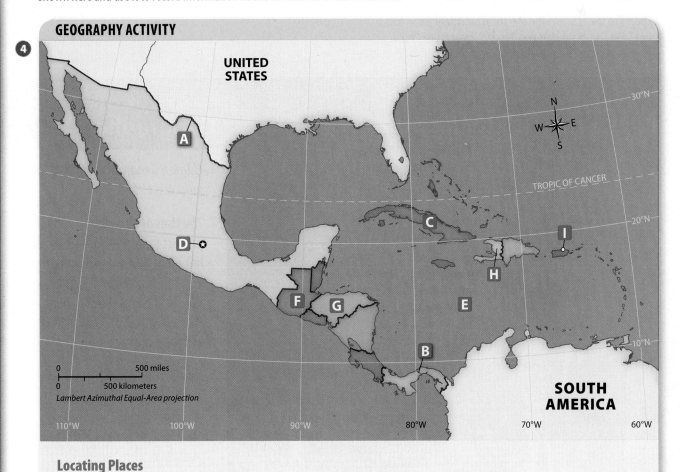

GEOGRAPHY ACTIVITY

4

Locating Places
Match the letters on the map with the numbered places listed below.

1. Panama Canal
2. Honduras
3. Mexico City
4. Rio Grande
5. Caribbean Sea
6. Haiti
7. Guatemala
8. San Juan, Puerto Rico
9. Cuba

REVIEW THE GUIDING QUESTIONS

Directions: Choose the best answer for each question.

1 What are Mexico's two most important natural resources?

A. gold and silver

B. oil and natural gas

C. iron ore and copper

D. bauxite and zinc

2 Which is one of the most important waterways in the world?

F. Lake Nicaragua

G. Río Bravo

H. Panama Canal

I. Caribbean Sea

3 A civilization that flourished in Southern Mexico, Belize, and Guatemala about 3,000 years ago and built pyramids like the Egyptians was the

A. Anasazi.

B. Olmec.

C. Aztec.

D. Maya.

4 Europeans established plantations and brought enslaved people to Cuba, Puerto Rico, and Hispaniola in order to grow

F. tobacco.

G. bananas.

H. sugar.

I. coffee.

5 Which country is Mexico's biggest trading partner?

A. Canada

B. China

C. Venezuela

D. the United States

6 What is the most serious economic challenge facing Central American countries?

F. high rate of population growth

G. fluctuating oil prices

H. food shortages

I. debt

DBQ ANALYZING DOCUMENTS

7 IDENTIFYING Read this passage about the Maya.

"*About six million Maya live in Central America. Like their ancestors, many of them survive by growing maize (Indian corn) or other crops on their land, or by producing woven textiles for sale. In some villages, the men have to leave their families to find work in the cities, or on coffee and cotton plantations.*"

—from *How People Live,* DK Publishing

Which Maya activity today is similar to one from ancient times?

A. working in tourism

B. working in factories

C. growing maize

D. working in cities

8 DETERMINING CENTRAL IDEAS Which best explains why some men have to leave their villages?

F. They leave to seek wives elsewhere.

G. They're forced to do so by the government.

H. The villages are overcrowded.

I. They face a lack of jobs within the villages.

SHORT RESPONSE

"*At the beginning of the 17th century the sweet crystal [sugar] transformed the Caribbean Islands into the Sugar Islands, though the islands did not turn sweet themselves. . . . Entire jungles were leveled; a slave or, later, cheap work force was massively imported from Africa and Asia; [and] a huge wave of European settlers arrived to stay.*"

—from Alonso Silva Lee, *Natural Cuba/Cuba Natural*

9 DESCRIBING In what ways were the Caribbean islands transformed by the spread of sugar farming?

10 DETERMINING WORD MEANINGS What does the author of the passage mean by the phrase "the islands did not turn sweet themselves"?

EXTENDED RESPONSE

11 INFORMATIVE/EXPLANATORY WRITING Research and then write a brief report comparing and contrasting the cotton and rice plantations of the American South with the sugar plantations of the Caribbean islands.

Need Extra Help?

If You've Missed Question	❶	❷	❸	❹	❺	❻	❼	❽	❾	❿	⓫
Review Lesson	1	1	2	2	3	3	3	3	2	2	3

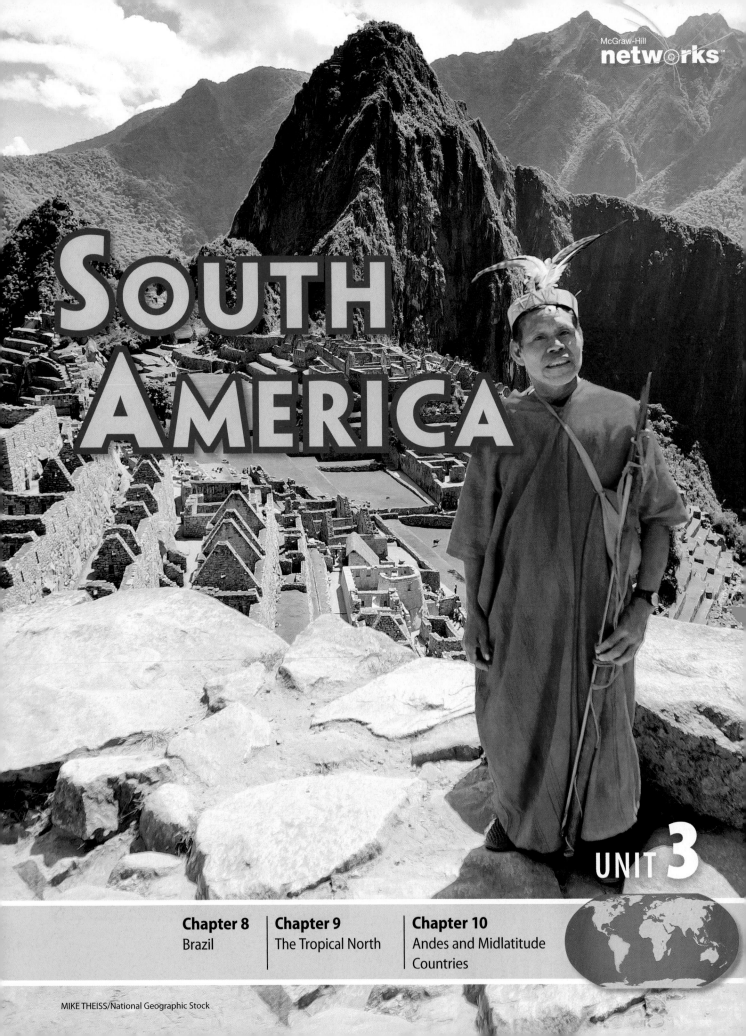

McGraw-Hill
networks™

SOUTH AMERICA

UNIT 3

MIKE THEISS/National Geographic Stock

EXPLORE the CONTINENT

SOUTH AMERICA

At nearly 7 million square miles (18 million sq. km) in area, South America is the fourth-largest continent in the world. Two great rivers—the Orinoco and the Amazon—flow through Brazil and the Tropical North. The most distinctive landform in the region is the Andes mountain ranges. The Andes, 4,500 miles (7,242 km) long, is the world's longest continental mountain range.

1 NATURAL RESOURCES Farmers grow a variety of potatoes in plots located in the Peruvian Andes. El Parque de la Papa, also known as "Potato Park," is a bio-reserve that is managed by the local communities surrounding it.

2 BODIES OF WATER Oil tankers travel near the port of Maracaibo, Venezuela. Like a circulatory system, the region's many waterways serve as arteries that transport people and goods throughout the region and to the world.

③ LANDFORMS Los Glaciares National Park in Argentina is an area of rugged mountains and many glacial lakes. Its name refers to the glaciers that are part of the Patagonian ice field. The ice field is the largest ice mantle outside Antarctica. Los Glaciares is located along Argentina's border with Chile.

FAST FACT

Earth's driest place is in South America.

(bkgd) ©Frank Lukasseck/Corbis; (l) JIM RICHARDSON/National Geographic Stock; (c) JUAN BARRETO/AFP/Getty Images; (r) Reto Stockli; NASA Earth Observatory

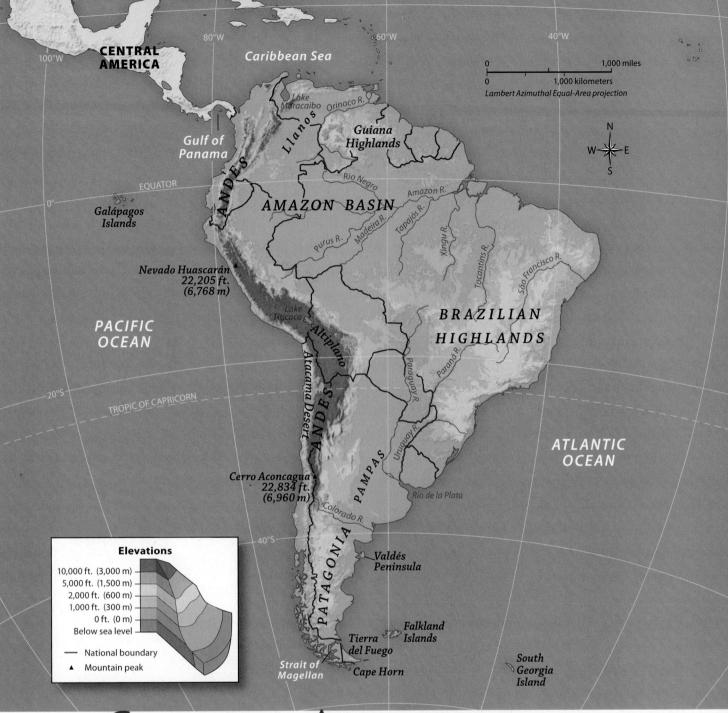

Elevations

10,000 ft. (3,000 m)
5,000 ft. (1,500 m)
2,000 ft. (600 m)
1,000 ft. (300 m)
0 ft. (0 m)
Below sea level

— National boundary
▲ Mountain peak

SOUTH AMERICA

PHYSICAL

MAP SKILLS

1 THE GEOGRAPHER'S WORLD What is the easternmost river in Brazil?

2 PLACES AND REGIONS Describe the elevation differences between western South America and eastern South America.

3 PLACES AND REGIONS Explain why elevation of land in the Amazon Basin makes travel easier.

POLITICAL

MAP SKILLS

1 PLACES AND REGIONS What is the capital of Uruguay?

2 PHYSICAL GEOGRAPHY Which two countries in South America do not have coastlines?

3 THE GEOGRAPHER'S WORLD Which country in South America shares its border with the most countries?

Cities
(Statistics reflect metropolitan areas.)

- ■ Over 5,000,000
- ☐ 2,000,000–5,000,000
- ◉ 1,000,000–2,000,000
- ● 500,000–1,000,000
- ○ Under 500,000

POPULATION

Per sq. mi.	Per sq. km
1,250 and over	500 and over
250–1,249	100–499
63–249	25–99
25–62	10–24
2.5–24	1–9
Less than 2.5	Less than 1

SOUTH AMERICA

POPULATION DENSITY

MAP SKILLS

1 HUMAN GEOGRAPHY Where do most people in South America live?

2 HUMAN GEOGRAPHY About how many people live in Lima?

3 PLACES AND REGIONS What are the largest cities on South America's eastern coast?

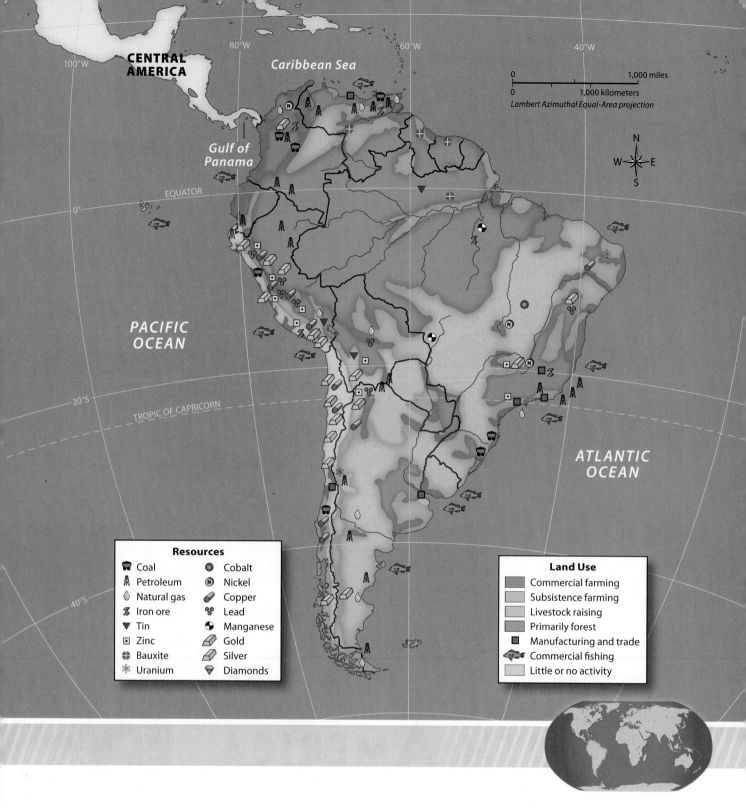

Resources

🪨 Coal	◉ Cobalt
⛏ Petroleum	Ⓝ Nickel
💧 Natural gas	🔋 Copper
⚡ Iron ore	🎱 Lead
▼ Tin	◒ Manganese
⊡ Zinc	▱ Gold
✚ Bauxite	▱ Silver
✳ Uranium	▽ Diamonds

Land Use

- Commercial farming
- Subsistence farming
- Livestock raising
- Primarily forest
- ■ Manufacturing and trade
- 🐟 Commercial fishing
- Little or no activity

ECONOMIC RESOURCES

MAP SKILLS

1 HUMAN GEOGRAPHY Is there more commercial farming or livestock raising in South America?

2 PHYSICAL GEOGRAPHY Where is the greatest concentration of minerals and ores?

3 ENVIRONMENT AND SOCIETY Is South America a manufacturing center?

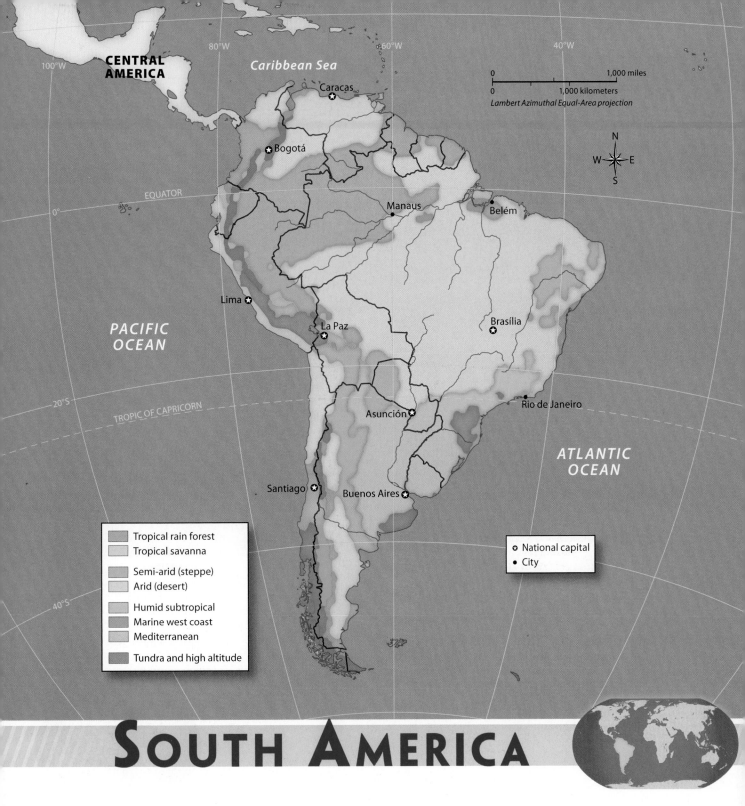

CENTRAL
AMERICA

Caribbean Sea

Caracas

Bogotá

PACIFIC
OCEAN

EQUATOR

Lima

La Paz

Manaus

Belém

Brasília

Rio de Janeiro

ATLANTIC
OCEAN

TROPIC OF CAPRICORN

Asunción

Santiago

Buenos Aires

0   1,000 miles
0   1,000 kilometers
Lambert Azimuthal Equal-Area projection

N
W   E
S

Tropical rain forest
Tropical savanna
Semi-arid (steppe)
Arid (desert)
Humid subtropical
Marine west coast
Mediterranean
Tundra and high altitude

National capital
City

SOUTH AMERICA

CLIMATE

MAP SKILLS

1 PHYSICAL GEOGRAPHY What is the most prevalent climate in South America?

2 PHYSICAL GEOGRAPHY Where is South America's desert climate?

3 PLACES AND REGIONS In which type of climate is Santiago, Chile, located?

BRAZIL

AP Photo/Silvia Izquierdo

ESSENTIAL QUESTIONS • *How does geography influence the way people live?*
• *How do governments change?* • *What makes a culture unique?*

Soccer ("football") player Robinho has many fans in Brazil and around the world.

networks
There's More Online about Brazil.

CHAPTER 8

The Story Matters...

Brazil is located in the eastern half of South America. Brazil's vast land area makes it the giant of South America. Water is also important in defining the country. The great Amazon River flows through Brazil for more than 2,000 miles (3,219 km) and carries as much as one-fourth of the world's freshwater. This river drains the Amazon Basin, which stretches across the northern half of Brazil and contains the world's largest remaining tropical rain forest.

FOLDABLES
Study Organizer

Go to the Foldables® library in the back of your book to make a Foldable® that will help you take notes while reading this chapter.

Valuable Natural Resources

Urban Population

BRAZIL (CCSS)

Brazil is the largest country in South America with almost 3.3 million square miles (8.5 million sq. km) of land. It accounts for most of the eastern coast of South America. Brazil contains more than 4,665 miles (7,508 km) of coastline along the Atlantic Ocean. The Equator and the Tropic of Capricorn run through the country. As you study the map, look for the geographic features that make this area unique.

Step Into the Place

MAP FOCUS Use the map to answer the following questions.

1 **PHYSICAL GEOGRAPHY** What is the main river in Brazil?

2 **PLACES AND REGIONS** How many countries share a border with Brazil?

3 **THE GEOGRAPHER'S WORLD** Why is it significant that the Equator and the Tropic of Capricorn both run through Brazil?

4 **CRITICAL THINKING**
ANALYZING Use the scale bar on the map to measure the distance between the cities of Brasília and Rio de Janeiro.

RIO, AERIAL VIEW The huge "Christ the Redeemer" statue overlooks Rio de Janeiro. Set between beautiful mountains and the Atlantic coast, Rio de Janeiro was Brazil's capital from 1763 to 1960.

BRAZIL'S CAPITAL Brasília is a planned city, built in Brazil's central wilderness area. Brasília has been the country's capital since 1960.

Step Into the Time

ANALYZING Select at least two events on the time line and explain how they illustrate the importance of the Amazon Basin to Brazil's development, as well as the environmental concerns caused by that development.

1500 Cabral is first European to reach Brazil's coast

1800

1822 Brazil gains independence from Portugal

Brazil

Caribbean Sea

VENEZUELA

GUYANA

FRENCH GUIANA
(France)

SURINAME

COLOMBIA

ATLANTIC OCEAN

ECUADOR

EQUATOR

Amazon R.

Madeira R.

Tapajós R.

Purus R.

BRAZIL

Xingu R.

Tocantins R.

São Francisco R.

Recife

PERU

B

Salvador

Brasília

BOLIVIA

A

PACIFIC OCEAN

Parand R.

São Paulo

Rio de Janeiro

TROPIC OF CAPRICORN

CHILE

PARAGUAY

ARGENTINA

URUGUAY

○ National capital
• City

N
W—E
S

0   500 miles
0   500 kilometers
Lambert Azimuthal Equal-Area projection

1889
Brazil is proclaimed
a republic

2010
Dilma Rousseff
elected president

1960 Capital moves from
Rio de Janeiro to Brasília

1900

1888 Slavery is
abolished in Brazil

2000

2009 Rio de Janeiro chosen
to host 2016 Olympic Games

(l to r) SambaPhoto/Milton Carelo/Getty Images; E. Hanazaki Photography/Flickr/Getty Images; Rodrigo Baleia/LatinContent/Getty Images; Kevin Schafer/Photographer's Choice/Getty Images; Benjamin Lowy/Getty Images News/Getty Images

Reading **HELP**DESK **CCSS**

Academic Vocabulary

- **area**
- **occur**

Content Vocabulary

- **tributary**
- **basin**
- **rain forest**
- **canopy**
- **plateau**
- **escarpment**
- **pampas**
- **Tropics**
- **temperate zone**

TAKING NOTES: *Key Ideas and Details*

Summarize As you read, use a graphic organizer to write a summary sentence about each topic.

Topic	Summary
Waterways	
Climate	
Resources	

Lesson 1
Physical Geography of Brazil

ESSENTIAL QUESTION • *How does geography influence the way people live?*

IT MATTERS BECAUSE
Brazil is the world's fifth-largest country in size and population.

Waterways and Landforms

GUIDING QUESTION *What are Brazil's physical features?*

Brazil is the largest country in South America. It occupies about half the continent. Rolling lowland plains and flat highland plateaus cover most of the country.

The Amazon

The Amazon River is one of Brazil's amazing natural features as well as a great natural resource. It begins high in the Andes of Peru and flows east across northern Brazil to the Atlantic Ocean. The river is the Western Hemisphere's longest river and the world's second longest, after the Nile River in Africa.

The Amazon is the largest river in terms of the amount of freshwater it carries. It moves more than 10 times the water volume of the Mississippi River. Of all the water that Earth's rivers empty into the oceans, about 25 percent comes from the Amazon. Its massive flow pushes freshwater more than 100 miles (161 km) out into the Atlantic Ocean. The river's depth allows oceangoing ships to travel more than 2,000 miles (3,219 km) upstream to unload or pick up cargo.

The Amazon Basin

One reason the Amazon carries so much water is that it has more than 1,000 **tributaries**. These smaller rivers feed into the Amazon as it flows from the Andes to the Atlantic Ocean. Several tributaries are more than 1,000 miles (1,609 km) long.

The **area** that a river and its tributaries drain is called a **basin**. The Amazon Basin covers more than 2 million square miles (5.2 million sq. km). Nearly half of Brazil's land lies within this vast region. Its wet lowlands cover most of the country's northern and western areas.

Much of the Amazon Basin is covered by the world's largest **rain forest**. A rain forest is a warm woodland that receives a great deal of rain each year. Tall evergreen trees form a **canopy**, or an umbrella-like covering. The Amazon rain forest is called the Selva. It is the world's richest biological resource. The Selva is home to several million kinds of plants, insects, birds, and other animals.

Only about 6 percent of Brazil's population live in the Amazon Basin. Most of the region contains fewer than two people per square mile. Some are Native Americans who live in small villages and have little contact with the outside world.

Brazilian Highlands

South and east of the Amazon Basin are the Brazilian Highlands. This is mainly a region of rolling hills and areas of high, flat land called **plateaus**. These highlands are divided into western and eastern parts.

(t) Manfred Gottschalk/Workbook Stock/Getty Images; (b) altrendo travel/Getty Images

Visual Vocabulary

Tributary A tributary is a smaller river or stream that flows into a larger one, or into a lake.

Academic Vocabulary

area a geographic region

South America's Amazon River and North America's Mississippi River cross vast distances and carry enormous amounts of water.

▶ **CRITICAL THINKING**

Comparing How are the Amazon and Mississippi Rivers similar? How are they different?

The western part of the highlands is largely grassland that is partly covered with shrubs and small trees. Farming and ranching are the major economic activities in this part of the highlands. Farther west is the Mato Grosso Plateau, a flat, sparsely populated area of forests and grasslands that extends into Bolivia and Peru.

Low mountain ranges form much of the eastern Brazilian Highlands, although some peaks rise above 7,000 feet (2,134 m). In other places, highland plateaus plunge to the Atlantic coast, forming **escarpments**, or steep slopes. These escarpments, rising from coast to highlands, have hindered development of inland areas.

Brazil's third-largest city, Brasília, is located in the Brazilian Highlands. It was built in the 1950s as Brazil's new capital to encourage settlement in the country's interior. Some 3.5 million people live in and around the city.

About 600 miles (966 km) south of Brasília is São Paulo. This huge city is located on a plateau at the highland's eastern edge, just 30 miles (48 km) from the Atlantic coast. With more than 17 million people, São Paulo is the largest city in the Southern Hemisphere. It is also South America's most important industrial city.

Farther south are grassy, treeless plains called **pampas**. The grass and fertile soil make the pampas one of Brazil's most productive ranching and farming areas.

Atlantic Lowlands

Brazil has one of the longest strips of coastal plains in South America, wedged between the Brazilian Highlands and the Atlantic Ocean. This narrow plains region, called the Atlantic lowlands, is just 125 miles (201 km) wide in the north; it becomes even narrower in the southeast. The rural parts of this region are another important area for farming.

An escarpment slopes down to an Atlantic Ocean beach near the city of São Paulo.

▶ **CRITICAL THINKING**

Describing How have escarpments affected Brazil's development?

SambaPhoto/Milton Carelo/Getty Images

Although the coastal lowlands cover only a small part of Brazil's territory, most of the nation's people live here. More than 12 million live in and around Rio de Janeiro, Brazil's second-largest city. Rio's beautiful beaches and vibrant lifestyle make it Brazil's cultural and tourist center.

☑ **READING PROGRESS CHECK**

Analyzing Why do many Brazilians live in the Brazilian Highlands?

Ranchers herd cattle on the Mato Grosso Plateau of west-central Brazil.

▶ **CRITICAL THINKING**
Describing What are the main features of the Mato Grosso Plateau?

A Tropical Climate

GUIDING QUESTION *What are Brazil's climate and weather like?*

Most of Brazil is located in the **Tropics**. This is the zone along Earth's Equator that lies between the Tropic of Cancer and the Tropic of Capricorn. Brazil's climate varies. In fact, the huge country has several different climates.

Wet Rain Forests
The area along the Equator in northern Brazil has a tropical rain forest climate. In this climate, every day is warm and wet. Daytime temperatures average in the 80s Fahrenheit (27°C to 32°C). It feels hotter than this because the wet rain forest makes the air humid.

During periods of drought, the Amazon River carries less water, which exposes sandbars in the river and low-lying areas along the shoreline.

▶ **CRITICAL THINKING**

Identifying What type of climate is found in areas along the Amazon River?

Academic Vocabulary

occur to happen or take place

Areas along the Amazon River have a tropical rain forest climate. They experience winds called monsoons that bring a huge amount of rain—120 inches to 140 inches (305 cm to 356 cm) per year. During the monsoon season, flooding swells the Amazon River in some places to more than 100 miles (161 km) wide. These areas also have a dry season when little rain **occurs**. During the dry season, forest fires are a danger, even in a rain forest.

Tropical Wet/Dry Climate

Tropical wet/dry climates usually exist along the outer edges of tropical rain forest climates. Most of the northern and central Brazilian Highlands has a tropical wet/dry climate. This climate has just two seasons—summer, which is wet, and winter, which is dry. Daily average temperatures change very little. Summers average in the 70°F range (21°C) and winters in the 60°F range (16°C). But even this slight difference is enough to change wind patterns, which affect rainfall. Between 40 inches and 70 inches (102 cm to 178 cm) of rain fall during the summer months. Winters get almost no rain.

Dry and Temperate Climates

The northeastern part of the Brazilian Highlands has a semiarid climate. This region is the hottest and driest part of the country. The daily high temperature during the summer often reaches 100°F (38°C). Frequent and severe droughts have caused many of the region's farms to fail. Even so, the desertlike plant life supports some light ranching.

Southeastern Brazil, including São Paulo and Rio de Janeiro, is located in the **temperate zone**—the region between the Tropic of Capricorn and the Antarctic Circle. It has a temperate climate called humid subtropical. It is the same type of climate that the southeastern United States experiences.

Temperatures vary according to location and elevation in this part of Brazil. Summers are generally warm and humid, and winters are mild. Rainfall occurs year-round. In the southern parts of this climate zone, snow can fall.

☑ **READING PROGRESS CHECK**

Identifying What factors make farming in the northeastern part of Brazil difficult?

Natural Resources

GUIDING QUESTION *What resources are most plentiful and important in Brazil?*

Brazil has some of the world's most plentiful natural resources. Many of the resources have been developed for years, especially in the south and southeast. Recent transportation improvements have made the resources in Brazil's vast interior available to its growing industries and population. Agriculture, mining, and forestry have been important for centuries. The natural riches of Brazil attracted European settlers to the region. They found abundant trees, rich mineral resources, and fertile farmland.

River floodwaters surge through an area of the Amazon rain forest in northwestern Brazil.

▶ **CRITICAL THINKING**
Identifying What yearly natural event causes flooding in the Amazon Basin?

Think Again?

Summer and winter occur at about the same time everywhere.

Not true. While American teens enjoy their summer vacation, young people in Brazil are going to school! That's because south of the Equator, the seasons are reversed. The summer months in the United States are winter months in Brazil.

Kevin Schafer/Photographer's Choice/Getty Images

A worker on a Brazilian coffee plantation picks ripe coffee berries. After picking, the coffee berries are separated for quality and packed in sacks to send to market.

▶ **CRITICAL THINKING**

Identifying Where are Brazil's major coffee-growing areas?

Abundant Forests

Forests cover about 60 percent of Brazil, accounting for about 7 percent of the world's timber resources. Most of the forests in the northeast and south were cleared long ago. Heavy logging continues in the Atlantic lowlands.

Logging in the Amazon Basin is increasing as more roads are built and settlement grows. The rain forest's mahogany and other hardwoods are highly desirable for making furniture. The rain forest is also a source of natural rubber, nuts, and medicinal plants. Logging, mining, and other development have become a major environmental issue. However, the rate of deforestation, or clearing land of forests or trees, has declined in recent years.

Minerals

Brazil has rich mineral resources that are only partly developed. They include iron ore, tin, copper, bauxite, gold, and manganese. At one time, most mining was done in the Brazilian Highlands. Recently, major deposits of minerals have been found in the Amazon basin. The new deposits might make Brazil the world's largest producer of many of the minerals. Brazil also has huge potential reserves of petroleum and natural gas deep under the ocean floor off its coast. Getting to the oil is a challenge, however.

Productive Farmland

Brazil is the world's largest producer of coffee, sugarcane, and tropical fruits. The country also produces great amounts of soybeans, corn, and cotton.

Brazilian farmers produce most of their country's food supply. Agriculture is also important in trade, accounting for more than one-third of Brazil's exports. It is a leading exporter of coffee, oranges, soybeans, and cassava. Cassava is used to make tapioca.

Major Crops

Production of coffee throughout the world was estimated to set an all-time high in 2012–2013, up 10 million bags from the previous year. Brazil and Vietnam accounted for most of the increase. The eastern Brazilian Highlands and the Atlantic lowlands are the main coffee-growing areas. Coffee was once Brazil's main export. Today, soybeans provide more income for the country. China is increasing its soybean imports, mostly for animal feed, and much of it comes from Brazil.

Most soybeans are grown in the south, but they are an important crop in the Brazilian Highlands, too. Farming has become easier in the highlands as farmers have begun using tractors and fertilizer to work the savanna soils.

Brazil grows one-third of the world's oranges, making it the world's leading supplier of the citrus fruit. Brazil is also the largest beef exporter in the world. Most of the country's grazing land is in the south and southeast.

In a recent year, Brazil's sugarcane production was more than two and a half times that of India, the second-leading producer. Brazilian sugarcane is used to make ethanol, which is mixed with gasoline and used as fuel for cars and trucks. For many years, the government has required cars to use ethanol. The country's car manufacturers make flexible-fuel vehicles that can use fuel with high levels of ethanol.

✔ **READING PROGRESS CHECK**

Identifying Which two regions are Brazil's most important agricultural areas?

Include this lesson's information in your Foldable®.

LESSON 1 REVIEW (CCSS)

Reviewing Vocabulary

1. How does Brazil's location in the Tropics affect its climate?

Answering the Guiding Questions

2. *Determining Central Ideas* Why is the Amazon Basin a unique region?

3. *Analyzing* How do a tropical rain forest climate and a tropical wet/dry climate differ?

4. *Describing* What resources are important Brazilian exports?

5. *Informative/Explanatory Writing* In which of Brazil's physical regions would you most like to live? Write a paragraph to explain why.

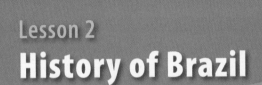

There's More Online!

☑ **MAP** Colonization of Brazil

☑ **SLIDE SHOW** Brazil's Natural Products

☑ **VIDEO**

Reading **HELP**DESK (CCSS)

Academic Vocabulary

- **comprise**
- **extract**

Content Vocabulary

- **indigenous**
- **slash-and-burn agriculture**
- **emancipate**
- **compulsory**

TAKING NOTES: *Key Idea and Details*

Sequencing As you read about Brazil's history, use the graphic organizer below to note how Brazil became a modern democratic republic.

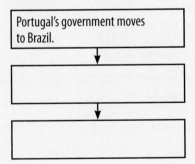

Portugal's government moves to Brazil.

↓

↓

Lesson 2
History of Brazil

ESSENTIAL QUESTION • *How do governments change?*

IT MATTERS BECAUSE
Brazil is one of the world's leading industrial powers.

Early History

GUIDING QUESTION *How did Brazil's early peoples live?*

In 1493 Christopher Columbus returned to Spain with news of his explorations and of new lands. The Spanish worried that neighboring Portugal, a powerful seafaring rival, would try to claim these lands for itself. So they asked the pope to find a solution. The pope decided that all new lands west of a certain line should belong to Spain. Lands east of the line would belong to Portugal. The two countries agreed to this division in 1494 by signing the Treaty of Tordesillas.

Almost nothing was known of the region's geography, so neither side realized how unequal the division was. Almost all of the Americas lay west of the line, which became Spanish territory. The only exception was the eastern part of South America, which became Portuguese territory. Today, this part of South America is Brazil. That is why Brazil is the only South American country that has a Portuguese heritage.

Indigenous Populations

The first Portuguese ships stopped in Brazil in 1500. Their destination was India, so they did not stay in Brazil for long. They had peaceful encounters with some of the **indigenous**, or native, peoples who lived along the coast. The Portuguese commander, Pedro Cabral, claimed the land for Portugal. After just 10 days, the Portuguese left. They had no idea of the vast region and many peoples included in Cabral's claim.

The people the Portuguese met were the Tupi. They lived along the coast and in the rain forests south of the Amazon River, where they grew cassava, corn, sweet potatoes, beans, and peanuts. They hunted fish and other water animals with arrows and harpoons from large log canoes, but they did little hunting on land.

Brazil's native peoples had lived there for more than 10,000 years when the Portuguese arrived. Estimates are that the population was between 2 million and 6 million by 1500. Besides the Tupi, it included the Arawak and Carib people of the northern Amazon and coast, and the Nambicuara in the drier grasslands and highlands. These are not the names of native peoples; they were Brazil's four main language groups. Each group **comprised** many different peoples.

Academic Vocabulary

comprise to be made up of
extract to remove or take out

Daily Life

Like the Tupi, Brazil's other lowland and rain forest peoples were mainly farmers. They lived in permanent, self-governing villages and practiced **slash-and-burn agriculture**. This is a method of farming in forests that involves cutting down trees and burning away underbrush to create fields for growing crops. Farther south, most of the Nambicuara of the Brazilian Highlands were nomads, people who move from place to place and have no permanent home. In the dry season, they lived as hunter-gatherers, people who get their food by hunting, fishing, and collecting seeds, roots, and other parts of trees and wild plants. In the wet season, they built temporary villages and practiced slash-and-burn agriculture.

An Ashaninka family fishes from a boat in Brazil's Amazon rain forest.
▶ **CRITICAL THINKING**
Describing How did indigenous peoples make a living when the first Europeans arrived in Brazil?

Europeans Arrive

For more than 30 years after Cabral's visit, the Portuguese did not pay much attention to Brazil. Their main focus was on their colonies and trade in Asia. Their trading ships sailed south and east around Africa on their way to Asia. Portuguese sailors established a few trading posts along Brazil's coast and collected brazilwood. The red dye **extracted** from this wood was highly valued in Europe. It was because of this trade that the Portuguese named the region Brazil.

Mike Goldwater/Alamy

The church of São Miguel das Missões was built about 1740 as the center of a Jesuit mission village in southern Brazil.

▶ **CRITICAL THINKING**

Explaining Why did the Jesuits build mission villages in Brazil and other parts of South America?

The valuable brazilwood trade made other Europeans more interested in Brazil. French traders began collecting the wood and shipping it to France. To bring Brazil under tighter Portuguese control, Portugal's King John III established a permanent colony and government there. The first Portuguese settlers arrived in 1533.

✓ **READING PROGRESS CHECK**

Determining Central Ideas Why did the Portuguese colonize Brazil?

Colonial Rule

GUIDING QUESTION *How did the Portuguese colony in Brazil develop?*

Portugal's rule of Brazil lasted more than 300 years. During that time, Portuguese settlements spread all along the coast. Explorers and others traveled up rivers and deep into Brazil's interior. The expansion brought wealth to Portugal, though much of it came at great cost to Brazil's indigenous peoples.

The Portuguese Conquest

King John III gave wealthy supporters huge tracts of land in Brazil. These tracts extended west from the coast about 150 miles (241 km) inland. In return, the people who received a land grant were responsible for developing it. They founded cities and gave land to colonists to farm.

Because the colonists could not do all the work that was required, they soon began enslaving nearby native peoples as laborers. Many of them resisted and were killed. Thousands more died from exposure to European diseases to which they had no natural resistance. Others fled into Brazil's interior. These conditions and other complaints caused King John to end the land-grant system in 1549. He put Brazil under royal control and sent a governor from Portugal to rule the colony.

Spread of Christianity

The new governor brought more colonists with him. They included a number of Jesuit Catholic priests who belonged to a missionary group called the Society of Jesus. The king asked the Jesuits to go to Brazil to help the native peoples and convert them to Christianity. Those who converted were settled in special Jesuit villages and were protected from slavery.

Those Portuguese colonists who held enslaved people complained to the king about the Jesuits' work. In 1574 he ruled that native peoples who did not live in Jesuit villages could be enslaved only if they were captured in war. This ruling sent Jesuits into Brazil's interior to protect and convert peoples there. Slave hunters also moved into the interior to attack and enslave the native peoples. Cattlemen and prospectors followed, slowly spreading development inland.

Sugar and Gold

As Brazil's sugar industry expanded, cattlemen needed new land. The rise of large sugarcane plantations, mainly in the northeast, pushed ranching westward.

Plantation workers carry sugarcane into a Brazilian mill, 1845.
▶ **CRITICAL THINKING**
Identifying Besides sugarcane, what else did large plantations grow?

In the 1600s, sugar became Brazil's main export and Portugal's greatest source of wealth. Coffee and cotton plantations also developed. The discovery of gold in the eastern highlands in the 1690s further boosted the development of the interior. Towns sprang up as thousands of colonists rushed to the area. Large numbers of new colonists arrived from Europe, as well. The discovery of diamonds in the region in the 1720s added to the population boom.

Plantation agriculture and mining required large numbers of workers. This increased the need for enslaved workers. When native populations could not fill the need, the Portuguese began importing large numbers of enslaved Africans. By the 1780s, more than 150,000 enslaved Africans worked in the mining districts. This was twice the size of the Portuguese population. By 1820, some 1.1 million enslaved people accounted for nearly one-third of Brazil's total population.

✓ **READING PROGRESS CHECK**

Determining Central Ideas Why did King John III send Jesuits to Brazil?

MAP SKILLS

1 **PLACES AND REGIONS** Where did the Portuguese settle in South America?

2 **HUMAN GEOGRAPHY** Why was the division of South America between Spain and Portugal so unequal?

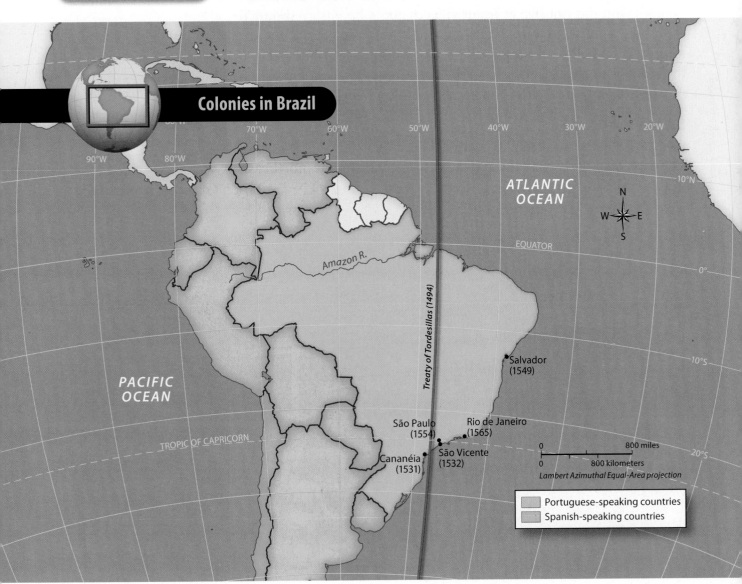

Colonies in Brazil

ATLANTIC OCEAN

EQUATOR

Amazon R.

Treaty of Tordesillas (1494)

PACIFIC OCEAN

Salvador (1549)

São Paulo (1554)

Rio de Janeiro (1565)

São Vicente (1532)

Cananéia (1531)

TROPIC OF CAPRICORN

0   800 miles
0   800 kilometers
Lambert Azimuthal Equal-Area projection

Portuguese-speaking countries
Spanish-speaking countries

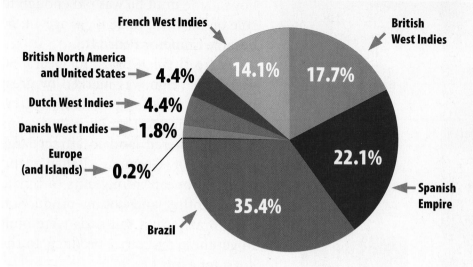

French West Indies

British North America and United States ➤ **4.4%**

Dutch West Indies ➤ **4.4%**

Danish West Indies ➤ **1.8%**

Europe (and Islands) ➤ **0.2%**

14.1%

British West Indies

17.7%

22.1%

Spanish Empire

35.4%

Brazil

THE SLAVE TRADE
More than one of every three enslaved Africans who were transported to the Americas were brought to Brazil.

▶ **CRITICAL THINKING**
1. *Identifying* What percentage of enslaved Africans were transported to the Spanish Empire?
2. *Integrating Visual information* What is the combined percentage of enslaved Africans brought to the West Indies?

Independent Brazil

GUIDING QUESTION *How did Brazil gain independence and become a democracy?*

Brazil gained independence from Portugal in an unusual way. It came gradually, fairly easily, and with little bloodshed. It was also the indirect result of the actions of the French emperor Napoleon Bonaparte.

Independence and Monarchy

In 1805, Britain joined by its allies Russia, Austria and Sweden, went to war with France to crush Napoleon. Instead, Napoleon defeated them and conquered much of Europe. In 1807 Napoleon invaded Portugal. As the French army closed in on Portugal's capital city of Lisbon, ruler Dom João, the royal family, and other government leaders fled to Brazil. Rio de Janeiro became the new capital of the Portuguese Empire. Brazil's status within the empire changed from a colony to a kingdom. This action gave Brazil equal status with Portugal within the empire.

After Napoleon was defeated, the Portuguese people wanted their king back. In 1821 Dom João and the rest of the government returned to Portugal. He left his son Pedro to rule Brazil. In 1822 Portugal's legislature restored Brazil's status as a colony and ordered Pedro to return. Pedro refused to give up the Brazilian throne. He declared independence and crowned himself Emperor Pedro I. Most other independent American nations became republics, but independent Brazil became a constitutional monarchy. In this form of government, a king, a queen, or an emperor acts as head of state.

Most Brazilians had supported independence from Portugal, but they soon tired of Pedro's harsh rule. In 1831 he was forced to turn over the throne to his five-year-old son.

As in the United States, voters in Brazil elect a president every four years.

▶ **CRITICAL THINKING**
Identifying Which group controlled the election of Brazil's president in the early republic?

A series of advisers ruled in the boy's name until he was old enough to rule on his own. In 1840, at age 14, he became Emperor Pedro II.

Pedro II ruled Brazil for nearly 50 years. His reign was marked by great progress. Brazil's population grew from 4 million to 14 million during his rule. He offered land to attract large numbers of Germans, Italians, and other European immigrants to Brazil. Sugar, coffee, and cotton production rose. Brazil's first railroads were built to get these and other products to the coast for export.

In 1850 Brazil stopped importing enslaved people from Africa. In the 1860s, a new movement began to **emancipate**, or free, the enslaved. Pedro II opposed slavery, but he thought it should be ended gradually. An 1871 law granted freedom to all children born to people in slavery. An 1885 law freed enslaved people who were over age 60. Finally, in 1888, all remaining enslaved people were freed.

The Brazilian Republic

Brazil's powerful plantation owners were angered by the loss of their enslaved workers. In 1889 they supported Brazil's army in overthrowing Pedro II. A new government was established, with a constitution based on the Constitution of the United States. Brazil became a republic, a system in which the head of state is an elected ruler instead of a king, a queen, or an emperor. In this republic, the right to vote was limited to wealthy property owners. In 1910, for example, out of a population of 22 million, only 627,000 people could vote.

Most of the power in the early republic was held by the governors of Brazil's southeastern states. Governors were elected by their state's wealthy voters. State governors controlled the election of Brazil's president, who usually came from the highly populated, coffee-rich states of São Paulo and Minas Gerais (General Mines).

These presidents followed economic policies that benefited southeastern Brazil. Coffee became Brazil's main export. By 1902, Brazil was supplying 65 percent of the world's coffee. São Paulo, Minas Gerais, and Rio de Janeiro also became the country's industrial and commercial centers. Over time, some people became unhappy with government policies that continued to favor the coffee growers and other rich Brazilians. In 1930 Getúlio Vargas overthrew the newly elected "coffee president" and seized power. He ruled for the

next 15 years. Vargas's reforms made him a hero to most Brazilians. He raised wages, shortened work hours, and let workers form labor unions. Yet for much of his rule, Vargas governed as a dictator. He dissolved the legislature and banned political parties. In 1945 military leaders forced Vargas to resign.

Brazil Under Military Rule

Vargas was elected president again in 1950, but again was forced from office by the military in 1954. For over 30 years, government in Brazil alternated between dictators and elected leaders. Manufacturing thrived throughout this period. Foreign investments brought rapid growth in the steel, auto, and chemical industries.

Industrial growth was accompanied by changes and unrest in Brazilian society. As a result, the military took control of Brazil in 1964, and a series of generals became the heads of government. An elected legislature was allowed, but the army controlled the elections. People who opposed the government were arrested. Many others were frightened into silence. The military gave up power in 1985 and allowed the election of a civilian president.

Modern Brazil

Today Brazil is a democratic republic in which people elect a president and other leaders. In Brazil, voting is **compulsory**. This means that citizens have no choice in deciding whether or not to vote. People from ages 18 to 70 are required by law to vote.

Because Brazil has a high number of well-supported political parties, coalition governments are common. A coalition government is one in which several political parties cooperate to do the work of government. In 2003 a democratically elected president replaced another democratically elected president for the first time in more than 40 years. In 2010 voters elected Dilma Vana Rousseff as the thirty-sixth president of Brazil. She is the first woman president in the country's history.

Include this lesson's information in your Foldable®.

☑ **READING PROGRESS CHECK**

Identifying Central Ideas Why did Brazil's monarchy come to an end?

LESSON 2 REVIEW (CCSS)

Reviewing Vocabulary

1. What kind of agriculture did some *indigenous* farmers practice?

Answering the Guiding Questions

2. *Analyzing* How were the Nambicuara similar to and different from the other main indigenous peoples of early Brazil?

3. *Identifying* Why did African slavery increase in Brazil before it was abolished completely in 1888?

4. *Describing* What were the main steps in Brazil's transition from a colony to a democratic country?

5. *Argument Writing* Take the role of a Brazilian living in 1889. Write a letter to the editor of your local newspaper supporting or opposing the establishment of the republic. Be sure to state the reasons for your opinion.

Reading **HELP**DESK

Academic Vocabulary

- **diverse**
- **unique**

Content Vocabulary

- **hinterland**
- **metropolitan area**
- **central city**
- **favela**

TAKING NOTES: *Key Ideas and Details*

Organize As you read the lesson, use the graphic organizer below to organize information about Brazil by adding one or more facts to each box.

Modern Brazil

Population — Culture — Challenges

Lesson 3
Life in Brazil

ESSENTIAL QUESTION · *What makes a culture unique?*

IT MATTERS BECAUSE
Brazil's cultures have influenced many people around the world.

People and Places

GUIDING QUESTION *What cultures are represented by Brazilians?*

With some 200 million people, Brazil is the world's fifth-largest country in population. Only China, India, the United States, and Indonesia are home to more people. About half of all South Americans live in Brazil.

Brazil's Diverse Population

Brazil is a mix of several cultures. Many people have a combination of European, African, and native American ancestry. Many are of Portuguese origin or immigrants from Germany and Italy. To a lesser degree, people came from Russia, Poland, and Ukraine. São Paulo, in particular, has a **diverse** population, including a large Japanese community.

Nearly 40 percent of Brazilians have mixed ancestry. This is largely because marriages between people of different ethnic groups have been more acceptable in Brazil than in many other countries. The largest group of multiethnic Brazilians are persons with European and African ancestors. People of European and Native American ancestry are a smaller group.

The smallest multiethnic group is persons of African and Native American descent. About 4 million Africans had been enslaved and brought to Brazil by the 1800s. Many escaped into the **hinterland**, the often remote inland regions, far from the coasts. The Africans lived there with the indigenous Native Americans or formed their own farming communities.

Today, about 80 percent of Brazilians live within 200 miles (322 km) of the Atlantic coast. After slavery ended, many formerly enslaved people left their homes and settled in other agricultural areas or towns. The northeast, however, still has Brazil's highest African and mixed populations. They also form the major population groups in coastal cities and towns north of Rio de Janeiro.

Most Brazilians of European descent live in southern Brazil. Indigenous Native Americans live in all parts of the country. The Amazon rain forest holds the greatest number, but about half of Brazil's Native Americans now live in cities.

Crowded Cities

For most of Brazil's history, the majority of Brazilians lived in rural areas, mainly on plantations, on farms, or in small towns. In the 1950s, millions of people began migrating to cities to take jobs in Brazil's growing industries. By 1970, more Brazilians lived in urban areas than in rural ones. Today, 89 percent of Brazilians live in and around cities.

São Paulo, Brazil's industrial center, is one of the world's largest cities. Some 17 million people live in its **metropolitan area**, or the city and built-up areas around the central city. The **central city** is the largest or most important city in a metropolitan area. São Paulo and Brazil's other large cities look much like cities in the United States. Skyscrapers line busy downtown streets. Cars and trucks jam highways in the mornings and evenings as people travel to and from their jobs. People work in office buildings, shops, and factories. Many own small businesses.

Favelas

Many middle-class urban dwellers live in apartment buildings. Others live in small houses in the suburbs, which are largely residential communities on the outskirts of cities. Wealthy Brazilians live in luxury apartments and mansions.

Most of Brazil's large cities also have shantytowns called **favelas**. Favelas are makeshift communities located on the edges of the cities.

Stuart Dee/Photographer's Choice RF/Getty Images

Academic Vocabulary

diverse differing from one another; varied

Sugarloaf Mountain looms above Rio de Janeiro's Copacabana Beach.
▶ **CRITICAL THINKING**
Explaining What has led to the growth of Brazil's cities since the 1950s?

The Estaiada Bridge, opened in 2008, is one of São Paulo's landmarks. It is known for its curved appearance and X-shaped tower.

▶ **CRITICAL THINKING**

Describing What role does São Paulo play in Brazil's economy?

Favelas arose as millions of poor, rural Brazilians with few skills and little education migrated to cities to seek better lives. These people could not afford houses or apartments. Instead, they settled on land they did not own and built shacks from scraps of wood, sheet metal, cinder blocks, and bricks. Some favelas lack sewers and running water. In many, disease and crime are widespread.

São Paulo and Rio de Janeiro have the most and largest favelas. Rio has about 1,000 of them. About one of every three of the city's residents live in a favela. Rio officials have tried to deal with this problem by offering favela dwellers low-cost housing in the suburbs. Many do not want to move because the long commute from the suburbs to jobs in the city can take hours.

✔ **READING PROGRESS CHECK**

Analyzing Why does Brazil have such a large percentage of people with multiethnic ancestry?

People and Cultures

GUIDING QUESTION *What is it like to live in Brazil?*

Brazilians get along well for a country whose population includes such a variety of racial and ethnic groups. This is largely due to Brazilians' reputation for accepting other people's differences. Personal warmth, good nature, and "getting along" are valued in Brazilian culture. These attitudes and behaviors are an important part of what is known as the "Brazilian Way."

Tensions exist in Brazilian society, but they involve social and economic issues more than ethnic or cultural ones. Ethnicity still plays a factor, though, because Brazilians of European origins have often had better educational opportunities. They hold many of the better jobs as a result.

Ethnic and Language Groups

Until the late 1800s, nearly all European immigrants to Brazil were from Portugal. After slavery ended, large numbers of Italians arrived to work on the coffee plantations.

During the same period, settlers from Germany started farming colonies in southern Brazil. In the early 1900s, the first Japanese arrived to work in agriculture in the Brazilian Highlands. Many of their descendants moved to cities. The first Middle Easterners, mainly Lebanese and Syrians, arrived at about the same time. They became involved in commerce in cities and towns around the country.

The diversity of Brazil's people has given the country a **unique** culture. Portuguese is Brazil's official language. Almost all Brazilians speak it. Brazilian Portuguese is quite different from the language spoken in Portugal. In fact, many Brazilians find it easier to understand films from Spanish-speaking countries in South America than films from Portugal. This is because Brazil's many ethnic groups have introduced new words to the language. Thousands of words and expressions have come from Brazil's indigenous peoples. Dozens of Native American languages are still spoken throughout Brazil.

Religion and the Arts

About two-thirds of Brazilians are Roman Catholics, but only about 20 percent attend services regularly. Women go to church more often than men, and older Brazilians are more active in the Church than the young.

Academic Vocabulary

unique unlike anything else; unusual

Most of the rest of Brazil's population follows the Protestant faith. Those who practice Islam and Eastern religions such as Buddhism are growing in numbers. Many Brazilians blend Christian teachings with beliefs and practices from African religions.

Other African influences on Brazilian culture include foods, popular music, and dance, especially the samba. Brazilians blended samba rhythms with jazz to introduce the world to music called bossa nova. Several Brazilian writers have gained world fame for their books exploring regional and ethnic themes. Brazilian movies and plays also have gained worldwide attention.

Each February, Brazilians celebrate a four-day holiday called Carnival. Millions of working-class and middle-class Brazilians spend much of the year preparing for it by making costumes and building parade floats. Nearly all city neighborhoods are strung with lights. Rio de Janeiro's Carnival is the largest and is world famous. Elaborately costumed Brazilians ride equally elaborate floats in dazzling parades. They are accompanied by thousands of costumed samba dancers moving to the lively music.

Rural Life

Family ties are strong in Brazil. Family members usually live close to one another. They hold frequent reunions or gather at a family farm or ranch on weekends and holidays. Life in rural Brazil has changed little over the years. Most rural families are poor. They work on plantations or ranches or own small farms. They live in one- or two-room houses made of stone or adobe—clay bricks that are dried and hardened in the sun. Their chief foods are beans, cassava, and rice. A stew of black beans, dried beef, and pork is Brazil's national dish.

Urban Life

Many city dwellers are poor, too, and they eat a similar diet. For those who can afford it, U.S. fast-food chains are rapidly expanding in larger Brazilian cities. In general, people in the industrial cities of southern Brazil have a better life than people in the more rural northeast.

Life in Brazil's cities moves at a faster pace. Government services and modern conveniences are available there. Many workers have good jobs and enjoy a decent quality of life. Most middle-class families have cars. Poor families rely on buses to get to work and to the beach or countryside on weekends.

Soccer ("football") is Brazil's most popular sport. It is played nearly everywhere on a daily basis. Matches between professional teams draw huge crowds in major cities. Brazil's national team is recognized as one of the best in the world.

☑ READING PROGRESS CHECK

Describing Describe one element of Brazil's culture. Explain why that element of culture is important to Brazilians.

Contemporary Brazil

GUIDING QUESTION *What challenges does Brazil face?*

Brazil has the world's seventh-largest economy. It ranks among the leaders in mining, manufacturing, and agriculture. These activities have produced great wealth for some people and a growing middle class. However, only 10 percent of Brazilians receive about half the country's income, while the bottom 40 percent receive only 10 percent of the total income. At the same time, 1 in 10 Brazilians is forced to live on less than $2 a day. About 1 in 5 workers is employed in agriculture, mainly on large farms and ranches owned by corporations or wealthy Brazilians.

Brazil is a member of several organizations designed to promote free trade. MERCOSUR, established in 1991, is South America's leading trading bloc. In 2008 the leaders of 12 South American nations created the Union of South American Nations (UNASUR).

Education and Earning a Living

Education is an important key to success in Brazil. College graduates earn twice as much as high school graduates do, and high school graduates earn four times as much as those with little or no schooling.

Soccer ("football") players scramble for the ball during a match at a Rio de Janeiro stadium.

▶ **CRITICAL THINKING**

Describing How important is football to Brazilians?

Boys read in front of the class at a public school in Brazil's Amazon area.

▶ **CRITICAL THINKING**

Describing How well educated are most Brazilians?

School is free up to age 17. Yet 60 percent of Brazilians have only four years of schooling or less. These people have a hard life. They work long hours for low pay. In 2011 the government launched "Brazil Without Poverty," a program aimed at raising the standard of living and improving access to education and health care.

Seeking to create a skilled workforce, Brazil's government is trying to improve education at all levels. It has increased funds to build better primary and secondary schools. At the university level, Brazil has introduced the "Science Without Borders" program, which aims to send thousands of students to universities abroad, including to colleges in the United States.

Connections and Challenges

Improving citizens' quality of life is just one of the challenges facing Brazil. The government is sponsoring a program to colonize the country's sparsely populated interior. Several highways have been built across the country. The most important is the Transamazonica Highway, from the coastal city of Recife to the border with Peru. To relieve poverty and overcrowding, poor rural Brazilians have been offered free land in the Amazon if they will develop it. Thousands have followed new roads into the Amazon Basin to take advantage of this offer.

Brazilians also have worked to develop the energy resources the country needs for continued economic development. Large power

plants along several major rivers use water power to produce most of Brazil's electricity. In the 1970s, the high cost of oil caused the government to develop a program that substitutes ethanol, a fuel made from sugarcane, for gasoline. Recent discoveries of oil and natural gas off Brazil's coast provide the country with the energy it needs.

Environmental Concerns

Programs to develop Brazil's interior have resulted in great concern for the future of the Amazon rain forest. Logging has long been a problem, as trees are cut down to sell as wood. The Transamazonica Highway and other new roads have increased this destruction by making it easier to get into the rain forest and to get the logs out.

The farmers, ranchers, miners, and other settlers the roads have brought into the region have become cause for even greater concern. About 15 percent of the rain forest is already gone, and the rate of its destruction has attracted worldwide attention.

It is easy to think that good soils must lie underneath tropical rain forests. However, this is often not true. The heat and moisture of the area keep the nutrients in the biosphere, that is, in the living organisms, particularly the plants. As a result, the soil is poor. When the forest is cleared for farming, the soil cannot support crops.

A highway cuts through Brazil's Amazon rain forest.

▶ **CRITICAL THINKING**

Explaining How do new roads benefit and harm Brazil's development, especially in rain forest areas?

Include this lesson's information in your Foldable®.

✔ **READING PROGRESS CHECK**

Identifying What are reasons for allowing development in the rain forest?

©Paulo Fridman/Sygma/Corbis

LESSON 3 REVIEW **CCSS**

Reviewing Vocabulary

1. What is Brazil doing to develop some of its *hinterlands*?

Answering the Guiding Questions

2. *Identifying* In what parts of Brazil do most of its population live?

3. *Determining Central Ideas* How has Brazil's African heritage affected its culture today?

4. *Analyzing* How do education issues contribute to economic inequalities in Brazil?

5. *Argument Writing* Choose one challenge Brazil faces today and write a short essay suggesting how to solve it.

Rain Forest Resources

Many medicines that we use today come from plants found in rain forests. From these plants, we derive medicines to treat or cure diabetes, heart conditions, glaucoma, and many other illnesses and physical problems.

Largest Rain Forests The world's largest rain forests are located in the Amazon Basin in South America, the Congo Basin in Africa, and the Indonesian Archipelago in Southeast Asia. The Amazon rain forest makes up more than half of Earth's remaining rain forest.

The Planet's Lungs Rain forests are often called the "lungs of the planet" for their contribution in producing oxygen, which all animals need for survival. Rain forests also provide a home for many people, animals, and plants. Rain forests are an important source of medicine and foods.

> **"Every year, less and less of the rain forest remains. Human activity is the main cause of this deforestation."**

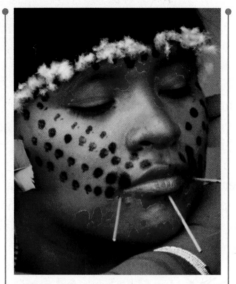

The Yanomami People
An ancient indigenous people, the Yanomami live in the Amazon rain forest regions of Brazil and Venezuela. For many years, the Yanomami lived in isolation. They rely on their environment for their food, shelter, and medicine.

Deforestation Every year, less and less of the rain forest remains. Human activity is the main cause of this deforestation. Humans cut rain forests for grazing land, agriculture, wood, and the land's minerals. Deforestation harms the native peoples who rely on the rain forest. The loss of rain forests also has an extreme impact on the environment because the rich biological diversity of the rain forest is lost as the trees are cut down.

Preserving Rain Forests More and more people realize that keeping the rain forests intact is critical. Groups plant trees on deforested land in the hope that forests will eventually recover. More companies are operating in ways that minimize damage to rain forests.

More Research Thirty years ago, very little research on the medicines of the rain forest was being done. Today, many drug companies and several branches of the U.S. government, including the National Cancer Institute, are taking part in research projects to find medicines and cures for viruses, infections, cancer, and AIDS.

Ashaninka children are at play in the rain ▶ forest. The Ashaninka comprise one of the largest indigenous groups in South America.

THERE'S MORE ONLINE

HEAR why the rain forest is important • **SEE** the loss of the rain forest • **WATCH** plants become medicine

These numbers and statistics can help you learn about the resources of the rain forest.

1.4 Billion Acres

The Amazon rain forest covers 1.4 billion acres (2,187,500 sq km). If the rain forest were a nation, it would be the 13th-largest country in the world.

OVER SEVEN PERCENT

Tropical rain forests make up about 7 percent of the world's total landmass. But found within the rain forest are half of all known varieties of plants.

40 Years

In 1950 rain forests covered about 14 percent of Earth's land. Rain forests cover about 7 percent today. Scientists estimate that, at the present rate, all rain forests could disappear from Earth within 40 years.

80%

About 80 percent of the diets of developed nations of the world originated in tropical rain forests. Included are such fruits as oranges and bananas; corn, potatoes, and other vegetables; and nuts and spices.

120

Today, 120 prescription drugs sold worldwide are derived from rain forest plants. About 65 percent of all cancer-fighting medicines also come from rain forest plants. An anticancer drug derived from a special kind of periwinkle plant has greatly increased the survival rate for children with leukemia.

50,000 Square Miles

When rain forests are cleared for land, animal and plant life disappears. Almost half of Earth's original tropical forests have been lost. Every year, about 32 million acres—50,000 square miles (129,499 sq. km)—of tropical forest are destroyed. That's roughly the area of Nicaragua or the state of Alabama.

EIGHTY PERCENT

For centuries, people who live in rain forests have used the plants and trees to meet their health needs. The World Health Organization (WHO) estimates that about 80 percent of the indigenous peoples still rely on traditional medicine.

ONE PERCENT

Although ingredients for many medicines come from rain forest plants, less than 1 percent of plants growing in rain forests have been tested by scientists for medicinal purposes.

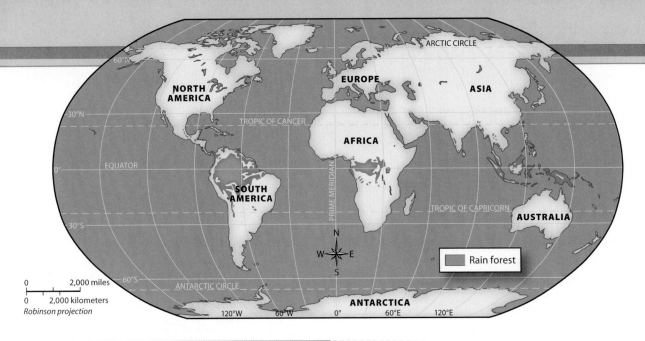

The map shows rain forest regions of the world with labels: ARCTIC CIRCLE, NORTH AMERICA, EUROPE, ASIA, 60°N, 30°N, TROPIC OF CANCER, AFRICA, EQUATOR, SOUTH AMERICA, PRIME MERIDIAN, 30°S, TROPIC OF CAPRICORN, AUSTRALIA, 60°S, ANTARCTIC CIRCLE, ANTARCTICA, 120°W, 60°W, 0°, 60°E, 120°E

0 — 2,000 miles
0 — 2,000 kilometers
Robinson projection

Legend: Rain forest

GLOBAL IMPACT

THE WORLD'S RAIN FORESTS Rain forests are located in a belt around Earth near the Equator. Abundant rain, relatively constant temperatures, and strong sunlight year-round are ideal conditions for the plants and animals of the rain forest.

Rain forests cover only a small part of Earth's surface. The Amazon Basin in South America is the world's largest rain forest area.

Rain Forest Research

Laboratories provide a research base for scientists to conduct environmental research. This laboratory in Mumbai attracts rain forest scientists from around the world.

Thinking Like a
Geographer

1. *Environment and Society* Why do you think scientists only know about a small fraction of potential medicines from the rain forest?

2. *Environment and Society* How do you think native doctors in the Amazon rain forest discovered medical uses for plants?

3. *Human Geography* List two reasons to explain why some people support saving rain forests. List two reasons to explain why some people support cutting down rain forests. Write a paragraph to state which position you support. Include facts to support your position.

Directions: Write your answers on a separate piece of paper.

1 Use your FOLDABLES to explore the Essential Question.

INFORMATIVE/EXPLANATORY Write an essay explaining how the Brazilians' conversion of rain forest land to farmland may affect the environment of the rest of the world.

2 21st Century Skills

IDENTIFYING POINT OF VIEW Given what you have learned about the benefits of rain forests, do you think Brazil has an obligation to maintain what remains of them? Write two or three paragraphs to explain your viewpoint.

3 Thinking Like a Geographer

DESCRIBING On a graphic organizer, note important differences between Brazilians who live in the major cities and those who do not. Add other categories you think are important in describing the differences.

	City dwellers	Country dwellers
Wealth		
Housing		
Work		

4 GEOGRAPHY ACTIVITY

Locating Places

Match the letters on the map with the numbered places below.

1. Brasília

2. Atlantic Ocean

3. Amazon River

4. São Paulo

5. Recife

6. Pacific Ocean

7. São Francisco River

8. Amazon Basin

REVIEW THE GUIDING QUESTIONS

Directions: Choose the best answer for each question.

1 Brazil is the world's largest exporter of

    A. beef.

    B. clocks.

    C. peanut butter.

    D. tropical plants.

2 Most of Brazil's population lives in

    F. the rain forest.

    G. the coastal lowlands.

    H. the Amazon Basin.

    I. northeastern Brazil.

3 The first Portuguese explorer to lay claim to Brazil was

    A. Ferdinand Magellan.

    B. a Jesuit priest.

    C. Getúlio Vargas.

    D. Pedro Cabral.

4 When Napoleon invaded Portugal in 1807 and the Portuguese royal family and government leaders fled to Brazil, which city became the new capital of the Portuguese Empire?

    F. São Paulo

    G. Campinas

    H. Rio de Janeiro

    I. Brasília

5 Brazil's largest metropolitan area, or city and surrounding suburbs, is

    A. Brasília.

    B. Buenos Aires.

    C. São Paulo.

    D. Rio de Janeiro.

6 The official language of Brazil is

    F. Spanish.

    G. Brazilian.

    H. Portuguese.

    I. English.

DBQ ANALYZING DOCUMENTS

❼ IDENTIFYING POINT OF VIEW Read the following news report:

"*Brazilian farmers meanwhile have been demanding the country's Congress ease environmental laws in the Amazon region. They support a bill that would let them clear half the land on their properties in environmentally sensitive areas. Current law allows farmers to clear just 20 percent of their land in the Amazon zone.*"

—from Marco Sibaja, "Amazon Deforestation in Brazil Increases"

Why do Brazilian farmers want to be able to clear more land?

A. They oppose any environmental protection laws.

B. They want more land for crops to increase their profits.

C. They plan to sell the cleared land for new housing developments.

D. They hope to drive Native Americans from the land.

❽ ANALYZING What is likely to happen if the bill the farmers support becomes law?

F. Brazil's economy will suffer from too much emphasis on agriculture.

G. Agriculture in Brazil will decline because the land is unproductive.

H. Deforestation in the Amazon will increase at a faster rate.

I. Environmentalists will stop fighting over deforestation.

SHORT RESPONSE

"*Exploiting vast natural resources and a large labor pool, [Brazil] is today South America's leading economic power. . . . Highly unequal income distribution and crime remain pressing problems.*"

—from *CIA World Factbook*

❾ ANALYZING How has Brazil's economy benefited from the nation's large size?

❿ IDENTIFYING Why is unequal income distribution in Brazil a problem?

EXTENDED RESPONSE

⓫ INFORMATIVE/EXPLANATORY WRITING Compare and contrast the cultures of the United States and Brazil, including political, economic, and social factors.

Need Extra Help?

If You've Missed Question	❶	❷	❸	❹	❺	❻	❼	❽	❾	❿	⓫
Review Lesson	1	1	2	2	1	3	3	3	1	3	3

THE TROPICAL NORTH

ESSENTIAL QUESTIONS • *How does geography influence the way people live?*
• *Why does conflict develop?* • *What makes a culture unique?*

Farmer Juan Lucas harvests palms in Ecuador's tropical coastal lowlands.

©Pablo Corral V/Corbis

The Story Matters...

The countries of the Tropical North are home to some of the most ethnically diverse populations in the world. Native Americans, Europeans, Africans, and Chinese are among those who live in this subregion of South America. It is also home to some of the most diverse environments in the world. Landscapes include jungles, towering mountain ranges, broad river plains, plunging waterfalls, and an archipelago renowned for its unique animal life.

FOLDABLES®
Study Organizer

Go to the
Foldables®
library in the
back of your
book to make
a Foldable® that will help you take
notes while reading this chapter.

Trade
Foreign Influences
and Resources
Geography
The Tropical North

Ecuador, Colombia, Venezuela, Guyana, Suriname, and French Guiana are the lands that make up South America's Tropical North.

Step Into the Place

MAP FOCUS Use the map to answer the following questions.

1 **THE GEOGRAPHER'S WORLD** Which country in the Tropical North is connected to Central America?

2 **THE GEOGRAPHER'S WORLD** In which direction would you go if you were traveling from French Guiana to Suriname?

3 **PLACES AND REGIONS** Why do you think the Galápagos Islands belong to Ecuador?

4 **CRITICAL THINKING** **DESCRIBING** Use the map to help you describe how Guyana and Suriname are similar geographically.

RIVER TRAVEL Indigenous peoples, such as the Makushi, live in small villages on the banks of rivers that wind their way through Guyana's rain forests.

URBAN CENTER With about 4 million people, Caracas is the capital and largest city of Venezuela.

Step Into the Time

DESCRIBING Choose one event from the time line and write a paragraph describing the social, political, or environmental effect that event had on the region and the world.

1821 Simón Bolívar frees Venezuela from Spanish rule

1667 The Netherlands acquires Suriname from Britain

1800

1835 English naturalist Charles Darwin arrives on Galápagos Islands

The Tropical North

TROPIC OF CANCER

20°N

ATLANTIC OCEAN

Caribbean Sea

GUATEMALA
HONDURAS
EL SALVADOR
NICARAGUA
COSTA RICA
PANAMA

ANTIGUA AND BARBUDA
ST. KITTS AND NEVIS
DOMINICA
ST. LUCIA
BARBADOS
GRENADA
ST. VINCENT AND THE GRENADINES
TRINIDAD AND TOBAGO

10°N

Barranquilla
B
Maracaibo
Valencia
Caracas
Maracay

Lake Maracaibo

Orinoco R.

Georgetown
Paramaribo
Cayenne
FRENCH GUIANA (Fr.)

Gulf of Panama

Atrato R.
Cauca R.
Magdalena R.

Medellín

Bogotá

Cali

VENEZUELA

GUYANA

SURINAME

Galápagos Islands (Ecuador)

EQUATOR

COLOMBIA

A

BRAZIL

0°

Quito
ECUADOR
Guayaquil — *Guayas R.*

⊛ National capital
○ Department capital
● City

PACIFIC OCEAN

PERU

N
W — E
S

0   400 miles
0   400 kilometers

10°S

Lambert Azimuthal Equal-Area projection

BOLIVIA

90°W   80°W   70°W

1935
Ecuador declares part of Galápagos Islands a wildlife sanctuary

1998
Hugo Chávez is elected president of Venezuela

1966 Guyana gains independence from Britain

1900

2000

1978 UNESCO adds the Galápagos Islands to the World Heritage List

1990s Ecuadoran Indians protest for rights

networks

There's More Online!

☑ **GRAPHIC ORGANIZER**

☑ **MAP** Tropical North

☑ **SLIDE SHOW** Emeralds

☑ **VIDEO**

(l to r) Fabio Filzi/Vetta/Getty Images; Tips Images/Tips Italia Srl a socio unico/Alamy; ©Last Refugee/Robert Harding World Imagery/Corbis

Reading **HELP**DESK (CCSS)

Academic Vocabulary

- **exceed**
- **despite**

Content Vocabulary

- **elevation**
- **trade winds**
- **cash crop**

TAKING NOTES: *Key Ideas and Details*

Identify As you read the lesson, use a graphic organizer like this one to record the important resources of each of these countries.

Country	Resources
Ecuador	
Colombia	
Venezuela	
Guyana	
Suriname	

Lesson 1
Physical Geography of the Region

ESSENTIAL QUESTION · *How does geography influence the way people live?*

IT MATTERS BECAUSE

The land and waters of the Tropical North provide oil, bauxite, and emeralds, along with shrimp and other food products that people and industries in the United States and around the world need or want.

Landforms and Waterways

GUIDING QUESTION *What are the major physical features of the Tropical North?*

South America's Tropical North consists of five countries and a colony. From west to east, they are Ecuador, Colombia, Venezuela, Guyana, Suriname, and French Guiana.

Colombia is the Tropical North's largest country, and Venezuela is the second largest. Each is more than twice the size of California. Ecuador and Guyana, the third and fourth largest, are about the size of Colorado and Kansas, respectively. Suriname is about the size of Washington State; French Guiana, the smallest, is the size of Maine. Together, the countries of the Tropical North total only about one-third the size of nearby Brazil.

Landforms of the Tropical North

Ecuador, Colombia, and Venezuela have the region's most diverse physical geography. The Andes mountain ranges, which extend the length of western South America, run through each country. Some of the peaks have **elevations**, or height above the level of the sea, that **exceed** 18,000 feet (5,486 m)—almost 3.5 miles (5.6 km) high. Many peaks are covered with snow year-round. About 40 peaks are volcanoes.

Cotopaxi in Ecuador, at 19,347 feet (5,897 m), is the world's highest active volcano. In Colombia, the Sierra Nevada de Santa Marta mountains along the Caribbean coast are the world's highest coastal range.

Colombia is the only country in South America with coastlines on both the Pacific Ocean and the Caribbean Sea. The mountains make travel between the coasts difficult. So does the Darién, a wilderness region of deep ravines, swamps, and dense rain forest along Colombia's border with Panama.

West of the Andes, Colombia and Ecuador have narrow lowlands that border their Pacific coasts. East of the mountains, more lowlands extend into Peru, Brazil, and Venezuela. The southern half of the lowlands is part of the Amazon Basin. The northern half is a grassy plain called the Llanos. This plain also covers most of northern Venezuela.

Southern Venezuela contains a heavily forested region of rolling hills, low mountains, and plateaus called the Guiana Highlands. Along the border with Brazil, groups of forest-covered mesas called *tepuis* rise to heights of 9,000 feet (2,743 m) in places. The Guiana Highlands extend east into Guyana, Suriname, and French Guiana. Rain forest covers most of this region except for a narrow band of low and sometimes swampy plains along the Atlantic coast.

Abundant Waterways

Rivers flow across much of northern South America. The 1,300-mile-long (2,092 km) Orinoco River is the continent's third-longest river. Its more than 400 tributaries form the north's largest river system. The Orinoco crosses Venezuela in a giant arc, dropping from the Guiana Highlands through the Llanos to the Atlantic Ocean. One of its tributaries flows over Angel Falls, the world's highest waterfall. Angel Falls is more than 20 times higher than Niagara Falls. From the top of a *tepui,* the water plunges more than a half-mile to the fall's base.

Academic Vocabulary

exceed to be greater than; to go beyond a limit

Visual Vocabulary

Mesa A mesa is a small, elevated area of land that has a flat top and sides that are usually steep cliffs.

A shepherd tends alpaca below the western slope of the Cotopaxi volcano in Ecuador.
► **CRITICAL THINKING**
Analyzing Why is there snow on Cotopaxi even though the volcano lies close to the Equator?

Academic Vocabulary

despite in spite of

The waters of Venezuela's Angel Falls drop from such a height that they are vaporized by the wind and turn into mist before reaching the ground.

▶ **CRITICAL THINKING**

Identifying Angel Falls is part of what major river system?

The Tropical North region has coastlines on three bodies of ocean water. Ecuador and western Colombia lie along the Pacific Ocean. Northern Colombia and Venezuela lie along the Caribbean Sea. The Atlantic Ocean washes the shores of Guyana, Suriname, and French Guiana.

Colombia's two main rivers, the Magdalena and the Cauca, flow north across Andes plateaus and valleys to the Caribbean Sea. These rivers form important routes into the country's agricultural and industrial interior. Both can be navigated by commercial ships for much of their length.

Other rivers that begin in the Andes flow west to the Pacific. Of these, Ecuador's Guayas River is the most important because it has made Guayaquil the country's largest city and a major port.

Rivers in Guyana, Suriname, and French Guiana flow north and empty into the Atlantic. Most are shallow, slow moving, and responsible for the region's swampy coastline. They are not useful for long-distance transportation into the interior.

Galápagos Islands

The Galápagos Islands lie in the Pacific, about 600 miles (966 km) west of Ecuador. They consist of 13 major islands, six smaller ones, and many tiny islands called islets. These rocky islands, which were formed by underwater volcanoes, are owned by Ecuador. Most have no human population.

The islands' isolation makes them home to many unusual animals, such as lizards that swim and birds with wings although they do not fly. In the 1800s, British scientist Charles Darwin studied the islands' animals to develop his theory of evolution. Today, the islands are tourist attractions. Many are protected as national parks.

☑ **READING PROGRESS CHECK**

Analyzing How do Colombia's rivers help the nation's economy?

Climates

GUIDING QUESTION *How and why do climates vary in the Tropical North?*

South America's Tropical North lies along the Equator. **Despite** its location, the region has a variety of climates. Many of the variations result from differences in elevation and location, and from the influence of ocean currents and winds.

Tropical Climates

The region's coasts, interior lowlands, plains, and highlands all have some type of tropical climate. This means warm temperatures throughout the year.

Much of the coastal and eastern lowlands of Ecuador and Colombia have a tropical monsoon climate, with a short, dry season and a long, wet season of heavy rainfall. In Colombia's coastal Chocó region, which includes the rugged Darién, it rains more than 300 days per year. This produces more than 400 inches (1,016 cm)—about 33 feet (10 m)—of rainfall each year, making it one of the wettest places on Earth.

The Llanos of Colombia and Venezuela have a tropical wet-dry climate, with an annual rainfall of 40 inches to 70 inches (102 cm to 178 cm). Most rain falls between May and October. Average daily temperatures are above 75°F (24°C) throughout the year. The Guiana Highlands have a tropical monsoon climate in some places. In other areas, a tropical rain forest climate (which has no dry season) is normal.

Guyana, Suriname, and French Guiana have the same climate as Venezuela's highlands. Yearly rainfall ranges from 70 inches to 150 inches (178 cm to 381 cm). Their coasts are not as hot as might be expected because of the **trade winds**, steady winds that blow from higher latitudes toward the Equator. The Caribbean coast of Venezuela and Colombia is also cooler. It has a semiarid climate, receiving less than 20 inches (51 cm) of rain per year.

Cooler Highlands

Mountain climates depend on elevation. From 3,000 to 6,500 feet (914 m to 1,981 m) is the *tierra templada,* or "temperate land." This zone has moderate rainfall and temperatures with daily averages between 65°F (18°C) and 75°F (24°C). Next is the *tierra fria,* or "cold land," reaching to about 10,000 feet (3,048 m). A colder zone called the *páramo* begins at about 10,000 feet (3,048 m); daily average temperatures in this zone are below 50°F (10°C). Wind, fog, and light drizzle are common in this zone. Vegetation is mainly grasses and hardy shrubs. Above 15,000 feet (4,572 m), the ground is permanently covered with snow and ice.

☑ READING PROGRESS CHECK

Identifying How do the climates of the Pacific coast, the Atlantic coast, and the Caribbean coast differ?

Natural Resources

GUIDING QUESTION *Which natural resources are most important to the economies of the Tropical North's countries?*

Tropical rain forests cover much of the North, but lack of roads and the region's physical geography have made it difficult for any of its countries to exploit this natural resource. The North's largest countries, Venezuela and Colombia, are its richest and most diverse in other resources, as well.

Think Again?

Angel Falls was named after a pilot.

True. The falls are called Salto Ángel in Spanish. They are named for Jimmie Angel, a Missouri-born pilot who was the first person to fly over the falls in a plane in 1933. In 2009, however, Venezuelan president Hugo Chávez declared that the falls should be known as Kerepakpai Merú, which means "waterfall of the deepest place" in the language of the local Pemón people. He believed that Venezuela's most famous natural wonder should have an indigenous name. At the time of Chávez's death in early 2013, the name of the falls remained in dispute.

EMERALD MINING
Emeralds are one of the most valued gems in the world.

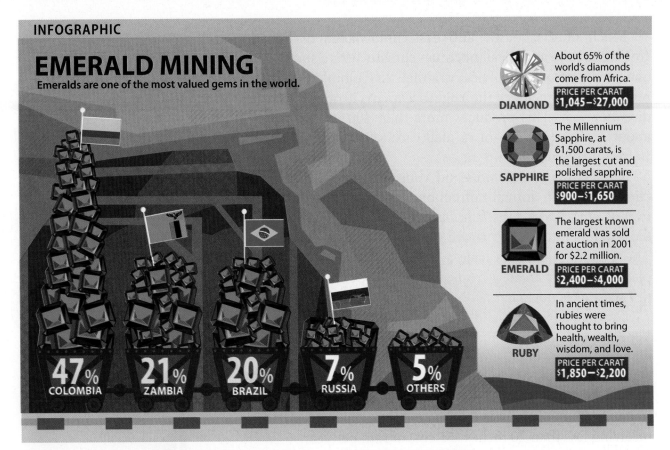

47% COLOMBIA
21% ZAMBIA
20% BRAZIL
7% RUSSIA
5% OTHERS

DIAMOND About 65% of the world's diamonds come from Africa.
PRICE PER CARAT
$1,045–$27,000

SAPPHIRE The Millennium Sapphire, at 61,500 carats, is the largest cut and polished sapphire.
PRICE PER CARAT
$900–$1,650

EMERALD The largest known emerald was sold at auction in 2001 for $2.2 million.
PRICE PER CARAT
$2,400–$4,000

RUBY In ancient times, rubies were thought to bring health, wealth, wisdom, and love.
PRICE PER CARAT
$1,850–$2,200

Emeralds were first mined in South America by indigenous peoples centuries ago. Later, the Spanish mined emeralds and shipped them to Europe as part of the valuable treasure from their American empire.

▶ **CRITICAL THINKING**

Analyzing Why do you think emeralds, gold, and diamonds are considered to be valuable?

Fossil Fuels

Oil is found across much of the Tropical North. Venezuela is South America's top producer of oil and ranks eleventh in the world. Some of the world's largest known reserves are in the Llanos, at the mouth of the Orinoco River, and offshore in the Caribbean. Large amounts also exist around Lake Maracaibo, South America's largest lake, along the country's northwestern coast. Venezuela has some of the world's largest natural gas deposits and is South America's second-largest coal producer. Most of the coal lies along the country's southwestern border with Colombia.

Colombia is South America's largest coal producer, with major deposits in its lowlands. It is also South America's third-largest oil producer (Brazil is second), with deposits in the Amazon lowlands, the Llanos, and the Magdalena River valley.

Ecuador produces less oil than Colombia, but it accounts for 40 percent of Ecuador's exports. It is piped over the Andes from oil fields in the east. Suriname and Guyana also have oil resources, but they do not produce enough to even meet their needs.

Minerals and Gems

Gold is found throughout the Tropical North. The largest deposits are in Colombia's mountains, eastern Ecuador, and Venezuela's

Guiana Highlands. In Ecuador, thousands of miners live in remote jungle regions and do dangerous work in tunnels that sometimes collapse in heavy rains.

Diamonds are mined from Colombia to Suriname, but Colombia is better known for high-quality emeralds and is the world's leading emerald producer. Guyana is one of the world's largest producers of bauxite, a mineral used to make aluminum. Venezuela and Suriname also have major bauxite deposits. In addition, the four countries have important deposits of copper, iron ore, and other minerals. Except for gold, Ecuador's mineral resources are limited, and French Guiana has no important mining industries.

Agriculture and Fishing

The differing elevations and climates in Ecuador and Colombia allow farmers to grow a variety of crops. Both countries export bananas from their tropical lowlands and coffee from the *tierra templada*. Ecuador's agriculture, however, is not well developed. The amount of farmland is limited, and most rural Ecuadorans grow only enough to feed their families. Corn, potatoes, beans, and cassava are common crops in both countries. Colombia produces rice, wheat, sugarcane, and cattle for sale, as well as cotton for the country's large textile industry.

Coffee is Venezuela's main **cash crop**, a product raised mainly for sale. Venezuela's main food crops are corn and rice. Most farming takes place in the northwest, and most ranching happens on the Llanos. Only about 10 percent of Venezuelans are farmers or ranchers, and much the same is true of Venezuela's neighbors to the east. Guyana, Suriname, and French Guiana have little farming because much of the land is covered by rain forest. Any farming takes place mainly along the coast.

Fishing is not a major economic activity in the Tropical North, which is unusual for countries that border the sea. The region's people do not eat much fish. The major catch of its small fishing industry is shrimp, most of which is exported.

☑ **READING PROGRESS CHECK**

Identifying Which fossil fuel, mineral, and gem are most widespread in the Tropical North?

Include this lesson's information in your Foldable®.

LESSON 1 REVIEW **CCSS**

Reviewing Vocabulary

1. What *cash crops* are important to the economy of the Tropical North?

Answering the Guiding Questions

2. *Analyzing* Why are Venezuela's Orinoco and Colombia's Magdalena rivers so important?

3. *Identifying* How are climate and elevation related in the Tropical North?

4. *Analyzing* Why is agriculture more important in Colombia than elsewhere in the Tropical North?

5. *Informative/Explanatory Writing* Which of the Tropical North's countries would you most like to visit? Write a paragraph to explain why.

Reading **HELP**DESK

Academic Vocabulary

- **conflict**
- **stable**

Content Vocabulary

- *immunity*
- *encomienda*
- **hacienda**

TAKING NOTES: *Key Ideas and Details*

Analyze As you read the lesson, write summary sentences about five important events in the history of the Tropical North on a graphic organizer like the one below.

Important Events
• Native Americans settle in villages along the region's coast.
•
•
•
•

Lesson 2
History of the Countries

ESSENTIAL QUESTION · *Why does conflict develop?*

IT MATTERS BECAUSE
The countries of the Tropical North export products that are sought after and highly valued by the rest of the world.

Early History and Colonization

GUIDING QUESTION *How did Europeans colonize the Tropical North?*

The Tropical North's indigenous peoples lived there for thousands of years before encountering Spanish explorers. These explorers invaded the region in the early 1500s. Less than 50 years later, the Spanish had conquered and colonized most of the region.

Early Peoples of the Tropical North

The Native Americans of the Tropical North included Carib, Arawak, and other hunter-gatherer peoples. They settled in villages along the Caribbean and Atlantic coasts.

To the west, the Cara and other peoples built fishing villages along the Pacific coast. Over time, groups like the Chibcha and Quitu moved inland to mountain valleys in the Andes. There they created advanced societies that farmed, made cloth from cotton and ornaments of gold, and traded with the Inca, an advanced civilization that developed to the south. In the late 1400s, some of the groups were conquered by the Inca and became part of the Inca Empire.

Arrival of the Europeans

In the early 1500s, Spanish adventurers landed on the Caribbean and Atlantic coasts, seeking gold and enslaving native peoples. When they met resistance and found no gold, they lost interest. The first Spanish settlements did not appear

(l to r) DEA/M. SEEMULLER/De Agostini Picture Library/Getty Images; ©Bettmann/Corbis; Danita Delimont/Gallo/Getty Images

on the Caribbean coast—in Venezuela and Colombia—until 1523 and 1525. The Spanish made no effort to colonize east of Venezuela.

On the Pacific coast, the Spanish conquered the Inca in 1530 and seized their silver and gold. Driven by hunger for more wealth, they invaded Ecuador in 1534. By the mid-1500s, the conquest of the area that is now Ecuador, Colombia, and Venezuela was complete.

Spanish Colonies

To control their new colonies, the Spanish set up governments. Bogotá, which the Spanish founded in 1538, became the capital of Colombia in 1549. The Spanish placed Ecuador's government at the native town of Quito in 1563. Caracas, which the Spanish founded in 1567, eventually became Venezuela's capital. The Spanish located these cities where Native Americans already had settlements. Most were located inland, in the higher elevations where climates are milder than on the tropical coasts. For many years, Venezuela was ruled from Peru. In the 1700s, Spain placed Venezuela, Ecuador, and Colombia under a single government located at Bogotá.

Native American peoples suffered greatly under the Spanish. As in Brazil, thousands died from European diseases to which they had no natural **immunity**, or protection against illness. Others found themselves forced to work for the Spanish under a system called ***encomienda***. This system allowed Spanish colonists to demand labor from the Native Americans who lived in a certain area.

The *encomienda* provided workers for Spanish mines and for the large estates, called **haciendas,** that developed in some rural areas. Native Americans in remote regions, such as Venezuela's Llanos and the rain forests of eastern Ecuador and Colombia, came under the control of Roman Catholic missionaries who were trying to convert them to Christianity.

Under Spanish rule, the Native American village of Teusaquillo became known as Bogotá. Life in Spanish Bogotá focused on the main plaza, or square, surrounded by a cathedral and government buildings.

▶ **CRITICAL THINKING**
Describing What role did location play in the selection of Bogotá and other cities as centers of government in northern South America?

Most haciendas became plantations that grew coffee, tobacco, sugarcane, or other cash crops. Others, mostly on the Llanos, were cattle ranches. As the hacienda system grew, the Spanish brought in thousands of enslaved Africans to provide more labor. African slavery was most common in Venezuela.

European Colonization

The French, British, and Dutch fought over and colonized Guyana, Suriname, and French Guiana. The British and the Dutch established sugar plantations and brought the first enslaved Africans to the area. Control of these colonies changed hands several times in the 1600s and 1700s. Eventually, what is now Guyana became British Guiana. Suriname was called Dutch Guiana, and French Guiana became a colony of France.

☑ **READING PROGRESS CHECK**

Identifying Which European nations founded colonies in the Tropical North, and which countries did each nation colonize?

Simón Bolívar, also known as "the Liberator," led the movement that won freedom for several countries in the Americas.

Independence

GUIDING QUESTION *How did Spain's colonies become independent countries?*

By the late 1700s, many Spanish colonists who were born in the Americas wanted independence from their Spanish rulers. Their chance came in 1808, when the French ruler Napoleon invaded and conquered Spain. Spain found it difficult to fight the French in Europe and to rule its colonies. Some of the colonists in the Americas took this opportunity to fight for independence from Spain.

Overthrow of Colonial Rule

Ecuadorans rose up against Spanish rule in 1809. Colombians and Venezuelans soon followed. A long war began, at first mainly between groups who remained loyal to Spain and those who favored independence. After the Spanish expelled the French from Spain in 1814, Spain's king sent troops to South America to try to restore Spanish control. In the south, resistance to the Spanish was led by Argentine general José de San Martín. In the north, Venezuela's Simón Bolívar led the revolt.

Spanish forces were not finally defeated until 1823. In 1819, however, Bolívar united Venezuela, Colombia, Panama, and Ecuador to form an independent republic called Gran Colombia. He became its first president.

Independent Countries

Gran Colombia broke apart after Bolívar's death in 1830. Ecuador and Venezuela formed independent countries. Colombia and Panama remained united as one country. In the early 1900s, Panama separated from Colombia and became independent.

Independence and self-government did not bring democracy and peace. Wealthy landholders competed with wealthy city businesspeople for control of the government. **Conflict** over the Catholic Church's role in society added to the unrest. The tensions resulted in civil wars in Colombia and Venezuela. Throughout the history of Ecuador, Colombia, and Venezuela, military or civilian leaders often ruled as dictators.

Labor and Immigration

While Ecuador, Colombia, and Venezuela struggled with self-government, British, Dutch, and French Guiana remained colonies. The British abolished slavery in their colony in 1838. The French and the Dutch followed in 1863.

To replace the once-enslaved workers, British and Dutch plantation owners recruited laborers from India and China. The Dutch also imported workers from their colony in Indonesia. The immigrants had to work on their colony's sugar, rice, coffee, or cacao plantations for a required length of time. At the end of their contract, they were free. Many stayed in the colony and, like the formerly enslaved people they replaced, founded towns along the coast.

In 1852 France began sending convicted criminals to its colony. More than 70,000 convict laborers arrived between 1852 and 1939. The worst convicts were imprisoned off the coast on notorious Devil's Island.

✔ **READING PROGRESS CHECK**

Determining Central Ideas How did British, Dutch, and French colonists find workers after slavery ended in their colony?

British and Dutch colonial rulers recruited foreign workers from India, China, and other parts of Asia to harvest various tropical crops in their South American colonies. The prison on Devil's Island (above) housed convict laborers.

▶ **CRITICAL THINKING**

Describing How did the arrival of foreign workers in British Guiana and Dutch Guiana change these territories in a way that made them different from other parts of South America?

Academic Vocabulary

conflict a serious disagreement

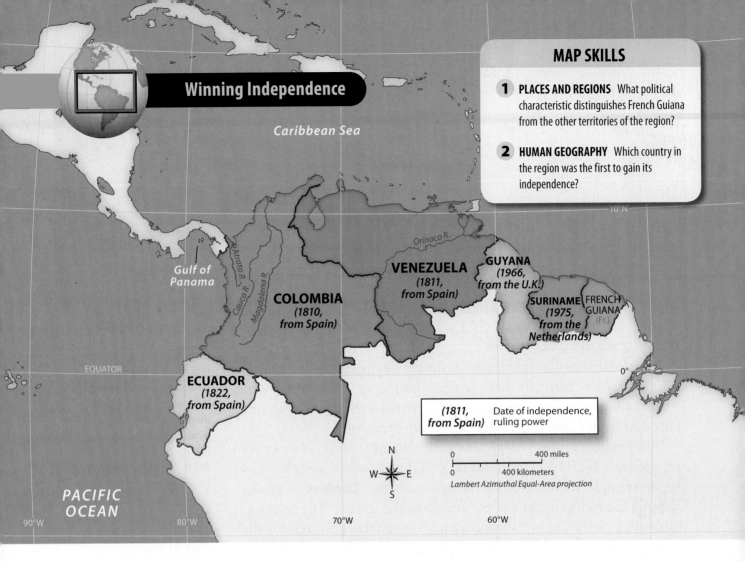

Caribbean Sea

MAP SKILLS

1 **PLACES AND REGIONS** What political characteristic distinguishes French Guiana from the other territories of the region?

2 **HUMAN GEOGRAPHY** Which country in the region was the first to gain its independence?

10°N

Orinoco R.

VENEZUELA
(1811, from Spain)

GUYANA
(1966, from the U.K.)

Gulf of Panama

Atrato R.

Cauca R.

Magdalena R.

COLOMBIA
(1810, from Spain)

SURINAME
(1975, from the Netherlands)

FRENCH GUIANA *(Fr.)*

EQUATOR

0°

ECUADOR
(1822, from Spain)

(1811, from Spain) Date of independence, ruling power

N
W E
S

0   400 miles
0   400 kilometers
Lambert Azimuthal Equal-Area projection

PACIFIC OCEAN

90°W   80°W   70°W   60°W

Challenges and Change

GUIDING QUESTION *What challenges do the countries of the Tropical North face?*

The political and social problems that plagued Ecuador, Colombia, and Venezuela after independence continued through most of the twentieth century. Venezuela, for example, did not achieve a peaceful transfer of power between opposing groups until 1969. Meanwhile, the region's other countries, which gained independence in the twentieth century, experienced similar issues and challenges.

Gaining Independence

Independence came slowly for Guyana and Suriname. The British granted their colony limited self-government in 1891. In 1953 all colonists were given the right to vote and allowed to elect a legislature. Guyana finally gained independence in 1966.

Colonists in Dutch Guiana obtained the right to vote in 1948 and self-government in 1953. The colony became the independent country of Suriname in 1975.

The people of French Guiana became French citizens and gained the right to vote in 1848. In 1946 French Guiana's status changed from

a colony to an overseas department, or district, of the country of France. French Guiana remains part of France and has representatives in France's national legislature.

Revolutions and Borders

The Tropical North's lack of strong, **stable** governments has resulted in major unrest in its countries, as well as conflicts between them. In Colombia, assassinations and other violence between feuding political groups took as many as 200,000 lives between 1946 and 1964. In the 1960s and 1970s, small rebel groups began making attacks throughout the country in hopes of overthrowing the government.

Ecuador's government has not maintained control over its remote region, which lies in the Amazon Basin, to the east of the Andes. In the 1940s, Peru seized some of this land. The two countries often clashed, until a settlement was finally reached in 1968. In 2008 tensions between Ecuador and Colombia were strained after Colombian forces attacked a Colombian rebel camp in Ecuador's territory. In 2010 Colombia accused Venezuela of allowing Colombian rebels to live in its territory. War was narrowly avoided.

Guyana's independence renewed an old border dispute with Venezuela that arose when Guyana was a British colony. The dispute was not settled until 2007. Another dispute arose on Guyana's eastern border after Suriname gained independence in 1975. Several clashes took place before that boundary was settled in 2007. Guyana also experienced years of social and political unrest as its African and South Asian populations competed for power.

Like Guyana, Suriname has faced internal unrest since independence. The military removed civilian leaders in 1980 and again in 1990. Meanwhile, rebel groups of Maroons, the descendants of escaped slaves, disrupted the country's bauxite mining in an effort to overthrow the government. The army responded by killing thousands of Maroon civilians. Thousands more fled to safety in French Guiana.

 READING PROGRESS CHECK

Identifying Which of the region's nations have experienced serious internal unrest since gaining independence?

Include this lesson's information in your Foldable®.

LESSON 2 REVIEW (CCSS)

Reviewing Vocabulary
1. How were the *encomienda* and the *hacienda* related?

Answering the Guiding Questions
2. ***Analyzing*** Why were the Spanish more interested in colonizing Ecuador, Colombia, and Venezuela than Guyana, Suriname, and French Guiana?

3. ***Identifying*** How did conflicts in society lead to independence for Spain's colonies and cause unrest afterward?

4. ***Determining Central Ideas*** Why do the Tropical North's nations have a history of tense relations and internal unrest?

5. ***Argument Writing*** Write a short speech calling for or opposing independence for French Guiana. Support your view.

Reading **HELP**DESK (CCSS)

Academic Vocabulary

- **ratio**
- **migrate**

Content Vocabulary

- **Creole**
- **tariff**

TAKING NOTES: *Key Ideas and Details*

Organize As you read the lesson, use a graphic organizer to list information about life in the Tropical North.

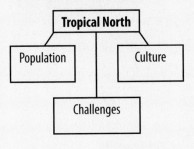

Tropical North
- Population
- Culture
- Challenges

Lesson 3
Life in the Tropical North

ESSENTIAL QUESTION · *What makes a culture unique?*

IT MATTERS BECAUSE
Many nations of the world, including the United States, have important trade relations with countries of the Tropical North.

People and Places

GUIDING QUESTION *What ethnic groups populate the Tropical North, and where do they live?*

People of European, African, Native American, and mixed descent are the major population groups of the countries that border the Pacific and Caribbean coasts. African, South Asian, and ethnically mixed peoples form the majority in the Atlantic coast countries.

Population Groups

Ecuador has the Tropical North's greatest indigenous population. About one in four Ecuadorans is Native American. If mestizos, or people of white and Native American descent, are added, the **ratio** becomes 9 of every 10 Ecuadorans.

Venezuela and Colombia have the opposite distribution. Some 20 percent of their populations are white, and 1 to 2 percent are Native American. Colombia's native population is the lowest of any Andean country. Mestizos are the largest group, accounting for more than two-thirds of Colombians and Venezuelans. The African populations of Venezuela, Ecuador, and Colombia are small, although some 15 percent of Colombians have mixed African and European ancestry.

The descendants of contract laborers from India are Suriname's largest group, making up nearly 40 percent of the population. An equal number are people of African and

mixed-African descent. A large Indonesian population is also present. Whites and Native Americans total less than 5 percent of Surinam's population.

Neighboring Guyana is home to more Native Americans; this group makes up almost 10 percent of the country's population. Ethnic Africans make up one-third of the population, and East Indians account for more than 40 percent. The country has no significant white population. About one in six Guyanese is of mixed ancestry.

People of mixed descent make up most of French Guiana's population. Small groups of French, Native Americans, Chinese, East Indians, Laotians, Vietnamese, Lebanese, Haitians, and Africans also live in French Guiana.

Where People Live

Guyana's population remains largely rural. Elsewhere in the Tropical North, most people live in cities. Bogotá, Colombia's capital on a high Andes plateau, is home to almost 5 million people. It is the north's largest city and the fifth largest in South America.

Colombia's Caribbean lowlands are home to about 20 percent of its people, mainly in Cartagena and other port cities along the coast.

John Coletti/AWL Images/Getty Images

Academic Vocabulary

ratio the relationship in amount or size between two or more things

The Iglesia de San Francisco, built by Catholic missionaries in about 1560, is the oldest restored church in Bogotá, Colombia.

▶ **CRITICAL THINKING**
Describing How important is the city of Bogotá in the Tropical North region today?

The country's Pacific coast is sparsely settled. Most of the people there are descendants of enslaved Africans who worked on plantations near the Caribbean Sea. As they were freed or they escaped, they migrated into remote areas in western Colombia. The Llanos, where cattle ranching is the main activity, is another area with few people.

Quito, Ecuador, is another mountain capital city, with nearly 2 million people. Most of Ecuador's Native Americans live in or around Quito, or they farm rural mountain valleys nearby. Most other Ecuadorans live along the coast. Guayaquil, the country's largest city and major port, is located there.

Most Venezuelans live along the coast. As in Colombia and Ecuador, Venezuelans began **migrating** to cities in the mid-1900s for the jobs and opportunities they offered. Today, more than 90 percent of the country's people live in Caracas, the capital city of 3 million, and other cities on or near the coast.

The countries of Guyana, Suriname, and the territory of French Guiana are sparsely populated. The population of the three combined totals only about half the population of Caracas. The interior of French Guiana has few roads and is largely uninhabited. In Suriname, small groups of Native Americans live in the Guiana Highlands. Nearly everyone else lives along the coast. Suriname's capital, Paramaribo, a city of 260,000, is home to more than half the country's population.

Most Guyanese also live on the coast, mainly in small farm towns. Each town's farmlands extend inland for several miles. The country's interior is home to a few groups of Native Americans and scattered mining and ranching settlements.

☑ **READING PROGRESS CHECK**

Determining Central Ideas Where do the greatest number of people in the Tropical North live?

Academic Vocabulary

migrate to move from one place to another

People and Cultures

GUIDING QUESTION *What is the Tropical North's culture like?*

Despite its largely Spanish heritage, no one culture unifies the Tropical North. Instead, its culture can be defined by the wide variety of ethnic groups that populate the region.

Language Groups

Spanish is the official language of Ecuador, Colombia, and Venezuela. There are differences in Ecuadoran Spanish because of the influence of Native American languages in each region of the country. More than 10 native languages are spoken in Ecuador. More than 25 native languages are spoken in Venezuela and some 180 in Colombia. Colombians, however, have taken great care to preserve the purity of the Spanish language.

Languages in Guyana, Suriname, and French Guiana reflect their colonial heritage as well as their ethnic populations. **Creole**, a group of languages that enslaved people from various parts of Africa developed to communicate on colonial plantations, is widely spoken. Most people in Guyana speak English. In Suriname, the official language, Dutch, is spoken only as a second language. Native American languages, Hindi, and other South Asian languages are heard in both countries.

Religion, Daily Life, and the Arts

Ecuador, Colombia, and Venezuela are overwhelmingly Roman Catholic. No more than 10 percent of the people in these countries practice other religions. The religions practiced in Guyana, French Guiana, and Suriname reflect the variety of ethnic groups that live there. Suriname's population is made up of about equal numbers of Roman Catholics, Protestants, Hindus, and Muslims. Guyana's population is largely Protestant and Hindu, with sizable Catholic and Muslim minorities. In all countries, some Native Americans practice indigenous religions.

Each country's foods, music, and other cultural elements reflect its ethnic and religious makeup. Venezuela, Colombia, and Ecuador celebrate Carnival, though the festivities are not as colorful or as lively as those in Brazil. Regional religious festivals are celebrated in many Andes communities.

Whether they come from the region's rural or urban areas, many people enjoy a celebration called Carnival. This festival is celebrated just before the beginning of Lent, the Christian holy season that comes before Easter.

Kristin Piljay/Lonely Planet Images/age fotostock

A girl in the Andes of Ecuador plays a pan flute, one of the most popular instruments of Andean music.

▶ **CRITICAL THINKING**
Describing In addition to music, what other cultural traditions are practiced by people in the region?

Culture often differs by geographic area. Native Americans in mountain regions weave baskets and cloth using designs that are hundreds of years old. They play Andean music using drums, flutes, and other traditional instruments. Along the coast of Colombia and Venezuela, a dance called the *cumbia* blends the region's Spanish and African heritages. Other Venezuelan coastal music and dance, such as salsa and merengue, show Caribbean island influences. Maracas and guitars are used to perform the music of the Llano.

☑ **READING PROGRESS CHECK**

Analyzing What language and religion are most common in the region?

Ongoing Issues

GUIDING QUESTION *What challenges do the countries of the Tropical North face?*

Many people who live in the Tropical North are poor, although natural resources in the region are plentiful. For generations, those resources have mostly benefited only a wealthy few. This situation has created tensions within and between countries.

Trade Relations

Many South American leaders believe that one way to strengthen their countries' economies is to expand trade. In 2008 the countries in the Tropical North joined with the rest of South America to form the Union of South American Nations (UNASUR). One of the organization's goals is ending **tariffs**—taxes on imported goods—on trade between member nations. Another goal is adopting a uniform currency, similar to the euro.

A Northern Neighbor

Another challenge is improving the region's relationship with the United States. The relationship has sometimes been rocky in the past, as when the United States helped Panama gain independence

Kymri Wilt/DanitaDelimont.com "Danita Delimont Photography"/Newscom

from Colombia in the early 1900s. Relations between the United States and Colombia have improved greatly. The United States and the Colombian government are working together to stop the flow of illegal drugs.

Challenges in Venezuela

In 1998 Venezuelans elected Hugo Chávez, a former military leader, as president. Chávez frequently criticized the United States and became friendly with anti-U.S. governments in Cuba and Iran.

After his election, Chávez promised to use Venezuela's oil income to improve conditions for the country's poor. Among other actions that angered U.S. leaders, in 2009 he seized control of U.S. companies that were developing oil resources in Venezuela. His strong rule split Venezuela into opposing groups. Working-class people supported Chávez, but middle-class and wealthy Venezuelans opposed his policies.

Struggles in Colombia and Ecuador

Colombia has undergone a long and bitter struggle between the country's government and a Colombian organization called the Revolutionary Armed Forces of Colombia (FARC). One of FARC's goals is to curtail the role of foreign governments and businesses in Colombia's affairs. Another goal is to provide help and support for the nation's poor farmers. FARC is funded through various means, including the production and sale of illegal drugs.

In Ecuador, indigenous peoples protested for rights and blamed President Rafael Correa for not keeping his promises. Correa had promised to rewrite Ecuador's constitution. Among other things, he pledged to extend the rights of the people. Disappointed when Correa did not act, indigenous peoples organized to win rights for access to land, basic services, and political representation.

FOLDABLES®
Study Organizer

Include this lesson's information in your Foldable®.

☑ **READING PROGRESS CHECK**

Determining Central Ideas How did Hugo Chávez increase tensions between Venezuela and the United States?

LESSON 3 REVIEW **CCSS**

Reviewing Vocabulary
1. Why did enslaved Africans create *Creole*?

Answering the Guiding Questions
2. *Identifying* Why is the Tropical North home to so many ethnic groups?

3. *Analyzing* Why are there Hindu and Muslim populations in northern South America?

4. *Analyzing* How and why is UNASUR likely to affect the economies and people of the Tropical North's countries?

5. *Informative/Explanatory Writing* Choose one of the challenges the Tropical North faces, and write a short essay suggesting how to solve it.

Directions: Write your answers on a separate piece of paper.

1 Use your FOLDABLES to explore the Essential Questions.
INFORMATIVE/EXPLANATORY WRITING Choose one of the region's countries or colonies. Compare the physical and population maps found at the beginning of the unit. Then write at least two paragraphs explaining how the physical geography affects where people live and work.

2 21st Century Skills
ANALYZING Working in small groups, choose one of the countries or the colony found in the region and research the most common occupations practiced by its people. Are any of those jobs unique to that country and its culture? Present your findings to the class in a slideshow or a poster.

3 Thinking Like a Geographer
IDENTIFYING Create a two-column chart. List the name of the Tropical North country or colony in the first column. List the primary languages spoken in the second column.

4 GEOGRAPHY ACTIVITY

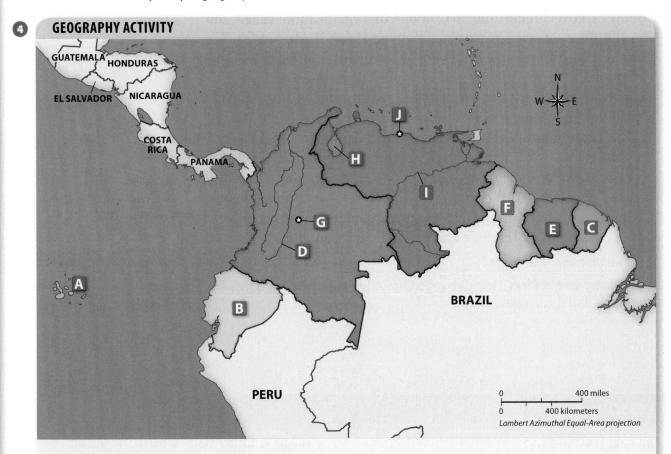

Locating Places
Match the letters on the map with the numbered places listed below.

1. Lake Maracaibo
2. Ecuador
3. French Guiana
4. Caracas
5. Orinoco River
6. Guyana
7. Bogotá
8. Galápagos Islands
9. Suriname
10. Magdalena River

REVIEW THE GUIDING QUESTIONS

Directions: Choose the best answer for each question.

1 The Galápagos Islands were the site of
   A. Christopher Columbus's second landing in the Americas.
   B. an outpost of the Incas.
   C. Charles Darwin's study that resulted in the theory of evolution.
   D. Ecuador's largest volcanic eruption.

2 The world's leading producer of emeralds is
   F. Venezuela.
   G. Colombia.
   H. Ecuador.
   I. French Guiana.

3 South America's native populations were forced into laboring for the Spanish under a system called
   A. *encomienda.*
   B. immunity.
   C. hacienda.
   D. Quito.

4 The transition from colonial governments to independence in the Tropical North of South America
   F. happened suddenly in 1550.
   G. took place slowly over a period of more than 100 years.
   H. began in Brazil.
   I. was a result of the War of 1812.

5 The most populous spot in the Tropical North is
   A. Ecuador.
   B. Bogotá.
   C. the Llanos.
   D. Cartagena.

6 Of the languages used in the Tropical North countries, Spanish is used in its purest form in
   F. Ecuador.
   G. French Guiana.
   H. Venezuela.
   I. Colombia.

DBQ ANALYZING DOCUMENTS

7 **CITING TEXT EVIDENCE** Read the following passage:

"*With one of the highest deforestation rates in Latin America, Ecuador is losing 200,000 hectares (494,211 acres) of forest every year . . . Although most of the forests of the country are public lands, an important percentage of what is left is in the hands of indigenous people and farmers, among the country's poorest citizens.*"

—from Steve Goldstein, "A Grand Plan: Ecuador and 'Forest Partners'" (2008)

What group in Ecuador controls the forests on public lands?

A. indigenous people   C. the government

B. farmers   D. businesses

8 **ANALYZING** Which most likely explains why the indigenous people and the farmers might be willing to sell their land?

F. to enjoy a better way of life

G. so they can buy more productive land

H. so they can move to the city

I. to rid themselves of the burden of the land

SHORT RESPONSE

"*Latin American pop superstar Shakira will be at this weekend's gathering of the Western Hemisphere's leaders advocating for [promoting] her favorite issues: early childhood development and universal education.*"

—from Gregory M. Lamb, "Shakira Advocates for Children at the Summit of the Americas" (2012)

9 **IDENTIFYING POINT OF VIEW** How would improving child development and education benefit the countries of the region?

10 **ANALYZING** What step do you think nations in this region should take to improve child development? Why?

EXTENDED RESPONSE

11 **INFORMATIVE/EXPLANATORY WRITING** What is school like for children who live in the Tropical North of South America? Research and then describe the educational system in this part of the world in a detailed report. Find out how many months of the year school is in session, how long the typical day is, and what kinds of subjects are taught. You might also want to find out how education is funded. Then compare the results with your school experience.

Need Extra Help?

If You've Missed Question	**1**	**2**	**3**	**4**	**5**	**6**	**7**	**8**	**9**	**10**	**11**
Review Lesson	1	1	2	2	3	3	1	3	3	3	3

ANDES AND MIDLATITUDE COUNTRIES

Peter Adams/Photographer's Choice/Getty Images

ESSENTIAL QUESTIONS • *How does geography influence the way people live?*
• *Why do civilizations rise and fall?* • *What makes a culture unique?*

Woman from the town
of Tarabuco in south
central Bolivia

Lesson 1
*Physical Geography
of the Region*

Lesson 2
History of the Region

Lesson 3
Life in the Region

The Story Matters...

Running the length of the Pacific
coast of South America, the
Andes define the countries of
Peru, Bolivia, and Chile. For the
people who have made the Andes
their home, the grandeur of the
high mountain peaks often
contrasts sharply with the
challenges of life in such a
rugged location. And yet, it was
in this very location of
mountains, high plateaus, plains,
and deserts that the ancient Inca
built their powerful and highly
developed civilization.

FOLDABLES®
Study Organizer

Go to the
Foldables® library
in the back of
your book to
make a Foldable®
that will help you
take notes while
reading this chapter.

Geography	Culture	Economy

ANDES AND MIDLATITUDE COUNTRIES

The world's longest mountain system runs parallel to the Pacific coast of South America. The Andes stretch about 4,500 miles (7,242 km) and include many high mountain peaks. As you study the map, look for other geographic features that make this area unique.

Step Into the Place

MAP FOCUS Use the map to answer the following questions.

1 PLACES AND REGIONS Which country has two capitals? Name them.

2 THE GEOGRAPHER'S WORLD Which two bodies of water does the Strait of Magellan connect?

3 THE GEOGRAPHER'S WORLD What is Uruguay's capital city?

4 CRITICAL THINKING ANALYZING Why do you think Chile has such a unique shape?

A

RUGGED TERRAIN Gigantic cacti dot the hilly landscape on Bolivia's Incahuasi Island. The island is in the middle of the world's largest salt flats.

B

HANDICRAFTS The people of the small island of Taquile on Lake Titicaca are known for making some of Peru's highest-quality handwoven clothing.

Step Into the Time

TIME LINE Which event on the time line discusses natural resources? Write a paragraph explaining positive and negative effects on a country's colony that holds valuable resources.

1533 Spanish conquer Inca Empire

1811 Paraguay gains independence from Spain

1800

1545 Silver is discovered in Potosí, Bolivia

1808 Rebellion against Spanish rule grows

Andes and Midlatitude Countries

ECUADOR

PERU

Lima
Cuzco

Ucayali R.

Beni R.

Mamoré R.

Lake Titicaca

La Paz

BOLIVIA

Sucre

B

A

BRAZIL

PARAGUAY

Paraná R.

EQUATOR 0°

N
W · E
S

CHILE

Asunción

Paraguay R.

Elqui R.

ARGENTINA

URUGUAY

Uruguay R.

TROPIC OF CAPRICORN 20°S

PACIFIC OCEAN

ATLANTIC OCEAN

Santiago

Buenos Aires

Montevideo

Río de la Plata

Bío-Bío R.

Colorado R.

0   1,000 miles
0   1,000 kilometers
Albers Equal-Area Conic projection

✪ National capital
○ Territorial capital
• City

40°S

Falkland Islands (U.K.)
Stanley

South Georgia Island (U.K.)

Strait of Magellan

2006
Michelle Bachelet elected first woman president of Chile

1946 Juan Perón becomes president of Argentina

2007 Earthquake in southwest Peru leaves 200,000 people homeless

1900

2000

1900s Foreign companies run mining operations in the region

2009 Bolivia's new constitution empowers indigenous peoples

Reading **HELP**DESK CCSS

Academic Vocabulary

• **isolate**

Content Vocabulary

• **cordillera**
• **altiplano**
• **pampas**
• **estuary**
• **altitude**

TAKING NOTES: *Key Ideas and Details*

Integrate Visual Information
As you read, use a graphic organizer like this one to identify significant physical features of the region.

Physical Features

Lesson 1
Physical Geography of the Region

ESSENTIAL QUESTION · *How does geography influence the way people live?*

IT MATTERS BECAUSE
Much of the terrain of the southern and western part of South America is extremely rugged. The geography of the area presents unique challenges to the people who live there.

Andes Countries

GUIDING QUESTION *What are the physical features of the Andean region?*

Three countries make up the bulk of the Andean region in South America. They are Peru, Bolivia, and Chile. From north to south, these countries span from the Equator to the southern tip of the continent of South America. The physical landscape includes towering mountains, sweeping plains, and significant waterways.

The Andes

On a map of South America, one of the first features you might notice is the system of mountain ranges running parallel to the continent's Pacific coast. These are the Andes, the longest continuous group of mountain ranges in the world and the tallest in the Western Hemisphere. The Andes include high plateaus and high plains, with even higher mountain peaks rising above them. The entire series of the Andes range stretches for 4,500 miles (7,242 km).

The peaks that make up the Andes are not arranged in one neat line. Instead, they form a series of parallel mountain ranges. The parallel ranges are called **cordilleras**. The rugged terrain of the cordilleras makes travel difficult. These ranges **isolated** human settlements from one another for centuries.

In Peru and Bolivia, the two main branches of the Andes border a high plain called the **altiplano**. In fact, *altiplano* means "high plain" in Spanish. About the size of Kentucky, the altiplano has an elevation of 11,200 feet to 12,800 feet (3,414 m to 3,901 m) above sea level.

The Andes mountain ranges are the result of collisions between tectonic plates. This kind of geologic activity comes as no surprise. After all, the Andes are part of the Ring of Fire. All around the rim of the Pacific Ocean, plates are colliding, separating, or sliding past each other. Those forces make earthquakes and volcanic eruptions a part of life throughout much of the Andes.

Plains and Deserts

The Andes run parallel to the Pacific coast but lie 100 miles to 150 miles (161 km to 241 km) inland from the coast. The land between the Andes and the coast averages more than 3,500 feet (1,067 m) above sea level. In most places, the land rises steeply from the ocean. The area has tall cliffs and almost no areas of coastal plain. In Peru and northern Chile, the area between the Pacific and the Andes is a coastal desert. On the Atlantic side of South America, broad plateaus and valleys spread across Uruguay and eastern Argentina.

Academic Vocabulary

isolate to separate

Spectacular mountains are part of Torres del Paine National Park in southern Chile.

▶ CRITICAL THINKING

Describing How were the Andes mountain ranges formed?

The Illimani mountains tower over a small village in the plains of western Bolivia.

▶ **CRITICAL THINKING**

Describing What are the main economic activities in the area?

This plain is called the **pampas**. Its thick, fertile soils come from sediments that have eroded from the Andes. The pampas, like North America's Great Plains, provide land for growing wheat and corn and for grazing cattle.

Coastal Peru and Chile and most of southern Argentina have deserts. Wind patterns, the cold Peru Current, and high elevations are the causes of the low precipitation. The Atacama Desert in Peru and northern Chile is so arid that in some places no rainfall has ever been recorded. The Patagonia Desert in Argentina lies in the rain shadow of the Andes.

Waterways

The Paraná, the Paraguay, and the Uruguay rivers combine to create the second-largest river system in South America, after the Amazon. This river system drains much of the eastern half of South America. The system is especially important to Paraguay, because Paraguay is a landlocked country. The river system provides transportation routes and makes possible the production of hydroelectric power. Along the Paraguay River is the Pantanal, one of the world's largest wetlands. This area produces a diverse ecosystem of plants and animals.

Steve Allen Travel Photography/Alamy

The Paraná-Paraguay-Uruguay river system flows into the Río de la Plata (Spanish for "river of silver"). This river then empties into the Atlantic Ocean on the border of Argentina and Uruguay. The Río de la Plata meets the ocean in a broad estuary. An **estuary** is an area where the ocean tide meets a river current.

South America has few large lakes. The largest lake in the Andean region is Lake Titicaca. It lies on the border between Bolivia and Peru. Lake Titicaca is on the altiplano at 12,500 feet (3,810 m) above sea level. It is the world's highest lake that is large enough and deep enough to be used by small ships.

☑ **READING PROGRESS CHECK**

Analyzing How do you think the geography of the Andean region affects the lives of the people who live there?

Climate Diversity

GUIDING QUESTION *How does climate affect life in the Andean region?*

Climate is part of a region's physical geography. The varying mountains, plains, and other landforms in the Andean region and midlatitude countries of South America mean that the region's climate is extremely diverse.

The Effect of Altitude

The main factor that determines climate in the Andes is **altitude**, or height above sea level. The higher the altitude, the cooler the temperatures are. This is true even in the warm tropics. Conditions in the region can range from hot and humid at lower elevations to freezing in the mountain peaks.

Farming is a challenge in the rugged Andean region. Farmers have successfully terraced the hillsides to grow crops such as potatoes, barley, and wheat.

Visitors to the Andes may find the altitude at the higher elevations hard to handle. Oxygen is thin. This results in heavy breathing and tiring easily. The region's inhabitants are adapted to the thinner air, as are various native species of plants and animals.

Think Again?

The Atacama Desert is unpopulated.

Not true. More than a million people live in this region. They live in mining towns, fishing villages, and coastal cities. Farmers use irrigation to grow olives and tomatoes. The Atacama is also a favorite place for teams of astronomers. They take advantage of the area's crystal-clear night skies to probe the secrets of the universe.

Midlatitude Variety

Climates of the midlatitude countries of South America are quite different from the Andean region. These countries enjoy a generally temperate, or moderate, climate. In Uruguay, for example, the average daily temperature in the middle of winter are a mild 50°F to 54°F (10°C to 12°C). In mid-summer, the average daily temperatures reach a comfortable 72°F to 79°F (22°C to 26°C). There is no wet or dry season—rainfall occurs throughout the year. Inland areas, however, are drier than the coast.

Argentina is much larger than Uruguay and includes a greater variety of landforms. As you might expect, Argentina's climate is extremely diverse. It varies from subtropical in the north to tundra in the far south. Northern Argentina has hot, humid summers. Southern Argentina has warm summers and cold winters with heavy snowfall, especially in the mountains.

Paraguay presents yet a different climate. Paraguay is a landlocked country. The climate is generally temperate or subtropical. Strong winds often sweep the pampas in Paraguay because the country lacks mountain ranges to serve as wind barriers.

INFOGRAPHIC

EFFECTS OF ALTITUDE

30,000 ft
Average height of commercial aircraft; cabins are pressurized and oxygenized

29,527 ft
Highest elevation considered possible for human survival without an oxygen supply

20,000 ft
Less than half the oxygen available at sea level; physical activity becomes taxing; breathing is difficult

10,000 ft
Shortness of breath from simple movements, such as walking uphill

1,000 ft above sea level
No effects felt

SEA LEVEL

At all latitudes, altitude, or the height above sea level, influences climate. Earth's atmosphere thins as altitude increases. Thinner air retains less heat.

▶ **CRITICAL THINKING**

Describing What happens to temperatures as altitude increases? How do higher elevations affect humans?

Climate Extremes

Extremes of climate can be experienced in the Andean countries without changing latitude. Altitude is all that has to change. The climate changes tremendously in the Andes from the lower to the higher elevations. The *tierra caliente,* or "hot land," is the land near sea level. The hot and humid conditions do not change much from month to month. In this zone, farmers grow bananas, sugarcane, cacao, rice, and other tropical crops.

From 3,000 feet to 6,000 feet (914 m to 1,829 m), the air is pleasantly cool. Abundant rainfall helps forests and a great variety of crops grow. This zone is the *tierra templada,* or "temperate land." It is the most densely populated area. Here, farmers grow a variety of crops, such as corn, coffee, cotton, wheat, and citrus fruits.

Higher up, the climate changes. This is the *tierra fría,* or "cold land." It extends from 6,000 feet to 10,000 feet (1,829 m to 3,048 m). The landscape is a combination of forests and grassy areas. Farmers here grow crops that thrive in cooler temperatures, including potatoes, barley, and wheat.

The land at the highest altitude is the *tierra helada,* or "frozen land." Here, above 10,000 feet (3,048 m), conditions can be harsh. The winds blow cold and icy, and temperatures fall well below freezing. Vegetation is sparse, and few people live in this zone.

Mining (left), especially copper mining, is an important part of the Chilean economy. Mining can be dangerous. Miners celebrate their rescue on October 12, 2010 (right), after being trapped underground for 69 days.

Identifying Besides minerals and metals, what natural resources are abundant in the Andean and midlatitude countries?

Llamas are camelids, relatives of camels. They are bred and raised for food, wool, and as pack animals for pulling carts.

Identifying What other camelids are common in the region?

El Niño and La Niña

Every few years, changes in wind patterns and ocean currents in the Pacific Ocean cause unusual and extreme weather in some places in South America. One of these events is called El Niño. During an El Niño, the climate along the Pacific coast of South America becomes much warmer and wetter than normal. Floods occur in some places, especially along the coast of Peru.

El Niños form when cold winds from the east are weak. Without these cold winds, the central Pacific Ocean grows warmer than usual. More water evaporates, and more clouds form. The thick band of clouds changes wind and rain patterns. Some areas receive heavier-than-normal rains. Other areas, however, have less-than-normal rainfall.

Scientists have found that El Niños occur about every three years. They also found that in some years, the opposite kind of unusual weather takes place. This event is called La Niña. Winds from the east become strong, cooling more of the Pacific. When this happens, heavy clouds form in the western Pacific.

☑ **READING PROGRESS CHECK**

Identifying Why is the *tierra templada* the most populated climate zone by altitude in the Andean region?

Imàgenes del Perú/Flickr/Getty Images

Natural Resources

GUIDING QUESTION *Which natural resources are important to the region?*

The Andean and midlatitude countries are rich in natural resources. Energy sources are especially important. Bolivia holds the second-largest reserves of natural gas in South America, trailing behind only Venezuela. Bolivia also has extensive deposits of petroleum. Paraguay's hydroelectric power plants produce nearly all the country's electricity. The governments want to use these resources to develop and strengthen their economies.

Minerals and Metals

Besides energy resources, the region has a number of mineral resources. Most of the area's mines are in the Andes. Chile leads the world in exports of copper. Tin production is important to the Bolivian economy. Bolivia and Peru have deposits of silver, lead, and zinc, and Peru also has gold.

Wildlife

The region's varied geography and climate support a variety of wildlife, including many species of birds and butterflies. The ability of plants and animals to thrive in the region varies with altitude.

A group of mammals called camelids is especially important in this region. Camelids are relatives of camels, but they do not have the typical humps of camels. Two kinds of camelids are the llama and the alpaca. The llama is the larger of the two. Llamas serve as pack animals and are a source of food, wool, and hides. Native Americans throughout the Andes tend herds of llamas. These animals are used to carry goods or pull carts. They are also raised for food, hides, and wool.

Alpacas are found only in certain parts of Peru and Bolivia. The animal's thick, shaggy coat is an important source of wool. Alpaca wool is strong yet soft and repels water. It is used for all kinds of clothing and as insulation in sleeping bags.

Include this lesson's information in your Foldable®.

☑ **READING PROGRESS CHECK**

Analyzing What metal is important to Chile's economy?

LESSON 1 REVIEW (CCSS)

Reviewing Vocabulary

1. Why is *altitude* an important feature in the Andean region?

Answering the Guiding Questions

2. *Determining Central Ideas* Why do earthquakes and volcanoes occur in the Andes?

3. *Analyzing* Why is the climate wet on the western slopes of the Andes in southern Argentina but dry on the eastern slopes?

4. *Describing* How does the climate of the Andes countries compare with that of the midlatitude countries?

5. *Identifying* Give a specific example of how a family living in the Andes might use llamas.

6. *Narrative Writing* You are living with a relative for a month somewhere in the Andes. Choose the country and the area where you are staying. Then write a letter to a friend describing where you are and what you did yesterday.

Lesson 2
History of the Region

ESSENTIAL QUESTION • *Why do civilizations rise and fall?*

IT MATTERS BECAUSE
In the Andean and midlatitude countries of South America, history and government have developed in very different ways.

Early History and Conquest

GUIDING QUESTION *How has history influenced the region?*

Native Americans and European colonizers made major contributions to the history of this region of South America. The actions and achievements of both groups continue to influence life today. Almost all the countries in the region have been independent for nearly two centuries. Still, their history continues to influence their culture.

Rise and Fall of the Inca Empire

Before the rise of the Inca in the 1100s, the Andean region was the home of small Native American societies such as the Moche, the Mapuche, and the Aymara. These societies were based primarily on agriculture. The Moche settled on the arid coastline of northern Peru. Archaeological finds show that they were talented at engineering and irrigation. They used a complex irrigation system to grow corn (maize), beans, and other crops.

The Inca developed a highly sophisticated civilization. They first settled in the Cuzco Valley in what is now Peru. It was not until their fifth emperor, Capac Yupanqui, that they began to expand outside the valley. In the 1400s, under the rule of Pachacuti Inca Yupanqui, the Inca made extensive conquests. By the early 1500s, the Inca ruled a region stretching from northern Ecuador through Peru and then southward into Chile. Historians estimate that the area was

home to 12 million people. This population included dozens of separate cultural groups, who spoke many different languages.

The Inca state was called Tawantinsuyu. The name means "the land of the four quarters." The imperial capital was located where the four quarters, or provinces, of the Inca Empire met, at Cuzco. Inca society was highly structured. At the top of the **hierarchy** were the emperor, the high priest, and the commander of the army. The nobility served the emperor as administrators. At the bottom of the social pyramid were farmers and laborers.

Inca technology and engineering were highly advanced. The Inca built extensive irrigation systems, roads, tunnels, and bridges that linked regions of the empire to Cuzco. Today you can still see the remains of Inca cities and fortresses. Some of the most impressive ruins are at Machu Picchu, located about 50 miles (80 km) northwest of Cuzco.

The Inca had no written language. Instead, they created a counting system called quipu for record keeping. A quipu was a series of knotted cords of various colors and lengths.

Messengers carrying quipu could travel as far as 150 miles (241 km) per day on the roads. The Inca became extremely wealthy because of their vast natural resources of gold and silver.

Spain Conquers Peru

Unfortunately for the Inca, however, their advanced culture could not turn back the invasion that led to the empire's downfall. The Spanish conquests in Mexico encouraged them to move into South America. In 1532 a Spanish adventurer named Francisco Pizarro landed in Peru with a small band of soldiers.

(tr) ©Werner Forman/Corbis; (b) HUGHES Hervé©/hemis.fr/Getty Images

The ruins of the city of Machu Picchu are high in the Andes.
▶ **CRITICAL THINKING**
Describing How did the Inca build a large empire?

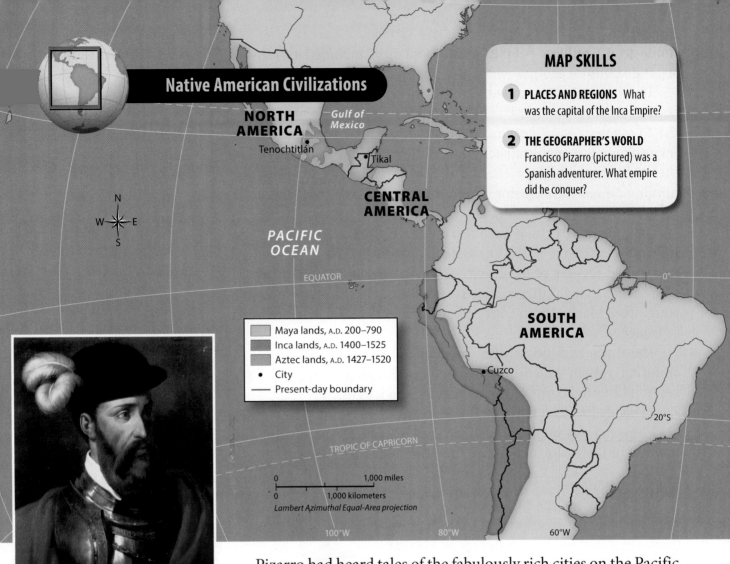

Native American Civilizations

NORTH AMERICA

Gulf of Mexico

Tenochtitlán

Tikal

CENTRAL AMERICA

PACIFIC OCEAN

EQUATOR

SOUTH AMERICA

Cuzco

0°

20°S

TROPIC OF CAPRICORN

	Maya lands, A.D. 200–790
	Inca lands, A.D. 1400–1525
	Aztec lands, A.D. 1427–1520
•	City
—	Present-day boundary

0   1,000 miles

0   1,000 kilometers

Lambert Azimuthal Equal-Area projection

100°W   80°W   60°W

MAP SKILLS

1 PLACES AND REGIONS What was the capital of the Inca Empire?

2 THE GEOGRAPHER'S WORLD Francisco Pizarro (pictured) was a Spanish adventurer. What empire did he conquer?

Spanish conquistador Francisco Pizarro conquered the empire of the Inca.

Pizarro had heard tales of the fabulously rich cities on the Pacific coast of South America. He also learned that the Inca empire had been badly weakened by a civil war from which the new emperor, Atahualpa, had emerged victorious. When the Spanish confronted Atahualpa, the Inca army, unprepared to face Spanish artillery, was nearly destroyed. Within a few years, the Spanish controlled the entire empire. The Spanish also seized control of the region's precious metals. The road system that the Inca developed became an important transportation route for Spanish goods. The Inca are the ancestors of the Native Americans who live in the Andes region today.

The Spanish also branched out from Peru to create colonies in Argentina, Chile, and other parts of South America. The Spanish military victors were called conquistadors, from the Spanish word for "conqueror." The Spanish colonies became sources of wealth for Spain. Some Spanish settlers prospered from gold and silver mining. Spanish rule of the Andean and midlatitude areas in the region continued for nearly 300 years.

After the Inca lost their empire to the Spanish, they and other Native Americans in the region endured great hardships. Their numbers declined drastically as a result of **smallpox**, a highly

infectious disease introduced by the Europeans. The introduction of epidemic diseases also affected numerous Native American groups in North America.

☑ READING PROGRESS CHECK

Analyzing How were the Spanish, under conquistador Francisco Pizarro, able to conquer such a mighty empire as the Inca?

Independent Countries

GUIDING QUESTION *How did the countries of the Andean region gain their independence?*

By the early 1800s, most of South America had been under Spanish control for nearly 300 years. History was about to change once again. The reasons for this shift were local as well as international.

Overthrow of Spanish Rule

In the early 1800s, revolution and liberation movements were occurring around the world. The United States threw off British rule, and the French replaced the monarchy with a republic. Struggles for independence also occurred in Mexico and the Caribbean. People in South America were encouraged by these events. It was exactly the right time for two South American revolutionary leaders to lead the fight against Spanish rule. These two leaders were Simón Bolívar and José de San Martín.

The two leaders were able to rally support for independence. San Martín pioneered many elements of **guerrilla** warfare—the use of troops who know the local landscape so well that they are difficult for traditional armies to find. By the mid-1800s, many South American countries had gained independence.

Power and Governance

After the Spanish left South America, several different countries formed. Their borders mostly followed the divisions set in place by the Spanish colonizers. But despite gaining independence, political and economic hardships continued on the continent.

In contrast to the United States, there was no strong momentum for unity in South America. In fact, the rulers of many of the newly independent states were wealthy aristocrats, powerful landowners, or military dictators. Their mindset was more European than South American. In addition, communication between countries was difficult because of the mountainous terrain.

The new countries lacked a tradition of self-government. The British colonies in North America had elected representatives in their colonial legislatures. The new states of South America, however, did not have a structure in place for a government to function. The newly independent countries drafted constitutions.

Thinking Like a Geographer

A Land of Two Capitals

A map of Bolivia reveals an unusual feature. Bolivia has two capitals: Sucre and La Paz. When Bolivia gained its independence, the question of which city would serve as the nation's capital was never resolved. Today, La Paz is the home of the country's executive and legislative branches, and Sucre serves as the center of the country's judicial system.

Eva Perón was married to Juan Perón, president of Argentina. She won admiration for her efforts to support the woman suffrage movement and to improve the lives of the poor.

▶ **CRITICAL THINKING**
Identifying How was Juan Perón removed from office? Who ruled Argentina after Perón?

The enormously uneven distribution of wealth between rich and poor, however, resulted in social and economic instability. Several countries engaged in bloody conflicts over boundary disputes and mineral rights. These conflicts led to much loss of life and weakened economies.

☑ **READING PROGRESS CHECK**

Describing Why are Simón Bolívar and José de San Martín important in the history of the region?

History of the Region in the Modern Era

GUIDING QUESTION *What challenges did the countries of the region face in the late 1800s and 1900s?*

The Andean and midlatitude countries continued to face challenges during the late 1800s and 1900s. With military backing, dictators seized power, and they ignored democratic constitutions. Economies were still dependent on outside powers.

Economic Challenges

The countries of the region faced economic challenges. Among the challenges were developing and controlling resources, building roads and railroads, and establishing trade links. Before independence, the countries of the region depended economically on Spain and Brazil. After independence, the economies of the region remained tied to countries outside South America.

Rapidly industrializing countries in Europe exploited the region for its raw materials. Wealthy landowners, cattle grazers, and mining operators refused to surrender their ties to European investors. Beginning in the early 1900s, large U.S. and European **multinational** firms—companies that do business in several countries—started mining and smelting operations in the region. As the economies expanded, profits grew for wealthy landowners and multinational companies. But many workers and farmers and their families remained mired in poverty.

Political Instability

Economic woes led to calls for reform. Political leaders promised changes for the better. In 1946 Argentinians elected General Juan Perón as the nation's president. Perón and his wife, Eva, were popular with the people. The new government enacted economic reforms to benefit the working people. However, the Perón government limited

free speech, censored the press, and added to the country's debt. After Perón was overthrown in 1955, the military government ruled Argentina.

The new government moved to put an end to unrest. The rulers imprisoned thousands of people without trial. Some were tortured or killed. Others simply "disappeared." Argentina was also troubled by conflict over the Falkland Islands. Argentina and Great Britain both claimed the Falklands. After a brief war in 1982, Argentina was defeated, and the Falklands remain a British territory.

Significant changes were also taking place in the country of Chile. In the presidential election of 1970, Chileans elected a socialist candidate named Salvador Allende. Allende took action to redistribute wealth and land. The government took over Chile's copper industry and banking system. Allende's economic reforms were popular with workers but angered the upper classes. In 1973 Chilean military officers staged a **coup**, an illegal seizure of power, and killed Allende. A military dictatorship, headed by General Augusto Pinochet, ruled Chile for the next 16 years.

Movements for Change

In recent years, democracies have replaced dictatorships. Yet the countries in the region are still struggling to end corruption in government, shrink the gap between rich and poor, provide jobs, and protect human rights.

Voters also have elected new leaders. In 2005 Bolivians elected Evo Morales, the country's first indigenous president. Morales introduced a new constitution and land reforms, brought industries under government ownership, and moved to limit U.S. corporate involvement in the country's politics. In 2006 Chileans elected the country's first female president, Michelle Bachelet. A year later, Cristina Fernández de Kirchner became Argentina's first elected female president. Both female leaders started efforts to improve human rights and equal opportunity.

Include this lesson's information in your Foldable®.

☑ **READING PROGRESS CHECK**

Determining Central Ideas After independence, why did the countries in this region continue to experience economic hardship?

LESSON 2 REVIEW (CCSS)

Reviewing Vocabulary

1. What is one advantage of *guerrilla* warfare?

Answering the Guiding Questions

2. *Identifying* What were some of the strengths and achievements of the Inca culture?

3. *Describing* What were two events in North America and Europe that set the stage for the independence movement in the Andes and midlatitude countries?

4. *Determining Central Ideas* What can you infer about democratic government based on what you learned about Juan Perón in Argentina and Salvador Allende in Chile?

5. *Argument Writing* You are either Simón Bolívar or José de San Martín. The year is 1822, when the two men met face to face in Guayaquil (now located in Ecuador). Write a paragraph or two in which you urge your fellow leader to pursue the struggle for independence from Spain.

There's More Online!

- ☑ **IMAGES** Cultures of the Andean Region
- ☑ **MAP** Comparing Highway Systems
- ☑ **SLIDE SHOW** Traditional and Modern Lifestyles
- ☑ **VIDEO**

— Inca road system
— Pan-American highway

Lesson 3
Life in the Region

ESSENTIAL QUESTION · *What makes a culture unique?*

Reading **HELP**DESK

Academic Vocabulary

- **impact**
- **contemporary**

Content Vocabulary

- **pueblo jóven**

TAKING NOTES: *Key Ideas and Details*

Describe As you read this lesson, use a graphic organizer like this one to write an important fact about each topic.

Topic	Fact
People and Places	
People and Cultures	
Ongoing Issues	

IT MATTERS BECAUSE
The population of the Andean and midlatitude countries is ethnically diverse. People, places, and cultures have been shaped by physical geography, urban growth, migration, and immigration.

People and Places

GUIDING QUESTION *What are the major population patterns in the Andean region?*

The population of the Andean region is not evenly distributed. Population patterns in this region reflect the changing **impact** of politics, economics, and the availability of natural resources and jobs.

Population Density and Distribution

The people of the region came from many different places. Three centuries of colonization by Spain have left their mark on the population. Enslaved Africans were brought as laborers, especially in Peru; this was less common in other countries of the region. After independence, immigrants from many European countries traveled to settle in South America. Immigrants also came from Asia.

Today, Bolivia and Peru have large Native American populations. Argentina and Chile have many people of European ancestry, including Spanish, Italian, British, and German backgrounds. Peru has descendants of people from Europe, but also people with origins in Japan and Southwest Asia.

In the Andean and midlatitude countries, as with the whole continent, the population is densest in the coastal areas. This area is sometimes called the "population rim." The rugged,

mountainous areas and the tropical rain forest have discouraged settlement in many inland areas. Transportation is difficult, and communication can be slow or nonexistent. Coastal regions offer fertile land, favorable climates, and easy transportation.

Large Cities and Communities

The largest city in the region is Buenos Aires, the capital of Argentina. This is a bustling port and cultural center. It resembles a European city with its parks, buildings, outdoor cafes, and wide streets. About 2.8 million people live in the central city of Buenos Aires, but the metropolitan area includes 11.5 million people. This is more than one-fourth of Argentina's entire population. Although Argentina as a whole is not densely populated, the area in and around Buenos Aires is.

Buenos Aires and many other large cities in the region have shantytowns. These makeshift communities often spring up on the outskirts of a city. Poor people migrate here from remote inland areas to seek a better life. They cannot afford houses or apartments in the city or suburbs, so they settle in the shantytowns. They build shacks from scraps of sheet metal, wood, and other materials. Shantytowns often lack sewers, running water, and other services. They tend to be dangerous places with widespread crime.

In Lima, the capital and largest city in Peru, the shantytowns are called **pueblos jóvenes**. The name means "young towns." One pueblo jóven was home to María Elena Moyano. She worked to improve education, nutrition, and job opportunities in the pueblo jóven. She refused to give in to the demands of the government or to the communist rebels, who assassinated her in 1992. She has become recognized as a national hero for her courage.

☑ **READING PROGRESS CHECK**

Analyzing What limits the population in many inland areas?

WINFIELD PARKS/National Geographic Stock

Academic Vocabulary

Academic Vocabulary

impact an effect

Shacks made of metal, wood, and other materials stand in front of high-rise apartments in Buenos Aires.

▶ **CRITICAL THINKING**
Describing What challenges do people living in the shantytowns face?

People and Cultures

GUIDING QUESTION *How do ethnic and religious traditions influence people's lives?*

The Andean and midlatitude regions are home to a wide range of ethnic groups. Although many people trace their ancestry back to Europe, Asia, and Africa, Native American groups still thrive in parts of the region.

Ethnic and Language Groups

The Guarani is a Native American group that lives mainly in Paraguay, but people of Guarani descent also can be found in Argentina, Bolivia, and Brazil. The Guarani lived in tropical forests and practiced slash-and-burn agriculture—cutting and burning small areas of forest to clear land for farming. After a few years, the soil's nutrients were used up, and the people moved to a new area.

Today, Guarani customs and folk art are an important part of the culture in Paraguay. Guarani is one of the country's official languages. A related language is Sirionó. The Sirionó live in eastern Bolivia. In Peru, Quechua, a surviving language of the central Andes, is still widely spoken, along with Spanish.

Traditional Medicine

In Bolivia, the custom of Kallawaya medicine is widespread. The word *Kallawaya* might come from the Aymara word for "doctor." Kallawaya healers use traditional herbs and rituals in their cures. Many Bolivians seek the help of Kallawayas when they get sick, either because they prefer these healers or because they cannot afford other doctors. In fact, 40 percent of the Bolivian population relies on the natural healers, who travel from place to place.

Religion and the Arts

During the centuries of Spanish colonization, the Roman Catholic Church was one of the region's most important institutions. The influence of the Catholic Church continues. Millions of people practice mixed religions. Many of the native peoples of the Andes combine their indigenous rituals and beliefs with Roman Catholicism. Others have adopted Protestant religions.

Traditional arts, crafts, music, and dance thrive in the Andean and midlatitude countries. In literature, two Chilean poets, Gabriela Mistral and Pablo Neruda, have won the Nobel Prize for Literature. The works of writers from Argentina,

Kallawaya healers use traditional methods in efforts to cure the sick. The customs and languages of the indigenous peoples are an important part of the culture of the region.

▶ CRITICAL THINKING
Identifying What are some of the languages that are spoken in the region?

©David Mercado/Reuters/Corbis

including Jorge Luis Borges and Manuel Puig, are popular with readers around the world. Isabel Allende from Chile, a cousin of the country's former president Salvador Allende, is a **contemporary** writer of great distinction. Many writers have been praised for their use of magic realism. This style combines everyday events with magical or mythical elements. It is especially popular in Latin America.

Academic Vocabulary

contemporary living or happening now

Daily Life

In large cities and towns and in wealthier areas, family life revolves around parents and children. In the countryside, extended families are more common. In the region, the *compadre* relationship is still valued. This relationship is a strong bond between a child's parents and other adults who serve as the child's godparents.

In a megacity like Buenos Aires, people can wander through large, modern shopping malls. They may dine in outdoor cafes or fancy restaurants. They may work in modern office buildings. Traditional Andean foods of the countryside include *pachamanca* in Peru. This is a mixture of lamb, pork, and chicken baked in an earthen oven. Pachamanca dates back to the time of the Inca Empire.

The Larcomar Shopping Mall in Lima, Peru, is a popular attraction for international tourists.

Soccer, or football, is the most popular sport in the region. Football is Argentina's national game, and its teams have won several World Cup titles. Equestrian sports, or sports featuring riders on horseback, are popular in the region. Argentina's polo teams have long dominated international competition. The Argentina team won the first ever Olympic polo gold medal in 1924.

Football is also the national sport for many of the other countries in the region. The top professional league, the American Football Confederation, is made up of teams from 10 South American nations. League teams are eligible for the World Cup and the America Cup. Other popular sports include basketball, golf, boxing, and rugby. Social life focuses on family visits, patriotic events, religious feast days, and festivals.

✓ **READING PROGRESS CHECK**

Analyzing How is Kallawaya medicine different from the modern medicine that is practiced in most Western countries?

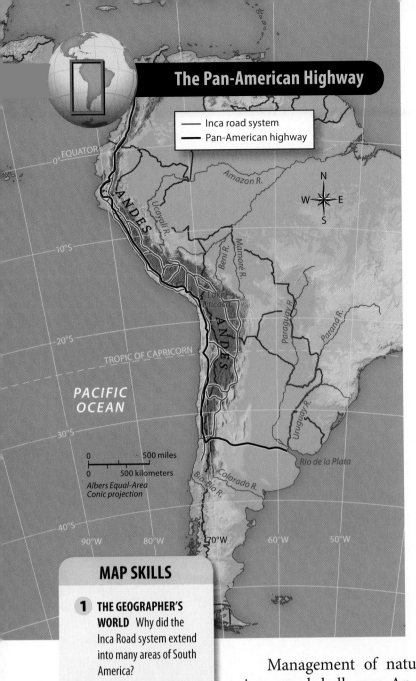

The Pan-American Highway

Legend:
- Inca road system
- Pan-American highway

ANDES

Amazon R.

Ucayali R.

Marañón R.

Mamoré R.

Beni R.

Lake Titicaca

ANDES

Paraguay R.

Paraná R.

Uruguay R.

PACIFIC OCEAN

TROPIC OF CAPRICORN

Rio de la Plata

Bío-Bío R.

Colorado R.

0°EQUATOR

10°S

20°S

30°S

40°S

90°W   80°W   70°W   60°W   50°W

0   500 miles
0   500 kilometers
Albers Equal-Area
Conic projection

MAP SKILLS

1 **THE GEOGRAPHER'S WORLD** Why did the Inca Road system extend into many areas of South America?

2 **THE GEOGRAPHER'S WORLD** Why do you think the Pan-American Highway links many capital cities?

Ongoing Issues

GUIDING QUESTION *How are economic and environmental issues affecting the region?*

The population growth rate in the Andean and midlatitude countries is generally not high. However, population is growing enough in some places to add to today's challenges of earning a living.

Earning a Living

It is difficult to build strong economies in the Andean region largely because of the rugged terrain. Many countries rely heavily on agriculture, which is limited and difficult in the mountains. About one-third of Peruvians, for example, are farmers. They grow potatoes, coffee, and corn on terraces built into the mountain slopes. Farms in the valleys along the coast produce sugarcane, asparagus, mangoes, and many other crops.

Mining and fishing are other important economic activities in Peru. Mines in the mountains produce silver, zinc, copper, and other minerals. Peru also produces oil and natural gas. The country's coastal waters are rich fishing grounds. Much of the fish catch is ground into fishmeal for animal feed and fertilizer.

Management of natural resources presents many important issues and challenges. An example is conflict between countries over gas reserves. Bolivia has the second-largest reserve of natural gas in Latin America. Bolivia is landlocked. So, to export the gas, it must move through Peru or Chile.

In 2003 the Bolivian government proposed moving the natural gas through Chile, because it would be cheaper than an alternate plan to go through Peru. The Bolivian people turned out in huge numbers to protest. In Bolivia, suspicion and anger against Chile are widespread. These feelings date back to the Pacific War of the early 1880s, when Chile took over Bolivia's former coastal lands.

As the economies in the region develop, the primary economic activities of agriculture, mining, and fishing remain important. Other activities have become important as well. About 20 percent of the workers in the region are employed in the secondary, or manufacturing, sector. They work in factories making products.

About 65 percent find jobs in the tertiary, or services, sector. They work in a wide variety of occupations, ranging from transportation and retail sales to banking and education.

Transportation and Trade

This region has many geographic and regional barriers. The Andes limit construction of roads and railroads. Yet highways do link large cities. The Pan-American Highway, for example, runs from Argentina to Panama, then continues northward after a break in the highway. A trans-Andean highway connects cities in Chile and Argentina. Peru and Brazil are building the Transoceanic Highway. Parts of it opened in 2012. Eventually, this road will link Amazon River ports in Brazil with Peruvian ports on the Pacific Ocean. Unlike other countries, Argentina has an effective railway system.

Trade also connects countries. In 1991 a trade agreement, known as MERCOSUR, was signed by Argentina, Paraguay, Uruguay, and Brazil. In 2011 MERCOSUR merged into a new organization—the Union of South American Nations (UNASUR). The Union set up an economic and political zone modeled after the European Union (EU). Its goals are to foster free trade and closer political unity.

Addressing Challenges

Looking toward the future, the Andean and midlatitude countries must address many challenges. Environmental issues are among the most important. Air and water pollution is a major problem, especially in the shantytowns of urban areas, where the lack of sewage systems and garbage collection increases disease.

Disputed borders have presented challenges for years. For example, Bolivia and Paraguay long disputed rights to a region thought to be rich in oil. In 1998 Peru and Ecuador finally settled a territorial dispute after years of tensions marked by episodes of armed conflict. Border wars use up resources of people, money, time, and brainpower that could be used to address economic development and environmental concerns.

☑ **READING PROGRESS CHECK**

Determining Central Ideas How does the physical landscape hamper transportation? What actions are being taken to improve transportation?

Include this lesson's information in your Foldable®.

LESSON 3 REVIEW

Reviewing Vocabulary

1. Why might someone live in a *pueblo jóven*?

Answering the Guiding Questions

2. ***Describing*** How are the populations of Peru and Bolivia different from the populations of Argentina and Chile?

3. ***Analyzing*** Why might the Kallawaya of Bolivia have an important influence on people's lives?

4. ***Identifying*** What are two important industries in Peru today?

5. ***Informative/Explanatory Writing*** Write a paragraph or two to explain what you think is the most pressing problem facing the people who live in shantytowns.

What Do You Think?

Is Globalization Destroying Indigenous Cultures?

Globalization makes it easier for people, goods, and information to travel across borders. Customers have more choices when they shop. Costs of goods are sometimes lower. However, not everyone welcomes these changes. Resistance is particularly strong among indigenous peoples. They see the expansion of trade and outside influences as a threat to their way of life. Is globalization deadly for indigenous cultures?

No !

PRIMARY SOURCE

❝ Indigenous people have struggled for centuries to maintain their identity and way of life against the tide of foreign economic investment and the new settlers that often come with it. … But indigenous groups are increasingly assertive. Globalization has made it easier for indigenous people to organize, raise funds and network with other groups around the world, with greater political reach and impact than before. The United Nations declared 1995–2004 the International Decade for the World's Indigenous People, and in 2000 the Permanent Forum on Indigenous Issues was created. … Many states have laws that explicitly recognize indigenous people's rights over their resources. … Respecting cultural identity [is] possible as long as decisions are made democratically—by states, by companies, by international institutions and by indigenous people. ❞

—Report by the United Nations Development Programme (UNDP)

A Quechuan family shops at an open air market in a Peruvian village. The Quechua is the term for several native groups who speak a common language. They live in many countries throughout the region.

TEXT: Jerry Mander is Distinguished Fellow of International Forum on Globalization. This quote is from Web site: http://www.ifg.org/
PHOTO: Martin Mejia/AP Images

A woman carries water home in a mining town in central Peru. Protesters contend that aggressive mining practices contaminate the environment and destroy the way of life of indigenous people.

Yes !

PRIMARY SOURCE

" Globalization . . . is a multi-pronged attack on the very foundation of [indigenous people's] existence and livelihoods. . . . Indigenous people throughout the world . . . occupy the last pristine [pure and undeveloped] places on earth, where resources are still abundant [plentiful]: forests, minerals, water, and genetic diversity. All are ferociously sought by global corporations, trying to push traditional societies off their lands. . . . Traditional sovereignty [control] over hunting and gathering rights has been thrown into question as national governments bind themselves to new global economic treaties. . . . Big dams, mines, pipelines, roads, energy developments, military intrusions all threaten native lands. . . . National governments making decisions on export development strategies or international trade and investment rules do not consult native communities. . . . The reality remains that without rapid action, these native communities may be wiped out, taking with them vast indigenous knowledge, rich culture and traditions, and any hope of preserving the natural world, and a simpler . . . way of life for future generations. "

—International Forum on Globalization (IFG), a research and educational organization

What Do You Think? DBQ

1 ***Citing Text Evidence*** According to the IFG, why are indigenous people at risk?

2 ***Describing*** According to the United Nations report, how has globalization given indigenous people more power?

Critical Thinking

3 ***Identifying*** One effect of globalization is that more tourists are visiting remote places such as rain forests in South America and wildlife areas in Africa. How do you think indigenous peoples feel about the growth of tourism in their communities?

Directions: Write your answers on a separate piece of paper.

❶ Use your **FOLDABLES** to explore the Essential Questions.

INFORMATIVE/EXPLANATORY WRITING Write a short essay to answer the question: Why are people who live in the Andes more likely to follow a traditional way of life than those who live in cities?

❷ 21st Century Skills

DESCRIBING Write a radio script for a two-minute segment about what people do for recreation in one of this region's countries. Research the topic, and outline the information you want to include in your script. Share your notes with an adult, and ask: (a) Is my outline clear and well-organized? (b) Does the outline have too much or too little information? Revise your outline as needed, and then use it to write the script. Exchange scripts and compare your script with a classmate's. Discuss strong and weak points in each script.

❸ Thinking Like a Geographer

ANALYZING Think about why industrialization requires good transportation and communication systems. Then, describe obstacles that slow the development of these systems.

❹ **GEOGRAPHY ACTIVITY**

Locating Places

Match the letters on the map with the numbered places below.

1. Rio de la Plata	**3.** Lima	**5.** Santiago	**7.** Paraguay	**9.** Uruguay River
2. Bolivia	**4.** Falkland Islands (Malvinas)	**6.** Chile	**8.** Uruguay	**10.** Buenos Aires

REVIEW THE GUIDING QUESTIONS

Directions: Choose the best answer for each question.

1 The ocean tide meets a river current at a(n)

A. cordillera.

B. altiplano.

C. estuary.

D. *tierrra templada.*

2 Chile leads the world in exports of

F. emeralds.

G. copper.

H. lead.

I. gold.

3 The Inca became extremely wealthy because of

A. the system called quipu.

B. rich farmland.

C. huge deposits of gold and silver.

D. a lucrative fur trade.

4 The broad plain that spreads across Uruguay and eastern Argentina is called

F. the pampas.

G. the Atacama.

H. the Rio Blanco.

I. the Río de la Plata.

5 The group of South American animals called camelids includes

A. horses and donkeys.

B. sheep and goats.

C. llamas and alpacas.

D. mules and llamas.

6 The Spanish explorer who took the Inca emperor Atahualpa hostage was

F. Pizarro.

G. Columbus.

H. Magellan.

I. Fernandez.

DBQ ANALYZING DOCUMENTS

7 CITING TEXT EVIDENCE Read the following passage:

"*[Argentina's] President [Cristina] Kirchner . . . announce[d] the nationalization of the Argentine oil company. . . . The move . . . raised concerns that this may be the first of many expropriations [government takeovers] of privately run companies.*"

—from Jonathan Gilbert, "The Next Venezuela?" (2012)

What does the writer say that Argentina's government might do in the future?

A. break the nation's dependence on imported oil

B. take over other companies

C. replace nuclear power with oil as the main source of energy

D. end high unemployment in Argentina

8 IDENTIFYING What impact will Argentina's action probably have on foreign companies doing business there?

F. They probably will try to sell their businesses to the government.

G. They will expect to have lower costs when oil prices fall.

H. They are likely to fear that their companies will be taken over, too.

I. They will seek to buy the oil company from the government.

SHORT RESPONSE

"*Visitors to modern Cuzco frequently marvel at the exquisite [very fine] workmanship of its many Inca walls. The . . . impression is that a great deal of Inca Cuzco has survived. . . . [Actually,] new streets have been created, ancient ones lost, and the bulk of the city's former palaces, halls, temples, and shrines [holy places] have been demolished.*"

—from Brian S. Bauer, *Ancient Cuzco: Heartland of the Inca* (2004)

9 DETERMINING CENTRAL IDEAS Is the impression visitors have that much of Inca Cuzco has survived correct? Why or why not?

10 IDENTIFYING Who do you think was responsible for these changes to ancient Cuzco? Why?

EXTENDED RESPONSE

11 INFORMATIVE/EXPLANATORY WRITING Think about what you have read about the physical geography of the Andes and midlatitude countries. What geographic factors influence where people have settled in the region?

Need Extra Help?

If You've Missed Question	**1**	**2**	**3**	**4**	**5**	**6**	**7**	**8**	**9**	**10**	**11**
Review Lesson	1	1	2	1	1	2	3	3	2	2	1

Jonathan Gilbert, BA (Hons), PgDip. This article first appeared in THE CHRISTIAN SCIENCE MONITOR. http://CSMonitor.com; From ANCIENT CUZCO: HEARTLAND OF THE INCA, by Brian S. Bauer, Copyright © 2004. By permission of the University of Texas Press.

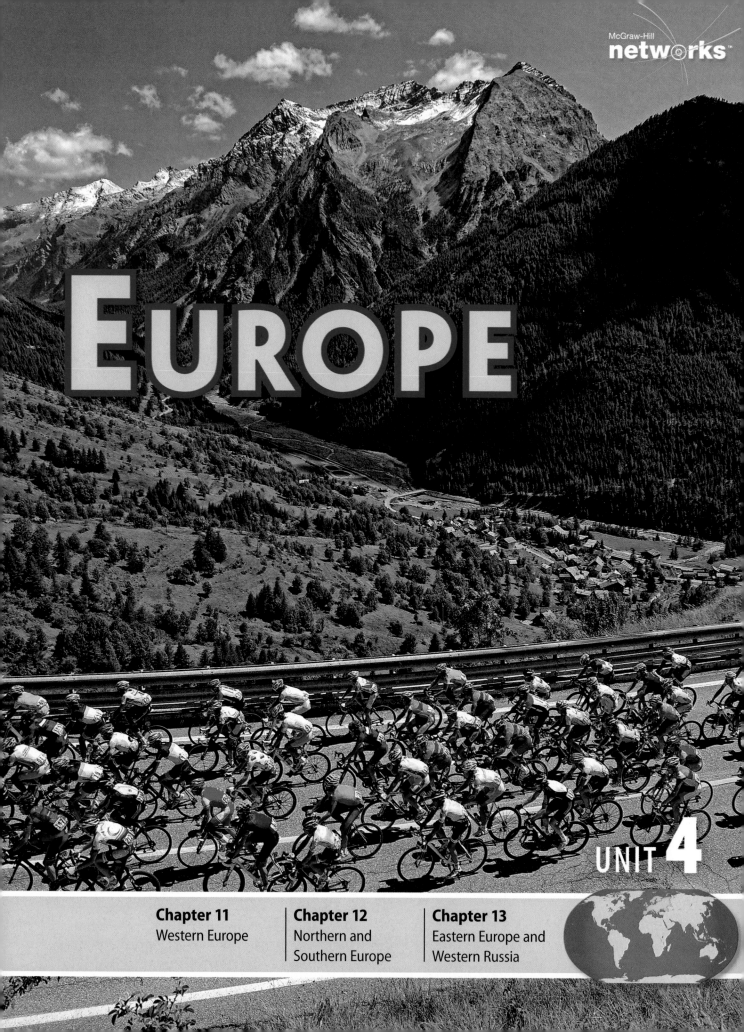

EUROPE

McGraw-Hill
networks™

UNIT 4

Chapter 11
Western Europe

Chapter 12
Northern and
Southern Europe

Chapter 13
Eastern Europe and
Western Russia

EXPLORE the CONTINENT

EUROPE

Europe is a relatively small continent with a long, jagged coastline. Fertile plains extend across much of northern Europe. Farther south, the plains become rugged hills, and then mountains. Great rivers wind their way through Europe's landscape, linking inland areas with the seas.

① **BODIES OF WATER** A fisher uses a net to catch salmon on Loughros Bay in northwest Ireland. Europe's deep bays and well-protected inlets shelter fine harbors. For centuries, the rivers of Europe have provided links between coastal ports and inland population centers.

② **LANDFORMS** A herder tends to cattle on the Great Hungarian Plain in the eastern part of Hungary. Farmers raise livestock and cultivate grains, fruits, and vegetables here. In the northern part of the region, the North European Plain stretches from France to Russia. The plain's fertile soil and wealth of rivers originally drew farmers to the area.

(3) NATURAL RESOURCES Located in west central Italy, the region of Tuscany is known for its natural resources as well as its beautiful landscape. In the mountains, iron and magnesium are produced, and marble is quarried. Carrara marble is in high demand for building and statues worldwide. Along the northern coast, pine forests are abundant.

FAST
FACT

Europe is only about 10 percent larger than the United States.

EUROPE

PHYSICAL

MAP SKILLS

1 **PHYSICAL GEOGRAPHY** How would you describe the landforms of central Europe?

2 **PLACES AND REGIONS** How high is the Meseta in the Iberian Peninsula?

3 **PLACES AND REGIONS** What body of water separates Great Britain from Europe's mainland?

Map labels:

ARCTIC OCEAN

Barents Sea

Murmansk

Europe/Asia boundary

Reykjavík · ICELAND

ARCTIC CIRCLE · PRIME MERIDIAN

Norwegian Sea

Faeroe Islands (Denmark)

FINLAND

SWEDEN

NORWAY

Helsinki

Oslo · Stockholm

Dvina R.

RUSSIA

St. Petersburg

Tallinn

EST.

LATVIA

Riga

Moscow

ATLANTIC OCEAN

IRELAND

Dublin

UNITED KINGDOM

North Sea

DENMARK

Copenhagen

Baltic Sea

LITH.

Vilnius

Kaliningrad

Minsk

BELARUS

KAZAKHSTAN

Ural R.

Thames R.

London

NETH.

Amsterdam

Brussels

BELG.

Elbe R.

Berlin

POLAND

Warsaw

Vistula R.

Kiev (Kyiv)

UKRAINE

Don R.

Volga R.

Volgograd

English Channel

Seine R.

Paris

LUX.

Luxembourg

Rhine R.

GERMANY

Prague

CZECH REP.

Oder R.

SLOVAKIA

Bratislava

Dniester R.

Dnieper R.

Sea of Azov

Caspian Sea

Loire R.

Bay of Biscay

FRANCE

Bern

SWITZ.

Vaduz

LIECH.

Vienna

AUST.

Budapest

HUNG.

SLOV.

Ljubljana

Zagreb

Chișinău

MOLDOVA

ROMANIA

Danube R.

Po R.

Rhône R.

PORTUGAL

Madrid

Andorra la Vella

ANDORRA

MONACO

SAN MARINO

ITALY

CROATIA

B.&H.

Sarajevo

SERB.

Belgrade

KOS.

Priština

MONT.

Podgorica

Skopje

MAC.

Bucharest

BULGARIA

Sofia

Black Sea

TURKEY

Ebro R.

Lisbon

Tagus R.

SPAIN

Corsica (France)

Rome

Sardinia (Italy)

Balearic Islands (Spain)

VATICAN CITY (within Rome)

Tiranë

ALBANIA

GREECE

Adriatic Sea

Aegean Sea

Strait of Gibraltar

Sicily (Italy)

MALTA

Valletta

Nicosia

CYPRUS

Crete (Greece)

Mediterranean Sea

AFRICA

o National capital
• City

0   400 miles
0   400 kilometers
Lambert Azimuthal Equal-Area projection

30°N · 0° · 10°E · 20°E · 30°E

60°N · 50°N · 40°N · 70°N · 40°E · 10°W

Abbreviations

AUST.	Austria
B.&H.	Bosnia & Herzegovina
BELG.	Belgium
CZECH REP.	Czech Republic
EST.	Estonia
HUNG.	Hungary
KOS.	Kosovo
LIECH.	Liechtenstein
LITH.	Lithuania
LUX.	Luxembourg
MAC.	Macedonia
MONT.	Montenegro
NETH.	Netherlands
SERB.	Serbia
SLOV.	Slovenia
SWITZ.	Switzerland

POLITICAL

MAP SKILLS

1 **THE GEOGRAPHER'S WORLD** Where is Monaco located?

2 **PLACES AND REGIONS** What country lies between Belarus and Romania?

3 **PLACES AND REGIONS** Which countries border the Baltic Sea?

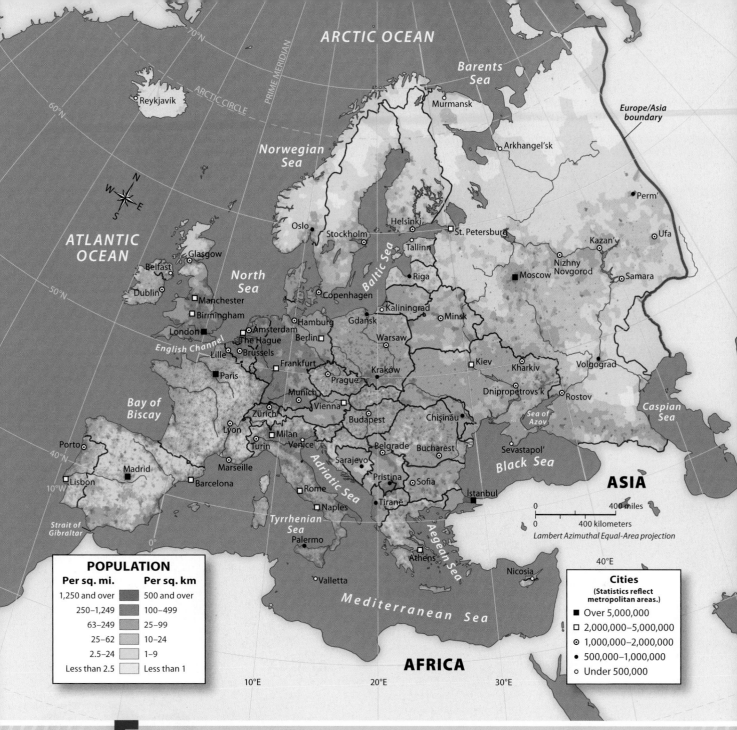

EUROPE

POPULATION DENSITY

POPULATION

Per sq. mi.	Per sq. km
1,250 and over	500 and over
250–1,249	100–499
63–249	25–99
25–62	10–24
2.5–24	1–9
Less than 2.5	Less than 1

Cities
(Statistics reflect metropolitan areas.)

- ■ Over 5,000,000
- ◻ 2,000,000–5,000,000
- ◉ 1,000,000–2,000,000
- • 500,000–1,000,000
- ◦ Under 500,000

MAP SKILLS

1 PLACES AND REGIONS How would you describe the population in the area north of St. Petersburg?

2 THE GEOGRAPHER'S WORLD What cities are located in Europe's most densely populated area?

3 PLACES AND REGIONS What part of Europe has a low population density?

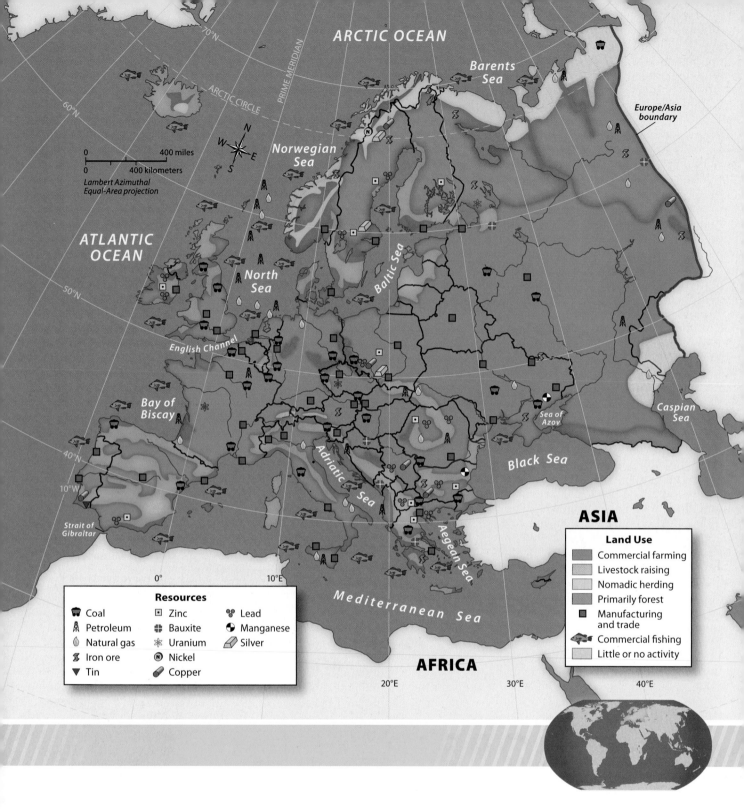

Resources

Coal	Zinc	Lead
Petroleum	Bauxite	Manganese
Natural gas	Uranium	Silver
Iron ore	Nickel	
Tin	Copper	

Land Use

- Commercial farming
- Livestock raising
- Nomadic herding
- Primarily forest
- Manufacturing and trade
- Commercial fishing
- Little or no activity

ECONOMIC RESOURCES

MAP SKILLS

1 PLACES AND REGIONS In what part of Europe is bauxite found?

2 ENVIRONMENT AND SOCIETY How is land used in the far northern areas of Europe?

3 PLACES AND REGIONS What mineral resources are located under the North Sea?

ARCTIC OCEAN

Barents Sea

RUSSIA

Europe/Asia boundary

Reykjavík

Norwegian Sea

ATLANTIC OCEAN

Oslo Stockholm Helsinki

Baltic Sea

Riga ⊙ Moscow

North Sea

Copenhagen

London Berlin Warsaw

English Channel Kiev

Paris

Bay of Biscay

Bern Vienna Budapest

Madrid Belgrade *Sea of Azov* *Caspian Sea*

Lisbon *Adriatic Sea* *Black Sea*

Strait of Gibraltar Rome **ASIA**

Aegean Sea

Athens

Mediterranean Sea

AFRICA

⊙ National capital

| 0 | 400 miles |
| 0 | 400 kilometers |

Lambert Azimuthal Equal-Area projection

40°E

Semi-arid (steppe)
Humid subtropical
Marine west coast
Mediterranean
Humid continental
Subarctic
Tundra and high altitude

EUROPE

CLIMATE

MAP SKILLS

1 **THE GEOGRAPHER'S WORLD**   Where are Europe's driest areas located?

2 **PLACES AND REGIONS**   What type of climate do people in Paris experience?

3 **PLACES AND REGIONS**   What area of Europe has a largely humid continental climate?

WESTERN EUROPE

ESSENTIAL QUESTIONS · *How does geography influence the way people live?*
· *Why do civilizations rise and fall?* · *How do governments change?*

©Alex Masi/Corbis

Coal mine worker in South Wales

The Story Matters...

Western Europe has always been a crossroad of cultures. Today, this is reflected in the diversity of cultures represented in the population, particularly in large cities, such as Paris, France. Long before modern times, however, the geography and resources of Western Europe influenced early civilizations and their rise to economic and political power, shaping the history and governments of this region.

FOLDABLES
Study Organizer

Go to the Foldables® library in the back of your book to make a Foldable® that will help you take notes while reading this chapter.

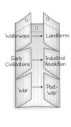

WESTERN EUROPE CCSS

Islands and peninsulas branch out from Western Europe into the various oceans, seas, bays, and channels that border the region. In the past, these landforms and waterways isolated groups of people. As a result, many different cultures developed. As you study the map, look for other geographic features that make this area unique.

Step Into the Place

MAP FOCUS Use the map to answer the following questions.

1 **THE GEOGRAPHER'S WORLD** Which two countries does the English Channel separate?

2 **PLACES AND REGIONS** Which island country in Western Europe lies farthest west?

3 **THE GEOGRAPHER'S WORLD** What is the capital city of Switzerland?

4 **CRITICAL THINKING**
ANALYZING In what ways might the large number of rivers and waterways be important to the people of Western Europe?

A

SWISS ALPS A train begins its climb to the summit. The railway, completed in the early 1900s, makes this scenic trip popular with visitors.

B

BRANDENBURG GATE Completed in the late 1700s, the Brandenburg Gate is one of the well-known landmarks in Berlin, Germany.

Step Into the Time

TIME LINE Choose an event from the time line and predict its long-term political, cultural, or geographical consequences.

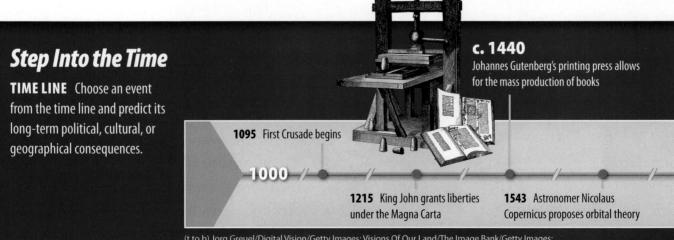

c. 1440
Johannes Gutenberg's printing press allows for the mass production of books

1095 First Crusade begins

1000

1215 King John grants liberties under the Magna Carta

1543 Astronomer Nicolaus Copernicus proposes orbital theory

Western Europe

National capital
City

NORWAY
SWEDEN
ESTONIA
LATVIA
LITHUANIA
RUSSIA

North
Sea
DENMARK
Baltic Sea

B

Dublin
UNITED
KINGDOM
POLAND

IRELAND

NETHERLANDS
Amsterdam
Berlin

Thames R.
London
GERMANY

Brussels
BELGIUM
Rhine R.
CZECH
REPUBLIC
SLOVAKIA

ATLANTIC
OCEAN
Luxembourg
LUXEMBOURG
Danube R.
Vienna

Paris
Seine R.
LIECHTENSTEIN
AUSTRIA
HUNGARY

Loire R.
Bern
Vaduz
SLOVENIA

0 400 miles
0 400 kilometers
Lambert Azimuthal Equal-Area projection
FRANCE
SWITZERLAND
CROATIA

A
ITALY
BOSNIA &
HERZEGOVINA

Bay of
Biscay
MONACO
MONTENEGRO

PORTUGAL
SPAIN
Ajaccio
Corsica
(France)

Mediterranean
Sea

PRIME MERIDIAN

1804
Napoleon crowned
emperor of France

1993
European Union created

1900

2000

1918 World War I ends

1945 Japan surrenders,
ending World War II

1989 Berlin Wall torn down as
East and West Germany reunite

Reading **HELP**DESK

Academic Vocabulary

- **adapt**

Content Vocabulary

- **dike**
- **polder**
- **estuary**
- **Westerlies**
- **deciduous**
- **coniferous**

TAKING NOTES: *Key Ideas and Details*

Identify Use a graphic organizer like this one to list important resources in Western Europe.

Resources

Lesson 1
Physical Geography of Western Europe

ESSENTIAL QUESTION · *How does geography influence the way people live?*

IT MATTERS BECAUSE
The geography of Western Europe has provided the people who live there with several advantages: closeness to the sea, abundant resources, and temperate climates.

Landforms and Waterways

GUIDING QUESTION *How do the physical features of Western Europe make the region unique?*

You probably have seen images of Europe on television or in photographs. Maybe you have seen a movie showing a Dutch windmill on a windswept plain. You might have watched news clips showing cyclists racing through the rolling French countryside during a yearly sports event called the Tour de France. These images show just a part of Western Europe's varied landscape.

Western Europe comprises the following nations: Ireland, the United Kingdom, France, Luxembourg, Germany, the Netherlands, Belgium, Austria, and Switzerland. Belgium, the Netherlands, and Luxembourg are often referred to as the Benelux Countries. Western Europe also includes the tiny countries of Monaco and Liechtenstein, which have a combined population of fewer than 70,000 people.

Low-Lying Plains
Western Europe's landscape consists of plains with mountains cutting through some places. Shaped by wind, water, and ice, these landforms have influenced the lives of the region's peoples. Much of Western Europe lies within an area called

(l to r) TADAO YAMAMOTO/a.collectionRF/Getty Images; Willfried Gredler/age fotostock; Arterra Picture Library/Alamy; Philippe Huguen/AFP/Getty Images

the Northern European Plain. France, Belgium, the Netherlands, Luxembourg, and most of Germany are located on this plain.

Massive sheets of ice shaped the Northern European Plain during the last ice age, which ended about 11,000 years ago. In some places, the melting glaciers left behind fertile soil, but also thick layers of sand and gravel. Ocean waves, currents, and winds have eroded these deposits into sand dunes along some of the North Sea coastline. The glaciers also left behind areas of poorly drained wetlands along the North Sea and Atlantic coasts in Ireland and the United Kingdom.

Mountains and Highlands

Two mountain ranges—the Pyrenees (PIR•eh•NEES) and the Alps—separate Western Europe from neighboring nations in Southern Europe. Both mountain ranges divide the cooler climates of the north from the warm, dry climate of the Mediterranean region to the south. To the southwest, the Pyrenees form a natural barrier between France and Spain. This mountain range stretches 270 miles (435 km) from east to west. Pico de Aneto—the tallest mountain in the Pyrenees—reaches a height of 11,169 feet (3,404 m). The average height of the Pyrenees, however, is only about 5,300 feet (1,615 m).

To the east of the Pyrenees lie the Alps. Like the Pyrenees, the Alps were created by the folding of rocks as a result of plate tectonics. Then these mountains were further shaped by glaciers. The Alps extend about 750 miles (1,207 km) along the southeastern border of France through Switzerland, Austria, and Germany. The Alps are much larger and higher than the Pyrenees. The tallest mountain in the Alps is Mont Blanc, at 15,771 feet (4,807 m).

The Alps and Pyrenees are geologically younger than other mountainous areas in Western Europe that have been worn down by glaciers. The most extensive is a plateau called the Middle Rhine Highlands in Germany, Belgium, Luxembourg, and France. Its highest elevations are about 3,000 feet (914 m).

Seas, Islands, and Waterways

Western Europe has long, irregular coastlines that touch many bodies of water, including the Atlantic Ocean and the North, Baltic, and Mediterranean seas. This closeness to the sea has long shaped European life.

The North Sea is part of the Atlantic Ocean but shallower. Separating the island of Britain from the rest of Europe, the North Sea is a rough, dangerous body of water to navigate. It is also a rich fishing ground for the Netherlands and the United Kingdom. The North Sea has long been important for trade between the United Kingdom and the rest of Europe. It is also the location of large oil and natural gas reserves.

Windmills are used to pump water into a reservoir for this village in the plains of the Netherlands.

▶ **CRITICAL THINKING**
Identifying What is the major plain in the region?

TADAO YAMAMOTO/a.collectionRF/Getty Images

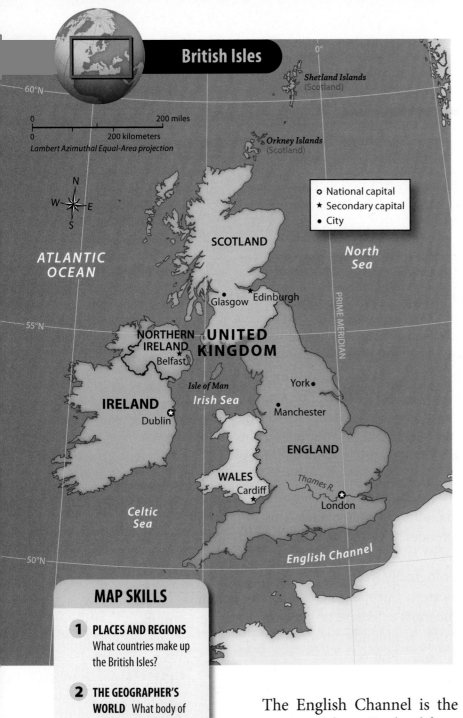

Shetland Islands
(Scotland)

Orkney Islands
(Scotland)

| National capital
★ Secondary capital
• City

ATLANTIC
OCEAN

SCOTLAND

North
Sea

Glasgow Edinburgh

NORTHERN
IRELAND
Belfast

UNITED
KINGDOM

PRIME MERIDIAN

Isle of Man
Irish Sea

York

Manchester

IRELAND

Dublin

ENGLAND

WALES

Thames R.

Celtic
Sea

Cardiff

London

English Channel

0   200 miles
0   200 kilometers
Lambert Azimuthal Equal-Area projection

N
W   E
S

MAP SKILLS

1 **PLACES AND REGIONS**
What countries make up
the British Isles?

2 **THE GEOGRAPHER'S
WORLD** What body of
water lies west of the
United Kingdom?

Academic Vocabulary

adapt to change a trait
to survive

The North Sea has helped but also hindered the Dutch, the people of the Netherlands. The Netherlands sits at a low elevation—25 percent of the country is below sea level. To **adapt**, the Dutch have built **dikes**, which are walls or barriers to hold back the water. The Dutch call the land they reclaim from the sea **polders**. This land is used for farming and settlement. Stormy seas, however, have broken dikes and caused flooding in recent times.

Just off the northern coast of France lie the British Isles, or the Atlantic Archipelago—the islands of Britain and Ireland. You might think each main island is its own country, but that is not the case. There are four parts of the United Kingdom: England, Wales, Scotland, and Northern Ireland. Three parts—England, Wales, and Scotland—occupy the island of Britain. The fourth part, Northern Ireland, is the northeastern section of Ireland. The country of Ireland, or Eire, occupies all but the northeastern section of the island of Ireland.

The English Channel is the part of the Atlantic Ocean that separates southern England from northern France. The channel is a busy sea route connecting the North Sea with the Atlantic Ocean. In addition to the ships that travel the route, trains pass beneath this waterway. A tunnel runs through the rock under the water. Called the Chunnel, it houses a high-speed train that connects Britain to mainland Europe.

Western Europe has a wealth of rivers and smaller waterways. The region's rivers have played an important role in how Europe developed over the centuries. Rivers determined the locations of cities, such as London, Paris, and Hamburg. Rivers provide transportation routes for goods and people. They also form some political borders. Rivers linked by canals provide transportation

This seaport in Dover, England, is one of the busiest in the world.
▶ **CRITICAL THINKING**
Analyzing What are some of the important waterways in Western Europe?

networks from inland areas to the sea. They also provide water for farming and electric power.

The Thames River in southern England is one of the most well-known rivers in the world. The Thames flows for 205 miles (330 km) from the Cotswold Hills to the city of London. When it reaches London, the Thames becomes an **estuary** and extends for another 65 miles (105 km) before it enters the North Sea. An estuary is where part of the sea connects to the lower end of a river.

The Rhine River is the busiest waterway in Europe. It begins high in the Swiss Alps and empties into the North Sea. Along its course, it plunges over waterfalls, cuts deeply into the Middle Rhine Highlands, and connects industrial areas to the port in Rotterdam. The Rhine meanders lazily across the plains of the Netherlands to the North Sea. The river serves as part of the political boundary between France and Germany. It runs through the most populated region of Europe.

Another important waterway in Germany is the Elbe River, which also empties into the North Sea. Historically, the Elbe formed part of the border between West Germany and East Germany when that country was split into two from 1945 to 1990 (after World War II). Another important river in Germany, the Danube, is different from the other rivers because it flows toward the east. It passes through southern Germany and Austria on its way into Eastern Europe. The Danube is Eastern Europe's most important waterway. The Main (MINE) River, a tributary of the Rhine, is linked to the Danube by the Main-Danube Canal. The canal links the North Sea with the Black Sea at the southeastern end of Europe.

Think **Again**?

London has black fogs that can kill people.

Not true. At one time, London had "pea soupers," thick, black and greenish fog. The fogs were caused in part by the burning of soft coal in homes and industry, which produced soot and poisonous sulfur dioxide. This mixed with the mist and fogs of the Thames Valley. In 1952 the Great Smog killed 4,000 London residents. Coal burning has been outlawed in London, but fumes from automobiles still produce smog in the city.

Medioimages/Photodisc

Roadways run along the Danube River in Vienna, Austria.

▶ **CRITICAL THINKING**

Describing How is the Danube different from other rivers in the region?

In France, the longest river is the Loire River, at a length of 634 miles (1,020 km). It passes through the Loire Valley, the most important agricultural region in France. It is the Seine River, however, which runs through the capital city of Paris, that carries far more of the country's inland water traffic.

☑ **READING PROGRESS CHECK**

Analyzing How did the rivers in Western Europe affect its economic development?

Climate

GUIDING QUESTION *Why is the climate mild in Western Europe?*

Western Europe is located at northern latitudes, but it has a milder climate than other places at the same latitudes. For example, southern France is at roughly the same latitude as Halifax, Nova Scotia, in Canada, but southern France experiences a milder climate. Why? Western Europe is located near the Atlantic Ocean. Warm winds off the ocean are the primary factor that shapes the region's climate.

Temperate Lowlands

Most of Western Europe lies in the path of the **Westerlies**, strong winds that travel from west to east. The Westerlies blow a constant stream of relatively warm air from the sea to the land. Why is the air so warm this far north? The answer is found on the other side of the Atlantic Ocean, in the tropical waters of the Caribbean Sea. Here, an

Willfried Gredler/age fotostock

ocean current called the Gulf Stream moves warm tropical water north along the eastern coast of North America. The Gulf Stream then flows across the Atlantic Ocean, where its eastern extension, the North Atlantic Current, approaches the European coast. The current's warm water heats the air above it. This warm, moist air moves inland on the Westerlies and brings mild temperatures and rain to most of Western Europe throughout the year. Summers are cool, and winters are mild. This climate is known as a marine west coast climate.

Most places that have a marine west coast climate have mountain ranges running north-south along the coast. The mountains block the warm, moist air from moving farther inland. Western Europe, however, does not have coastal mountain ranges, so the Westerlies blow farther across the European continent.

Mediterranean Climate Area

Other areas of Western Europe, such as southern France, have a drier climate. A high-pressure system called the Azores High travels north over the Atlantic Ocean during the summer months. This pressure system pushes moist air northward. As a result, summers in southern France are hot and dry. In winter, the nearby Mediterranean Sea moderates the climate so that winters are mild or cool. Most of the rainfall occurs in spring and autumn. This is called a Mediterranean climate.

☑ READING PROGRESS CHECK

Describing How do the Westerlies affect the climate of Western Europe?

Walkers take an autumn stroll through a forest in southern Belgium. In this area, autumns start mild but soon become cool. Winters are cold and snowy.

Natural Resources

GUIDING QUESTION *How do the people of Western Europe use the region's natural resources?*

Western Europe has many important natural resources. Layers of coal lie beneath rich, fertile soils. Beneath the North Sea are large pockets of oil and natural gas. Western Europeans use these natural resources to support their economies and populations.

Energy Sources

Deposits of coal are plentiful throughout Britain, Belgium, the Netherlands, France, and Germany. Coal was used to fuel machines invented during the Industrial Revolution of the 1800s. Today, coal production is declining in Northern Europe. Much less coal is available than in the past, so the region is importing more coal. Coal has also become less important as a source of energy as people rely more on other energy sources to meet their needs. In places where other fuels are scarce, Europeans burn peat. A peat bog is a wetland in which large masses of vegetable matter decay in the poorly drained soil. Peat—the name for the decaying vegetable matter—can be used as fuel for heating. The peat is dug up, cut into blocks, and dried so that it can be burned.

In 1959 oil and natural gas were discovered under the North Sea. Since then, the North Sea has become the region's most important source for these resources. The United Kingdom and the Northern European nation of Norway are the leading producers of oil and natural gas from North Sea oil fields. The Netherlands and Germany also produce oil and natural gas from the North Sea.

Other countries in Western Europe use their rivers to supply energy. Switzerland, for example, uses its fast-flowing rivers to produce electricity. Hydroelectricity supplies more than half of Switzerland's electricity needs.

Rich Soils

The Northern European Plain has some of the richest soils in Europe so people can farm there. Soils of the Northern European Plain contain humus. Humus is decomposed plant and animal material that makes soils rich and fertile, and is good for growing crops and raising livestock.

France is Western Europe's leading agricultural producer. France devotes more surface area to agriculture than any other country in the region. Large wheat fields stretch across northern France. Orchards and vineyards are common in the central and southern parts of the country. On the Northern European Plain, farmers grow a variety of crops and raise cattle and hogs. In the Netherlands, dairy farming is important for the country's economy.

Plants and Wildlife

The moderate climate and abundance of rainfall in most of Western Europe support a wide variety of plant and animal life. The British Isles have dense forests, grasslands, scrublands, and wetlands. The natural vegetation in most of the British Isles is **deciduous** forest, which includes trees such as oak, maple, beech, and chestnut, that lose their leaves in autumn. The climate on the mainland of Europe is more diverse than the climate of the British Isles. As a result, mainland Europe has a wider variety of plant life.

Farther inland, in Germany, Austria, and Switzerland, the drier climate as well as highlands and mountain ranges support other kinds of plants. Coniferous forests have become more common. **Coniferous** trees, such as fir and pine trees, have cones and needle-shaped leaves, and they keep their leaves during the winter. Many peaks in the Alps and Pyrenees lie above the tree line. Here, grasses and shrubs are the most common plants. Much of Western Europe was once covered in forests. Most of these forests were destroyed to make towns, cities, and roads.

France is one of the largest producers of wheat in the world; it is the largest European producer.

▶ **CRITICAL THINKING**

Analyzing What makes the soils of the Northern European Plain good for growing crops?

Wildlife

Animals have had to adapt to these changes. Deer, wild boars, hare, and mice are common. Wildcats, lynx, and foxes roam the forests. Small populations of brown bears live in the Pyrenees. In the British Isles, the population of large mammals, such as wolves, reindeer, and boars, has decreased. The islands have more than 200 kinds of birds, many of which have adapted to life in towns and cities.

☑ **READING PROGRESS CHECK**

Determining Central Ideas What effect did coal have on the Industrial Revolution?

Include this lesson's information in your Foldable®.

Philippe Huguen/AFP/Getty Images

LESSON 1 REVIEW **CCSS**

Reviewing Vocabulary

1. How do the *Westerlies* affect the climate in Western Europe?

Answering the Guiding Questions

2. *Identifying* What landmasses make up the British Isles, and what countries do they form?

3. *Describing* Why is the North Sea important to Western Europe?

4. *Analyzing* How does the marine west coast climate in Western Europe differ from marine west coast climates in other parts of the world?

5. *Describing* Why has the use of coal dwindled in Western Europe? What discovery in the North Sea helped bring about that change?

6. *Informative/Explanatory Writing* Write a paragraph describing how Europe's rivers contribute to the region's industry.

netw*o*rks

There's More Online!

☑ **GRAPHIC ORGANIZER**
☑ **SLIDE SHOW** WWII
☑ **VIDEO**

Reading **HELP**DESK

Academic Vocabulary

- **theory**

Content Vocabulary

- **smelting**
- **feudalism**
- **Middle Ages**
- **pilgrimage**
- **Parliament**
- **industrialized**
- **Holocaust**

TAKING NOTES: *Key Ideas and Details*

Summarize Use a chart like this one to describe the different ways nationalism affected people in Britain, France, and Germany.

Britain	France	Germany

Lesson 2
History of Western Europe

ESSENTIAL QUESTION · *Why do civilizations rise and fall?*

IT MATTERS BECAUSE
Even though the nations of Western Europe are not geographically large, their culture and technology have had a worldwide impact.

History of the Region Through 1800

GUIDING QUESTION *How did Western Europe change from a land controlled by loose-knit tribes to a region of monarch-ruled nations?*

Western Europeans were not the first Europeans to begin farming, but the rich soil and moderate climate drew many early people to the region. As the forests were cleared for farmland, people began a long struggle to control the land.

Beginnings

Modern humans have lived in Europe for about 40,000 years. The early people were hunters and gatherers, but over time the practice of agriculture was introduced in Western Europe. Populations began to grow, and settlements became towns. People began to make tools from metal, especially bronze. To make bronze, people needed to know how to melt and fuse tin and copper, a process called **smelting**.

Roman Empire

Meanwhile, the Romans were spreading throughout Southern Europe and advancing into Western Europe. By A.D. 14, all of France and most of Germany were under Roman control. Within 100 years, Rome also controlled most of the island of Britain.

The Romans did not just conquer people and territories. They brought their beliefs, their language (Latin), and their technologies with them. They built concrete roads and bridges throughout the empire. They also built aqueducts, which carried water long distances, from remote areas to cities and towns. Some of these structures are still visible today.

Over the centuries, Rome's empire in Western Europe began to weaken. The Huns, a warrior people from Asia, invaded from the east, driving invading groups of Germanic peoples, such as the Visigoths, westward. Rome could no longer protect its colonies in Western Europe. Some Germanic groups settled there and created kingdoms. The Franks ruled what is now France, and the Angles and Saxons ruled what is today England.

Christianity and Western Europe

Christianity, which became Europe's major religion, began in the eastern Roman Empire. It gradually spread throughout the empire. Once Emperor Constantine converted to Christianity in A.D. 312, Christianity began to spread quickly. By the time Rome's western empire fell in A.D. 476, Christianity was common throughout most of Western Europe. The Christian Church played a key role in education and developed religious communities called monasteries. The Roman Catholic Church became a major force in Western European life.

The Middle Ages

As time went on, invaders threatened the region. No strong governments existed to help Western Europeans fight off invasion. To bring order, a system called **feudalism** arose. Under feudalism, kings gave land to nobles. The nobles in turn gave kings military service. Many nobles became knights, or warriors on horseback. Today, we call this period of time the **Middle Ages**, or the Medieval Age. This term describes a period of transition between ancient and modern times.

Conflicts also arose over religious beliefs. One of the most important rituals in medieval European society was the religious **pilgrimage**, a visit to lands that were important to the history of Christianity. Jerusalem was the most important destination for pilgrims, but in the late 1000s, Muslims controlled the city. Pope Urban II, leader of the Catholic Church, called for a crusade to conquer Jerusalem for Christianity. The kings and noblemen of Western Europe formed great armies to meet the pope's demand. They won Jerusalem in the First Crusade. More crusades followed, but they were not successful. Muslims regained control of Jerusalem, and Muslim power continued to grow.

Thinking Like a Geographer

The Channel in History

Because the English Channel is the most direct route from continental Europe to the island of Britain, it has served as the route of many invasions. When Julius Caesar first invaded Britain in 55 B.C., the Roman army crossed the English Channel. In World War II, American, British, and Canadian troops invaded from the opposite direction, from Britain to the beaches of Normandy in France.

This iron and bronze helmet dates from the 300s B.C.

▶ **CRITICAL THINKING**
Identifying What warrior people from Asia invaded Western Europe?

Many of the crusaders returned to their European homes with changed ideas. They had seen a richer, more powerful, more modern world in the east. These ideas from the east began to spread across Northern Europe.

The economy in Western Europe was changing. Villages grew into towns. Traders and merchants began to play a bigger role in town life. Work became more specialized. People with important skills—metalsmiths, butchers, carpenters—began to organize into guilds. Guilds were not as powerful as the noblemen or the Church, but they helped the towns grow stronger.

Hundred Years' War

The threat of war between France and England flared throughout the 1200s and early 1300s. When war finally broke out between the two countries in 1337, the fighting lasted for a total of more than 100 years. England won important battles early on, gaining land in France. By the end of the Hundred Years' War, France had won all that land back. Several truces were agreed to during the war, some of them lasting many years. One of the most important developments of the war though was not a truce. It was the rapid spread of a terrible disease called a plague.

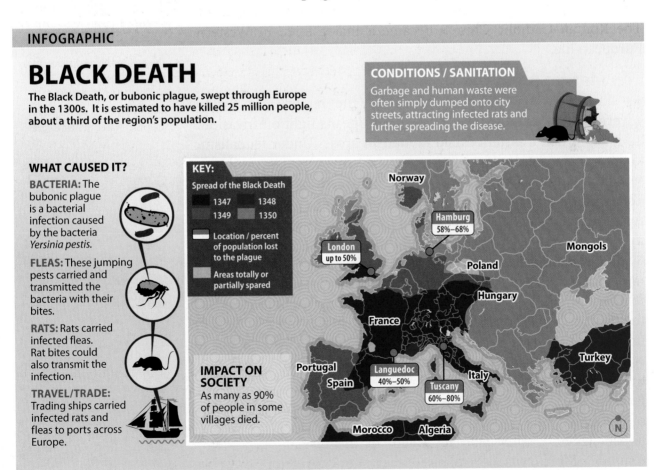

INFOGRAPHIC

BLACK DEATH

The Black Death, or bubonic plague, swept through Europe in the 1300s. It is estimated to have killed 25 million people, about a third of the region's population.

CONDITIONS / SANITATION
Garbage and human waste were often simply dumped onto city streets, attracting infected rats and further spreading the disease.

WHAT CAUSED IT?

BACTERIA: The bubonic plague is a bacterial infection caused by the bacteria *Yersinia pestis*.

FLEAS: These jumping pests carried and transmitted the bacteria with their bites.

RATS: Rats carried infected fleas. Rat bites could also transmit the infection.

TRAVEL/TRADE: Trading ships carried infected rats and fleas to ports across Europe.

KEY:
Spread of the Black Death
1347   1348
1349   1350

Location / percent of population lost to the plague

Areas totally or partially spared

IMPACT ON SOCIETY
As many as 90% of people in some villages died.

Norway
Hamburg 58%–68%
Mongols
London up to 50%
Poland
Hungary
France
Portugal
Spain
Languedoc 40%–50%
Tuscany 60%–80%
Italy
Turkey
Morocco   Algeria
N

With the great loss of population, the production of food and goods also decreased. Workers demanded higher wages and farmers increased the price of their goods.

▶ **CRITICAL THINKING**

Analyzing Explain the effect of supply and demand on wages.

In 1347 the plague, called the Black Death, reached Western Europe, where it raged for four years. Whole towns were wiped out. Four more outbreaks struck Europe by the end of the century. Victims of the Black Death often stayed in monasteries and hospitals run by Roman Catholic officials.

Early Modern Europe

The Roman Catholic Church was wealthy and had power over numerous aspects of society. Many people wanted to reform, or change, some Church teachings and practices. For example, most people did not speak Latin. Yet, the Bible was largely available only in Latin. People began to demand translations of the Bible in their languages so that they could read and interpret it on their own.

People also began questioning the moneymaking practices of Church officials. This included the sale of indulgences, pardons from the Church for a person's sins. A German priest named Martin Luther protested this practice. He declared that only trust in God could save people from their sins. In 1517 Luther wrote the Ninety-Five Theses, a document that attacked the practice of selling indulgences. The Church expelled Luther for his beliefs, but his ideas spread quickly. His followers became known as Lutherans, and his efforts spurred a religious movement called the Protestant Reformation.

As the Catholic Church's power weakened, England's kings also were being forced to share power with a new government institution called **Parliament**. This lawmaking body was made up of two houses. The House of Lords represented the wealthy, powerful nobles. The "lower" house, or House of Commons, represented the common citizens, usually successful guild members and business owners.

The Enlightenment

A wave of discovery and scientific observation swept over Europe in the 1600s and 1700s. During this time, European explorers were traveling and mapping the world, and European astronomers were mapping the solar system.

In 1543 the Polish astronomer Nicolaus Copernicus proposed the **theory** that Earth and the other planets orbit the sun instead of the sun and other planets orbiting Earth, as was then believed. Philosophers began to consider ways of improving society. People began to use reason to observe and describe the world around them. Reason transformed the way people thought about how to answer questions about the natural world. This period is called the Enlightenment.

English philosophers John Locke and Thomas Hobbes used reason to study society itself. Locke believed that the best form of government was a contract, or agreement, between the ruler and the people. People began to question the authority of kings and of the Church.

Think Again

Johannes Gutenberg invented the printing press.

Not true. The earliest printing was probably done with stamps that were pressed into clay or soft wax to create an image. The Chinese started printing on paper using wooden blocks with text carved into the wood. The Chinese also invented movable type, which used carved blocks of individual characters. Gutenberg did not invent the first printing press. His design for a movable-type printing press, however, is the one that was adopted and used going forward.

Academic Vocabulary

theory an explanation of why or how something happens

During the French Revolution in July 1789, a huge mob stormed the Bastille, an old fort used as a prison and a weapons armory.

▶ **CRITICAL THINKING**

Analyzing What was the result of the French Revolution?

Reform

In 1789 France was a powerful country, ruled by a king. Most of the people in France were peasants, living in poverty. But a growing, successful middle class resented not having a voice in government. In July of that year, a revolution limited the king's power and ended the privileges of nobles and church leaders. An important document was written: *The Declaration of the Rights of Man and of the Citizen*. It stated that government's power came from the people, not the king. A few years later, the king was removed and executed.

Not everyone in France supported the revolution. Violence raged in France for the next 10 years. Finally, in 1799 a young French general named Napoleon Bonaparte quickly took military and political control of the country. With a powerful army, he brought much of Europe under French control. By 1814, the combined might of France's enemies in conquered lands led to Napoleon's defeat and removal from power.

☑ **READING PROGRESS CHECK**

Identifying What roles did the Reformation and the Enlightenment play in changing the balance of power in Western Europe?

Change and Conflict

GUIDING QUESTION *How did the industrial system change life in Western Europe?*

During the 1800s, some Western European nations **industrialized**, or changed from an agricultural society to one based on industry. As a result, many people moved from the countryside to the city to

©Stefano Bianchetti/Corbis

find work in factories. The urban population grew, and the cities became powerful. At the same time, some Europeans began to feel strong loyalty to their country. A new, national spirit was rising.

The Industrial Revolution

A big change took place in Britain in the period from 1760 to 1830. People began to use steam-powered machines to perform work that had been done by humans or animals. For example, weavers in small villages once wove cloth on looms in their own homes, but new machines were invented to weave more cloth at greater speed for lower cost.

Machines of the Industrial Revolution did not affect only urban populations; they improved farm labor so much that fewer people were needed to work the land. People began to leave farms and villages for industrial cities where they could work in the factories.

As nations industrialized, loyalties shifted. Former enemies Great Britain and France grew closer as Germany gained military strength. As the possibility of war increased, alliances formed between countries.

World War I

Rivalries among European powers for new territory and economic power helped lead to World War I. Political changes also contributed as monarchies and empires were being replaced by modern nation-states.

A steam hammer in an English factory molds steel and iron into engine and machine parts.

▶ **CRITICAL THINKING**
Describing How did the Industrial Revolution affect rural populations?

Axis Control in World War II

Maximum extent of territory under Axis control

MAP SKILLS

1 PLACES AND REGIONS Which country in central Europe did not fall under Axis control?

2 THE GEOGRAPHER'S WORLD Why would control of Southern Europe and northern Africa be important?

Fought between 1914 and 1918, World War I resulted in millions of deaths and great destruction. Germany lost the war, and the victorious countries—led by Great Britain, France, Italy, and the United States—demanded that Germany pay for damages.

The defeat nearly wrecked the German economy. Germans believed they were being punished too harshly for their role in the war. A political radical named Adolf Hitler used the people's anger to build a political party called the Nazi Party. By 1933, he was the dictator, or absolute ruler, of Germany. The Nazis believed the Germans were a superior race. They carried out the **Holocaust**, the government-sponsored murder of 6 million Jews. Other minorities also suffered at the hands of the Nazis. Hitler and his Nazi Party envisioned a new German empire.

World War II

War came when Hitler's armies began seizing other countries. World War II stretched far beyond Western Europe. Germany allied with Italy and Japan to form the Axis Powers. Great Britain, the United States, and the Soviet Union formed the Allied Powers. The war was fought in Western and Southern Europe, in Africa, and in the Pacific. A combination of American, British, and Canadian troops invaded

France in June 1944 and liberated it from the Germans.

After Hitler's death and Germany's surrender in May 1945, the war continued in East Asia and the Pacific for another three months. The fighting ended after the United States used atomic bombs on the cities of Hiroshima and Nagasaki in Japan. Worldwide, between 40 million and 60 million people died in World War II. More civilians died than military forces.

The Cold War

Before World War II, Britain, France, and Germany were among the most powerful nations in the world. However, World War II had weakened them. After the war, the United States and the Soviet Union emerged as the leading world powers. Both superpowers were interested in Europe's fate. The Soviet Union took control of most of Eastern Europe. The United States was a strong ally to nations in Western Europe. Germany was split in half, with Britain, the United States, and France occupying western Germany, and the Soviet Union controlling the eastern half.

For more than 40 years, the United States and the Soviet Union engaged in a cold war, a conflict that never erupted into war, but the threat of war always existed. Both sides stockpiled nuclear weapons. In the 1980s, Soviet influence began to weaken. Protest movements spread in European countries under Soviet control. The Cold War ended when the government of the Soviet Union collapsed in 1991.

☑ **READING PROGRESS CHECK**

Determining Central Ideas How were the causes of World War I and World War II similar? How were the causes different?

Nazi soldiers post a notice urging Germans not to buy from this establishment because it is owned by a Jew. Nazi actions against Jews became violent.
▶ CRITICAL THINKING
Describing What was the Holocaust?

FOLDABLES Study Organizer

Include this lesson's information in your Foldable®.

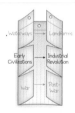

Hulton Archive/Getty Images

LESSON 2 REVIEW (CCSS)

Reviewing Vocabulary
1. Describe how *feudalism* worked.

Answering the Guiding Questions
2. *Identifying* What was the economic result of the plague?

3. *Determining Central Ideas* How did thinking change during the historical period known as the Enlightenment?

4. *Analyzing* How did the Industrial Revolution change life in Western Europe?

5. *Identifying* What factors led to World War I?

6. *Informative/Explanatory Writing* Write a paragraph to discuss this statement: The printing press was one of the greatest inventions in history. Explain why you agree or disagree with the statement.

netw⊙rks

There's More Online!

☑ **GRAPHIC ORGANIZER**

☑ **IMAGE** Dublin, Ireland

☑ **MAP** Empires: Western Europe

☑ **VIDEO**

Reading **HELP**DESK CCSS

Academic Vocabulary

- **cooperate**
- **regulate**
- **diverse**

Content Vocabulary

- **postindustrial**

TAKING NOTES: *Key Ideas and Details*

Identify Use a graphic organizer like the one here to identify three characteristics of major European cities.

Major European Cities

Lesson 3
Life in Western Europe

ESSENTIAL QUESTION · *How do governments change?*

IT MATTERS BECAUSE
Western European nations recognize that in a global economy, they need to work together if their region is to prosper.

People, Places, and Cultures

GUIDING QUESTION *What are contributions of Western Europe to culture, education, and the arts?*

Western Europe's great cities are major population centers. They are also historical landmarks and tourist attractions. National capitals such as London, Dublin, Berlin, and Paris are among the world's most famous cities.

A Changing World

Political events in the 1900s threatened all of Europe. In order to survive and compete in a changing world, the nations of Western Europe needed to learn to work together.

When Western Europe began to rebuild after World War II, countries made efforts to **cooperate**. In April 1951, the Treaty of Paris called for an international agency to supervise the coal and steel industries in France, West Germany, Belgium, the Netherlands, Luxembourg, and Italy.

Those six nations then created the European Economic Community, or EEC, in 1958 to make trade among its member nations easier. The spirit of cooperation among these countries continued when they created the European Commission, or EC, in 1967. Two more Western European nations, the United Kingdom and Ireland, joined the EC in 1971. By the late 1980s, Denmark, Greece, Spain, and Portugal had also joined.

Forming the European Union

Those 12 nations formed the European Union, or EU, in 1993 with one goal being to strengthen trade among the countries of Europe. Member nations have control over their own political and economic decisions, but they also follow EU laws to **regulate** the use of natural resources and the release of pollutants. They also have agreements on law enforcement and security.

When the Soviets lost control of Eastern Europe in the late 1980s, those nations began forming their own governments. With the Soviet threat gone, East Germany and West Germany reunited. A united Germany became a strong voice in the EU.

The European Union now has 27 members. Eight of those nations lie in Western Europe: Austria, Belgium, France, Germany, Ireland, Luxembourg, the Netherlands, and the United Kingdom.

Ethnic and Language Groups

Celts, Saxons, Romans, Vikings, Visigoths, and others fought for dominance in ancient Western Europe. Those traditional ethnic divisions faded as the modern nations of Europe began to take shape. The people of a nation share a common language and a common history. Ethnic groups such as the French, the Germans, and the British rule entire countries. Their languages are the main languages of those nations.

Paul M O'Connell/Flickr/Getty Images

Academic Vocabulary

cooperate to act or work with others

regulate to adjust or control something according to rules or laws

O'Connell Street is Dublin, Ireland's, main thoroughfare. Dublin is among the world's most famous cities.

Employees prepare a food order at a fast-food restaurant that caters to the city's large Islamic population.

Identifying From what areas did many Muslims come to Western Europe?

Western Europe is home to significant numbers of other ethnic groups who are minorities in the country. Many are immigrants. They often speak the language of their homeland and continue their own culture and way of life.

Most people in the region speak one of the Indo-European languages. Indo-European is a family of related languages. It includes languages spoken in most of Europe, parts of the world that were colonized by Europeans, Persia, India, and some other parts of Asia.

Two major divisions of Indo-European languages spoken in Western Europe are Romance and Germanic. Romance languages are based on Latin, the language of the Roman Empire. The most common Romance language in Western Europe is French. The Germanic languages spoken in Western Europe include German, Dutch, and English, although about half of the English vocabulary comes from the Romance languages. Not all European languages are Indo-European, however. For example, Basque, a language spoken in the Pyrenees region of France and Spain, is unrelated to any other language spoken today. It is common for Western Europeans to speak more than one language—their native language in addition to English, French, or German.

Religion in Western Europe

Romans accepted Christianity, and Christian missionary-monks spread their religion during the Middle Ages. Christianity continues as Europe's major religion. Germany was the birthplace of the Protestant Reformation, and Western Europe was the first place that Protestantism took hold. Today, most Western European Christians are either Catholic or Protestant. The Roman Catholic faith is strongest in France, Ireland, and Belgium. Protestant churches are strongest in the United Kingdom and Germany.

Immigration from Africa and Asia has brought many Muslims to Western Europe, especially to the United Kingdom, France, Germany, Austria, the Netherlands, and Switzerland. Muslims follow the religion of Islam.

World War II and the Holocaust nearly wiped out Europe's Jewish population. Today, Europe's Jewish communities are growing in Western Europe, especially in France, the United Kingdom, and Germany.

©JACKY NAEGELEN/Reuters/Corbis

Literature, Music, and the Arts

For centuries, Western Europe has been a world leader in the arts and culture. As European explorers spread European culture to other parts of the globe, the names of their greatest artists became known worldwide. England's William Shakespeare is one of the most famous playwrights in the world, nearly 400 years after his death. The music of German and Austrian composers such as Bach, Mozart, Beethoven, and Schubert is among the most important in all of classical music. The paintings of great artists from France, the Netherlands, and Belgium are among the most treasured in the world.

The arts are an important part of Western European culture. Museums and cultural institutions celebrate each nation's art and history, and national governments support the arts. The German government, for example, funds hundreds of theaters, and concerts and plays attract large audiences. Most important is the influence Western European culture has had on the rest of the world. German architects from the Bauhaus School influenced buildings in cities throughout the 1900s. British popular music and television have had an impact, especially on American culture.

MAP SKILLS

1 **THE GEOGRAPHER'S WORLD** The original members of the EU are from what part of Europe?

2 **HUMAN GEOGRAPHY** From a geographical standpoint, would you predict that the EU in the future will grow, stay the same, or decrease in members? Explain.

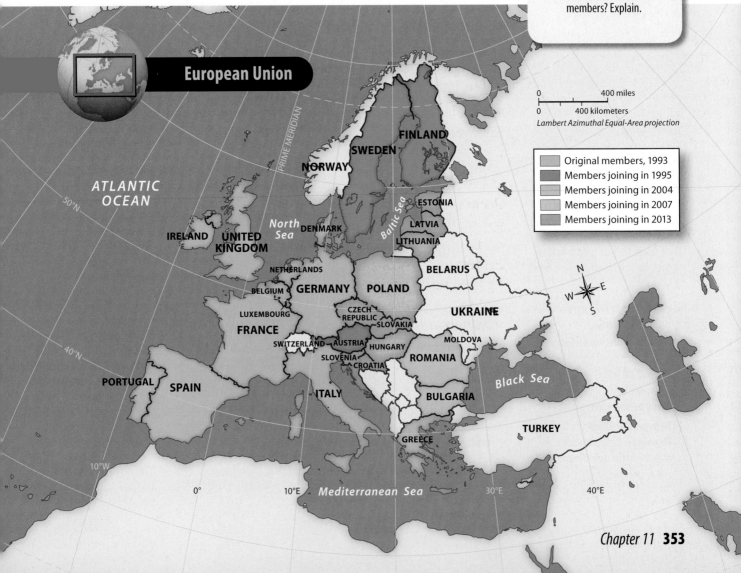

European Union

Original members, 1993
Members joining in 1995
Members joining in 2004
Members joining in 2007
Members joining in 2013

Lambert Azimuthal Equal-Area projection

Daily Life

The most popular team sport across Western Europe is football—what Americans call soccer. Professional leagues have formed throughout Western Europe. In the United Kingdom, cricket and rugby are popular team sports. Switzerland and Austria's rugged Alps and plentiful winter snow make mountain climbing, skating, downhill skiing, and cross-country skiing popular in both countries.

Because so much of the population of Western Europe lives in cities, roads are crowded. Automobile traffic and pollution are extensive in parts of the region. In Switzerland, traffic congestion has created serious air pollution in the Alpine valleys. To relieve congestion and address problems with pollution, much of Europe turned to high-speed rail travel.

In many areas, tradition is part of their everyday lives. The people of Scotland and Wales, for example, take pride in their ancient languages—Scottish Gaelic in Scotland and Welsh in Wales. These languages are taught in schools to keep the old cultures alive.

Railways and Highways

Europeans first began riding high-speed rail lines in France in 1981. France went on to build a high-speed line connecting all of its major cities. These trains travel at speeds of up to 185 miles (298 km) per hour. In the 1990s, the French high-speed rail lines began connecting to other high-speed rail lines: from Paris to London via a tunnel beneath the English Channel, from Paris to the Netherlands, and from Paris to Brussels, Belgium.

CHART SKILLS >

ENGLISH WORDS FROM OTHER LANGUAGES

Many of the words we use every day are derived from French or German. Many words that are used in fields of science, law, and medicine have Greek or Latin roots.

▶ CRITICAL THINKING

1. **Determining Word Meanings** The word *telephone* is derived from the Greek prefix *tele* (far) and the suffix *phone* (voice). Determine the meaning of the words: *telescope* and *telecast*.

2. **Determining Word Meanings** Determine the meaning of the words: *microphone* and *megaphone*.

French	German	Latin	Greek
ballet	kindergarten	a.m. /p.m.	atmosphere
denim	blitz	census	comedy
garage	poltergeist	millennium	democracy
infantry	noodle	lunar	geography
salon	hamster	solar	pediatrician
	pretzel		

Here are other English words derived from other languages:

African: banana, cola, jazz, zebra

Arabic: algebra, chemistry

Spanish: breeze, canyon, mesa

Japanese: anime, tycoon, tsunami

Norwegian: fjord, ski, slalom

The Louvre in Paris, France, is one of the world's great museums and the most-visited art museum in the world.

A well-developed highway system also links Europe's major cities. Germany's superhighways, called autobahns, are among Europe's best roads. Many European countries are participating in the Forever Open Roads project. By combining efforts to develop innovative technology, the planners are working to transform the way roads are designed, built, and maintained in the twenty-first century.

Education

Western Europe is one of the wealthiest, most urban, and well-educated regions in the world. In most of Western Europe, school is mandatory until students reach the age of 16, but many students then attend college.

Western Europe contains some of the oldest and most renowned universities. Oxford University in England and the University of Paris opened their doors to students before 1200. Many universities started at this time after Pope Gregory VII issued a ruling calling for the creation of schools of education for the clergy. Hundreds of secular colleges—those without religious affiliation—were established by the 1400s.

✔ **READING PROGRESS CHECK**

Describing In what ways do nations of Western Europe support art and culture?

Buena Vista Images/The Image Bank/Getty Images

Current Challenges

GUIDING QUESTION *Why is Western Europe considered a postindustrial region?*

France, Germany, and the United Kingdom are not the military giants they were in 1900, but they still have some of the biggest economies in the world. A global financial crisis, however, has hurt the entire region since the early 2000s.

Earning a Living

Since the Industrial Revolution, improvements in agriculture have made it possible for fewer people to cultivate larger areas of land. Today, more than half the population of Western Europe lives and works in cities. Even in France, Western Europe's leading agricultural nation, less than 4 percent of the workforce works in agriculture.

In the past few decades, the number of industrial workers has also declined. Only about 25 percent of Western Europeans work in the industrial, or secondary sector, of the economy. Many more people work in the tertiary sector, which is service industries. This sector includes government, education, health care, banking and financial services, retail, computing, and repair of mechanical equipment. The United Kingdom was the birthplace of modern industry. Yet today, only 18.2 percent of the workforce in the United Kingdom works in industry.

MAP SKILLS

1 **THE GEOGRAPHER'S WORLD** Which empire controlled the largest territory?

2 **THE GEOGRAPHER'S WORLD** What effect do you think world wars had on the British, French, and German empires?

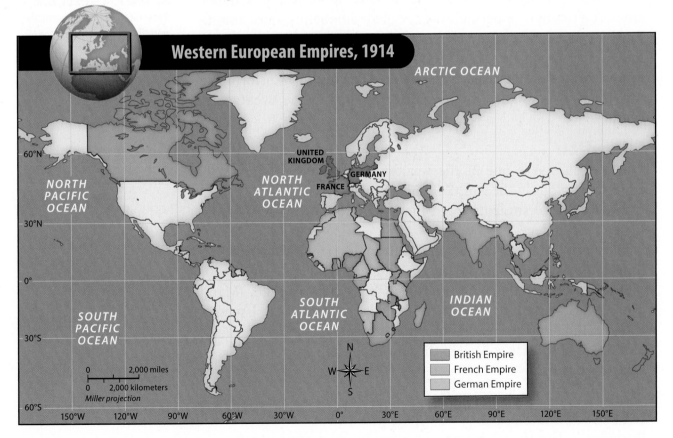

Western European Empires, 1914

British Empire
French Empire
German Empire

When the economy of a country depends more on services than it does on industry, that country is said to be **postindustrial**. Every nation in Western Europe has a postindustrial economy.

Challenges

For hundreds of years, the nations of Western Europe were among the most powerful in the world. In 1900 Great Britain, France, and Germany ruled over empires that extended beyond Europe to Asia, Africa, the Americas, and the Pacific Islands. The 1900s was hard on Western Europe. The two world wars did extensive damage to nearly the entire region. Then the Cold War kept Western Europe on the brink of war for more than 40 years.

Even so, Germany, France, and the United Kingdom have been economically strong for a long time and remain among the seven biggest economies in the world. The cooperation made possible by the European Union helps Western European nations compete with larger economies, such as the United States, China, and Japan. For that to continue, the economies of all the EU member nations must be healthy. The global financial crisis of 2008, however, had an impact on all of Europe. Governments of the EU disagreed about how to deal with ongoing financial problems.

Immigration Brings Changes

The population of Western Europe is changing. Most population growth in Western Europe is caused by immigration. Many people come to Western Europe from Africa, Asia, and Eastern Europe looking for job opportunities or trying to escape political oppression. When they immigrate, they bring parts of their culture with them, including their religions. Germany, France, the Netherlands, and the United Kingdom each have large Muslim populations. The mix of European and immigrant cultures creates a richer, more **diverse** culture, but it also creates racial and religious tensions. To avoid these problems, some countries have attempted to restrict immigration.

✓ READING PROGRESS CHECK

Determining Central Ideas What challenges do the nations of Western Europe face?

Academic Vocabulary

diverse different from each other

 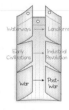

Include this lesson's information in your Foldable®.

LESSON 3 REVIEW (CCSS)

Reviewing Vocabulary

1. Define *postindustrial* as it relates to industry and services.

Answering the Guiding Questions

2. *Identifying* Name one advantage and one disadvantage resulting from the creation of the European Union.

3. *Analyzing* How did the collapse of the Soviet Union affect the European Union?

4. *Determining Central Ideas* Why is Western Europe considered a postindustrial region?

5. *Argument Writing* Write an open letter to the nations of Western Europe explaining the need to form the European Union.

What Do You Think?

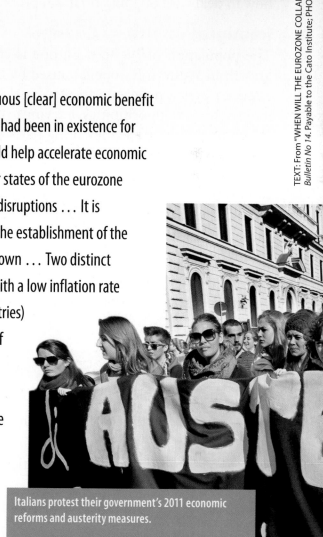

Is the European Union an Effective Economic Union?

The European Union was formed to build peace among its member states, but it also aims to build prosperity. In the 1990s, member states agreed to do away with trade barriers, create a single market, and make the euro their common currency. But the global financial crisis of 2008 hurt the EU greatly. Greece, Ireland, and Portugal needed bailouts to cover their huge debts. Italy and Spain struggled economically, too. EU leaders disagreed over how to respond to the crisis, and tensions still exist.

No!

PRIMARY SOURCE

❝ The creation of the eurozone was presented as an unambiguous [clear] economic benefit to all the countries willing to give up their own currencies that had been in existence for decades or centuries. ...[S]tudies promised that the euro would help accelerate economic growth and reduce inflation and stressed ... that the member states of the eurozone would be protected against all kinds of unfavorable economic disruptions ... It is absolutely clear that nothing of that sort has happened. After the establishment of the eurozone, the economic growth of its member states slowed down ... Two distinct groups of countries have formed within the eurozone—one with a low inflation rate and one (Greece, Spain, Portugal, Ireland and some other countries) with a higher inflation rate. ... [T]he economic performance of individual eurozone members diverged [went in different directions] and the negative effects of the 'straight-jacket' of a single currency over the individual member states have become visible. ... [A]s a project that promised to be of considerable economic benefit to its members, the eurozone has failed." ❞

—Václav Klaus, president of the Czech Republic

Italians protest their government's 2011 economic reforms and austerity measures.

TEXT: From "WHEN WILL THE EUROZONE COLLAPSE?" by Václav Klaus. Published May 26, 2010, Cato Institute Economic Bulletin No 14. Payable to the Cato Institute; PHOTO: Giorgio Cosulich/Getty Images/Getty Images News/Getty Images

TEXT: Reproduced and adapted from IT'S A BETTER LIFE. HOW THE EU'S SINGLE MARKET BENEFITS YOU ©European Communities, http://ec.europa.eu, 2002. Responsibility for the adaptation lies entirely with McGraw-Hill Education; PHOTO: Lester Lefkowitz/Iconica/Getty Images

Cargo containers at Felixstowe in the United Kingdom are prepared for transfer to cargo ships.

Yes!

PRIMARY SOURCE

" [T]he single market has transformed for the better many aspects of European life. . . . People move freely across most borders. . . . Going to work in another Member State is much easier. . . . Goods are no longer delayed for hours or days at borders by heavy paperwork: this makes delivery times shorter, allowing manufacturers to save money and reduce prices for customers. . . . Consumer choice is vast: the range of products on sale across the EU is wider than ever and in most cases prices are easily compared thanks to the euro. Manufacturers have to keep prices down because they are selling into one huge competitive market. . . . Capital—the investment that businesses need to start and to grow—flows easily within the single market, sustaining companies and generating jobs. "

—The European Commission

What Do You Think? DBQ

1. **Determining Central Ideas** How has the EU's single market benefited travelers, workers, consumers, and manufacturers?

2. **Identifying Point of View** Why does Václav Klaus consider the EU's single currency to be a "straight-jacket"?

Critical Thinking

3. **Analyzing** Some people think the eurozone will dissolve. Others, including Václav Klaus, believe that it will survive, especially if new rules force member countries to behave responsibly. What factors might affect what happens in the future?

Directions: Write your answers on a separate piece of paper.

1 Use your **FOLDABLES** to explore the Essential Question.
INFORMATIVE/EXPLANATORY WRITING Write an essay explaining how the physical geography and the climate of Western Europe influenced the development of agriculture in the region.

2 **21st Century Skills**
INTEGRATING VISUAL INFORMATION Working in small groups, choose any country in Western Europe and create a slide show depicting the country's landforms, waterways, and natural resources. Include images of the country's capital city. Describe at least one cultural tradition or holiday that is not familiar to you or to most Americans.

3 **Thinking Like a Geographer**
INTEGRATING VISUAL INFORMATION Fill in this graphic organizer to help you remember the climates of Western European countries.

Climate	Characteristics

4 **GEOGRAPHY ACTIVITY**

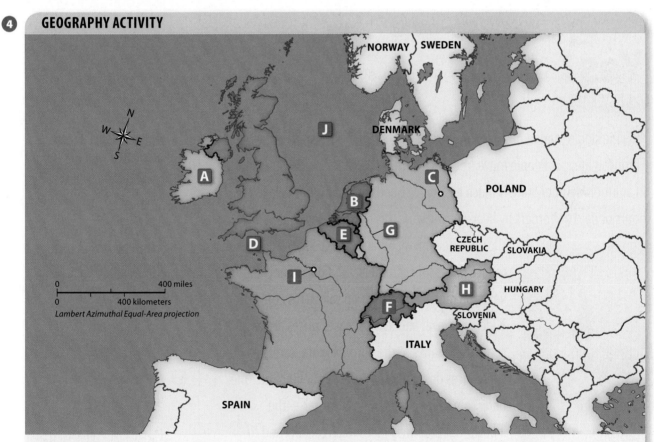

Locating Places
Match the letters on the map with the numbered places below.

1. Ireland
2. English Channel
3. Paris
4. Switzerland
5. the Netherlands
6. Belgium
7. Austria
8. Rhine River
9. Berlin
10. North Sea

Chapter 11 ASSESSMENT CCSS

REVIEW THE GUIDING QUESTIONS

Directions: Choose the best answer for each question.

1. People in the Netherlands build dikes and reclaim land from the sea that they call
   - A. peat bogs.
   - B. estuaries.
   - C. polders.
   - D. reserves.

2. Which two mountain ranges separate Western Europe from Southern Europe?
   - F. Alps and Urals
   - G. Pyrenees and Alps
   - H. Middle Rhine Highlands and Carpathians
   - I. Tian Shan and Apennines

3. In which Western European country did the Industrial Revolution begin?
   - A. Germany
   - B. France
   - C. the Netherlands
   - D. Great Britain

4. The system of land ownership and farming during the Middle Ages was called
   - F. pilgrimage.
   - G. channeling.
   - H. feudalism.
   - I. slavery.

5. The Industrial Revolution resulted in
   - A. a shortage of workers on farms.
   - B. a great migration of people from rural areas to cities.
   - C. more time for people to spend with their families.
   - D. handmade goods.

6. Many Europeans now work in the tertiary sector, which includes
   - F. factory work.
   - G. construction.
   - H. retail sales.
   - I. mining.

DBQ ANALYZING DOCUMENTS

7 **DETERMINING CENTRAL IDEAS**   Read the following passage:

"*Intense green vegetation . . . covers most of [Ireland]. . . . Ireland owes its greenness to moderate temperatures and moist air. The Atlantic Ocean, particularly the warm currents in the North Atlantic Drift, gives the country a more temperate climate than most others at the same latitude.*"

—from NASA, Earth Observatory Web Site (2011)

Based on this passage, what is unusual about Ireland's climate?

A. warmer than expected given its latitude

B. colder than expected given its nearness to the ocean

C. wetter than expected given that it is an island

D. drier than expected given its northerly location

8 **ANALYZING**   Which economic activities in Ireland most likely benefit most from this climate?

F. manufacturing and transportation

G. agriculture and tourism

H. finance and retailing

I. mining and fishing

SHORT RESPONSE

"*The European Union [EU] . . . was created in the aftermath of the Second World War. The first steps were to foster economic cooperation: the idea being that countries who trade with one another become economically interdependent and so more likely to avoid conflict.*"

—from European Union, "Basic Information" (2012)

9 **IDENTIFYING**   What was the original purpose of the European Union?

10 **CITING TEXT EVIDENCE**   Why did World War II convince European leaders to form the European Union?

EXTENDED RESPONSE

11 **INFORMATIVE/EXPLANATORY WRITING**   Explain why more countries have joined the European Union since its founding and whether you think it has been successful.

Need Extra Help?

If You've Missed Question	**1**	**2**	**3**	**4**	**5**	**6**	**7**	**8**	**9**	**10**	**11**
Review Lesson	1	1	2	2	2	3	1	3	3	3	3

http://europa.eu ©European Union, 1995–2012

NORTHERN AND SOUTHERN EUROPE

ESSENTIAL QUESTIONS • *How do people adapt to their environment?*
• *Why do civilizations rise and fall?* • *How do new ideas change the way people live?*

Norwegian fisherman from a
Lofoten Island fishing village

networks

There's More Online about Northern and
Southern Europe.

CHAPTER 12

Lesson 1
*Physical Geography of
the Regions*

Lesson 2
History of the Regions

Lesson 3
*Life in Northern and
Southern Europe*

The Story Matters...

The people of Northern Europe
live in a land of cold, harsh
winters. In contrast, the people
of Southern Europe enjoy the
warm Mediterranean climate.
What do these people have in
common? The countries in this
subregion have long histories,
dating back to the Vikings in the
north and the ancient Greeks and
Romans in the south. These
ancient peoples influenced many
aspects of culture—from
exploration to philosophy to the
discovery of new ideas.

FOLDABLES
Study Organizer

Go to the Foldables® library in the back
of your book to make a Foldable® that
will help you take notes while reading
this chapter.

Northern and Southern Europe

Geography | History | Culture

NORTHERN AND SOUTHERN EUROPE (CCSS)

Despite the distance between Northern and Southern Europe, many of the regions' countries are members of the European Union. Their histories show conflict, but today they are unified as members of the European continent, with similar economic goals.

Step Into the Place

MAP FOCUS Use the map to answer the following questions.

1 THE GEOGRAPHER'S WORLD Northern and Southern Europe have several large peninsulas. Identify two peninsulas in Northern Europe and three in Southern Europe, and name the countries of each.

2 PLACES AND REGIONS With so many countries bordering water, such as the North Sea, the Atlantic Ocean, and the Mediterranean Sea, what activity would you expect to find in Northern and Southern Europe?

3 PLACES AND REGIONS Name the capital cities of Spain, Italy, Norway, and Sweden.

4 CRITICAL THINKING
Analyzing With the difference in latitude between Northern and Southern Europe, how would you expect their climates to differ?

A

SOUTHERN COASTLINE Warm, bright sunshine floods the ruins of an ancient temple that overlooks the Mediterranean Sea in Greece.

B

NORTHERN COASTLINE A red fox walks across rocks in a wilderness area of northern Finland. The rugged land in far northern areas of Europe was shaped by retreating glaciers at the end of the last Ice Age.

Step Into the Time

TIME LINE Choose an event from the time line and write a journal entry from the point of view of a teenager living during that time. Describe how that event has changed your perception of the world.

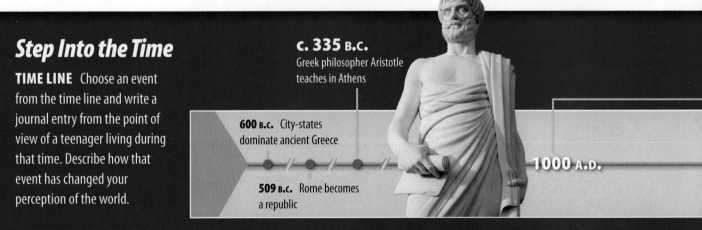

c. 335 B.C. Greek philosopher Aristotle teaches in Athens

600 B.C. City-states dominate ancient Greece

509 B.C. Rome becomes a republic

1000 A.D.

(t to b) Philip Coblentz/Brand X Pictures/PictureQuest; FLPA/Harri Taavetti/FLPA/age fotostock; Dimitris Tavlikos/Alamy

Northern and Southern Europe

ICELAND
Reykjavík

ARCTIC CIRCLE

PRIME MERIDIAN

60°N

Faeroe
Islands
(Denmark)

Norwegian
Sea

B

FINLAND

SWEDEN

NORWAY

Glama R.

Oslo   Stockholm

Helsinki

WESTERN
RUSSIA

50°N

North
Sea

DENMARK

Baltic Sea

Copenhagen

- National capital
- City

400 miles

400 kilometers

Lambert Azimuthal Equal-Area projection

ATLANTIC
OCEAN

WESTERN
EUROPE

EASTERN
EUROPE

40°N

PORTUGAL

Ebro R.

Andorra
la Vella

Milan

Po R.

SAN
MARINO

Black Sea

Madrid

ANDORRA

Florence

ITALY

İstanbul

Lisbon

Tagus R.

Rome

Naples

TURKEY

ASIA

SPAIN

Sardinia
(Italy)

VATICAN CITY
(within Rome)

GREECE

Aegean Sea

10°W

Balearic
Islands
(Spain)

Sicily
(Italy)

A

Athens

Nicosia

Strait of
Gibraltar

30°N

AFRICA

MALTA   Valletta

Mediterranean Sea

Crete
(Greece)

CYPRUS

0°

10°E

30°E

---

c. 1000
Leif Eriksson reaches
North America

1436 Gutenberg
invents printing press

1492 Columbus sets
sail to the Americas

1995 Sweden joins the
growing European Union

1945 World War II ends

1400

1900

2000

c. 1400 During Renaissance, astonishing
developments in arts and literature occur

1610 Italian astronomer Galileo
discovers moons orbiting Jupiter

1992 European Union forms

2008 Financial crisis leads
to worldwide recession

(l to r) Maremagnum/Photographer's Choice/Getty Images; ©Doug Pearson/JAI/Corbis; Katja Kreder/age fotostock; Arnulf Husmo/Stone/Getty Images

Reading **HELP**DESK ⓒⓒˢˢ

Academic Vocabulary

• **uniform**

Content Vocabulary

• **glaciation**
• **fjord**
• **tundra**
• **scrubland**
• **trawler**

TAKING NOTES: *Key Ideas and Details*

Identify Choose one of the countries in the region. List important resources of that country on a graphic organizer like the one below.

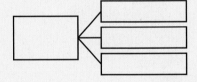

Lesson 1
Physical Geography of the Regions

ESSENTIAL QUESTION • *How do people adapt to their environment?*

IT MATTERS BECAUSE
Northern Europe and Southern Europe have different landforms, climates, and resources. The geography of the regions presents unique challenges to the people who live there.

Landforms and Waterways

GUIDING QUESTIONS *How are the landforms in Northern and Southern Europe similar? How are they different?*

Much of Northern Europe is a land of rugged mountains, rocky soils, and jagged coasts. A map of Northern Europe would show Denmark, Sweden, Norway, Finland, and Iceland. Together, these five far-northern lands are often called the Nordic countries.

The Mediterranean Sea dominates the coast of much of southern Europe, affecting the climate and the movement of people. Southern Europe is made up of Spain, Portugal, Italy, and Greece, as well as the tiny countries of Andorra, San Marino, and Vatican City. It also includes the island countries of Malta and Cyprus, the westernmost part of Turkey, and the tiny British territory of Gibraltar.

A Land of Peninsulas

A peninsula is an area of land surrounded on three sides by water. In the United States, Florida is a good example of a peninsula.

Two peninsulas make up most of Northern Europe. Jutland is a peninsula that extends northward from Germany and includes most of Denmark. The Scandinavian Peninsula is made up of Norway and Sweden. The large landmass east

of Sweden is Finland. Much of the land of Norway, Sweden, and Finland lies north of 60° N latitude.

Northern Europe also includes many islands. Iceland is a large island in the northern Atlantic Ocean near the Arctic Circle. Denmark has about 400 islands. Its capital, Copenhagen, is on the largest of the islands.

Southern Europe also has several peninsulas. Spain and Portugal form the Iberian Peninsula. Most of Italy is the long, boot-shaped Italian peninsula. East of Italy, the larger Balkan Peninsula includes several Eastern European nations, with Greece at its southern tip.

In Southern Europe, the large Mediterranean islands of Sicily and Sardinia are part of Italy. The nearby islands of Malta form an independent country. The island of Crete is part of Greece. Farther east, the island of Cyprus contains the largely Greek but independent nation of Cyprus, as well as North Cyprus, a Turkish territory.

Mountains and Plains

The Scandinavian Peninsula has a spine of rugged mountains, formed when two tectonic plates collided. **Glaciation**, or the weathering and erosion caused by moving masses of ice called glaciers, carved the land into the mountains and plateaus we see today.

Iceland also has rugged terrain, but it formed from volcanic activity. The island is part of a mountain range, the Mid-Atlantic Ridge, which is mostly underwater. At Iceland, it rises above sea level. Iceland is home to more than 200 volcanoes and many hot springs, as well as Europe's largest glacier.

Part of the boundary between Western Europe and Southern Europe is formed by two mountain ranges, the Pyrenees and the Alps. The Pyrenees mark the boundary between southern France and the Iberian Peninsula. The Alps form the northern border of Italy and separate the Italian peninsula from the rest of Europe.

Hikers pass beneath steep rock formations in the Dolomites, a branch of the Alps in northeastern Italy.

▶ **CRITICAL THINKING**

Describing Describe the geographical relationship of the Alps to the Italian peninsula.

Maremagnum/Photographer's Choice/Getty Images

Some countries of Southern Europe are mountainous. The Apennines extend along the length of Italy. They are volcanic and subject to earthquakes. Greece also has rugged highlands. The tallest and most famous of its mountains is Mount Olympus, which legend says was the home of the gods of Greek mythology.

Most of Spain lies on a plateau called the Meseta Central. It is a harsh landscape. To the north and south of the Meseta Central are mountain ranges, but its western side slopes gently toward the Atlantic Ocean. Valencia, along Spain's eastern coast, is a coastal plain of rolling hills.

Portugal is divided geographically by the Tagus River. South of the river, the landscape is characterized by extensive rolling plains. Throughout the area are abundant trees and plants, including evergreen oak trees, olive trees, figs, and vineyards. The northeastern coast features a landscape of wide valleys and steep hills.

The Mediterranean Sea

The most important body of water in Southern Europe is the Mediterranean Sea. It stretches about 2,500 miles (4,023 km) from the southern coast of Spain in the west to the coasts of Greece, Turkey, and various countries of Southwest Asia. The Mediterranean is almost completely surrounded by land. In the west, it connects to the Atlantic Ocean through the Strait of Gibraltar. At the Strait's narrowest point, just 8 miles (13 km) separates the southern tip of Spain from Africa.

Waterways

Among the important rivers of Southern Europe is Italy's Po. Many smaller rivers drain into the Po as it travels from the Alps to the Adriatic Sea. The Ebro is the longest river that lies entirely in Spain. It originates in the Cantabrian Mountains in the north and ends at Spain's Mediterranean coast. The longest river on the Iberian Peninsula is the Tagus, which crosses Spain and Portugal on its way to the Atlantic Ocean.

Northern Europe has few important rivers, but it has a long coastline, indented by many seas and bays. Norway is surrounded on three sides by water, and much of Norway's west coast is dotted by narrow, water-filled valleys called **fjords**.

The Baltic Sea borders Sweden's southern and southeastern coasts. Sweden has major ports on the Baltic, but the sea is more likely to freeze than other bodies of salt water. This is because the Baltic is shallow and has a fairly low concentration of salt. Finland also borders the Baltic Sea. Its interior is better known for its many lakes. Glaciers have carved as many as 56,000 lakes in Finland.

✓ **READING PROGRESS CHECK**

Analyzing Which landforms best characterize Northern Europe?

©Doug Pearson/JAI/Corbis

Visual Vocabulary

Fjord A fjord is a deep, narrow, sea-filled valley at the base of steep cliffs.

Contrasting Climates

GUIDING QUESTION *How is the climate of Northern Europe different from the climate of Southern Europe?*

Northern Europe has a cool or cold climate. Yet, the conditions in some places are much harsher than in others. Landforms, distance from the sea, and ocean currents play a part in determining climate.

Chilly Northern Europe

The northern part of Norway is located as far north as Alaska, but it is not nearly as cold. The Norwegian Current, part of the Gulf Stream, flows past Norway carrying warm water from the tropics. As a result, western Norway has a marine climate, with mild winters and cool summers. The relatively mild climate does not extend far to the east. Mountains reduce the eastward flow of milder air. Eastern Norway thus has colder and snowier winters. Even so, the climate in eastern Norway and in Sweden is milder than in other parts of the world at the same latitude.

Finland's climate is considered continental because it receives little influence from the seas. It has cold winters and hot summers. Winters are harsh, especially north of the Arctic Circle. In the mountainous parts of northern Finland, the snow never melts. This land is mostly **tundra**, a region where subsoil is frozen and only plants such as lichens and mosses can survive.

The Gulf Stream brings warm water to the southern and western coasts of Iceland. This moderates the temperatures in these parts of the country. The effect does not reach northern Iceland, which just touches the Arctic Circle. There, drifting ice and fog are common in winter, and temperatures year-round are several degrees colder than they are in the southern and western parts of the country.

Warm Southern Europe

The most common climate in Southern Europe is Mediterranean. The climate features warm or hot summers and cool or mild winters. Spring and fall are rainy, but summers are dry. If you have ever been to the southern coast of California, you have experienced a Mediterranean climate.

Temperatures are not **uniform** across Southern Europe or even within countries. Northern Italy is nearer the Alps and has a cooler mountain climate, where winters are colder than in southern Italy and snow is heavy at higher elevations. The Meseta Central, a vast plateau in Spain, on the other hand, has a continental climate.

Katja Kreder/age fotostock

Academic Vocabulary

uniform the same or similar

Vacationers enjoy an attractive beach in Ibiza, one of the Balearic Islands. These scenic islands in the Mediterranean Sea are part of Spain.

▶ **CRITICAL THINKING**
Analyzing Why might many Northern Europeans vacation in Spain and other Mediterranean countries during the winter months?

The area's high elevation and mountain barriers cause dry winds and drought conditions year-round. The dryness causes temperature extremes, with cold winters and hot summers.

☑ **READING PROGRESS CHECK**

Identifying How do landforms and waterways affect the climates of Norway and Italy?

Natural Resources

GUIDING QUESTION *What natural resources are available to the people of Northern and Southern Europe?*

Northern and Southern Europe hold rich stores of resources. The sea also provides a variety of resources, from fish to oil and gas.

Vegetation

Plants need to be drought resistant in order to survive the dry summers in a Mediterranean climate. Because of the dry climate and poor soils in Southern Europe, many areas are **scrubland**, or places where short grasses and shrubs are the dominant plants. Trees such as olive, fig, and cypress are common. Two of the most important crops throughout Southern Europe are grapes and olives. Wine, which is made from grapes, is an important export for Spain, Portugal, Italy, and Greece. Italy and Greece are also major exporters of olive oil.

In widely forested Northern Europe, the main plant resource is wood. Forests cover nearly three-fourths of Finland, and wood from these forests is Finland's most important natural resource. Finland exports birch, spruce, and pine wood and paper products to Western Europe. Sweden's forests produce timber, paper, wood pulp, and furniture.

Northern Minerals and Energy

Norway might not be considered an energy powerhouse. But, thanks to the discovery of petroleum in the North Sea, Norway is now Europe's biggest exporter of oil and one of Europe's leading suppliers of natural gas.

Northern Europe also has rich mineral ore resources. With deposits of iron ore, copper, titanium, lead, nickel, and zinc, Norway remains one of the world's leading metal exporters. Sweden lacks fossil fuels, but it has rich mineral resources. These include iron ore, copper, gold, zinc, and lead.

The Sautso Dam in Norway spans northern Europe's largest canyon. The water reservoir behind the dam is 11.8 miles (19 km) long.

▶ **CRITICAL THINKING**

Identifying What are Norway's main exports?

Arnulf Husmo/Stone/Getty Images

Denmark has few natural mineral or energy resources. However, Denmark uses wind turbines to supply much of its electricity. Sweden and Norway get much of their electricity from hydroelectric power plants. Because of the volcanic nature of the island, Iceland has enormous reserves of geothermal energy. It provides energy for industries and all of the heating needs of Reykjavik, Iceland's major industrial area, as well as several other towns. Iceland's rivers also supply hydroelectric power.

Sea Resources

A rich variety of fish inhabit the North Sea, and the nations along its shores have long-standing fishing traditions. Norway has one of the biggest fishing industries in Europe. Some of the fish caught, such as sand eels and mackerel, are ground into fish meal, a powder that is used in animal feed and fertilizer. Today, fewer people work in the fishing industry. This is because ships that tow large nets behind them, called factory ships or **trawlers**, have increased catches.

At one time, whaling was an important industry for many countries by the sea, including Norway. Whales were hunted for their meat and oil, but by the mid-1900s, many of the largest species of whales were in danger of becoming extinct. Whaling is now limited to a few smaller species that are not believed to be endangered.

Spain and Portugal have extensive Atlantic Ocean coastlines, and fishing is an important industry in both countries. The cities of Vigo and La Coruña are Spain's biggest fishing ports. Spanish fishing fleets range far from their shores, however, leading to conflicts with other countries.

The Mediterranean Sea has long been an important fishing ground for the countries that border its shores, including Italy and Greece. However, overfishing and pollution have reduced fish populations. In addition, the Mediterranean Sea lacks the nutrients necessary to support large populations of fish. Commercial fisheries use fish hatcheries to cultivate and breed fish. After the fish mature, they are released into the sea. That way, they repopulate the fish population that has been overfished.

✓ READING PROGRESS CHECK

Analyzing How can a thriving fishing industry be a positive and a negative factor for a country?

Thinking Like a Geographer

A "Hot Water" Project

The building of the Sautso Dam and water plant sparked a heated debate in Norway when the original plans were presented. Opponents pointed out that the project threatened local villages of the indigenous Sami people, as well as an important river. Protests focused attention on the need to protect Sami culture and the environment. The dam was finally built, but the plans changed enough to meet many of the protesters' concerns.

FOLDABLES
Study Organizer

Include this lesson's information in your Foldable®.

Northern and Southern Europe
Geography | History | Culture

LESSON 1 REVIEW

Reviewing Vocabulary

1. How does *glaciation* affect the landscape?

Answering the Guiding Questions

2. *Describing* How are the landforms of Northern and Southern Europe alike and different?

3. *Describing* How would the climate of Norway be different without the Norwegian Current?

4. *Analyzing* What kinds of problems does the fishing industry in Southern Europe face?

5. *Informative/Explanatory* Describe what is unique about the island of Iceland.

Reading **HELP**DESK

Academic Vocabulary

- **achievement**
- **convert**
- **rational**

Content Vocabulary

- **city-state**
- **longship**
- **pagan**
- **Renaissance**

TAKING NOTES: *Key Ideas and Details*

Summarizing Choose one of the countries in the lesson. As you read, use a graphic organizer like the one below to describe three or more important events in that country's history.

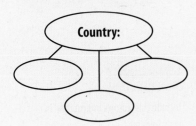

Country:

Lesson 2
History of the Regions

ESSENTIAL QUESTION • *Why do civilizations rise and fall?*

IT MATTERS BECAUSE
Some of the most influential civilizations in the world originated in these two regions. As you read, think about how the accomplishments of the past influence us today.

Early History of the Regions

GUIDING QUESTION *Why were early civilizations in Northern and Southern Europe important?*

Southern Europe produced two of the world's most influential civilizations: ancient Greece and ancient Rome. Greek and Roman, or classical, culture continue to affect our world. During the Middle Ages, the period between A.D. 500 and 1500, Christianity and other classical ideas helped build a new, orderly European civilization. But wars and invasions were still common. The Vikings of Northern Europe were seafarers and invaders. Their voyages changed the history of Western Europe and North America.

Ancient Greece

Greece's many mountains and seacoasts influenced the ancient Greeks to form separate communities called **city-states**. Each city-state was independent, but each one was linked to the other city-states by Greek language and culture. Powerful city-states, such as Athens and Sparta, were rivals, but they faced a common enemy to the east in mighty Persia. When the Persians invaded the Greek mainland in 490 B.C., the combined forces of Athens's navy and Sparta's army spent 40 years defeating them.

After the Persian Wars, Athens emerged as the most developed city-state. Its **achievements** were momentous. For

example, the ideas of philosophers such as Socrates, Plato, and Aristotle are still studied. Greek art set a standard for beauty that later influenced the Romans and inspires people to this day. Athens was also the first known democracy. The free citizens of Athens enjoyed a way of life that was unique in ancient times.

Wars weakened the Greek city-states, and Macedon, a kingdom north of Greece, took advantage. The Macedonian king, Alexander the Great, extended his rule over not only Greece, but also Asia Minor (now the Asian part of Turkey), Persia, and Egypt. Even though Alexander died at the age of 33, he spread Greek culture throughout an empire that lasted another 300 years.

Roman Empires

While the Greek city-states were at their height, another group was slowly gaining power to the west on the Italian peninsula. A series of small settlements built on hills along the Tiber River eventually merged into a single city that became Rome. Later, the Romans put government into the hands of consuls, who were elected to office annually. This was the birth of the Roman Republic.

The Romans had a talent for warfare, and they set out to conquer their neighbors. By 275 B.C., they controlled the Italian peninsula, inspiring Rome to add even more territory.

Academic Vocabulary

achievement the result gained by a great or heroic deed

MAP SKILLS

1 **PHYSICAL GEOGRAPHY** Why was Italy an ideal location to be the center of the Roman Empire?

2 **PLACES AND REGIONS** Which areas were part of Alexander's empire and later part of the Roman Empire?

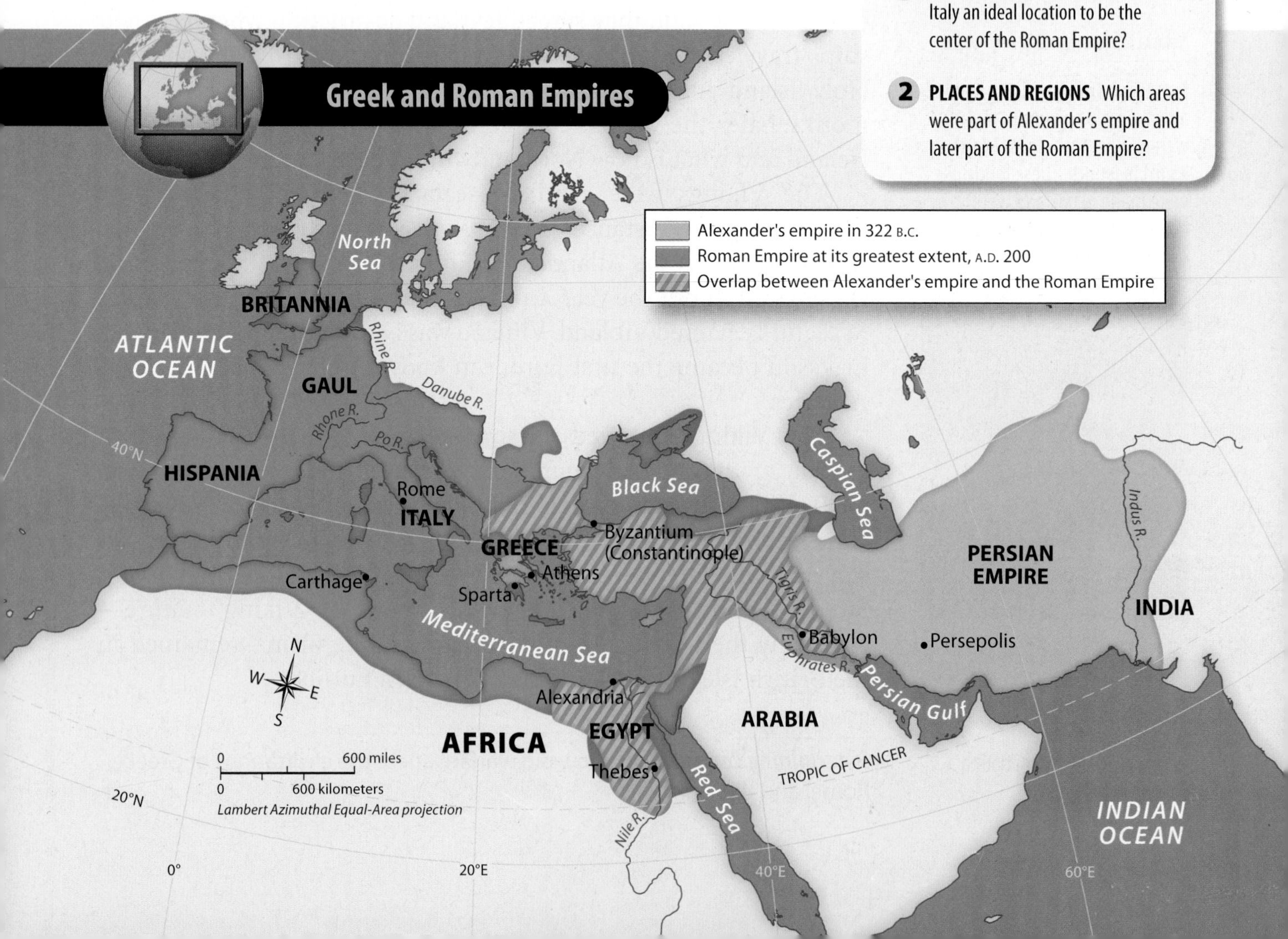

Greek and Roman Empires

Alexander's empire in 322 B.C.
Roman Empire at its greatest extent, A.D. 200
Overlap between Alexander's empire and the Roman Empire

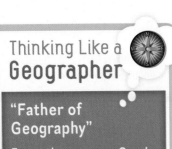

Spain, Sicily, Macedonia, Greece, and Asia Minor fell to Roman armies and were turned into Roman provinces. Eventually, a powerful military leader, Julius Caesar, seized control of Rome. After Caesar was assassinated, his great-nephew, Octavian, given the title Augustus, became the first of a series of emperors, and the Roman Republic was no more. The new Roman Empire expanded eastward to Egypt and westward to the British Isles.

In A.D. 330, Emperor Constantine moved the capital of the empire from Rome to the Greek city of Byzantium, in what is now Turkey. This location was closer to important trade routes to China and Southwest Asia. The new capital, renamed Constantinople, was also farther from the barbarians who were attacking the Roman Empire in the west. Repeated invasions continued to weaken the western empire. German invaders took control of Rome in A.D. 476. This was the end of the western Roman Empire. The eastern empire lasted for almost the next 1,000 years until it fell in 1453 to the Ottoman Turks. The Turks changed the name Constantinople to Istanbul.

The Viking Age

In the A.D. 700s, ships carrying warriors from Scandinavia began raiding the coasts of Western Europe. At home, these warriors were farmers or young men eager for adventure. At sea, they were pirates called Vikings, and they spread fear and destruction wherever their ships traveled. In A.D. 793, they raided and destroyed the abbey at Holy Island in northeastern England, killing and enslaving the monks. Later, the Vikings conquered other parts of Britain as well as Ireland and what is now Normandy in France.

The Vikings were excellent seafarers, and they sailed their **longships** great distances to explore and to trade. They sailed westward across the Atlantic, founding settlements in Iceland and Greenland. About the year A.D. 1000, Leif Eriksson led the Vikings to a land he named Vinland. Vinland was Newfoundland, in Canada. Eriksson became the first European known to have reached North America.

The Vikings followed a **pagan** religion, which was based on ancient myths and had a number of different gods. After about A.D. 1000, Viking groups throughout Scandinavia began to **convert** to Christianity. The Viking threat died out as more Scandinavians stayed home. They contributed to building the kingdoms of Norway, Sweden, and Denmark. However, traces of Viking culture—especially their epic tales of adventure and heroism—remained in the British Isles and other parts of Western Europe.

Visual Vocabulary

Longship The Vikings raided and explored on sturdy ships that were 45 feet to 75 feet (14 m to 23 m) in length. These longships were made of oak and were powered by wind and up to 60 oarsmen. The square sails were sometimes brightly colored.

Academic Vocabulary

convert to bring about a change in beliefs

☑ **READING PROGRESS CHECK**

Determining Central Ideas How did warfare affect the civilizations of Greece, Rome, and the Vikings?

Discovery and "Rebirth"

GUIDING QUESTION *How did the Renaissance pave the way for voyages of discovery?*

During the long period known as the Middle Ages, many of the ancient achievements were forgotten. Important manuscripts were lost or destroyed. Many of the writings that survived ended up in the East, where scholars could still read classical Greek. Beginning in the 1300s, a curiosity for Greek and Roman learning took hold in the Italian city of Florence, where poets such as Dante and Petrarch were inspired by ancient literature. To them, these works were freer, more **rational**, and more joyous than the works of their world.

Renaissance

When the Byzantine Empire fell in 1453, many scholars traveled west with ancient Greek manuscripts. At the same time, a practical printing press was invented in Germany. Suddenly, it was possible to print many copies of manuscripts that until then had to be lettered by hand. People could now own and read books.

These breakthroughs resulted in a period of artistic and intellectual activity known as the **Renaissance**. The Italian city of Florence became a center of learning and culture. Architects drew inspiration from the ancients and created new architectural styles. Painters and sculptors, such as Leonardo da Vinci, looked to nature for inspiration.

Curiosity about the natural world also led to the birth of modern science. In 1609 Italian astronomer Galileo designed a telescope to observe the moon and the planets. Galileo's observations helped prove Copernicus's theory that the planets, including Earth, orbit the sun. The Renaissance began in Italy, but Galileo's work influenced scientists throughout Europe.

Empires and Exploration

By the 1400s, Europeans wanted to do more business with China and India, but overland routes were long and dangerous. Prince Henry of Portugal inspired sailors and navigators to find a sea route to Asia by sailing around Africa. In 1488 Portuguese sea captain Bartholomeu Dias reached the Cape of Good Hope at the southern tip of Africa.

Academic Vocabulary

rational based on reason or logic

Hagia Sophia (or "Holy Wisdom") in Istanbul, at first, was a Christian church. It later became a Muslim place of worship, and today it is a museum.

▶ **CRITICAL THINKING**
Identifying What was the name of the city of Istanbul when Hagia Sophia was built?

Ten years later, Vasco de Gama rounded the Cape and sailed to India. The Portuguese established sea trade with South Asia.

Christopher Columbus, an Italian navigator, had a different idea: Why not reach Asia via a westward sea route? Spain agreed to finance the expedition, and Columbus left Spain with three ships on August 3, 1492. Columbus underestimated the size of Earth and overestimated the size of Asia. When he finally saw land, he assumed he had reached Asia. In fact, it was a Caribbean island. Columbus, like Leif Eriksson before him, had landed in the Americas.

Through its expeditions to the Americas, Spain became the most powerful country in Europe. The Spanish built an empire in Mexico, Central America, and South America. They conquered the Aztec of Mexico and the Inca Empire in South America, and they enslaved the native peoples of the Caribbean.

Contact between Europe and the Americas also resulted in the exchange of goods. Europeans brought wheat, olives, bananas, coffee, sugar, horses, sheep, pigs, and cattle to the Americas. In exchange, the Europeans received tomatoes, corn (maize), potatoes, squash, cacao (the source of chocolate), and hot peppers. This commerce is known as the Columbian Exchange.

Religion in the Regions

Meanwhile, Christianity had become identified with Europe. Rome considered itself the seat of Christianity as early as the A.D. 100s. When the Roman Empire was split into eastern and western empires, Christianity in the empire also split into eastern and western branches. The western branch evolved into the Roman Catholic Church, which was dominant in Italy and Spain and throughout Western Europe. The eastern branch became the Eastern Orthodox Church, centered in Greece and parts of Eastern Europe.

The rise of the religion of Islam threatened the power of the Christian churches. The Moors were Muslims, followers of Islam, who invaded Spain from Northern Africa in the A.D. 700s and ruled most of Spain for more than 700 years. The Byzantine Empire fell to the Ottoman Turks, another Muslim people, in 1453. Under Ottoman rule, Greek Christians were free to practice their religion, but their rights were limited compared to those of Muslims. Over the next few centuries, Greek Christians struggled to preserve their traditions and beliefs. Today, the vast majority of people in Greece still belong to the Greek Orthodox Church.

During the 1520s, the ideas of Martin Luther contributed to the spread of the Protestant Reformation. Kingdoms across Northern Europe broke away from the Roman Catholic religion. The countries adopted some form of Protestantism as their official state religion.

☑ **READING PROGRESS CHECK**

Determining Central Ideas Why did Christianity change as it took hold in Southern and Northern Europe?

History in the Modern Era

GUIDING QUESTION *What has been the relationship between Northern Europe and Southern Europe over the last 200 years?*

The 1800s brought sweeping changes to the regions. The Scandinavian countries saw their military glory vanish, but they became prosperous democracies. Spain and Portugal lost much of their overseas empires, followed by conflicts at home. Greece won freedom from the Turks. Italy's separate territories, except for San Marino and what later became Vatican City, united in 1870.

Conflict and War

During the 1900s, Northern and Southern Europe were involved in both world wars. After fighting a civil war in the 1930s, Spain stayed out of World War II. However, Italy, ruled by dictator Benito Mussolini, sided with Nazi Germany in that conflict. The Italians were defeated by allied U.S. and British forces in 1943, a year and a half before the war in Europe ended.

The period following World War II brought even more political changes. Italy became a democracy and began to rebuild its economy. Greece suffered a brutal civil war between Communists and opponents of communism. Spain, Portugal, and Greece joined Italy as democracies.

The Modern Era

Since 1945, Scandinavia has enjoyed a high standard of living as well as political and social freedoms. Its leaders worked for world peace and economic growth in the world's new nations. Beginning in the 1990s, the nations of Northern and Southern Europe developed closer ties with each other and with other European countries as members of the European Union (EU).

 READING PROGRESS CHECK

Identifying What country did Italy side with during World War II?

Swedish troops helped Finland in that country's conflict with the Soviet Union during World War II.

▶ **CRITICAL THINKING**

Describing How was Southern Europe involved in World War II?

Include this lesson's information in your Foldable®.

Keystone-France/Gamma-Keystone/Getty Images

LESSON 2 REVIEW

Reviewing Vocabulary

1. What changes did the *Renaissance* bring about in Europe?

Answering the Guiding Questions

2. *Identifying* Why were the ancient Greek and Roman civilizations important?

3. *Analyzing* How did the development of printing help promote voyages of discovery?

4. *Describing* How would you characterize the relationship among the countries of Northern Europe over the past 200 years?

5. *Argument Writing* Select a Renaissance thinker you believe was the most important. In a paragraph, explain why you believe this person's contributions were the most significant to the age and to people today.

Reading **HELP**DESK (CCSS)

Academic Vocabulary

- **contribution**

Content Vocabulary

- **homogeneous**
- **dialect**
- **welfare capitalism**
- **recession**

TAKING NOTES: *Key Ideas and Details*

Identify Use a graphic organizer like this one to list several artistic contributions of the countries of Northern and Southern Europe.

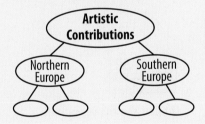

Lesson 3
Life in Northern and Southern Europe

ESSENTIAL QUESTION · *How do new ideas change the way people live?*

IT MATTERS BECAUSE

The countries of Europe are experiencing changes that will affect the lives of the people who live there. These changes will alter the relationships of the countries to each other and to the rest of the world.

People and Places

GUIDING QUESTION *What is the distribution of the populations in Northern Europe and in Southern Europe?*

Aging Populations

In most places in Northern and Southern Europe, improvements in the standard of living have helped reduce infant mortality—the number of babies who die in their first year of life. Improvements in the standard of living have also helped people live longer. You might think this would mean that populations in European countries are growing more quickly than in the past. That is not the case. Birthrates have declined as Europeans decided to have fewer children, and population growth has slowed in the past few decades. Older people have become a larger percentage of the population. Therefore, the population is aging.

Where People Live

As in other parts of Europe, the proportion of people who live in the rural areas of Southern Europe has decreased over the last 100 years. In 1900 Italy's three largest cities—Rome, Milan, and Naples—each had about 500,000 residents. The combined metropolitan area population of the three cities is

now about 10 million. About 60 percent of the population of Greece live in cities. In fact, 25 percent of the population live in the capital city of Athens. More than 75 percent of Spain's population live in cities and towns, making rugged rural areas such as the Meseta Central region seem nearly empty in comparison. Madrid, Spain's capital, has a population of more than 5 million, and the population density is 1,750 persons per square mile (675 per sq. km). Compare this to the population density for the country as a whole: 220 persons per square mile (85 per sq. km).

In Northern Europe, the population is even more concentrated in cities. Most people live in the southern parts of the countries because of the milder climates. More than half of Iceland's population lives in the capital city, Reykjavik. The capital cities of Northern European countries—Copenhagen (Denmark), Oslo (Norway), Stockholm (Sweden), and Helsinki (Finland)—have by far the largest populations of any cities in their countries. They are also their countries' primary cultural centers.

☑ **READING PROGRESS CHECK**

Analyzing Why is the population of Northern and Southern Europe growing older?

Bikers cross a historic square in the central area of Copenhagen, the capital of Denmark.

▶ **CRITICAL THINKING**

Describing What are major characteristics of capital cities in Northern European countries?

People and Cultures

GUIDING QUESTION *Why are most people in Northern Europe Protestant, whereas most people in Southern Europe are Catholic?*

Ethnic and Language Groups

The populations of the Northern European countries are relatively **homogeneous**, or alike, although they have some ethnic diversity due to immigration from Asia and Africa. The population of Norway is more than 90 percent Norwegian, and the population of Finland is more than 90 percent Finnish. Denmark and Sweden are similar. The original settlers of these lands were the Sami, or Lapps. Many of the Sami live by fishing and hunting, as they have for thousands of years. Most live in the northern parts of Sweden, Norway, and Finland. Iceland was first settled by Celts and later conquered by Norway. The bulk of the population in Iceland is a blend of the two ethnic groups.

The languages spoken in Denmark, Sweden, and Norway developed from a common German language base. Today, Danish, Swedish, and Norwegian are similar enough that the speakers can usually understand each other, even though each language is distinct. Finnish is unrelated to the other languages; it is closer to the Hungarian and Estonian languages.

The populations of the Southern European countries are relatively homogeneous but more diversified than it might seem at first. Spain is actually a nation of regions that constantly resist unifying pressures from the central government.

Greece also has small populations of Turks, Albanians, Macedonians, and Rom (gypsies). Most people speak Greek, which is closely related to the language spoken in ancient Greece. Most of Italy is ethnically Italian, but minority groups in northern Italy speak German, French, or Slovenian. The Italian language has many **dialects**, or regional variations, but most people speak and understand the Italian that originates around Florence and Rome. Most people in Spain speak a form of Spanish called Castilian. Other dialects are spoken in Spain as well. People in Catalonia, Valencia, and the Balearic Islands speak Catalan. Galicia is an area in northwestern Spain. People there speak Galician, which is closely related to Portuguese.

Religion and the Arts

The Protestant Reformation was successful in Northern Europe. To this day, more than three-quarters of all people living in Northern Europe belong to a Lutheran church. A small percentage are either Catholic or Muslim. In Finland, 15 percent of the people do not belong to any church.

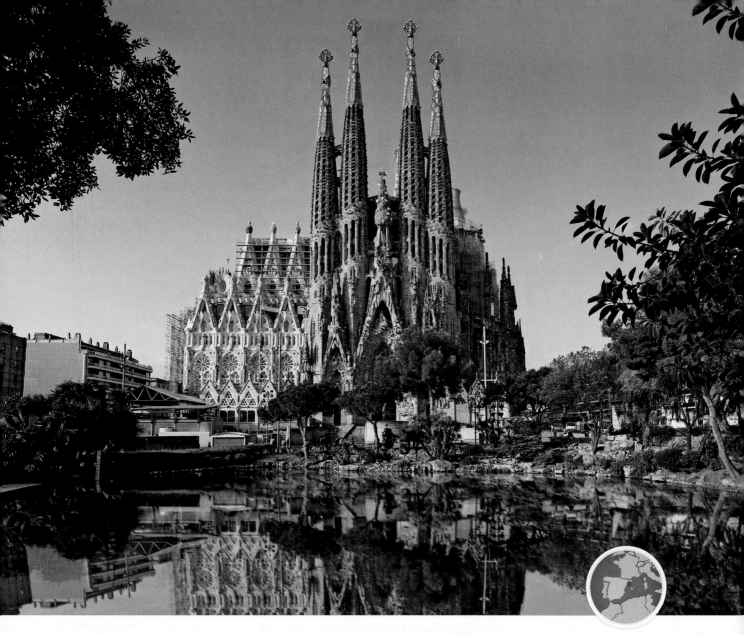

In Southern Europe, the Reformation had little lasting impact. Most people in Spain, Portugal, and Italy belong to the Roman Catholic Church. In Greece, nearly 98 percent of the population belongs to the Greek Orthodox Church. Immigrants from Muslim countries, however, have been slowly changing the religious makeup of Southern Europe.

As the birthplace of the Renaissance, Italy has made many contributions to painting, sculpture, and architecture. The Baroque artists who followed the Renaissance period also emerged from the Italian city-states. The genius of Renaissance and Baroque art can be seen in the palaces, churches, city squares, statues, and paintings of cities such as Rome and Florence. The art centers of Europe moved to Western Europe following the Baroque period—to countries such as France and the Netherlands. In the 1900s, the Italian futurist painters rebelled against traditional art. They painted scenes of modern life that celebrated the industrial age. Italy today is one of the world centers for architecture and fashion.

The Spanish Catalan architect Antoni Gaudí began to build the Sagrada Familia ("Holy Family") Church in Barcelona during the 1880s. This Roman Catholic place of worship, still being built, is due to be finished about 2026.

▶ **CRITICAL THINKING**

Analyzing Why are three different forms of Christianity practiced in Northern Europe and Southern Europe today?

Sylvain Sonnet/Photographer's Choice/Getty Images

A bin is filled with toy blocks at a factory in Billund, Denmark.

©David McLain/Aurora Photos/Corbis

Academic Vocabulary

contribution an important part played by a person, bringing about a significant result or advancement

The musical legacies of Northern and Southern Europe are rich and deep. Italian operas are among the most popular in the world. Operas are stage presentations that use music to tell a story. Among the masterworks of opera are *The Barber of Seville* by Rossini and *Tosca* by Puccini.

Spanish music has been influential all over Europe and the Americas. It was the Spanish, more than any other people, who advanced and promoted the guitar as a serious musical instrument. Northern European composers such as Jean Sibelius (Finland), Edvard Grieg (Norway), and Carl Nielsen (Denmark) made important **contributions** to the classical music tradition. Northern European pop music, such as that by the 1970s Swedish group ABBA, has also had an impact around the world. Iceland has produced original and critically acclaimed popular musicians, including Björk and Sigur Rós.

Denmark has made a major contribution to children's toys: the LEGO™. Based in the town of Billund, LEGO™ (a play on the Danish words for "play well") has been creating plastic building-block toys since the late 1940s. LEGO™ sets were introduced in the United States in the 1960s, and their popularity has remained strong.

Daily Life

One of Northern Europe's most notable achievements is their literacy rate. In Norway, Sweden, and Denmark, the literacy rate is nearly 100 percent. The educational system in Northern European countries has strong support from the government, and most schooling is free. Northern European citizens pay relatively high taxes. In return, they receive a variety of public services and social welfare benefits. Every citizen is covered by health insurance.

Winter sports, such as skating and skiing, are popular in Northern Europe. Many people also enjoy skiing in the mountainous regions of Spain. Soccer (or, as it is known in Europe, football) is popular in Southern Europe. Spain, Portugal, Italy, and Greece have outstanding national soccer teams. Basketball is also common, especially in Spain and Greece. Bullfighting is still popular in Spain, although it is controversial because of the violence of the sport and cruelty to the bull.

Northern and Southern Europe are affected by the spread of popular culture, especially youth culture, from place to place. Young people throughout Europe are familiar with the same kinds of music. American fast-food chains, television programs, and films are popular throughout these regions.

✓ **READING PROGRESS CHECK**

Analyzing How is life in Southern Europe similar to life in the United States? How is it different?

Skiing is the most popular activity at a winter resort on Spain's side of the Pyrenees mountains.

▶ **CRITICAL THINKING**
Identifying Point of View Why does the sport of bullfighting arouse controversy?

Students relax in front of the Parliament building in Oslo, Norway's capital. The governments in Northern Europe and Southern Europe are all democracies. Some, such as Norway and Spain, are constitutional monarchies, while others like Finland and Italy are republics.

▶ **CRITICAL THINKING**
Determining Central Ideas What is the purpose of welfare capitalism?

Issues in Northern and Southern Europe

GUIDING QUESTION *How do the financial problems of one country in Southern Europe affect other European countries?*

The natural resources of the North Sea are key to many people's employment in Norway. The petroleum resources in the North Sea provide the single most important industry.

Earning a Living

Many Northern Europeans work in service industries. The standard of living is high in these countries, but so are taxes. Northern Europeans practice a form of **welfare capitalism**, where the government uses tax money to provide a variety of services, such as health care and education, to all citizens. The intention of welfare capitalism is to ensure that all people have access to those aspects of life that are considered essential, even those people who might not otherwise be able to afford them.

Following World War II, Italy had one of the weakest economies in Europe. In the decades since, the country has built a strong industrial base. Still, the economy is much stronger in northern Italy. The government has tried to stimulate the economy in the south, but most attempts have failed.

Making Connections

In 1992 a group of European countries signed a treaty creating the European Union (EU). The European Union is an attempt to provide common social, economic, and security policies throughout the continent. Today, 27 countries belong to the EU. Three Northern European countries are EU members: Denmark, Sweden, and Finland. In Southern Europe, Spain, Portugal, Italy, and Greece are members.

Meeting Challenges

Although the EU has helped ease conflicts among member countries, there have been many stumbling blocks to unifying the member nations. Immigration is one issue. Immigrants from many non-European countries bring their culture with them, creating a changing cultural landscape that can lead to conflict and violence. The EU promotes cooperation among member countries in easing conflicts, as well as preventing illegal immigration.

In 2008 a financial crisis swept the world. Europe experienced a **recession**, or a period of slow economic growth or decline. The three biggest banks in Iceland failed, forcing Iceland's prime minister to resign. The financial crisis hit Spain and Greece hard. Greece's inability to pay its skyrocketing debts became a threat to the economies of all the European countries.

Another issue surrounds two Northern European countries that are not members of the EU: Iceland and Norway. Whaling was an important industry in both countries. As populations of large whales began to shrink to near-extinction levels, a ban was placed on whaling. The ban has been lifted several times, but only to allow for hunting small, toothed whales. For now, the three remaining countries that hunt for whales—Norway, Iceland, and Japan—are under the watch of the International Whaling Commission.

Include this lesson's information in your Foldable®.

☑ READING PROGRESS CHECK

Determining Central Ideas How does welfare capitalism work, and what advantages does it offer the people of Northern Europe?

LESSON 3 REVIEW

Reviewing Vocabulary

1. Why are the populations of Northern Europe considered *homogeneous*?

Answering the Guiding Questions

2. ***Determining Central Ideas*** Why do you think people in Northern and Southern Europe find living in cities more appealing than living in rural areas?

3. ***Analyzing*** Why do you think people in Norway, Denmark, and Sweden can easily understand each other's languages but find it more difficult to understand Icelandic?

4. ***Describing*** How has membership in the European Union changed the relationship among the countries of Northern and Southern Europe?

5. ***Informative/Explanatory*** Write a paragraph that compares the lives of people in Northern and Southern Europe.

Aging of Europe's Population

An aging population is defined as a population in which the number of elderly (65 years old and older) is increasing at a faster rate than younger age groups. Europe is the region with the highest proportion of older persons.

Median Age The median age of a population is the age that divides the population into two equal groups. In the world as a whole, the median age is 28.4 years old. For Europe, the median age is 40 years old. The median age of Europeans has continued to rise and is expected to increase to 46 years by 2050.

Standard of Living Europeans enjoy a higher standard of living than people in some of the other regions of the world. In general, Europeans are living longer thanks to improvements in health, diet, and preventive health care.

> **"Some senior citizens continue working because they cannot afford to retire."**

Living Longer As people live longer, many experience age-related health problems. That puts an added strain on the health care system and health care costs. Some senior citizens continue working because they cannot afford to retire.

Smaller Families Over the past 40 years, families in Europe have grown smaller. As a result, fewer young workers are entering the labor force, while the number of senior citizens continues to grow.

Age of Retirement Some European nations are seeking ways to cut costs. Several European nations have raised the age of retirement. Adjusting the retirement age upward means more people in the workforce and

fewer people depending on government pensions.

The Workforce European countries are trying to increase their workforce by attracting skilled workers from other countries. Some are trying to attract workers only from other parts of Europe. Some European nations will have to change their immigration laws if they want to attract immigrants from regions outside of Europe.

Hikers make their way on a trek through ▶ the Alps.

THERE'S MORE ONLINE

HEAR how the EU is funded • *SEE* a slide show on European health care • *WATCH* a video on immigration

These numbers and statistics highlight the problems associated with an aging population.

population in 2060

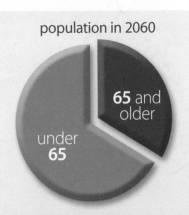

65 and older

under **65**

Up to
30%

The share of the population aged 65 and over will rise from 17% in 2010 to 30% in 2060. The percentage of people aged 80 or older will more than double: from 5% in 2010 to 12% in 2060.

Seventy Five Years Old

Why is Europe's population growing older? People are living longer because of improved medical care and a better standard of living. In 1950 the average European lived to be 65. Today, the length of life is 75 years. At the same time, a decreasing birthrate means that there are fewer young people. In 2010 the birthrate—the number of births per 1,000 population—was half the birthrate in 1950.

65 **and Older**

Which countries have the highest percentage of residents who are 65 years old or older? Except for Japan, the world's 15 oldest countries are European countries. The five countries with the largest percentage of older citizens are: Japan and Italy, both at 19.5%, followed by Germany, 18.6%; Greece, 17.8%; and Sweden, 17.3%.

Life Expectancy at
FIFTY

What problems are caused by an aging population? As the aging population grows and more people retire, great strain is put on the health care and pension systems. Life expectancy varies greatly, depending on the country or region. For example, a 50-year-old female can expect 10.4 more healthy life years in Estonia, but 24.1 more in Denmark. The life expectancy for a 50-year-old male is slightly less.

14
or younger

European countries have many more senior citizens than most other nations—and far fewer young people. More than one-quarter—26.3%—of the world's population is 14 years old or younger. In Europe, only 15% are in that age group.

AN AGING POPULATION

Percentage of population age 60 and older

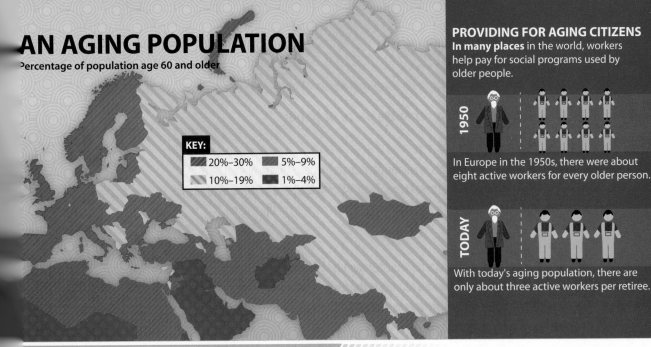

KEY:
- 20%–30%
- 10%–19%
- 5%–9%
- 1%–4%

PROVIDING FOR AGING CITIZENS

In many places in the world, workers help pay for social programs used by older people.

1950

In Europe in the 1950s, there were about eight active workers for every older person.

TODAY

With today's aging population, there are only about three active workers per retiree.

GLOBAL IMPACT

DEMOGRAPHIC TRENDS In many areas, the percentage of people who are older is increasing while the percentage of children is growing smaller.

The number of persons 60 years old and older has tripled between 1950 and 2000. By 2050, that number will more than triple again.

There is a higher proportion of older persons in Europe, but lower proportions in developing regions such as Africa. In 2050, 10 pecent of the population of Africa is projected to be 60 years old or older, up from 5 percent in 2000.

Europe, 2050

If projections hold, Europeans 60 years and older will comprise about 37 percent of the population by 2050. Persons younger than 15 will make up only about 14 percent of Europe's people.

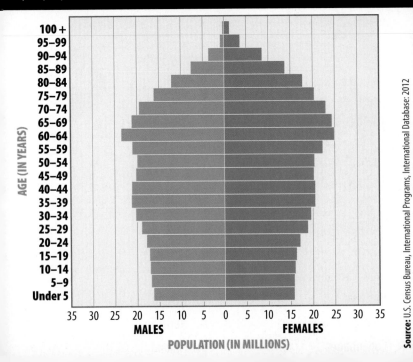

AGE (IN YEARS)

100 +
95–99
90–94
85–89
80–84
75–79
70–74
65–69
60–64
55–59
50–54
45–49
40–44
35–39
30–34
25–29
20–24
15–19
10–14
5–9
Under 5

35 30 25 20 15 10 5 0 5 10 15 20 25 30 35

MALES **FEMALES**

POPULATION (IN MILLIONS)

Source: U.S. Census Bureau, International Programs, International Database: 2012

Thinking like a
Geographer

1. ***Human Geography*** Why is the percentage of elderly growing in Europe?

2. ***Economic Geography*** What kinds of businesses will grow in a place where the population is aging? What kinds of businesses will be in less demand?

3. ***Human Geography*** You and your classmates have been appointed by the mayor of a city to help companies find skilled workers from other countries. Prepare a PowerPoint presentation for the owners of those companies about ways to attract foreign workers.

Directions: Write your answers on a separate piece of paper.

1 Use your FOLDABLES to explore the Essential Question.

INFORMATIVE/EXPLANATORY Students in Europe participate in sports that might not be as well-known where you live. Identify one of those sports. Then, write at least two paragraphs to answer this question: Is the popularity of the sport related to the physical geography of the region?

2 21st Century Skills

ANALYZING The nations of Northern Europe have made progress in developing and using renewable energy sources. Work in small groups to choose a country in the region, and research that country's efforts to ensure that its people have adequate energy sources well into the future. Present your findings to the class in a slide show presentation.

3 Thinking Like a Geographer

DETERMINING CENTRAL ISSUES You have read that some locales in the regions are densely populated, but others are not. Identify the three major factors that affect population density.

4 **GEOGRAPHY ACTIVITY**

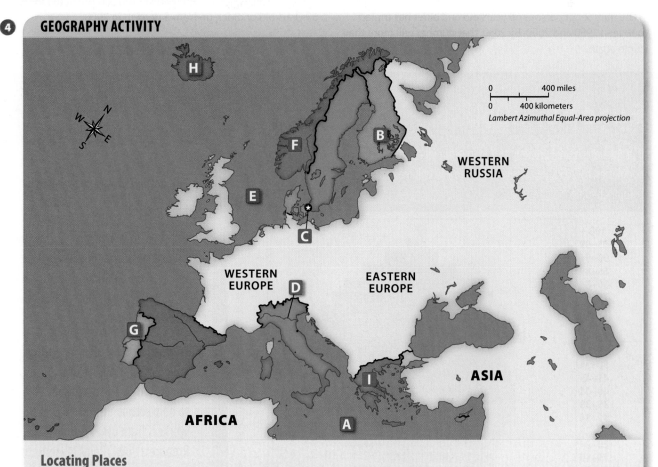

Locating Places
Match the letters on the map with the numbered places below.

1. Mediterranean Sea
2. Po River
3. Norway
4. Portugal
5. Greece
6. Copenhagen
7. Iceland
8. North Sea
9. Finland

Chapter 12 ASSESSMENT (CCSS)

REVIEW THE GUIDING QUESTIONS

Directions: Choose the best answer for each question.

1 What kind of energy heats all of the homes and most of the businesses in Reykjavík, Iceland?

A. nuclear

B. natural gas

C. geothermal

D. solar

2 Most of Southern Europe has a Mediterranean climate, which is much like the climate along the

F. southeastern coast of the United States.

G. Pacific northwest.

H. Chesapeake Bay.

I. coast of southern California.

3 What weakened Greece's city-states and eventually led to the fall of Greek civilization?

A. invasion by Vikings

B. years of war

C. Roman conquest

D. lack of trade routes

4 Who were the Scandinavian warriors and pirates who eventually became better known for exploring and trading?

F. Moors

G. Vikings

H. barbarians

I. Ottoman Turks

5 Most people in Northern and Southern Europe live in

A. rural areas.

B. small towns.

C. villages.

D. cities.

6 The Basques, one of the oldest minority groups in Europe, live in and around the Pyrenees mountains in

F. Spain and Portugal.

G. Italy and Greece.

H. France and Italy.

I. France and Portugal.

DBQ ANALYZING DOCUMENTS

❼ ANALYZING DOCUMENTS Read the following passage about the Iberian Peninsula:

"Its eastern seaboard forms part of the Mediterranean world, and in early times [it] was drawn successively [by turns] into the Carthaginian, Roman, and Muslim spheres. But much of the arid interior is drawn . . . towards the Atlantic."

—from Norman Davies, *Europe: A History* (1998)

What statement about Iberia it true based on the passage?

A. Iberians defeated the Carthaginians.

B. Iberia was isolated from all other regions.

C. Iberia was often under the control of other empires.

D. Iberians invaded Rome.

❽ IDENTIFYING What impact did Iberia's nearness to the Atlantic have on its history?

F. The area was often invaded by the British.

G. Iberian nations led the way in exploring the Atlantic.

H. Navies from the region could not fight Mediterranean navies.

I. Iberian nations were cut off from trade with Asia.

SHORT RESPONSE

"The expansion of most industry in Norway has largely been governed by private property rights and the private sector. Nevertheless, some industrial activities are owned or run by the state."

—from Aschehoug and Gyldendal's *Norwegian Encyclopedia*

❾ CITING TEXT EVIDENCE Which economic sector, private or public, has been responsible for the expansion of most industry in Norway?

❿ ANALYZING Why would economists label Norway's economy a "mixed economy"?

EXTENDED RESPONSE

⓫ INFORMATIVE/EXPLANATORY WRITING As you learned in your reading, the literacy rate in the Scandinavian countries is nearly 100 percent and education is free through college. How does that compare with literacy rates and the cost of education in Southern Europe and the United States? Research the issue, and report your findings.

Need Extra Help?

If You've Missed Question	❶	❷	❸	❹	❺	❻	❼	❽	❾	❿	⓫
Review Lesson	1	1	2	2	3	3	1	1	2	2	3

From EUROPE: A HISTORY, by Norman Davies, ©Norman Davies, 1996; Edited from Aschehoug and Gyldendal's Norwegian Encyclopedia, as seen on http://www.norway.org/aboutnorway/economy/economy/mixed/, official Web site in the USA of the Royal Norwegian Embassy in Washington, D.C.

EASTERN EUROPE AND WESTERN RUSSIA

Dancers with Moscow's world-famous Bolshoi Ballet rehearse for an upcoming performance.

ESSENTIAL QUESTIONS · *How does geography influence the way people live?*
· *How do governments change?*

networks

There's More Online about Eastern Europe and Western Russia.

CHAPTER 13

Lesson 1
Physical Geography

Lesson 2
History of the Regions

Lesson 3
Life in Eastern Europe and Western Russia

The Story Matters...

The people of Eastern Europe and Western Russia share an agricultural background, which has been important in their history. Over the centuries, however, the people in this subregion have faced many political, ethnic, and economic challenges. The rise and fall of communism in Russia has had a tremendous impact on the countries of Eastern Europe.

FOLDABLES
Study Organizer

Go to the Foldables® library in the back of your book to make a Foldable® that will help you take notes while reading this chapter.

Eastern Europe *Western Russia* *Empires* *Populations*

EASTERN EUROPE AND WESTERN RUSSIA

Rolling hills and fertile soil blanket much of Eastern Europe and Western Russia, two regions that merge just south of the Baltic Sea and north of the Black Sea and the Caspian Sea. As you study the map of Eastern Europe and Western Russia, look for geographic features that make this area unique.

Step Into the Place

MAP FOCUS   Use the map to answer the following questions.

1 **THE GEOGRAPHER'S WORLD**
Which Eastern European country bordering Russia is located farthest north?

2 **PLACES AND REGIONS**   Which river runs across the border of Russia and Eastern Europe?

3 **THE GEOGRAPHER'S WORLD**
Name four landlocked countries in Eastern Europe.

4 **CRITICAL THINKING**
Analyzing   If you were a farmer in Belarus who needed to ship crops overseas, from which sea would you transport your crops? How would you get them to that sea?

CHURCH ARCHITECTURE   St. Sophia Cathedral, with its 13 domes, is one of the oldest churches in Ukraine. It was built in 1037 to rival the church of the same name in Constantinople.

EUROPE'S LONGEST RIVER   Boats line the Volga River, which flows about 2,300 miles (3,701 km) through west central Russia.

Step Into the Time

TIME LINE   Using events on the time line, write a paragraph explaining how these events contributed to the rise of communism in Russia and Eastern Europe.

1762
Catherine the Great becomes empress of Russia

1700

1721   Peter the Great founds the Russian Empire

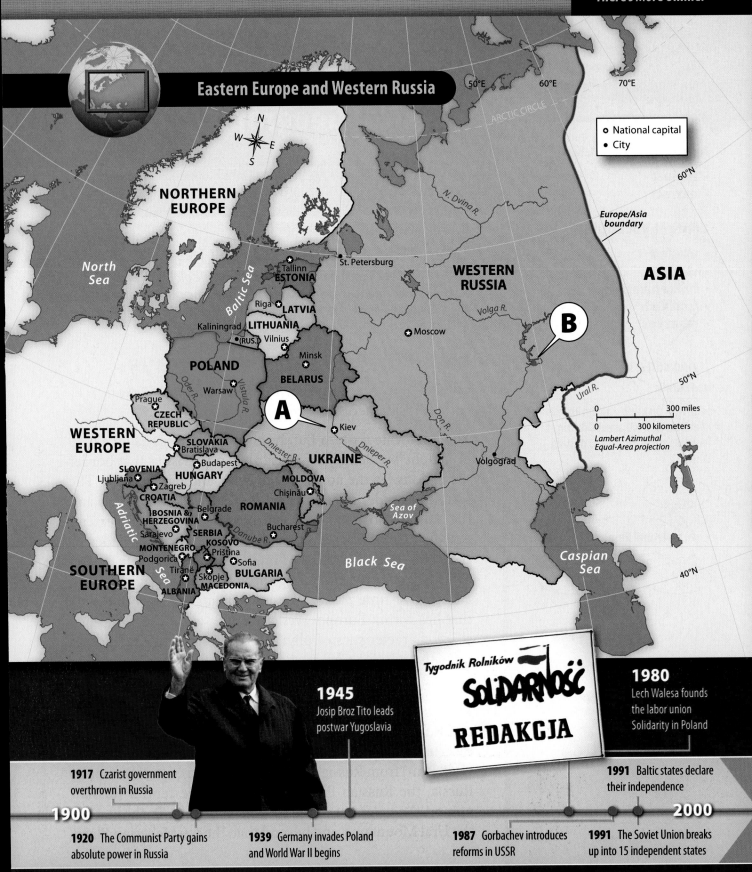

Eastern Europe and Western Russia

○ National capital
● City

NORTHERN EUROPE

North Sea

Baltic Sea

ARCTIC CIRCLE

Europe/Asia boundary

60°N

50°E   60°E   70°E

St. Petersburg

WESTERN RUSSIA

N. Dvina R.

ASIA

Tallinn
ESTONIA

Riga   LATVIA

Volga R.

Moscow

B

Kaliningrad   LITHUANIA
(RUS.)   Vilnius

POLAND   Minsk

Oder R.   BELARUS

Warsaw

Vistula R.

Ural R.

50°N

Prague

CZECH REPUBLIC

A

Kiev

WESTERN EUROPE

SLOVAKIA
Bratislava

Dniester R.

Dnieper R.

Don R.

0   300 miles
0   300 kilometers
Lambert Azimuthal Equal-Area projection

SLOVENIA   Budapest

UKRAINE

Volgograd

Ljubljana   HUNGARY

MOLDOVA

Zagreb   Chişinău

CROATIA

ROMANIA

Sea of Azov

Caspian Sea

BOSNIA & HERZEGOVINA   Belgrade

40°N

Sarajevo   SERBIA

Danube R.

Bucharest

MONTENEGRO   KOSOVO

Black Sea

Podgorica   Priština

SOUTHERN EUROPE

Tiranë   Sofia

Skopje   BULGARIA

ALBANIA   MACEDONIA

Adriatic Sea

1945
Josip Broz Tito leads postwar Yugoslavia

Tygodnik Rolników
SOLIDARNOŚĆ
REDAKCJA

1980
Lech Walesa founds the labor union Solidarity in Poland

1917 Czarist government overthrown in Russia

1991 Baltic states declare their independence

1900

2000

1920 The Communist Party gains absolute power in Russia

1939 Germany invades Poland and World War II begins

1987 Gorbachev introduces reforms in USSR

1991 The Soviet Union breaks up into 15 independent states

Reading **HELP**DESK (CCSS)

Academic Vocabulary

- impact
- extract

Content Vocabulary

- **upland**
- **steppe**
- **balkanization**
- **brackish**
- **reserves**

TAKING NOTES: *Key Ideas and Details*

Identify On a web diagram like this one, list at least four natural resources found in Eastern Europe and Western Russia.

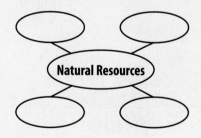

Natural Resources

Lesson 1
Physical Geography

ESSENTIAL QUESTION • *How does geography influence the way people live?*

IT MATTERS BECAUSE
Both the rugged mountains and the gentle plains of Eastern Europe and Western Russia have shaped the cultures of the people living there.

Landforms and Waterways

GUIDING QUESTION *In what way have the landforms in the Balkan Peninsula shaped the cultures of that region?*

You can locate Eastern Europe and Western Russia on a map or a globe by identifying physical characteristics that border the regions. To the north are the Baltic and Barents Seas. The southern border is defined by the Caucasus Mountains and the Adriatic, Black, and Caspian Seas. The regions extend eastward to the Ural Mountains. Eastern Europe includes 10 countries in the north and 11 on the Balkan Peninsula. Russia is a huge country, extending through Europe and Asia and covering 11 time zones. Western Russia is the part of Russia that lies within Europe. Western Russia and Eastern Europe share characteristics of physical and human geography that unite them into a single region.

Vast Plains

Eastern Europe and Western Russia rest mostly on a group of plains. The largest plain is the Russian Plain, which begins in Belarus and Ukraine and stretches east about 1,000 miles (1,609 km) from Russia's western borders. In central European Russia, the Russian Plain rises to form the central Russian **upland**. An upland is an area of high elevation. To the east are the Ural Mountains, and beyond that, the west Siberian Plain.

The Northern European Plain includes Poland in Eastern Europe, but it also extends into parts of Western Europe. South and southeast of the Northern European Plain is the Hungarian Plain, which includes parts of many different countries. Within Romania is the Transylvanian basin. A basin is an area of land that slopes gently downward from the surrounding land. Much of Ukraine is **steppe**, or vast, level areas of land that support only low-growing, vegetation-like grasses.

Bordering Mountains

To the south of the Russian Plain are two chains of mountains that make up the Greater and Lesser Caucasus Mountains. They extend from the northwest to the southeast with a valley between.

East of the Russian Plain, the Ural Mountains form a boundary between Europe and Asia. The Urals are up to 250 million years old. The northern mountains are covered in forests and some glaciers. Grasslands cover the southern Urals.

The Carpathian Mountains are much younger. On a map, the Carpathians appear almost as an eastward extension of the Alps. The Vienna basin in Austria separates the two mountain ranges.

The Balkan Peninsula is a mountainous region. In fact, *balkan* is a Turkish word for mountain. The Carpathian Mountains run through the peninsula's north and are linked to the Balkan Mountains. The region is so mountainous that human settlements are isolated from one another. This isolation results in cultural diversity among the people, but it is also the source of conflict among ethnic groups. Conflict among ethnic groups within a state, a country, or a region is known as **balkanization**.

Russia's Komi region borders the Urals and other mountain ranges. It is rich in coal, oil, natural gas, diamonds, gold, and timber.

▶ **CRITICAL THINKING**

Describing What characteristic of the Ural Mountains makes them unique?

Bulgaria, located in the Balkan Peninsula, has a mild climate and sandy beaches along its Black Sea coast.

▶ **CRITICAL THINKING**

Identifying What seas other than the Black Sea border the Balkan Peninsula?

Surrounding Seas

The Baltic Sea lies northwest of Russia and Eastern Europe. The Baltic is shallow and **brackish**, or somewhat salty, because it is seawater mixed with river water. In the southwest, the Adriatic, Ionian, and Black seas surround the Balkan Peninsula on three sides. The Black Sea borders the southern coast of Ukraine and southwestern Russia. The sea also separates Turkey from Ukraine and the Balkan Peninsula. At Europe's most southeastern point is the Caspian Sea. The Caspian Sea is the world's largest inland body of water, covering an area larger than Japan.

Rivers and Lakes

A vast number of rivers, canals, lakes, and reservoirs are found in Eastern Europe and Western Russia. The Volga River is the longest river in Europe and Russia's most important waterway. Originating northwest of Moscow, the Volga and its many tributaries carry more freight and passenger traffic than any other river in Russia. It provides hydroelectric power and water to many parts of Russia.

The Dnieper River also originates in Russia. It flows through Belarus and Ukraine before emptying into the Black Sea. Dams and reservoirs southeast of Kiev provide hydroelectric power. They also irrigate farmlands and help relieve water shortages in parts of Ukraine. Originating in the Carpathian Mountains and emptying into the Black Sea, the Dniester River is the second-longest river in Ukraine. The Dniester carries freight and passenger ships, and it serves as an important route to the Black Sea.

From its origins in southwestern Germany, the Danube flows toward the east through several countries before emptying into the Black Sea. The Danube provides transportation, hydroelectric power, fishing, and water for irrigation. Historically, the river transported traders as well as invading armies. Today, many cities are located along its banks, including three capital cities: Vienna, Austria; Budapest, Hungary; and Belgrade, Serbia. The Main River became connected to the Danube via the Main-Danube Canal, which linked the North Sea with the Black Sea.

☑ **READING PROGRESS CHECK**

Analyzing How is the location of the Black Sea strategic to the region?

©Dallas and John Heaton/Corbis

Climates

GUIDING QUESTION *How does climate affect plants that are grown and harvested in Eastern Europe and Western Russia?*

Several different types of climate are found in Eastern Europe and Western Russia, from the hot summers and rainy winters in Albania to the cold, polar reaches of northern Russia.

Humid Continental Climates

Much of Eastern Europe and Western Russia have a humid continental climate. These areas experience mild or warm summers and long, cold winters. Farther south, in places such as Croatia, Serbia, and Bulgaria, summers are hotter, and winter weather is similar to that of areas farther north.

Albania and Macedonia experience a more Mediterranean climate, especially in the western areas. Summers tend to be hot and dry, and winters are mild to cool and rainy.

Russia's Far North

North of 60°N latitude, Western Russia has a subarctic climate. Winters are very cold, with temperatures as low as −40°F (−40°C). The summers are short and cool, though temperatures can range from 50°F (10°C) to 86°F (30°C).

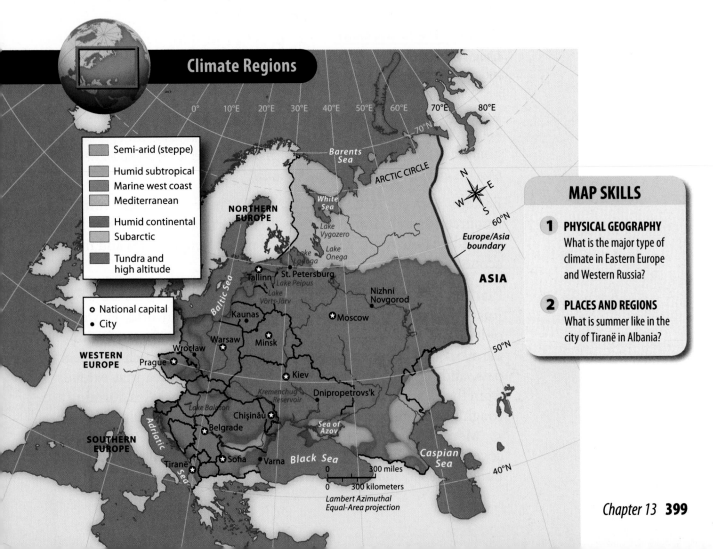

Climate Regions

Semi-arid (steppe)
Humid subtropical
Marine west coast
Mediterranean
Humid continental
Subarctic
Tundra and high altitude

○ National capital
● City

MAP SKILLS

1 PHYSICAL GEOGRAPHY
What is the major type of climate in Eastern Europe and Western Russia?

2 PLACES AND REGIONS
What is summer like in the city of Tiranë in Albania?

Lambert Azimuthal Equal-Area projection

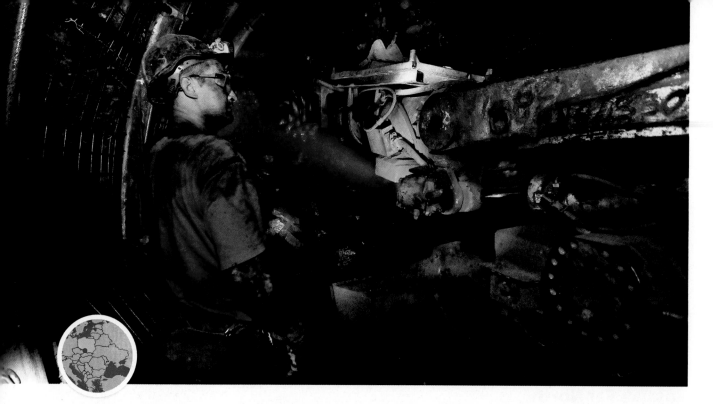

A miner watches as the arm of a drill machine draws out coal from a mine in southern Poland.

▶ **CRITICAL THINKING**

Identifying What other minerals are important to Poland's economy?

Farther north is Novaya Zemlya, an archipelago consisting of two large islands and several small islands. The climate here is polar, and a large part of Novaya Zemlya is covered in ice year-round. Only the southern island is inhabited by a small number of the indigenous Nenets, who are herders and fishers.

☑ **READING PROGRESS CHECK**

Identifying What area in Eastern Europe or Western Russia has a climate most similar to where you live?

Natural Resources

GUIDING QUESTION *What are three important challenges to the development of resources in Eastern Europe and Western Russia?*

Eastern Europe and Western Russia have abundant mineral resources, as well as dense forests, fertile farmlands, and rich fishing grounds. These resources play a vital role in people's lives and in the economy of the countries in which they live.

Forests and Agriculture

Russia is a vast country—by far the largest in the world. However, only about one-sixth of Russia's land is suitable for agriculture. Farmers grow a number of crops, including grains such as wheat, oats, and barley. Most agricultural land is in an area that extends from the western shores of the Baltic Sea to the Black Sea, forming a roughly triangular shape. This area is known as the fertile triangle.

More than one-fifth of the world's forests are in Russia. They cover an area almost the size of the continental United States. Lumber, paper, and cardboard are important products of the forestry

Bloomberg/Getty Images

industry. The long, cold winters of Western Russia's continental climate, however, cause forests to grow slowly. The intense harvesting of forests and the slow rates of growth threaten the forests and the forestry industry.

In 2010 Russia experienced the hottest summer in 130 years, with drought conditions and temperatures reaching 104°F (40°C). That summer, wildfires destroyed 37 million acres (about 15 million ha) of forests, agricultural crops, and other vegetation. The **impact** of these fires was tremendous, taking lives, destroying homes, and damaging Russia's forestry and agricultural industries.

Academic Vocabulary

impact a dramatic or forceful effect or influence on something

Energy and Minerals

Most of Russia's vast coal, oil, and natural gas **reserves** are in Siberia. Reserves are the estimated total amount of a resource in a certain area. Russia's coal and rich deposits of iron ore fuel the country's steel industry. Machines made from steel are used to build Russia's automobiles, railroads, ships, and many consumer products.

Poland's important mineral resources include aluminum, coal, copper, lead, and zinc. Poland is one of the world's major sources of sulfur. Romania has rich deposits of coal, and it **extracts** oil from the Black Sea. Hydroelectric and thermal power plants also support Romania's energy needs. Other important mineral resources include copper and bauxite, the raw material for aluminum.

Academic Vocabulary

extract to draw or pull out

Fishing Industry

Russia's fishing industry is an important part of the country's economy. Salmon, cod, herring, and pollack are among the most important commercial fish in Russia. Many of Russia's lakes and rivers are also used for freshwater fishing.

Romania's fishing industry is concentrated in the southeastern area of that country. The Danube River and lakes and rivers near the Black Sea provide much of the fish. The European Union's restrictions on overfishing has hurt Romania somewhat, but fishing remains important to the country's economy.

☑ **READING PROGRESS CHECK**

Analyzing Why are Russia's mineral industries so important to its economy?

FOLDABLES®
Study Organizer

Include this lesson's information in your Foldable®.

LESSON 1 REVIEW

Reviewing Vocabulary

1. What caused the *balkanization* on the Balkan Peninsula?

Answering the Guiding Questions

2. *Describing* In what way have the landforms in the Balkan Peninsula shaped the cultures of the region?

3. *Analyzing* Why do few people live on the archipelago of Novaya Zemlya?

4. *Identifying* What are two important challenges to Russia's forestry industry?

5. *Informative/Explanatory Writing* Explain how the mineral resources in Russia are important to its industry.

Reading **HELP**DESK CCSS

Academic Vocabulary

• **strategy**
• **inevitable**

Content Vocabulary

• **czar**
• **serf**
• **genocide**
• **communism**
• **collectivization**

TAKING NOTES: *Key Ideas and Details*

Identify Use a graphic organizer like the one shown here to identify two ways that Joseph Stalin fashioned the Soviet Union into a communist dictatorship.

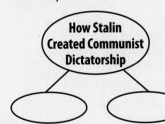

How Stalin Created Communist Dictatorship

Lesson 2
History of the Regions

ESSENTIAL QUESTION • *How do governments change?*

IT MATTERS BECAUSE
For most of the late 1900s, the USSR was one of the two most powerful countries in the world. The Soviets ruled Russia, nearly all of Eastern Europe, and much of Central Asia.

Early History

GUIDING QUESTION *How did Peter I and Catherine II change Russia?*

For the last 1,000 years, the people of Eastern Europe and Western Russia have been part of great empires that struggled against each other—and sometimes against their own people.

Early Slavic States

Many different ethnic groups settled in the regions of Eastern Europe and Western Russia long before modern national borders were set. Most of the people in the region are Slavs. Slavs are an ethnic group that includes Poles, Serbs, Ukrainians, and other Eastern Europeans.

Early Slavs migrated from Asia and settled in the area that now includes Ukraine and Poland. In the A.D. 400s and A.D. 500s, Slavs moved westward and southward, coming into contact with migrating Celtic and Germanic groups.

In the A.D. 800s, Slavic groups in the present-day Czech Republic formed Great Moravia, an empire covering much of central Europe. Other Slav people settled in the Balkans, eventually coming under the rule of the Ottoman Empire.

Another early Slav group settled in the forest and plains of present-day Ukraine and Belarus. The people of a settlement called Kiev organized the Slav communities into a union of city-states known as Kievan Rus. The leaders controlled the area's trade, using Russia's western rivers as a link between the

(l to r) DEA/A. DAGLI ORTI/De Agostini Picture Library/Getty Images; ©Dean Conger/Corbis; ©Peter Turnley/Corbis

Baltic Sea and the Black Sea. Kievan Rus prospered from trade with the Mediterranean world and Western Europe. Later, non-Slavic people also settled in the region. Besides ethnic Russians who make up the majority of the population today, there are Hungary's Magyars, Romanians, Slavs, Ukrainians, and many others.

Throughout Russia's history, the Russian Slavs have dominated the country's politics and culture. Most Slavs practice Eastern Orthodoxy, a form of Christianity brought to Russia from the eastern Mediterranean area. By the A.D. 1000s, the ruler and people of Kievan Rus had accepted Eastern Orthodox Christianity. It remains Russia's largest religion today.

Imperial Russia

During the later 1200s, the warrior armies of the Mongols of Central Asia invaded Russia. For the next 250 years, they controlled most of Russia. Near the end of the Mongol reign, the princes of Muscovy (now the city of Moscow) rose to power.

MAP SKILLS

1 HUMAN GEOGRAPHY
Why did early Slav communities develop in the area of Western Russia?

2 PLACES AND REGIONS
In what time period did Russia gain control of the area around the Baltic Sea?

Expansion of Russia

Kievan Territory	— Boundary of the Soviet Union in 1945
1360–1533	— Present-day Russian boundary
1533–1689	
1689–1917	

0 — 1,000 miles
0 — 1,000 kilometers
Two-Point Equidistant projection

FEUDALISM IN EUROPE & RUSSIA

As in other parts of medieval Europe, Russia's feudal system depended on a large number of laborers. In exchange for a serf's labor, the lord or noble provided a place to live and protection.

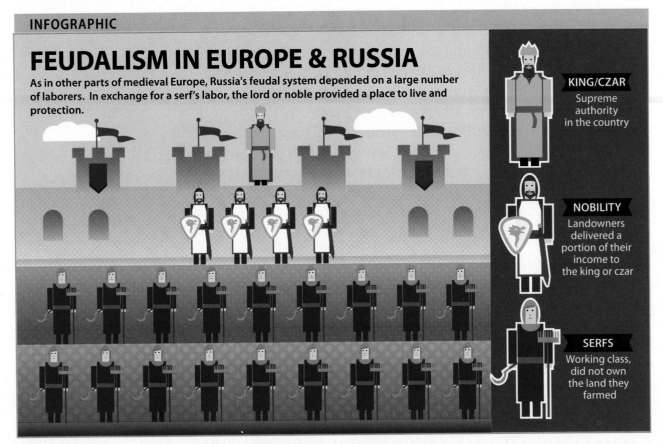

KING/CZAR
Supreme authority in the country

NOBILITY
Landowners delivered a portion of their income to the king or czar

SERFS
Working class, did not own the land they farmed

Under feudalism, monarchs gave land to nobles and lords in exchange for military protection. Serfs farmed the land of the nobles in exchange for the use of the land and protection.

▶ **CRITICAL THINKING**

Explaining How did constant warfare lead to the development of feudalism?

The most powerful of these princes, Ivan IV, defeated the Mongols and declared himself the **czar** of Russia. *Czar* is Russian for Caesar, or powerful ruler. The Russian nobility, dissatisfied with the czars who ruled after Ivan, looked for a young noble to lead the country. In 1613 they elected 16-year-old Michael Romanov as czar. The Romanovs ruled for the next 300 years.

Powerful Czars

Later, a czar now known as Peter the Great attempted to turn Russia into a major power. After Peter's death in 1725, Russia endured a string of weak czars. During the late 1700s, Empress Catherine the Great came to power. Catherine encouraged the development of Russian education, journalism, architecture, and theater. During her reign, Russia expanded its empire and took possession of the entire northern coast of the Black Sea.

Plight of the Serfs

The czars and nobles enjoyed rich, comfortable lives. At the bottom of society, however, were the great masses of people. Most were **serfs**, or farm laborers who could be bought and sold along with the land. These people lived hard lives. In 1861 Czar Alexander II abolished serfdom. The new law, however, did little to help the serfs.

They had no education and few ways to earn a living. Industrialization drew some serfs to cities, where they worked long hours for low wages.

☑ **READING PROGRESS CHECK**

Describing How did Catherine the Great expand the Russian Empire?

Conflict and Communism

GUIDING QUESTION *How did the Russian Communist Party plan to transform Russia into an industrial giant?*

In the early 1900s, discontent with the rule of the czars spilled into the streets. Strikes and demonstrations in 1905 nearly ended the reign of Czar Nicholas II. One event, called Bloody Sunday, began with workers marching toward the czar's palace in St. Petersburg to demand better working conditions. The march ended when soldiers fired into the marchers, killing nearly 1,000 people. Another much larger conflict was brewing—one that would involve millions of people, military and civilian.

Wars and Revolution

The threat of war in Europe had been brewing for many years. The major powers had already formed alliances. Austria-Hungary, Germany, and the Ottoman Empire made up the Central Powers. Great Britain, France, and Russia were called the Allies. An assassination triggered World War I. A Bosnian terrorist named Gavrilo Princip assassinated Archduke Francis Ferdinand of Austria-Hungary on June 28, 1914, in Bosnia. By August, nearly all of Europe was at war.

At first, the Russian people supported the war effort, but as military failures, high casualties, and food shortages began mounting, public opinion turned against the war and against the czar. Russia encouraged Armenians in Turkish-controlled lands to fight alongside them. The Turks responded by deporting 1.75 million Armenians to Syria and Mesopotamia. During this mass deportation, about 600,000 Armenians starved or were murdered by Turkish soldiers and police. The mass murder of vast numbers of an ethnic or cultural group is called **genocide**.

In 1917 food shortages in Russia triggered riots in the capital. Soldiers began deserting, joining civilians in their protests against the war. Even though the Allies won the war, Russia emerged as a weakened nation.

The killing of Archduke Ferdinand, heir to Austria-Hungary's throne, lit the fuse of World War I.

▶ **CRITICAL THINKING**
Determining Central Ideas Why did this killing lead to a war involving many nations?

Czar Nicholas was forced to step down. A new government was installed, but it could not maintain power. By the end of 1917, a group of revolutionaries known as the Bolsheviks had seized control of the government.

Rise of Communism

The Bolsheviks had strong support all over Russia. Inspired by the writings of Karl Marx, they remade Russia into a communist state. **Communism** is an economic system built on the idea that all property should belong to the community or the state, not to private individuals. The Bolsheviks, who had become the Russian Communist Party, took control of all land and industry. Their leader, Vladimir Lenin, became the first premier of the new Union of Soviet Socialist Republics (also known as the Soviet Union).

When Lenin died in 1924, the secretary of the Central Committee of the Communist Party, Joseph Stalin, became leader of the Soviet Union. Stalin used terror and brute force to fashion the Soviet Union into a communist dictatorship. He forced the **collectivization** of all agriculture, so that all farmland was owned and controlled by the government. Peasants and landowners protested, especially in Ukraine. The clash between agricultural workers and the government resulted in a famine that killed millions.

May Day was an official holiday in the Soviet Union and Soviet satellite countries. Held on May 1, celebrations, like this one in Moscow, typically included military parades.

©Dean Conger/Corbis

By the early 1930s, the Soviet Union was on its way to becoming one of the world's industrial giants. Stalin wanted something more, however. His vision was to spread a Soviet-style communist government throughout the world.

The USSR and Its Satellites

In 1941 Nazi Germany invaded the Soviet Union, drawing the country into World War II. During the conflict, the Soviets joined with Great Britain and the United States to defeat the Germans. At the end of World War II, the fate of Europe was left to the victors—the United States, Great Britain, and the USSR. The Soviet army already occupied Czechoslovakia, Poland, Romania, Hungary, and Bulgaria. Stalin agreed to allow elections in those countries but soon installed communist governments. Germany was split in two. The United States, Great Britain, and their allies set up West Germany as a democracy under their guidance, and East Germany became a communist state. Countries under Soviet rule came to be known as satellite countries, meaning they were under the economic and political domination of a more powerful country.

☑ **READING PROGRESS CHECK**

Analyzing How did the USSR come to control most of Eastern Europe?

The Regions in the Modern Era

GUIDING QUESTION *How is a "cold war" different from other kinds of war?*

After World War II, the Soviet Union shared superpower status with the United States. Both superpowers possessed weapons of unimaginable destructive force. Would they dare to use those weapons against each other?

The Cold War

The Cold War was the rivalry and conflict between the USSR and the United States and their allies. During the next four decades, the Soviet Union and the United States engaged in a struggle for world influence and power.

Although both superpowers built destructive weapons, they also used other **strategies**, such as the threat of force and providing military and financial aid to their allies. At times, however, an outcome of nuclear warfare seemed **inevitable**.

The United States and its allies created the North Atlantic Treaty Organization (NATO) in 1949. Any attack on a member country would be considered an attack on all of them, and NATO countries agreed to respond as a group. The original NATO countries included many of the non-Eastern European nations. When NATO admitted West Germany in 1955, the USSR responded by creating the Warsaw Pact. Member countries were Albania, Bulgaria, Czechoslovakia, East Germany, Hungary, Poland, Romania, and the USSR.

Academic Vocabulary

strategy a plan to solve a problem

inevitable sure to happen

Eastern Bloc

(1949) Date joined Soviet Union/Eastern Bloc
— Border of the Soviet Union
— Border of the Eastern Bloc states

Barents Sea

ARCTIC CIRCLE

RUSSIA

ESTONIA
(1940)
LATVIA
(1940)
LITHUANIA
(1940)
EAST
GERMANY
(1949)
POLAND
(1947)
BELARUS
(1922)
CZECHOSLOVAKIA
(1948)
UKRAINE
(1922)
HUNGARY
(1947)
MOLDOVA
(1940)
ROMANIA
(1947)
YUGOSLAVIA
(1945)
BULGARIA
(1946)
ALBANIA
(1945)

Baltic Sea

Black Sea

Caspian Sea

0   600 miles
0   600 kilometers
Lambert Azimuthal Equal-Area projection

MAP SKILLS

1 HUMAN GEOGRAPHY
What countries became tied to the Soviet Union during the early part of World War II?

2 PHYSICAL GEOGRAPHY
What about the location of Eastern Europe made this region important to the Soviet Union?

The two superpowers came close to war during the Cuban Missile Crisis. In October 1962, after learning that the Soviets were sending missiles to Cuba, the U.S. set up a naval blockade around Cuba to prevent the shipment of missiles. Both sides seemed prepared to go to war. As tensions grew, Soviet premier Nikita Khrushchev agreed to stop shipping the missiles to Cuba. Another crisis was brewing, however. The Soviet satellite countries in Eastern Europe began to rebel against Soviet control.

Unrest in the Soviet Satellites

In 1968 Czechoslovakia's leader Alexander Dubček announced sweeping reforms. He wanted to give the press more freedom and to guarantee citizens' civil rights. The Czech people welcomed the reforms, but the Soviets removed Dubček from power.

In 1980 dozens of Polish trade unions joined together to form Solidarity. Solidarity used strikes to put pressure on the government. The Polish government responded by declaring Solidarity illegal and putting its leaders in jail. Solidarity became an underground, or secret, organization.

Changes Under Gorbachev

Then in the 1980s, a new Soviet leader, Mikhail Gorbachev, came to power and implemented new policies. *Glasnost*, which means "openness," was an attempt to allow the people in the USSR and its

satellite countries to have more social and political freedoms. *Perestroika,* which means "restructuring," was an attempt to reform the Soviet economy.

Change came quickly in Eastern Europe. Solidarity was legalized in Poland in 1989, then the Communists were voted out of power. By 1990, Hungary, Czechoslovakia, Bulgaria, and Romania had new governments. In Germany, the Berlin Wall that separated West and East Berlin was torn down. By the next year, East and West Germany were reunited. Soviet control of Eastern Europe was broken.

In 1991 a group of Soviet officials, who thought Gorbachev's policies meant the downfall of the Soviet Union, staged a coup and arrested Gorbachev. Gorbachev's allies resisted, the people protested, and the military turned against the coup leaders. Gorbachev was released, and the coup leaders were arrested. Communist control of the USSR was at an end. By the end of 1991, the Soviet Union was dissolved, and all the republics had become independent countries.

Divisions and Conflict

When Eastern Europe shook free of Soviet domination in 1989, ethnic tensions flared in the Balkan Peninsula. The former Yugoslav republics used to be one large country called Yugloslavia. In the early 1990s, disputes among ethnic groups tore the country apart. Croatia, Slovenia, Macedonia, and Bosnia and Herzegovina broke free of Yugoslavia and became separate countries. Serbia and Montenegro each became its own country in 2006.

Kosovo, which was considered part of Serbia, has a mostly Albanian Muslim population. When Yugoslavia broke apart, many people in Kosovo decided they wanted to break free of Serbian control. When the Kosovo Liberation Army began an armed rebellion in 1998, Serbs responded with military force. NATO intervened to end the bloodshed, and the United Nations began governing Kosovo. Finally, in 2008 Kosovo declared itself independent, though Serbia, Russia, and other countries refused to recognize this.

✔ **READING PROGRESS CHECK**

Determining Central Ideas How did glasnost and perestroika affect the USSR?

Soviet leader Mikhail Gorbachev worked to improve the Soviet Union's relations with the United States and other Western countries.
▶ **CRITICAL THINKING**
Citing Text Evidence How did Gorbachev's policy of glasnost affect Eastern Europe?

Include this lesson's information in your Foldable®.

LESSON 2 REVIEW

Reviewing Vocabulary

1. Why did landowners protest *collectivization* of agriculture?

Answering the Guiding Questions

2. ***Describing*** How did Peter I and Catherine II change Russia?

3. ***Determining Central Ideas*** What were Stalin's main goals for the Soviet Union?

4. ***Analyzing*** How is a "cold war" different from other kinds of war?

5. ***Argument Writing*** Write a speech encouraging Czar Alexander II to abolish serfdom and to grant rights and liberties to all peasants.

Academic Vocabulary

- **decline**
- **unique**
- **factor**

Content Vocabulary

- **inflation**
- **oligarch**
- **devolution**

TAKING NOTES: *Key Ideas and Details*

Identify Use a graphic organizer like this one to identify three challenges faced by Russia and the countries of Eastern Europe.

Challenges for Russia and Eastern Europe Today

Lesson 3
Life in Eastern Europe and Western Russia

ESSENTIAL QUESTION • *How does geography influence the way people live?*

IT MATTERS BECAUSE
It has been more than 20 years since the collapse of the USSR. The countries that the USSR once ruled in Eastern Europe have all moved forward, some more easily than others.

People and Places

GUIDING QUESTION *What were some of the challenges Russia faced after the fall of the Soviet Union?*

Life has changed in Eastern Europe and Western Russia. Some people have benefited from the changes; others have not. The attempt to change from a communist state to a free market economy has not been easy or particularly successful.

Economic Changes

Russia faced enormous challenges following the collapse of the USSR. The economy that was centrally controlled by the communist government had been in **decline** for years. Most of its industry had centered on military hardware and heavy industrial machinery. The country was not prepared to transform into a producer of consumer products that are the real engine of free market economies.

Inflation, or the rise in prices for goods and services, increased, while production slowed. The transfer of industry to private ownership was a great deal for wealthy individuals who had connections in government. However, these changes did not improve the living conditions for most Russians. People refer to the new owners of industrial Russia as **oligarchs**. An oligarch is one of a small group of people who control the government and use it to further their own goals.

Social and Political Changes

When the Soviet Union existed, the central government kept tight control over its many ethnic groups. Some groups wanted to form their own countries. Among them are the Chechens, who live in Chechnya near the Caspian Sea and Caucasus Mountains in southern Russia. The region has many oil reserves, and its oil pipelines transport fuel to major Russian cities. Russian troops fought Chechen rebels to keep Chechnya a part of Russia. When Russia finally pulled out in 1996, the Chechen rebellion was still not over. President Boris Yeltsin was widely blamed for being unable to solve these problems.

At the end of 1999, Yeltsin resigned and was replaced by Vladimir Putin, who was elected president in 2000. Putin, a former officer in the KGB, the country's secret police, was viewed as someone who wanted to keep a tight rein on government power. Putin launched reforms to reduce the power of the oligarchs and encouraged economic development. Although Putin helped stabilize the Russian economy, he dealt harshly with those who opposed him.

He was reelected for a second term in 2004. In 2008 Dmitry Medvedev was elected president, and he appointed Putin to be prime minister. In 2012 Putin ran for the presidency a third time and won. Soon after, new restrictive laws were passed strengthening penalties against demonstrators, blocking some Internet sites, and restricting free speech.

Where People Live

The two largest cities in Western Russia are Moscow and St. Petersburg. In addition to being the political capital of Russia, Moscow is the cultural, educational, and scientific capital. It has also been the spiritual home of the Russian Orthodox Church for more than 600 years. St. Petersburg was founded by Peter the Great in 1703.

The biggest population centers in Eastern Europe are the capital cities, such as Kiev, Ukraine; Minsk, Belarus; Budapest, Hungary; Warsaw, Poland; and Prague in the Czech Republic. Each of these cities is a center of national culture.

At one time, most Eastern Europeans lived in rural areas. Now more people live in urban areas. The urban population of Albania, Macedonia, and Croatia, for example, is above 50 percent. In some countries, urbanization is even higher. In Poland and Montenegro, the urban population is about 60 percent. In Hungary, the Czech Republic, and Belarus, urbanization is 70 percent or more.

Academic Vocabulary

decline a gradual deterioration into a weakened condition

A family sits in front of its farmhouse in Senj, a seaside town in Croatia. Although parts of Eastern Europe remain rural, most people now live in urban areas.

During the industrial age, people began moving from Eastern Europe to Western Europe and North America; that trend continues. Eastern Europeans have moved to escape political oppression and to seek better economic opportunities. Countries such as Romania have lost population since the lifting of Soviet travel restrictions.

☑ **READING PROGRESS CHECK**

Determining Central Ideas Why have so many Eastern Europeans emigrated to other parts of Europe or to the United States?

People and Cultures

GUIDING QUESTION *How did geographical barriers affect the development of Slavic culture in Eastern Europe?*

The history of Western Russia and Eastern Europe has created a rich mix of cultures and people. People take great pride in their folk and religious traditions, most of which were frowned upon by Soviet authorities.

Ethnic and Language Groups

At one time, a single Slavic language, understood by most Slavic people, was spoken. As Slavic people settled in different parts of Eastern Europe, geographical barriers separated and isolated them. These groups developed distinct languages and cultures.

Slavs generally belong to one of three categories. East Slavs are represented by the Slavic ethnic groups in Russia, Ukraine, and Belarus. West Slavs include ethnic Slavs in Poland, the Czech Republic, Slovakia, and parts of eastern Germany. The most diverse group are the South Slavs, who live in Bulgaria and other countries

GRAPH SKILLS >

POPULATION OF RUSSIA

The graph shows the changes in the population of Russia from 1940 to 2010.

▶ **CRITICAL THINKING**

1. ***Describing*** In what decades did the population of Russia increase?

2. ***Analyzing*** What is a possible explanation for Russia's sharp drop in population during the 1940s?

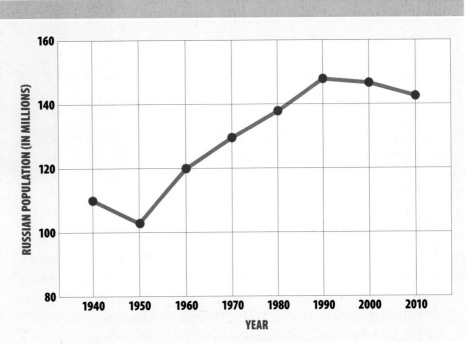

RUSSIAN POPULATION (IN MILLIONS)

YEAR

East Slavs
West Slavs
South Slavs

MAP SKILLS

1 HUMAN GEOGRAPHY
What three large divisions make up the Slavic people of Eastern Europe and Western Russia?

2 PLACES AND REGIONS
What countries in the region do not have majority Slavic populations?

of the Balkans. Each of these groups speaks its own language. Russia is made up of more than 120 ethnic groups, although almost 80 percent of the population is ethnic Russian.

The people of Albania are a distinct ethnic group that has been living in that region for about 4,000 years. Albanian is the last surviving language of an entire Indo-European language group. It is the ancestor of the language spoken by present-day Albanians, and it has survived thousands of years of conquest and cultural change.

Religion

For most of the 1900s, religious practice was strongly discouraged throughout Eastern Europe and Western Russia. In some countries in Eastern Europe, a sizable percentage of the population does not practice any religion. Less than half the population of the Czech Republic belongs to any church. In the Baltic region, nearly two-thirds of Latvians and one-third of Estonians are not affiliated with any church. In most of Eastern Europe, however, Soviet repression strengthened religious faith. The dominant religion in most of these countries is the Eastern Orthodox Church. Many different churches exist within the Orthodox faith.

Most of these churches are affiliated with a specific ethnic group or country. The majority of people in Belarus, Bulgaria, Moldova, Montenegro, Romania, Serbia, and Ukraine worship at an Eastern Orthodox Church. The majority of people in Croatia, Hungary, Lithuania, Poland, Slovakia, and Slovenia are Roman Catholics. Most of these countries also support minority populations of

Shoppers walk along the boulevard past a mall in Plovdiv, Bulgaria. Plovdiv is the country's second-largest city. Only the capital city of Sofia is larger.

Muslims, Roman Catholics, Eastern Orthodox, Protestants, and Jews. Nearly 70 percent of Albanians are Muslim, as are a sizable number of people in Bosnia and Herzegovina.

The Arts

In the 1800s and early 1900s, Russians produced some of the most important cultural works in all of Europe. The novels of Tolstoy and Dostoyevsky; the music of Mussorgsky, Tchaikovsky, and Rimsky-Korsakov; and the plays and short stories of Chekhov and Gogol are still considered among the world's finest.

People in the countries of Eastern Europe are proud of the great art produced by their people. In many cases, those works are an important symbol of their national character. Although a small amount of literary work was written in the Czech language, Czech literature did not became internationally important until Czechoslovakia became an independent country in 1918. Karel Čapek was a Czech writer who was famous for his plays and novels. His most well-known contribution to world literature is a word he coined—*robot*—in his 1921 play *R.U.R.* Eastern European composers, such as Béla Bartók and Zoltán Kodály, celebrated the traditional music of Hungary and Romania by using it in their compositions. Bulgaria also has a rich tradition in folk and choir music.

Western popular culture has had a huge impact on the art of Eastern Europe and Western Russia. Russian and Polish filmmakers can follow their national traditions, but they also can see how well-liked and influential American films and television are. Rock and pop music from the United States and Western Europe are extremely popular. Young artists in this part of the world are creating international popular culture while trying to bring something uniquely Eastern European to it.

Daily Life

For much of the 1900s, the people of Eastern Europe and Western Russia lived under communist governments that attempted to control their private lives. The collapse of the USSR brought about **devolution** in Russian government and in governments throughout Eastern Europe. Devolution occurs when a strong central government surrenders its powers to more local authorities.

One of the results of this change is the return of national traditions and identity. Most of these countries have strong cultural and religious traditions. These traditions were never really lost, but the lack of strong Soviet control has made it possible for people to live and speak more freely about their beliefs and interests. Such freedoms emphasize **unique** aspects of these countries and their people.

The other result of the loss of Soviet control is the rising influence of international popular culture. Soviet authorities did not trust the music, films, and television programs coming from capitalist countries such as the United States, but they could not effectively outlaw them. Young Russians and Eastern Europeans are now having the same cultural experiences that young people are having in the rest of Europe and in the Americas. This shared culture emphasizes those things that all these people have in common with each other.

One issue in Russia is the generation gap that exists between people who grew up in communist USSR or are old enough to remember it, and those people who have lived most or all of their lives in the post-Soviet era. One big question is how to teach the history of the USSR to young people who never experienced life under the Soviet system.

Bruce Yuanyue Bi/Lonely Planet Images/Getty Images

Academic Vocabulary

unique unusual or distinctive

Teenagers stroll on the main street of Minsk, the capital and largest city of Belarus.
▶ CRITICAL THINKING
Analyzing How has the end of Communist rule affected young people in Eastern Europe and Western Russia?

In the 1990s, historians took a critical look at the Russian Revolution, the leadership of Lenin and Stalin, and the excesses of the Soviet government. They took a more positive approach when looking back at the era of the Romanov czars.

☑ **READING PROGRESS CHECK**

Determining Central Ideas How has daily life changed in Russia since the fall of the USSR?

Academic Vocabulary

factor something that actively contributes to a result

Issues in Eastern Europe and Western Russia

GUIDING QUESTION *What are the economic advantages and disadvantages of Eastern Europe's location between continents?*

Eastern Europe and Western Russia are still in the process of change. They are trying to modernize their industries and governments during difficult economic times.

Earning a Living

Western Russia and the countries of Eastern Europe cover a vast area. The landforms, the soil, the mineral resources, the climate, the economies, and the national traditions are different throughout the region. These **factors** combine to determine how people earn a living in these places. Nearly half the working population in Albania is employed in agriculture. Romania, Serbia, and Bosnia and Herzegovina also have many people employed in agriculture.

Russia is one of the world's leading suppliers of oil and natural gas. Most of it comes from western Siberia or the region between the Volga River and the Ural Mountains. Pipelines link these regions to the rest of the country. Russia supplies oil as well as natural gas to European countries, especially the countries in Eastern Europe.

Russia is also a major supplier of iron ore and other metals. About 1 million people work in Russia's forestry industry. The majority of people in Russia and in most of Eastern Europe now work in the service industries, or businesses that provide services to individuals as well as other businesses.

Many of the industries in Eastern Europe fell on hard times after the collapse of the Soviet Union. Many industries suffered big losses, leading to high unemployment, especially in the

During Communist rule and shortly after its end, shoppers in Eastern Europe had to wait in long lines outside of stores. Food and other products often were in short supply and had to be rationed, or divided equally.

▶ **CRITICAL THINKING**

Describing What benefit has come to Eastern European workers in recent years?

©Shepard Sherbell/Corbis SABA

Balkans. One positive move is the number of Eastern European countries that have joined the European Union. Since 2004, the following countries have joined the EU: the Czech Republic, Hungary, Latvia, Lithuania, Poland, Slovakia, Slovenia, Bulgaria, and Romania. The EU has sought to protect employment, improve workers' living and working conditions, and create a strong European trading bloc that can compete effectively with the United States.

Connections: Europe and Asia

Developments in Europe and Asia have affected Russia. Russians lived under Mongol rulers for centuries and later were next door to the Ottoman Turks, one of the most powerful Islamic empires to exist. Russia was influenced by developments in Europe, and, in turn, made contributions to European culture. Russian culture has always been a mix of European and Asian influences.

Today, oil and natural gas extracted from the Siberian oil fields in Central Asia are delivered via pipeline to all of Russia, Eastern Europe, and as far west as Italy and Germany. Geographically and culturally, Russia plays a key role in the relationship between Europe and Asia.

Addressing Challenges

Russia and countries that were formerly part of the Soviet republic continue to discuss agreements on the borders of the countries. The countries involved are Estonia, Latvia, Lithuania, Ukraine, and Kazakhstan (in Central Asia). Even though fighting died down in Chechnya, occasional outbreaks of violence still occur in that republic, and rebels continue to call for independence.

The 2008 financial crisis hit Eastern Europe hard. The countries there struggled as they transformed into free market economies. Joining the EU should have been a great benefit to the economies of its new members, but all members of the EU have suffered as a result of the financial crisis.

FOLDABLES®
Study Organizer

Include this lesson's information in your Foldable®.

✅ **READING PROGRESS CHECK**

What have been two setbacks in the economies of Eastern Europe since the collapse of the USSR?

LESSON 3 REVIEW

Reviewing Vocabulary

1. Why were the *oligarchs* unpopular with some Russians?

Answering the Guiding Questions

2. *Analyzing* How did the rebellion in Chechnya affect the presidency of Boris Yeltsin?

3. *Describing* How did geographical barriers affect the development of Slavic culture in Eastern Europe?

4. *Determining Central Ideas* What are the economic advantages and disadvantages of Eastern Europe's location between continents?

5. *Narrative Writing* Imagine that you live in an Eastern European country and are writing a letter to your cousin who has lived in the United States for many years. Try to get your cousin to come visit you. Be sure to remind him or her of all the positive changes that have occurred since the collapse of the Soviet Union.

Directions: Write your answers on a separate piece of paper.

1 Use your FOLDABLES to explore the Essential Question.

INFORMATIVE/EXPLANATORY WRITING Explain the meaning of the word *balkanization* and how it acquired its meaning.

2 **21st Century Skills**

ANALYZING With a partner, research to find one primary source and one secondary source about the Cuban missile crisis. Then, discuss and answer these questions: Which source provided a clearer picture of the event? Why? Did either source seem to support or favor one side over the other?

City	Country
Moscow, St. Petersburg	Russia
Kiev	Ukraine

3 **Thinking Like a Geographer**

IDENTIFYING On a chart like the one shown, list the most populous cities of Western Russia and Eastern Europe with their countries.

4 **GEOGRAPHY ACTIVITY**

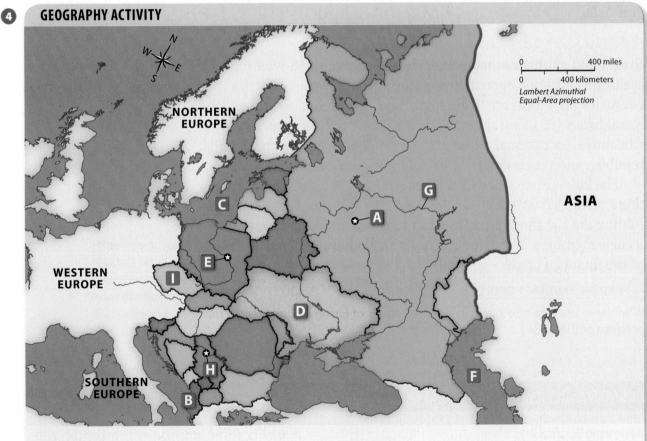

Identifying

Match the letters on the map with the numbered places listed below.

1. Baltic Sea
2. Moscow
3. Volga River
4. Caspian Sea
5. Ukraine
6. Belgrade
7. Albania
8. Warsaw
9. Czech Republic

REVIEW THE GUIDING QUESTIONS

Directions: Choose the best answer for each question.

1 Which mountains form a boundary between Europe and Asia?
 A. the Carpathians
 B. the Urals
 C. the Balkans
 D. the Alps

2 Although Russia is the largest country in the world, it has a relatively small percentage of land available for farming. Its main crop is
 F. tobacco.
 G. sunflowers.
 H. grains.
 I. lavender.

3 What did Czar Alexander II abolish in Russia in 1861?
 A. the Russian Orthodox Church
 B. diamond mining
 C. the Russian army and navy
 D. serfdom

4 The North Atlantic Treaty Organization (NATO), created in 1949, is an alliance of which countries?
 F. the Balkan states
 G. Western Russia and Eastern Europe
 H. the United States and its allies
 I. the USSR and Cuba

5 The dominant religion in most of Eastern Europe and Western Russia is
 A. Islam.
 B. Eastern Orthodox.
 C. Protestantism.
 D. Moldavan.

6 The difference in experience and viewpoints between those who grew up in the Communist USSR and Russians of the post-Soviet era is referred to as
 F. a generation gap.
 G. the diaspora.
 H. cultural dissonance.
 I. capitalism.

DBQ ANALYZING DOCUMENTS

7 IDENTIFYING Read the following passage about Russian culture:

"*As one looks at the history of Russian culture, it may be helpful to think of the forces rather than the forms behind it. Three in particular—the natural surroundings, the Christian heritage, and the Western contacts of Russia—hover bigger than life.*"

—from James H. Billington, *The Icon and the Axe* (1970)

Which theme of geography is represented by the influence of Western contacts on Russia?

A. human-environment interaction

B. location

C. movement

D. place

8 ANALYZING Which of these aspects of natural surroundings is most likely to have affected Russian culture?

F. sense of vast space

G. warm climate

H. fertile farmland

I. nearness to neighbors

SHORT RESPONSE

"*The southern half of Eastern Europe is referred to as the Balkans or Balkan Peninsula, after the name of a mountain range in Bulgaria. Balkanization [means] the recurrent division and fragmentation of this part of Eastern Europe, and it is now applied to any place where such processes take place.*"

—from H.J. de Blij and Peter O. Muller, *Geography* (2006)

9 ANALYZING What characteristics of the Balkans and the people who live there led to these frequent divisions?

10 IDENTIFYING POINT OF VIEW If you were a leader of one of the peoples of the Balkans, how would you try to bridge the divisions separating your group from others?

EXTENDED RESPONSE

11 INFORMATIVE/EXPLANATORY WRITING In the 1990s, Russia began to make a transition from a communist economy to a more capitalistic economy. Discuss how this transition has worked so far. What are some of the positive and negative factors in the change? You will want to examine some outside sources to explain this economic transition.

Need Extra Help?

If You've Missed Question	1	2	3	4	5	6	7	8	9	10	11
Review Lesson	1	1	2	2	3	3	3	1	3	3	3

ASIA

UNIT **5**

©Keren Su/Corbis

EXPLORE the CONTINENT

ASIA is the largest continent on Earth. It extends from the Mediterranean Sea in the west to the Pacific Ocean in the east. Asia is home to about 60 percent of the world's population. Yet, at least two-thirds of the land area is too cold or too dry to support a large population. Winds called monsoons affect climate in much of Asia. They bring cool, dry weather in winter and heavy rainfall and floods in summer.

① NATURAL RESOURCES Asia's numerous resources include petroleum, copper, rice, and fish. Growing populations and industries in cities such as Shanghai, China, demand more and more resources. Asian countries are quickly trying to meet this need.

② LANDFORMS Asia's landforms are varied. Mountain ranges, plateaus, and plains dominate western and central areas. Hundreds of islands dot coastlines in the east. Plate movements formed many of these islands. The island country of Indonesia has about 130 active volcanoes, more than any other country on Earth.

③ BODIES OF WATER Water is plentiful in eastern and southern Asia, where many people depend on it for a living. In Sri Lanka, fishers use baitless hooks to snare mackerel and herring. In parts of western Asia, however, water is scarce. The lack of water is a major issue for countries in these areas.

FAST FACT

Asia comprises 30 percent of the world's land area.

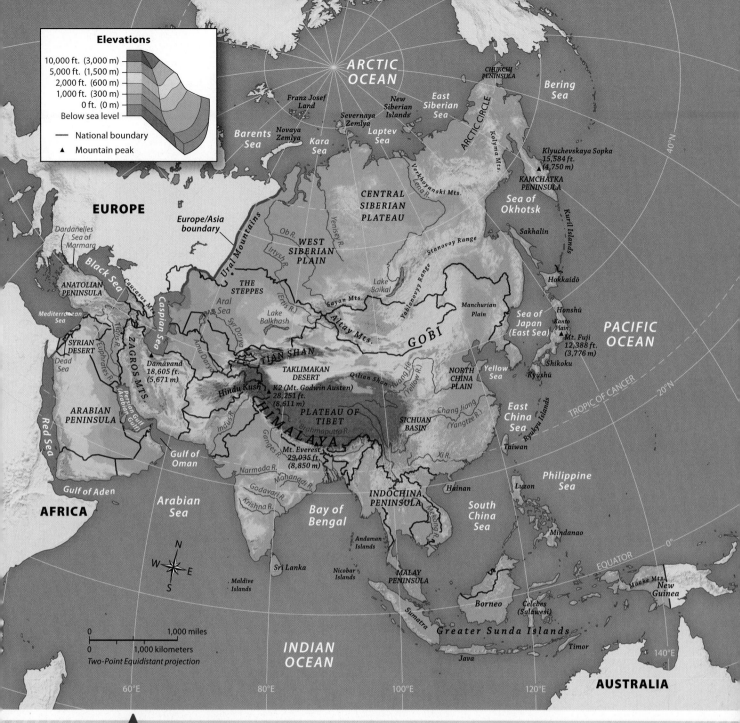

Elevations

10,000 ft. (3,000 m)
5,000 ft. (1,500 m)
2,000 ft. (600 m)
1,000 ft. (300 m)
0 ft. (0 m)
Below sea level

— National boundary
▲ Mountain peak

ARCTIC OCEAN

CHUKCHI PENINSULA

Bering Sea

Franz Josef Land

New Siberian Islands

East Siberian Sea

Severnaya Zemlya

Klyuchevskaya Sopka
15,584 ft.
(4,750 m)

Barents Sea

Novaya Zemlya

Kara Sea

Laptev Sea

ARCTIC CIRCLE

Kolyma Mts.

KAMCHATKA PENINSULA

EUROPE

Europe/Asia boundary

Ural Mountains

CENTRAL SIBERIAN PLATEAU

Verkhoyanski Mts.

Lena R.

Sea of Okhotsk

Sakhalin

Kuril Islands

Dardanelles
Sea of Marmarg

Black Sea

ANATOLIAN PENINSULA

Caucasus Mts.

THE STEPPES

Ob R.

Irtysh R.

WEST SIBERIAN PLAIN

Yenisey R.

Stanovoy Range

Hokkaidō

Mediterranean Sea

Aral Sea

Lake Balkhash

(Irtis R.)

Lake Baikal

Sayan Mts.

Yablonovy Range

Manchurian Plain

Sea of Japan (East Sea)

Honshū

PACIFIC OCEAN

SYRIAN DESERT

ZAGROS MTS.

Caspian Sea

Syr Darya

Altay Mts.

GOBI

Kanto Plain

Mt. Fuji
12,388 ft.
(3,776 m)

Dead Sea

Tigris R.

Euphrates R.

Persian Gulf (Arabian Gulf)

Amu Darya

Damávand
18,605 ft.
(5,671 m)

TIAN SHAN

TAKLIMAKAN DESERT

Qilian Shan

Huang He

NORTH CHINA PLAIN

Yellow Sea

Shikoku

Kyūshū

ARABIAN PENINSULA

Hindu Kush

K2 (Mt. Godwin Austen)
28,251 ft.
(8,611 m)

PLATEAU OF TIBET

Yellow R.

SICHUAN BASIN

Chang Jiang
(Yangtze R.)

East China Sea

Ryukyu Islands

Red Sea

Indus R.

HIMALAYA

Brahmaputra R.

Xi R.

Taiwan

TROPIC OF CANCER

Gulf of Oman

Ganges R.

Mt. Everest
29,035 ft.
(8,850 m)

20°N

Gulf of Aden

Narmada R.

Godavari R.

Mahanadi R.

INDOCHINA PENINSULA

Hainan

Philippine Sea

AFRICA

Arabian Sea

Krishna R.

Bay of Bengal

Mekong R.

South China Sea

Luzon

N
W E
S

Andaman Islands

Sri Lanka

Nicobar Islands

MALAY PENINSULA

Mindanao

Maldive Islands

0   1,000 miles

0   1,000 kilometers

Two-Point Equidistant projection

INDIAN OCEAN

Borneo

Celebes
(Sulawesi)

Maoke Mts.

New Guinea

EQUATOR

0°

Sumatra

Greater Sunda Islands

Java

Timor

140°E

AUSTRALIA

60°E   80°E   100°E   120°E

ASIA

PHYSICAL

MAP SKILLS

1 PHYSICAL GEOGRAPHY   What part of Asia has the highest elevation?

2 THE GEOGRAPHER'S WORLD   Which ocean borders the Central Siberian Plateau?

3 PLACES AND REGIONS   How would you describe the region of Southeast Asia?

POLITICAL

MAP SKILLS

1 **PLACES AND REGIONS** Which country in the region is the largest?

2 **THE GEOGRAPHER'S WORLD** Which country is located between China and Russia?

3 **THE GEOGRAPHER'S WORLD** What body of water lies between Kazakhstan and Iran?

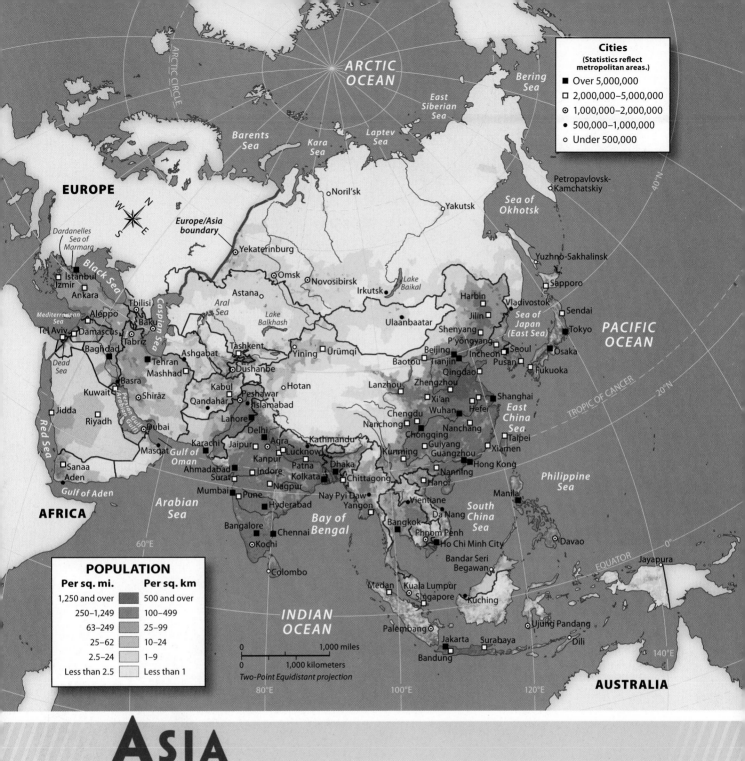

Cities
(Statistics reflect metropolitan areas.)

- ■ Over 5,000,000
- □ 2,000,000–5,000,000
- ◉ 1,000,000–2,000,000
- • 500,000–1,000,000
- ○ Under 500,000

POPULATION

Per sq. mi.	Per sq. km
1,250 and over	500 and over
250–1,249	100–499
63–249	25–99
25–62	10–24
2.5–24	1–9
Less than 2.5	Less than 1

0 1,000 miles
0 1,000 kilometers
Two-Point Equidistant projection

ASIA

POPULATION DENSITY

MAP SKILLS

1 **THE GEOGRAPHER'S WORLD** What parts of Asia are the most densely populated?

2 **THE GEOGRAPHER'S WORLD** Which part of the region has the lowest population density?

3 **THE GEOGRAPHER'S WORLD** In general, what population pattern do you see in Southwest Asia?

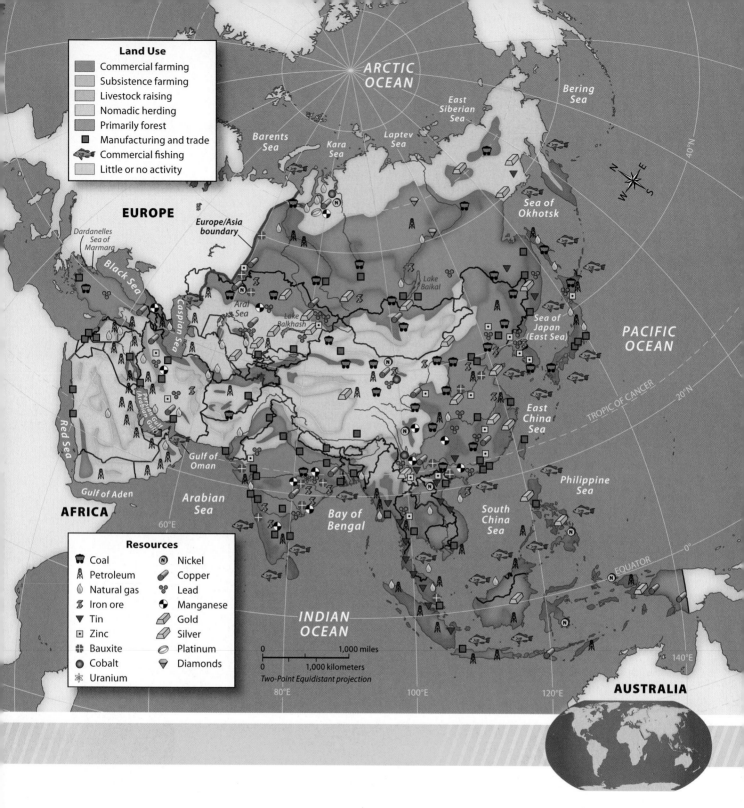

Land Use

- Commercial farming
- Subsistence farming
- Livestock raising
- Nomadic herding
- Primarily forest
- Manufacturing and trade
- Commercial fishing
- Little or no activity

Resources

- Coal
- Petroleum
- Natural gas
- Iron ore
- Tin
- Zinc
- Bauxite
- Cobalt
- Uranium
- Nickel
- Copper
- Lead
- Manganese
- Gold
- Silver
- Platinum
- Diamonds

0 — 1,000 miles
0 — 1,000 kilometers
Two-Point Equidistant projection

ECONOMIC RESOURCES

MAP SKILLS

1 PLACES AND REGIONS What mineral resources can be found around the Persian Gulf?

2 HUMAN GEOGRAPHY What economic activity takes place in the South China Sea?

3 THE GEOGRAPHER'S WORLD In what part of Russia are gold and silver deposits found?

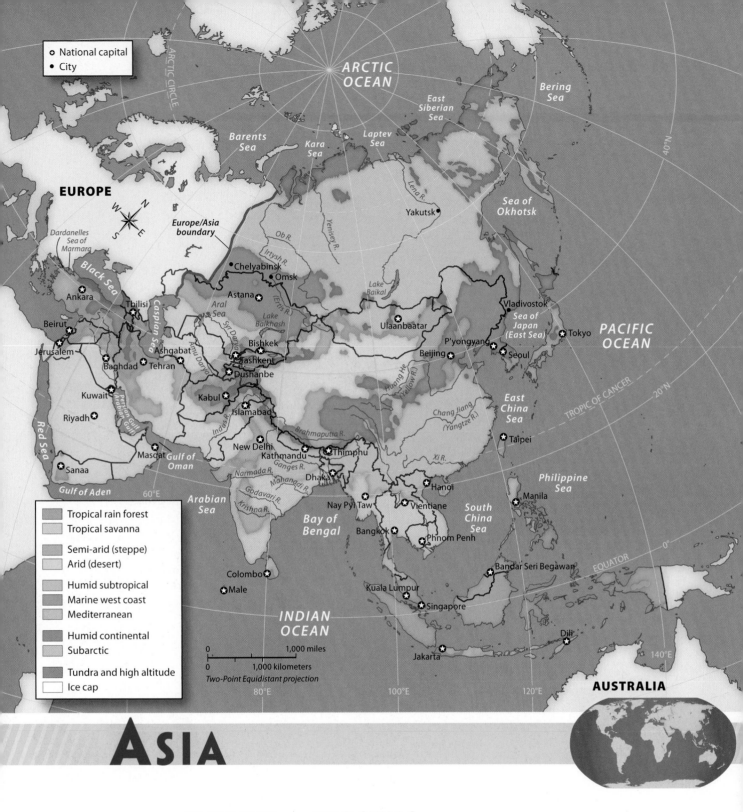

Legend:
- ⊛ National capital
- • City

ARCTIC OCEAN

EUROPE

Europe/Asia boundary

Dardanelles
Sea of Marmara

Black Sea

Ankara ⊛
Tbilisi ⊛
Beirut •
Jerusalem ⊛
Baghdad ⊛
Kuwait ⊛
Riyadh ⊛
Sanaa ⊛
Masqat ⊛

Ashgabat ⊛
Tehran ⊛
Kabul ⊛
Islamabad ⊛

Chelyabinsk •
Omsk •
Astana ⊛
Aral Sea
Lake Balkhash
Bishkek ⊛
Tashkent ⊛
Dushanbe ⊛

Caspian Sea
Amu Darya
Syr Darya
Irtysh R.
(Ertis R.)

Ob R.
Yenisey R.
Lena R.

Lake Baikal

Ulaanbaatar ⊛

Yakutsk •

Sea of Okhotsk

Bering Sea

East Siberian Sea

Laptev Sea

Kara Sea

Barents Sea

Vladivostok •
Sea of Japan (East Sea)
Tokyo •
P'yongyang ⊛
Beijing ⊛
Seoul ⊛

PACIFIC OCEAN

40°N
20°N
TROPIC OF CANCER

Huang He (Yellow R.)
Chang Jiang (Yangtze R.)
Xi R.

East China Sea

Taipei •

New Delhi ⊛
Kathmandu ⊛
Thimphu ⊛
Dhaka ⊛
Ganges R.
Brahmaputra R.
Narmada R.
Mahanadi R.
Godavari R.
Krishna R.
Indus R.

Nay Pyi Taw ⊛
Hanoi ⊛
Vientiane ⊛
Bangkok ⊛
Phnom Penh ⊛

Manila ⊛

Philippine Sea

South China Sea

Bandar Seri Begawan ⊛

Red Sea
Gulf of Aden
Gulf of Oman
Arabian Sea
Persian Gulf (Arabian Gulf)

Bay of Bengal

Colombo ⊛
Male ⊛

Kuala Lumpur ⊛
Singapore ⊛

Jakarta ⊛

Dili ⊛

INDIAN OCEAN

EQUATOR
0°
140°E
120°E

Climate Legend:
- Tropical rain forest
- Tropical savanna
- Semi-arid (steppe)
- Arid (desert)
- Humid subtropical
- Marine west coast
- Mediterranean
- Humid continental
- Subarctic
- Tundra and high altitude
- Ice cap

0 — 1,000 miles
0 — 1,000 kilometers
Two-Point Equidistant projection

80°E   100°E   120°E

AUSTRALIA

ASIA

CLIMATE

MAP SKILLS

1 PHYSICAL GEOGRAPHY How would you describe the climate zones found along the Equator?

2 PLACES AND REGIONS Which city do you think receives more rain each year—Beijing or Ulaanbaatar? Why?

3 PHYSICAL GEOGRAPHY In general, how does the climate of India's coastal areas differ from the climate of the Ganges River area?

EAST ASIA

ESSENTIAL QUESTIONS · *How does geography influence the way people live?*
· *What makes a culture unique?* · *How do cultures spread?* · *Why do people trade?*
· *How does technology change the way people live?*

Performers with the Beijing Opera wear facial paint and colorful costumes.

The Story Matters...

East Asia can trace many of its cultural features to an ancient civilization that arose in China thousands of years ago. In the centuries that followed, powerful dynasties ruled China, creating an enormous empire that influenced the entire region. Today, migration and trade have paved the way for an exchange of ideas and practices between East Asia and other parts of the world. Japan, Taiwan, and South Korea have become modern industrial nations.

FOLDABLES®
Study Organizer

Go to the Foldables® library in the back of your book to make a Foldable® that will help you take notes while reading this chapter.

Economic Growth
Cultural Influences
Mainland and Islands
East Asia

Jon Arnold Alamy

429

EAST ASIA (CCSS)

East Asia borders Central Asia , South Asia, and Southeast Asia and extends to the Pacific Ocean. As you study the map of East Asia, look for the geographic features that make this region of Asia unique.

Step Into the Place

MAP FOCUS Use the map to answer the following questions.

1 **THE GEOGRAPHER'S WORLD** What country lies south of Mongolia?

2 **PLACES AND REGIONS** What seas border China?

3 **THE GEOGRAPHER'S WORLD** What bodies of water lie between Japan and the Koreas?

4 **CRITICAL THINKING**
Analyzing If you traveled east along 40° N latitude, which East Asian countries would you travel through?

A **BUSY CITY MARKET** One of the main attractions in Hong Kong, China, is the Temple Street Night Market. Each night, crowds visit the market to buy a variety of goods, such as clothes and household appliances.

B **REMOTE MYSTERIOUS HIGHLAND** A nomadic sheep herder crosses the vast spaces of the Plateau of Tibet in western China.

Step Into the Time

TIME LINE Choose an event from the time line and write a paragraph predicting the effect of that event on the future of East Asia.

c. 550 B.C. Confucius is born

1766 B.C. Rule of China's Shang dynasty begins

1000 B.C.

East Asia

National capital

City

RUSSIA

CENTRAL ASIA

MONGOLIA

⊙Ulaanbaatar

NORTH KOREA

Sea of Japan (East Sea)

JAPAN

Beijing⊙

P'yŏngyang⊙

Tokyo⊙

⊙Seoul

B

CHINA

Huang He (Yellow R.)

SOUTH KOREA

Hiroshima

Yellow Sea

SOUTH ASIA

Brahmaputra R.

Chengdu•

Chang Jiang (Yangtze R.)

Shanghai•

East China Sea

PACIFIC OCEAN

TROPIC OF CANCER

A

•Taipei

Xi R.

TAIWAN

•Hong Kong

Macao•

Bay of Bengal

SOUTHEAST ASIA

90°E

South China Sea

Philippine Sea

0   500 miles
0   500 kilometers
Lambert Azimuthal Equal-Area projection

100°E

110°E

120°E

130°E

50°N
40°N
30°N
20°N
10°N

N W E S

c. 200 B.C.
Travel on the Silk Road begins

1945
Atomic bombs dropped on Hiroshima, Nagasaki; Japan surrenders, ending World War II

935 A.D. Wank Kon renames kingdom Koryo (Korea)

1192 Shoguns begin rule of Japan

2011 Earthquake, tsunami strike Japan

1000 A.D.

2000 A.D.

1001 Japan's Murasaki Shibubi writes *The Tale of Genji*

1949 Communists seize power in China

Reading **HELP**DESK **CCSS**

Academic Vocabulary

• **dominate**

Content Vocabulary

• **de facto**
• **archipelago**
• **tsunami**
• **loess**

TAKING NOTES: *Key Ideas and Details*

Summarize As you study the lesson, use a graphic organizer like the one below to write one summary sentence for each topic.

Topic	Summary
Landforms	
Waterways	
Resources	

Lesson 1
Physical Geography of East Asia

ESSENTIAL QUESTION • *How does geography influence the way people live?*

IT MATTERS BECAUSE
East Asia is where some of the most densely populated areas in the world are found. It also has vast areas where few people live. To understand why these extreme differences exist, it is necessary to understand the region's physical geography.

Landforms and Waterways

GUIDING QUESTION *What are the main physical features and physical processes in East Asia?*

The landscapes and physical features of East Asia are varied and sometimes unusual. East Asia is home to the world's highest mountain range, as well as a vast plateau that sits miles high. It is also the site of islands created by volcanoes and fertile plains where hundreds of millions of people live.

A Regional Overview
East Asia covers much of the eastern half of the Asian continent. Its eastern boundary stretches along the Pacific Ocean. Bordering East Asia to the north are Russia and Central Asia. To the south are the regions of South Asia and Southeast Asia.

East Asia is made up of six countries: China, Japan, Mongolia, North Korea, South Korea, and the de facto country of Taiwan. A **de facto** country is one not legally recognized by other countries. China is the largest country in the region. It has more than four-fifths of the region's total land area. Slightly smaller than the United States, China is the world's fourth-largest country in land area.

At the opposite extreme is Taiwan, the region's smallest country. It is roughly the size of Massachusetts and Connecticut combined. China and Mongolia sit on the Asian mainland, while the other four countries occupy islands or a peninsula.

The Mainland

Geographers often divide mainland East Asia, which consists of China and Mongolia, into three broad geographic subregions. Because of differences in elevation, the subregions are like steps in a staircase. The highest step is the Plateau of Tibet, a vast area of mountains and uplands in southwestern China. The plateau is sometimes called "the roof of the world" because of its extremely high elevation. Much of the land sits more than 2.5 miles (4 km) above sea level.

Lofty mountains circle the Plateau of Tibet. The Kunlun Shan range runs along the northern edge, and the Himalaya—the tallest mountains in the world—rise along the southern edge. The Himalaya reach their highest point at Mount Everest, which soars higher than any other mountain in the world. When mountain climbers reach Everest's peak, they stand nearly 5.5 miles (8.9 km) above sea level.

North and east of Tibet, land elevation drops sharply to the second step in the staircase. Mountains and plateaus **dominate** this subregion, too, but they are generally much lower than those of Tibet. Some of the mountain ranges lie along the edges of enormous basins, or natural land depressions. Much of the land in the northern part of the subregion is desert or near desert, with little or no vegetation.

Land along the southern part of the subregion is forested. Some of the steepest and deepest canyons in the world lie where the land descends from Tibet.

The third and lowest step in the staircase covers most of the eastern third of China. The main landforms in this subregion are low hills and plains. Most Chinese live on these plains.

Academic Vocabulary

dominate to have the greatest importance

A train crosses a bridge spanning a deep valley in western China. The Qingzang, the world's highest rail system, links newly developing areas in the west to densely populated plains areas in eastern China.

▶ **CRITICAL THINKING**
Describing Why is western China's Plateau of Tibet called "the roof of the world?"

©Fritz Hoffmann/In Pictures/Corbis

A Peninsula and Many Islands

In addition to the mainland, East Asia includes a large peninsula and, running parallel to the eastern coast, a long string of islands. The Korean Peninsula is a thumb of land that juts southward from the mainland between the Yellow Sea and the Sea of Japan (East Sea). It is home to two countries: North Korea and South Korea. The peninsula is mountainous, especially in the northeast. In the south and west, broad plains stretch between the mountains and the coast.

Along the eastern edge of the Sea of Japan, an arc of islands stretches for roughly 1,500 miles (2,414 km). The islands—four large ones and thousands of much smaller ones—form the **archipelago** (ahr kuh PEH luh goh) of Japan. An archipelago is a group or chain of islands.

The islands of Japan are part of the Ring of Fire, which nearly encircles the Pacific Ocean. In this area, huge sections of Earth's crust grind against each other and cause earthquakes and volcanic eruptions. The islands of Japan were formed by volcanic eruptions millions of years ago.

Forested mountains cover nearly three-fourths of the land on the islands. Plains are generally small and isolated. On Honshū, Japan's largest island, a beautiful, cone-shaped volcano called Mount Fujiyama or Mount Fuji rises about 12,400 feet (3,780 m) above the Kanto plain. Snow-covered Mount Fuji is a well-known symbol of Japan. Although it has not erupted in nearly 300 years, scientists believe it could.

Japan is one of the most earthquake-prone countries in the world. More than 1,000 small earthquakes shake the country every year. Major quakes occur less often, but they can cause tremendous damage and loss of life. When an earthquake occurs below or close to the ocean, it can trigger a **tsunami** (soo NAH mee). This is a huge wave that gets higher as it approaches the coast. Tsunamis can wipe out coastal cities and towns. In 2011 the most powerful earthquake in Japan's history produced massive tsunamis that devastated areas along the northeastern coast.

Hundreds of miles southwest of Japan's main islands lies another large island, Taiwan. Like Japan, Taiwan was formed as a result of volcanic activity. Mountains form a rugged spine that stretches the length of this sweet-potato-shaped

Mount Fuji, on Japan's Honshū island, soars above blooming cherry trees. Each spring, people come to the five lakes around Mount Fuji to view the cherry blossoms with the beautiful mountain in the distance.

▶ **CRITICAL THINKING**

Describing What physical features make up the islands of Japan?

Peerapat Tandavanitj/Flickr/Getty Images

island. The spine is actually the edge of a huge, tilting block of Earth's crust. The western face of the block slopes much more gradually than the steep eastern face. Broad plains spread across the western part of the island.

Bodies of Water

Bodies of water in East Asia provide food for its people, give them a place to live, move their goods, and power factories to light homes. Fish and other seafood caught in the seas and ocean make up an important part of the diet of many East Asians.

Four large seas sit along East Asia's eastern edge. The largest is the South China Sea. It is partly enclosed by Taiwan, southeastern China, and islands of Southeast Asia. Because the South China Sea lies between important ports in the Pacific and Indian oceans, it has some of the busiest shipping lanes in the world.

The East China Sea lies between China and Japan's Ryukyu Islands, a long archipelago extending southwestward toward Taiwan. In the north, this sea meets the Yellow Sea, which is shaped by the Korean Peninsula and the northeastern coast of China. Farther north, Japan, the Korean Peninsula, and the Asian mainland together are shaped like a corral that almost entirely encircles the Sea of Japan (East Sea).

Rivers in China

The water of East Asia's two most important rivers, the Huang He (Yellow River) and the Chang Jiang (Yangtze River), flow across China. Both of these rivers begin high on the Plateau of Tibet in southwestern China and flow down the eastern slope of the plateau. The Huang He flows by twists and turns to the Yellow Sea far to the east. Along the way, it picks up a tremendous amount of yellow-brown silt called **loess** (LEHS). This silt gives the river and the sea their name and color.

In eastern China, silt deposited by floods over millions of years has created a broad, fertile plain called the North China Plain. This is one of China's most productive farming areas. Throughout history, however, floods have regularly destroyed homes and crops and have drowned many people. For this reason, the Huang He is sometimes called "China's sorrow."

Like the Huang He, the Chang Jiang begins on the Plateau of Tibet. From its headwaters, it flows about 3,450 miles (5,552 km) to its mouth at the port city of Shanghai on the East China Sea. It is the longest river in Asia and the third longest in the world. Only the Nile and the Amazon are longer.

The Chang Jiang is China's principal waterway. It also provides water for a fertile farming region where more than two-thirds of the country's rice is grown. Nearly one-third of China's people live in the river's basin.

Stocktrek Images/Getty Images

Peninsula A peninsula is an area of land that juts into a lake or an ocean and is surrounded on three sides by water. The term comes from the Latin words *paene,* meaning "almost," and *insula,* meaning "island."

The Huang He flows peacefully past irrigated farmland in eastern China on its way to the Yellow Sea.

▶ **CRITICAL THINKING**

Describing Why is the Huang He sometimes called "China's sorrow?"

Rivers in Japan and Korea

Japan's major rivers are relatively short, steep, and swift. They flow down from the mountains in the interior of the islands to low plains along the coast. Most of the rivers, including the two longest, the Shinano and the Tone, generate hydroelectric power.

The main rivers of the Korean Peninsula flow from inland mountains westward toward the Yellow Sea. The Han River flows through South Korea's capital, Seoul. North Korea's longest river, the Yalu (or Amnok), forms the country's border with China.

☑ **READING PROGRESS CHECK**

Identifying What are some ways the people of East Asia depend on rivers?

Climate

GUIDING QUESTION *What are the main factors that affect climate in different parts of East Asia?*

A traveler to East Asia would encounter a great range of climates, from hot and rainy to cold and dry. This is partly due to the area's vast size and partly because of its range of elevation.

Climate Factors

East Asia spans a tremendous distance north to south. The region's southernmost lands lie within Earth's hot tropical zone, while the northernmost lands sit closer to the frigid North Pole than to the Equator. Much of the climate variation results from these differences in latitude.

Another important factor is land elevation. Higher areas are generally colder than lower areas. Two areas at the same latitude can have very different climates if one area is much higher than the other. This situation is found in many parts of East Asia.

Thinking Like a
Geographer

Rice as a Staple Food

Rice is the main source of food for over half the world's people. Most East Asians depend on it as a staple food. Rice is such an important part of Asian cultures that the word for *rice* in some languages is the same as the word for *food*. What foods are considered staples of the American diet?

©George Steinmetz/Corbis

Air masses also play a major role in shaping East Asia's climate zones. A cold, dry, polar air mass spreads southward from northern Asia during the colder months of the year. A warm, moist, tropical air mass spreads northward and eastward from the Pacific Ocean during the warmer months.

Climates in the East

Southeastern China stays hot and rainy through much of the year. Vegetation is lush, and conditions are ideal for growing grain, especially rice. Farther north in eastern China, more seasonal variation occurs. Summers are hot and rainy, but winters are cold and fairly dry. China's capital, Beijing, lies in this part of the country. In the summer, temperatures in Beijing often soar above 90°F (32°C). In the winter, icy winds from the northwest whip through the city, and temperatures can drop below 0°F (−18°C).

Being surrounded or nearly surrounded by water affects the climates of East Asia's island and peninsula countries: Taiwan, Japan, and North and South Korea. These countries are generally wetter and experience milder temperatures than mainland areas at the same latitudes.

Climates in the North, Northwest, and Southwest

Across north-central and northwestern China and neighboring Mongolia, the climate ranges from semiarid (somewhat dry) to arid (very dry). Winters are bitterly cold. This sparsely populated area includes the vast, rocky Gobi Desert and the sandy Taklimakan Desert. It also has great expanses of treeless grasslands.

The Plateau of Tibet in southwestern China also has a dry climate. This is mainly because the towering Himalaya block moist air flowing northward from the Indian Ocean. Because the plateau sits at such a high elevation, the weather is cold and windy throughout the year.

✔ READING PROGRESS CHECK

Identifying How do the Himalaya affect the climate of the Plateau of Tibet?

Natural Resources

GUIDING QUESTION *What mineral resources are most abundant in East Asia?*

East Asia is rich in minerals, forests, and other natural resources. But the resources are not evenly distributed.

The nomadic Tsaatan herd reindeer and hunt for gold in mountainous areas of northern Mongolia.
▶ **CRITICAL THINKING**
Describing What is the climate like in Mongolia?

Minerals

China, which covers a large portion of East Asia, holds by far the greatest share of the region's resources. China is a world leader in the mining of tin, lead, zinc, iron ore, tungsten, and other minerals. Manufacturers use tungsten to make high-quality steel, lightbulbs, rockets, and electrical equipment.

Japan is one of the world's leading industrial countries, but the islands of Japan have few mineral resources. Japan has coal, copper, some iron ore, and a few other minerals, but it must import a variety of raw materials. Taiwan, like Japan, is a major industrial country, but it has limited mineral reserves. As a result, Taiwan also has to import various minerals to meet increasing demand.

Cultured pearls are harvested in the seas surrounding the Japanese islands. They range in color from white to silver and pink. They are known for their beautiful, smooth, shiny appearance.

Energy Resources

East Asia has a variety of energy resources, including coal, oil, and natural gas. The largest deposits of these fossil fuels are in China. China is the world's largest producer of coal. Its deposits are larger

Workers haul coal to barges on the Chang Jiang River in China. The barges transport the coal down stream to power plants.

©Bob Sacha/Corbis

than those of any other country except the United States and Russia. China also has substantial oil and natural gas reserves under the South China Sea and in the Taklimakan Desert in the far west.

Despite these fossil fuel resources, China still cannot meet all of its energy needs. The country's economy is growing rapidly. As a result, China is turning to energy-rich countries in Central and Southwest Asia for supplies of oil and natural gas. China also imports coal from Southeast Asia and Australia.

Several East Asian countries use hydroelectric power to help meet their energy needs. China produces electricity from the Three Gorges Dam on the Chang Jiang. This massive dam also helps control floods and provides water for irrigation. Hydroelectric dams in Japan harness the power of the country's swift-flowing rivers.

Forests

Much of western and northwestern China is so dry that trees cannot grow. Forestland once blanketed the eastern part of the country, but over the centuries, forests became smaller as people cut down trees for heating, building, and to create farmland. Today, forested areas cover less than one-sixth of the country.

Other East Asian countries, however, still have extensive forested areas. More than half of Taiwan's rugged landscape is covered by forests. The country used to be a major exporter of timber and wood products. Now much of the forested land is protected, so the country relies on imports to meet its need for wood products.

Thick forests cover the steep hillsides of Japan's inland areas. Almost two-thirds of the country is forested. Logging is limited in parts of the country, however, because the Japanese consider many forest areas to be sacred.

In the Korean Peninsula, many trees have been cleared for farmland, but forests still cover mountainous areas. About three-fourths of North Korea's rugged landscape is forested.

☑ **READING PROGRESS CHECK**

Analyzing Why is it necessary for people in Taiwan and Japan to import wood products?

Include this lesson's information in your Foldable®.

LESSON 1 REVIEW **CCSS**

Reviewing Vocabulary
1. How did Japan's *archipelago* form?

Answering the Guiding Questions
2. *Citing Text Evidence* How do the climates of East Asia's island and peninsula areas differ from climates of the mainland areas?

3. *Analyzing* How has the Huang He helped and hurt the people of China?

4. *Identifying* How does Japan make use of its short, swift rivers?

5. *Describing* Why must China import petroleum, natural gas, and coal?

6. *Informative/Explanatory Writing* Think about how elevation affects climate. Write a paragraph that explains how mountains and plateaus affect East Asia's climate.

Reading **HELP**DESK **CCSS**

Academic Vocabulary

• intertwine

Content Vocabulary

• **dynasty**
• **shogun**
• **samurai**
• **sphere of influence**
• **communism**

TAKING NOTES: *Key Ideas and Details*

Describe Use a chart like the one below to describe two key events in the histories of China, Japan, and Korea.

	Key Events
China	1.
	2.
Japan	1.
	2.
Korea	1.
	2.

Lesson 2
History of East Asia

ESSENTIAL QUESTIONS • *What makes a culture unique?* • *How do cultures spread?*

IT MATTERS BECAUSE
East Asia has a long, rich, and fascinating history. Learning about the events, innovations, and ideas that shaped the region will give you a better understanding and appreciation of its countries, cultures, and peoples.

Early East Asia

GUIDING QUESTION *What important inventions from East Asia spread across the rest of the world?*

China's civilization is more than 4,000 years old. Throughout its history, Chinese civilization influenced the development of other East Asian countries.

Early China

For many centuries until the early 1900s, rulers known as emperors or empresses governed China. A **dynasty**, or line of rulers from a single family, would hold power until it was overthrown. Then a new leader would start a new dynasty. Under the dynasties, China built a highly developed culture and conquered neighboring lands.

As their civilization developed, the Chinese tried to keep out foreign invaders. In many ways, this was easy. On most of China's borders, natural barriers such as seas, mountains, and deserts already provided protection. Still, invaders threatened from the north. To defend this area, the Chinese began building the Great Wall of China about 2,200 years ago. Over the centuries, the wall was continually rebuilt and lengthened. In time, it snaked more than 4,000 miles (6,437 km) from the Yellow Sea in the east to the deserts of the west. It remains in place today.

China's Dynasties

The Shang was the first dynasty to leave written records. From the writings, we learned that the Shang took power about 1600 B.C. in the North China Plain. Like all succeeding dynasties, the Shang faced rebellions by local lords, attacks by Central Asian nomads, and natural disasters such as floods. When the government was stable, it could defend its people against some of these problems. Eventually, however, the dynasty weakened and fell.

After the Shang, the Zhou dynasty ruled for about 800 years, beginning around 1045 B.C. Under Zhou rule, Chinese culture spread, trade grew, and the Chinese began making iron tools.

After the Zhou, powerful dynasties expanded China's territory. In the 200s B.C., the Han united all of China and started building the Great Wall that exists today. Under the Han and Tang dynasties, traders and missionaries spread Chinese culture to all of East Asia.

In the early 1400s, under the Ming dynasty, the naval explorer Zheng He reached as far as the coast of East Africa. The last dynasty of China, the Qing, ruled from the mid-1600s to the early 1900s.

Achievements and Ideas

China underwent many changes during the Zhou dynasty. During this period, laws were recorded for the first time, the first coins were created, and farmers began to use plows pulled by oxen. It was also an age of great thinkers. One of these thinkers was a man named Confucius. He thought people should be morally good and loyal to their families. He also believed that a ruler should lead his people as though he were the head of a family.

Another of the thinkers was Laozi, who founded an important belief system called Daoism. Laozi thought that people should live in harmony with nature.

Think Again?

The Great Wall of China is a single, continuous wall.

Not true. The Great Wall, which runs for about 5,500 miles (8,851 km) across northern China, is actually made up of many different walls. Some of the walls run parallel to each other. The best-preserved sections of wall were constructed of bricks and stones during the Ming dynasty (1368–1644). Some walls are made of dirt, gravel, stone, and wood. Long stretches of wall have been destroyed by erosion or buried by blowing sand.

To build the Great Wall, Chinese rulers used teams of workers that included border guards, peasant farmers, and prison convicts.

▶ **CRITICAL THINKING**
Describing Why was the Great Wall built?

©Image Source/Corbis

DYNASTIES OF CHINA

Dynasty	Time Span
Xia	2200–1700 B.C.
Shang	1766–1080 B.C.
Zhou	1045–221 B.C.
Qin	221–206 B.C.
Han	206 B.C.–A.D. 221
Sui	A.D. 581–618
Tang	A.D. 618–907
Song	A.D. 960–1279
Yuan (Mongol)	1279–1368
Ming	1368–1644
Qing	1644–1911

For centuries, rulers of various dynasties used their power to expand and unite China.

Identifying Which dynasties lasted 500 years or more?

Later, another important belief system called Buddhism was introduced to China from India. Confucianism, Daoism, and Buddhism have been major influences on China and the rest of East Asia for many centuries.

The Han dynasty (202 B.C.–A.D. 220) had such an impact that many of China's people today call themselves the People of Han. With Han rule came unity and stability. The arts and sciences flourished. Papermaking was invented, and government officials began using paper for keeping records. Han rulers encouraged trade along the Silk Road. This was a caravan route that stretched 4,000 miles (6,437 km) between China and Southwest Asia, and then extended into Europe and South Asia. The Chinese sent goods such as silk, tea, spices, paper, and fine porcelain (POHR•suh•luhn) as far west as ports along the Mediterranean Sea in exchange for wool, gold, and silver.

The Han dynasty was followed by several centuries of decline and disunity as individual states fought to gain power. When China was eventually reunified, the stage was set for a long period of stability and cultural advancement under the Tang dynasty (A.D. 618–907) and Song dynasty (A.D. 960–1279).

During the Tang dynasty, probably in the A.D. 800s, a new type of printing emerged. The Chinese developed a way to use blocks of wood and clay to print characters on paper. This invention, known as woodblock printing, made it possible to print large numbers of books quickly, which allowed ideas to spread more rapidly. Another important invention was gunpowder, which was used in explosives and fireworks. The Chinese also invented the magnetic compass, which helped sailors find their direction at sea.

Early Korea and Japan

Korea was first populated thousands of years ago by people who migrated from northern Asia. By the first century A.D., the peninsula was divided among three rival kingdoms. Six centuries later, one of these kingdoms—the Silla—conquered the other two and unified the peninsula. After enduring for three centuries, the Silla kingdom was succeeded in A.D. 935 by the Koryo.

In the 1200s, the Mongols, a people from the steppes of Central Asia, had conquered North China, parts of Asia, and the northern half of the Korean peninsula. The vast territory became part of the Mongol's Yuan dynasty. At the end of the 1300s, the Mongols were driven out of Korea and a new Korean dynasty called the Choson came to power. It would stay in power until modern times.

Korea went through religious changes during this period. Buddhism had spread from India to China to Korea in the A.D. 300s and became popular during the Koryo dynasty. Later, during the Chosun dynasty, Confucianism became Korea's dominant religion. Chinese characters came to be used for Korean writing. Korean artists and writers were inspired by the art and literature of China. Korean rulers also adopted Confucianism as a basis for government. In some periods, China provided Korea with military protection. In other periods, Koreans lived in fear of Chinese invasion.

The history of Japan is **intertwined** with that of its neighbors to the west. The Japanese islands were settled by people from Korea and China. By A.D. 500, the clans and tribal kingdoms of Japan had close ties to Korea. Eventually, the ties extended to China, beginning a flow of ideas and culture that transformed Japan. Japanese people began using the Chinese calendar and the Chinese system of writing.

Academic Vocabulary

intertwine to twist or twine together

MAP SKILLS

1 **THE GEOGRAPHER'S WORLD** What was the Silk Road?

2 **PLACES AND REGIONS** Which areas of the world were linked by the Silk Road?

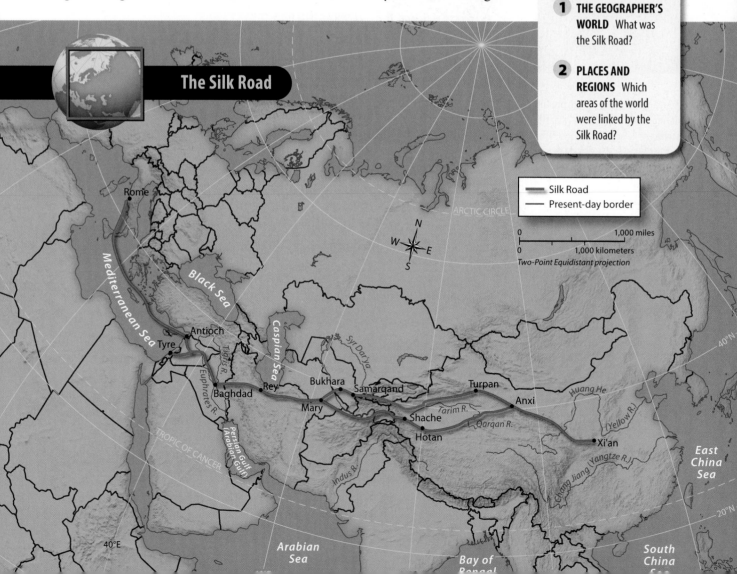

The Silk Road

In addition, the Japanese adopted Chinese technology and Buddhism spread to the islands from Korea and mixed with Shinto, a Japanese religion. Shinto, or "Sacred Way," stressed that all parts of nature—humans, animals, plants, and rivers—have spirits.

Japan was ruled by emperors, but over time, they lost power. Eventually landowning families set up a feudal system. Under the system, high nobles gave land to lesser nobles in return for their loyalty and military service. At the bottom of the social ladder, peasants farmed nobles' estates in exchange for protection. By the 1100s, armies of local nobles had begun fighting for control of Japan. Minamoto Yoritomo (mee•nah•moh•toh yho•ree•toh•moh) became Japan's first **shogun**, or military leader. Landowning warriors, who were called **samurai**, (SA•muh•ʀʏ) supported the shogun. Although the emperor kept his title, the shoguns held the real power.

☑ **READING PROGRESS CHECK**

Describing What are some ways in which China influenced Japan?

A Japanese folding screen, produced during the 1700s, displays a battle scene.

▶ **CRITICAL THINKING**
Citing Text Evidence How was early Japan ruled?

Change in East Asia

GUIDING QUESTION *How did increased contact with the West influence the region?*

Throughout much of its history, East Asia was mostly isolated from the rest of the world. High mountains, harsh deserts, and vast distances limited the flow of ideas and goods between the region and other parts of Eurasia. From the 1500s onward, however, increasing trade brought East Asian countries into greater contact with other cultures, especially Europe.

Spheres of Influence

By the early 1800s, internal problems had weakened China. Meanwhile, European countries were growing more powerful and making stronger claims. By the 1890s, European governments and Japan had claimed large areas of China as spheres of influence. A **sphere of influence** is an area of a country where a single foreign power has been granted exclusive trading rights.

The foreign intrusion in their country made many Chinese people angry. Their anger fueled a revolution in 1911, and the new government could not control the country. By 1927, a military leader named Chiang Kai-shek had formed the Nationalist government.

Two Chinas

Meanwhile, Chiang's rival, Mao Zedong (MOW dzuh•DUNG), gained support from Chinese farmers. Mao believed in **communism**, a system in which the government controls all economic goods and services. After years of civil war, the Communists won power in 1949. They set up the People's Republic of China on the Chinese mainland. The Nationalists fled to the island of Taiwan. There, they set up a government called the Republic of China.

Rise of Japan

Around 1542, a Portuguese ship heading to China was blown off its course and landed in Japan. The traders on the ship became the first Europeans to visit Japan. Soon, more traders began arriving, along with Christian missionaries. By the early 1600s, Japan's rulers had begun to fear that European powers were planning a military conquest of the islands. They decided to isolate Japan by forcing all foreigners to leave, banning European books, and blocking nearly all relations with the outside world.

Japan's isolation lasted for roughly two centuries. In 1854 U.S. naval officer Matthew C. Perry sailed to Japan with four warships. He pressured the Japanese to end their isolation and open their country to foreign trade. Not long afterward, rebel samurai forced the shoguns to return full power to the emperor.

Recognizing that European countries were far more advanced and powerful, Japan set out to transform itself by learning everything it could about the West. The country soon became an industrial and military power, and it began developing an empire.

By 1940, Japanese forces had gained control of Taiwan, Korea, parts of mainland Asia, and some Pacific islands. This expansion was one factor that led Japan to fight the United States and its allies in World War II.

☑ **READING PROGRESS CHECK**

Analyzing How was Korea affected by Japanese expansion?

Modern East Asia

GUIDING QUESTION *What conflicts divided East Asian countries?*

After World War II, East Asia saw substantial changes in its governments and economies. Some of the region's countries developed into important economic powers.

U.S. naval officer Matthew Perry met representatives of the Japanese shogun at the port of Yokohama.

▶ **CRITICAL THINKING**
Describing How did Perry's visit change Japan?

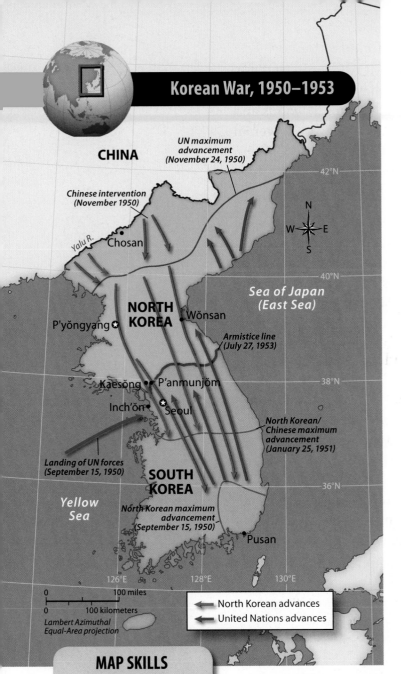

Korean War, 1950–1953

CHINA

UN maximum advancement (November 24, 1950)

Chinese intervention (November 1950)

Yalu R.

Chosan

42°N

N
W · E
S

40°N

Sea of Japan (East Sea)

NORTH KOREA

Wŏnsan

P'yŏngyang

Armistice line (July 27, 1953)

Kaesŏng · · P'anmunjŏm

38°N

Inch'ŏn · · Seoul

North Korean/ Chinese maximum advancement (January 25, 1951)

Landing of UN forces (September 15, 1950)

SOUTH KOREA

36°N

Yellow Sea

North Korean maximum advancement (September 15, 1950)

Pusan

126°E   128°E   130°E

0   100 miles
0   100 kilometers

Lambert Azimuthal Equal-Area projection

→ North Korean advances
→ United Nations advances

MAP SKILLS

1 THE GEOGRAPHER'S WORLD When and where did the North Koreans make their maximum advance?

2 HUMAN GEOGRAPHY What is the armistice line?

Modern China

After 1949, China became "two Chinas"— one was mainland China ruled by the Communists, and the other was the Nationalist government on the island of Taiwan. In the 1950s, the Communist mainland government took control of businesses and industry. It also took land and created state-owned farms.

In the late 1950s, Mao Zedong introduced the Great Leap Forward. This program's goal was to increase China's industrial output. Many peasants left the fields and began working in factories. Cities grew rapidly. The program failed, however. Poor planning, natural disasters, and a drop in food production led to widespread famine.

During China's Cultural Revolution in the late 1960s, intellectuals such as doctors and teachers were ordered to work on farms. Students also were taken from school and sent to the countryside to work. In this way, Mao hoped to get rid of any cultural elements that did not support his idea of communism.

Taiwan, on the other hand, pursued a goal of "one China"— two parts of one nation moving toward reunification. Taiwan wanted the mainland Communist government to negotiate with Taiwan as an equal, but Communist leaders said no.

At first, the Taiwanese government limited the freedom of its people. By 1970, however, Taiwan's leaders had instituted democratic reform and developed an economy based on capitalism. Prosperity transformed the island into an economic powerhouse.

By comparison, China's Communist government was stagnant. After Mao's death in 1976, though, Chinese leaders started to open China to the West. Economic reforms helped China become a rising global power. Chinese leaders gradually allowed a market economy to develop alongside the communist economy by letting people own businesses and sell products and services freely.

A Divided Korea

After World War II ended, Korea was divided into two countries: South Korea and North Korea. South Korea was supported by the United States, and Communist North Korea had strong ties to China

and to the Soviet Union, a powerful communist country that stretched across northern Europe and Asia.

Both Koreas claimed the entire peninsula. In 1950 North Korea invaded South Korea. United Nations forces and the United States rushed to support South Korea, and China helped North Korea. The Korean War ended in 1953 without a peace treaty or a victory for either side. A buffer zone, called the demilitarized zone, was established to separate the two countries.

Stark differences developed between the two Koreas after the war. Over several decades, South Korea followed the path of capitalism and the country's economy began to grow rapidly. In contrast, North Korea's economy, which is strictly controlled by its Communist government, has struggled. North Koreans face many hardships because most resources go to the military.

Modern Japan

After being defeated by the United States and its allies in World War II, Japan was stripped of its overseas territories and military might. The country adopted a democratic constitution, and women and workers gained more rights. During the Korean War, the United States needed Japanese factories to provide supplies for its war effort. Japanese shipbuilders, manufacturers, and electronics industries benefited from giving this assistance.

The Japanese government worked closely with businesses to plan the country's economic growth. Both invested in research and development of electronics products for the home. A highly skilled workforce and the latest technology helped Japan develop its industries. Within a few decades, the Japanese were leading producers of ships, cars, cameras, and computers. By the 1990s, world demand for Japanese-made goods had turned it into a global economic power. Despite some economic ups and downs in the 2000s, Japan's economy remains one of the world's strongest. Based on gross national product, Japan ranks third in the world, trailing only the United States and China.

Include this lesson's information in your Foldable®.

☑ **READING PROGRESS CHECK**

Determining Central Ideas What led to the growth of China's economy beginning in the 1970s?

LESSON 2 REVIEW **CCSS**

Reviewing Vocabulary
1. How can a new *dynasty* form?

Answering the Guiding Questions
2. *Describing* How did Buddhism spread across East Asia?

3. *Identifying* Who ruled during the Yuan dynasty?

4. *Determining Central Ideas* Why did Europeans want access to China and Japan?

5. *Analyzing* What led to the creation of "two Chinas"?

6. *Argument Writing* Write a one-page essay explaining which early Chinese invention had the greatest effect on the rest of the world and why.

networks

There's More Online!

☑ **IMAGE** Baseball in Japan

☑ **ANIMATION** Population Pyramid of East Asia

☑ **SLIDE SHOW** Influence of Japanese Anime

☑ **VIDEO**

Reading HELPDESK CCSS

Academic Vocabulary

• structure

Content Vocabulary

• **urbanization**
• **megalopolis**
• **trade deficit**
• **trade surplus**

TAKING NOTES: *Key Ideas and Details*

Organize Information As you read the lesson, use a diagram like the one below to record four key facts about the people of East Asia.

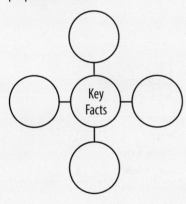

Key Facts

Lesson 3
Life in East Asia

ESSENTIAL QUESTIONS • *Why do people trade?* • *How does technology change the way people live?*

IT MATTERS BECAUSE
East Asia is one of the largest population centers on Earth. China alone holds about one-fifth of the world's people. The powerful economies of the East Asian countries affect global trade and manufacturing.

The People

GUIDING QUESTIONS *Which areas in East Asia have the highest population densities?*

Most people in East Asia live crowded together in river valleys, basins, and deltas, or on coastal plains. These lands and climates are favorable to agriculture and industry and are among the most densely populated places on Earth. In contrast, vast areas in the west and northwest are sparsely populated.

Population Patterns

Throughout much of history, China has had a large population. Two thousand years ago, the number of inhabitants had already reached an astonishing 59 million. Even today, only about two dozen countries have this many people.

For many centuries, China's population growth was slowed by epidemics, famines, warfare, and other factors. After the 1300s, however, growth began to increase steadily, and in the middle of the 1900s, it became explosive. Because the uncontrolled growth was causing many problems, the government enacted a policy in 1979 that required families to have no more than one child. This "one-child" policy helped to slow China's growth. China's 2010 Census showed a population of 1.37 billion people on the mainland and in

Taiwan and Hong Kong. Far more densely populated in the east than in the west and northwest, China has an average population density of about 140 persons per square mile (54 per sq. km). In contrast, its northern neighbor, Mongolia, has a population density of less than 4 persons per square mile (1.6 persons per sq. km).

Population growth in some other parts of East Asia slowed at the end of the 1900s. Japan's low birthrate means that the average age of its population is increasing. Nearly one-quarter of the population is age 65 or older. Since the mid-1990s, Japan has encouraged more births by providing programs such as child care. Soon the country could face a shortage of workers. A shortage might encourage leaders to allow more foreign workers into the country.

Where People Live

Throughout China's history, most of its people lived off the land as farmers. Economic reforms in the late 1970s, however, caused a surge of **urbanization**. Millions of peasants left their farms and moved to booming cities in eastern China. Nearly half of the country's people now live in cities. Shanghai has about 11 million people, making it the largest city in China. Just over 7 million people live in Beijing, the capital. Hong Kong, which China regained from Britain in 1997, has more than 5 million people. Dozens of other Chinese cities have populations greater than 1 million.

In other East Asian countries, urbanization began earlier and is farther advanced than in China. In Japan, two-thirds of the people live in cities. The cities of Tokyo, Osaka, Nagoya, and Yokohama form a **megalopolis**, or supersized urban area, along the coast. Greater Tokyo, Japan's capital and its largest city, is home to about 32 million people.

As South Korea industrialized, more people moved to cities. Now, 83 percent of South Koreans live in urban areas. The country's capital city, Seoul, has more than 10 million people. Across East Asia, the standard of living for people in cities is generally higher than that of people in rural areas.

✓ **READING PROGRESS CHECK**

Determining Central Ideas How can a country's growth rate influence its economy?

Because of the lack of building space, Hong Kong's urban skyline is dominated by many high-rise office and apartment buildings. As a result, Hong Kong is described as "one of the most vertical places on Earth."

▶ **CRITICAL THINKING**
Describing What major change has affected East Asia's cities in recent decades?

Culture in East Asia

GUIDING QUESTION *What are some of the cultural differences among East Asian countries?*

The people of East Asia have a rich cultural heritage. Traditions, beliefs, art, literature, and other elements that make up their heritage have been shaped by contributions from many different groups. The rise of a global culture has also brought change.

Ethnic and Language Groups

In each East Asian country, people tend to be ethnically similar. In Japan, about 99 percent of the population is ethnic Japanese and speaks the Japanese language. Nearly all the people in North and South Korea are ethnic Koreans who speak the Korean language. About 95 percent of Mongolia's people are ethnic Mongolian, and almost all of them speak the Khalkha Mongolian language.

In China, the Han ethnic group makes up about 92 percent of the population. The other 8 percent belong to more than 50 different ethnic groups. The official language in China is Mandarin Chinese, but many dialects are spoken. In Taiwan, Mandarin Chinese is the official language. Most of the people of Taiwan are Taiwanese, and many also speak the Taiwanese, or Min, language. The official languages of Hong Kong are Cantonese and English.

Worshipers often leave flowers and light candles at street-corner religious shrines in Macau. The territory was under Portuguese rule until becoming part of China in 1999.

▶ **CRITICAL THINKING**

Describing What role does China's government play in setting religious policy?

Christian Goupi/age fotostock

Religion and the Arts

The people of East Asia have many religions and belief systems. Many Chinese practice a mix of Buddhism, Daoism, and Confucianism. The Communist government that took over in 1949 believed that religion had no place in a communist country, so it began limiting religious practice. In recent decades, however, antireligious policies have been relaxed somewhat.

Buddhism has a large following in Korea and Japan, although North Korea's government also limits religious practice. In Japan, many people combine Buddhism with Shinto, the country's traditional religion. In South Korea, Christianity has a strong presence.

A number of art forms have long been popular in East Asia. In China, Korea, and Japan, many artists have painted the rugged landscapes of their countries. Their works reflect a reverence for nature that is part of Daoism and Shinto. Ceramics and pottery have been important parts of East Asian art since ancient—even prehistoric—times. Craftspeople in East Asia are also skilled at weaving, carving, and lacquerwork.

Calligraphy, the art of turning the written word into beautiful, expressive images, is considered one of the highest forms of art in China and Japan. Chinese characters are visually interesting and complex, so they lend themselves well to calligraphy. It is common for East Asians to display works of calligraphers in their homes.

East Asians also have strong literary and theatrical traditions. Japanese poets often write haiku, brief poems that follow a specific **structure**. Japan is famous for its traditional forms of theater. Today, Japan is also known for anime, a type of animation. Comic books and cartoons using this style have become popular all over the world. Jingxi, known to English-speakers as Peking opera, is a popular type of musical theater in China. In Jingxi performances, actors and actresses wear colorful costumes, sing in high-pitched voices, and use gestures, postures, and steps to reveal the attitudes of their characters.

Along with art and literature, East Asians have developed new forms of expression. South Korean "K-pop" is a popular type of music that young people in Japan and other East Asian countries enjoy. It has its roots in dance and electronic music from the West. Communication technology, the Internet, and travel increase the reach of new art forms. At the same time, different countries adopt and change them in unique ways.

For centuries, East Asian artisans have used lacquer, a clear or colored wood finish, to protect and add luster to jewelry boxes, furniture, and other art objects.

Academic Vocabulary

structure organization

Yueliang Yao/Kalium/age fotostock

Daily Life

Traditionally, the family is the center of social life in East Asia. In rural areas of East Asia, for example, different generations of one family may share the same home. In crowded cities, tall apartment buildings provide housing for many families. As more people have moved into urban areas, some traditional attitudes have begun to change. For example, many women in the region now work outside the home.

East Asian cultures place a high value on education. Teachers are greatly respected, and children are expected to work hard. At a young age, children begin taking important exams that can determine whether they will get into top colleges.

Rice and noodles are staples in the diets of most East Asians, but otherwise cuisine varies widely across the region. For example, China's Sichuan (or Szechuan) Province is known for its bold, spicy dishes. Cantonese cuisine from the Guangdong Province is generally milder, with a careful balance of flavors. In Japan, meals are often built around seafood and soybean-based foods such as tofu. In Mongolia, where many people raise livestock, the cuisine features meat and dairy products.

People in East Asia enjoy many different pastimes, some traditional and some modern. Millions of people young and old practice martial arts such as tai chi (TY CHEE) and tae kwon do (TY KWAHN DOH), both of which originated in the region centuries ago. Baseball, introduced from North America, is widely

Billboards of anime characters form a backdrop to young fans lining up to attend an exhibition of animation, comics, toys, and games in Hong Kong.

Identifying What other new forms of expression have recently developed in East Asia?

Philippe Lopez/AFP/Getty Images

popular in Japan, Taiwan, and South Korea. Many children play on Little League baseball teams, and Japan has its own major baseball leagues. Basketball, also imported from America, has become one of the top sports in China.

Holidays are also important in East Asia. The biggest holiday of the year in China is Spring Festival, also known as Chinese New Year. It is celebrated with dances, fireworks, traditional foods, and religious ceremonies to honor ancestors. In Japan, people observe New Year's Day by visiting temples and shrines. South Korea's most popular holiday is the Harvest Moon Festival, which is something like Thanksgiving Day in the United States. During this festival, South Koreans celebrate the fall harvest and hold special ceremonies in honor of their ancestors.

☑ **READING PROGRESS CHECK**

Analyzing Why is religious activity limited in China?

Members of a Little League team in Tokyo, Japan, prepare for a baseball game. In 2012 a Japanese team won the Little League World Series Championship held in South Williamsport, Pennsylvania.

Identifying What other American sport has become popular in East Asia?

Current Issues in East Asia

GUIDING QUESTION *How do East Asian economies affect economies around the world?*

In the last half-century, rapid economic growth has transformed East Asia. China and Japan now have larger economies than any other country in the world except the United States. With this growth, however, have come problems, including environmental damage.

Economies and Environments

In China, factories, coal-burning power plants, and the growing numbers of cars and trucks have led to dramatic increases in air pollution. Factory waste, sewage, and farm chemicals have poisoned water. Rapidly growing urban areas have eaten up valuable farmland. Many cities face a constant shortage of water.

Japan has similar issues. Polluted air from power plants has produced acid rain and other problems. But as a more developed country with a longer history of industrialization, Japan has engaged in stronger environmental protection efforts. In addition, Japan faces a constant threat of earthquakes.

A boy runs across a rubbish-covered beach in Hainan, a tropical island in the South China Sea. Hainan, a part of China, has seen its economy grow rapidly in recent years.

▶ CRITICAL THINKING

Determining Central Ideas What have been the effects of rapid economic growth in East Asia?

In 2011 the strongest earthquake ever recorded in Japan killed thousands of people and damaged several nuclear power plants. It also disrupted trade and manufacturing around the world.

Trade

Many of the goods manufactured in East Asia are shipped to the United States and Europe. China's exports to the United States include electronic goods, toys and games, clothing, and shoes. From the United States, China imports soybeans, cotton, automobiles, and many other goods. Trade between the two countries, however, is not balanced. In 2010 the U.S. trade deficit with China rose to more than $273 billion. The **trade deficit** means the United States imports more goods from China than it sells to China. A **trade surplus** occurs when a country exports more products than it imports.

Challenges Facing East Asia

Japan and China are dealing with the challenges of population growth. Ever-growing populations put a strain on limited resources and services. In 1979 China began a policy that allowed each family to have no more than one child. The policy did slow population growth. However, with fewer children and better living standards, the percentage of elderly has grown substantially. Economic growth and productivity depend on a labor force of young adult workers. To increase the population of young people, some Chinese are demanding that the government change or repeal the law that penalizes families that have more than one child.

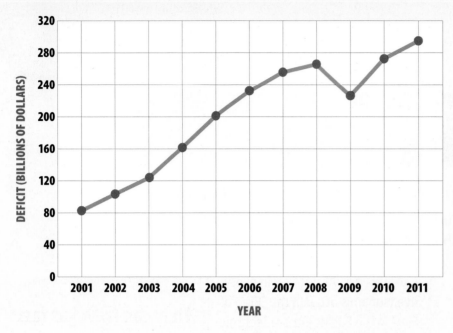

U.S. TRADE DEFICIT WITH CHINA, 2001–2011

The United States depends on imports from China. Except for the 2009 recession, the U.S. trade deficit with Chinas has risen dramatically.

 CRITICAL THINKING

1. *Integrating Visual Information* In what years since 2001 has the United States had a trade surplus with China?

2. *Identifying* In what years was the trade deficit greater than $280 billion?

Source: U.S. Department of Commerce; U.S. International Trade Commission (ITC)

Political differences in East Asia are another challenge. Japan is in dispute with Russia over a long chain of islands known as the Kuril Islands, which lie north of Japan. Russia claims the entire archipelago, but Japan claims the southernmost islands. North Korea's efforts to develop nuclear weapons have drawn harsh criticism from several countries.

Both China and North Korea face questions about human rights in their countries, and China continues to receive international pressure for its views on Tibet and Taiwan. China's economic boom has drawn many people out of poverty, but the country faces a growing income gap between people in its cities and the countryside.

☑ **READING PROGRESS CHECK**

Drawing Conclusions How might an earthquake in Japan affect the economies of other parts of the world?

FOLDABLES
Study Organizer

Include this lesson's information in your Foldable®.

| Economic Growth |
| Cultural Influences |
| Mainland and Islands |
| East Asia |

LESSON 3 REVIEW **CCSS**

Reviewing Vocabulary

1. What parts of Japan make up its *megalopolis*?

Answering the Guiding Questions

2. *Describing* How is Japan trying to compensate for the rise in the average age of its people?

3. *Determining Central Ideas* Today, many East Asian families are becoming scattered as people move to cities. Also, more women now work outside the home. How might these changes affect the structure and role of the family?

4. *Analyzing* In recent decades, hobbies and sports have become more important in the lives of people in China and other East Asian countries. What is the connection between this change and economic growth?

5. *Determining Central Ideas* What are some of the negative results of economic growth in East Asia?

6. *Argument Writing* Would you rather live in urban or rural areas in East Asia? Write a paragraph explaining your answer.

The Fury of a
TSUNAMI

A tsunami is a series of ocean waves generated by earthquakes, landslides, volcanic activity, or other disturbances below or on the seafloor. Scientists say that earthquakes cause about 90 percent of all tsunamis.

Warning Signs A tsunami can strike quickly. There are a few warning signals that typically occur. If an earthquake strikes, do not stay in low areas or near water. Earthquakes often trigger a tsunami. Sometimes, the ocean appears to drain away before an approaching tsunami hits. This is a warning sign that a tsunami is approaching.

4 of every 5 About four in every five tsunamis occur in the "Ring of Fire" in the Pacific Ocean. Tectonic shifts make volcanoes and earthquakes common there.

500 MPH Some tsunamis are 100 miles (161 km) long or longer. They can travel as fast as 500 miles (805 km) per hour—as fast as a commercial jet plane. At that speed, they can cross the Pacific Ocean in less than a day.

Read about two significant tsunamis in recent times:

July 9, 1958 The largest recorded tsunami in modern times luckily struck the isolated region around Lituya Bay, Alaska. Waves rose 1,700 feet (518 m). That is taller than the Empire State Building.

> **They can travel as fast as 500 miles (805 km) per hour—as fast as a commercial jet plane.**

Dec. 26, 2004 The deadliest tsunami in recorded history, the Indian Ocean tsunami killed an estimated 230,000 people. The tsunami swept through a wide area, including Indonesia, India, Madagascar, and Ethiopia. Many people died in the weeks after the tsunami hit because of lack of water and medical treatment.

Waters flood the city of Miyako shortly after the March 2011 earthquake struck northern Japan. ▶

Did You Know ?

The Richter scale is used to rate the magnitude—or amount of energy released—of an earthquake. Most earthquakes that occur today rate less than 3 on the Richter scale and do not produce much damage. On average, at least one earthquake with a magnitude of 8.0 occurs each year.

THERE'S MORE ONLINE

HEAR stories from tsunami survivors   •   *SEE* before and after images   •   *WATCH* an animation of a tsunami

These numbers and statistics can help you understand the full effects of the earthquake and tsunami that struck Japan in March 2011.

8.9
MAGNITUDE

Japan was hit by an 8.9 magnitude earthquake on March 11, 2011, that triggered a deadly, 23-foot tsunami in the northern part of the country. In comparison, the 1998 earthquake that caused massive damage in California was only a magnitude 6.9 on the Richter scale.

40 Years | The tsunami in

Japan recalled the 2004 disaster in the Indian Ocean. On December 26, a 9.0 magnitude earthquake—the largest earthquake in 40 years—ruptured in the Indian Ocean, off the northwestern coast of the Indonesian island of Sumatra. The earthquake stirred up the deadliest tsunami in world history. It was so powerful that the waves caused loss of life on the coast of Africa and were even detected on the East Coast of the United States.

100,000

At a news conference on March 13, Prime Minister Naoto Kan, who later gave the disaster the name "Great East Japan Earthquake," emphasized the gravity of the situation: "I think that the earthquake, tsunami, and the situation at our nuclear reactors makes up the worst crisis in the 65 years since the war. If the nation works together, we will overcome." The government called in 100,000 troops to aid in the relief effort.

200,000
Evacuated

Cooling systems in one of the reactors at the Fukushima Daiichi nuclear power station in the Fukushima prefecture on the east coast of Japan failed shortly after the earthquake, causing a nuclear crisis. This initial reactor failure was followed by an explosion and eventual partial meltdowns in two reactors, then by a fire in another reactor that released radioactivity directly into the atmosphere. The nuclear troubles were not limited to the Daiichi plant; three other nuclear facilities also reported problems. More than 200,000 residents were evacuated from affected areas.

15,839

According to the official toll, the disasters left 15,839 dead and 3,647 missing. One year after the tsunami, about 160,000 people had not returned to their homes.

LEVEL 7 | On April 12, 2011, Japan

increased its assessment of the situation at the Fukushima Daiichi nuclear power plant to Level 7, the worst rating on the international scale, putting the disaster on par with the 1986 Chernobyl explosion.

450 MPH

The waves travel in all directions from the epicenter of the disturbance. The waves may travel in the open sea as fast as 450 miles (724 km) per hour. As they travel in the open ocean, tsunami waves are generally not particularly large—hence the difficulty in detecting the approach of a tsunami. But as these powerful waves approach shallow waters along the coast, their velocity is slowed and they consequently grow to a great height before smashing into the shore.

Map (Global Impact)

December 26, 2004, Sumatra, Indonesia: A 9.1–9.3 earthquake triggered a devastating tsunami in the Indian Ocean. More than 200,000 people were estimated to have been killed.

October 25, 2010, Sumatra, Indonesia: A 7.7 earthquake and resulting tsunami caused the loss of 509 lives when it struck off the coast of Sumatra.

July 17, 2006, Java, Indonesia: 668 lives were lost when a 7.7 earthquake caused a tsunami on the island of Java.

March 11, 2011, Japan: A 9.0 earthquake caused a tsunami on Japan's northeast coast. More than 20,000 people were confirmed killed or missing.

JAPAN

PACIFIC OCEAN

CHINA

TAIWAN

BANGLADESH

MYANMAR (BURMA) LAOS

THAILAND VIETNAM
CAMBODIA

PHILIPPINES

MALAYSIA

SINGAPORE

INDONESIA

EAST TIMOR (TIMOR-LESTE)

INDIAN OCEAN

EQUATOR

0 1,000 miles
0 1,000 kilometers
Two-Point Equidistant projection

N E S W

20°N
0°
20°S
100°E 120°E 140°E

GLOBAL IMPACT

DESTRUCTIVE FORCES Hundreds of small earthquakes shake Japan every year. Major quakes occur less often, but they may cause disastrous damage and loss of life. When an undersea earthquake generates a tsunami—a huge tidal wave that gets higher and higher as it approaches the coast—many lives may be lost. Because earthquakes and tsunamis are difficult to predict, some parts of the region, especially in high-population areas, rely on special building methods and emergency preparedness to help reduce casualties.

Natori, Japan

Natori (left) is shown before the March 2011 earthquake and tsunami. A photograph of the city (right) shows the effects afterwards.

BEFORE

AFTER

Thinking Like a Geographer

1. **Physical Geography** How are tsunamis created?

2. **Physical Geography** Research to find information about the 2004 Indian Ocean tsunami. Create a map that identifies the nations that suffered deaths and great destruction.

3. **Environment and Society** Research online to find out what happened when the Fukushima Daiichi nuclear power plant was hit by the 2011 tsunami. Create a PowerPoint® presentation that includes diagrams and photos to show the chain of events that occurred when the nuclear plant was damaged. Present your slides to

Directions: Write your answers on a separate piece of paper.

1 **Use your FOLDABLES to explore the Essential Question.**

INFORMATIVE/EXPLANATORY WRITING Why do people in China live in densely populated areas in river valleys and along the coastal plains when the country has so much sparsely populated land in its western regions?

2 **21st Century Skills**

DESCRIBING Working in small groups, choose an East Asian country and research how its schools operate. Find out what kinds of classes students take, how they are graded, what sports and activities are offered, and what a typical school day is like. Prepare a report or a slide show and share results with another class in your school.

3 **Thinking Like a Geographer**

IDENTIFYING The United States and China are about the same size and lie within the Northern Hemisphere at similar latitudes. Make a T-chart on a sheet of paper. On one side list similarities between the two countries, and on the other side list differences.

4 **GEOGRAPHY ACTIVITY**

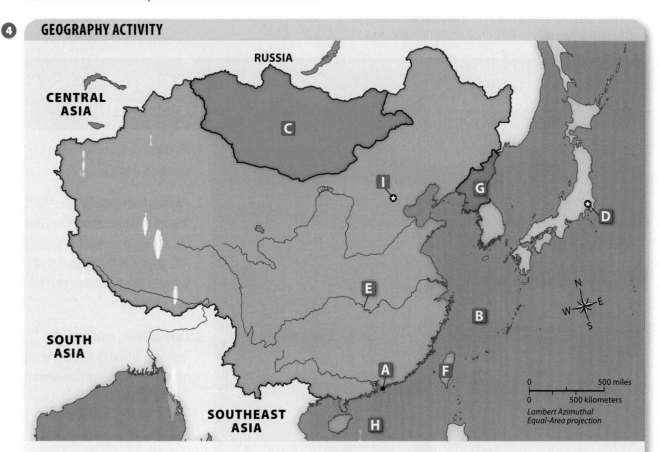

Locating Places

Match the letters on the map with the numbered places below.

1. Taiwan
2. East China Sea
3. South China Sea
4. Chang Jiang (Yangtze River)
5. Mongolia
6. Hong Kong
7. Beijing
8. North Korea
9. Tokyo

REVIEW THE GUIDING QUESTIONS

Directions: Choose the best answer for each question.

1 Why do most Japanese live within a few miles of the coast?

A. The interior area of the country consists of steep, heavily forested mountains.

B. The weather is better.

C. Huge corporate farms take up most of the interior space.

D. People like living near the beach.

2 China's major river systems

F. flow through Beijing.

G. begin on the Plateau of Tibet.

H. empty into the South China Sea.

I. are not navigable.

3 What product invented during China's Han dynasty is still in use around the world?

A. fireworks

B. paper

C. gunpowder

D. soap

4 Why did the United States send warships to Japan in 1854?

F. to conquer the island

G. to establish a naval base

H. to pressure the Japanese to open their country to foreign trade

I. to protect Japan from a Chinese invasion

5 During the Korean War, which country sent its troops to help North Korea and fight against American and United Nations forces?

A. Taiwan

B. Japan

C. China

D. South Korea

6 Which two East Asian countries have the largest economies?

F. China and North Korea

G. China and Taiwan

H. China and Japan

I. China and South Korea

DBQ ANALYZING DOCUMENTS

7 **ANALYZING**   Read this passage about the earthquake and tsunami that struck Japan in 2011:

"*Very little of the devastation resulting from this earthquake was from the initial shaking. This is partly because of Japan's stringent [tough] building codes. But mainly because any damage from the seismic waves that sent skyscrapers in Tokyo swaying was dwarfed by the impact of the 10 metre tsunami that hit the Japanese coast less than an hour later.*"

—from Chris Rowan, "Japan Earthquake" (2011)

What does Rowan mean when he calls the tsunami a 10-meter (33-foot) tsunami?

A.   how far the wave traveled from the point of the earthquake

B.   how far along the coast the tsunami hit

C.   how far inland the wave traveled

D.   how high the tsunami was above sea level

8 **CITING TEXT EVIDENCE**   What detail in the passage explains that the Japanese are aware of the dangers of living along the Ring of Fire?

F.   strict building codes that limit earthquake damage

G.   the height of the tsunami

H.   the short duration of the initial shaking caused by the quake

I.   heavy damage from the tsunami

SHORT RESPONSE

"*But for many [Chinese], the growing gap between rich and poor is the most pressing issue, especially in Beijing's slums. . . . Li Yulan, 78, runs a small shop. . . . She says the rich are too rich. The poor are too poor. Of eight people in her family, just two have income, she says.*"

—from Shannon Van Sant, "China Struggles to Bridge Gap Between Rich, Poor" (2012)

9 **IDENTIFYING POINT OF VIEW**   Would China's middle class agree that the income gap is the country's most pressing issue? Why or why not?

10 **DISTINGUISHING FACT FROM OPINION**   Is the example of Li Yulan convincing evidence of the income gap problem? Why or why not?

EXTENDED RESPONSE

11 **INFORMATIVE/EXPLANATORY WRITING**   Every year America's trade deficit with China increases by billions of dollars. How do trade deficits affect the American economy? How might they affect you and your family personally?

Need Extra Help?

If You've Missed Question	❶	❷	❸	❹	❺	❻	❼	❽	❾	❿	⓫
Review Lesson	1	1	2	2	2	3	1	1	3	3	3

"China Struggles to Bridge Gap Between Rich, Poor," by Shannon Van Sant, March 13, 2012. *Voice of America*, http://voanews.com

SOUTHEAST ASIA

ESSENTIAL QUESTIONS • *How does geography influence the way people live?* • *What makes a culture unique?* • *Why does conflict develop?*

©ADREES LATIF/X90022/Reuters/Corbis

Royal guard at the Grand Palace in Bangkok, Thailand

The Story Matters...

Southeast Asia is made up of a mainland and hundreds of islands. The mainland has several mountain ranges, coastal plains, and major rivers. The islands also have varied landforms, but their rivers are short and steep. Shaped by tectonic plate movements, these islands have many active volcanoes. Today, plate movements cause dangerous earthquakes, sometimes generating tsunamis. Recovering from such natural disasters presents many challenges to the people of the region.

FOLDABLES
Study Organizer

Go to the Foldables® library in the back of your book to make a Foldable® that will help you take notes while reading this chapter.

SOUTHEAST ASIA

Two peninsulas form Southeast Asia's mainland—the Indochina and the Malay Peninsulas. Like the mainland, most of the region's islands are mountainous.

Step Into the Place

MAP FOCUS Use the map to answer the following questions.

1 THE GEOGRAPHER'S WORLD
What is the capital of Cambodia?

2 THE GEOGRAPHER'S WORLD
In what country is the city of Manila located?

3 THE GEOGRAPHER'S WORLD
In which direction would you travel to go from Hanoi to Mandalay?

4 CRITICAL THINKING
Describing What does it mean to say a country is "landlocked"? Does Southeast Asia have any landlocked countries?

URBAN SKYLINE Skyscrapers dominate Jakarta, the capital and leading economic center of Indonesia. With more than 11 million people, Jakarta has the largest population of any city in Southeast Asia.

FLOATING MARKET Fruit and vegetable dealers sell their wares from boats on Inya Lake in Myanmar (Burma). The artificial lake was created in the 1880s to provide water to Yangon (Rangoon), the capital.

Step Into the Time

TIME LINE Choose an event from the time line and write a paragraph predicting the social, political, or economic causes and effects of that event.

802 A.D.
Khmer empire, based in what is now Cambodia, rules much of region

1565 The Spanish establish their first settlement in the Philippines

1000 **1100** **1500**

939 Vietnam becomes semi-independent from China after 1,000 years of domination

c. 1150 King Suryavarman II has the Angkor Wat temple built in Cambodia

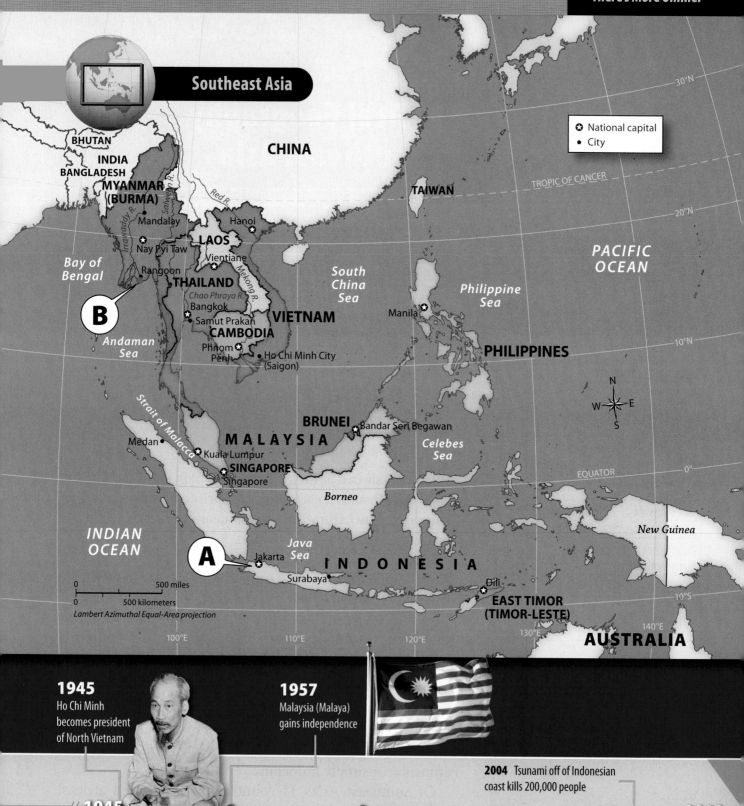

Southeast Asia

National capital
City

BHUTAN
INDIA
BANGLADESH
MYANMAR
(BURMA)

CHINA

TAIWAN

TROPIC OF CANCER

Mandalay
Hanoi

LAOS

Nay Pyi Taw
Vientiane

Bay of
Bengal

Rangoon

THAILAND

Chao Phraya R.

Bangkok
Samut Prakan

CAMBODIA

VIETNAM

South
China
Sea

Manila

Philippine
Sea

PACIFIC
OCEAN

Red R.
Salween R.
Irrawaddy R.
Mekong R.

B

Andaman
Sea

Phnom
Penh

Ho Chi Minh City
(Saigon)

PHILIPPINES

30°N

20°N

10°N

BRUNEI

MALAYSIA

Bandar Seri Begawan

Celebes
Sea

Medan

Kuala Lumpur

SINGAPORE

Singapore

Strait of Malacca

Borneo

New Guinea

INDIAN
OCEAN

A

Jakarta

Java
Sea

INDONESIA

Surabaya

Dili

EAST TIMOR
(TIMOR-LESTE)

EQUATOR

0°

10°S

AUSTRALIA

0   500 miles
0   500 kilometers
Lambert Azimuthal Equal-Area projection

100°E   110°E   120°E   130°E   140°E

1945
Ho Chi Minh
becomes president
of North Vietnam

1957
Malaysia (Malaya)
gains independence

2004 Tsunami off of Indonesian
coast kills 200,000 people

1945

2010

1965–1973 U.S. fights in
Vietnam War

2002 East Timor
gains independence

2009 Trade between Vietnam and the
United States exceeds $15 billion annually

netw⊘rks

There's More Online!

- ☑ **IMAGE** Indonesian Tsunami
- ☑ **MAP** Malaysia/Indonesian Islands
- ☑ **SLIDE SHOW** Wildlife of Southeast Asia
- ☑ **VIDEO**

Reading **HELP**DESK   **CCSS**

Academic Vocabulary

- **commodity**

Content Vocabulary

- **insular**
- **flora**
- **fauna**
- **endemic**

TAKING NOTES: *Key Ideas and Details*

Identify As you read the lesson, use a graphic organizer like this one to identify the countries that occupy or share the region's main peninsulas and islands.

Peninsula or Island	Country or Countries
Indochinese and Malay Peninsulas	
Borneo	
New Guinea	
Java	
Timor	
Luzon	

Lesson 1
Physical Geography of Southeast Asia

ESSENTIAL QUESTION • *How does geography influence the way people live?*

IT MATTERS BECAUSE

Southeast Asia is like a challenging puzzle. Some of its main pieces, its countries, are divided into smaller pieces. Learning how the region's pieces fit together will help you answer a question geographers ask: What makes it a region?

Landforms and Resources

GUIDING QUESTION *How are the landforms of Southeast Asia's mainland different from the landforms of its islands?*

Southeast Asia can be divided into two main parts: a mainland area and an **insular** area, an area comprised of islands. (*Insular* comes from the Latin word *insula*, meaning "island." Another term that comes from *insula* is *peninsula*, which means "almost an island.")

The mainland sits at the southeastern corner of the Asian continent, bordering the world's two most populous countries: China and India. In this area, where the Indian Ocean meets the Pacific, thousands of islands stretch across miles of tropical waters.

Peninsulas and Islands

Most of the mainland occupies a large peninsula that juts southward from the Asian continent. Located between India and China, this extension of land is known as the Indochinese Peninsula, or simply Indochina.

Of Southeast Asia's 11 countries, 6 are located at least partly on the mainland peninsulas. These countries are Cambodia, Laos, Malaysia (the western region of this divided

country), Myanmar (also known as Burma), Thailand, and Vietnam. The tiny island country of Singapore sits just off the southern tip of the Malay Peninsula. A bridge connects Singapore's main island to the peninsula.

The region's other four countries, along with the eastern part of Malaysia, occupy islands in the vast Malay Archipelago. Larger in area than any other island group in the world, it contains more than 24,000 islands, stretching from mainland Southeast Asia to Australia. Among the islands are 7 that rank among the 20 largest in the world: New Guinea, Borneo, Sumatra, Sulawesi, Java, Luzon, and Mindanao.

More than 17,000 of the islands in the Malay Archipelago are part of Indonesia, which is by far the largest country in Southeast Asia. Indonesia shares the islands of Timor, New Guinea, and Borneo with other countries. East Timor, one of the world's newest countries, occupies the eastern half of Timor. Malaysia's eastern region spreads across northern Borneo and surrounds the small country of Brunei.

The northernmost islands in the Malay Archipelago are the Philippines, which form their own archipelago east of the mainland. The Philippines are a country of more than 7,000 islands and islets.

Mountains and Volcanoes

Much of the land in Southeast Asia is rugged and mountainous. On the mainland, mountain ranges generally have a north-south orientation. They include a group of ranges along the western edge of Myanmar that follow the border between Myanmar and Thailand and form the backbone of the Malay Peninsula. The Annamese Cordillera, a range that stretches through Laos and Vietnam, runs parallel to the mainland's eastern coast.

Many of the mountains on Southeast Asia's islands have volcanic origins. The islands lie along the Ring of Fire, a seismically active zone that encircles much of the Pacific Ocean. Most of the world's earthquakes and volcanic eruptions occur within this belt.

This aerial view of Komodo National Park in Indonesia reveals that mountainous islands are a major part of the landscape in Southeast Asia.

Identifying What Southeast Asian island group has a larger area than any other island group in the world?

A farmer tends to his rice crop near Mount Mayon, the most active volcano in the Philippines.

▶ **CRITICAL THINKING**

Analyzing Despite the dangers, how do farmers in the region benefit from volcanic eruptions?

Academic Vocabulary

commodity a material, resource, or product that is bought and sold

In Southeast Asia, four major plates meet: the Eurasian Plate, the Indo-Australian Plate, the Pacific Plate, and the Indian Plate. The pressures and tensions produced by the meeting of these plates have fractured Earth's crust into many smaller plates across the region. This action has also produced the many islands and the fractured geography, as well as the volcanoes.

Indonesia has more than 100 active volcanoes—more than any other country in the world. Most are in a long arc along the country's southern edge. One of the most famous volcanoes, Krakatoa, lies between the islands of Sumatra and Java. In 1883 Krakatoa erupted and collapsed into the sea, triggering tsunamis that claimed about 36,000 lives.

An even deadlier tsunami occurred in Indonesia in 2004. A powerful undersea earthquake off the coast of Sumatra produced huge waves that slammed into coastal areas of Southeast and South Asia, causing more than 230,000 deaths.

Natural Resources

Southeast Asia possesses a rich variety of mineral resources, including tin, copper, lead, zinc, gold, and gemstones such as rubies and sapphires. Indonesia, Malaysia, and Thailand rank among the world's top tin producers, with Indonesia accounting for roughly a fourth of the total world production of this **commodity**.

Teak, mahogany, ebony, and other hardwood trees that grow in Southeast Asia's tropical forests have long been in high demand. Many of the region's countries export wood and wood products. To combat deforestation, some countries have restricted logging.

Southeast Asia is also rich in fossil fuels. Indonesia and Malaysia rank among the top 30 countries in the world in oil reserves and production. They rank among the top 15 in natural gas reserves and production.

☑ **READING PROGRESS CHECK**

Analyzing What effect has the Ring of Fire had on the formation of the region?

Bodies of Water

GUIDING QUESTION *Why does Southeast Asia have so many different seas?*

Bodies of water are key parts of Southeast Asia's geography and identity. The region encompasses about 5 million square miles (13 million sq. km), but only a third of the area is land.

Per-Andre Hoffmann/Picture Press/Getty Images

Oceans and Seas

The Malay Peninsula and Indonesia's Sunda Isles represent the boundary between two oceans. To the west and south lies the Indian Ocean, and to the north and east lies the Pacific Ocean. The region's largest seas are the South China Sea and the Philippine Sea. West of the Malaysian Peninsula lies the Andaman Sea.

Some of the busiest lanes in the world pass through Southeast Asia's seas and their waterways. About a quarter of the world's trade, including half of all sea shipments of oil, goes through the Strait of Malacca, which links the South China and Andaman Seas. As a result, Singapore, which controls the Strait of Malacca, has become one of the most important ports in the world.

River Systems

Southeast Asia's longest and most important rivers are on the mainland. Most rain flows into one of five major rivers. From west to east, the rivers are the Irrawaddy, Salween, Chao Phraya, Mekong, and Red. Each river generally flows from highlands in the north to lowlands in the south before emptying into the sea.

The Irrawaddy, Salween, and Mekong begin high on the Plateau of Tibet. The Irrawaddy flows almost straight south through a valley in Myanmar's center. The river plays a major role in the region's transportation system. Its vast fertile delta is important for farming.

The Salween runs southward through the eastern part of Myanmar, forming part of that country's border with Thailand. Like the Irawaddy, it drains into the Andaman Sea. The Mekong is Southeast Asia's longest river. It twists and turns for about 2,700 miles (4,345 km) through or near Myanmar, Thailand, Laos, Cambodia, and Vietnam. The river's drainage basin is twice the size of California, and its enormous delta is one of the world's most productive agricultural regions.

☑ **READING PROGRESS CHECK**

Analyzing Why do you think Southeast Asia's longest rivers are found on the mainland and not on islands?

The port city of Banda Aceh in Indonesia suffered great damage from a tsunami that struck on December 26, 2004.

Eugene Hoshiko/AP Images

A plantation worker in Indonesia gathers latex, a milky liquid tapped from rubber trees. The latex is eventually refined into useable rubber.

▶ **CRITICAL THINKING**
Analyzing Why do rubber trees and plants grow so well in Indonesia and other parts of Southeast Asia?

Climate, Vegetation, and Wildlife

GUIDING QUESTION *In what ways does Southeast Asia's location shape its climate?*

Climates in Southeast Asia are generally hot and humid. Much of the region receives more than 60 inches (152 cm) of rain fall each year. Because of these weather conditions and the variety of habitats in Southeast Asia, the region is home to an astonishing abundance of plant and animal life.

A Tropical Region

Latitude and air currents play major roles in shaping Southeast Asia's climates. Nearly the entire region lies within the Tropics—the zone that receives the hottest, most direct rays of the sun. From November to March, the direct rays of the sun are south of the Equator in the Southern Hemisphere. This produces areas of low pressure that draw monsoon winds that blow across the region from the northeast to the southwest. These winds bring cooler, drier air to much of the mainland but deliver heavy rains to the southern Malay Peninsula and to the islands. From May to September, the pattern reverses. The direct rays of the sun and the associated low pressure are north of the Equator. This causes monsoon winds to blow from the southwest to the northeast, bringing warm air and rain to the islands and the mainland.

Most of the region's land is surrounded or nearly surrounded by the sea, which has a moderating effect on air temperatures. Elevation also affects weather conditions. Temperatures in the highland areas are generally cooler than in the lowland areas. At the top of Indonesia's highest peak, Puncak Jaya (PUHN-chock JAH-yuh), which is close to the Equator, glaciers are visible.

JUSTIN GUARIGLIA/National Geographic Stock

The sea, the elevation, and air currents combine to create four climate zones. The southern Malay Peninsula, the southern Philippines, and most of Indonesia have a tropical rain forest climate. A tropical monsoon climate, with rainy and dry seasons, prevails in the northern Philippines, the northern Malay Peninsula, and coastal areas of the mainland. Most inland areas fall within a tropical savanna climate zone with less distinct rainy and dry seasons. The northernmost mainland has a humid subtropical climate in which summers are hot and wet and winters are mild and dry.

Weather in Southeast Asia sometimes turns deadly. Intense tropical storms called typhoons form over warm Pacific Ocean waters and slowly spin westward, often targeting the Philippines. With winds that can exceed 150 miles (241 km) per hour and torrential rains that sometimes continue for days, typhoons can wipe out homes and buildings, cause devastating floods, destroy crops, and kill large numbers of people.

Plants and Animals

Few places in the world rival Southeast Asia in diversity of **flora**, or plant life. Indonesia, for example, has more than 40,000 species of flowering plants, including about 5,000 species of orchids and more than 3,000 species of trees.

Blanketing much of the region are either tropical rain forests or forests with a mix of evergreen and deciduous trees. In coastal areas, forests of mangrove trees, which have aboveground roots, form a border between land and sea.

Southeast Asia's **fauna**, or animal life, consists of a wide variety of animals. Many species, including mammals, birds, fish, and insects, are **endemic**, or found nowhere else in the world. Fires, logging, mining, agriculture, and poaching have reduced the habitat of many animals.

☑ **READING PROGRESS CHECK**

Identifying What are three factors that affect climate in Southeast Asia?

Think Again?

Hurricane, *typhoon*, and *cyclone* refer to different types of storms.

Not true. All three terms refer to a large, powerful, circular storm that originates over warm tropical ocean waters and is characterized by strong winds and heavy rain. These storms have different names in different parts of the world.

FOLDABLES
Study Organizer

Include this lesson's information in your Foldable®.

LESSON 1 REVIEW **CCSS**

Reviewing Vocabulary

1. What peninsula forms a long bridge between mainland and *insular* Southeast Asia?

Answering the Guiding Questions

2. Identifying What are some of Southeast Asia's most important mineral resources?

3. Analyzing Why are most of Indonesia's volcanoes located along its southern edge?

4. Analyzing What is the connection between Singapore's status as one of the most important ports in the world and its location at the southern end of the Strait of Malacca?

5. Describing How do monsoon winds cause wet and dry seasons on Southeast Asia's mainland?

6. Argument Writing Write a persuasive letter urging the government leaders of Southeast Asia to take urgent action to protect the region's tropical forests from uncontrolled logging. Discuss the importance of the forests and the consequences of deforestation.

networks

There's More Online!

☑ **IMAGE** Angkor Wat

☑ **MAP** History of Southeast Asian Civilizations

☑ **VIDEO**

Reading **HELP**DESK **CCSS**

Academic Vocabulary

- **ultimate**

Content Vocabulary

- **sultan**
- **plantation**
- **absolute monarchy**
- **constitutional monarchy**

TAKING NOTES: *Key Ideas and Details*

Describe Select one of the countries from the lesson. As you read, describe three important events in that country's history on a web diagram like the one below.

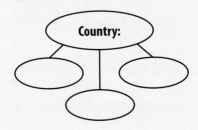

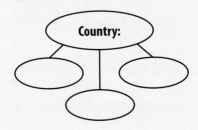

Country:

Lesson 2

History of Southeast Asia

ESSENTIAL QUESTION · *How does geography influence the way people live?*

IT MATTERS BECAUSE

Southeast Asia sometimes gets less attention than its two giant neighbors, China and India. The region has played an important role in world history in recent centuries, however, and it is likely to play an even larger role in the future.

Kingdoms and Empires

GUIDING QUESTION *What role has trade played in Southeast Asia's history?*

Southeast Asia is known as "the Crossroads of the World" because it is located along important maritime trade routes. Trade has exposed the region to many different cultural influences. It also has made the region prey for foreign powers seeking to increase wealth and power by controlling the routes.

Prehistoric Cultures

Humans have lived in Southeast Asia for at least 40,000 years. For much of this period, the region looked quite different from today. Earth was in the grip of an ice age, so sea levels were lower. Much of the continental shelf lay above water as dry land. This land connected the mainland to many areas that are now islands, including Borneo, Sumatra, and Java. The mainland area, therefore, was much larger than it is now, and the island area was much smaller. As the ice age waned, the seas began to rise. They reached their present levels about 8,000 years ago.

Throughout the prehistoric period, Southeast Asia's inhabitants survived by hunting and gathering. They used stone tools and weapons. Around 6,000 years ago, people had begun practicing agriculture, growing rice in fertile river

(l to r) ©Jose Fuste Raga/Corbis; ©badz/age fotostock; Manfred Bail/Alamy; Keystone/Hulton Archive/Getty Images

valleys and deltas. Agriculture allowed for a more settled existence and led to the development of complex societies.

Powerful Societies Emerge

Several thousand years ago, early metalworking societies arose in Southeast Asia. Using sophisticated techniques, these cultures produced bronze tools, weapons, ornaments, and ceremonial drums. The most famous of these cultures, the Dong Son, was centered in northern Vietnam.

In the middle of the 100s B.C., China and India began to exert a powerful influence. China conquered the Red River delta in the northeast and made Vietnam a province of the Han empire. For the next 1,000 years, Vietnam remained under Chinese control. Chinese culture became embedded in the cultures of Vietnam and its neighbors.

People from India came to Southeast Asia as traders and missionaries. India's culture and its religions, Hinduism and Buddhism, spread along trade routes from India into Southeast Asia. Societies with Indian roots began to develop throughout the region.

Trading Societies

Funan, one of the first important trade-based states in the region, was established in the A.D. 100s. It covered parts of what are now Cambodia, Thailand, and Vietnam. Its people traded with China and India, but its cultural influences were mostly Indian.

Around the A.D. 600s, a kingdom called Srivijaya arose on the island of Sumatra and gained control of the Strait of Malacca. This waterway connecting the Pacific and Indian oceans was important because of the many trade goods that had passed through it by boat.

The main Hindu temple at Angkor Wat displays ancient building styles of India and Southeast Asia. Angkor Wat forms the largest single religious site in the world. Today, it is the national symbol of Cambodia, and the main temple appears on that country's flag.

Identifying How did Hinduism spread to Southeast Asia?

©Jose Fuste Raga/Corbis

This modern mosque built on stilts is located near the port city of Malacca in Malaysia. When the tide is high and the water level rises, the building appears to be floating.

▶ CRITICAL THINKING

Describing How did the religion of Islam reach Southeast Asia?

This included much of the spice trade, which connected China and Southeast Asia with India, Southwest Asia, Africa, and Europe. By controlling trade, Srivijaya grew into an empire that dominated commerce in the region for 600 years. Its capital, Palembang, became an important center of Buddhism.

Agricultural Societies

Southeast Asia's major agricultural societies developed where rice could be grown. The Pagan kingdom sprang up in Myanmar's Irrawaddy delta; Vietnamese society took root in the Red River delta; and the Khmer empire was centered near a large lake called Tonle Sap in Cambodia. The Khmer invented a water-management system to improve crops. The empire is also known for architecture, especially the temple complexes Angkor Wat and Angkor Thom. Built 800 years ago, the complexes still stand, drawing millions of visitors annually.

Islamic States

Islam, the religion of Muslims, could have reached Southeast Asia as early as the A.D. 800s or A.D. 900s when traders from the Middle East or western India traveled the sea route to China. By the 1200s, two Islamic kingdoms had been established in northern Sumatra. From there, Islam gradually spread eastward. By the 1400s, other Islamic kingdoms had risen near ports located along the main trade route through the region's waterways. By the 1600s, Islam had become the dominant religion across the Malay Archipelago.

The most important Islamic kingdom was centered on the port of Malacca on the Malay Peninsula. Founded in about 1400, Malacca capitalized on its location along the Strait of Malacca and developed into a powerful trading empire. Its **sultans**, or kings, ruled over much of the peninsula and the island of Sumatra across the strait.

☑ **READING PROGRESS CHECK**

Identifying What are some ways in which India influenced Southeast Asia?

Western Colonization

GUIDING QUESTION *How did European colonization change Southeast Asia?*

During the period known as the Age of Discovery, European explorers made long voyages in search of gold, silver, spices, and other sources of wealth. They also sought to spread Christianity and to map the world. The arrival of European ships in Southeast Asia marked the beginning of conquest and colonization that would leave the region dramatically changed.

European Traders

From ancient times through the Middle Ages, spices from Southeast Asia had reached Europe through Chinese, Indian, and Arab traders. Europeans used spices such as ginger, cinnamon, cloves, nutmeg, and mace to flavor food, to preserve meat, and to make perfumes and medicines. The demand was great and supplies were limited, so traders were able to charge high prices. Some spices were worth more than gold. The fabulous wealth of the spice trade led European powers to seek ocean routes to the source of the spices, and to gain control of the trade.

These spices on display on the Indonesian island of Bali include nutmeg, lemon grass, and cinnamon.

▶ **CRITICAL THINKING**

Describing Why did Europeans value Southeast Asian spices?

Manfred Bail/Alamy

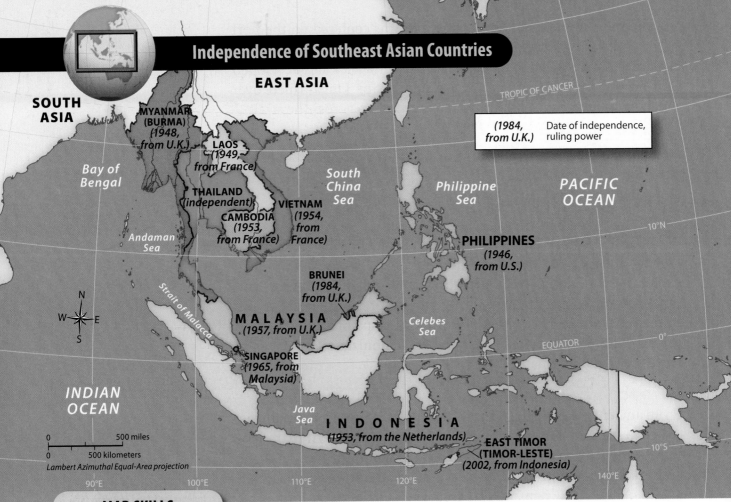

Independence of Southeast Asian Countries

EAST ASIA

SOUTH ASIA

(1984, from U.K.) Date of independence, ruling power

MYANMAR (BURMA) *(1948, from U.K.)*

Bay of Bengal

LAOS *(1949, from France)*

THAILAND *(independent)*

VIETNAM *(1954, from France)*

CAMBODIA *(1953, from France)*

South China Sea

Philippine Sea

PACIFIC OCEAN

TROPIC OF CANCER

Andaman Sea

PHILIPPINES *(1946, from U.S.)*

10°N

BRUNEI *(1984, from U.K.)*

Strait of Malacca

MALAYSIA *(1957, from U.K.)*

Celebes Sea

SINGAPORE *(1965, from Malaysia)*

EQUATOR

0°

INDIAN OCEAN

Java Sea

INDONESIA *(1953, from the Netherlands)*

EAST TIMOR (TIMOR-LESTE) *(2002, from Indonesia)*

10°S

0 ___ 500 miles
0 ___ 500 kilometers
Lambert Azimuthal Equal-Area projection

90°E   100°E   110°E   120°E   140°E

MAP SKILLS

1 PLACES AND REGIONS
Which Southeast Asian country does not have an independence date? Why?

2 PHYSICAL GEOGRAPHY
Why might Indonesia find it difficult to achieve a sense of national unity?

In the early 1500s, Portuguese navigators discovered they could reach India and Southeast Asia by sailing around the southern tip of Africa. In 1511 the Portuguese conquered Malacca. That same year, they discovered the sources of cloves, nutmeg, and mace: the Moluccas and the Banda Islands, which became known as the Spice Islands. Through much of the 1500s, wealth generated by spices enriched Portugal's monarchy.

Meanwhile, other European powers sought alternate routes that would allow them to profit from the spice trade. The explorer Ferdinand Magellan commanded five Spanish ships that reached the Philippines by sailing across the Pacific Ocean from Mexico. Soon the Philippines became a Spanish colony.

At the beginning of the 1600s, Holland (known today as the Netherlands) jumped into the spice trade. Because the Portuguese controlled the Strait of Malacca, the Dutch charted a new route to the Spice Islands. By the middle of the century, they replaced the Portuguese as the dominant trading power.

Colonial Rule

During the 1800s and early 1900s, European countries gained control over other parts of Southeast Asia. Burma and Malaysia became colonies of Great Britain, and Vietnam, Laos, and Cambodia

became colonies of France. In addition to the spice trade, European countries developed tin and coal mines and built factories to process the raw materials. They also established **plantations**, or large farms on which a single crop is grown for export. Plantation crops included tea, coffee, tobacco, and rubber trees. Thousands of laborers were brought in from China and India to work in the mines and on plantations. Many of them stayed on as permanent residents of the colonies.

Thailand (Siam)

Thailand, then known as Siam, was the only state in Southeast Asia that Europeans did not colonize. It was ruled by an **absolute monarchy** from the mid-1300s until 1932. In an absolute monarchy, one ruler has **ultimate** governing power over the country. The rulers of Siam had resisted threats from Burma since the 1500s. When the British declared war there, Siam's leaders agreed to allow free trade with Great Britain and other Western countries. This admission allowed the country to remain independent, acting as a buffer state between British and French possessions.

Academic Vocabulary

ultimate most extreme or greatest

☑ **READING PROGRESS CHECK**

Identifying In what ways did European powers gain wealth from their colonies in Southeast Asia?

Independent Countries

GUIDING QUESTION *What events ended the colonial era in Southeast Asia?*

In the early 1900s, nearly all of Southeast Asia was under the control of foreign powers. Colonial rule was often harsh, unjust, and exploitive, and sometimes it was met with violent resistance by the region's people. It was not this resistance, however, that ended colonialism. Instead, it was dramatic events that unfolded elsewhere in the world.

Dawn of Freedom

The first Southeast Asian colony to glimpse independence was the Philippines, which had been ruled by Spain since the 1500s. The United States took control of the colony in 1898 after defeating Spain in the Spanish-American War. During World War II, Japan sent its military forces to take control of Southeast Asian lands.

A Buddhist monk (right) in Cambodia casts his vote in a local election held in 2012. Today, Cambodia is struggling to build a democracy after years of civil war.

▶ **CRITICAL THINKING**
Describing Why were the 1970s a difficult decade for Cambodia's people?

TANG CHHIN SOTHY/AFP/Getty Images

Viet cong fighters patrol a waterway in the Mekong Delta during the Vietnam War. The Viet cong carried out armed attacks on U.S. and South Vietnamese forces. Their goal was to unite all of Vietnam under Communist rule.

▶ **CRITICAL THINKING**

Predicting What was the outcome of the Vietnam War?

Japan's nearness to Southeast Asia was a great advantage. The Japanese hoped to use the region's oil, rubber, and timber to help its economy and enable it to continue the war. Japan's rule lasted four years and caused the region's people great hardship.

The war ended in 1945, and the United States granted independence to the Philippines in 1946. Soon, other Southeast Asian colonies began to break free of European rulers. Myanmar negotiated its independence from Britain in 1948, and Indonesia was freed from the Netherlands in 1949. In the 1950s, Vietnam, Laos, and Cambodia freed themselves from French rule, and Malaysia and Singapore became independent of Britain.

The last country to gain its freedom was East Timor, sometimes called by its Portuguese name, Timor-Leste (TEE-mor LESS-tay). After being ruled by Portugal since the 1500s, East Timor declared independence in 1975 but was invaded by Indonesia. Indonesia's harsh, violent rule lasted there until 2002.

Regional Conflicts

The newly independent countries of Southeast Asia faced many challenges. Wars, revolutions, and dictatorships tested the new leaders. In Vietnam, Communist forces defeated the French in 1954 and ruled the northern part of the country. The United States supported leaders in the south. Fighting led to the Vietnam War, which lasted until 1975 and took more than 2 million lives. After the war ended, North Vietnam united the country under Communist rule.

Keystone/Hulton Archive/Getty Images

The Vietnam War affected other parts of Southeast Asia. In Laos, a conflict pitted the government of Laos against nationalists allied with Vietnamese communists. In 1975 the government of Laos collapsed, and many people fled to other countries.

In 1970, Cambodia's military took over the government. Five years later, a rural communist movement, called the Khmer Rouge, overthrew this government. The country's new leaders began a brutal campaign of terror and destruction. Between 1975 and 1979, at least 1.5 million people died or were killed. In 1978 Vietnamese forces invaded Cambodia and installed a new government controlled by Vietnam, sparking almost 13 years of civil war.

Modern Southeast Asia

The economic growth of East Asian countries such as China and Taiwan in the late 1900s helped some countries in Southeast Asia. Thailand, Malaysia, Indonesia, the Philippines, and Vietnam became centers of manufacturing in the region. Textiles and tourism are important parts of Cambodia's economy. Singapore, which gained independence from Great Britain in 1959 and then from Malaysia in 1965, is one of the world's most prosperous countries.

Thailand has been a **constitutional monarchy** since 1932. A constitutional monarchy's ruler is legally bound by a constitution and government. Thailand's government has often been threatened by the military. Numerous military coups took place in the 1900s. By the 1980s, Thailand had combined a tradition of monarchy with democratic reforms, and it enjoyed high rates of economic growth through the 1990s.

Myanmar has struggled since gaining independence. In 1962 the military seized power and established a socialist government. Since then, leaders have closed off the country to outside influences. Even a cyclone in 2008 that killed more than 100,000 people did not persuade the government to open its borders to allow aid from abroad.

Include this lesson's information in your Foldable®.

☑ **READING PROGRESS CHECK**

Identifying What colonial power ruled Vietnam before 1954?

LESSON 2 REVIEW

Reviewing Vocabulary

1. What crops were grown on *plantations* in Southeast Asian colonies?

Answering the Guiding Questions

2. *Describing* How did the geography of Southeast Asia change as a result of a rise in sea level?

3. *Analyzing* What made the Strait of Malacca such an important waterway?

4. *Determining Central Ideas* Why did European countries want to control the spice trade?

5. *Analyzing* How could the resources of Southeast Asia help Japan's economy and its ability to continue fighting in World War II?

6. *Informative/Explanatory Writing* Knowing about the history of colonization in Southeast Asia, write a paragraph in which you answer these questions: What are the dangers of resisting colonization? What are the consequences of not resisting?

Reading **HELP**DESK

Academic Vocabulary

- **exploit**

Content Vocabulary

- **primate city**
- **minority**
- **Pacific Rim**
- **subsistence farming**
- **ecotourism**

TAKING NOTES: *Key Ideas and Details*

Identify As you read, use a graphic organizer like this one to note key facts about the people of Southeast Asia.

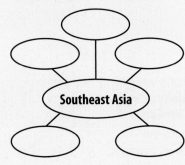

Southeast Asia

Lesson 3
Life in Southeast Asia

ESSENTIAL QUESTIONS • *What makes a culture unique?* • *Why does conflict develop?*

IT MATTERS BECAUSE

Southeast Asia is one of the most culturally diverse regions in the world. Learning about this diversity will lead to a better understanding of the region.

People and Places

GUIDING QUESTION *How is Southeast Asia's population shifting?*

In population size, Southeast Asia has fewer people than its neighbors, China and India. Compared to other areas, however, Southeast Asia's population is high. This lesson describes where the region's people live. The lesson also discusses how the population is expanding and changing.

Population Profile

Southeast Asia is home to about 625 million people. Indonesia has a population of approximately 250 million, which is 40 percent of the regional total.

Many Southeast Asian countries experienced rapid population growth in the 1900s, but today the region's growth rate is only slightly above the world average. Population is not evenly distributed. The highest densities generally are found in areas where good soil and abundant water allow agriculture to thrive. These areas include coastal plains, river valleys, and deltas. One of the greatest population densities is on the island of Java, whose volcanic soils are exceptionally fertile.

People on the Move

Historically, Southeast Asian societies were mostly rural. Since World War II, however, people have moved steadily from rural areas to cities. This movement is called

urbanization. Urbanization is occurring most rapidly in countries that recently were or are currently the most rural.

Urbanization has produced explosive population growth in some cities. Manila, the capital of the Philippines, is home to more than 11 million people. Manila is a **primate city**—a city so large and influential that it dominates the rest of the country. Like many large cities, Manila suffers from overcrowding. More than a third of its inhabitants live in slums, where people are desperately poor and most dwellings are crudely built shacks.

Rapid population growth in and around Jakarta, the capital of Indonesia, has created a continuous urban chain of cities that were once separated. This megalopolis, as such areas are called, is one of the largest in the world, with more than 26 million people.

☑ **READING PROGRESS CHECK**

Identifying What types of geographical areas in Southeast Asia have the highest population densities?

People and Cultures

GUIDING QUESTION *How have China and India influenced Southeast Asian cultures?*

Peoples from other regions have migrated to Southeast Asia for more than 2,000 years for many different reasons. As a result, Southeast Asia has a rich mix of peoples and cultures.

Southeast Asia's cities provide contrasts in the ways people live. In Manila, the Philippines (left), residents crowd into slum areas not far from the city's modern business center. Singapore's Chinatown district (right) is known for its restored historic buildings and numerous shops and restaurants.

▶ **CRITICAL THINKING**

Describing How have Southeast Asia's cities changed since World War II?

A woman of the Padaung ethnic group drives a car in Myanmar near Yangon (Rangoon). Many Southeast Asians today blend traditional and modern ways.

▶ **CRITICAL THINKING**

Describing What is a major characteristic of peoples and cultures in Southeast Asia?

Ethnic and Language Groups

Southeast Asia's population consists of many ethnic groups. Five main groups dominate the mainland: the Burmese in Myanmar, the Siamese in Thailand, the Malay in Malaysia, the Mon-Khmer in Laos and Cambodia, and the Vietnamese in Vietnam. The largest group in Indonesia is the Javanese, and the largest group in the Philippines is the Tagalog. Many smaller groups, sometimes called **minorities**, also live in each country.

The people of Southeast Asia speak many languages. Traveling throughout Indonesia, one could encounter several hundred different languages. Most of the region's languages are indigenous, or native to the region. Others, including English, Spanish, and French, arrived with trade and colonialism.

Religion and the Arts

Buddhism is the most widely practiced religion across most of the mainland. Islam is dominant on the southern Malay Peninsula and across the Indonesian islands. In the Philippines and East Timor, most people are Roman Catholic. In some remote areas, people practice animist religions. Animism is based on the belief that all natural objects, such as trees, rivers, and mountains, have spirits.

When people from India brought Buddhism and Hinduism to the region, they also brought rich traditions in architecture, sculpture, and literature. Some of the greatest architecture and sculpture are Hindu and Buddhist temples built centuries ago.

The wide variety of art forms that flourish in Southeast Asia reflect the region's great diversity of cultures. In Thailand, plays known as *likay* feature singing, dancing, and vibrantly colored costumes. Actors improvise song lyrics, dialogue, and plots. In the popular shadow-puppet theater in Indonesia, Malaysia, and Cambodia, one puppeteer sings, chants, and controls the puppets behind a screen that is lit from the back. An orchestra of metal instruments provides music.

Daily Life

Although Southeast Asian cities are growing rapidly, three-fourths of the region's people live in rural areas. Many rural people move to the cities, but some leave the region to work in other countries. The money they send home helps their families survive.

Education and literacy rates vary across the region. The highest literacy rate is in Vietnam, where 94 percent of the people can read and write. At the other extreme are East Timor, Laos, and Cambodia, all of which have literacy rates below 75 percent. Roughly 95 percent of the schools in East Timor were destroyed during the country's fight for independence from Indonesia. The government of the new country is working to build new schools.

Many sports are popular in Southeast Asia. Some are traditional and generally unknown outside the region, while others—including soccer, badminton, martial arts, and volleyball—are international. All 11 countries in the region participate in the Southeast Asian Games that take place every two years. The countries take turns hosting the games, which include Olympic sports and traditional regional sports like *sepaktakraw*, a kind of volleyball played with the feet, knees, head, and chest.

☑ **READING PROGRESS CHECK**

Identifying Which religions are most widespread in Southeast Asia?

Issues in Southeast Asia

GUIDING QUESTION *In what ways have human activities affected the environment in Southeast Asia?*

Southeast Asia is a region in transition. In recent decades, the rapid economic growth along much of the **Pacific Rim**—the area bordering the Pacific Ocean—has brought great changes to some of the region's countries. With these changes, however, have come new challenges and problems.

Residents of Manila, the Philippines, celebrate a religious festival that has its roots in the culture of Spain. The Philippines was part of the Spanish Empire from the 1500s to the late 1800s.

A young American visitor poses with a produce seller on a busy street in Hanoi, Vietnam. Tourism is a growing industry in Vietnam and many other Southeast Asian countries.

Identifying What other new industries have developed in Southeast Asia in recent decades?

Earning a Living

Farming remains the most common livelihood in most countries. Many rural villages depend on rice, the most important food staple in the region, as a cash crop. Rice is by far the most widely grown crop. Two of the region's countries, Thailand and Vietnam, lead the world in rice exports. Plantations located in Malaysia, Indonesia, and southern Thailand produce natural rubber and palm oil, while coconuts and sugar are important crops in the Philippines. Southeast Asia's other top export crops include cacao, coffee, and spices.

Many farmers grow food only to feed themselves and their families. This is called **subsistence farming**. Some practice subsistence farming but also work seasonally to earn money.

Since the end of the colonial period, many countries have focused on industry. The tiny country of Singapore has developed into a major industrial center. Mining contributes to the economies of some countries in the region. Indonesia, Malaysia, and Thailand together produce more than half of the world's tin. Large deposits of copper and gold are mined on the Indonesian portion of the island of New Guinea.

Fishing is an important livelihood in Thailand, Indonesia, Malaysia, and the Philippines. Tourism is a growing industry in countries such as Cambodia, Thailand, and Vietnam. An important part of this industry is **ecotourism**, or touring natural environments such as rain forests and coral reefs.

Finance, communication, and information technology have also improved greatly in the region. Many people in the cities use cell phones and have access to the Internet. Due to its large and generally inexpensive labor force, many U.S., Japanese, and European industries employ workers in the region.

Making Connections

In 1968, Indonesia, Malaysia, the Philippines, Singapore, and Thailand joined together to form the Association of Southeast Asian Nations (ASEAN). The organization seeks to increase economic development, social progress, and cultural development in the region. All the countries in the region except East Timor are members. Although many countries have experienced economic growth since the late 1960s, Cambodia, Laos, Myanmar, and Vietnam rank among the poorest countries in the world.

Economic and Environmental Challenges

Southeast Asia faces many challenges. The gap between rich and poor has widened in recent decades. Economic growth has helped some people out of poverty, but many still struggle to meet daily needs. In Manila, Jakarta, and other big cities, rapid urbanization has led to overcrowding and water shortages.

Protecting the environment is another challenge. Industries provide an economic boost, but they sometimes **exploit** and harm the environment. Tin mining has created huge wastelands in Malaysia, Thailand, and Indonesia. Commercial logging and farming destroy the tropical forests at an alarming rate. Indonesia and Malaysia recently introduced laws to slow deforestation while still promoting economic growth.

Dams along the Mekong River create hydroelectric power, but that also has created problems for the fishing industry. Vietnam, Cambodia, Laos, and Thailand established the Mekong River Commission to encourage safe management and usage of the river and its resources. Conflict over control of oil and gas resources in the South China Sea is another challenge. Vietnam, the Philippines, and Malaysia surround the sea. The countries' claims on the resources are in dispute.

Political Challenges

Since the end of the colonial era, economic and social progress have been slowed by political instability. After the military took over the government of Myanmar in 1962, the country slid into isolation, poverty, and brutality. The government has refused democratic elections and limits any protest by its people. Some people in Myanmar have tried to bring democracy to the country. In 2011 the military government began to allow more freedom for its people. In 2012 the United States reestablished diplomatic ties with Myanmar and assigned an ambassador to the country for the first time in more than 20 years.

☑ **READING PROGRESS CHECK**

Determining Central Ideas What is the mission of the Association of Southeast Asian Nations (ASEAN)?

Academic Vocabulary

exploit to use a person, a resource, or a situation unfairly and selfishly

Include this lesson's information in your Foldable®.

LESSON 3 REVIEW (CCSS)

Reviewing Vocabulary

1. *Identifying* What are some of the problems that a *primate city* might experience?

Answering the Guiding Questions

2. *Analyzing* Why might the literacy rate in East Timor soon begin to rise?

3. *Determining Central Ideas* What are some of the environmental problems facing Southeast Asia?

4. *Analyzing* What might happen to agricultural production in Southeast Asia if the urbanization trend continues?

5. *Determining Word Meanings* What is *animism*, and where is it practiced in Southeast Asia?

6. *Argument Writing* Imagine that your class is trying to pick a destination for an overseas field trip. Write a letter to your classmates encouraging them to select Southeast Asia. Give specific reasons why this region would make a good destination.

Directions: Write your answers on a separate piece of paper.

1 Use your **FOLDABLES** to explore the Essential Question.

INFORMATIVE/EXPLANATORY WRITING Explain in a paragraph where Southeast Asia's agricultural societies began and why they originated in those locations.

2 21st Century Skills

ANALYZING Research information to write about the economies of Vietnam and Cambodia. Discuss which country you think will undergo stronger economic development in the next decade.

3 Thinking Like a Geographer

IDENTIFYING The population of the Greater Jakarta, Indonesia, metropolitan area reached 26 million people. How many large U.S. cities, including the largest city in your state, would have to be combined to reach a population of 26 million? Use the chart to list several American cities and their respective populations. Add the totals until you reach approximately 26 million.

City	Population

4 **GEOGRAPHY ACTIVITY**

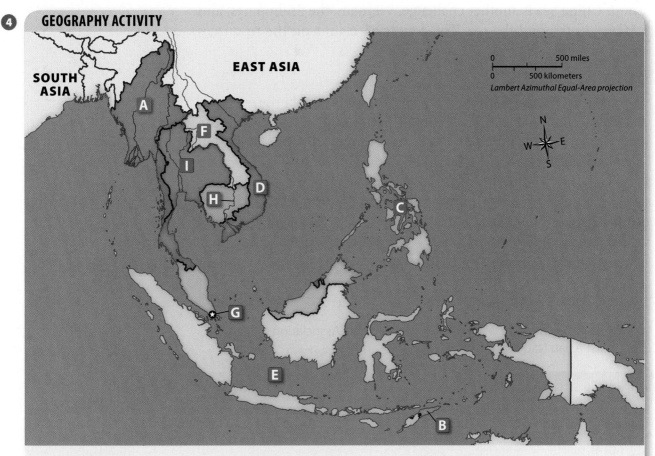

Locating Places

Match the letters on the map with the numbered places below.

1. Philippines
2. Thailand
3. Laos
4. Java Sea
5. Myanmar
6. Singapore
7. Mekong River
8. East Timor (Timor-Leste)
9. Vietnam

Chapter 15 ASSESSMENT CCSS

REVIEW THE GUIDING QUESTIONS

Directions: Choose the best answer for each question.

1. The region of Southeast Asia consists mainly of
   A. constitutional monarchies.
   B. Singapore.
   C. peninsulas and islands.
   D. war-ravaged countries.

2. Which country has the greatest number of active volcanoes in the world?
   F. Indonesia
   G. Vietnam
   H. Krakatoa
   I. Thailand

3. Why did foreign powers begin to colonize Southeast Asian countries?
   A. to defeat local uprisings
   B. to control the profitable spice trade
   C. to test new oceangoing navigation instruments
   D. to enslave the Southeast Asian peoples

4. In which country did the Khmer Rouge arise?
   F. the Philippines
   G. Laos
   H. Myanmar
   I. Cambodia

5. Forty percent of the population of Southeast Asia live in
   A. Manila.
   B. Brunei.
   C. Indonesia.
   D. Malaysia.

6. What is one of the greatest challenges Southeast Asian countries face?
   F. loss of trade
   G. environmental damage
   H. loss of cottage industries
   I. devaluation of their monetary systems

DBQ ANALYZING DOCUMENTS

7 IDENTIFYING Read the following passage:

"*The opening of the Suez Canal in 1869 and the advent of steamships launched an era of prosperity for Singapore as . . . trade expanded. . . . In the 20th century, the automobile industry's demand for rubber from Southeast Asia and the packaging industry's need for tin helped make Singapore one of the world's major ports.*"

—from U.S. State Department, "Singapore: History" (2011)

What kind of change was the initial cause of Singapore's growth?

A. political

B. technological

C. cultural

D. geographical

8 ANALYZING Where was the tin mentioned in the passage most likely mined?

F. southern Africa

G. Southeast Asia

H. East Asia

I. Oceania

SHORT RESPONSE

"*The Chinese conquerors referred to Vietnam as Annam, the 'pacified south.' But it was not peaceful. . . . Revolts recurred chronically [regularly], and . . . [leaders] stressed that Vietnam's customs, practices, and interests differed from those of China.*"

—from Stanley Karnow, *Vietnam: A History* (1983)

9 IDENTIFYING POINT OF VIEW Why would leaders of Vietnamese revolts emphasize that Vietnam's culture was different from China's?

10 ANALYZING How were the events that took place early in Vietnam's history, described here, similar to its later history?

EXTENDED RESPONSE

11 INFORMATIVE/EXPLANATORY WRITING Vietnam has one of the highest literacy rates in Southeast Asia and one of the highest poverty rates. With this in mind, research online and at the library to write about this seeming contradiction. In your essay, explore the effects of poverty on child labor. What impact does the purchase of high-priced consumer goods made in factories staffed by children have on the practice of child labor?

Need Extra Help?

If You've Missed Question	**1**	**2**	**3**	**4**	**5**	**6**	**7**	**8**	**9**	**10**	**11**
Review Lesson	1	1	2	2	3	3	3	1	2	2	3

From VIETNAM: A HISTORY, by Stanley Karnow, Copyright © 1983 by WGBH Educational Foundation and Stanley Karnow. Used by permission of Viking Penguin, a division of Penguin Group (USA) Inc. Reprinted by permission of SLL/Sterling Lord Literistic, Inc. Copyright by Stanley Karnow

SOUTH ASIA

ESSENTIAL QUESTIONS • *How does geography influence the way people live?*
• *How do governments change?* • *What makes a culture unique?*

A man in traditional clothes celebrates at a Hindu festival in northern India.

The Story Matters...

The landscape of South Asia is one of contrasts—from lowlands only a few feet above sea level to the Himalaya, the highest mountains in the world. Many of the rivers of South Asia begin in the Himalaya and flow down onto valleys and plains, providing the fertile soils needed for farming. Many of the countries of South Asia have their roots in ancient civilizations and religions that developed in these river valleys.

FOLDABLES®
Study Organizer

Go to the Foldables® library in the back of your book to make a Foldable® that will help you take notes while reading this chapter.

Geography | History | Economy

The northern part of South Asia is separated from the rest of Asia by mountains. The southern part of South Asia juts out of the Asian continent and into the waters of the Arabian Sea, Indian Ocean, and Bay of Bengal.

Step Into the Place

MAP FOCUS Use the map to answer the following questions.

1 PLACES AND REGIONS What is the capital of Bangladesh?

2 THE GEOGRAPHER'S WORLD What are the two island countries of South Asia?

3 PHYSICAL GEOGRAPHY Which South Asian countries are landlocked?

4 CRITICAL THINKING
Analyzing Use the scale bar to estimate the distance between Mumbai and New Delhi.

MOUNTAIN GATEWAY A road winds through the Khyber Pass at the border between Pakistan and Afghanistan. Invaders and traders of long ago used the pass to enter South Asia from the northwest.

HISTORIC MONASTERY Built in the 1600s, this Buddhist monastery overlooks Thimphu, the capital of Bhutan. Today, the monastery holds government offices and the throne room of Bhutan's king.

Step Into the Time

TIME LINE Choose two events from the time line and explain the connection between those events and their effects on the people of South Asia today.

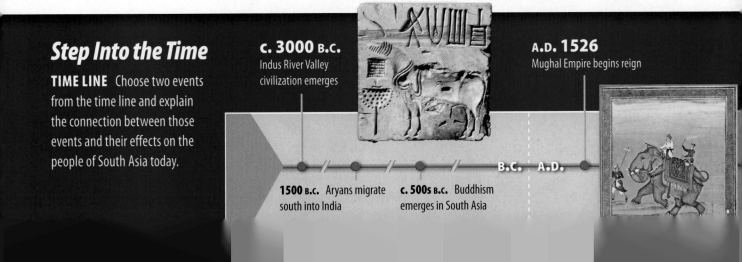

c. 3000 B.C. Indus River Valley civilization emerges

A.D. 1526 Mughal Empire begins reign

1500 B.C. Aryans migrate south into India

c. 500s B.C. Buddhism emerges in South Asia

B.C. A.D.

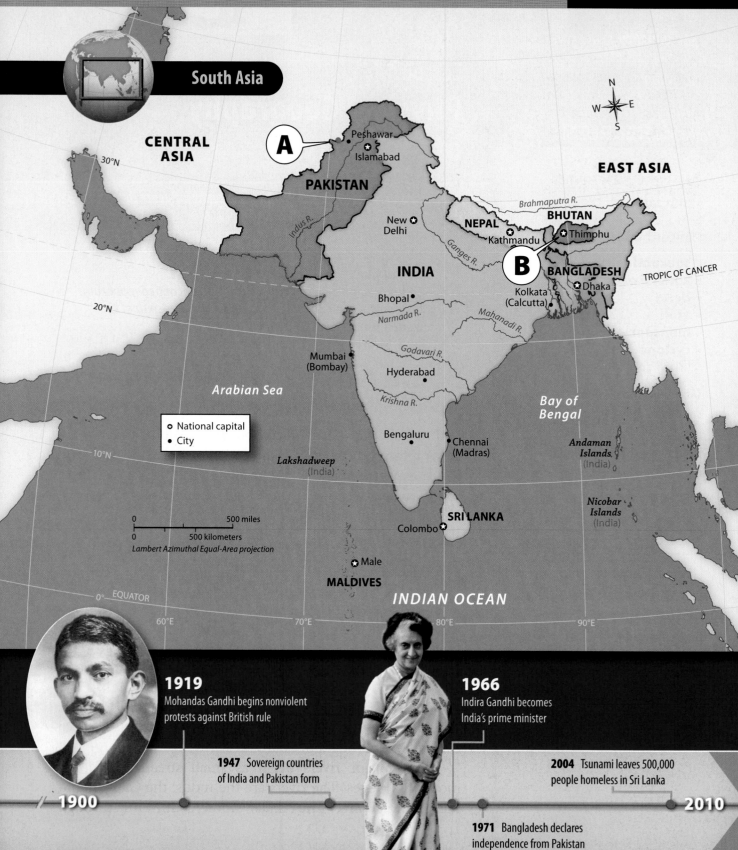

South Asia

CENTRAL ASIA

30°N

A Peshawar

Islamabad

PAKISTAN

EAST ASIA

Brahmaputra R.

New Delhi

NEPAL

BHUTAN

Kathmandu

Thimphu

Indus R.

Ganges R.

B

BANGLADESH

TROPIC OF CANCER

INDIA

Kolkata (Calcutta)

Dhaka

20°N

Bhopal

Narmada R.

Mahanadi R.

Godavari R.

Mumbai (Bombay)

Hyderabad

Arabian Sea

Krishna R.

Bay of Bengal

○ National capital
• City

Bengaluru

Chennai (Madras)

Andaman Islands (India)

10°N

Lakshadweep (India)

Nicobar Islands (India)

0   500 miles

0   500 kilometers

Lambert Azimuthal Equal-Area projection

SRI LANKA

Colombo

Male

MALDIVES

INDIAN OCEAN

0° EQUATOR

60°E

70°E

80°E

90°E

1919
Mohandas Gandhi begins nonviolent protests against British rule

1966
Indira Gandhi becomes India's prime minister

1947 Sovereign countries of India and Pakistan form

2004 Tsunami leaves 500,000 people homeless in Sri Lanka

1900

2010

1971 Bangladesh declares independence from Pakistan

networks

There's More Online!

☑ **CHART/GRAPH** Water Wells

☑ **MAP** Monsoons

☑ **ANIMATION** Rivers in South Asia

☑ **VIDEO**

Reading HELPDESK

Academic Vocabulary

- **annual**

Content Vocabulary

- **subcontinent**
- **alluvial plain**
- **delta**
- **atoll**
- **monsoon**
- **cyclone**

TAKING NOTES: *Key Ideas and Details*

Organize As you read the lesson, identify at least three facts about the physical geography of South Asia and write them in a graphic organizer like the one below.

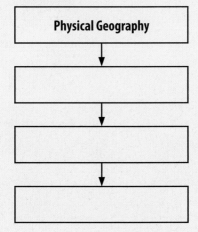

Physical Geography
↓
↓
↓

Lesson 1
Physical Geography of South Asia

ESSENTIAL QUESTION · *How does geography influence the way people live?*

IT MATTERS BECAUSE
South Asia is a land of contrasts, with snowcapped mountains towering over parched deserts. More than one-sixth of the world's people live in this region.

South Asia's Physical Features

GUIDING QUESTION *What physical features make South Asia unique?*

South Asia forms a **subcontinent**. A subcontinent is a geographically or politically unique part of a larger continent. Seven countries make up the region of South Asia. Of these, India is by far the largest. The other six countries are Pakistan, Nepal, Bhutan, Bangladesh, Maldives, and Sri Lanka.

Northern Mountains and Plains

Three mighty mountain ranges form South Asia's northern border. They are the Hindu Kush, the Karakoram, and the Himalaya. The Himalaya range includes the highest mountain in the world: Mount Everest, at 29,028 feet (8,848 m). The mountain ranges form a physical barrier. Invaders and traders could enter South Asia through only a few openings, such as the Khyber Pass between Afghanistan and Pakistan. Plate tectonics created the ranges millions of years ago. Today, the mountains are still rising. Plate movements also cause earthquakes throughout South Asia.

Three major rivers begin as small streams from the mountain ranges. The rivers are the Indus, the Ganges, and the Brahmaputra. The Indus flows southward through

Pakistan to the Arabian Sea. The Ganges and the Brahmaputra flow east and southeast to the Bay of Bengal.

The three rivers cross a vast plains area, where their annual flooding has deposited rich soil for growing crops. One-tenth of the world's people now live in the **alluvial plain** created by the Ganges River. This alluvial plain, or area of fertile soil deposited by floodwaters, is the world's longest.

The Brahmaputra and the Ganges come together and form the largest **delta** on Earth. Deltas are places where rivers deposit soil at the mouth of a river. The Brahmaputra/Ganges delta has some of the world's richest farmland.

Central and Southern Highlands

Mountains and rivers also dominate central and southern parts of South Asia. They physically and culturally separate India into northern and southern parts. Much of southern India is a high, flat area called the Deccan Plateau. Two low mountain ranges, the Western and Eastern Ghats, form the plateau's edges. A narrow coastal plain lies between each mountain range and the seacoast. The soils of the plain are rich and fertile.

Islands of South Asia

Sri Lanka and Maldives are the two island countries of South Asia. Sri Lanka, shaped like a teardrop, lies off the southeastern tip of India. Maldives lies southwest of India's tip and is made up of numerous islands. Many of the islands are small, ring-shaped islands called **atolls**.

☑ **READING PROGRESS CHECK**

Describing Describe the main physical regions of South Asia.

People wash clothes along the banks of the Ganges River in Varanasi, India. The Ganges is important for daily activities, but to Hindus it is also sacred. The city of Varanasi draws religious pilgrims from all over India.

▶ CRITICAL THINKING
Describing How does the Ganges help India's economy?

©Frederic Soltan/Sygma/Corbis

Atoll An atoll is a small, ring-shaped island formed when coral builds up around the edges of an ancient, underwater volcano.

South Asia's Climates

GUIDING QUESTION *How does climate affect people's lives in South Asia?*

Climate is closely related to the physical features of the region. Because the physical features of South Asia are so diverse, the region's climate is diverse, too. About half of South Asia has a tropical climate, with different kinds of plant life. Much of the northern half of the region enjoys a warm temperate climate. You can also find cool highlands in the north and scorching deserts to the west.

Monsoons

Much of South Asia's climate is a result of seasonal wind patterns called **monsoons**. Most of the region has little or no rainfall for eight months of the year. Then, beginning in May and early June, temperatures begin to rise sharply. Heated air causes a change in wind direction. Winds from the Indian Ocean carry moisture inland, bringing heavy rains and flooding. Most areas along the coast get at least 90 inches (229 cm) of rain per year. Millions of family farms depend on rain for survival.

The annual monsoon rains support South Asia's large population. Without the rains, the region could not grow enough food for its people. The floods come at a cost, though—they damage property and can cause loss of life.

Other natural hazards include tropical **cyclones**. These large, swirling storms often slam into the coast along the Bay of Bengal. Their violent winds and heavy rains can cause devastation. The strong winds push water from the Bay of Bengal to the shore, flooding low-lying areas far inland. One cyclone can kill tens of thousands of people. The delta lands of the Brahmaputra and Ganges rivers are especially vulnerable to flooding.

Tropical and Dry Areas

Much of South Asia has a tropical wet/dry climate with just three seasons—hot, wet, and cool. The hot and cool seasons are dry. The three seasons are a result of the monsoon wind patterns.

Tropical wet climates are found along the western coast of India, southern Sri Lanka, and the Ganges Delta in Bangladesh. These areas get plenty of rain year-round and have thick, green vegetation.

Not all of South Asia gets drenched by seasonal monsoons. Some places are dry. Parts of the Deccan Plateau, for example, get little rain, because the Western Ghats block the winds and rains of the wet-season monsoons.

Northwestern South Asia has the region's driest climate. The Thar Desert straddles the border between Pakistan and India. The area gets relatively little rain; **annual** rainfall is less than 20 inches (51 cm). The vegetation is mostly low, thorny trees and parched grasses.

Academic Vocabulary

annual yearly or each year

©Westend61 GmbH/Alamy

Highland and Temperate Climates

The tops of the huge mountain ranges on South Asia's northern border are covered year-round in snow. The mountains affect the climate of lower-lying areas. In winter, the Himalaya block the cold winds sweeping down from Central Asia. This forms a large temperate zone that stretches across Nepal, Bhutan, northern Bangladesh, and northeastern India. Farther south, the elevation of the Deccan Plateau combined with the wind-blocking effect of the Western and Eastern Ghats creates another temperate climate area.

☑ **READING PROGRESS CHECK**

Analyzing What positive and negative effects do the monsoons have on the lives of people in South Asia?

South Asia's Natural Resources

GUIDING QUESTION *Which natural resources are most important to South Asia's large population?*

South Asia has many natural resources, but they are not evenly distributed. As South Asia's largest country, India has the most productive land, as well as water and mineral resources.

MAP SKILLS

1 **PHYSICAL GEOGRAPHY**
What time of year brings wet weather to much of South Asia? What time of year brings drier weather? Why?

2 **PLACES AND REGIONS**
Which parts of South Asia receive the least amounts of rainfall during the summer period?

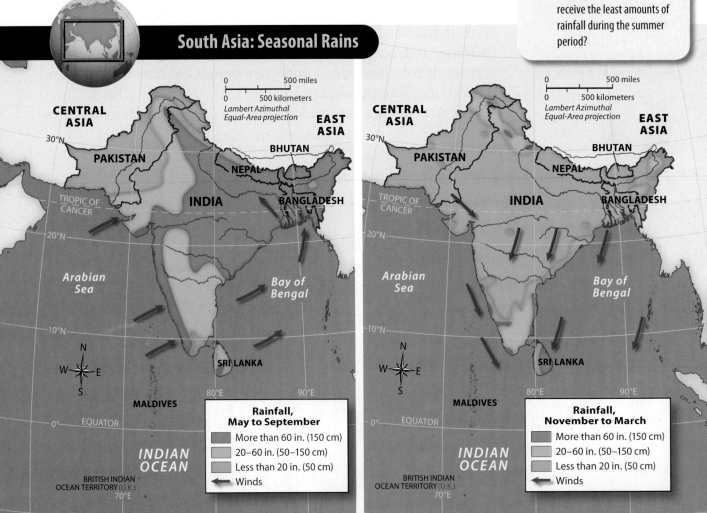

South Asia: Seasonal Rains

Rainfall, May to September
- More than 60 in. (150 cm)
- 20–60 in. (50–150 cm)
- Less than 20 in. (50 cm)
- Winds

Rainfall, November to March
- More than 60 in. (150 cm)
- 20–60 in. (50–150 cm)
- Less than 20 in. (50 cm)
- Winds

Water Resources

South Asians depend on rivers for irrigation, drinking and household water, and transportation. Water in rivers is also considered sacred in Hinduism, the principal religion in India. Hindus revere the Ganges, named for the goddess Ganga.

Today, water is an important source of energy for South Asia. Mountains provide swift-flowing rivers that can be used to generate electricity. Several dams, such as the Narmada River project, are being built, but hundreds more are planned. The Indian government argues that the projects will provide water for drinking, irrigation, and electricity. Opponents point out that areas must be flooded to build dams. This will displace millions of people and destroy ecosystems. They favor smaller-scale projects and traditional ways to manage water needs.

Mineral and Energy Resources

India has most of South Asia's mineral resources. These include iron ore, manganese, and chromite, all used in making steel. India also has large quantities of mica, a rock used to manufacture electrical equipment.

Nepal's natural resources include mica and copper. To the south, Sri Lanka boasts some of the world's finest gemstones, including sapphires and rubies. Sri Lanka also has large quantities of graphite. This is the "lead" that is used in pencils. Graphite is also used in batteries and as a lubricant.

South Asia has several important petroleum reserves. They are located in northern Pakistan and near the Ganges Delta. Exploration in the Arabian Sea has yielded some oil. One offshore oil field was

Villagers in northwestern India try to get water from a large well. When this region is affected by drought, dams, wells, and ponds often go dry.

▶ **CRITICAL THINKING**

Describing What are the physical characteristics of northwestern India?

discovered in the mid-1970s about 100 miles (161 km) west of the Mumbai (Bombay) coast. The field accounted for a large portion of India's domestic oil production. Overall, though, South Asia depends on imported oil. Natural gas fields are found in southern Pakistan and in Bangladesh. India also has an important deposit of uranium north of the Eastern Ghats. The uranium is used in the country's nuclear power plants.

Forests and Wildlife

Like rivers, forests have greatly influenced the history of South Asia. In colonial times, when the British ruled much of the subcontinent, forests were admired for their beauty but exploited for their commercial value. Important timber resources then included teak, sal, and sandalwood. The woods are still valuable today. There is much debate about how they should be used or whether they should be conserved.

Each kind of wood has qualities that make it valuable. Teak is a strong, attractive wood used to make high-quality furniture. Sal is a hardwood used for construction. Sandalwood, with its sweet scent, is often used to make decorative objects.

Forests are, however, more than resources for wood products. They take in carbon dioxide—a greenhouse gas—and release oxygen. Tree roots hold soil in place, reducing erosion. People live in the forests and depend on leaves and fruits for food. Forests also provide habitat for much of South Asia's unique wildlife. Indian forests, for example, are home to three of Earth's most endangered mammals: the tiger, the Asian elephant, and the one-horned rhinoceros. South Asians are working to reverse some of the region's wildlife losses. The creation of wildlife reserves and laws controlling hunting and logging have started to make a difference.

✔ **READING PROGRESS CHECK**

Analyzing Think about how people use rivers in South Asia. How is it similar to how rivers are used in other parts of the world?

A porter in Nepal uses a headband to carry a heavy load in the snowy, high altitudes of the Himalaya.

▶ **CRITICAL THINKING**

Explaining Why are Himalaya snows important to people living in the plains areas of South Asia?

Include this lesson's information in your Foldable®.

LESSON 1 REVIEW **CCSS**

Reviewing Vocabulary

1. Why is a *delta* often used as an agricultural area?

Answering the Guiding Questions

2. *Analyzing* Why was the Khyber Pass so important to South Asia for centuries?

3. *Determining Central Ideas* What might happen in South Asia if there were no monsoons?

4. *Analyzing* What might be the consequences of cutting down a forest in South Asia?

5. *Informative/Explanatory Writing* Take notes about the physical features of the countries of South Asia. Use your notes to write about the features. Use descriptive terms to contrast the mountains, deserts, plains, and rivers of the region.

Reading **HELP**DESK ⓒⓒⓢⓢ

Academic Vocabulary

• **policy**

Content Vocabulary

• **caste**
• **reincarnation**
• **Raj**
• **boycott**
• **civil disobedience**
• **nuclear proliferation**

TAKING NOTES: *Integration of Knowledge and Ideas*

Sequence As you read about the history of South Asia, use a time line like this one to order events from the beginning of the Indus Valley civilization to independence.

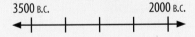

3500 B.C.   2000 B.C.

Lesson 2
History of South Asia

ESSENTIAL QUESTION • *How do governments change?*

IT MATTERS BECAUSE
South Asia is home to one of the oldest known civilizations. You will learn about the birth of a civilization and the struggle for independence by studying its rich history.

Early South Asia

GUIDING QUESTION *How did South Asia's early history lay the foundation for modern life in the region?*

The history of South Asia stretches back thousands of years. The region has experienced much change, yet many social structures and religious beliefs are steeped in tradition.

Early Civilizations

In the early 1920s, archaeologists in South Asia discovered the remains of one of the oldest known civilizations. Because of its location near the Indus River, this culture is known as the Indus Valley civilization. Its origins date back to 3500 B.C. The Indus Valley civilization formed about the same time as other river-valley civilizations around the world.

Two major centers of Indus Valley culture were located at Harappa and Mohenjo-Daro in modern-day Pakistan. These were large cities. Archaeologists have also discovered dozens of smaller settlements.

Written records by the Indus Valley people—mainly strings of symbols—have never been fully deciphered, but archaeological excavations indicate that the civilization was advanced. City streets were laid out in a grid-like pattern. Most houses were made of brick and had their own wells. Houses even had bathrooms and drains. Although most residents were farmers, craftsmanship flourished, as is evident

from toys and other artifacts. Some of the same artifacts show that people traded over long distances.

The Indus Valley culture lasted for about 1,000 years. What brought it to an end? No one knows for sure. A natural disaster such as an earthquake may have occurred. Disease or enemy invasions might have played a part in bringing down the civilization.

Equally mysterious are the beginnings of the Aryans, a group that swept into what is now India about 1500 B.C. They came from the northwest, probably from southern Russia and central Asia. Were they invaders, migrants, or wandering nomads? There are different theories about the Aryans. They likely herded sheep, cows, and other livestock in their homeland. Once in India, they settled down to become farmers.

The Aryan civilization lasted for about 1,000 years. It left behind two important legacies for South Asians. The first legacy was social. Aryans believed that society could be successful only if people followed strict roles and tasks. So, they established a system of *varnas,* or castes. **Castes** were social classes. The top class was made up of Brahmans, or priests. Next came the warriors. Third were the merchants. The bottom class consisted of laborers and peasants.

The caste system had a deep impact on South Asia for thousands of years, causing much inequality. If they were born into a lower caste, people could not move up in society, regardless of their talents. After India won its independence in 1947, the new country outlawed the caste system. Some effects of the system are still present though.

The second major legacy of the Aryans was literary. Recall that scholars have not been able to fully decipher the Indus Valley writing. In contrast, the Aryans composed long poetic texts, called the Vedas, in the ancient Sanskrit language. Sanskrit is the parent language of Hindi, one of the most important languages in modern India. Sanskrit also greatly influenced the development of ancient Greek and Latin.

A worker replaces broken bricks with new ones at the ruins of the ancient city of Mohenjo-Daro in Pakistan.
▶ **CRITICAL THINKING**
Describing What have archaeological digs revealed about the Indus Valley civilization and its large cities?

A steplike Hindu temple (top) and a wall statue of the Buddha (bottom) show the effects of Hinduism and Buddhism on South Asia's architecture and art.

The Vedas were religious hymns handed down orally for many centuries before they were written down. The *Rig Veda* [rihg vay•duh] likely took shape around 1200 B.C. This poem is a series of hymns in honor of Aryan deities. The hymns are full of vivid imagery and philosophical ideas. This poem laid the foundation for the growth of Hinduism.

Religious Traditions

South Asia is the birthplace of several major religions. The first of these is Hinduism. Often described as a way of life, Hinduism has no founder, no holy book, and no central set of core beliefs. Hindus usually pay respect to the Vedas and take part in religious rituals, either at home or in a local temple.

Hindus believe in **reincarnation**, or the rebirth of a soul in another body. Related to this idea is karma—the belief that actions in this life can affect your next life. After many lifetimes, an enlightened soul can be released from the reincarnation cycle of birth, death, and rebirth. Then the soul enters nirvana, a state of eternal bliss.

Around 500 B.C., two new religions arose in South Asia in response to Hinduism and its emphasis on the caste system. The first was Jainism. This religion was based on the Hindu principle of ahimsa, or noninjury. Followers of Jainism turned from farming to commerce so they would not have to kill or injure any living creature.

The second new religion began in northeastern India. It was founded by a noble prince named Siddhārtha Gautama. When he was 29 years old, Siddhārtha gave up his wealthy lifestyle and traveled in poverty, searching for spiritual truth. He reached his goal at the age of 35 and became known as "the Buddha," or "the enlightened one." He passed on to his followers what he believed to be the Four Noble Truths:

- Life is full of suffering.
- The cause of suffering is selfish desire.
- Suffering can be stopped by conquering desire.
- Desire can be conquered by following the Eightfold Path: right view, right intention, right speech, right action, right way of living, right effort, right mindfulness, and right concentration.

Buddhists, like Jains, largely rejected the caste system. Hinduism remains the major religion throughout India today. Buddhism has spread to other Asian countries, while Jainism remains a small but vibrant religion in India.

Three Indian Empires

The Aryan civilization faded by 500 B.C. Around 200 years later, another power arose: the Mauryas. The Mauryas conquered much of South Asia. Their most famous ruler was named Ashoka. A highly successful warrior, he converted to Buddhism around 260 B.C. and adopted a life of nonviolence. His conversion influenced many people throughout the subcontinent. Trade and culture thrived under his rule.

Hundreds of years later, another empire, the Guptas, managed to unify much of northern India. Under the ruler Chandragupta I, science, medicine, mathematics, and the arts flourished. Gupta scholars developed the decimal system in mathematics that we still use today.

Finally, during the 1500s and 1600s, India witnessed the flowering of a third great empire: the Mughals. In contrast to earlier emperors, Mughal rulers were Muslim rather than Hindu. They were the first Indian emperors to be members of a minority religion. During this era, many South Asians converted to Islam.

Some of the Mughals were tolerant. Akbar the Great, who ruled from 1556 to 1605, was a devoted Muslim, but he encouraged freedom of religion. He regularly held discussions with religious scholars. As with the Mauryas and the Guptas, culture, science, and the arts flourished under the Mughals. The architectural monument known as the Taj Mahal was constructed by the fifth Mughal emperor, Shah Jahan, in memory of his beloved wife.

✔ READING PROGRESS CHECK

Identifying What two especially important legacies did the Aryans leave behind?

A reflecting pool leads to the magnificent Taj Mahal in Agra, northern India. Built of white marble, the Taj Mahal is considered the greatest example of Muslim architecture in South Asia.

▶ CRITICAL THINKING

Describing What was the purpose of the Taj Mahal?

Modern South Asia

GUIDING QUESTION *How has conflict in South Asia led to change?*

Since their early history, South Asians have been no strangers to conflict. Then, beginning in the late 1800s, a combination of internal and external factors led to great changes in the region.

The British in South Asia

Beginning in the 1500s, European countries used improved ships and maps in exploration. In the 1600s, British traders established settlements in India. With the decline of the Mughal Empire, the traders became a powerful presence in South Asia. The British were especially interested in textiles, timber, and tea.

After a bloody rebellion in northern India in 1857, the British government took direct control of most of South Asia. They ruled over what is today India, Pakistan, and Bangladesh. The British used the name India to refer to the entire area. By this time, India was a British colony rather than a trading partner. Although the British built railways, schools, and ports, Indians resented a foreign presence in their land. In the late 1800s, an independence movement began.

Achieving Independence

In 1885, less than 30 years after the rebellion of 1857, Indian supporters of independence formed the Indian National Congress. The British, however, were reluctant to give up the **Raj**, as their imperialist rule of India was called. The Congress responded by endorsing a **boycott**. A boycott means refusing to buy or use certain goods—in this case, Indians refused to buy imported British goods.

In the early 1900s, two members of the Congress became leaders. The first was Mohandas K. Gandhi. Often called "Mahatma," which

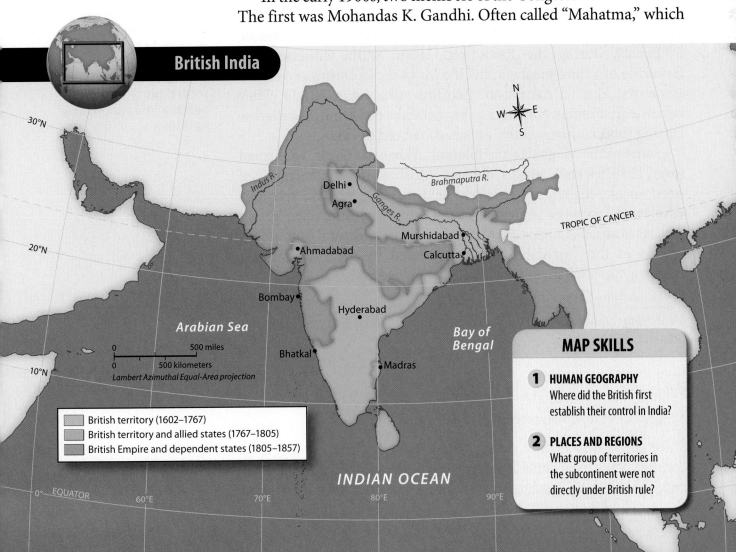

British India

British territory (1602–1767)
British territory and allied states (1767–1805)
British Empire and dependent states (1805–1857)

MAP SKILLS

1 HUMAN GEOGRAPHY
Where did the British first establish their control in India?

2 PLACES AND REGIONS
What group of territories in the subcontinent were not directly under British rule?

means "great soul," Gandhi studied law. He practiced law in South Africa, where the racist **policies** shocked and angered him. He returned to India determined to fight for independence from the British.

Gandhi was deeply opposed to violence. His most powerful weapon against British rule was **civil disobedience**, or nonviolent resistance. He was joined by a younger leader, Jawaharlal Nehru. Also trained as a lawyer, Nehru was the son of one of the original leaders of the Congress. Gandhi and Nehru finally succeeded in persuading the British to leave South Asia and surrender colonial rule. India became independent in August 1947. Conflicts between Hindus and Muslims, however, divided the subcontinent.

India and Pakistan

Since Mughal times, South Asia has been troubled by religious and cultural divisions between Hindus and Muslims. In 1947, as part of the independence settlement, the British negotiated a division of the subcontinent. Two countries were created: India (mainly Hindu) and Pakistan (mainly Muslim). To complicate matters, Pakistan was divided into western and eastern sectors. In the 1970s, East Pakistan achieved independence after a civil war and became the country of Bangladesh.

Tensions between India and Pakistan did not die down. The two countries have fought several wars and are involved in a dispute over the region of Kashmir, located in the Himalaya and Karakoram mountain ranges. In the late 1990s, both countries developed nuclear weapons. Because of this **nuclear proliferation**, or the spread of enormously powerful atomic weapons, conflict between the two countries could prove dangerous.

☑ **READING PROGRESS CHECK**

Describing What countries were created out of the South Asia subcontinent? What religions do the people of these countries follow?

Jawaharlal Nehru (left) and Mohandas K. Gandhi (right) were the main leaders of India's independence movement during the 1930s and 1940s.

▶ **CRITICAL THINKING**
Describing What method did Gandhi and Nehru use to get the British to leave India?

Academic Vocabulary

policy plan or course of action

Include this lesson's information in your Foldable®.

©Bettmann/Corbis

LESSON 2 REVIEW **CCSS**

Reviewing Vocabulary

1. How has the *caste* system influenced life in South Asia?

Answering the Guiding Questions

2. *Describing* How are the religions of Jainism and Buddhism alike? How are they different?

3. *Determining Central Ideas* Why are Hindu-Muslim conflicts in South Asia so significant to the history of the region?

4. *Narrative Writing* You are a time traveler whose machine can transport you to any one of the following: the Indus Valley cities of Harappa or Mohenjo-Daro; the empires of Ashoka or Akbar the Great; or the struggle for independence by the Indian National Congress under the leadership of Mohandas Gandhi. Write a brief story describing what you see and hear in the time and setting of your choice.

net*w***orks**

There's More Online!

☑ **CHART/GRAPH** Population of India

☑ **SLIDE SHOW** Culture Groups in South Asia

☑ **VIDEO**

Reading **HELP**DESK

Academic Vocabulary

- **establish**

Content Vocabulary

- **sitar**
- **green revolution**
- **cottage industry**
- **outsourcing**
- *dalit*

TAKING NOTES: *Key Ideas and Details*

Describe As you read the lesson, create a chart like the one below and write one key fact about each topic.

Urban growth	
Languages	
Religion	

Lesson 3
Life in South Asia

ESSENTIAL QUESTION · *What makes a culture unique?*

IT MATTERS BECAUSE

By 2030, India will become the most populous country on the planet. Among the challenges India faces are raising the standard of living of its people.

People and Places

GUIDING QUESTION *What are the population patterns of South Asia?*

South Asia is about half the size of the lower 48 states of the United States, but more than 1.5 billion people live in the region. That is about five times the size of the American population.

Population Profile

The largest countries in South Asia—in geographical area and in population—are India, Pakistan, and Bangladesh. The estimated population of India is 1.22 billion. This makes it the world's second-most-populous country, after China. India's birth rate, though, is higher than China's. By 2030, India will be the world's most populous country. Today, India alone has about 17.3 percent of all the world's people. The median age of Indians is 25 years old.

Pakistan's population is about 190 million. It is the world's sixth-largest country. The population of Bangladesh is about 160 million. By 2025, that number could grow to 205 million.

Where People Live

South Asians live mainly in areas that are good for farming. This means, for example, that a high percentage of Indians live in the fertile Ganges Plain in the north-central and northeastern parts of the country.

Agriculture plays a major role in India's economy. This is also true of other countries in South Asia. Although India has many large cities, about 7 of every 10 people live in the country's small villages. In contrast, the United States was already more urban than rural as early as 1920. Today, more than 7 of every 10 U.S. residents live in urban areas.

Population trends are changing, though, as more and more Indians leave their villages every year. The 2011 census showed more growth in urban areas than in rural areas—a first for India. People migrate to big cities, called "metros," in hopes of finding better jobs and a higher standard of living. Major cities in India include Mumbai, New Delhi (the capital city), Chennai (Madras), Kolkata (Calcutta), Bengaluru (Bangalore), and Hyderabad.

Census data show that Indian cities are experiencing rapid growth. For example, Mumbai is the business center of the country. The population of Mumbai and its surrounding urban area has grown from a few million around the time of independence to about 20 million. That makes Mumbai India's largest city and the fourth-largest city in the world. Bengaluru, India's third-largest city and a center for technology industries, is growing rapidly, as well.

GRAPH SKILLS >

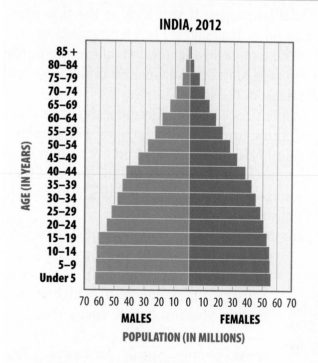

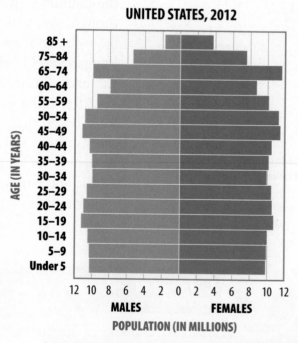

Source: U.S. Census Bureau, Statistical Abstract of the United States: 2012

POPULATION PYRAMIDS

Population pyramids use bar graphs to show a country's population by age and gender. A population pyramid shaped like India's indicates high birthrates and short life expectancy. A population pyramid shaped like the United States's indicates low birthrates and long life expectancy.

▶ **CRITICAL THINKING**

1. **Analyzing** How many males and females in India are between 10 and 14 years of age? How many in the United States?

2. **Analyzing** Compare the *Population in Millions* numbers at the bottom of each pyramid. What do the numbers tell you about the total population of India and the United States?

Population *density* is the number of people who live in a given unit of area. In Mumbai, the population density is 80,100 per square mile (30,900 per sq. km)—seven times the world's average.

The rapid growth of South Asia has put a strain on its resources. Cities struggle to provide essential services to all people, some of whom live in slums surrounding the city centers. Air and water pollution have increased with overcrowded conditions.

READING PROGRESS CHECK

Citing Text Evidence Why is India projected to overtake China as the most populous country by 2030?

People and Cultures

GUIDING QUESTION *How are the diverse cultures in South Asia rooted in ethnic and religious traditions?*

The cultures of South Asia are highly diverse. We think of the United States as a "melting pot" or a "nation of immigrants." The same might be said of South Asia.

Academic Vocabulary

establish to set up

Ethnic and Language Groups

If you look at an Indian currency note, you will see that all the important information appears in Hindi and in English. These are the country's two official languages. If you look at the fine print, though, you will find 15 other languages. Each language is spoken by millions of people. The languages include Punjabi in the north; Bengali and Oriya in the east; Marathi and Gujarati in the west; and Kannada, Tamil, and Telugu in the south.

After India won independence in 1947, boundaries for its states were based mainly on ethnic groups and languages. For example, the southern state of Tamil Nadu was **established** because most of its people spoke Tamil rather than Hindi. Indians are proud of their ethnic and language heritage. In fact, many who speak Tamil would rather talk to you in English than in Hindi.

Most Pakistanis are Muslim. The country's most common languages are Urdu and English. Urdu, like Hindi, developed from ancient Sanskrit. In Bangladesh, which is also mainly Muslim, people speak Bangla, a variation of Indian Bengali. In Sri Lanka, most people speak Sinhalese. This language has its own unique alphabet. An important minority in Sri Lanka speaks Tamil, mainly because Sri Lanka is so close to the southern state of India, called Tamil Nadu.

Indian currency note with portrait of Mohandas K. Gandhi

Identifying What two official languages appear on Indian currency notes?

©Jeremy Horner/Corbis

Religion and the Arts

South Asia's diversity is also apparent in religion and the arts. Six main religions are practiced in the region: Hinduism, Islam, Buddhism, Jainism, Sikhism, and Christianity. You have already learned about the first four religions. Sikhism developed in the 1500s, nearly 2,000 years after Buddhism and Jainism. Yet it, too, was a reaction to Hinduism—by and large, Sikhs reject the Hindu caste system. Sikhs deeply respect their original guru, or teacher. Like Muslims and Christians, they are *monotheists,* meaning they believe in only one God.

A South Asian family enjoys a meal of dishes of their region.

Identifying What foods make up a typical meal in South Asian homes?

Artistic expression is rich in South Asia. Two great epic poems of ancient India, the *Ramayana* and the *Mahabharata,* embody Hindu social and religious values. Their impact extends far beyond the Hindu community, though. South Asian children of all religions can tell you about the plots and characters of these epics. They are part of the region's heritage.

South Asia also is a center for classical dance. Music is a thriving art, and many of the region's musicians, singers, and composers are popular. Ravi Shankar was probably the best-known Indian musician and composer. He performed on the **sitar**, a stringed instrument. Motion pictures first arrived in India in 1896. The country now has the largest film industry in the world. The Hindi film industry, nicknamed "Bollywood," is based in Mumbai. Kolkata is also a filmmaking center.

Daily Life

Life in South Asia centers on family. Several generations of family members often live together. They share household chores and finances. Within the family, age and gender play important roles. Elders are respected, and females are often subordinate to males.

Arranged marriages are still common for many South Asians, but less so than in the past. Parents often introduce couples to each other but allow them to decide whether to marry. When couples meet independently of their parents, it is still important to get approval to marry from their parents.

A typical, South Asian family meal would include rice, legumes, and flatbreads. Curry is a combination of spices that is part of many dishes. South Asians do not eat a great deal of meat, in part because of religious guidelines.

✔ **READING PROGRESS CHECK**

Describing How do South Asian culture and American culture compare?

Thinking Like a
Geographer

What's In a Name?

Why have the names of many Indian cities been changed? National pride sparked Indians to discard names linked to colonial history. For example, the name Bombay came from the Portuguese *bom baia,* meaning "good bay." The city was renamed Mumbai in 1995 in honor of an ancient Hindu goddess.

Customers can apply for visas, check e-mail and Web sites, and receive computer instruction at a cybercafe in Bengaluru, India.

▶ **CRITICAL THINKING**

Describing How has South Asia been affected by the global revolution in communications?

Issues in South Asia

GUIDING QUESTION *What impact do economic and environmental issues have on life in South Asia?*

South Asia is a region of rapid growth. As the population grows, it will be ever more important for South Asians to work together to resolve their differences and jointly address the issues they face.

Earning a Living

One of the biggest challenges for South Asians is making a living. Many people are farmers, but good cropland is scarce. Yet South Asians have managed to grow enough food to feed their huge population. How do they do it?

Agricultural advances known as the **green revolution** have helped increase crop yields. The green revolution involves the use of irrigation, fertilizers, and high-yielding crops. Because of these improvements, India has not had to import food to feed its people since the 1970s. This situation might change, however. Green revolution methods are no longer yielding increases in crop productivity to meet the growing need.

In addition to farming, some South Asians make a living from mining or fishing. Others own or work for **cottage industries**— small businesses that employ people in their homes. Cottage industries include textile weaving, making jewelry and furniture, and wood carving. These small businesses help traditional crafts survive.

Advanced technology is a fast-growing part of the South Asian economy. Entire cities have grown up around the high-tech field. Indian computer specialists, engineers, and software designers are in high demand throughout the world, including the United States.

Another growing part of South Asia's economy is ecotourism. Ecotourism combines recreational travel and environmental awareness.

Making Connections

The efficiency of communication systems varies. Newspapers are thriving, cheap, and widely read, but landline phones are often unreliable. Cell phones have helped ease communication problems.

Because of its vast numbers of English speakers, India is well suited for **outsourcing**. Outsourcing occurs when a company hires an outside company or individual to do work. U.S. companies often outsource to foreign countries where workers may be more flexible or willing to work for lower wages. A U.S. resident who calls customer service with a laptop problem might speak to an agent in India.

©Steve Raymer/Corbis

Meeting Challenges

Despite South Asia's rich resources, the region faces a number of challenges. One challenge is the long-standing conflict between India and Pakistan over the largely Muslim territory of Kashmir. The area containing the headwaters of the Indus River is one source of conflict. Both countries want to control the source of the river. Today, control of Kashmir is divided between the two countries.

Sometimes, political conflicts have resulted in tragedy. Two of India's prime ministers have been assassinated in recent decades: Indira Gandhi (the daughter of Jawaharlal Nehru) in 1984, and her son Rajiv Gandhi in 1991. Both prime ministers were killed by extremists who opposed their views. Another issue in India involves oppression of the **dalits**, or so-called "untouchables." For centuries, these people have been discriminated against as outcasts, barely clinging to the lowest rung on the social ladder.

Other South Asian countries also face challenges. In Sri Lanka, Buddhists and Hindus engaged in a civil war that lasted several decades and cost more than 60,000 lives. In Nepal, a rebel group has been struggling since 1996 to overthrow democracy and establish a communist government. Tribal people control large parts of the Pakistan-Afghanistan border.

Other challenges involve health and environmental issues. Rapid population growth and lack of infrastructure have led to water pollution and air pollution in many parts of South Asia. Deforestation is also a major problem. Most of India's forests have been cut down for timber or development. Water quality, wildlife habitats, and climate have been affected as a result.

South Asia has achieved some success in meeting the challenges. The civil war in Sri Lanka has ended, restoring peace to that island country. India has implemented stricter controls on air and water pollution. Government expenditures for wildlife conservation have increased many times over.

Include this lesson's information in your Foldable®.

 READING PROGRESS CHECK

Analyzing Why might a U.S. company outsource its customer-service operations to a company in South Asia?

LESSON 3 REVIEW (CCSS)

Reviewing Vocabulary

1. Why has the *green revolution* been important for South Asians?

Answering the Guiding Questions

2. *Determining Central Ideas* Why might population growth in South Asia be a major problem in the years ahead?

3. *Determining Word Meanings* Why is South Asia sometimes called a religious and ethnic "melting pot"?

4. *Identifying* Give two examples of environmental issues confronting South Asia today.

5. *Informative/Explanatory Writing* Assume you are a seventh-grader in India, Pakistan, Bangladesh, Sri Lanka, or another South Asian country. Write a journal entry describing your typical day. Include some information about your family, as well as your friends and your favorite hobbies.

Directions: Write your answers on a separate piece of paper.

1 Use your **FOLDABLES** to explore the Essential Questions.

INFORMATIVE/EXPLANATORY WRITING Choose one of the countries that make up South Asia. Compare the physical and population maps of that country found in the front of the chapter. Explain in two or more paragraphs how the geographical features of that country influence where people live.

2 21st Century Skills

ANALYZING Use the Internet and print resources to find three or more facts about the economy of one of the South Asian countries. Present your facts in a chart. Then write at least two paragraphs explaining the effects physical geography has on the country's economy.

3 Thinking Like a Geographer

INTEGRATING VISUAL INFORMATION Use a web graphic organizer like the one shown here to list the six main religions practiced by people in South Asia, along with one fact you think is important to remember about each one.

4 **GEOGRAPHY ACTIVITY**

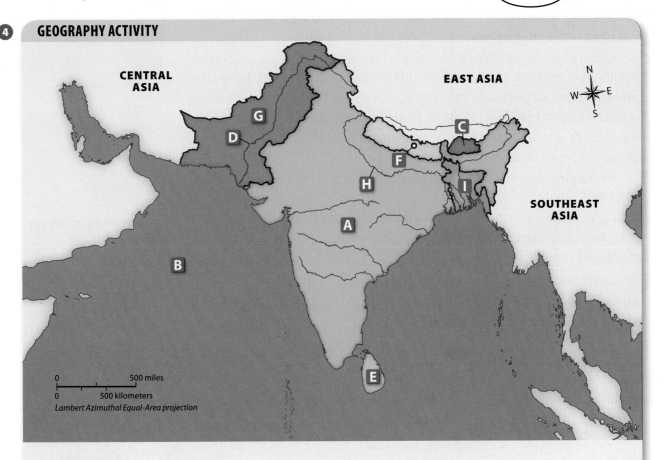

Locating Places

Match the letters on the map with the numbered places listed below.

1. Ganges River   **3.** Sri Lanka   **5.** Kathmandu   **7.** Indus River   **9.** India

2. Bhutan   **4.** Arabian Sea   **6.** Pakistan   **8.** Bangladesh

REVIEW THE GUIDING QUESTIONS

Directions: Choose the best answer for each question.

1 The subcontinent of South Asia is separated from the rest of Asia by the

   A. Bay of Bengal.

   B. Ganges River.

   C. Hindu Kush, the Karakoram Range, and the Himalaya.

   D. Thar Desert.

2 Much of South Asia's weather is directly influenced by wind patterns called

   F. cyclones.

   G. northerlies.

   H. doldrums.

   I. monsoons.

3 Which ancient Indian civilization was responsible for establishing the caste system?

   A. Mauryans

   B. Aryans

   C. Guptas

   D. Mughals

4 In what year did India finally obtain its independence from Great Britain?

   F. 1885

   G. 1918

   H. 1947

   I. 1960

5 Which South Asian nation is expected to become the most heavily populated country on Earth within the next 15 to 20 years?

   A. Pakistan

   B. India

   C. Bangladesh

   D. Sri Lanka

6 Which area has been the source of political conflict and violence between India and Pakistan?

   F. Nepal

   G. Kashmir

   H. Tibet

   I. Bhutan

DBQ ANALYZING DOCUMENTS

7 **ANALYZING** Read the following passage:

"*Because it cannot produce enough electricity to meet demand, [Pakistan's] government shuts off power for extended periods of time. These chronic blackouts, called load-shedding, sometimes last up to 18 hours a day and hamper economic activity, particularly affecting the country's textile industry, and leave people across a wide socio-economic spectrum in sweltering heat.*"

—from Azmat Khan, "You Aren't Hearing About Pakistan's Biggest Problems" (2011)

How is Pakistan's population related to this problem?

A. The large number of young people creates high demand for electricity.

B. The aging population relies on energy-demanding health care.

C. A growing population causes growing demand for electricity.

D. Consumers' demand for electric cars means an increase in energy needs.

8 **IDENTIFYING** Which of the following is an example of an economic problem that can result from a loss of electricity?

F. lack of power for cooking meals

G. people being unable to recharge electronics

H. lack of hot water for cleaning

I. factory machinery that cannot run

SHORT RESPONSE

"*Buddhism, like other faiths of India, believes in a cycle of rebirth. Humans are born many times on earth, each time with the opportunity to perfect themselves further. And it is their own karma—the sum total of deeds [actions in life], good and bad—that determines the circumstances of a future birth.*"

—from Vidya Dehejia, "Buddhism and Buddhist Art"

9 **ANALYZING** How does the Buddhist belief in rebirth fit in the broader culture of India?

10 **CITING TEXT EVIDENCE** In Buddhism, is a person always reborn into better circumstances in a new life? Why or why not?

EXTENDED RESPONSE

11 **INFORMATIVE/EXPLANATORY WRITING** What do you think is the biggest challenge facing South Asia in the twenty-first century? In several paragraphs, explain what you would do to meet that challenge if you were a leader in that region of the world.

Need Extra Help?

If You've Missed Question	**1**	**2**	**3**	**4**	**5**	**6**	**7**	**8**	**9**	**10**	**11**
Review Lesson	1	1	2	2	3	3	3	3	2	2	3

CENTRAL ASIA, THE CAUCASUS, AND SIBERIAN RUSSIA

ESSENTIAL QUESTIONS · *How does geography influence the way people live?*
· *How do governments change?*

©Strunin Anatoly/ITAR-TASS Photo/Corbis

A woman of
northwestern Siberia
in traditional dress

networks

There's More Online about Central Asia, the Caucasus, and Siberian Russia.

CHAPTER 17

The Story Matters...

Central Asia, the Caucasus, and Siberian Russia share a harsh land. Siberia is an enormous and cold land, but it contains vast and economically important oil and natural gas resources. The mountainous terrain, semiarid steppes, and vast deserts of Central Asia are home to many ethnic groups, some with roots in ancient cultures. Throughout history, people living in Central Asia have been affected by conflicts in the region, as well as expansion of the Russian Empire and the Soviet Union.

FOLDABLES®
Study Organizer

Go to the Foldables® library in the back of your book to make a Foldable® that will help you take notes while reading this chapter.

CENTRAL ASIA, THE CAUCASUS, AND SIBERIAN RUSSIA

Siberian Russia, the Central Asian countries, and the Caucasus region make up a large area. The regions were once part of the Soviet Union.

Step Into the Place

MAP FOCUS Use the map to answer the following questions.

1 **PHYSICAL GEOGRAPHY** The Aral Sea extends into which two countries?

2 **THE GEOGRAPHER'S WORLD** Which country borders Kyrgyzstan to the north?

3 **PLACES AND REGIONS** Yerevan is the capital city of which country?

4 **CRITICAL THINKING**
Analyzing In which direction do most Siberian rivers flow? Why do you think this is so?

Central Asia, the Caucasus, and Siberian Russia

Step Into the Time

TIME LINE Choose one event on the time line and describe what effects that event might have on the people of the region.

c. 1500s–1600s Russian trappers and traders settle in Siberia

c. A.D. 900s–1450s Mongol khans rule southern Siberia

c. 1600s Russia expands to Pacific coast

Bruno Morandi/Robert Harding World Imagery/Getty Images

ARCTIC
OCEAN

Bering
Sea

East
Siberian
Sea

Laptev
Sea

ARCTIC CIRCLE

Lena R.

Yakutsk

RUSSIA

Sea of
Okhotsk

Sakhalin

Kuril Islands

Lake
Baikal

Irkutsk

PACIFIC
OCEAN

0   500 miles
0   500 kilometers
Two-Point Equidistant projection

⊙ National capital
● City

ASIA

Vladivostok

Sea of
Japan
(East Sea)

100°E   20°E   140°E   160°E

1891
Construction of
Trans-Siberian
Railroad begins

1991
Soviet Union breaks up into
15 independent countries

1921–1922 Famine and
economic collapse in Russia

1979 Soviet Union invades Afghanistan

1900   **2000** **2010**

1917 Bolshevik Revolution
puts Communists in power

1922 Soviet Union
is formed

1947–1991 Soviet Union,
U.S. engage in Cold War

2008 Russia invades the Republic of Georgia

Reading **HELP**DESK (CCSS)

Academic Vocabulary

- **define**

Content Vocabulary

- **tundra**
- **taiga**
- **permafrost**
- **steppe**
- **deciduous**

TAKING NOTES: *Key Ideas and Details*

Summarize As you read the lesson, list important details for each topic of the lesson on a graphic organizer like the one below.

Topic	Details
Regions	
Landforms and Climates	
Waterways	
Natural Resources	

Lesson 1
Physical Geography of the Regions

ESSENTIAL QUESTION • *How does geography influence the way people live?*

IT MATTERS BECAUSE
Russian Siberia, the Central Asian countries, and the Caucasus make up a large area. The regions cover about one-ninth of the world's total land area.

Regions

GUIDING QUESTION *Which countries make up the regions?*

Siberian Russia, Central Asia, and the Caucasus are separate regions. They share certain characteristics, such as rugged terrain, harsh climate, and sparse population.

Siberian Russia

The eastern part of Russia is known as Siberia. It stretches from the Ural Mountains in the west to the Pacific Ocean in the east. Siberia is 25 percent larger than Canada, the world's second-largest country. The region is so vast that people living on Siberia's Pacific coast are farther from Moscow, Russia's capital, than they are from Maine in the United States.

Central Asia

Central Asia is made up of five countries: Kazakhstan, Turkmenistan, Uzbekistan, Kyrgyzstan, and Tajikistan. Kazakhstan is the northernmost and largest country in Central Asia and the ninth-largest country in the world. It is also the most sparsely settled country in Central Asia.

Kazakhstan is bordered by China on the east, the Caspian Sea on the west, and Russia to its north. The Central Asian countries of Turkmenistan, Uzbekistan, and Kyrgyzstan lie along Kazakhstan's southern border.

Turkmenistan and Uzbekistan are about the same size. Both are about twice the size of Minnesota. Turkmenistan has the smallest population in Central Asia, and Uzbekistan has the largest.

Kyrgyzstan and Tajikistan form Uzbekistan's eastern border. Both are small countries with small populations. Kyrgyzstan is about the size of South Dakota, and the smaller Tajikistan is about the size of Iowa.

Turkmenistan, Kyrgyzstan, and Tajikistan are Central Asia's southernmost countries. South of the countries lie China, Pakistan, Afghanistan, and Iran.

The Caucasus Region

Between the Caspian Sea and the Black Sea and south of the Caucasus Mountains is a region called the Caucasus. Its three small countries—Georgia, Azerbaijan, and Armenia—together total about the size of North Dakota. To their north lies Russia. Turkey and Iran border the Caucasus region to the south.

✅ **READING PROGRESS CHECK**

Analyzing Why is Kazakhstan, which has a larger population than Turkmenistan, more sparsely settled than Turkmenistan?

Landforms and Climates

GUIDING QUESTION *What are the major landforms and climates of Siberia, Central Asia, and the Caucasus region?*

The variety of landforms and climates that characterize the Caucasus, Central Asia, and Siberia make for one of the most geographically interesting regions of the world.

Mountains

The Caucasus Mountains **define** the Caucasus region. These mountains generally mark the border between Europe and Asia. Many of the mountains in this range are volcanoes. Although they have not erupted in thousands of years, the forces that formed them make the region prone to frequent earthquakes.

To the east, two high mountain ranges lie along Central Asia's border with China. The rugged Tian Shan cover most of Kyrgyzstan and extend into eastern Kazakhstan and Uzbekistan. On their southwestern edge is a smaller range called the Pamirs. They cover most of Tajikistan. Like the Caucasus, the Tian Shan and the Pamirs are an earthquake zone.

Academic Vocabulary

define to describe the nature or extent of something

Musicians perform in a mountain village in the Caucasus of southern Russia.
Identifying Where is the Caucasus located?

©Irina Popova/Corbis

The Ural Mountains, which run north and south through Russia, are yet another border range. These heavily forested mountains are much lower in elevation than the other ranges. Geographers consider the Ural Mountains to be a boundary between Europe and Asia. The Urals also mark the western border of Russia's Siberia region. The eastern third of Siberia is another mountainous region that extends south along Russia's border with Mongolia.

Plains and Deserts

Plains and deserts also characterize the regions. In Siberia, the west Siberian plain extends from the Ural Mountains east to the Yenisey River. Covering an area of almost 1 million square miles (2.6 million sq. km), the west Siberian plain is one of the world's largest and flattest plains. Its lowland areas are poorly drained, with many swamps and marshes. East of the Yenisey River, the land rises to form the central Siberian plateau, a rugged region of hills cut by deep river gorges.

Siberia's geography changes from north to south as well as from west to east. Extreme northern Siberia is mostly **tundra**—a treeless zone found near the Arctic Circle or at high mountain elevations. Northern Siberia is a harsh environment of bare, rocky ground with patches of small shrubs, mosses, and lichens.

South of the tundra lies a vast area of **taiga**—a zone of coniferous forest. Siberia's taiga is swampy, because, like the tundra, the region is covered in **permafrost**, a layer of permanently frozen ground that lies beneath the surface soil and rocks. Permafrost covers about two-thirds of Siberia. Southern portions of the western plains and central plateau contain dry grasslands called **steppe;** the steppe extends into Kazakhstan in Central Asia.

Central Asia is made up of lowland mountains and dune-covered deserts. Most of Kazakhstan consists of dry grassland plains and plateaus, with lowlands on its coast along the Caspian Sea. In eastern Kazakhstan, the Kyzyl Kum desert stretches south into Uzbekistan. Another desert, the Kara–Kum, extends over most of nearby Turkmenistan. Together, they form a harsh region, nearly the size of Texas, consisting of sand ridges, scattered grasses, and desert plants.

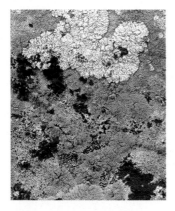

Visual Vocabulary

Lichen a plantlike organism that consists of algae and fungus

Nomadic people live in Siberia and Central Asia. A family, including reindeer and a dog, gather outside their home (left). A shepherd and his flock (right) cross the steppe near a Central Asian mountain range in Kazakhstan.

▶ CRITICAL THINKING

Describing What landscapes are found in much of Central Asia?

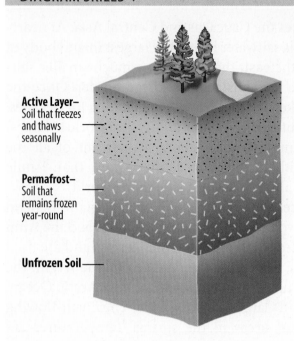

Active Layer– Soil that freezes and thaws seasonally

Permafrost– Soil that remains frozen year-round

Unfrozen Soil

BELOW GROUND IN SIBERIA

This cutaway diagram shows how permafrost penetrates below ground in Siberia.

▶ **CRITICAL THINKING**

1. Identifying What is the active layer?

2. Describing How might climate change affect permafrost?

A Variety of Climates

The regions have many climates, ranging from arctic climates in Siberia to desert climates in Central Asia. Dry conditions prevail across much of Central Asia, characterizing what is called an arid climate. The steppe of southwest Siberia and northern Kazakhstan receives a little more rain and has a semiarid climate. Moving north and east, summers become shorter and cooler, and the winters become colder. Summer in the Siberian tundra, for example, lasts only two months. Temperatures rarely rise above 50°F (10°C). Summers in the mountain valleys of eastern Siberia are longer and milder. Eastern and central Siberia have some of the coldest winters on Earth. There, the temperature has reached as low as –96°F (–71°C). The west Siberian plain experiences heavy snows, but elsewhere snowfall is light.

The mountain areas of Central Asia have a humid continental climate. Temperatures and precipitation in these areas vary according to location and elevation. Mountain valleys have hot, dry summers and cold winters. Rainfall varies from 4 inches to 20 inches (10 cm to 51 cm) per year. Mountain foothills are cooler and get more rain. The high elevations are even colder, with more rain and snow.

The Caucasus Mountains give Georgia, Azerbaijan, and Armenia a climate much like the mountains of eastern Central Asia. The presence, however, of two large bodies of water—the Caspian Sea and the Black Sea—makes the region's summers cooler and winters warmer. The Black Sea gives Georgia's coastal lowlands a humid subtropical climate, with up to 100 inches (254 cm) of rainfall in a year. Farther east, in the mountains and mountain valleys, the climate is much drier.

Waterways

The Caspian Sea separates the Caucasus and Central Asia. At nearly the size of California, this saltwater lake is the largest inland body of water in the world. To the east, the Aral Sea, a much smaller saltwater lake, straddles the Kazakhstan-Uzbekistan border. Once the world's fourth-largest lake, the Aral has been shrinking for decades.

Lake Baikal, in southeastern Siberia, is the world's largest freshwater lake by volume. It holds about 20 percent of all the freshwater on Earth. With a maximum depth of more than 1 mile (0.6 km), Lake Baikal is also the deepest lake in the world.

The dry climate forces many people in Central Asia to depend heavily on the region's two major rivers, the Syr Dar'ya and the Amu Dar'ya. The rivers flow from mountains across deserts and are used for irrigation. In Siberia, four great rivers—the Ob', the Irtysh, the Yenisey, and the Lena—flow north to empty into the Arctic Ocean. They are among the world's largest river systems. The north-flowing Siberian rivers flood vast areas in the spring. Temperatures are warmer where the rivers begin in the south than at their mouths in the north. Ice in the north blocks the rivers from emptying into the Arctic Ocean, resulting in floods.

Unlike other main Siberian rivers, the Amur River drains eastward. Some of it forms the border between Russia and China. Affected by summer winds, the Amur river valley is warmer than the rest of Siberia. It is Siberia's main food-producing area.

☑ **READING PROGRESS CHECK**

Citing Text Evidence What type of terrain is common in Siberia, Central Asia, and the Caucasus region?

Natural Resources

GUIDING QUESTION *Which natural resources of Siberia, Central Asia, and the Caucasus are economically important?*

Siberia holds some of the greatest wealth in natural resources on Earth. The Caucasus is also rich in natural resources. Central Asia is rich in natural resources too, but the area has little water.

Vast Forests

Siberia's vast taiga contains about 20 percent of all the world's trees. The economic value of this resource is limited, however. One reason is that a lack of roads makes logging

The nuclear-powered ship *Rossiya* breaks through Arctic Sea ice to open passageways for other ships. Nuclear-powered icebreakers like *Rossiya* can travel only in cold water because their reactors need to be kept cool.

▶ **CRITICAL THINKING**

Describing How are Siberia's rivers affected by the ice in the far north?

POCCHA

Patrick Landmann/Photo Researchers

in the region difficult. Another is that most of the taiga's trees are larch, one of the few species that will grow on permafrost. Because the quality of larch wood is poor, it is not in demand. Central Asia has few trees because of the arid climate. Parts of the Caucasus are forested. Oak and other **deciduous** trees are found in some lowlands and at lower mountain elevations. Higher in the mountains, pine, fir, and other coniferous trees grow.

Energy and Mineral Resources

All three regions are important producers of oil and natural gas. Fields in the tundra and taiga of Siberia's Ob' River basin make Russia a major provider of these fuels. The Central Siberian Plateau supplies most of Russia's coal. Other huge coal deposits exist in the Lena River valley in southeast Siberia. These areas are so remote, however, that most of their resources remain untapped. Eastern Siberia also holds most of Russia's gold, lead, and iron ore. The tundra region near the mouth of the Yenisey River is one of the world's leading producers of nickel and platinum.

Important oil and gas resources are found in Kazakhstan, Uzbekistan, and Turkmenistan in Central Asia. Large coal deposits are present in Kazakhstan, Tajikistan, and Uzbekistan. Kazakhstan is a major producer of uranium. The mountains of Tajikistan and Kyrgyzstan contain rich mineral resources, as do the eastern mountain areas of Kazakhstan and Uzbekistan. Gold, mercury, copper, iron, tin, lead, zinc, and other metals are mined there.

Most of the Central Asian countries have harnessed their rivers, especially in mountain areas, to produce electricity. The same is true of the Caucasus region, which does not have the rich energy resources of Siberia or Central Asia. Only Azerbaijan is a major oil and gas producer, mainly from fields near or in the Caspian Sea.

✔ **READING PROGRESS CHECK**

Identifying Which three resources are economically important in all three regions?

A worker welds a pipeline at the Urengoy gas field in northwestern Siberia. The Urengoy is one of the largest gas fields in the world. Its natural gas is sent by pipeline to customers as far away as Western Europe.

Identifying What minerals other than natural gas are found in Siberia?

FOLDABLES®
Study Organizer

Include this lesson's information in your Foldable®.

Dmitry Beliakov/Bloomberg/Getty Images

LESSON 1 REVIEW **CCSS**

Reviewing Vocabulary

1. How does the *permafrost* affect the geography of Siberian Russia?

Answering the Guiding Questions

2. *Identifying* What countries make up the Caucasus region and Central Asia?

3. *Identifying* Where are Siberia's lowlands and mountains?

4. *Identifying* What geographic feature separates the Caucasus and Central Asia?

5. *Analyzing* Why are so many of Siberia's vast resources undeveloped?

6. *Informative/Explanatory Writing* Write a paragraph to answer the question: How do you think climate affects the way people of the regions make a living?

Academic Vocabulary

- inhibit
- complex

Content Vocabulary

- pastoral
- collective
- irrigate

TAKING NOTES: *Key Ideas and Details*

Categorize As you read about the histories of Siberia, Central Asia, and the Caucasus, use a graphic organizer like this one to keep track of the invaders who conquered each region.

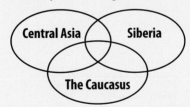

Central Asia — Siberia — The Caucasus

Lesson 2
History of the Regions

ESSENTIAL QUESTION · *How do governments change?*

IT MATTERS BECAUSE
Under communism, the Soviet Union was the main force in the regions. Today, the republics of Russia, the Caucasus, and Central Asia are running their own affairs.

Siberia

GUIDING QUESTION *How did the settlement of Siberia progress over time?*

Scientists are uncertain exactly when the first humans arrived in Siberia and whether they came from Europe or from central and eastern Asia. What is more certain is that thousands of years ago, some of them crossed from eastern Siberia to what is now Alaska to become the first people in the Americas.

Settlement, Invasion, and Conquest
After about 1000 B.C., Turkic, Iranian, Mongol, and Chinese people began migrating into southern Siberia from central and eastern Asia. These early people were nomads who lived in small groups. Some were hunter-gatherers. Others were **pastoral**; their lives were based on herding animals.

In the 200s B.C., invaders from Manchuria in northeast China drove many groups north onto the central Siberian plateau. The invaders from Manchuria were the first of several peoples to conquer parts of Siberia and add it to their empire. Other invaders included the Huns, Mongols, and Tartars. The Tartars were defeated by the Russians, who had been coming into Siberia to trade for furs.

Russian Siberia
As they pushed east, Russian traders built small forts and trading posts. Some of these sites eventually became towns.

Some Siberian groups welcomed the Russians. Groups that resisted eventually were subdued. By 1700, Russian control extended to the Pacific Ocean, and some 230,000 Russians were living in Siberia. The czars who were the rulers of Russia also used Siberia as a place of exile for political prisoners.

Lack of roads and other transportation links **inhibited** Siberia's development until the Trans-Siberian Railroad was built across the region between 1891 and 1905. To encourage settlement, the czar began offering settlers free land. By 1914, more than 3 million people had settled in Siberia.

The railroad allowed easier export of products from the region. Coal mines were opened in several locations. Modern farming methods were introduced, and the farmers began producing dairy products and large amounts of grain.

Revolution and Development

When the czar was overthrown and the Communists took control in 1917, some Siberian leaders resisted the new government. In 1922 the Communists brought Siberia under control. Siberia became part of the Union of Soviet Socialist Republics (also called the Soviet Union and USSR)—the country they formed from the Russian Empire.

©John Warburton-Lee Photography/Alamy

Academic Vocabulary

inhibit to restrict or prevent a process or an action

A richly decorated train station stands in the city of Irkutsk, a major stopping point on the Trans-Siberian Railroad.

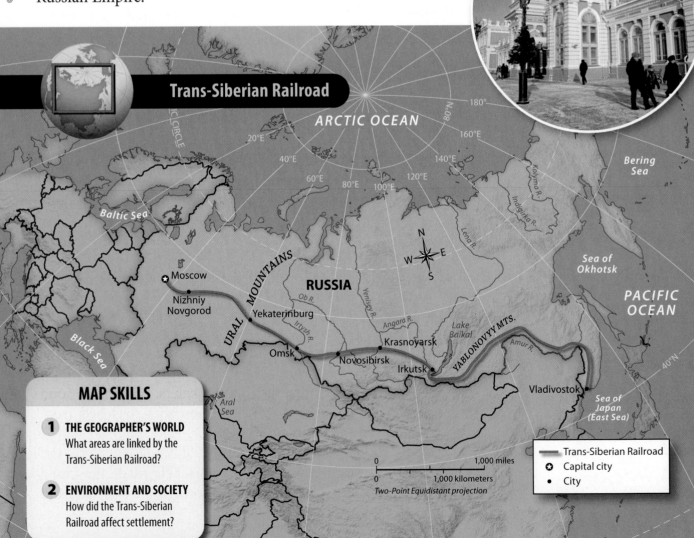

Trans-Siberian Railroad

ARCTIC OCEAN

Bering Sea

Baltic Sea

RUSSIA

Moscow

Nizhniy Novgorod

URAL MOUNTAINS

Yekaterinburg

Ob R.

Irtysh R.

Omsk

Novosibirsk

Krasnoyarsk

Angara R.

Yenisey R.

Lake Baikal

YABLONOVYY MTS.

Irkutsk

Amur R.

Lena R.

Kolyma R.

Indigirka R.

Sea of Okhotsk

PACIFIC OCEAN

Vladivostok

Sea of Japan (East Sea)

Black Sea

Aral Sea

Sea of Okhotsk

0   1,000 miles
0   1,000 kilometers
Two-Point Equidistant projection

— Trans-Siberian Railroad
✪ Capital city
• City

MAP SKILLS

1 THE GEOGRAPHER'S WORLD
What areas are linked by the Trans-Siberian Railroad?

2 ENVIRONMENT AND SOCIETY
How did the Trans-Siberian Railroad affect settlement?

Soviet leaders increased mining and industry in the region, especially along the Trans-Siberian Railroad. Much of this expansion came through the use of forced labor. Labor camps called gulags spread across Siberia in the 1930s. Millions of people who disagreed with Communist government policies were imprisoned in gulags. The Communists also combined the lands of small farmers into large, government-run farms called **collectives**. Farmers who resisted giving up their land were sent to the gulags.

When Germany invaded Russia during World War II, Soviet leaders relocated some factories to Siberia to keep them from being captured by German forces. Siberian industry continued to grow after the war. In the late 1990s, oil production increased but has not reached the levels of the late 1980s. Yet oil prices have risen, giving the Russian government a steady source of income. China's increased need for oil has led to calls to develop eastern Siberia's untapped reserves.

☑ **READING PROGRESS CHECK**

Determining Central Ideas How did control of Siberia change over time?

Central Asia

GUIDING QUESTION *How have invaders affected Central Asia's history and culture?*

For most of their history, the countries of Central Asia have been part of outside empires. In the 300s B.C., the region was the eastern border of the Greek empire created by Alexander the Great. In the centuries that followed, the Chinese, Huns, and other groups took turns ruling Central Asia. For many conquerors, the greatest prize was control of the Silk Road, the network of trade routes that linked China and Europe. Nomad groups that could not defeat the invaders often retreated into the region's deserts and dry plains, where they continued to live in freedom.

The Influence of Islam

The Arab conquest of Central Asia in the early A.D. 700s brought the religion of Islam to the region, where it remains the main religion today. The Arabs were soon replaced by the Persians, who ruled until the Mongols conquered the region in the early 1200s. In the mid-1300s, a Central Asian conqueror, Timur, overthrew the Mongol rulers. His armies soon conquered a huge region that stretched from the Caucasus to India.

The Silk Road city of Samarqand was the capital of Timur's empire. He made the city a center of culture by bringing in artists and

The city of Samarqand, in present-day Uzbekistan, was Central Asia's major political and cultural center during the 1300s and 1400s. Many buildings from this period, such as this mosque and burial site, still stand today.

▶ **CRITICAL THINKING**

Describing Why did Samarqand become an important center of the Islamic world during the 1300s and 1400s?

MARTIN GRAY/National Geographic Stock

scholars from other parts of the empire. His successors continued his support of the arts and sciences by establishing madrassas, Islamic centers of learning. Samarqand became a major center for the study of astronomy and mathematics.

Russian Rule and After

Czar Peter the Great began expanding the Russian Empire into Central Asia in the 1700s. By the mid-1800s, most of the region was under Russian control. The Russians' main interest was to grow cotton in Kazakhstan and Uzbekistan. To increase production, they began to **irrigate** the land. They also expanded the railroads to link the region with Russia.

When the Russian Revolution began in 1917, parts of Central Asia took advantage of the unrest to declare their independence. By 1920, Communist forces from Russia had regained control. Under a treaty of union, five of the Central Asian countries became part of the Soviet Union.

Soviet rule brought great change. The Soviets built dams on rivers to generate electric power and collect water for irrigation. They built one of the world's longest canals, the Kara-Kum Canal, to carry water from the Amu Dar'ya River more than 500 miles (805 km) to irrigate desert land near the Caspian Sea. They also increased industry by opening mines and factories to develop the region's mineral wealth.

The changes also had negative impacts. Soviet leaders resettled large numbers of Russians and other Europeans in the region. Local farmers and nomadic herders were forced onto collectives. The Soviets tried to uproot local cultures. They closed mosques to end the practice of Islam in the region.

The Sayano-Shushenskaya Dam, in south central Siberia, was built during the Soviet era. For many years, it was Russia's largest hydroelectric project. In 2009 an accident at the dam's power plant caused flooding that killed 78 people and destroyed much property.

▶ **CRITICAL THINKING**

Describing What economic changes did Soviet rule bring to Central Asia?

When the Soviet Union collapsed in 1991, the republics declared their independence. The region's present-day countries were formed. Most of the countries are still struggling with problems that arose after Soviet rule. Among the problems are modernizing their economies and dealing with ethnic tensions and dictatorial rulers.

✔ **READING PROGRESS CHECK**

Analyzing Why was Central Asia difficult to conquer and control completely?

After independence, Georgia ended government economic controls and allowed free enterprise. As a result, many inefficient, Soviet-built factories were abandoned and left to decay.

▶ **CRITICAL THINKING**
Determining Central Ideas Why did the Soviet economic system fail?

Academic Vocabulary

complex having interrelated parts that are difficult to understand or separate

The Caucasus

GUIDING QUESTION *How did the countries of Georgia, Armenia, and Azerbaijan develop?*

The Caucasus are a crossroads between Europe and Asia. This region has drawn conquerors since ancient times. It was part of the ancient Persian, Greek, and Roman empires. Later, Persians, Turks, and Russians dominated the region. This history of conquests and migrations makes the Caucasus one of the most ethnically **complex** places in the world today.

Early History

Georgia and Armenia have each had periods of self-rule during their long histories. Both existed briefly as powerful kingdoms in ancient times but came under the control of powerful neighbors. Around the year A.D. 300, each threw off Persian rule and established kingdoms again. They became two of the earliest countries to convert to the Christian religion. Beginning in the A.D. 600s, Muslim conquerors brought Islam to the Caucasus.

Later, much of the region was conquered by the Mongols in the 1200s and became part of Timur's Central Asian empire in the late 1300s. The Byzantines—Christians who controlled the eastern part of the old Roman Empire—and Persians also sought control of the region. By the late 1400s, the struggle had shifted to the Persians and a new power in Southwest Asia—the Ottoman Turks. The Ottomans and Persians competed for the Caucasus for the next 300 years.

Russian and Soviet Rule

The Christian Armenians and Georgians suffered under Persian and Turkish control. In the early 1700s, they turned to Russia for protection from their Muslim rulers. By the early 1800s, Russia had

annexed most of the region. Only part of Armenia remained in the Ottoman Empire.

During World War I, hundreds of thousands of Armenians died at the hands of Ottoman troops, and Russia took the rest of Armenia. Then, the Russian Revolution gave the entire Caucasus the chance to break free of Russian control. Georgia, Armenia, and Azerbaijan briefly formed independent states. Following the Communist victory in Russia's civil war, the Soviet Union annexed the Caucasus in 1922 and eventually created three Soviet socialist republics from the region—the present-day countries of Georgia, Armenia, and Azerbaijan.

The Caucasus benefited but also suffered under Soviet rule. Communists transformed the Caucasus from a largely agricultural area to an urban and industrial one. Soviet power finally ended centuries of invasion and instability, but crushed all opposition. Then during World War II, Germany invaded the Caucasus.

Some of the region's ethnic groups were accused of helping the German invaders. After the war, these groups were broken apart and resettled in various other republics of the USSR. The Soviets also punished other Caucasus people for showing pride in and loyalty to their ethnic identity and culture. Persecution eased after the death of the brutal Soviet dictator Joseph Stalin in 1953. Caucasus leaders took steps toward freedom when the Soviet Union began falling apart in the late 1980s. Non-Communists were already in power in Armenia and Georgia when the Soviet Union dissolved in 1991.

Like the Central Asian republics, Georgia, Armenia, and Azerbaijan declared their independence in 1991. As in Central Asia, though, their 70-some years under Soviet control continue to affect these countries. Since independence, these countries have struggled with economic changes, ethnic tensions, and border conflicts.

✓ **READING PROGRESS CHECK**

Citing Text Evidence Which Caucasus country did not have a history of independence before the end of Soviet rule?

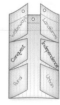

Include this lesson's information in your Foldable®.

LESSON 2 REVIEW (CCSS)

Reviewing Vocabulary

1. How did Soviet *collectives* change agriculture in Siberia and Central Asia?

Answering the Guiding Questions

2. *Describing* What factors contributed to Siberia's economic growth and development?

3. *Analyzing* How might Central Asia's culture and history have been different if Arabs had not been among the region's many conquerors?

4. *Determining Central Ideas* Why are the countries of the Caucasus so culturally complex?

5. *Argument Writing* Do you believe that the Soviet Union's persecution of groups for their ethnic loyalty and pride was a good policy? Why or why not?

networks

There's More Online!

- ☑ **IMAGES** Living in a Yurt
- ☑ **MAP** Understanding Time Zones
- ☑ **VIDEO**

Reading **HELP**DESK (CCSS)

Academic Vocabulary

- **significant**

Content Vocabulary

- **oasis**
- **homogeneous**
- **yurt**

TAKING NOTES: *Key Ideas and Details*

Summarize As you read the lesson, list important details for each region using a graphic organizer like the one below.

Siberian Russia	1. 2.
The Caucasus	1. 2.
Central Asia	1. 2.

Lesson 3
Life in the Regions

ESSENTIAL QUESTION · *How does geography influence the way people live?*

IT MATTERS BECAUSE
To live in these regions, people must adjust to a harsh and mountainous landscape.

People and Places

GUIDING QUESTION *Who are the peoples of Siberia, Central Asia, and the Caucasus, and where do they live?*

Siberia and most of Central Asia share much the same settlement pattern—the regions have a few large cities and sparsely settled remote areas. The Caucasus is a more urban region, and its rural areas are more densely populated.

Siberian Russia

More than half of Siberians live in the hundreds of cities and towns in the region. Most are alongside or south of the route of the Trans-Siberian Railroad, especially on the west Siberian plain. Novosibirsk, a city of nearly 1.5 million, is located there. It is Siberia's largest city and the third largest in Russia.

A few small cities and many towns also exist along the Ob', the Yenisey, and the Lena rivers and their tributaries, as well as along Siberia's eastern coast. Vladivostok, a city of 500,000 on the Sea of Japan in extreme southeastern Siberia, is located at the eastern end of the Trans-Siberian Railroad. It is a major port for shipping goods into and out of all of Russia.

The resettlement programs of the czars and Soviets have made Siberia's population overwhelmingly Russian. There are still populations of Mongol and Turkic groups, as well as other indigenous peoples. Most people have settled on farms or in cities and towns. A few in the north still live as pastoral nomads.

Central Asia

Kazakhstan is the only country in Central Asia with a largely urban population. About two-thirds of the people are ethnic Kazakhs. Another 25 percent are Russians. Most of the Russians and many Kazakhs live in Alma-Ata. With 1.4 million people, it is the country's largest city and industrial center. About 40 percent of Kazakhstan's people live in rural areas. Most live in the small towns and farm villages that are scattered across the lowlands and plateaus of the steppe.

Turkmenistan is about evenly divided between people who live in cities and large towns and those who live in rural settlements. The greatest number of people live along the Amu Dar'ya and in **oasis** areas in the south. An oasis is a green area by a water source in a dry region. Ethnic Turkomans make up about three-fourths of the population. Russians, at about 10 percent, form the second-largest ethnic group. Most Turkomans live in villages. Most Russians live in the capital, Ashkhabad, which has a population of about 850,000, and in Turkmenistan's two smaller cities.

Most people in Uzbekistan live in the eastern half of the country, almost two-thirds of them in rural areas. Heavily populated oases are covered with orchards, farm fields, and irrigation canals. Most rural people are Uzbeks. Most of the city dwellers are Russian and Kazakh. Uzbekistan's capital city, Tashkent, is home to more than 2 million people. It is Central Asia's largest city and its economic and cultural center.

Thinking Like a Geographer

World Time Zones

The world has 24 official time zones, each 15° of longitude apart. The Prime Meridian is the reference for measuring time at 0°. Traveling east from 0°, the time is one hour later in each time zone. Traveling west from 0°, the time is one hour earlier. When crossing the International Date Line at 180° longitude from west to east, one day is lost; when crossing from east to west, one day is gained.

MAP SKILLS

PLACES AND REGIONS How many time zones are in the area from the Caucasus to Siberia's easternmost point at the International Date Line?

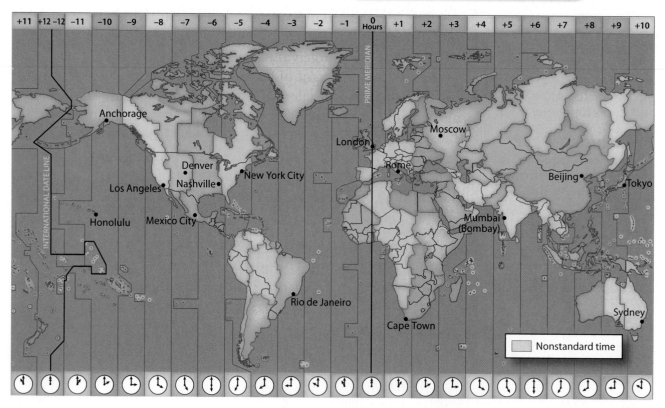

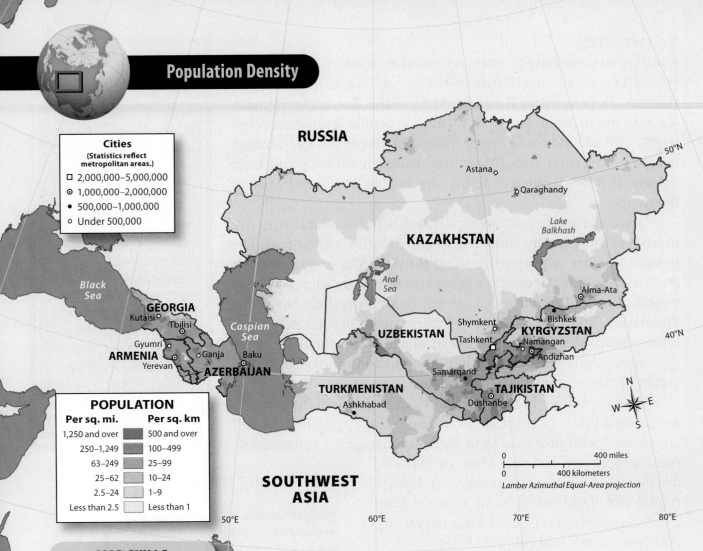

Cities
(Statistics reflect metropolitan areas.)

- □ 2,000,000–5,000,000
- ◉ 1,000,000–2,000,000
- ● 500,000–1,000,000
- ○ Under 500,000

POPULATION

Per sq. mi.	Per sq. km
1,250 and over	500 and over
250–1,249	100–499
63–249	25–99
25–62	10–24
2.5–24	1–9
Less than 2.5	Less than 1

0 — 400 miles
0 — 400 kilometers
Lamber Azimuthal Equal-Area projection

MAP SKILLS

This map shows the population density and major cities of Central Asia and the Caucasus.

1 HUMAN GEOGRAPHY
What city on the map has the largest population?

2 PLACES AND REGIONS
How does population density in Armenia compare with Turkmenistan's?

Academic Vocabulary

significant a noticeably large amount; important

About two-thirds of Kyrgyzstan's people are Kirghiz. The Kirghiz were once nomads who were settled onto collectives during Soviet rule. They remain largely rural today.

Less than 40 percent of the population live in cities and towns. Most of the town dwellers are Uzbeks and Russians. Tajikistan is also mainly rural, but the country's largely mountainous terrain causes its settled areas to be densely populated. Villages dot the foothills and mountain valleys. In dry regions, irrigation has created densely populated oases. About 80 percent of the people are ethnic Tajiks. Most of the rest are Uzbeks.

The Caucasus

Nearly all of Armenia's people are ethnic Armenians. Most of Azerbaijan's people are ethnic Azeris. Georgia is largely ethnic Georgian but has a population that contains **significant** minorities—mainly Armenians and Azeris. The Caucasus is also home to about 50 smaller ethnic groups.

About half the region's people are urban, and half are rural. Armenia, however, has a larger urban population overall. Two-thirds of Armenians live in cities and towns.

The region's rural population is not evenly distributed. Mountain valleys and foothills and some parts of the mountains are heavily settled. Few people live above 6,000 feet (1,829 m). The area along Georgia's Black Sea coast is also densely populated. At the same time, one of every four Georgians lives in Tbilisi, its capital city of 1.1 million.

Armenia's Ararat plain, near the country's border with Turkey and Iran, is its most heavily populated region. The plain and the surrounding foothills and mountains are Armenia's economic and cultural center. Yerevan, its capital city of 1.1 million people, is located there.

Azerbaijan's most densely populated area is around its capital, Baku, on a peninsula in the Caspian Sea. Baku, with about 2 million people, is the largest city and most important industrial center in the Caucasus. Azerbaijan's most heavily settled rural region is in the extreme southeast, between the Caspian Sea and its border with Iran.

☑ **READING PROGRESS CHECK**

Identifying Where does most of Central Asia's Russian population live?

People and Cultures

GUIDING QUESTION *How are the cultures of Central Asia and the Caucasus alike and different?*

Siberia, Central Asia, and the Caucasus not only have different population distributions, their cultures are also very different. In parts of Central Asia, Soviet rule left lasting Russian influences. In parts of the Caucasus, Russian influences are strong but less widespread.

Language and Religion

Russian is the main language in Siberia, and Christianity, mainly Russian Orthodox, is the dominant religion, along with Buddhist and Muslim minorities. Russian is widely spoken in Central Asia. In Kazakhstan and Kyrgyzstan, Russian is an official language, in addition to the language of the country's ethnic majority. Russian is often used in business. In Kazakhstan, 95 percent of the population speaks it, mostly as a second language.

About half of Kazakhstan's people are Christians, and half are Muslims. In the rest of Central Asia, between 75 percent and 90 percent of the people practice Islam, and most of the rest are Christians.

In the Caucasus, most Armenians and Georgians are Christians, and most Azerbaijanis are Muslims. Georgia has a significant Muslim minority as well. In each country, the language of its ethnic majority is the official language and is widely spoken.

Armenia was the first country in the world to adopt Christianity as its official religion—in A.D. 301. Today, most Armenians belong to the Armenian Apostolic Church.

Identifying Which country in the Caucasus has a large Muslim minority?

Yurt a round tent with a domed roof covered with heavy, waterproof wool felt or animal skins

Village women in Uzbekistan and other parts of Central Asia have traditionally woven carpets and rugs. After years of decline under the Soviets, the art of carpet weaving is now being taught at workshops in some urban areas.

▶ **CRITICAL THINKING**

Analyzing What economic change has affected the making of traditional goods by hand?

Ethnic Unrest

Compared to the rest of the Caucasus, Armenians are **homogeneous**, or of the same kind. The Turks and Azeris, who live in the countries bordering Armenia, historically have been the Armenians' enemies. In the 1890s and again during World War I, the Turks massacred millions of Armenians who were living under Turkish rule. Meanwhile, Christian Armenians were fighting the influence of Islam being spread by Azeris living in the Caucasus and Iran.

More recently, Armenians and Azeris have been fighting over Nagorno-Karabakh, a region with an Armenian majority that is part of Azerbaijan. In 1988 Armenia seized Nagorno-Karabakh and the territory that connected it to Azerbaijan. As tensions and violence increased, many Azeris fled the region. The dispute remains unsettled.

Georgia also has experienced much ethnic unrest. Its Armenian minority has long sought greater self-rule. In 1992 ethnic minority groups in Georgia's Abkhaz Republic region revolted for independence. In 2008 other minorities in its South Ossetia region did the same. Russia sent troops to each region to help the rebels and increase its influence over Georgia. Pressure from other countries caused the Russians to withdraw from South Ossetia, but tensions over both regions continue.

In Central Asia, Tajikistan was torn by civil war after Soviet rule ended. Tajik Communists, backed by Russian troops, fought Islamic groups and their allies. Some 10,000 people were killed and 500,000 were forced to flee before a peace agreement was reached in 1997.

Culture and Daily Life

Most urban Kazakhs live and dress like Europeans. Rural Kazakhs, women especially, continue to wear the heavily embroidered and colorful traditional Kazakh clothing. Many rural Kazakhs also live in **yurts**. Colorful embroidered rugs decorate traditional Kazakh homes. Such rugs are typical of most Central Asian cultures.

Some Kirghiz and other rural Central Asians also live in sturdy yurts. Rural Kirghiz women stay at home caring for their children while the men tend to the family's crops and livestock. Raising and racing horses are important in most Central Asian cultures.

Uzbek families are often large. Family relationships are close, and parents have great authority. Children help with tasks around the home. In cities and towns, most families live in

In a nomad camp, Kirghiz herders tend to their horses. Equestrian sports (sports with horses), such as racing and a form of polo, are important parts of the culture of Central Asia's nomadic people.

houses with a courtyard, where they spend much of their leisure time. Uzbeks also decorate their homes with colorful rugs, but most wear Western-style clothing.

Most Tajiks live in the western half of the country in the valleys that lie between the steep mountains. They live in villages of 200 to 700 flat-roofed houses along an irrigation canal or a river. A mud fence surrounds each house and the orchard or vineyard next to it. In mountain communities, the villages are smaller. On steep slopes, the houses' flat roofs are the yards for the houses above them.

✓ **READING PROGRESS CHECK**

Citing Text Evidence What ethnic unrest exists in the Caucasus today?

Relationships and Challenges

GUIDING QUESTION *What challenges lie ahead for the regions?*

Most Caucasus and Central Asian countries have struggled to establish stable, democratic governments since the end of Soviet rule. They are also trying to develop free-enterprise economies after years of Communist government controls.

The Soviet Legacy

The countries of the Caucasus and Central Asia adopted new constitutions after they gained independence. The constitutions created democratic governments, gave the people rights and freedoms, and set the stage for free market economies. In practice, however, the promises were hard to keep.

Old rusting ships lie in the sand where the Aral Sea used to be. The Aral Sea supported a major fishing industry before its water level dropped.

▶ **CRITICAL THINKING**

Analyzing How have Soviet policies of the past affected Central Asia's environment today?

In Azerbaijan and Kyrgyzstan, former Communists came to power and were reluctant to change their authoritarian ways. Uzbekistan's Communist party stayed in control by changing its name. In Tajikistan and Georgia, civil war and government corruption brought hardships to citizens.

Turkmenistan's first leader was not a Communist. He seized all power, however, declared himself president for life, and showed little concern for the people. The country's first real elections were not held until a year after his death in 2006. Most other countries in the two regions are now also moving slowly toward democracy.

Russians have been leaving Central Asia gradually since the end of Soviet rule. This emigration has deprived some countries of workers with the skills needed to run the mines, factories, and other businesses created during the Soviet era. Students who are graduating from the many universities, scientific and technical institutes, trade schools, and public schools the Soviets created are closing the gap. Another benefit of schooling is that nearly all adults in Central Asia and the Caucasus can now read and write.

Public health in Central Asia has suffered. The Soviets' carelessness in developing agriculture and industry has resulted in high levels of pollution and related health risks. The environmental disaster they created in the Aral Sea is a good example.

The Shrinking Aral Sea

The Aral Sea was once the fourth-largest inland lake in the world. Today it has lost about 90 percent of its water. Towns that were once important fishing ports are now dozens of miles from the sea.

The problem began in the 1960s, when the Soviets built huge farms in the desert and steppes; they dug long canals to irrigate them. Not enough water from the Amu Dar'ya and Syr Dar'ya reached the sea to replace the water lost to evaporation and irrigation. The sea's water level dropped, and its shoreline receded.

What was once the seafloor became dry land filled with salt and farm chemicals from decades of runoff from the irrigated fields. Winds whipping across this polluted wasteland carried salt and chemicals hundreds of miles away into people's lungs. Lung diseases and cancer rates rose as a result.

Kazakhstan is now trying to raise the sea's water level again; some improvement has resulted. Uzbekistan and Turkmenistan have not been able, however, to agree on ways to reduce their use of the Amu Dar'ya's water. Without a regional solution, restoring the sea might not be possible.

Territorial Issues

The dispute over the Amu Dar'ya is just one of several regional issues caused by Soviet rule. When the Soviets created the borders of their Central Asian and Caucasus republics, they paid little attention to where each region's ethnic groups lived. Ethnic-related disputes exist along every shared boundary in Central Asia.

Azerbaijan, Iran, Russia, Turkmenistan, and Kazakhstan also cannot agree on how to share the surface and seabed of the Caspian Sea. At stake is access to minerals such as oil and natural gas, fishing rights, and transportation for the landlocked countries of Azerbaijan and Turkmenistan. Russia and Japan have a long-running dispute over the Kuril Islands, an island chain that extends south from Siberia's Kamchatka peninsula. Japan still claims rights to the southernmost islands, which it lost to the Soviet Union during World War II.

Include this lesson's information in your Foldable®.

✓ **READING PROGRESS CHECK**

Determining Central Ideas Why were the Caucasus and Central Asia slow to establish democratic governments when the Soviet Union collapsed?

LESSON 3 REVIEW

Reviewing Vocabulary

1. Why are *oases* important in Central Asia?

Answering the Guiding Questions

2. ***Describing*** Why does most of Siberia's population live along the Trans-Siberian Railroad?

3. ***Identifying*** What cultural characteristics do many of Central Asia's main ethnic groups share?

4. ***Analyzing*** How have ethnic tensions affected relationships in the Caucasus and Central Asia?

5. ***Narrative Writing*** Create a journal entry describing a day on the Trans-Siberian Railroad or in a rural community in Central Asia.

Directions: Write your answers on a separate piece of paper.

1 Use your **FOLDABLES** to explore the Essential Questions.

INFORMATIVE/EXPLANATORY WRITING Review the population map at the beginning of this unit. Explain why some areas are heavily populated. Describe factors that influenced this pattern of settlement in three or more paragraphs.

2 **21st Century Skills**

INTEGRATING VISUAL INFORMATION Research to find a a primary source and a secondary source on the collapse of the Soviet Union. Read and discuss the two documents. Answer the questions: What kind of information is provided by the primary source? Is it different from information provided by the secondary source? How? Which source provides information that is more accurate? Why?

3 **Thinking Like a Geographer**

INTEGRATING VISUAL INFORMATION Create a time line. List five key events in the history of the region. Include the entries on your time line.

4 **GEOGRAPHY ACTIVITY**

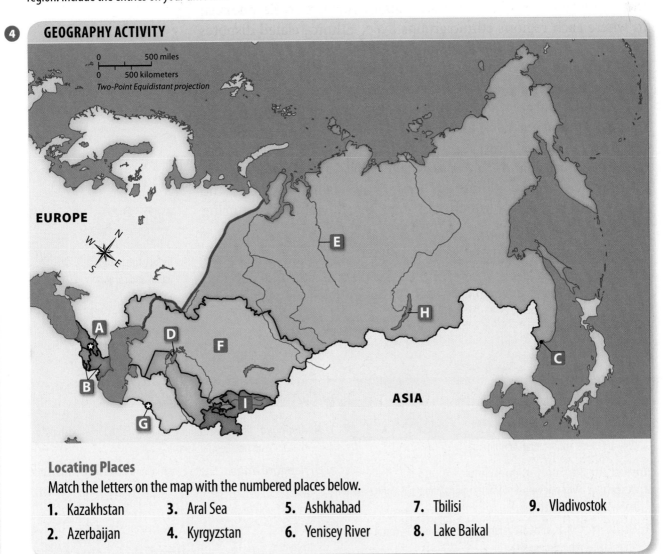

Locating Places
Match the letters on the map with the numbered places below.

1. Kazakhstan
2. Azerbaijan
3. Aral Sea
4. Kyrgyzstan
5. Ashkhabad
6. Yenisey River
7. Tbilisi
8. Lake Baikal
9. Vladivostok

REVIEW THE GUIDING QUESTIONS

Directions: Choose the best answer for each question.

1 Natural resources found in Siberian Russia and Central Asia include huge deposits of

   A. oil and natural gas.

   B. lead and antimony.

   C. copper and tungsten.

   D. bauxite and iron.

2 What is the name of the mountain chain that divides Europe from Asia?

   F. Ural Mountains

   G. Caucasus Mountains

   H. Syr Dar'ya

   I. Tien Shan

3 What technological development opened up Siberia for settlement and aided economic expansion?

   A. the telephone

   B. the steam engine

   C. the Trans-Siberian Railroad

   D. the Moscow to Novosibirsk Canal

4 What benefits did the Caucasus countries of Armenia, Georgia, and Azerbaijan gain from the years they were part of the Soviet Union?

   F. Roads and airports were built.

   G. They profited greatly from the Cold War arms race.

   H. Soviet military power reduced ethnic tensions and brought peace to the region.

   I. The Soviets encouraged the development of the area's natural resources.

5 Which large inland lake nearly dried up because of mismanagement and too much irrigation?

   A. Caspian Sea

   B. Aral Sea

   C. Black Sea

   D. Red Sea

6 The two most popular religions practiced by the people who live in Central Asia, the Caucasus, and Siberia are

   F. Protestantism and Catholicism.

   G. Islam and Hinduism.

   H. Hinduism and Christianity.

   I. Christianity and Islam.

DBQ ANALYZING DOCUMENTS

7 **IDENTIFYING** Read the following passage about the Sogdians, a group of people who were important during the early history of Central Asia:

"*The Sogdians were the major participants in the Silk Road caravans, ... their alphabet the source of later alphabets to the east, they carried with them such religions as Zoroastrianism . . . and Nestorian Christianity.*"

—from Albert Dien, "The Glories of Sogdiana"

What was the purpose of caravans along the Silk Road in ancient times?

A. to carry out trade of valuable goods

B. to achieve military conquest

C. to spread Sogdian culture to new areas

D. to enrich the Sogdian rulers

8 **ANALYZING** The Sogdians were important because they

F. built the silk road.

G. organized the caravans.

H. had a strong cultural impact.

I. ruled the region for many centuries.

SHORT RESPONSE

The following passage discusses the work of USAID, a U.S. government agency that provides humanitarian aid and assistance for development to other countries:

"*When USAID arrived in Kazakhstan in 1992, . . . the country's centrally controlled [economic] system lay in ruins. No one could have imagined that, less than 15 years later, this once-underdeveloped . . . republic would be co-financing USAID's economic development work in the country.*"

—from Geoff Minott and Leanne McDougall, "One Part U.S., Two Parts Kazakhstan" (2012)

9 **DETERMINING WORD MEANINGS** What do the authors mean when they say that now Kazakhstan is "co-financing" USAID projects?

10 **ANALYZING** What can you infer about Kazakhstan's economy from this report?

EXTENDED RESPONSE

11 **INFORMATIVE/EXPLANATORY WRITING** Research to find out about the construction of America's transcontinental railroad. In an essay, compare the building of that railroad with the building of the Trans-Siberian Railroad in Russia. In your writing, describe the effects each railroad had on settlement patterns and economic growth.

Need Extra Help?

If You've Missed Question	❶	❷	❸	❹	❺	❻	❼	❽	❾	❿	⓫
Review Lesson	1	1	2	2	3	3	2	2	3	3	2

SOUTHWEST ASIA

ESSENTIAL QUESTIONS · *How does geography influence the way people live?*
· *Why do civilizations rise and fall?* · *How does religion shape society?*

A woman in Baghdad, Iraq, displays her inked fingers, showing that she has just voted in the election.

Wathiq Khuzaie/Getty Images News/Getty Images

networks

There's More Online about Southwest Asia.

CHAPTER 18

The Story Matters...

The area between the Tigris and Euphrates rivers in Southwest Asia is known as the Fertile Crescent. Here, fertile soil and water for irrigation supported the growth of an early civilization—Mesopotamia. Three major world religions—Judaism, Christianity, and Islam—began in this region. From ancient to modern times, the region's physical features, resources, and cultures have greatly influenced how people live.

FOLDABLES®
Study Organizer

Go to the Foldables® library in the back of your book to make a Foldable® that will help you take notes while reading this chapter.

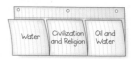

Southwest Asia lies where the continents of Asia, Africa, and Europe meet. Some of the world's earliest civilizations started here.

Step Into the Place

MAP FOCUS Use the map to answer the following questions.

1 **PLACES AND REGIONS**
What is the capital city of Syria?

2 **THE GEOGRAPHER'S WORLD**
What countries border Yemen?

3 **THE GEOGRAPHER'S WORLD**
What body of water lies west of Saudi Arabia?

4 **CRITICAL THINKING**
Describing In which direction would you travel from the Persian Gulf to the city of Ankara?

A

BEACH RESORT Vacationers gather at a beach along the Dead Sea. Because of the high salt content of the water, people who bathe in the Dead Sea can float on its surface without effort.

B

PORT CITY Modern buildings tower over the city of Jaffa, along Israel's Mediterranean coast. One of the world's oldest cities, Jaffa is known for its historic areas, fragrant gardens, and sprawling flea market.

Step Into the Time

TIME LINE Choose an event from the time line and write a paragraph explaining its effect on the region and the world.

c. 3000 B.C. The Sumerians develop early form of writing

c. A.D. 30 Jesus preaches in Jerusalem

B.C. **A.D.**

c. 4000 B.C. People settle along the Tigris and Euphrates rivers

1792–1750 B.C. King Hammurabi develops written laws, known as the Code of Hammurabi

Southwest Asia

RUSSIA

Black Sea

40°N

•Ankara

TURKEY

CENTRAL ASIA

Caspian Sea

Aral Sea

Mediterranean Sea

Beirut

SYRIA

•Arbil

Euphrates R.

Tigris R.

LEBANON

Damascus

•Tehran

Kabul•

AFGHANISTAN

Tel Aviv-Jaffa

•Baghdad

Jerusalem

30°N

A

IRAQ

IRAN

ISRAEL

JORDAN

B

Kuwait

SOUTH ASIA

KUWAIT

Persian Gulf (Arabian Gulf)

AFRICA

SAUDI ARABIA

BAHRAIN

Manama

TROPIC OF CANCER

QATAR

Doha

Abu Dhabi

Gulf of Oman

Riyadh

UNITED ARAB EMIRATES

•Masqat

20°N

•Makkah (Mecca)

OMAN

◇ National capital
• City

Red Sea

INDIAN OCEAN

0   500 miles

0   500 kilometers

Lambert Azimuthal Equal-Area projection

YEMEN

•Sanaa

30°E

40°E

Gulf of Aden

50°E

2003 The Iraq War begins

1290 The Ottoman Empire begins; lasts until 1922

1948 Israel is founded

1900

2010

A.D. 600s Muhammad begins preaching the teachings of Islam

1938 Oil is discovered in Saudi Arabia

1990 Persian Gulf War begins

2010 Arab Spring revolts take place in the Middle East

Reading **HELP**DESK

Academic Vocabulary

- **vary**

Content Vocabulary

- **alluvial plain**
- **oasis**
- **wadi**
- **semiarid**

TAKING NOTES: *Key Ideas and Details*

Identify As you read the lesson, complete the graphic organizer by listing the physical features found in this region.

Southwest Asia	Examples
Mountains	
Deserts	
Natural Resources	

Lesson 1
Physical Geography of Southwest Asia

ESSENTIAL QUESTION · *How does geography influence the way people live?*

IT MATTERS BECAUSE
Southwest Asia is characterized by a complex physical geography that influences its people, history, and importance in the world today.

Southwest Asia's Physical Features

GUIDING QUESTION *What are the main landforms and resources in Southwest Asia?*

Southwest Asia comprises 15 countries that lie in the area where Asia meets Europe and Africa. Similarities in physical geography help unite these countries into a single region. Mountains and plateaus formed by active plate tectonics can be seen throughout the region. Dry, desert climates are also widespread.

Mountains and Plateaus

Mountains and plateaus dominate the landscape of Southwest Asia. They have been created over the past 100 million years by collisions between four tectonic plates. This movement also caused earthquakes.

Southwest Asia's loftiest mountains rise in the Hindu Kush range, which stretches across much of Afghanistan and along Afghanistan's border with the South Asian country of Pakistan.

The Hindu Kush and neighboring ranges form natural barriers to travel and trade. As a result, mountain passes have been important in this area. One of the world's most famous

passes is the Khyber Pass. It links the cities of Kabul, Afghanistan, and Peshawar, Pakistan. The pass has served as a route for trade and invading armies for thousands of years.

A vast plateau, covering much of Iran, is encircled by high mountain ranges. The mountains of western Iran merge with those of eastern Turkey. Close to the border rises Turkey's highest peak, Ararat, a massive, snowcapped volcano that last erupted in 1840. An elevated area known as the Anatolian Plateau spreads across central and western Turkey.

The Arabian Peninsula consists of Saudi Arabia, Yemen, Oman, and several other countries. It is a single, vast plateau that slopes gently from the southwest to the northwest. Long mountain ranges that parallel the peninsula's southwestern, northwestern, and southeastern coasts are actually the deeply eroded edges of the plateau.

Bodies of Water

The region of Southwest Asia has thousands of miles of coastline. Turkey has coasts on the Mediterranean and Black seas. Syria, Lebanon, Jordan, and Israel have coasts on the Mediterranean Sea. Jordan, Saudi Arabia, and Yemen border the long, narrow Red Sea. The Red Sea has been one of the world's busiest waterways since Egypt's Suez Canal, connecting the Red Sea and the Mediterranean, was completed in 1869. To the southeast of the Arabian Peninsula lies a part of the Indian Ocean called the Arabian Sea. Yemen and its neighbor Oman have coasts along this sea.

©Yannis Kontos/Sygma/Corbis

A deadly earthquake in 1999 left widespread destruction in Turkey.
▶ **CRITICAL THINKING**
Describing What causes earthquakes throughout Southwest Asia?

Oil tankers pass through the Strait of Hormuz. The strait is the only sea passage from the Persian Gulf to the open ocean.

▶ **CRITICAL THINKING**
Analyzing Why is the Strait of Hormuz considered a strategic waterway?

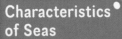

Thinking Like a Geographer

Characteristics of Seas

The word *sea* is most often used to describe a large body of salt water that is part of, or connected to, an ocean. The Black Sea and the Mediterranean Sea, both of which are connected to the Atlantic Ocean, meet this description, as does the Red Sea, a narrow extension of the Indian Ocean. The landlocked Caspian and Dead seas, however, do not. They best fit the description of *lake*. Large lakes with salty water, however, are sometimes called seas.

In the northeast, the Arabian Peninsula is shaped by the Persian Gulf, which is connected to the ocean by a strategic waterway called the Strait of Hormuz. The Persian Gulf has become tremendously important in world affairs since the middle of the 1900s.

Eight of Southwest Asia's 15 countries border the Persian Gulf: Oman, the United Arab Emirates, Saudi Arabia, Qatar, Bahrain, Kuwait, Iraq, and Iran. In the north, Iran also borders the landlocked Caspian Sea.

The Dead Sea, which lies between Israel and Jordan, is also landlocked. It is far smaller than the region's other seas. At 1,300 feet (396 m) below sea level, it ranks as the world's lowest body of water, and its shore represents the lowest land elevation.

Southwest Asia's two longest and most important rivers are the Tigris and the Euphrates, which are often considered parts of the same river system. The rivers begin within 50 miles (80 km) of each other in the mountains of eastern Turkey. In their lower courses, they flow parallel to one another across a broad **alluvial plain**, a plain created by sediment deposited during floods. The plain covers most of Iraq as well as eastern Syria and southeastern Turkey. This area has been known since ancient times as Mesopotamia, which is Greek for "land between the rivers." Thousands of years ago, one of the world's earliest civilizations took root in the fertile lands of Mesopotamia.

Deserts

Desert landscapes spread across most of Southwest Asia. The Arabian Desert, which covers nearly the entire Arabian Peninsula, is the largest in the region and one of the largest in the world. It is made up of rocky plateaus, gravel-covered plains, salt-crusted flats, flows of black lava, and sand seas, which are unbroken expanses of sand.

©Dean Conger/Corbis

In the southern part of the peninsula lies the largest sand sea in the world: the Rub' al-Khali, or Empty Quarter. Winds have sculpted its reddish-orange sands into towering dunes and long, winding ridges. The climate is so dry and hot that this starkly beautiful wilderness cannot support permanent human settlements. In some areas, nomadic people known as the Bedouin keep herds of camels, horses, and sheep.

The Arabian Desert is a harsh environment, but plants thrive in oases. An **oasis** is an area in a desert where underground water allows plants to grow throughout the year.

☑ **READING PROGRESS CHECK**

Analyzing How has tectonic activity—that is, movement of Earth's crustal plates—helped shape landforms in Southwest Asia?

Southwest Asia's Climates

GUIDING QUESTION *What are some ways that mountains, seas, and other physical features affect climate in Southwest Asia?*

A single type of climate dominates most of Southwest Asia. The only parts of the region with greater climatic variety lie in the northwest and northeast.

An Arid Region

Although this region is surrounded by seas and gulfs, water is a scarce resource here. Most of the region falls within an arid, or very dry, climate zone. Deserts—areas that receive less than 10 inches (25 cm) of annual rainfall—cover nearly the entire Arabian Peninsula as well as large parts of Iran. These deserts are part of a broad band of arid lands that stretch from western North Africa to East Asia. Southwest Asia's arid lands can be brutally hot in the summer. Temperatures in the Arabian Desert can soar as high as 129°F (54°C).

Mountain springs provide a constant flow of water through the Bani Wadi Khalid riverbed in northern Oman. Along the stream's course are large rock formations and shimmering pools of turquoise water.

▶ **CRITICAL THINKING**

Analyzing What feature makes the Bani Wadi Khalid area different from most other landscapes in the Arabian Peninsula?

OIL: RESERVES AND CONSUMPTION
KEEPING US ON THE MOVE

The countries of Southwest Asia have some of the largest reserves of oil in the world.

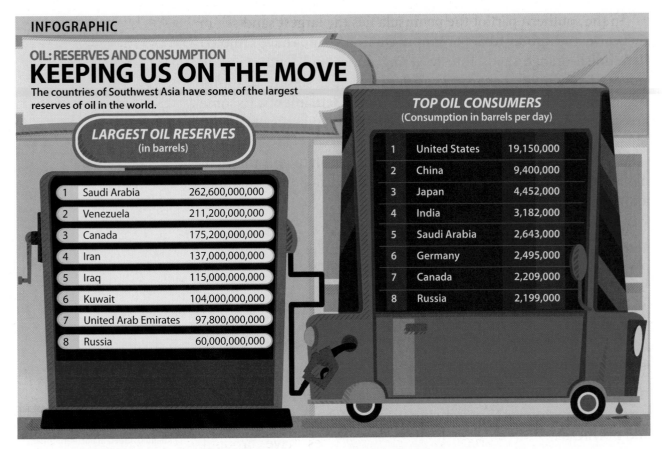

LARGEST OIL RESERVES
(in barrels)

1	Saudi Arabia	262,600,000,000
2	Venezuela	211,200,000,000
3	Canada	175,200,000,000
4	Iran	137,000,000,000
5	Iraq	115,000,000,000
6	Kuwait	104,000,000,000
7	United Arab Emirates	97,800,000,000
8	Russia	60,000,000,000

TOP OIL CONSUMERS
(Consumption in barrels per day)

1	United States	19,150,000
2	China	9,400,000
3	Japan	4,452,000
4	India	3,182,000
5	Saudi Arabia	2,643,000
6	Germany	2,495,000
7	Canada	2,209,000
8	Russia	2,199,000

Oil reserves are estimates of the amount of crude oil located in a particular economic region. Oil consumption is the amount of oil an economic region uses.

▶ **CRITICAL THINKING**

Describing Which Southwest Asian countries are among the countries with the largest oil reserves?

Although rain is scarce in this region, rainfall can quickly transform the desert landscapes. Torrents of water race through **wadis** (WAH-deez), or streambeds that are dry. Buried seeds sprout within hours, carpeting barren gravel plains in green.

At the margins of Southwest Asia's dry zones lie areas that are considered **semiarid** (seh-mee-AIR-id), or somewhat dry. These areas are found in the highlands and mountain ranges of the region.

A Mediterranean climate prevails along Southwest Asia's Mediterranean and Aegean coasts and across much of western Turkey. Winds blowing off the seas bring mild temperatures and moderate amounts of rainfall during the winter months. The summer months are warm and dry.

Mountainous areas of eastern Turkey, western Iran, and central Afghanistan have continental climates in which temperatures **vary** greatly between summer and winter. The mountains of the Hindu Kush range in far eastern Afghanistan fall within a highland climate zone, and glaciers are found among the soaring peaks.

Academic Vocabulary

vary to show differences between things

☑ **READING PROGRESS CHECK**

Identifying In what parts of Southwest Asia could farmers grow crops without irrigation?

Natural Resources

GUIDING QUESTION *How do natural resources influence the lives of people in Southwest Asia?*

Scarcity of water has shaped this region's human history and settlement patterns. Other natural resources, however, are found in abundance. The most important resources are two fossil fuels for which the world has a seemingly unquenchable thirst: oil and natural gas.

The gaseous form of petroleum is called natural gas, and the liquid form is called crude oil, or simply oil. Crude oil is refined to produce energy sources such as gasoline, diesel fuel, heating oil, and industrial fuel oil. Petroleum is also the basic raw material used to make many other products, such as plastics, bicycle tires, and cloth fibers.

The world's largest known deposits of petroleum are in Southwest Asia. Most are concentrated around and under the Persian Gulf. Together, five countries that border the gulf—Saudi Arabia, Iran, Iraq, Kuwait, and United Arab Emirates—hold more than half the oil that has been discovered in the world.

Most of the petroleum produced by these countries is exported to industrialized countries. Petroleum revenues have brought tremendous wealth to a few people in the exporting countries. But only in a relatively few areas has the wealth been used to improve the lives of the people or bring about modernization.

Southwest Asia also has a great variety of mineral resources. Large coal deposits are found in Turkey and Iran. Phosphates, used to make fertilizers, are mined in Iraq, Israel, and Syria. Between 2006 and 2010, American geologists conducting a survey of Afghanistan discovered enormous deposits of iron, copper, gold, cobalt, lithium, and other minerals such as rare earth elements used to make electronic devices.

Include this lesson's information in your Foldable®.

Water | Civilization and Religion | Oil and Water

✓ READING PROGRESS CHECK

Identifying Five countries that border the Persian Gulf hold more than half the oil that has been discovered in the world. Name three of the countries.

LESSON 1 REVIEW (CCSS)

Reviewing Vocabulary

1. Describe the difference between a *wadi* and an *oasis*.

Answering the Guiding Questions

2. *Identifying* What makes the Dead Sea distinct?

3. *Describing* What are the major physical geography features of the Arabian Peninsula?

4. *Describing* If you were to travel across the Arabian Desert, what are two types of landscapes or landforms you might see?

5. *Citing Text Evidence* The United Nations ranks Afghanistan as one of the world's poorest countries. How might recent discoveries change that situation?

6. *Narrative Writing* Imagine that you are spending a few days exploring the area of the Arabian Desert called the Rub' al-Khali. Write a one-paragraph journal entry describing the experience.

netw⊙rks

There's More Online!

☑ **CHART/GRAPH** Ziggurats
☑ **IMAGES** The Kurds
☑ **MAP** Islamic Expansion
☑ **VIDEO**

Reading **HELP**DESK

Academic Vocabulary

- expand
- collapse

Content Vocabulary

- polytheism
- millennium
- monotheism

TAKING NOTES: *Key Ideas and Details*

Sequence As you read about Southwest Asia's history, use a time line to put key events and developments in order.

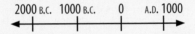

2000 B.C. 1000 B.C. 0 A.D. 1000

Lesson 2
History of Southwest Asia

ESSENTIAL QUESTION • *Why do civilizations rise and fall?*

IT MATTERS BECAUSE

Southwest Asia has played a large role in human history. The world's earliest civilization developed here, and three major religions were born. Great empires that arose in the region grew to cover parts of three continents.

Early Southwest Asia

GUIDING QUESTION *What are some of the most important advancements that occurred in Southwest Asia in ancient times?*

Mesopotamia

Throughout most of human history, people lived as hunter-gatherers. In small groups, they hunted wild animals and searched for wild fruits, nuts, and vegetables. They were nomadic, frequently moving from place to place. About 10,000 years ago, though, a dramatic change began to occur: People started practicing agriculture—raising animals and growing crops. One of the first places this agricultural revolution unfolded was in Mesopotamia. Mesopotamia was a fertile plain between the Tigris and Euphrates rivers in present-day Iraq.

With the shift to agriculture came a shift to a more settled lifestyle. Villages began to appear in Mesopotamia. Because food was plentiful, some villagers were freed up from farming and could undertake toolmaking, basket weaving, or record keeping. Over time, some villages grew into large, powerful cities that had their own governments and military forces. These cities represent the world's first civilizations.

Over thousands of years, Mesopotamian societies such as the Sumerians and the Babylonians invented sophisticated

irrigation and farming methods. They built huge, pyramid-shaped temple towers, and made advances in mathematics, astronomy, government, and law. Using a writing system called cuneiform (kew-NAY-ih-form), they produced great works of literature, including a poem known as the *Epic of Gilgamesh*. Mesopotamia's achievements helped shape later civilizations in Greece, Rome, and Western Europe.

Birthplace of World Religions

Southwest Asia is also a cradle of religion. Three of the world's major religions originated there. In ancient times, most people in the region worshiped many gods. This practice is known as **polytheism**. During the second millennium B.C., a new religion arose. A **millennium** is a period of a thousand years. This new religion was based on **monotheism**—the belief in just one God—and developed among a people called the Israelites. This religion came to be known as Judaism and its followers as Jews.

Jews believe the father of the Israelite people was Abraham. According to the Hebrew Bible, God called Abraham to leave his home in Mesopotamia and found a new nation in a land called Canaan, between the Jordan River and the Mediterranean Sea. The area is shared today by Israel, the Palestinian territories, and Lebanon.

About A.D. 30, a Jewish teacher named Jesus began preaching in this area. The teachings of Jesus led to the rise of Christianity. This new religion spread rapidly throughout the Mediterranean world and into Europe. It also spread across Southwest Asia.

Then, in the A.D. 600s, Islam—the religion of Muslims—arose in the Arabian Peninsula. Muhammad, regarded by Muslims as the last and greatest of the prophets, announced his message in the desert city of Makkah (Mecca). Many of the teachings of Islam are similar to those of Judaism and Christianity. For example, all three religions are monotheistic and regard Abraham as the messenger of God who first taught this belief.

The Ziggurat of Ur was a temple built in Mesopotamia during the 2000s B.C. It served as the political and religious center of the city of Ur.

▶ **CRITICAL THINKING**
Determining Word Meanings
Why is early Mesopotamia called a civilization?

©nik wheeler/Alamy

Southwest Asia is the birthplace of Judaism, Christianity, and Islam. The three religions still influence the region today. (left) A Jewish teenager carries a scroll of the Hebrew Bible at the Western Wall in Jerusalem. (center) Muslims circle the Kaaba, a cube-shaped shrine, at the Grand Mosque in Makkah, Saudi Arabia. (right) Christian clergy lead a procession held in Jerusalem shortly before Easter.

▶ **CRITICAL THINKING**

Describing How is the influence of Judaism, Christianity, and Islam reflected in Southwest Asia today?

Academic Vocabulary

expand to increase or enlarge

Islamic Expansion

The religion that Muhammad preached was relatively simple and direct. It focused on the need to obey the will of Allah, the Arab word for God. It obligated followers to perform five duties, which became known as the Pillars of Islam: promising faith to God and accepting Muhammad as God's prophet, praying five times daily, fasting during the month of Ramadan (RAHM-uh-don), aiding the poor and unfortunate, and making a pilgrimage to the holy city of Makkah.

In its first several years, Islam attracted few converts. By the time of Muhammad's death, however, in A.D. 632, it had **expanded** across the Arabian Peninsula. Under Muhammad's successors, known as caliphs (KAY-lifs), Arab armies began spreading the religion through military conquests. It was also spread by scholars, by religious pilgrims, and by Arab traders.

By about A.D. 800, Islam had spread across nearly all of Southwest Asia, including Persia (present-day Iran) and part of Turkey. It also extended into most of Spain and Portugal and across northernmost Africa. It later expanded to northern and eastern Africa, Central Asia, and South and Southeast Asia.

Islamic society was enriched by knowledge, skills, ideas, and cultural influences from many different peoples and areas. The influences contributed to a flowering of Islamic culture that lasted for centuries. During this period, great works of architecture were built, and centers of learning arose. Arab scholars made advances in math and science. This golden age was to have a lasting impact on every place it touched.

During the 1100s and 1200s, crusaders from Western Europe set up Christian states along Southwest Asia's Mediterranean coast. The Muslims fought back and gained control of these territories by 1300. However, in other areas, Muslim military power weakened.

In the middle of the 1200s, a Central Asian people known as the Mongols, led by the grandson of the famous leader Genghis Khan, conquered Persia and Mesopotamia. These areas became part of a vast Mongol empire that stretched across much of Eurasia.

As a result of the Mongol attacks, the Islamic world was fragmented and fell into decay. Soon, however, a new era of Islamic expansion began. At its heart were the Ottomans, a group of Muslim tribes who began building an empire on the Anatolian Peninsula in the early 1200s. By the mid-1300s, the Ottoman Empire had grown to include much of western Southwest Asia and parts of southeastern Europe and northern Africa. At its height, it was one of the world's most powerful states. It endured for six centuries before finally **collapsing** in the early 1900s.

✔ **READING PROGRESS CHECK**

Determining Central Ideas What are some ways in which Islam was spread?

Academic Vocabulary

collapse to break down completely

MAP SKILLS

1 **THE GEOGRAPHER'S WORLD** How far did Islam spread by A.D. 750?

2 **PLACES AND REGIONS** Why did Muslims from Arabia conquer lands such as Syria, Persia, and Egypt?

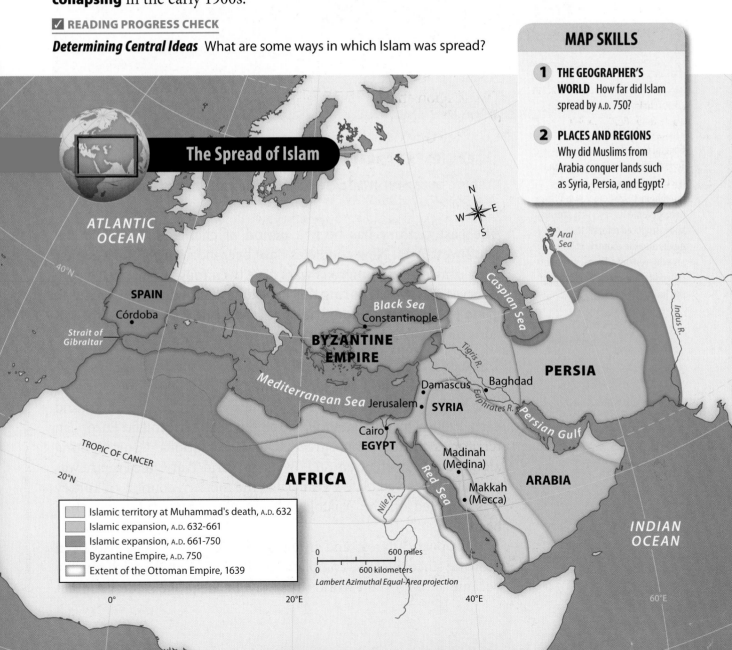

The Spread of Islam

ATLANTIC OCEAN

40°N

SPAIN
Córdoba

Strait of Gibraltar

Aral Sea

Black Sea
Constantinople

BYZANTINE EMPIRE

Caspian Sea

Mediterranean Sea

Tigris R.

PERSIA

Damascus • Baghdad

Jerusalem • SYRIA

Euphrates R.

Persian Gulf

Indus R.

Cairo•
EGYPT

TROPIC OF CANCER

20°N

AFRICA

Nile R.

Red Sea

Madinah (Medina)

Makkah • (Mecca)

ARABIA

INDIAN OCEAN

Islamic territory at Muhammad's death, A.D. 632
Islamic expansion, A.D. 632–661
Islamic expansion, A.D. 661–750
Byzantine Empire, A.D. 750
Extent of the Ottoman Empire, 1639

0   600 miles
0   600 kilometers
Lambert Azimuthal Equal-Area projection

0°   20°E   40°E   60°E

General Mustafa Kemal reviews Turkish troops during the war that led to the creation of a Turkish republic in 1923. Kemal, a military hero, became Turkey's first president and introduced reforms to modernize the country. In honor of his achievements, Kemal was later called Atatürk, meaning "Father of the Turks."

Modern Southwest Asia

GUIDING QUESTION *What present-day issues facing Southwest Asia have their roots in ancient times?*

The past century has been a period of change and conflict for Southwest Asia. New countries have been born, new borders have been drawn, and numerous wars have been fought. Vast petroleum reserves discovered during this period have brought great wealth to some of the region's countries but have also created new tensions and conflicts.

Independent Countries

After reaching the peak of its power in the 1500s, the Ottoman Empire began to decline. The decline worsened in the 1800s and early 1900s. During that time, the empire lost African and European territories through wars, treaties, and revolutions. After fighting alongside the losing Central Powers in World War I, the empire was formally dissolved. A few years later, the modern country of Turkey was founded on the Anatolian Peninsula, where the empire had been born.

European interest and influence in Southwest Asia had been growing since the 1869 completion of the Suez Canal, which quickly became an important world waterway. In the peace settlement that

General Photographic Agency/Hulton Archive/Getty Images

ended World War I, Britain and France gained control over the Ottoman Empire's former territories under a mandate system. In this arrangement, the people of these territories were to be prepared for eventual independence.

Dividing up their territories, the British and French created new political boundaries that showed little regard for existing ethnic, religious, political, or historical divisions. These boundaries would take on deep importance when the territorial units became independent countries and when new discoveries of petroleum deposits were made.

Long-simmering resentment toward the European colonial powers soon grew into strong nationalist movements among Arabs, Persians, Turks, and other groups. Between 1930 and 1971, one country after another won its independence, and the map of Southwest Asia began to take its present form.

Arab-Israeli Conflict

One of the mandates received by Britain after World War I was the territory called Palestine. It roughly corresponded to the Land of Israel, which was the area inhabited by the Jewish people in ancient times. Most of the people living in Palestine at the time of the mandate were Muslim Arabs. During the same period, growing numbers of Jewish immigrants seeking to escape persecution had been arriving from Europe and other parts of the world. As the Jewish population increased, tensions between Palestinian Arabs and Jews deepened.

Jewish nationalists called for the reestablishment of their historic homeland in Palestine. This movement gained support as a result of the Holocaust—the systematic murder of 6 million European Jews by Nazi Germany during World War II. Hundreds of thousands of Jews who survived the Holocaust were now refugees in search of a place to live.

In 1947 the United Nations decided on the issue of Palestine. The United Nations voted to divide the territory into two states, one Arab and one Jewish. The proposal was rejected by the Arabs.

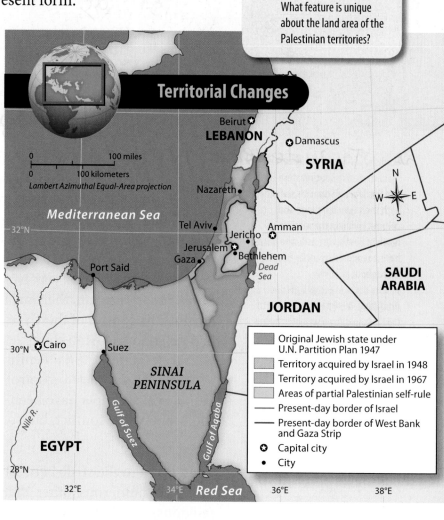

MAP SKILLS

1 HUMAN GEOGRAPHY
How does Israel's territory today compare with Israel's territory in 1967?

2 PHYSICAL GEOGRAPHY
What feature is unique about the land area of the Palestinian territories?

Territorial Changes

0   100 miles
0   100 kilometers
Lambert Azimuthal Equal-Area projection

Mediterranean Sea

Beirut
LEBANON
Damascus
SYRIA
Nazareth
Tel Aviv
Jericho   Amman
Jerusalem
Gaza   Bethlehem
Dead Sea
JORDAN
SAUDI ARABIA
Port Said
Cairo   Suez
SINAI PENINSULA
Gulf of Suez
Gulf of Aqaba
Nile R.
EGYPT
Red Sea

- Original Jewish state under U.N. Partition Plan 1947
- Territory acquired by Israel in 1948
- Territory acquired by Israel in 1967
- Areas of partial Palestinian self-rule
- Present-day border of Israel
- Present-day border of West Bank and Gaza Strip
- Capital city
- City

Kurdish families travel by cart over a modern road in southeastern Turkey. The Kurds are a Sunni Muslim people with their own language and culture. Living in the mountains north of Southwest Asia, the Kurds have been ruled by other people throughout history.

▶ **CRITICAL THINKING**
Analyzing Why has the demand of the Kurds for their own independent country been difficult to achieve?

The Arabs did not want to give up land. On the day in 1948 that Israel, the Jewish state, declared its independence, armies from neighboring Arab countries invaded. Hundreds of thousands of Palestinian Arabs became refugees after fleeing the violence. That war ended with a truce in 1948. Other major Arab-Israeli wars were fought, however, in the 1950s, 1960s, and 1970s.

During a brief 1967 war, Israel captured areas known as the West Bank, East Jerusalem, the Sinai Peninsula, the Gaza Strip, and Golan Heights. Its control of these areas was opposed by Palestinian Arabs and neighboring Arab countries, which led to further conflict. Israel withdrew from the Sinai Peninsula in 1982 and from the Gaza Strip in 2005. It continues to control the West Bank and East Jerusalem. Numerous attempts have been made to find a peaceful solution to the Arab-Israeli conflict, but so far none have been successful.

Civil Wars

In addition to the strife between Arabs and Israelis, Southwest Asia has seen numerous other conflicts since World War II. Ethnic, religious, and political differences have fueled many conflicts. So has the rise of Islamist movements that consider Islam to be a political system as well as a religion. The desire to control large oil fields has also caused, or contributed to, many of the conflicts.

Civil wars have torn apart Lebanon, Afghanistan, and Yemen. The Kurds, a fiercely independent people living in eastern Turkey, northern Iraq, and western Iran, have fought to gain their own

©Nik Wheeler/Corbis

country. A revolution in Iran in the late 1970s resulted in the overthrow of that country's monarchy and the establishment of an Islamic republic. Iraq invaded Iran in 1980, touching off an eight-year-long war. A decade later, Iraq invaded and annexed its small but oil-rich neighbor, Kuwait. This invasion triggered the Persian Gulf War, in which a coalition led by the United States quickly liberated Kuwait.

Conflict and Terrorism

On September 11, 2001, an Islamist organization called al-Qaeda carried out terrorist attacks on U.S. soil that killed nearly 3,000 people. The United States determined that Afghanistan's Islamist ruling group, the Taliban, was supporting al-Qaeda and sheltering its leaders. In October, forces led by the United States and the United Kingdom invaded Afghanistan and removed the Taliban from power.

Two years later, the Second Persian Gulf War began when forces from the United States and the United Kingdom invaded Iraq and overthrew the government of Saddam Hussein. Hussein was accused of possessing weapons of mass destruction, a suspicion that eventually was proved to be untrue.

Looking to the Future

Despite the many conflicts, there is hope for a more peaceful and brighter future in Southwest Asia. Revenue from petroleum has brought prosperity and modernization to oil-rich countries of the Persian Gulf. In 2010 and 2011, a popular uprising in Tunisia inspired democratic movements in Yemen, Bahrain, and Syria.

On the other hand, militant Islamic political movements limit the growth of democracy and civil rights. In addition, throughout the region, major gaps still exist in standards of living between the oil-rich countries and poorer countries.

☑ **READING PROGRESS CHECK**

Determining Central Ideas What event in Europe helped spur the creation of a Jewish state in Southwest Asia?

Think Again?

The Middle East and Southwest Asia are two names for the same region.

Not true. According to most authorities, the Middle East includes all or part of North Africa as well as all or most of Southwest Asia. Some authorities also consider the countries of Central Asia to be part of the Middle East.

FOLDABLES
Study Organizer

Include this lesson's information in your Foldable®.

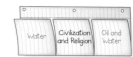

LESSON 2 REVIEW (CCSS)

Reviewing Vocabulary

1. What is the difference between *monotheism* and *polytheism*?

Answering the Guiding Questions

2. *Identifying* What is one of the Pillars of Islam?

3. *Describing* What change in Mesopotamia around 10,000 years ago resulted in a less nomadic lifestyle?

4. *Identifying* What are two developments that occurred during Islam's golden age?

5. *Citing Text Evidence* What empire represented a second period of Islamic expansion, and where did that empire begin?

6. *Determining Central Ideas* How did the 2001 terrorist attacks on the United States lead to a U.S. invasion of Afghanistan?

7. *Informative/Explanatory Writing* Some conflicts in Southwest Asia relate to the struggle for a homeland by groups such as the Jews, the Palestinians, and the Kurds. Write a short essay discussing what a homeland is and why groups are willing to fight for one.

Reading **HELP**DESK (CCSS)

Academic Vocabulary

- **widespread**

Content Vocabulary

- **hydropolitics**
- **fossil water**

TAKING NOTES: *Key Ideas and Details*

Determine the Main Idea As you read the lesson, write the main idea for each section on a graphic organizer like the one below.

Main Ideas

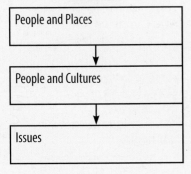

| People and Places |
| People and Cultures |
| Issues |

Lesson 3
Life in Southwest Asia

ESSENTIAL QUESTION · *How does religion shape society?*

IT MATTERS BECAUSE
Because of its strategic location at the convergence of three continents, its huge petroleum reserves, and the deep-rooted conflicts that divide its people, Southwest Asia occupies a central place in world affairs.

People and Places

GUIDING QUESTION *In what parts of Southwest Asia do most people live?*

Southwest Asia's population is slightly greater than that of the United States, although the region is only about three-fourths as large as the United States in area. Throughout history, population patterns in Southwest Asia have been shaped largely by the availability of water. In recent times, another resource—petroleum—has also played an important role.

Population Profile

Southwest Asia is home to about 330 million people. Iran and Turkey, its most populous countries, each have about 80 million people. Some oil-rich countries around the Persian Gulf are experiencing population booms as their fast-growing economies attract foreign workers. Qatar has had one of the world's highest population growth rates in recent years.

Today, many countries of Southwest Asia are highly urbanized. In Israel, Saudi Arabia, and Kuwait, for example, more than four of every five people live in cities. In Afghanistan and Yemen, however, more than two-thirds of the people live in rural areas. However, these countries have the region's highest annual urbanization rate as people move to the cities. As a whole, the region has a rapidly growing population and a high percentage of people below 15 years of age.

Where People Live

Population is not evenly distributed across Southwest Asia. The highest densities are in the region's northern and western parts and in its southern tip. These areas include parts of Turkey, Iraq, Iran, and Afghanistan; the countries along the coast of the Mediterranean Sea; and the highlands of southern Saudi Arabia and southwestern Yemen. Most of these areas have relatively higher rainfall.

Areas with dry or somewhat dry climates are more sparsely populated. These areas include the Arabian Desert and the desert lands that spread across central and eastern Iran. Some desert areas are almost completely uninhabited. One exception is Mesopotamia, the land between the Tigris and Euphrates rivers in Iraq. Although its climate is relatively dry, the area supports high population density because the rivers provide abundant water for irrigating crops.

Southwest Asia has metropolises, such as Istanbul, Damascus, Tehran, and Baghdad, that are home to millions of people. Gleaming modern cities, such as Dubai, Abu Dhabi, and Riyadh, rise from the sands of oil-rich Persian Gulf countries. Tel Aviv, Israel's largest city after Jerusalem, is a thriving urban center. These cities stand in sharp contrast to ancient rural villages that seem untouched by the passing of time. In some of the region's desert areas, nomads, known as Bedouins, sleep in tents and raise herds of camels, sheep, goats, and cattle.

☑ **READING PROGRESS CHECK**

Citing Text Evidence Why do some countries around the Persian Gulf have rapidly growing populations?

At a height of more than 2,700 feet (823 m), the Burj Khalifa (left) is the tallest building in the world. The building is located in Dubai, United Arab Emirates. (right) Adobe storage buildings stand along Al-Assad Lake, a reservoir on the Euphrates River in Syria. A network of canals carries water from the lake to irrigate land on both sides of the Euphrates.

(l) ©Charles Bowman/Robert Harding World Imagery/Corbis; (r) Oliver Gerhard/age fotostock

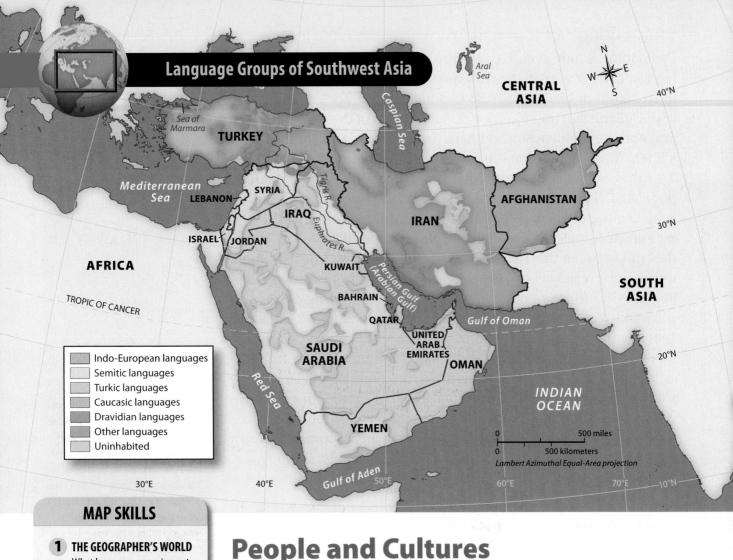

Language Groups of Southwest Asia

Legend:
- Indo-European languages
- Semitic languages
- Turkic languages
- Caucasic languages
- Dravidian languages
- Other languages
- Uninhabited

500 miles
500 kilometers
Lambert Azimuthal Equal-Area projection

MAP SKILLS

1 THE GEOGRAPHER'S WORLD
What language group is most common in Syria?

2 THE GEOGRAPHER'S WORLD
What two language groups are most prevalent in Iran?

Academic Vocabulary

widespread spread out

People and Cultures

GUIDING QUESTION *What cultural differences are found across Southwest Asia?*

Southwest Asia is often thought of as an Arab or an Islamic realm. The reality, however, is more complex. The region, which has always been a crossroads of humanity, is home to many different people.

Ethnic and Language Groups

Arabs represent the largest group in Southwest Asia. In Saudi Arabia, Syria, Jordan, and other countries, 9 out of 10 people are Arab. The two most populous countries in the region, however, have only small Arab populations. In Turkey, Turks form the majority. In Iran, which once was the historical region called Persia, most people are Persian.

In Israel, which was founded as a Jewish state, Jews account for about three-fourths of the population. Kurds, who have no country of their own, represent significant minorities in Turkey, Iran, and Iraq. The region they inhabit is traditionally known as Kurdistan.

Arabic, spoken by Arabs, is the most **widespread** language in Southwest Asia. Other important languages include Turkish and Farsi, the language of Persians. Hebrew is the official language of Israel, and Kurdish is spoken by Kurds.

Some of the region's countries have complex ethnic and linguistic makeups. Afghanistan, for example, is home to Pashtuns, Tajiks, Hazaras, Uzbeks, Aimaks, Turkmen, and Balochs. In addition to the official languages of Afghanistan—Pashto and Afghan Persian—the Afghani people speak Uzbek and more than 30 other languages.

The presence of so many ethnic and language groups in one country presents a challenge to national unity. Many people in Southwest Asia identify with their ethnic group more strongly than with the country they live in. This is clearly evident in countries such as Afghanistan, where people identify themselves as Pashtun or Hazari rather than as Afghani. Even in countries that are mostly Arab, such as Syria and Iraq, people identify with tribes that are based on family relationships. Tribal identity is often stronger than national identity.

Religion and the Arts

From its birthplace in the cities of the Arabian Peninsula, Islam spread across Southwest Asia some 1,300 years ago. It remains the region's dominant religion, helping to unite people of different ethnicity and languages. It is practiced by Arabs, Turks, Persians, Kurds, and many other groups.

Islam has two main branches, Sunni and Shia. Most of Southwest Asia's Muslims are Sunnis. In Iran, however, Shias—Muslims of the Shia branch—outnumber Sunnis nine to one.

Judaism is practiced by about three-fourths of the people in Israel. Christians represent about 40 percent of the population in Lebanon and 10 percent in Syria.

An Iranian woman copies the tile design on a wall at a mosque in Eşfahān, Iran. Islam discourages showing living figures in religious art, so Muslim artists often work in colorful geometric patterns, floral designs, and calligraphy. Passages from the Quran decorate the walls of many mosques.

▶ **CRITICAL THINKING**
Analyzing Why would Muslim artists use passages from the Quran in calligraphy?

A family in Iraq eats a pre-dawn meal before fasting on the second day of the Muslim holy month of Ramadan. Muslims believe that fasting will help people focus on God and on living better lives. According to the Quran, Muhammad first received teachings from God during the month of Ramadan.

▶ **CRITICAL THINKING**

Describing How does religion affect daily life in Southwest Asia?

Religion and art have been closely tied in Southwest Asia throughout history. Some of the region's most magnificent works of architecture are mosques, temples, and other religious structures. Sacred texts such as the Hebrew and Christian Bibles and Islam's Quran stand as works of literature as well as guides to their followers.

The region also has other rich artistic traditions, including calligraphy, mosaics, weaving, storytelling, and poetry. Colorful, handwoven carpets from Persia, or present-day Iran, have been famous for centuries, as has the collection of folktales known as *The Thousand and One Nights*.

Daily Life

Across Southwest Asia, daily life varies greatly. Some people live in cities, some live in villages, and a few live as nomads. Throughout the region's history, most people practiced traditional livelihoods such as farming, raising livestock, or fishing. In recent times, more people have been leaving the land to work in petroleum production, food processing, auto manufacturing, textiles, and construction.

Religion plays a central role in the daily lives of many people in Southwest Asia. Islam is a complete way of life, with rules regarding diet, hygiene, relationships, business, law, and more. To Muslims, families are the foundation of a healthy society; maintaining family ties is an important duty.

Ramadan, the ninth month of the Muslim calendar, is a holy month of fasting. Between dawn and dusk, Muslims are obligated to refrain from eating and drinking. After ending their fast with prayer each evening, people enjoy festive meals. The end of Ramadan is marked by a three-day celebration called *Eid al-Fitr*, which translates as Festival of Breaking Fast.

✔ **READING PROGRESS CHECK**

Identifying What is the major ethnic group in Iran, and what language does that group speak?

Issues

GUIDING QUESTION *How have oil wealth and availability of natural resources created challenges for countries of Southwest Asia?*

The period since World War II has brought a great deal of change and conflict to Southwest Asia. Looking to the future, the region faces many difficult issues. Some of them relate to resources and others to ethnic, religious, and cultural divisions. Some are new, and others are rooted in the distant past.

Oil Dependency and Control

The discovery in the mid-1900s of vast petroleum deposits in Southwest Asia had a strong impact on the region. Exports of petroleum products have brought great wealth to countries around the Persian Gulf, where the largest deposits are found. With this wealth came modernization in some countries. In other countries, little has changed, especially for the average person.

Petroleum has brought new challenges. Many people living in modern cities in oil-producing countries grew up living in tents and practicing traditional farming and herding. Some Muslims believe that increased exposure to Western ways is corrupting the region's people. Another issue is the growing gap between rich and poor countries. Qatar and Kuwait, for example, rank among the wealthiest countries in the world; Afghanistan ranks among the poorest. The struggle to control oil has led to tension and wars. It has also resulted in increased intervention in Southwest Asia by foreign powers.

In 1991 U.S.-led forces pushed Iraqi invaders out of Kuwait. As Iraqi troops left, they set fire to more than 600 oil wells. Tons of thick oil smoke filled the air, and unburned oil spilled into the Persian Gulf. This environmental disaster took place in the area around Kuwait. The air and the soil were polluted, and animal and sea life were destroyed.

©Peter Turnley/Corbis

Because of irrigation, farming is possible in some dry areas of Saudi Arabia. This wheat farm in Saudi Arabia's Nejd region is supported by center-pivot irrigation. This method involves a long pipe of sprinklers that moves in a circle around a deep well that supplies the water.

▶ **CRITICAL THINKING**
Determining Central Ideas Why is fossil water valuable to Saudi Arabia?

Oil dependency is also an issue. Exporting countries thrive when oil commands high prices, but they suffer when worldwide prices drop. Further, oil is not a renewable resource, and the countries have already depleted some of their reserves. To lessen their dependency on oil, exporting countries have invested money in other industries. Countries that import oil are investigating alternatives to oil.

Changing Governments

More than six decades after it began, the Arab-Israeli conflict continues as one of the biggest issues facing Southwest Asia. At the heart of the conflict are the Gaza Strip and West Bank territories, which Israel captured in 1967. Eruptions of violence have hampered progress toward a peaceful solution.

The years 2010 and 2011 marked the beginning of the Arab Spring, a wave of pro-democracy protests and uprisings in North Africa and Southwest Asia. Protests against authoritarian rulers broke out in Tunisia, Egypt, Libya, Yemen, Bahrain, Syria, Jordan, and Oman in early 2011.

By the end of the year, leaders in Tunisia, Egypt, Libya, and Yemen were overthrown. Protests in Bahrain were quashed by security forces, but the government later agreed to implement reforms. Syria fell into upheaval when the government used armed force to stop protests. More peaceful reform efforts are underway in Jordan and Oman.

Water Concerns

Scarcity of freshwater has plagued Southwest Asia throughout history. Dramatic population growth has produced greater demand for this precious resource, making the situation more dire and increasing the importance of **hydropolitics**, or politics related to water usage and access.

Water from the saltwater seas that surround Southwest Asia can be made into freshwater through desalination, or the removal of salt. Unfortunately, this process is expensive and therefore not practical for meeting the region's water needs.

Saudi Arabia, which has no rivers that flow year-round, has tapped into **fossil water**. This term refers to water that fell as rain thousands of years ago, when the region's climate was wetter, and is now trapped between rock layers deep below ground. By pumping the water to the surface for irrigation, Saudi Arabia has transformed areas of barren desert into productive farmland. Fossil water is not a renewable resource, however, and the underground reservoirs could soon run dry.

The region's greatest source of freshwater is the Tigris-Euphrates river system. From their sources in the mountains of eastern Turkey, the Tigris and Euphrates rivers flow southeastward through the desert plains of Syria and Iraq.

The three countries depend heavily on the rivers and their tributaries. In recent decades, all have built dams to control flooding, to generate electricity, and to capture water for irrigation. Syria and Iraq, which are downstream from Turkey, have bitterly opposed an ambitious, decades-long dam-building project in Turkey that threatens to reduce river flow.

☑ **READING PROGRESS CHECK**

Describing What was the Arab Spring? What countries in Southwest Asia were involved?

Include this lesson's information in your Foldable®.

LESSON 3 REVIEW (CCSS)

Reviewing Vocabulary
1. What is *fossil water*?

Answering the Guiding Questions
2. *Identifying* What are Southwest Asia's two most populous countries, and approximately how many people live in each country?

3. *Determining Word Meanings* What is hydropolitics?

4. *Analyzing* How might dams built on the Tigris and Euphrates rivers in Turkey affect agriculture in Syria and Iraq?

5. *Identifying* What are the two main branches of Islam, and to which branch do most Muslims in Southwest Asia belong?

6. *Identifying Point of View* How might Persian Gulf countries be affected if oil-importing countries begin turning to alternate energy sources?

7. *Describing* Who are the Bedouin?

8. *Narrative Writing* Imagine that have you spent your whole life in a poor village somewhere in Southwest Asia. Then, one day you visit Dubai, a bustling, modern city of skyscrapers and shopping malls. Write a letter to a friend or a family member back in your village describing your experience in Dubai.

What Do You Think?

CCSS

Are Trade Restrictions Effective at Changing a Government's Policies?

Sometimes, one country restricts trade with another country as a way to force it to change its policies. For example, if the U.S. government wants a country to give its citizens more democratic rights, it might not allow that country to sell goods in the United States. Although the U.S. government often applies trade restrictions on countries, opinions differ about their effectiveness.

Yes!

PRIMARY SOURCE

" Sanctions aimed at achieving major policy objectives have the strongest chance of success if applied by a multilateral coalition [many countries]; it helps that the multilateral approach also shares the cost. ... Smart sanctions have their use in tightening the screws on a recalcitrant [stubborn] or defiant [openly disobedient] regime [government] without inflicting collateral [additional] damage on the population. ... And even when sanctions don't achieve stated objectives, they nonetheless signal resolve [determination], and may in fact be essential to prepare the political ground at home and abroad for military action against the target."

—Gordon Kaplan, "Making Economic Sanctions Work"

TEXT: From "Making Economic Sanctions Work," ©2010 by Gordon Kaplan. Published by UT San Diego, August 20, 2010; PHOTO: (t) ©Justin Guariglia/Corbis; (b) ©Monty Rakusen/cultura/Corbis

Rows of cargo containers await shipment at Singapore, a city-state in Southeast Asia. Singapore depends on international trade for its survival. Its port is one of the busiest cargo shipping centers in the world.

No !

PRIMARY SOURCE

❝ Imposing sanctions. . .are not only an act of war according to international law, they are most often the first step toward a real war starting with a bombing campaign. We should be using diplomacy rather than threats and hostility. Nothing promotes peace better than free trade. Countries that trade with each other generally do not make war on each other, as both countries gain economic benefits they do not want to jeopardize. Also, trade and friendship applies much more effective persuasion to encourage better behavior, as does leading by example. ❞

—Ron Paul, "The Folly of Sanctions"

A port security worker checks cargo at a warehouse in London, England. Ships from all over the world pass through major international ports, such as London.

What Do You Think? DBQ

❶ *Identifying Point of View* According to Kaplan, why are sanctions effective even if they do not achieve their objectives?

❷ *Identifying Point of View* Why does Paul think that free trade is better than economic sanctions for promoting peace?

❸ *Analyzing* Who do you think makes the stronger argument, Kaplan or Paul?

Directions: Write your answers on a separate piece of paper.

1 Use your **FOLDABLES** to explore the Essential Question.
INFORMATIVE/EXPLANATORY WRITING Choose one of the countries in this region and learn about its natural resources. Then write an essay to answer: Is the country using its resources wisely?

2 **21st Century Skills**
INTEGRATING VISUAL INFORMATION Working in small groups, identify one problem facing the countries of Southwest Asia. Research its effects on the people of the region and offer possible solutions. Produce a slide show or build a visual display to share your findings.

3 **Thinking Like a Geographer**
IDENTIFYING Identify five countries of Southwest Asia. List the countries and their capital cities. Then write one interesting fact about each country.

4 **GEOGRAPHY ACTIVITY**

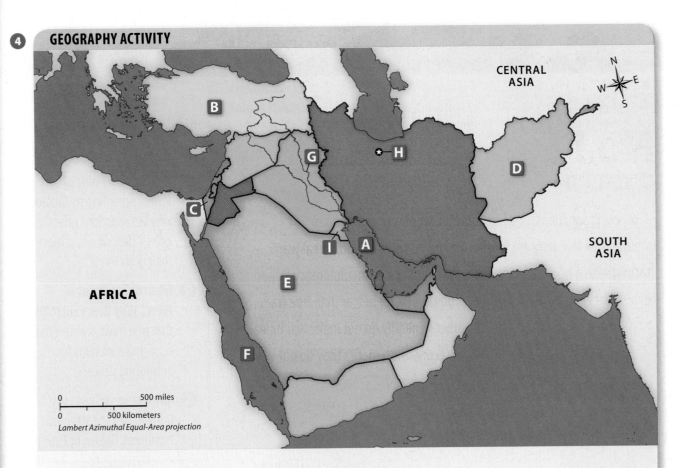

Locating Places
Match the letters on the map with the numbered places below.

1. Saudi Arabia
2. Tehran
3. Afghanistan
4. Israel
5. Red Sea
6. Kuwait
7. Persian Gulf
8. Tigris River
9. Turkey

REVIEW THE GUIDING QUESTIONS

Directions: Choose the best answer for each question.

1 Which scarce resource has most directly shaped Southwest Asia's history and settlement patterns?
   A. rich soil
   B. natural gas
   C. water
   D. forests

2 Southwest Asia's physical geography can be described as one of
   F. high mountains.
   G. extremes.
   H. little variety.
   I. sandy deserts.

3 How long ago did humans convert from living as hunter-gatherers to living in settlements and practicing agriculture?
   A. a million years ago
   B. 10,000 years ago
   C. 50,000 years ago
   D. 100,000 years ago

4 Which three major world religions originated in Southwest Asia?
   F. Sikhism, Hinduism, Judaism
   G. Christianity, Judaism, Hinduism
   H. Islam, Judaism, Christianity
   I. Judaism, Islam, Hinduism

5 What twentieth-century discovery brought great wealth to some countries in Southwest Asia?
   A. the cell phone
   B. vast petroleum deposits
   C. gold
   D. rubies and emeralds

6 The population of Southwest Asia is
   F. growing rapidly.
   G. declining.
   H. aging.
   I. leaving to find work in India.

DBQ ANALYZING DOCUMENTS

7 **CITING TEXT EVIDENCE** Read the following passage about Israel's economy:

"*Israel has a diversified, technologically advanced economy. . . . The major industrial sectors include high-technology electronic and biomedical equipment, metal products, processed foods, chemicals, and transport equipment. . . . Prior to the violence that began in September 2000, [Israel] was a major tourist destination.*"

—from U.S. State Department Background Notes, "Israel"

Which detail supports the idea that Israel's economy is technologically advanced?

A. biomedical equipment industry

B. processed food industry

C. transport equipment industry

D. tourism industry

8 **ANALYZING** What inference can you draw about the decline of tourism to Israel after 2000?

F. Economic hard times reduced tourism to all locations.

G. Other places became more fashionable as tourist destinations.

H. Tourism declined because people were worried about their safety.

I. Lower prices for air travel would revive tourism to Israel.

SHORT RESPONSE

"*One challenge the Saudis face in achieving their strategic vision to add production capacity is that their existing fields experience 6 to 8 percent annual 'decline rates' on average . . . , meaning that the country needs around 700,000 [billion barrels per day] in additional capacity each year just to [make up] for natural decline.*"

—from Energy Information Administration, *Saudi Arabia*

9 **DETERMINING WORD MEANINGS** Based on this passage, what do the "decline rates" refer to?

10 **ANALYZING** What could cause these decline rates?

EXTENDED RESPONSE

11 **INFORMATIVE/EXPLANATORY WRITING** Write an essay explaining why you think conflict in this part of the world since the end of World War II has increased so dramatically. Consider how conflict and wars in this part of the world affect you and your family.

Need Extra Help?

If You've Missed Question	**1**	**2**	**3**	**4**	**5**	**6**	**7**	**8**	**9**	**10**	**11**
Review Lesson	1	1	2	2	3	3	3	3	1	1	2

McGraw-Hill
netw⊕rks™

AFRICA

UNIT 6

EXPLORE the CONTINENT

AFRICA The continent of Africa covers about one-fifth of Earth's total land area. Africa is the world's second-largest continent, after Asia. For the 54 independent countries located in Africa, physical features such as the Sahara, the Congo River, and the Great Rift Valley have greatly influenced how people live in this environment.

1 **NATURAL RESOURCES** The region has an abundance of minerals. South Africa, Botswana, and the Democratic Republic of the Congo are among the countries that have valuable deposits of gold, uranium, diamonds, and other minerals. These miners work in one of the world's deepest gold mines in South Africa.

2 **BODIES OF WATER** Many of Africa's lakes lie in huge basins formed millions of years ago by the uplifting of the land. The Nile, the world's longest river, flows from the East African highlands through Egypt to the Mediterranean Sea.

(3) LANDFORMS Africa's geography is made up of a variety of landforms. Step-like plateaus rise from the coasts to inland mountains. Tropical grasslands with scattered trees cover almost one half of the continent. These giraffes are among the millions of animals—zebras, gazelles, lions, and cheetahs—that roam East Africa's Serengeti, one of the world's largest plains.

FAST FACT

Africa is three times the size of the United States.

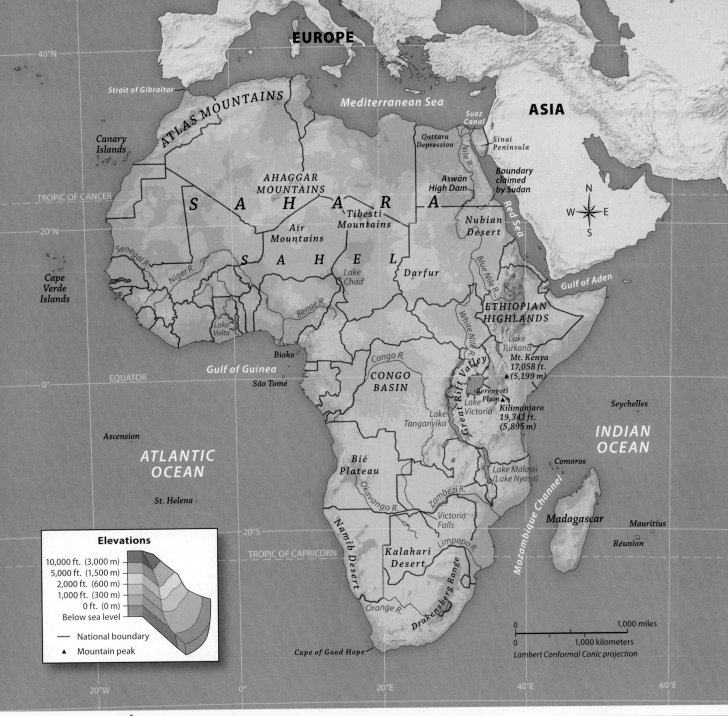

EUROPE

Strait of Gibraltar

ATLAS MOUNTAINS

Mediterranean Sea

Suez Canal

ASIA

Canary Islands

Qattara Depression

Sinai Peninsula

TROPIC OF CANCER

AHAGGAR MOUNTAINS

S A H A R A

Aswān High Dam

Nile R.

Boundary claimed by Sudan

Nubian Desert

Red Sea

Tibesti Mountains

Aïr Mountains

S A H E L

Senegal R.

Niger R.

Lake Chad

Darfur

Blue Nile R.

Gulf of Aden

Cape Verde Islands

Benue R.

ETHIOPIAN HIGHLANDS

Lake Volta

Lake Turkana

Bioko

Congo R.

White Nile R.

Mt. Kenya 17,058 ft. ▲(5,199 m)

Gulf of Guinea

CONGO BASIN

Serengeti Plain ▲

São Tomé

EQUATOR

Great Rift Valley

Lake Victoria

Kilimanjaro 19,341 ft. (5,895 m)

Seychelles

Lake Tanganyika

Ascension

ATLANTIC OCEAN

Bié Plateau

Lake Malawi (Lake Nyasa)

Comoros

INDIAN OCEAN

St. Helena

Okavango R.

Mozambique Channel

Madagascar

Mauritius

Zambezi R.

Réunion

Namib Desert

Victoria Falls

Limpopo R.

Kalahari Desert

Drakensberg Range

Orange R.

Cape of Good Hope

Elevations

10,000 ft. (3,000 m)
5,000 ft. (1,500 m)
2,000 ft. (600 m)
1,000 ft. (300 m)
0 ft. (0 m)
Below sea level

— National boundary
▲ Mountain peak

0   1,000 miles
0   1,000 kilometers
Lambert Conformal Conic projection

AFRICA

PHYSICAL

MAP SKILLS

1 PLACES AND REGIONS What is the major mountain range of North Africa?

2 THE GEOGRAPHER'S WORLD Which of Africa's largest lakes lies farthest to the south?

3 PHYSICAL GEOGRAPHY What is the highest mountain in Africa?

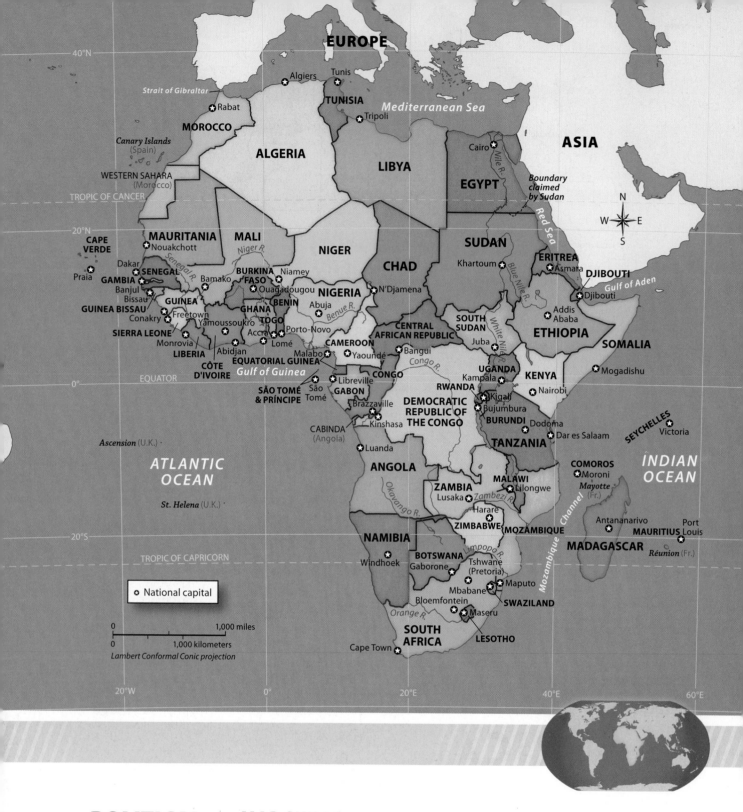

EUROPE

Strait of Gibraltar

Algiers · Tunis
Rabat ·

MOROCCO

Canary Islands (Spain)

WESTERN SAHARA (Morocco)

TROPIC OF CANCER

ALGERIA

TUNISIA
Tripoli ·

Mediterranean Sea

Cairo ·

LIBYA

EGYPT

Boundary claimed by Sudan

ASIA

Red Sea

N
W · E
S

20°N

CAPE VERDE

Praia ·

Nouakchott ·

MAURITANIA

Dakar ·

SENEGAL

Bissau

GUINEA BISSAU

Banjul

GAMBIA

Senegal R.

MALI

Bamako ·

BURKINA FASO

Ouagadougou ·

Niger R.

Niamey ·

NIGER

Abuja ·

NIGERIA

CHAD

N'Djamena ·

Benue R.

Khartoum ·

SUDAN

Blue Nile R.

Asmara ·

ERITREA

DJIBOUTI
Djibouti ·

Gulf of Aden

Addis Ababa ·

Conakry ·

GUINEA

Freetown ·

SIERRA LEONE

Monrovia ·

LIBERIA

GHANA

Yamoussoukro ·

Accra ·

TOGO

Lomé ·

Abidjan ·

CÔTE D'IVOIRE

BENIN

Porto-Novo ·

CAMEROON

Yaoundé ·

CENTRAL AFRICAN REPUBLIC

Bangui ·

SOUTH SUDAN

Juba ·

White Nile R.

ETHIOPIA

SOMALIA

Mogadishu ·

EQUATOR

Malabo

EQUATORIAL GUINEA

Gulf of Guinea

SÃO TOMÉ & PRÍNCIPE

São Tomé ·

Libreville ·

GABON

CONGO

Brazzaville ·

Kinshasa ·

Congo R.

DEMOCRATIC REPUBLIC OF THE CONGO

RWANDA

Kigali ·

BURUNDI

Bujumbura ·

UGANDA

Kampala ·

KENYA

Nairobi ·

Dodoma ·

Dar es Salaam

SEYCHELLES

Victoria ·

0°

Ascension (U.K.) ·

CABINDA (Angola)

Luanda ·

ATLANTIC OCEAN

St. Helena (U.K.) ·

ANGOLA

Okavango R.

TANZANIA

MALAWI

Lilongwe ·

COMOROS

Moroni ·

Mayotte (Fr.)

Mozambique Channel

INDIAN OCEAN

20°S

ZAMBIA

Lusaka ·

Zambezi R.

Harare ·

ZIMBABWE

MOZAMBIQUE

Antananarivo ·

MAURITIUS

Port Louis

TROPIC OF CAPRICORN

NAMIBIA

Windhoek ·

BOTSWANA

Gaborone ·

Limpopo R.

MADAGASCAR

Réunion (Fr.)

✪ National capital

Tshwane (Pretoria) ·

Maputo ·

Mbabane ·

SWAZILAND

0   1,000 miles

0   1,000 kilometers

Lambert Conformal Conic projection

Orange R.

Bloemfontein ·

SOUTH AFRICA

Maseru ·

LESOTHO

Cape Town ·

20°W   0°   20°E   40°E   60°E

POLITICAL

MAP SKILLS

1. **PLACES AND REGIONS** Which country in the region of North Africa is the largest in area?

2. **PLACES AND REGIONS** Name three island countries of Africa.

3. **THE GEOGRAPHER'S WORLD** Which countries border Namibia in Southern Africa?

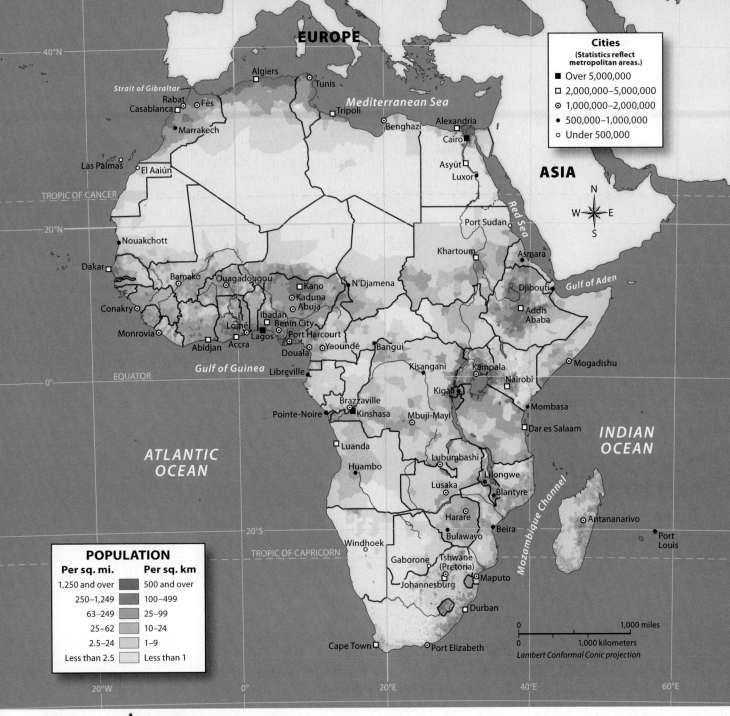

POPULATION

Per sq. mi.	Per sq. km
1,250 and over	500 and over
250–1,249	100–499
63–249	25–99
25–62	10–24
2.5–24	1–9
Less than 2.5	Less than 1

Cities
(Statistics reflect metropolitan areas.)
- ■ Over 5,000,000
- ▢ 2,000,000–5,000,000
- ◉ 1,000,000–2,000,000
- ● 500,000–1,000,000
- ○ Under 500,000

0   1,000 miles
0   1,000 kilometers
Lambert Conformal Conic projection

AFRICA

POPULATION DENSITY

MAP SKILLS

1 PLACES AND REGIONS Which areas of Africa have the lowest population density?

2 THE GEOGRAPHER'S WORLD Generalize about the areas of Africa with the highest population densities. What can you say about these places?

3 PLACES AND REGIONS What are the largest cities in Africa?

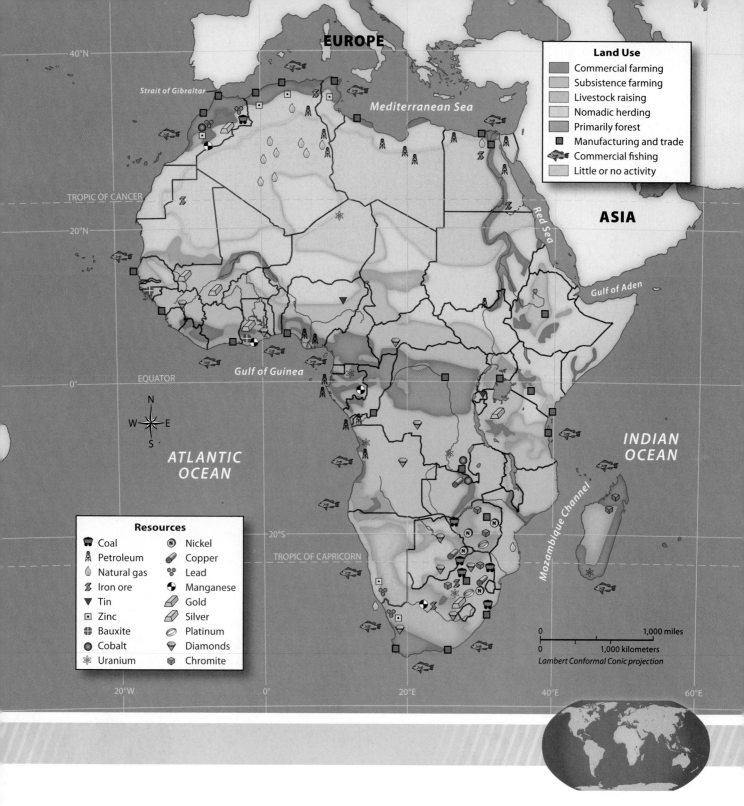

Land Use
- Commercial farming
- Subsistence farming
- Livestock raising
- Nomadic herding
- Primarily forest
- Manufacturing and trade
- Commercial fishing
- Little or no activity

Resources
- Coal
- Petroleum
- Natural gas
- Iron ore
- Tin
- Zinc
- Bauxite
- Cobalt
- Uranium
- Nickel
- Copper
- Lead
- Manganese
- Gold
- Silver
- Platinum
- Diamonds
- Chromite

EUROPE

Strait of Gibraltar

Mediterranean Sea

ASIA

TROPIC OF CANCER

Red Sea

Gulf of Aden

EQUATOR

Gulf of Guinea

ATLANTIC OCEAN

INDIAN OCEAN

Mozambique Channel

TROPIC OF CAPRICORN

0   1,000 miles
0   1,000 kilometers
Lambert Conformal Conic projection

40°N

20°N

0°

20°S

20°W   0°   20°E   40°E   60°E

ECONOMIC RESOURCES

MAP SKILLS

1 **ENVIRONMENT AND SOCIETY** Where are precious minerals mined in different areas of Africa?

2 **HUMAN GEOGRAPHY** What is the most common type of farming in Africa?

3 **PLACES AND REGIONS** Why do you think the Nile River valley specializes in commercial farming?

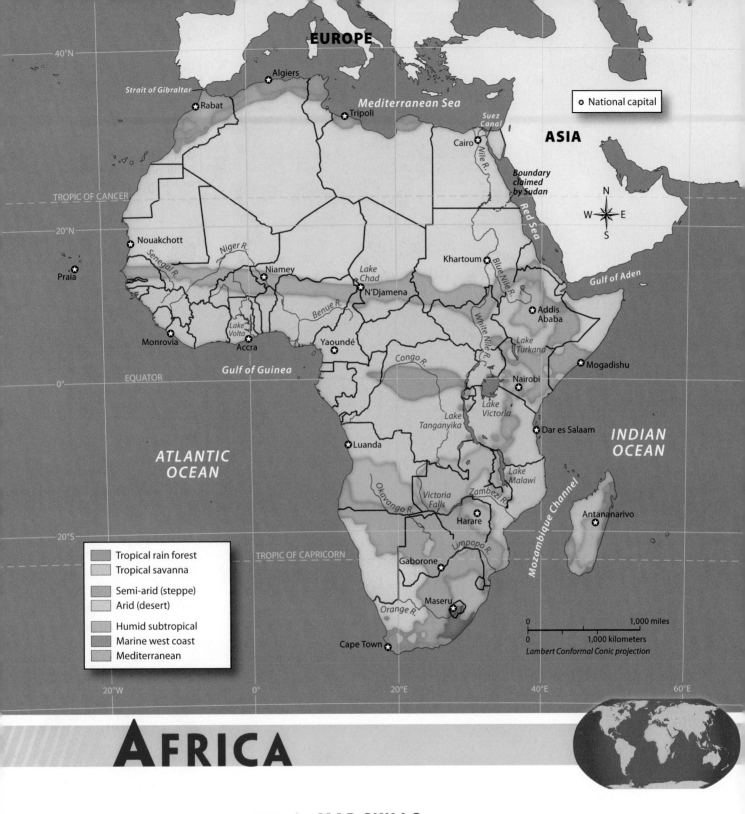

AFRICA

CLIMATE

Legend:
- Tropical rain forest
- Tropical savanna
- Semi-arid (steppe)
- Arid (desert)
- Humid subtropical
- Marine west coast
- Mediterranean

○ National capital

MAP SKILLS

1 PLACES AND REGIONS In the area of Africa south of the Sahara, what is the predominant climate?

2 PHYSICAL GEOGRAPHY What climate types are in the area around the Congo River?

3 PLACES AND REGIONS What is the climate of Cape Town, South Africa?

NORTH AFRICA

ESSENTIAL QUESTIONS • *How do people adapt to their environment?*
• *How does religion shape society?* • *Why do conflicts develop?*

The Story Matters...

Because of the fertile soil in the Nile River valley, ancient Egyptians created a farming society that developed into an empire thousands of years ago. Their great achievements had a tremendous impact on later civilizations in the region. Other waterways influenced the development of trade and communications among Africa, Asia, and Europe. Waterways and trade routes were also important to the spread of ideas and religion, including Islam.

FOLDABLES®
Study Organizer

Go to the Foldables® library in the back of your book to make a Foldable® that will help you take notes while reading this chapter.

	Know	Learned
Geography		
History		
Economy		

Tips Images/SuperStock

Souks, or markets, such as this one in Esna, Egypt, carry a wide assortment of goods.

NORTH AFRICA CCSS

The countries of North Africa border the Mediterranean Sea. They make up one region of the African continent. To their south, the Sahara crosses Africa, and to the south of the Sahara lie the other regions of the second-largest continent.

Step Into the Place

MAP FOCUS Use the map to answer the following questions.

1. **PLACES AND REGIONS** What is the capital of Morocco?

2. **THE GEOGRAPHER'S WORLD** What body of water flows to the east of Egypt?

3. **THE GEOGRAPHER'S WORLD** Which North African country has the largest land area?

4. **CRITICAL THINKING Analyzing** What do the capitals of the North African countries have in common geographically?

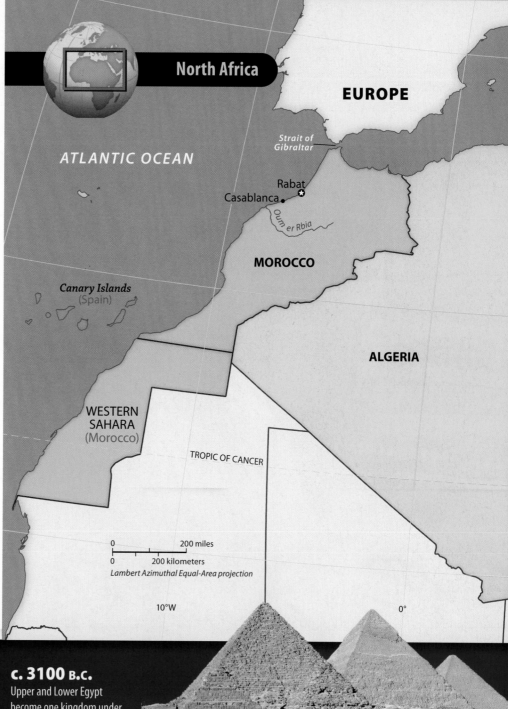

North Africa

EUROPE

ATLANTIC OCEAN

Strait of Gibraltar

Rabat
Casablanca

Oum er Rbia

MOROCCO

Canary Islands (Spain)

ALGERIA

WESTERN SAHARA (Morocco)

TROPIC OF CANCER

0   200 miles
0   200 kilometers
Lambert Azimuthal Equal-Area projection

10°W   0°

Step Into the Time

TIME LINE Choose an event from the time line and write a paragraph explaining the social, economic, or environmental effects of that event on the region and the world.

c. 3100 B.C. Upper and Lower Egypt become one kingdom under the pharaoh Menes

c. 5000 B.C. People begin farming along the Nile River

A.D. 644 Islam spreads to Egypt and other parts of North Africa

B.C. | A.D.   **1800**

c. 3200 B.C. Ancient Egyptians develop hieroglyphic writing

40°N

SOUTHWEST ASIA

Tunis

Medjerda R.

Mediterranean Sea

TUNISIA

Tripoli

● Benghazi

N
W   E
S

○ National capital
● City

Alexandria

Suez Canal

30°N

Cairo

LIBYA

EGYPT

Nile R.

Red Sea

Aswān High Dam

Lake Nasser

10°E

20°N

1830
Algeria, Tunisia, and Morocco become part of the French Empire

1969
Muammar al-Qaddafi seizes power in Libya

2010–2011 Pro-democracy revolts known as the Arab Spring take place in Tunisia, Egypt, and Libya

1956 Oil is discovered in Libya

1900

2000

1859–1869 The Suez Canal is built, linking the Mediterranean and Red seas

1970 The Aswān High Dam is completed

2003 Earthquake in northern Algeria leaves 200,000 people homeless

2011 Egyptian president Mubarak leaves office

Reading **HELP**DESK CCSS

Academic Vocabulary

- **margin**
- **channel**

Content Vocabulary

- **delta**
- **silt**
- **wadi**
- **erg**
- **nomad**
- **phosphate**
- **aquifer**

TAKING NOTES: *Key Ideas and Details*

Organize Information As you read about North Africa's physical geography, take notes using the graphic organizer below. Add rows to list more features and their characteristics.

Feature	Characteristic
Atlas Mountains	

Lesson 1
The Physical Geography of North Africa

ESSENTIAL QUESTION · *How do people adapt to their environment?*

IT MATTERS BECAUSE
The Sahara, in North Africa, is the world's largest hot desert. The desert extends over almost the entire northern one-third of the continent of Africa.

Landforms and Waterways

GUIDING QUESTION *How have physical features shaped life in the region?*

Hassan is an Egyptian farmer. In winter, when temperatures are milder than during the summer, he grows wheat. Rainfall is scarce in Egypt, however. How does Hassan get the water he needs to grow his wheat? He draws it from a canal. Canals carry water from the Nile River to the country's farms around the river. Just as they did thousands of years ago, Egypt's farmers still depend on the waters of the Nile.

Countries of the Region

Egypt is the easternmost country in North Africa. The Sinai Peninsula, a triangle of land across the Red Sea from Africa, belongs to Egypt but it is considered a part of Southwest Asia.

North Africa includes five countries. All, like Egypt, sit on the southern shore of the Mediterranean Sea. Libya is to Egypt's west. Tunisia and Algeria are west of that country. Farthest west is Morocco, which has a small Mediterranean coast and a longer coast along the Atlantic Ocean. South of Morocco lies an area called Western Sahara. Morocco claims this area, although the United Nations does not recognize its ownership of this land.

North Africa is a large region. If you placed North Africa over the 48 connected states of the United States, it would reach from Maine to Washington state and cover the northern half of the country.

Coastal Plains and Mountains

In North Africa, low, narrow plains sit on the **margins**, or edges, of the Mediterranean and Atlantic coasts. In the west, the high Atlas Mountains rise just behind this coastal plain. These mountains extend about 1,200 miles (1,931 km) across Morocco and Algeria into Tunisia. They form the longest mountain chain in Africa and greatly influence the region's climate.

The Atlas Mountains are actually two sets of mountains that run alongside each other. A high plateau sits between them. The southern chain is generally higher than the one to the north. It includes Mount Toubkal in Morocco. At 13,665 feet (4,165 m), it is the highest peak in North Africa.

South of these mountains is a low plateau that reaches across most of North Africa. The land rises higher in a few spots formed by isolated mountains. In Egypt, the southern reaches of the Nile River cut through a highland area to form a deep gorge, or valley. Southeastern Egypt has low mountains on the shores of the Red Sea. The southern part of Egypt's Sinai Peninsula is also mountainous. This area includes Egypt's highest point, Gebel Katherina. It reaches 8,652 feet (2,637 m) high. Another set of low mountains lies in northeastern Libya, near the coast.

Lowlands

Northwestern Egypt has a large area of lowland. Called the Qattara Depression, it sinks 440 feet (134 m) below sea level. This area is nearly the size of New Jersey. Marshes and lakes prevent cars and trucks from passing through it.

©Amar Grover/JAI/Corbis

Academic Vocabulary

margin an edge

A village stands at the foot of a large granite formation in the Atlas Mountains of Morocco.

▶ **CRITICAL THINKING**
Describing How many sets of mountains make up the Atlas Mountains? What landform separates the sets of mountains?

Lush green farmland contrasts sharply with the vast desert areas that stretch for hundreds of miles on either side of the Nile River.

▶ **CRITICAL THINKING**

Analyzing How is the relationship of Egyptians to the Nile River today different from the relationship of ancient Egyptians to the river?

Academic Vocabulary

channel course

Waterways

For centuries, North Africa has been linked by the Mediterranean Sea to other lands. The sea has brought trade, new ideas, and conquering armies.

Next to the Mediterranean, the most important body of water in the region is the Nile River. At 4,160 miles (6,695 km), the mighty Nile is the longest river in the world. It begins far south of Egypt at Lake Victoria in East Africa. That lake sits on the border of Uganda and Tanzania. The river flows northward, joined by several tributaries. The most important of them is the Blue Nile, which begins in the highlands of Ethiopia.

The Nile has a massive delta at its mouth. A **delta** is an area formed by soil deposits that build up as river water slows down. Many deltas form where a river enters a larger body of water. The Nile delta is found where the Nile meets the Mediterranean Sea. Here, at the mouth of the river, the Nile's delta covers more than 9,500 square miles (24,605 sq. km)—larger than the size of New Hampshire. The river once took seven different **channels**, or courses, to reach the sea. Today, only two remain. The others have been filled with soil.

The Nile brings life to dry Egypt. In ancient times, filled by rains to the south, the Nile flooded each year. These floods left **silt**—a fine, rich soil that is excellent for farming—along the banks of the river and in the delta. Farmers used the soil to grow crops. Because they could grow large amounts of food, they were able to support the growth of a great civilization. Ancient Egypt was called "the gift of the Nile."

©B. Anthony Stewart/National Geographic Society/Corbis

Today, several dams control the floods. The largest is Aswān High Dam. These dams hold back the high volume of water produced in the rainy season. The water can then be released during the year. An important benefit is that Egypt's farmers today can grow crops year-round. This is the water the farmer Hassan uses to grow his wheat. Another benefit of the dams is that people in Egypt have security from floods. One negative consequence of the dams is that the silt no longer settles on the land and enriches the soil.

Egypt controls another important waterway. This one, the Suez Canal, is human-made. The canal connects the Mediterranean Sea to the Red Sea. As a result, it links Europe and North Africa to the Indian and Pacific oceans. International trade depends on this canal. Using it enables ships traveling between Asia and Europe to avoid going all the way around Africa. The Suez Canal saves many days of travel time and much costly fuel.

☑ **READING PROGRESS CHECK**

Citing Text Evidence Why was ancient Egypt called "the gift of the Nile"?

Climate

GUIDING QUESTION *How do people survive in a dry climate?*

What would it be like if it hardly ever rained? That is the situation that many North Africans face. Large areas of the region receive only a few inches of rainfall each year—if that much.

A container ship passes through Egypt's Suez Canal, one of the world's most heavily used shipping lanes. Opened in 1869, the canal has been enlarged over the years to handle much bigger ships.

▶ **CRITICAL THINKING**
Describing What advantages does the Suez Canal provide for ship travel?

Causes of North Africa's Climates

The Atlas Mountains play a major role in controlling the climate in the western part of North Africa. These mountains create the rain shadow effect. Moist air blows southward from the Atlantic Ocean and the Mediterranean Sea toward the mountains. As the air rises up the northern slopes, it cools and releases rain. By the time it passes over the mountains, the air is dry. This dry air reaches the interior. Inland areas, then, remain arid.

The vast inland area of North Africa is dry for another reason. High-pressure air systems descend over areas to the south of the region for much of the year. They send hot, dry air blowing to the north. This air mass dries out the land. On the rare occasions when it does rain in the desert, the southern winds soon follow. They dry the land and leave behind **wadis**, or dry streambeds.

Desert and Semiarid Areas

Much of North Africa, then, is covered by a desert: the Sahara. Imagine a vast expanse of space, like an ocean, but covered in sand and rock. That is what the Sahara looks like. Spreading across more than 3.5 million square miles (9.1 million sq. km), the Sahara is as large as the entire United States. It covers most of North Africa and spills into three other regions of Africa, as well.

The Sahara's vast stretches of sand are called **ergs**. Strong winds blow the sand about, creating huge dust storms that choke people and animals that are caught outside. The winds also build towering sand dunes. When new winds blow, they can change the shape and size of those dunes.

Landscapes in the Sahara include rugged mountains, stony plains, and large sand dunes. A traveler (left) leads a camel caravan past towering dunes in the Moroccan part of the Sahara. Farther east, Egypt's part of the Sahara (right)—called the Libyan Desert—has rocky surfaces.

▶ **CRITICAL THINKING**

Describing How are sand dunes formed?

Ergs cover only about a quarter of the Sahara. In other areas, rocky plateaus called *hamadas* and rocks eroded by wind are common. Some areas contain oases, areas fed by underground sources of water. Plants can grow in oases and trade caravans that cross the desert stop at them for needed water. **Nomads**, people who move about from place to place in search of food, rely on these oases during their travels. They use the plants to graze herds of sheep or other animals. Some people live on oases and grow crops.

In the North African part of the Sahara, temperatures soar during the day in the summer. They can reach as high as 136°F (58°C). During the winter, though, daytime temperatures can drop as low as 55°F (13°C).

Mediterranean and Other Climates

North of the desert are different climate zones. A band of steppes encircle the desert immediately to the north. Temperatures here are high, and rainfall is slightly greater than in the desert. This band extends to the eastern coast. Coastal cities in Libya receive only 10 inches to 15 inches (25 cm to 38 cm) of rain per year. Alexandria, near Egypt's coast, generally receives only 7 inches (18 cm) of rainfall per year.

A Mediterranean climate dominates the western coast. This climate gives the region warm, dry summers and mild, rainy winters. More rain falls along the coast than in the dry interior. Rain amounts are higher in the west than in the east. In the west, they are higher on the mountain slopes than along the coast. Coastal areas of Morocco receive 32 inches (81 cm) or less of rain per year.

Coastal areas of North Africa have a Mediterranean climate that is well suited for growing cereal crops, citrus fruits, grapes, olives, and dates.

Think Again

Was the Sahara always a desert?

No. Thousands of years ago, the Sahara received more rainfall than it does now. Over time, though, the climate changed. The Sahara became dry—as it remains today.

As North Africa's population grows, the demand for water increases. This pump provides water from deep underground to people living in a Sahara environment.

▶ **CRITICAL THINKING**

Describing Why is so much water available underground in parts of the Sahara?

Mountain areas with highland climates also receive more rainfall—as much as 80 inches (203 cm) per year. Highland climates are found within the mountains. Morocco's Atlas Mountains often are covered by snow in the winter. As hard as it might be to believe, just a few hundred miles north of the Sahara, people can snow ski.

☑ **READING PROGRESS CHECK**

Analyzing Where do you think most people in North Africa live? Explain why this might be so.

Resources

GUIDING QUESTION *What resources does North Africa have?*

Oil and natural gas are resources that we use to power our cars and trucks and to generate electricity and heat. Some countries of North Africa have these resources in large quantities. All five countries in the region, though, struggle to get enough of another precious resource—water.

Oil, Gas, and Other Resources

Libya is the most oil-rich country in North Africa. Its oil reserves are ranked ninth in the world and it exports more oil than all but 15 other countries. Libya also has natural gas, but in lesser amounts. The money Libya earns from oil has fueled its economy.

Algeria has large reserves of natural gas—more than all but nine other countries. It also has large supplies of oil. These two resources make up nearly all of its exports.

Like Algeria, Egypt has larger reserves of natural gas than oil. Still, it has enough oil to supply most of what it consumes each year. Egypt even sells a small amount to other countries.

Tunisia's main resources are iron ore and phosphates. **Phosphates** are chemical compounds that are often used in fertilizers. These products are important in Morocco, as well. In addition, rich fishing grounds off Morocco's coast are a vital resource. Fish is one of that country's leading exports.

Water

Limited rainfall and high temperatures in this region leave little freshwater on the surface. Rains can be heavy when they come, but the sandy soil soon absorbs the water. Dry winds evaporate the rest. Only the Nile is a reliable source of water for farming throughout the year.

How vital is the Nile? Ninety-five out of every 100 Egyptians live within 12 miles (19 km) of the Nile River or its delta. Yet this narrow river valley and the large delta make up only a small part of Egypt's total area. Without the waters of the Nile, Egypt's people could not survive.

Outside of the Nile valley, most of the region's water needs are met with water that comes from oases and aquifers. **Aquifers** are underground layers of rock in which water collects. People use wells to tap into this water. Libya, for instance, relies on aquifers to meet almost all of its water needs. However, nearly half of Libya's people have no access to water that has been treated to be sure it meets health standards.

A growing population in this region poses problems for the future. Demand for the water in an aquifer shared by Algeria, Libya, and Tunisia has increased ninefold in recent years. In North Africa, aquifers take a long time to refill. If people continue to take water out at a high rate, the aquifers might not be able to refill quickly enough and the region's water problem will become much worse.

Include this lesson's information in your Foldable®.

☑ **READING PROGRESS CHECK**

Analyzing Why would aquifers take a long time to fill up in North Africa?

LESSON 1 REVIEW (CCSS)

Reviewing Vocabulary

1. In which desert feature can people live year-round, a *wadi* or an oasis? Why?

Answering the Guiding Questions

2. *Describing* How has the Mediterranean Sea affected the region?

3. *Analyzing* Does the northern or the southern chain of the Atlas Mountains receive more rainfall? Why?

4. *Determining Central Ideas* Which nations in the region are likely to import energy resources? Why?

5. *Analyzing* How can governments in the region prevent aquifers from being used up?

6. *Informative/Explanatory Writing* Write a paragraph comparing and contrasting the climates of Egypt and Morocco.

networks

There's More Online!

☑ **IMAGES** Islam

☑ **MAP** The Punic Wars

☑ **ANIMATION** How the Pyramids Were Built

☑ **SLIDE SHOW** Egyptian Artifacts

☑ **VIDEO**

Reading **HELP**DESK CCSS

Academic Vocabulary

- **project**
- **demonstrate**

Content Vocabulary

- **pharaoh**
- **myrrh**
- **hieroglyphics**
- **convert**
- **monotheism**
- **caliph**
- **regime**
- **fundamentalist**
- **civil war**

TAKING NOTES: *Key Ideas and Details*

Summarize As you read about the history of North Africa, note key events and their importance using a graphic organizer like the one below.

Year, Event	Importance

Lesson 2
The History of North Africa

ESSENTIAL QUESTION · *How does religion shape society?*

IT MATTERS BECAUSE
One of the world's first civilizations arose in North Africa thousands of years ago.

Ancient Egypt

GUIDING QUESTION *Why was ancient Egypt important?*

Egypt, in North Africa, was one of the earliest known civilizations. Egyptian civilization arose along the Nile River, and Egyptians depended on the Nile for their livelihood. They built cities, organized government, and invented a writing system to keep records and create literature.

The Rise of Egypt

People have been living along the banks of the Nile River for thousands of years. As many as 8,000 years ago, people settled in the area to farm. The rich floodwaters of the Nile allowed farmers to produce enough food to support a growing population. Over time, some members of this early society began to do other things besides farming. Some made pottery. Others crafted jewelry. Some became soldiers. A few became kings.

About 5,000 years ago, two kingdoms along the Nile were united into one. For most of the next 3,000 years, kings called **pharaohs** ruled the land. The great mass of people farmed the land. They paid a share of their crops to the government. The government's leaders also made them work on important **projects**, or planned activities. These projects included building temples and other monuments. Sometimes the people had to fight in the pharaoh's armies.

The Expansion of Egypt

For centuries, Egypt traded with nearby lands. Merchants carried Egyptian grain and other products to the south. There they traded for luxury goods like gold, ivory, and incense. They also traded to the east for wood from what is now Lebanon.

Around 1500 B.C., the Egyptians decided to expand their area. They took control of lands to the south that held gold and seized areas along the Red Sea that had **myrrh**. This plant substance gives off a pleasing scent. Priests burned it in religious ceremonies. Egypt also conquered the eastern shores of the Mediterranean. That gave them control of the timber there. Egypt's kings gained wealth by taxing conquered peoples.

Religion and Culture in Ancient Egypt

The pharaoh was the head of Egyptian society. He was seen as more than a man. He was thought to be the son of the sun god. The Egyptians practiced polytheism, which is the belief in many gods. The sun god was one of the most important of their gods. His daily journey through the sky brought the warmth needed to grow crops. The pharaoh, Egyptians believed, connected them to the gods. He made sure that they would flourish as a people.

KENNETH GARRETT/National Geographic Stock

Academic Vocabulary

project a planned activity

One of the most famous Egyptian pharaohs was the boy-king Tutankhamen. At 10 years of age, Tutankhamen became ruler of Egypt, but he died unexpectedly nine years later.

▶ **CRITICAL THINKING**
Describing Based on the map, describe the area controlled by ancient Egypt.

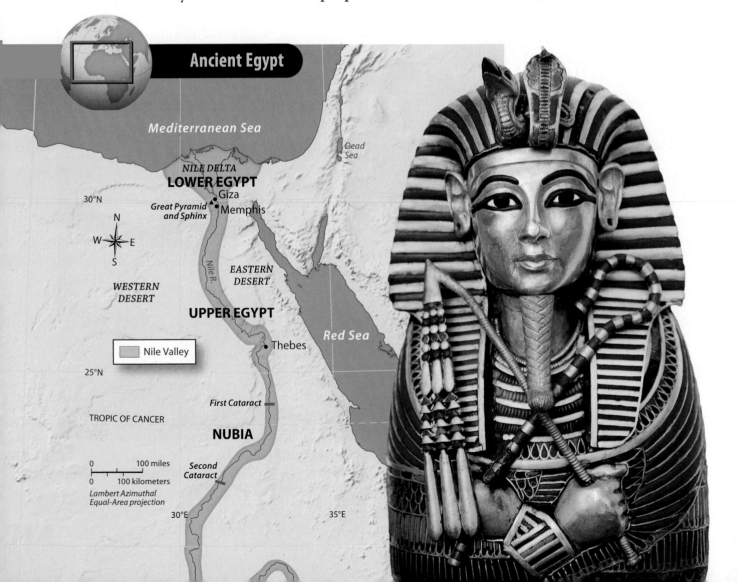

Ancient Egypt

Mediterranean Sea

Dead Sea

NILE DELTA
LOWER EGYPT
Giza
Great Pyramid and Sphinx
Memphis

30°N

N
W   E
S

WESTERN DESERT

EASTERN DESERT

Nile R.

UPPER EGYPT

Red Sea

Thebes

Nile Valley

25°N

First Cataract

TROPIC OF CANCER

NUBIA

0   100 miles
0   100 kilometers
Lambert Azimuthal Equal-Area projection

Second Cataract

30°E   35°E

Egyptians believed in life after death. Because of this belief, the pharaohs had vast tombs built for themselves. The tombs were filled with riches, food, and other goods. These goods were meant to support the pharaohs in the afterlife. When the pharaoh died, his body was preserved as a mummy and placed in the tomb.

At first the tombs were low structures built of bricks. Around 2600 B.C., the first pyramid was built as a tomb. These huge tombs, made of rock, were built by thousands of workers. Later, the pharaohs stopped building pyramids. Instead, workers carved their tombs out of rocky cliffs.

Historians know much about ancient Egypt because the Egyptians had a system of writing. The system, called **hieroglyphics**, used pictures to represent sounds or words.

Influence of Ancient Egypt

The Egyptians made many advances in mathematics and science. They used mathematics to measure farm fields and to figure out taxes. Their studies of the stars and planets led to advances in astronomy. They were masters of engineering as **demonstrated** by their great pyramids and temples.

Academic Vocabulary

demonstrate to show

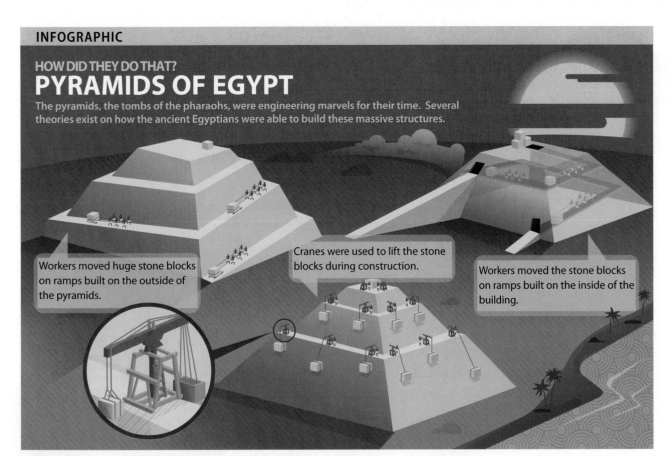

INFOGRAPHIC

HOW DID THEY DO THAT?
PYRAMIDS OF EGYPT
The pyramids, the tombs of the pharaohs, were engineering marvels for their time. Several theories exist on how the ancient Egyptians were able to build these massive structures.

Workers moved huge stone blocks on ramps built on the outside of the pyramids.

Cranes were used to lift the stone blocks during construction.

Workers moved the stone blocks on ramps built on the inside of the building.

BUILDING THE PYRAMIDS
Thousands of people were involved in building a pyramid. Much of the work was done by farmers during the Nile floods, when they could not tend their fields. Surveyors, engineers, carpenters, and stonecutters also lent their skills.

▶ **CRITICAL THINKING**

Analyzing How might the building of the pyramids have led to advances in science and mathematics?

Some of this knowledge was spread to other areas through trade and conquest. Later, Egypt had one of the world's earliest libraries. It was built in the 200s B.C., when Greece conquered and ruled Egypt. The library stored many important works of ancient literature.

☑ **READING PROGRESS CHECK**

Determining Central Ideas Why is it important to know about ancient Egypt?

The Middle Ages

GUIDING QUESTION *How was North Africa connected to other areas?*

Today, people use the Internet to contact each other anywhere in the world. In ancient times, people had to make contact in person. The people of North Africa used the Mediterranean Sea to make this contact with other peoples. Sometimes they were joined by trade. Other times they were joined by conflict.

Carthage and Rome

Western North Africa was first visited by other Mediterranean peoples in the 600s B.C. At that time, traders from what is now Lebanon sailed southwest across the Mediterranean. They built new settlements in many areas. One was a city in what is now Tunisia. They called it Carthage. Within about 200 years, the city had grown powerful. It controlled North Africa from modern Tunisia to Morocco. It also ruled parts of modern Spain and Italy.

In the 200s B.C. and 100s B.C., Carthage fought three wars with the Roman Empire. In the last war, Rome defeated Carthage and destroyed the city. Rome, then, came to control western North Africa. Eventually, Rome conquered Egypt, as well.

The waters of a Roman bath reflect the ruins of the city of Leptis Magna in Libya. The Romans made Leptis Magna one of the most beautiful cities in North Africa during the A.D. 100s.

Identifying What ancient city fought Rome for control of much of North Africa?

Guenter Fischer/Getty Images

Religion plays a central role in the lives of most North Africans today. These Muslim women gather for prayer in the main square of El Mansûra, a city in Egypt's Nile delta.

▶ **CRITICAL THINKING**

Describing How did Islam develop during the century after Muhammad?

During Roman times, many North Africans **converted**, or changed, religions. Because the Roman Empire had adopted Christianity, many North Africans converted to this religion. Others followed their native religions. Except for religion, Roman rule had little effect on native North Africans. Most people continued to live as before. Millions of Berbers who live in western North Africa today are descended from these native people.

Rise of Islam

The Roman Empire fell in the A.D. 400s. Afterward, several local kingdoms formed in North Africa. In the A.D. 600s, though, a new influence emerged in the region. The religion of Islam was founded on the Arabian Peninsula by the prophet Muhammad in A.D. 632. Followers of this religion—called Muslims—began to conquer other lands. By A.D. 642, they had conquered Egypt. By A.D. 705, they ruled all of North Africa. Islam, like Judaism and Christianity, is a monotheistic religion. **Monotheism** means belief in just one god.

Islamic Rule

The Muslim empire was ruled by the **caliph**. This figure had political and religious authority. Caliphs had trouble keeping control over North Africa, however. By the A.D. 800s, separate Berber kingdoms had arisen in parts of the region. These kingdoms often fought one another. Some gained control of most of North Africa. Others only ruled parts of the area.

An Islamic group known as the Fatamids arose in Egypt in the A.D. 1000s. Its rulers expanded Cairo and made it their capital. The city became a center of Muslim learning and trade.

Islamic Culture

At first, Berbers and Egyptians resisted the Islamic religion. By the A.D. 1000s, though, most of them had converted. They also adopted the Arabic language. This language and Islamic learning linked North Africa to the Muslim world. It also helped unite the cultures and people of North Africa and Southwest Asia. Considerable similarities between the regions exist to this day, more than 1,000 years later.

☑ **READING PROGRESS CHECK**

Identifying Point of View Did the Roman or the Islamic empire have more impact on North Africa? Why do you think so?

Khaled Desouki/AFP/Getty Images

The Modern Era

GUIDING QUESTION *What leads people to revolt against a government?*

North Africans formed their own countries in the late 1900s. In recent decades, these countries have changed in far-reaching ways. Often, unrest accompanied the changes.

Foreign Rule

In the 1500s, North Africa began to fall under the rule of foreign armies. The Portuguese and Spanish captured parts of Morocco. The Ottoman Empire, based in modern Turkey, took the rest.

The 1800s saw Ottoman power weaken and Europeans move into North Africa. France began to conquer Algeria in 1830. Although it took several decades, by the late 1800s France controlled that area and Tunisia, too. Some Europeans who settled in these areas grew wealthy. Muslim natives, though, were largely poor. In the early 1900s, France and Spain split control of Morocco. At about the same time, Italy seized Libya.

Egypt kept its independence for much of the 1800s. Its kings tried to build a more modern state. One of the accomplishments was completing construction of the Suez Canal in 1869.

MAP SKILLS

1 PLACES AND REGIONS
Which North African country was the first to become independent?

2 HUMAN GEOGRAPHY
How was Libya governed before independence?

North African Independence

EUROPE

ASIA

ATLANTIC OCEAN

Strait of Gibraltar

Rabat

Oum er Rbia

Algiers

Tunis

Medjerda R.

TUNISIA *(1956, from France)*

Tripoli

Mediterranean Sea

Suez Canal

Cairo

Nile R.

Red Sea

MOROCCO *(1956, from France)*

ALGERIA *(1962, from France)*

LIBYA *(1951, from United Nations trusteeship, administered by British and French governors)*

EGYPT *(1922, from U.K.)*

WESTERN SAHARA *(Morocco)*

TROPIC OF CANCER

Lake Nasser

40°N

30°N

20°N

10°W 0° 10°E 20°E 30°E

0   500 miles
0   500 kilometers
Lambert Azimuthal Equal-Area projection

AFRICA

| *(1962, from France)* | Date of independence, ruling power |

In 1987 soldiers marched in a parade in Tripoli, Libya's capital, to celebrate the rule of Muammar al-Qaddafi. Opponents finally overthrew the military dictator in 2011.

▶ **CRITICAL THINKING**

Analyzing Why was Qaddafi able to rule Libya for so long? Why was he finally overthrown?

The Suez Canal quickly became a vital waterway. Because of the canal's importance, though, other nations wanted to control Egypt. In 1882 Britain sent troops to Egypt. Kings continued to rule, but the British were the real power in the country.

Independence

Many North Africans resented European control. Independence movements arose across the region in the early 1900s. They gained strength after World War II. Italy had been defeated in the war, and France and Britain were severely weakened.

Egypt broke free of foreign control first. In 1952 a group of Egyptian army officers revolted against the king and the British. They created an independent republic, and they put the government in charge of the economy.

Algerians had to fight long and hard for independence. They rebelled against French rule starting in 1954. Not until 1962 did they succeed in ousting the French. Many Europeans fled the country after independence was achieved.

Military leaders also took control of Libya in 1969. They were led by Muammar al-Qaddafi. He remained in control of the nation— and its oil wealth—for more than 40 years. Tunisia and Morocco have avoided military rule. Tunisia has been a republic since gaining independence in 1959. Morocco has had a monarchy since gaining freedom from France in 1956.

Recent Decades

Independence has not always led to success for the countries of North Africa. Algeria has been plagued by unrest among Islamic

Thomas Hartwell/Time & Life Pictures/Getty Images

political groups. Tunisia's government was often accused by the U.S. government of neglecting the rights of the nation's people. Libyan leader Qaddafi had a harsh **regime**, or style of government. Dissent was suppressed, and the government controlled all aspects of life. Qaddafi angered other nations by supporting terrorist groups.

Meanwhile, other problems built up in these nations. High population growth strained their economies. Corrupt governments fueled unrest. In recent years, Muslim **fundamentalists** have led a movement for the people and government to follow the strict laws of Islam. They also reject Western influences on Muslim society.

These problems came to a head in late 2010 in a series of revolts called the Arab Spring. The revolts began in Tunisia, where widespread unrest succeeded in convincing the longtime president to step down from power early in 2011. Tunisians celebrated as a new government took office.

Emboldened by this success, many Egyptians took to the streets. For more than two weeks, thousands of Egyptians turned out every day in Cairo and other cities to protest the government. This revolt also succeeded. In February 2011, Egypt's longtime president Hosni Mubarak gave up power. A group of officers took control and promised to create a new government run by civilians. In 2012 Egyptians voted in the first free presidential election in the country's history.

Unrest also arose in Morocco. There, the king agreed to several reforms that would give more power to the people.

The Arab Spring revolt also reached Libya. The government cracked down on protests. That response angered more Libyans. A **civil war**, or a fight for control of the government, broke out. After months of fighting, the rebels succeeded in taking control of the country. In October of 2011, they killed Qaddafi, and his remaining supporters gave up.

✓ **READING PROGRESS CHECK**

Determining Central Ideas How did the people of North Africa react to European control of the region? Compare that reaction to how North Africans reacted to rule by the Islamic Empire.

FOLDABLES
Study Organizer

Include this lesson's information in your Foldable®.

	Know	Learned
Geography		
History		
Economy		

LESSON 2 REVIEW (CCSS)

Reviewing Vocabulary

1. How were the *pharaohs* of ancient Egypt and the *caliphs* of the Muslim empire similar? How were they different?

Answering the Guiding Questions

2. *Identifying Point of View* Why did the people of Egypt not revolt against the pharaoh even though they had to pay high taxes and work on major building projects?

3. *Integrating Visual Information* Look at a map of the world. What routes do you think the people of North Africa traveled to trade with the people of Southwest Asia in the Middle Ages?

4. *Determining Central Ideas* What has caused unrest in North Africa in recent years?

5. *Informative/Explanatory Writing* Write a summary of the events and results of the Arab Spring.

Lesson 3
Life in North Africa

(l to r) Franz Marc Frei/Lonely Planet Images/Getty Images; Andrew Woodley/Alamy; ©Adam Reynolds/Corbis; ©Claudia Wiens/Corbis; ©Nichole Sobecki/Corbis

---

Reading HELP DESK (CCSS)

Academic Vocabulary

- **emphasis**
- **factor**

Content Vocabulary

- **souk**
- **fellaheen**
- **couscous**
- **diversified**
- **constitution**

TAKING NOTES: *Key Ideas and Details*

Summarize As you read about daily life, culture, and society in the region, take notes using the graphic organizer below.

Daily Life	Culture	Society
•	•	•
•	•	•

ESSENTIAL QUESTION · *Why do conflicts develop?*

IT MATTERS BECAUSE
North Africa is experiencing political changes.

Culture of North Africa

GUIDING QUESTION *What is daily life like in North Africa?*

The vast majority of people in North Africa practice the Islamic religion. Five times a day, the call to prayer rings out from mosques across North Africa, and devout Muslims stop what they are doing to say prayers. Each week on Friday, millions assemble in the mosques for Friday prayer and to hear a sermon. Once a year during Ramadan, the ninth month of the Islamic calendar, Muslims fast (do not eat) from dawn to dusk.

The People

Three main groups—Egyptians, Berbers, and Arabs—make up the population of North Africa. The region has a varied culture. Egypt's ancient heritage looms over that nation just as the pyramids tower over some of its cities. French influence can be seen from Morocco to Tunisia. Although Arab Muslim culture dominates, some Berber traditions continue.

Although most people are Muslims, some Christians and Jews also live in the region. One in 10 of Egypt's people are Christians. Most of them belong to the Coptic Christian church, which formed in the A.D. 400s.

Of the North African nations, Libya has the highest rate of urbanization. More than three of every four Libyans live in an urban area. Only about half of Egypt's people are city dwellers.

Daily Life

Patterns of daily life differ between the cities and the countryside. The region's cities tend to be busy, bustling centers of industry and trade. They also are a blend of traditional cultures and modern life.

Towns and cities of North Africa show no signs of having been planned. Instead, they have grown steadily over the centuries. Streets are narrow and curving. Some built-up areas extend into the surrounding rural farming areas.

Cairo, Egypt, is by far North Africa's largest city, with more than 9.3 million people. The next three largest cities are Algiers, Algeria; Casablanca, Morocco; and Tunis, Tunisia. Combined they have fewer people than Cairo.

Cairo's buildings reflect its more than 1,000-year history. The waterfront along the Nile River boasts gleaming modern skyscrapers and parks. Throughout the city are historic mosques—Islamic places of worship. Tourists flock to the city's famous museums, though they have to endure traffic jams to get there. A jumble of old apartment buildings spreads to the west. Beyond them, a million or so people live in mud huts in a massive poor neighborhood called "the City of the Dead."

An important feature of North African cities is the **souk**, or open-air market. Here, businesspeople set up stalls where they sell food, craft products, and other goods. Singers and acrobats perform here and there in the markets, especially at night.

Life in rural areas follows a different pattern. Farming villages in rural Egypt can be as small as 500 people. Families live in homes built of mud brick with few windows. Each morning, the **fellaheen**— poor farmers of Egypt—walk to work in the fields outside the village.

Spices are among the many products sold at the Khan el-Khalili, the largest souk in Cairo, Egypt. Founded in 1382, the marketplace is a network of streets lined with shops, coffeehouses, and restaurants.

▶ **CRITICAL THINKING**

Describing How did cities in North Africa develop?

A family enjoys a meal at home in the town of Matmata in southern Tunisia. The Berber town is known for its dwellings that are built underground in cave-like structures. To create a home, a resident digs a wide pit in the ground and then hollows out caves around the pit's edge. The caves serve as rooms, which are connected by trench-like corridors.

Many use hand tools and rely on muscle power or animal power. At day's end, they return home.

Farms in Libya are clustered around oases. These communities are small because so little land can be farmed. In Morocco, many farmers live in the well-watered highland areas. They build terraces on steep hillsides to plant their crops.

Some rural dwellers still live like nomads. This is the same kind of life Berbers have followed for centuries. They tend herds of sheep, goats, or camels. They move from place to place in search of food and water for their herds. Some settle in one area for part of the year to grow grains.

Food

Moroccan food has gained fame around the world for its rich and complex flavors. The base of many Moroccan meals is **couscous**, small nuggets of semolina wheat that are steamed. Rich stews of meat and vegetables are poured over it. This style of cooking is also common in Algeria and Tunisia.

Sandwiches in this region are often made with flat pieces of pita bread. They might include grilled pieces of lamb, chicken, or fish. Falafel is made from ground, dried beans and formed into cakes and fried. Pigeon is also popular in Egypt and Morocco.

Arts

The arts in North Africa reflect the influence of Islam. The Islamic religion forbids art that shows the figures of animals or humans. Folk art, like weaving and embroidery, has intricate patterns but no figures. These patterns are also used to decorate buildings.

Many young people in North Africa are attracted to Western

Andrew Woodley/Alamy

music and movies. This has provoked an angry response among some strict Muslims. In Algeria, some artists have left the country because of harsh criticism. Egypt has long been a center of television and film production. Its shows and movies are seen throughout the Arab world.

Languages and Literature

Arabic is the official language of all five countries in North Africa. French is prominent in Morocco, Algeria, and Tunisia. French and English are most often heard in the region's cities, but Berber languages are more common in rural areas.

As the largest Arabic-speaking country, Egypt has played an important part in the literature of the region. Egyptian writers have explored themes like the impact of influences from Western culture. Novelist Naguib Mahfouz, who wrote more than 30 novels and hundreds of stories, achieved worldwide recognition when he won the Nobel Prize for Literature in 1988.

✓ **READING PROGRESS CHECK**

Identifying What is an example of the influence of Islam on daily life in North Africa?

Challenges in North Africa

GUIDING QUESTION *What challenges face North Africa?*

Standards of living vary widely across the region and even within countries. In addition to economic issues, the region faces significant social challenges.

Economic Issues

When oil was discovered in Libya, Muammar al-Qaddafi, the leader of the country, said that a major goal was to provide social benefits to everyone. That did not happen. The income gained from selling oil did not reach most of the country's people. When Qaddafi fell from power in 2011, Libyans hoped that their lives would improve, but progress started slowly.

Algeria has tried to shift its economy away from the **emphasis** on the sale of oil and natural gas. The government keeps tight control of businesses, however. As a result, companies from other countries are not willing to invest there.

Academic Vocabulary

emphasis importance

A craftsperson in Cairo, Egypt, uses copper thread to embroider Arabic writing onto fabric. Muslims prize the art of beautiful writing, which they use to express the words of the Quran, the Islamic holy book.

▶ **CRITICAL THINKING**
Determining Central Ideas
How has Islam influenced the arts of North Africa?

Morocco's economy is the most **diversified**. A diversified economy includes a mix of many different economic activities. The people of the country engage in mining, some manufacturing, farming, and tourism. Poverty and unemployment are widespread in Morocco, however.

In recent years, thousands have left the region for Europe. They move mostly to Spain and France looking for jobs. Morocco, Algeria, and Tunisia have lost the most people.

Social Issues

High population growth is a major concern in Libya and Egypt. This growth rate contributes to crowding and inadequate health care, as well as poverty. A large share of the population in the region is 14 years old or younger. This is especially true in Egypt and Libya. These countries will have to work hard to develop their economies so that today's young people can find jobs in the future.

In February 2012, U.S. Secretary of State Hillary Clinton addressed young people in Tunisia and across the region. She cited the work they did to bring about the massive changes of the Arab Spring. Clinton warned, though, that it would take a long time and hard work to build the country's economy and increase jobs for young people. The U.S. government has pledged money to several countries to help them accomplish these goals.

Another issue is literacy. Libya has the highest literacy rate in the region: 89 percent of Libyans can read and write. The literacy rate is

Workers load freight wagons at a phosphate mine in southern Tunisia. North Africa has rich deposits of phosphates, which are mineral salts used to make fertilizer.

▶ **CRITICAL THINKING**

Analyzing What social and economic challenges does North Africa face?

mediacolor's/Alamy

much lower in the other North African countries. Literacy is most serious in Morocco, where little more than one-half of Moroccans can read and write. A very low literacy rate among women is a major **factor**, or cause, for this trend. More than 65 percent of Moroccan men can read and write; less than 40 percent of that nation's women can. Literacy among women is about 20 percent lower than among men in the other four countries of the region, as well. This gap hinders the ability of the countries to build strong economies.

☑ READING PROGRESS CHECK

Identifying Point of View What might happen in North Africa if young people grow impatient with the slow rate of economic growth? Why?

North Africa's Future

GUIDING QUESTION *How will North Africa address the problems it faces?*

Powerful new social movements have swept through the region of North Africa in recent years. They have led to major political changes in three countries and put pressures on the governments of the other two.

Political Issues

Two political forces are strong in the region of North Africa. One is a push for democracy. Many North Africans have grown more and more frustrated with their leaders. They think the leaders focused more on building their own power than on building the economy and improving their countries. Many question the government's harsh treatment of people who criticize their countries' leaders. Some leaders are calling for the different groups to learn to work together to avoid the conflicts that pull societies apart.

Women students meet for class at Cairo, Egypt's Al-Azhar University, the world's chief center for Islamic learning. In Egypt, women may work outside the home, attend universities, vote, and run for office. However, opportunities for women still lag behind those for men in education and the labor market.

▶ CRITICAL THINKING

Analyzing How does the issue of women's literacy affect economies in North Africa?

Academic Vocabulary

factor a cause

©Claudia Wiens/Corbis

Young people in Benghazi, Libya, yell protest slogans against dictator Muammar al-Qaddafi. Clashes between street demonstrators and armed government forces in February 2011 led to much bloodshed. A civil war broke out, and rebel groups eventually overthrew Qaddafi.

▶ **CRITICAL THINKING**

Describing What happened in Libya after the fall of Qaddafi?

The second force was an increase in Islamic fundamentalism. Some strict Muslims want laws changed to conform to the rules of Islam. They want to see an end to Western influences on their culture. The political party of the Muslim Brotherhood gained a majority in Egypt's parliament in the 2011 elections. It also won a majority in Morocco and a large share of seats in Tunisia. These forces helped bring about the Arab Spring of 2010 and 2011. They have left conditions across the region uncertain.

Egypt began writing a new **constitution** in 2012. A constitution is a set of rules for a nation and its government. Egypt's new government could give more power to the parliament, the lawmaking body. It is not clear how well this new government will work or what groups will control it, though.

For nearly 20 years, Algeria has undergone brutal conflict between Islamist groups and the government and its forces. As many as 100,000 people have died in the fighting. As in Morocco, the government was able to keep power after the Arab Spring, but it had to promise to reform the political system.

By 2012, Libya's victorious rebels were working on making a new government. They also faced the need to rebuild much of the country after the civil war. In 2012, leaders in eastern Libya said they wanted self-rule in their part of the country. Although they said that they did not wish to divide the country or to keep their area's oil wealth for themselves, the move raised the possibility of continued conflict in Libya.

Islam in the Modern World

Many Muslims in the region worry about the impact of Western culture on their lands. They think that Western entertainment conflicts with Islamic values. They also disagree with Western ideas about women's rights.

Women in North Africa generally have more rights than those in other Muslim lands. In Tunisia, for instance, they can own businesses and have their own bank accounts. About half of all university students in Tunisia are women. Women may lose some of these rights if extreme Muslim leaders take control of the governments.

Several million of Egypt's Coptic Christians have grown more worried about their position in recent years as well. Some Muslim extremists have attacked them and bombed churches. Early in 2012, the longtime head of the Coptic church died. He had led the church for nearly 40 years in relative peace until near the end of his life. His death increased the uncertainty for Copts in that area.

Relations with Other Nations

Egypt broke ranks with other Muslim nations in 1979 when it signed a peace treaty with Israel. It has also developed close ties with the United States since then. That friendship has come under increasing criticism from Muslim fundamentalists. Morocco has also had close relations with the United States. Its government has been criticized for this as well.

These situations raise more questions about what will happen if Muslim conservatives gain power. Will the new governments reject close ties with the United States? Will they take steps against Israel?

The situations in Algeria and Libya also are uncertain. Will new governments there be less willing to sell oil to the United States? For what purposes will they use the money they earn from selling oil? The answers to these questions will help to shape the future of North Africa and the world.

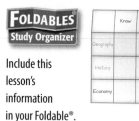

Include this lesson's information in your Foldable®.

☑ **READING PROGRESS CHECK**

Analyzing Why were the results of the Arab Spring different in Algeria and Morocco compared with the other countries of the region?

LESSON 3 REVIEW

Reviewing Vocabulary

1. Is it important for an economy to be *diversified*? Why or why not?

Answering the Guiding Questions

2. *Determining Central Ideas* Why do you think many Muslims worry about the impact of Western culture on their lands?

3. *Describing* How is the relatively young population connected to the economic issues in these nations?

4. *Analyzing* About half of Egypt's people live in rural areas. Most of them are farmers. What impact does that have on Egypt's economy? Why?

5. *Identifying Point of View* Why is the political situation in North Africa important to the United States?

6. *Argument Writing* Do you think the most serious issues facing North Africa are political, social, or cultural? Write a paragraph explaining why.

Directions: Write your answers on a separate piece of paper.

1 Use your FOLDABLES to explore the Essential Question.
INFORMATIVE/EXPLANATORY WRITING Briefly describe the population distribution in the region of North Africa and suggest a likely reason for the distribution.

2 21st Century Skills
INTEGRATING VISUAL INFORMATION With a partner, create a set of flash cards showing an outline of the five North African countries combined and outlines of the individual countries. Make five separate cards with the names of the capital cities. Devise a game using the flash cards, and exchange games with another pair of classmates. After playing both games, work with the other pair to turn the flash card games into a computer game.

3 Thinking Like a Geographer
INTEGRATING VISUAL INFORMATION On an outline map of North Africa, indicate the region's climates. Include the rain shadow areas on the map key.

4 GEOGRAPHY ACTIVITY

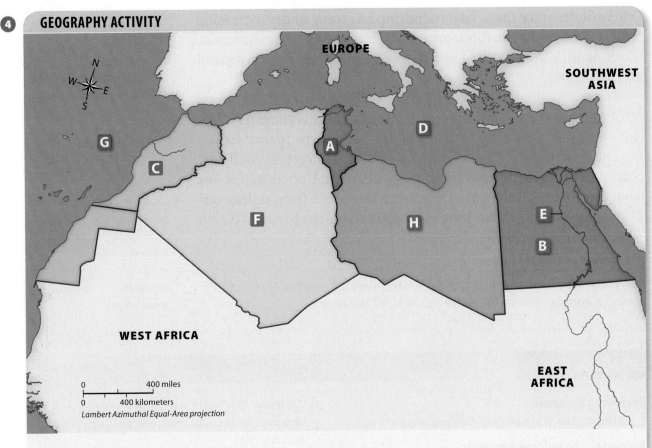

Locating Places
Match the letters on the map with the numbered places below.

1. Mediterranean Sea
2. Egypt
3. Algeria
4. Libya
5. Nile River
6. Morocco
7. Tunisia
8. Atlantic Ocean

REVIEW THE GUIDING QUESTIONS

Directions: Choose the best answer for each question.

1 One nickname for ancient Egypt was

  A. serpent of the sea.

  B. wadi of the floods.

  C. gift of the Nile.

  D. delta dawn.

2 Of the North African countries, Libya has the most

  F. olives.

  G. oil.

  H. water.

  I. cedarwood.

3 Hieroglyphics were

  A. equipment used for building pyramids.

  B. Muslim political and religious leaders.

  C. pictures that represented sounds or words.

  D. spices burned for religious ceremonies.

4 The Berbers and Egyptians became linked to the Muslim world in the A.D. 1000s by

  F. the defeat at Carthage.

  G. the power of the caliphs.

  H. the Arabic language and Islamic learning.

  I.   the pharaoh's desire for more territory.

5 The official language of the five North African nations is

  A. Coptic.

  B. French.

  C. Afrikaner.

  D. Arabic.

6 Today's Islamic fundamentalists in North Africa want to

  F. return to their families' farms.

  G. convert the citizens of Israel.

  H. end Western influence on the Islamic culture.

  I. turn around the Tunisian economy.

DBQ ANALYZING DOCUMENTS

7 **CITING TEXT EVIDENCE** Read the following passage about recent changes in Libya's government:

"*In March 2011, a Transitional National Council (TNC) was formed . . . with the stated aim of overthrowing the Qaddafi regime and guiding the country to democracy. . . . Anti-Qaddafi forces in August 2011 captured the capital, Tripoli. In mid-September, the [United Nations] General Assembly voted to recognize the TNC as the legitimate interim governing body of Libya.*"

—from CIA World Factbook, "Libya"

How did the Transitional National Council view its role in Libya?

A. as an ally of Qaddafi

B. as supporters of Libya's former king

C. as a temporary government

D. as social reformers

8 **ANALYZING** What happened to the Qaddafi government in 2011?

F. It remained in control of the country.

G. It relocated to a new capital.

H. It forged an alliance with the TNC.

I. It fell out of power.

SHORT RESPONSE

"*Morocco, on account of the invasions of Arabs and the exterior adventures of Moorish kings, was strongly influenced by Middle Eastern culture and the culture of the Andaluz [Muslim-ruled Spain]. The Arabs learned [cooking] secrets from the Persians and brought them to Morocco; from Senegal and other lands south of the Sahara came caravans of spices. Even the Turks made a contribution.*"

—from Paula Wolfert, *Couscous and Other Good Food From Morocco* (1973)

9 **DETERMINING CENTRAL IDEAS** What is the main idea of this passage?

10 **DESCRIBING** What kinds of contact by different groups led to these influences on Moroccan culture?

EXTENDED RESPONSE

11 **INFORMATIVE/EXPLANATORY WRITING** Research the Arab Spring of 2011. Write an essay contrasting the current situation in the countries that were involved in the Arab Spring with their situation before the upheaval. Did the "spring" last? Did the citizens of these countries gain or lose what they were trying to achieve? Are their lives better or worse today than they were before 2011? Do any of them now have a stable, democratic government?

Need Extra Help?

If You've Missed Question	**1**	**2**	**3**	**4**	**5**	**6**	**7**	**8**	**9**	**10**	**11**
Review Lesson	1	1	2	2	3	3	2	2	2	2	2

EAST AFRICA

Teenage girl from the East African country of Somalia

Ariadne Van Zandbergen/Lonely Planet Images/Getty Images

ESSENTIAL QUESTIONS · *How does geography influence the way people live?* · *Why do people trade?* · *Why does conflict develop?*

The Story Matters...

Some of Africa's earliest kingdoms developed in East Africa, where trade in gold and ivory brought great wealth. Since ancient times, thriving trade has fostered interaction among different cultures, influencing language and religion and creating much ethnic diversity across the region. The landscape of East Africa also has great diversity—from the Serengeti Plain and the Great Rift Valley to the highlands in Ethiopia and Kilimanjaro in Kenya.

FOLDABLES®
Study Organizer

Go to the Foldables® library in the back of your book to make a Foldable® that will help you take notes while reading this chapter.

607

EAST AFRICA CCSS

Some of Africa's important early civilizations flourished in East Africa. Many of the countries have been scarred by conflict in recent years.

Step Into the Place

MAP FOCUS Use the map to answer the following questions.

1 **THE GEOGRAPHER'S WORLD** Which three East African countries share Lake Victoria?

2 **PLACES AND REGIONS** What is the capital city of Kenya?

3 **THE GEOGRAPHER'S WORLD** The Tekeze is a major river in what country?

4 **CRITICAL THINKING** Integrating Visual Information What country is cut off from the sea by Eritrea, Djibouti, and Somalia?

A

INACTIVE VOLCANO Snowcapped Kilimanjaro looms over savanna plains near the border of Tanzania and Kenya. The mountain is made up of three volcanic cones, all inactive.

B

WAR-TORN CITY Ruined buildings line an Indian Ocean beach in Mogadishu, the capital of Somalia. Since the early 1990s, various armed groups have fought over Somalia.

Step Into the Time

TIME LINE Using at least two events on the time line, write a paragraph describing how trade influenced the development of East Africa.

800 B.C. Kingdom of Kush develops along the Nile River

A.D. 400 Kingdom of Aksum prospers from trade

30,000–20,000 B.C. Ancient people live in what is now Sudan

B.C. | A.D.

1100s Muslim settlements multiply in East Africa

EAST AFRICA

NORTH AFRICA

ASIA

TROPIC OF CANCER

20°N

SUDAN

Nile R.

Omdurman
Khartoum

Atbara R.

ERITREA

Red Sea

Asmara

DJIBOUTI

Gulf of Aden

Blue Nile R.

Tekeze R.

Lake
Tana

Djibouti

10°N

Dire Dawa

Hargeysa

SOUTH
SUDAN

White Nile R.

Omo R.

Addis Ababa

ETHIOPIA

Shabeele R.

SOMALIA

CENTRAL
AFRICA

Juba

Lake
Turkana

Jubba R.

Mogadishu

EQUATOR

0°

UGANDA

Lake
Albert

Kampala

KENYA

INDIAN
OCEAN

RWANDA

Lake
Victoria

Kigali

A

Tana R.

Nairobi

B

Bujumbura

BURUNDI

TANZANIA

Mombasa

○ National capital
• City

N
W   E
S

0   500 miles
0   500 kilometers
Lambert Azimuthal Equal-Area projection

Lake
Tanganyika

Dodoma

Dar es Salaam

10°S

SOUTH
AFRICA

20°E   30°E   40°E

Lake
Malawi

1896
Ethiopian troops
defeat Italian troops
at Battle of Adwa

1961
Tanganyika becomes
independent; changes
name to Tanzania in 1964

1800   **1900**   **2000**

1400s Arab conquests bring Islam
to the area of present-day Sudan

1800s Swahili language
spreads inland

1880s Germany, Britain, and
France take control of the region

2000s Civil war in Darfur region of
Sudan kills hundreds of thousands

netw⚡rks

There's More Online!

- ☑ **IMAGES** Glaciers in East Africa
- ☑ **MAP** Desertification of the Sahel
- ☑ **SLIDE SHOW** The Nile River's Source
- ☑ **VIDEO**

Reading **HELP**DESK

Academic Vocabulary

- **consist**

Content Vocabulary

- **rift**
- **desertification**
- **hydroelectric power**
- **geothermal energy**

TAKING NOTES: *Key Ideas and Details*

Identifying As you study the lesson, use a web diagram like this one to list information about the land and water features of the region.

Lesson 1
Physical Geography of East Africa

ESSENTIAL QUESTION · *How does geography influence the way people live?*

IT MATTERS BECAUSE
East Africa offers a rugged, beautiful landscape and different climates. The region provides variety, potential, and considerable challenges for economic development.

Land and Water Features

GUIDING QUESTION *What makes the ecosystem of East Africa diverse?*

The region of East Africa **consists** of 11 countries. Sudan and South Sudan dominate the northern part of the region. Eritrea, Djibouti, Somalia, and Ethiopia are located in the northeast. This area is called the Horn of Africa because it is a horn-shaped peninsula that juts out into the Arabian Sea. Three countries occupy the central and southern parts of the region: Kenya, Tanzania, and Uganda. Finally, in the western sector lie the landlocked countries of Rwanda and Burundi. East Africa offers a rugged, beautiful landscape that has great variety.

Landforms

The Great Rift Valley is the most unusual feature of East Africa's physical geography. Sometimes it is called the Great Rift system because it is not one single valley. Rather, it is a series of large valleys and depressions in Earth's surface. These are formed by long chains of geological faults. The Great Rift started forming about 20 million years ago when tectonic plates began to tear apart from one another. Africa was once connected to the Arabian Peninsula. But as the two **rifted** apart, or separated from one another, the land in between

(l to r) ©Hemis/Alamy; ©Michael Freeman/Corbis; ©John Warburton-Lee Photography/Alamy; Bruno Zanzottera/Parallelozero/Aurora Photos; ©Nigel Pavitt/JAI/Corbis

sank and was filled by the Red Sea. Eventually, all of East Africa will separate from the rest of Africa, and the Red Sea will fill the rift.

The Great Rift system's northern end is in Jordan in Southwest Asia. From Jordan, it stretches about 4,000 miles (6,437 km) to its southern end in Mozambique in southeastern Africa. The rift has an average width of 30 miles to 40 miles (48 km to 64 km).

The rift system has an eastern and western branch in East Africa. The eastern Rift Valley—the main branch—runs from Southwest Asia along the Jordan River, Dead Sea, and Red Sea. It continues through the Danakil plain in Ethiopia. It is one of the hottest and driest places on Earth, and earthquakes and volcanic activity occur here regularly. Long, deep cracks develop in Earth's surface as the tectonic plates rift apart.

As the eastern Rift Valley continues south from the Danakil plain, the conditions are not as severe. It takes the form of deep valleys as it extends into Kenya and Tanzania, and down to Mozambique. The shorter western Rift Valley stretches from Lake Malawi in the south through Uganda in the north through a series of valleys. A chain of deep lakes that includes Lake Tanganyika, Lake Edward, and Lake Albert marks the western rift's northward path.

Along the branches of the Great Rift Valley, much volcanic and seismic activity occurred. The largest volcanoes are located on the eastern Rift. These include Mount Kenya and Kilimanjaro. Kilimanjaro is on the border between Kenya and Tanzania. With a summit of 19,341 feet (5,895 m), Kilimanjaro is the tallest mountain in Africa. Its summit is covered with snow year-round, even though the mountain is near the Equator.

Sudan is home to vast plains and plateaus. The northern part of the country is desert covered in sand or gravel. Somalia lies in the eastern part of the region, along the Indian Ocean.

©Hemis/Alamy

This aerial view shows a section of the floor of the eastern Rift Valley in Kenya. Many fault lines appear in the valley. Hardened lava from volcanoes and openings in the ground also mark the landscape.

▶ **CRITICAL THINKING**

Describing How will East Africa eventually be affected by the Rift's tectonic plate activity?

The palace of Sudan's president in Khartoum stands near where the Blue Nile joins the White Nile. The White Nile is named for the light-colored clay sediment found in its waters. The Blue Nile's name comes from the river's appearance during flood season when the water level is high.

▶ **CRITICAL THINKING**

Describing Where do each of the two Nile tributaries begin?

Somalia is also an extremely dry area. The country is made up largely of savanna and semidesert. To the north of Somalia lies the small country of Djibouti. Located on the coast between the Red Sea and the Gulf of Aden, Djibouti displays a highly diverse landscape. It has rugged mountains and desert plains.

South of Sudan, at the western edge of Uganda, the Ruwenzori Mountains divide that country from the Democratic Republic of the Congo. These peaks are sometimes called the "Mountains of the Moon." Mountains give way to hills in small, landlocked Rwanda. It is known as the "land of a thousand hills" for its beautiful landscape.

Bodies of Water

The longest river in the world is the Nile (4,132 miles or 6,650 km). The Nile Basin includes parts of many countries in the East African region: Tanzania, Burundi, Rwanda, Kenya, Uganda, Ethiopia, South Sudan, and the Sudan. Beginning in the 1800s, European explorers made numerous expeditions in attempts to find the source of the Nile River. The great river was discovered to have two sets of headwaters. One of them, the Blue Nile, rises in the northern highlands of Ethiopia. The other source, the White Nile, begins in Lake Victoria and runs through Lake Albert. The White Nile then passes through the swampy wetlands of central South Sudan, a huge area called the Sudd.

In northern Sudan, the Blue Nile and the White Nile meet at the city of Khartoum. The great river then runs northward through Egypt and empties into the Mediterranean Sea. Other than the Nile,

©Michael Freeman/Corbis

East Africa has few important rivers. This is due to the intermittent rainfall and the high temperatures in many areas of the region.

In the late 1970s, the swampy Sudd was the focus of a huge construction project called the Jonglei Canal. This channel was designed to avoid the Sudd. The goal was to allow the headstreams of the White Nile to flow more freely. Instead of the water spreading across the Sudd and slowly moving through it, the canal would allow more water to flow downstream and reach Sudan and Egypt. That would support more agriculture and better city services in those countries. But it would also damage the wetland environment of the Sudd. Fisheries could collapse and go extinct. Construction was suspended in 1983. The project could not continue because civil war in Sudan made it too dangerous.

Many of the lakes in East Africa are located near the Great Rift Valley. The largest lake on the continent of Africa is Lake Victoria. This lake lies between the western and the eastern branches of the Great Rift. The lake stretches into three countries: Uganda and Kenya in the north and Tanzania in the south. With an area of 26,828 square miles (69,484 sq. km), Lake Victoria is the second-largest freshwater lake in the world, after Lake Superior in the United States. For such a large body of water, Lake Victoria is relatively shallow. Its greatest known depth is about 270 feet (82 m). The lake is home to more than 200 species of fish. Of these, tilapia has the most economic value.

Another important lake in the region is Lake Tanganyika. This long, narrow body of water is located south of Lake Victoria, between Tanzania and the Democratic Republic of the Congo. The lake is only 10 to 45 miles (16 km to 72 km) wide, but very long. Measuring 410 miles (660 km) north to south, it is the world's longest freshwater lake. With a maximum depth of 4,710 feet (1,436 m), it is also the second deepest. Only Lake Baikal in Russia is deeper than Lake Tanganyika.

Farther south is Lake Malawi. It is the third-largest lake in the East African Rift Valley. The lake lies mainly in Malawi and forms part of that country's border with Tanzania and Mozambique.

Fishers leave the eastern shore of Lake Victoria by boat early in the morning to fish for tilapia and Nile perch. With its many fish species, Lake Victoria supports Africa's largest inland fishery.

☑ **READING PROGRESS CHECK**

Identifying What caused the striking physical features of the Great Rift Valley in East Africa?

Climates of East Africa

GUIDING QUESTION *How does climate vary in East Africa?*

Climate varies widely in the East African region. Temperature and rainfall can be quite different from one local area to another. The major factors explaining these variations include latitude, altitude, distance from the sea, and the type of terrain, such as mountains, highlands, desert, or coastal plains.

Temperatures

The diverse physical features of East African geography are matched by an extremely varied climate. In general, temperatures tend to be warmer toward the coast and cooler in the highlands. Sudan, Djibouti, and Somalia have high temperatures for much of the year. High mountains such as Kilimanjaro and the peaks of the Ruwenzori Range have had glaciers for thousands of years. Due to climate change, however, these glaciers are melting. Some experts predict that the glaciers of Kilimanjaro will completely disappear over the next 20 years.

The climate is always spring-like in the highlands of Kenya and Uganda. As a whole, however, Kenya and Uganda display considerable variations in climate. These variations depend on factors such as latitude, elevation, wind patterns, and ocean currents.

Rainfall

In many parts of East Africa, rainfall is seasonal. This is especially true close to the Equator. Wet seasons alternate with dry ones. For example, on the tropical grasslands, or savannas, of Kenya and

The Savoia glacier is located along the border of Uganda and the Democratic Republic of the Congo. Many scientists are concerned about the effects of climate change on the Ruwenzori glaciers. Around 1900, some 43 glaciers were distributed over 6 mountains in the range. Today, fewer than half of these glaciers still exist, on only 3 of the mountains. The rest have melted.

▶ **CRITICAL THINKING**

Analyzing Why do temperatures tend to be cool in inland East Africa despite the region's closeness to the Equator?

Bruno Zanzottera/Parallelozero/Aurora Photos

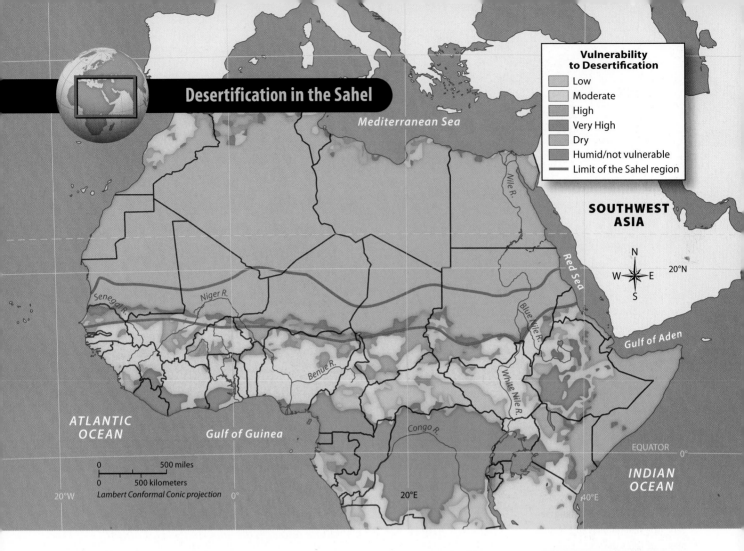

Desertification in the Sahel

Mediterranean Sea

SOUTHWEST
ASIA

20°N

Red Sea

Gulf of Aden

Nile R.

Blue Nile R.

White Nile R.

Senegal R.

Niger R.

Benue R.

Congo R.

ATLANTIC
OCEAN

Gulf of Guinea

EQUATOR   0°

INDIAN
OCEAN

0   500 miles
0   500 kilometers
Lambert Conformal Conic projection

20°W   0°   20°E   40°E

Vulnerability to Desertification

- Low
- Moderate
- High
- Very High
- Dry
- Humid/not vulnerable
- Limit of the Sahel region

N W E S

Tanzania, two rainy seasons occur in most years. These are the "long rains" of April and May and the "short rains" of October and November. The months in between these periods are dry, with little or no rainfall.

Rainfall in the region, however, can be unpredictable. Sparse rainfall can result in severe drought. In 2011, for example, Somalia suffered one of the worst droughts in its history. Political instability in that country made the effects of the drought especially severe. Observers estimated that 13 million people struggled to survive in the countries of Somalia, Ethiopia, Djibouti, and Kenya.

Another urgent issue in the region is **desertification**, or the process by which agricultural land is turned into desert. This process occurs when long periods of drought and unwise land use destroy vegetation. The land is left dry and barren. During the past half century, desertification has affected much of the Sahel. The Sahel is the "edge," or border area, between the Sahara and the countries farther to the south. Two such border nations in East Africa are Sudan and South Sudan.

☑ **READING PROGRESS CHECK**

Determining Central Ideas What generalization can you make about the variations in temperature in East Africa?

MAP SKILLS

1 **PLACES AND REGIONS**
Based on the legend, most of the Sahel is at what level of desertification?

2 **THE GEOGRAPHER'S WORLD** What causes desertification?

Chapter 20 **615**

Workers collect salt at Lake Assal in Djibouti. Salt covers everything, so very little vegetation is able to grow along the lake's shoreline. Located in the hot desert, the lake's area has summer temperatures as high as 126°F (52°C).

Identifying What other mineral resources are found in East Africa?

Resources of East Africa

GUIDING QUESTION *Which natural resources are important in East Africa?*

The natural resources of a region are closely linked to its economy and people's way of life. Settlement patterns in a geographical area have often been shaped by that area's natural resources. Important resources in East Africa are minerals, energy sources, landscapes, and wildlife. The ability of some countries to exploit these resources, however, has been hampered by political issues.

Mineral Resources

Mineral resources in East Africa include small gold deposits along the rifts in Kenya, Uganda, and Tanzania; gemstones like sapphires and diamonds in Tanzania; and tin in Rwanda. Ethiopia and Uganda produce lumber. Lake Assal in Djibouti, located about 500 feet (152 m) below sea level, is the world's largest salt reserve, with more than 1 billion tons of salt. This lake is located at the lowest point in Africa.

Energy Resources

Energy resources in East Africa include coal in Tanzania, as well as petroleum in Uganda, South Sudan, and northwestern Kenya. East Africa's energy potential has yet to be realized, though. For example, Sudan has the opportunity to develop **hydroelectric power**, or the production of electricity through the use of falling water. Hydroelectric power is already used in Kenya and Tanzania.

Likewise, Kenya and Djibouti are favorable locations for the development of **geothermal energy**. This type of energy comes from underground heat sources, such as hot springs and steam. In Kenya, an international group of companies is working with the government to develop geothermal energy sources. If they are successful, 30 percent of the country's energy needs could be met by geothermal energy by the year 2030. In Djibouti, geothermal energy production is expected to begin by the year 2014.

In East Africa, management of energy resources and energy use often has been inconsistent and uneven. Major cities gobble up much of the energy that is produced. Energy is often unavailable in rural areas.

Land and Wildlife

Besides mineral and energy resources, East Africa's land and wildlife are important assets. The soils in the region are not especially rich for agriculture, and farming is challenging. The breathtaking scenery of the Great Rift Valley, however, is an important tourist resource.

East Africa is also home to the greatest assemblage of wildlife in the world. Many national parks and wildlife sanctuaries are found in the region. Perhaps the most well-known wildlife reserves are located in Kenya and Tanzania. An outstanding example is the Serengeti Plain; this vast area, larger than the state of Connecticut, consists of tropical savanna grasslands. Two internationally famous national parks are located in East Africa—Serengeti National Park in Tanzania and the Masai Mara National Reserve in Kenya. These parks harbor lions, leopards, cheetahs, giraffes, zebras, elephants, and dozens of species of antelope.

Every year, thousands of tourists pour in from all over the world to see the marvel of the Great Migration. In this mass movement, more than 1 million animals travel hundreds of miles in search of fresh grazing land. The spectacular wildlife of East Africa makes an important contribution to the economy of the region.

☑ READING PROGRESS CHECK

Identifying What two promising alternatives might help improve energy supplies in the East African region?

Think Again?

Animals involved in the Great Migration on the Serengeti Plain travel together.

Not True. Nature employs a more sophisticated system. The three major migrating species are zebras, wildebeests, and Thomson's gazelles. These species migrate in a succession. First come the zebras. They consume crude, coarse, high grasses. Then the wildebeests follow, grazing on the lower shoots exposed by their predecessors. Last are the smaller Thomson's gazelles, antelopes that eat tender, fine shoots close to the ground.

FOLDABLES
Study Organizer

Include this lesson's information in your Foldable®.

Reviewing Vocabulary
1. What causes the process of *desertification*?

Answering the Guiding Questions
2. *Describing* What are the differing characteristics that make Lake Victoria and Lake Tanganyika noteworthy bodies of water, both in East Africa and in the world as a whole?

3. *Analyzing* How might desertification affect the economy in a region?

4. *Identifying* How are energy supplies distributed in East Africa?

5. *Informative/Explanatory Writing* Write a letter to a friend or a relative explaining why you want to visit East Africa to see the region's wildlife.

networks

There's More Online!

☑ **IMAGE** British at Omdurman

☑ **MAP** African Trade Routes and Goods

☑ **SLIDE SHOW** Ancient Africa

☑ **VIDEO**

Reading **HELP**DESK (CCSS)

Academic Vocabulary

- **impact**

Content Vocabulary

- **tribute**
- **imperialism**
- **genocide**
- **refugee**

TAKING NOTES: *Key Ideas and Details*

Organizing As you study the lesson, use a chart like this one to list important facts about the places.

Place	Facts
Nubia/Kush	
Aksum	
Coastal City-States	

Lesson 2
History of East Africa

ESSENTIAL QUESTION · *Why do people trade?*

IT MATTERS BECAUSE
East Africa has been a center of trade since ancient times. Throughout much of its history, East Africa has attracted people from many other continents.

Kingdoms and Trading States

GUIDING QUESTION *How has the history of trade impacted the region?*

Trade was important in the ancient kingdoms in East Africa. Contact between East Africa and other areas brought together people from different civilizations. Trade also resulted in the spread of Christianity and Islam into the region.

Ancient Nubia

The ancient region of Nubia was located in northeastern Africa, below ancient Egypt. The region stretched southward along the Nile River valley almost to what is now the Sudanese city of Khartoum. The region was bounded by the Libyan Desert in the west and by the Red Sea in the east. The Nile River was the pathway by which Nubia and the powerful empire of Egypt interacted.

In about 1050 B.C., a powerful civilization arose in Nubia. This was known as Kush. The Egyptians traded extensively with the Kushites, purchasing copper, gold, ivory, ebony, slaves, and cattle. The Kushites, in turn, adopted many Egyptian customs and practices. For example, they built pyramids to mark the tombs of their rulers and nobles.

During the final centuries of their civilization, the Kushites were isolated from Egypt. As a result, they turned increasingly to other African people south of the Sahara for trade and cultural contact. Around A.D. 350, Kush was

conquered by Aksum, a powerful state in what is now northern Ethiopia.

Aksum

The date of Aksum's establishment is uncertain but it might have been around 1000 B.C. The people of Aksum derived their wealth and power primarily from trade. Aksum was strategically located, and it controlled the port city of Adulis on the Red Sea. At its height of power, Aksum was the most important trading center in the region. Its trading connections extended all the way to Alexandria on the Mediterranean Sea. Aksum traders specialized in sea routes that connected the Red Sea to India.

Through the port of Adulis flowed gold and ivory, as well as raw materials. It is possible that Aksum sold captives for the slave trade. Aksum traded glue, candy, and gum arabic, a substance from acacia trees that today is used in the food industry. Christianity spread from its origin in Jerusalem along the trade routes. The Aksum kings adopted Christianity as their religion.

Trade Cities

Beginning around the A.D. 900s, after the decline of Aksum, Arabs settled on the East African coast of the Indian Ocean. The religion of Islam grew steadily more important in the region. At the same time, the Arabic and Bantu languages mingled to create a new language. This language is known as Swahili. The name comes from an Arabic word meaning "coast dwellers." Swahili is widely spoken today in Tanzania and Kenya, as well as in some other countries.

Gradually, the coastal settlements formed independent trading states. From coastal Somalia southward, along the shores of Kenya and Tanzania, these city-states prospered. They included Mogadishu, Lamu, Malindi, and Mombasa.

Many of the pyramids of ancient Kush still stand in present-day Sudan. Near the pyramids, the Kushites built a capital city called Meroë. Archaeologists have uncovered some of the remains of Meroë, including a royal palace, temples, and mud-brick homes.

▶ **CRITICAL THINKING**
Determining Central Ideas What do the pyramids of Meroë reveal about Kushite culture?

Nigel Pavitt/AWL Images/Getty Images

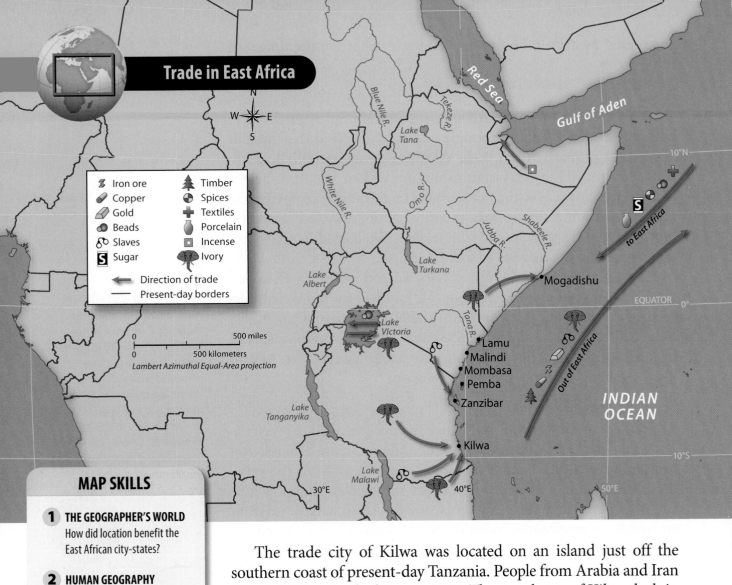

Trade in East Africa

Iron ore
Copper
Gold
Beads
Slaves
Sugar

Timber
Spices
Textiles
Porcelain
Incense
Ivory

→ Direction of trade
— Present-day borders

0   500 miles
0   500 kilometers
Lambert Azimuthal Equal-Area projection

MAP SKILLS

1 **THE GEOGRAPHER'S WORLD**
How did location benefit the East African city-states?

2 **HUMAN GEOGRAPHY**
What part did inland Africa play in the region's trade?

The trade city of Kilwa was located on an island just off the southern coast of present-day Tanzania. People from Arabia and Iran founded Kilwa in the late A.D. 900s. The merchants of Kilwa dealt in copper, iron, ivory, and gold. They exchanged these goods for products from many lands, including Chinese porcelain and Indian cotton.

Kilwa was a walled city. Its ruler lived in an impressive palace. For two centuries, the city was probably the wealthiest trading center in East Africa. The fourteenth-century traveler Ibn Battuta praised Kilwa as a beautiful city. At the time of Ibn Battuta's visit, Kilwa was ruled by Abu al-Mawahib. The sultan was so generous that people called him "the father of gifts."

☑ **READING PROGRESS CHECK**

Identifying Compare the economies of the coastal city-states in East Africa to those of the kingdom of Aksum.

The Colonial Era

GUIDING QUESTION *What was the effect of colonization on East Africa?*

Until the late 1800s, most Europeans knew little or nothing about Africa. Two of the continent's most famous explorers were Henry Morton Stanley and David Livingstone. In 1878 Stanley published a

popular travel book about his adventures in Africa. The book's title was *Through the Dark Continent*. The goal of Stanley's journey was to locate Livingstone, a medical missionary. Livingstone had traveled to Africa in the hope of locating the source of the Nile River.

European Traders

Just before 1500, the European age of discovery began to **impact** East Africa. Among the European countries, Portugal took the lead in overseas exploration. Along with other Europeans, the Portuguese established a sea route to India. From Europe, they sailed south along the west coast of Africa and then along the east coast of Africa. Then, they sailed along the coast of Arabia and on to India. This was a much easier and less expensive way to trade with India than any of the overland trade routes. In this way, the Portuguese were able to bring back many valuable spices from India.

As trade increased, the Portuguese began to demand **tribute**, or a regular tax payment, from the East African trading cities. The Portuguese had religious as well as economic motives; they believed that Christianity should replace Islam as the region's religion. Portuguese influence in the region did not last long, however. The Portuguese could not withstand attacks by African groups in the region. Other European countries became interested in colonizing Africa.

European Colonial Rule

In the late 1800s, European leaders set out a plan to dominate and control the continent of Africa. The action by which one nation is able to control another smaller or weaker nation is known as **imperialism**.

©Lebrecht Music & Arts/Lebrecht Music & Arts/Corbis

The Battle of Omdurman was fought in Sudan in 1898. In this battle, British and Egyptian forces—equipped with modern guns—defeated a much larger Mahdist army that used older weapons.
▶ **CRITICAL THINKING**
Integrating Visual Information
How does Hale's painting present the battle scene? What view of imperialism does it seem to support?

A painting in traditional Ethiopian style shows King Menelik II receiving ammunition for his army. Menelik worked to bring modern ways to Ethiopia. He especially wanted to prepare his army to successfully resist European invaders.

Africa was carved up into colonies. The reasons for colonization included economic profit, access to raw materials, and the opening of new markets. These reasons also included national pride, the protection of sea routes, the maintenance of the balance of power, and a quest to convert Africans to Christianity.

Occasional rebellions challenged European colonial rule. An especially bloody rebellion occurred against British and Egyptian domination in Sudan. Muhammad Ahmad, a religious and military leader, declared that he was the Mahdi, or redeemer of Islam. Mahdist forces succeeded in capturing Khartoum, the Sudanese capital. They established a new state there. In 1898 the British succeeded in reasserting their control of the region.

Independent Ethiopia

The revolt against foreign influence in Sudan eventually resulted in failure. In Ethiopia, however, the desire for independence prevailed. Italy had colonized the neighboring territory of Eritrea along the Red Sea coast. In 1889 the Italians signed a treaty with the Ethiopian emperor, Menelik II. Over the next few years, Italy claimed that, according to one provision of this treaty, it had the right to establish a "protectorate" in Ethiopia.

Menelik firmly denied these claims. He rejected the treaty in 1893. The Italian governor of Eritrea finally launched a major military attack in response in 1896. At the Battle of Adwa on March 1 of that year, Menelik defeated the Italian army. This conflict was

Sabena Jane Blackbird/Alamy

one of the most important battles in African history. After the Battle of Adwa, the European powers had no choice but to recognize Ethiopia as an independent state. Physical geography played an important role in Ethiopia's ability to remain independent. Rugged mountains with difficult terrain provided a barrier that was difficult for attacking forces to overcome.

☑ **READING PROGRESS CHECK**

Explaining What was the significance of Menelik II's victory at the Battle of Adwa in 1896?

Independence

GUIDING QUESTION *How did the countries of East Africa gain their independence?*

After the end of World War II in 1945, a movement ensued to end colonialism in Africa, Asia, and Latin America. In East Africa, particularly, Europeans were seen as disrupting traditional life. In addition, European countries were weakened by the fighting in World War II. Because of these pressures, Europeans granted East African colonies their independence in the 1960s. However, many of the former colonies faced difficulties in establishing their own countries.

New Nations Form

The early 1960s was a turning point for East Africa. During the period from 1960 to 1963 alone, six East African countries obtained independence: Somalia, Kenya, Uganda, Tanzania, Rwanda, and Burundi.

The achievement of independence in Kenya and Tanzania was especially important. Kenya had been a British colony for about 75 years. British plantation owners dominated the economy. They disrupted the traditional East African agricultural system. Local village agriculture was replaced by the production of cash crops, such as coffee and tea, on a large scale. Native people, such as the Kikuyu, were driven off the land. The British also controlled the government.

A nationalist named Jomo Kenyatta led the political protest movement in Kenya and negotiated the terms of independence for his country. In late 1963, Kenya became independent. Jomo Kenyatta served as the country's first prime minister and later as its president.

Tanzania also sought independence. Before independence, the country was called Tanganyika.

As independent Kenya's first leader, Jomo Kenyatta brought stability and economic growth to the country. When appearing in public, Kenyatta often carried a fly whisk, a symbol of authority in some traditional African societies.

▶ **CRITICAL THINKING**
Describing How did Kenya win its independence from British rule?

Villagers in South Sudan try to put out fires after warplanes from neighboring Sudan raided the area in early 2012. A year earlier, South Sudan had gained independence from Sudan following years of civil war. However, tensions remained high and conflict continued.

▶ **CRITICAL THINKING**

Describing Why have some African countries after independence faced civil wars and conflicts with neighboring countries?

When Germany was defeated in World War I, Tanganyika came under British control. Independence was the ultimate goal for Tanganyika—a goal it reached in late 1961. Three years later, the country merged with Zanzibar, and its name was changed to Tanzania.

Highland Countries

The Highland areas had a difficult road to independence. Many ethnic groups in the former colonies were often in conflict with one another. Ethnic tensions have long simmered in Rwanda and Burundi. These countries are home to two rival ethnic groups. The Hutu are in the majority there, and the Tutsi are a minority. In the 1990s, the Hutu-dominated government of Rwanda launched an attack on the Tutsi that amounted to **genocide**—the slaughter of an entire people on ethnic grounds. Hundreds of thousands of people were killed.

Bloodshed also stained the history of Uganda after independence. From 1971 to 1979, the country was ruled by the military dictator Idi Amin. Cruelty, violence, corruption, and ethnic persecution marked Amin's regime. Human rights groups estimate that hundreds of thousands of people lost their lives under his rule. Amin was finally forced to flee into exile. He died in 2003.

The Horn of Africa

The history of Somalia since independence in 1960 offers another example of the problems East African countries have faced. Since the 1970s, Somalia has been scarred by civil war. Border disputes with Ethiopia have also increased instability. Rival clan factions have

Michael Onyiego/AP Images

engaged in bitter feuds. Drought has brought famine to much of the country. In late 1992, the United States led a multinational intervention force in an effort to restore peace to the country. The civil war in Somalia, however, remained unresolved.

The instability, misery, and violence in Somalia also have affected neighboring countries. Thousands of **refugees**, for example, have made their way into Kenya. A refugee is a person who flees to another country for safety.

Elsewhere in the Horn of Africa, more than 30 years of fighting have marked the recent history of Eritrea. This country achieved independence in 1993 after a long struggle with Ethiopia. Access to the sea was an important territorial issue in this conflict. In the years since independence, Eritrea has undertaken military conflicts with Yemen and resumed attacks on Ethiopia. The country is unable to provide enough food for its people. Furthermore, economic progress has been limited because many Eritreans serve in the army rather than in the workforce.

A New Nation

Africa's newest country emerged as a result of civil war. Sudan won independence from Egyptian and British control in 1956. Leaders in southern Sudan were angered because the newly independent Sudanese government had failed to carry out its promise to create a federal system. Southern leaders also feared that the new central government would try to establish an Islamic and Arabic state.

Religion was also an issue that generated conflict. Most people in Sudan are Muslim, but in the southernmost 10 provinces, most people follow traditional African religious practices or the Christian religion. Economic issues are also a problem. The southern provinces hold a large share of the area's petroleum deposits. As a result of the civil war, the country of South Sudan became independent from Sudan in 2011.

Include this lesson's information in your Foldable®.

☑ **READING PROGRESS CHECK**

Determining Central Ideas How has civil war played an important part in the recent history of East Africa?

LESSON 2 REVIEW

Reviewing Vocabulary

1. What were some of the factors that led European nations to practice *imperialism* in Africa?

Answering the Guiding Questions

2. *Identifying* Discuss two important events that occurred in the history of the Ethiopian kingdom of Aksum.

3. *Identifying* Which two countries took the lead in the European colonization of East Africa in the late 1800s?

4. *Describing* What have been some of the major problems that East African countries have faced in building their nations after achieving independence?

5. *Narrative Writing* You are a modern-day Ibn Battuta, traveling through East Africa. Write a series of journal or diary notes telling about the people you meet and the sights you see there.

Reading **HELP**DESK

Academic Vocabulary

• **diverse**

Content Vocabulary

• **population density**
• **clan**
• **subsistence agriculture**
• **oral tradition**
• **poaching**

TAKING NOTES: *Key Ideas and Details*

Summarizing As you read about East African populations, daily life, culture, and challenges today, use a web diagram like the one here to list facts and details about each important idea.

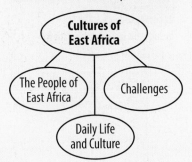

Cultures of East Africa

The People of East Africa

Challenges

Daily Life and Culture

Lesson 3
Life in East Africa

ESSENTIAL QUESTION · *Why does conflict develop?*

IT MATTERS BECAUSE
East Africa is a region of great diversity in ethnicity, religion, and language—not only across the region, but also within individual countries.

The People of East Africa

GUIDING QUESTION *What ethnic groups contribute to the diversity of the population?*

East African countries typically are home to many ethnic groups. Another striking feature in this region is the split between urban and rural populations. Languages and religions make up a mosaic of many different elements.

Where People Live

The population of East Africa is split between large cities and rural areas. Many large cities are on or near the coast of the Indian Ocean (for example, Mogadishu in Somalia, Mombasa in Kenya, and Dar es Salaam in Tanzania). Some large cities, however, developed from important trading centers. Such cities include Nairobi, the capital of Kenya, and Addis Ababa, the capital of Ethiopia.

Of the 11 countries in the region, Ethiopia has the largest population (about 80 million), and Djibouti has the smallest (about 1 million). People are distributed unevenly in East Africa. **Population density** measures how many people live in a given geographical area. A thickly settled area has a high population density. In thinly settled areas, the density is low. In Tanzania, population density varies greatly from one area to another. Overall, Rwanda has the highest population density in the region. Somalia has the lowest.

(l to r) Tom Cockrem/age fotostock; ©Cameron Davidson/Corbis; Harry Hook/Stone/Getty Images; Charles O. Cecil/Alamy; Zhao Yingquan/Newscom

In Ethiopia, the majority of people live in the central highlands. The warmer and drier areas of lower elevations are thinly inhabited. In Sudan, most people live along the Nile River. Arid parts of the country are thinly populated. In Somalia, most people are nomadic or seminomadic.

Ethnic Groups

The populations of Kenya, Tanzania, and Ethiopia are **diverse** in terms of ethnicity. Sometimes competition among different ethnic groups has led to political and economic conflict. Ethnic identity is closely linked to language and also to geography.

In Kenya, for example, the Kikuyu, Kamba, Meru, and Nyika people inhabit the fertile highlands of the Central Rift. The Luhya live in the Lake Victoria basin. The rural Luo people are located in the lower parts of the western plateau. The Masai people tend their herds of cattle in the south, along the Kenya-Tanzania border. Like the Masai, the Samburu and the Turkana are pastoralists. They live in the arid northwestern region of Kenya.

Another type of ethnic identity is the **clan**. A clan is a large group of people sharing a common ancestor in the far past. A group of related clans is called a clan family. Smaller groups of related people within a clan are called subclans. In Somalia, the basic ethnic unit is the clan.

Tom Cockrem/age fotostock

Academic Vocabulary

diverse having or exhibiting variety

Nairobi, the capital of Kenya, was founded in 1899 as a railway stop between plantations in Uganda and ports on the Kenyan coast. Today, Nairobi is one of East Africa's largest cities, with a population of about 3 million.

▶ CRITICAL THINKING
Describing Why are most East African cities located either along the Indian Ocean coast or in inland, highland areas?

In the A.D. 1100s, an Ethiopian king had the Church of St. George carved from solid red volcanic rock. Today, St. George and 10 similar churches in the town of Lalibela attract Ethiopian Christian worshippers as well as tourists from around the world.

Identifying What are the major religions in East Africa today?

In countries that have many diverse ethnic groups, building a sense of national identity is difficult. People often feel a stronger attachment and allegiance to their ethnic group than to their country. A Somali, for example, might feel a greater attachment to his or her clan than to the country of Somalia.

Languages

East Africa is a region where many African languages are spoken. For example, Ethiopians speak about 100 distinct languages. Kenya also has a wide variety of spoken languages. Swahili and English are used by large numbers of people to communicate. Those two languages are the official languages of the Kenyan legislature and of the courts.

Swahili is almost universal in Tanzania. The geographical location and colonial history of East African countries have often made an impact on the languages spoken there. For example, in Somalia the official language is Somali. However, Arabic is widely spoken in the northern area of the country, and Swahili is widespread in the south. In Somalia's colleges and universities, it is not uncommon to hear people speaking English or Italian. In Djibouti, Arabic and French are important languages.

Religion

Most people of East Africa follow either the Christian or Muslim faith. However, a number of traditional African religions also thrive in the region. Traders and missionaries from the Mediterranean region brought Christianity to Ethiopia in the A.D. 300s. The Ethiopian Orthodox Church is one of the world's oldest Christian churches. Today, about 60 percent of Ethiopians are Christians.

©Cameron Davidson/Corbis

In Kenya, the constitution guarantees freedom of religion. Christianity first arrived in Kenya with the Portuguese in the 1400s. But the religion was not practiced for several hundred years, until colonial missionaries arrived in Kenya in the late 1800s. Muslims are an important religious minority in Kenya. Today, Christianity is practiced by more than two-thirds of Kenya's population.

Tanzania is evenly split among Christianity, Islam, and traditional African religions. About one-third of the population follows each one of these three religious traditions.

☑ **READING PROGRESS CHECK**

Analyzing In a region with such diverse languages, how do you think East Africans can communicate with people outside their own language group?

Life and Culture

GUIDING QUESTION *What is daily life like for people in East Africa?*

In East Africa, traditional customs, as well as the impact of modernization, can be seen in daily life and culture. Culture in East Africa often displays a blend of African and European ways of life.

Daily Life

The rhythms of daily life are varied in East Africa. One factor is where people live: in cities or in rural areas. Most East Africans live in the countryside. But cities are growing rapidly, due to the economic opportunities they provide.

Nairobi is Kenya's capital and most important industrial city. The city is home to more than 3 million people. This makes Nairobi the most populous city in East Africa. It is a city of contrasts. High-rise business and apartment buildings sit near slums built of scrap material.

Daily life in rural areas is quite different from life in the cities. A rural family's housing, for example, might consist of a thatched-roof dwelling with very little in the way of modern or sanitary conveniences. Often, no electricity is available. Some rural people practice **subsistence agriculture**, growing crops to feed themselves and their families. Other rural people grow cash crops to sell.

A Masai mother and son (top) stand outside their home built of mud, sticks, and grass. The Masai people herd cattle on the inland plains of Kenya and Tanzania. A mosque and Islamic-style buildings (bottom) crowd the harbor of Mombasa, a city on Kenya's Indian Ocean coast.

▶ **CRITICAL THINKING**

Describing How do ways of life differ in East Africa depending on location and culture?

A tarab orchestra performs in Zanzibar, an Indian Ocean island that is part of Tanzania. Tarab is a form of music that began in Zanzibar and spread to other areas. The musician (left) plays a *qanun*, a stringed instrument believed to have been first used in Islamic Persia during the A.D. 900s.

▶ **CRITICAL THINKING**
Determining Central Ideas
What does a form of music like tarab reveal about East African culture?

In Tanzania, groups such as the Sukuma farm the land south of Lake Victoria. The Chaggas grow coffee in the plains around Kilimanjaro.

The Masai are a nomadic people who live in Tanzania and Kenya. They wander from place to place throughout the year as they tend herds of cattle. Their cattle provide the Masai with most of their diet.

The Masai have developed a unique way of living. Groups of four to eight families build a kraal, or a circular thornbush enclosure. The kraal shelters their herds of livestock. The families live in mud-dung houses inside the kraal.

The governments of Kenya and Tanzania have set up programs to persuade the Masai to abandon their nomadic lifestyle. The governments want to conserve land and protect wildlife, but the Masai have resisted. They want to preserve their way of life.

Arts and Culture

East African culture is deeply influenced by **oral tradition**. This means that stories, fables, poems, proverbs, and family histories are passed by word of mouth from one generation to the next. Folktales and fables offer good examples of oral tradition. In Kenya, the oral tradition functioned in a political way. Hymns of praise were passed on to support independence.

The small country of Djibouti is well known for its colorful dyed clothing. This includes a traditional piece of cloth that men wear around their waist like a skirt. It is common clothing for herders.

A leading novelist in East Africa is Kenya's Ngugi wa Thiong'o. His novel *Weep Not, Child* (1964) is considered the first important English-language novel written by an East African. This book is a story about the effects of conflict on families in Kenya. He also has authored works in the Bantu language of Kenya's Kikuyu people.

In Tanzania, an appealing and popular form of music is *tarab*. This type of music combines African, Arab, and Indian elements and instruments. Tarab has developed an international following. In Kenya, a popular musical style is *benga*. This pop style emerged in the 1960s in the area near Lake Victoria, which is inhabited by the Luo ethnic group.

East Africa is also linked to important findings in the fields of anthropology and ecology. Evidence indicates that East Africa is where human beings originated. The earliest known human bones come from Kenya and Ethiopia. The fossil beds of Olduvai Gorge in northern Tanzania have furnished us with an important record of 2 million years of human evolution.

In the domain of ecology, the national park systems of East Africa have no equal in the world. Protected areas like the Masai Mara National Reserve and Samburu National Reserve in Kenya, the Serengeti National Park in Tanzania, Queen Elizabeth National Park in Uganda, and Volcanoes National Park in Rwanda are preserving a precious inheritance.

☑ **READING PROGRESS CHECK**

Describing Compare and contrast urban and rural daily life in East Africa.

Challenges

GUIDING QUESTION *How do economic, environmental, and health issues affect the region today?*

Today, the people of East Africa face many complex, challenging issues. Some of the most important challenges involve economic development, the environment, and health.

Economic Development

Agriculture is the main economic activity in East Africa. Farmers in the region, however, face difficult challenges. First, the soils in East Africa are not especially fertile. Second, climate conditions are often unpredictable. Rainfall can be intermittent. Drought can severely damage crops.

Government policies in some countries of East Africa also favor the production of cash crops such as coffee for export. Such policies harm subsistence farmers who attempt to produce enough food to meet local needs. Much of this pattern of growing cash crops results from colonialism. Even after the countries of East Africa gained independence, the practice of growing cash crops for sale continued.

Self-sufficiency is a challenge in East Africa. The region is one of the poorest in the world. In addition, the population of many countries there is growing at a faster rate than the world's average. Industrialization has come slowly for East Africa.

Students view an exhibit of African wildlife at the Kenya National Museum in Nairobi. The museum's purpose is to collect, preserve, study, and present Kenya's cultural and natural heritage.

▶ **CRITICAL THINKING**

Describing How do East Africans protect living wildlife?

Scores of elephant tusks, seized from illegal poachers, are burned in Kenya. The purpose of the burning was symbolic: to point out the need to keep ivory from reaching international markets and to stop the illegal killing of elephants for their tusks.

Necessary resources such as trained workers, new facilities, and equipment have been lacking. In Ethiopia, for example, manufacturing amounts to only about 10 percent of the economy. Most of Ethiopia's exports are agricultural products. Its most important export is coffee.

The emphasis on primary industries that harvest or extract raw material, such as farming, mining, and logging, is also derived from colonialism. Colonial powers developed their colonies to provide products for the powers. Even after independence, the former colonies continue to produce the same products.

In Tanzania, the economy is mostly agricultural. Many farmers practice subsistence agriculture. Corn (maize), rice, millet, bananas, barley, wheat, potatoes, and cassava are among the important crops. Coffee and cotton are the most important cash crops. Gold is Tanzania's most valuable export.

In parts of East Africa, the economy has suffered because of civil war and political instability. The economy also is linked to the availability of transportation, communication, and education. One key indicator of progress in education is a country's literacy rate. Literacy rates across the region range from a low of 38 percent in Somalia to a high of 87 percent in Kenya.

Environmental Issues

East Africa faces challenging issues related to the environment. The region's lack of electric power has quickened the pace of deforestation. People are cutting down trees to meet their energy needs at an

alarming rate. They use the wood to cook food and heat their homes. Along with deforestation, desertification poses serious problems in countries like Sudan.

By setting up national parks and wildlife sanctuaries, the countries of East Africa are hoping that this will boost their economies and preserve their heritage. Ecotourism is tourism for the sake of enjoying natural beauty and observing wildlife. Revenue from ecotourism is important to the East African economy.

Wild animals such as elephants and lions also face the threat of **poaching**. Poaching is the trapping or killing of protected wild animals for the sake of profit in the illegal wildlife trade. African elephants are especially vulnerable to poaching; they are killed for their ivory tusks.

Health Issues

In East Africa, poor nutrition continues to be a difficult problem to overcome. One of the main causes of hunger and malnutrition in the region has been war. Since 1990, conflict in several East African countries has halted economic development and caused widespread starvation. Large numbers of refugees have poured across international borders.

HIV/AIDS is a serious and often fatal disease affecting people in the region. AIDS is an abbreviation that stands for "acquired immune deficiency syndrome." AIDS is caused by a virus that spreads from person to person. This disease continues to be a major health issue in Kenya, Tanzania, and Ethiopia. The resources required for medical education and treatment have put a further strain on East African economies.

Deaths from AIDS have cut the average life expectancy in East Africa. Drought and famine also have an impact on life expectancy. In East Africa, life expectancy at birth is 58 years in Rwanda and 62 years in Sudan. In Kenya, it is 63 years. By contrast, life expectancy in the United States is now about 78.5 years.

Include this lesson's information in your Foldable®.

 READING PROGRESS CHECK

Citing Text Evidence What is one major cause of deforestation in the region of East Africa?

LESSON 3 REVIEW (CCSS)

Reviewing Vocabulary
1. How might *poaching* affect the economies of some East African countries?

Answering the Guiding Questions
2. ***Determining Central Ideas*** What general statements can you make about the ethnic groups and where people live in East Africa?

3. ***Identifying*** Identify two ways in which trade has played a central role in the history of East Africa.

4. ***Describing*** What are two of the most important challenges confronting East Africa today?

5. ***Informative/Explanatory Writing*** Write a paragraph or two in which you explain some of the environmental issues that confront East Africa today.

Sudan Refugees and Displacement

Sudan has been involved in civil war for many years. Most people in the northern part of Sudan are Arab Muslim and live in cities. People in the southern part are African, rural farmers, and follow either African traditional religions or Christianity.

Geography Sudan is the sixteenth-largest country in the world in area and the third-largest country in Africa. It was the largest before South Sudan gained independence. Sudan's population is 33.4 million. South Sudan has approximately half that number. In land area, South Sudan ranks forty-fourth in the world.

Northern Control As an independent country, northern Sudan and its leaders controlled the government. They wanted to unify Sudan under Arabic and Islamic rule. In opposition were non-Muslims and the people of southern Sudan.

> **By the end of 2010, about 43.7 million people of the world did not have a home.**

Violence Continues When South Sudan became an independent country on July 9, 2011, many people hoped to start a new, peaceful life. However, several violent conflicts broke out, including continued conflict in Darfur.

Conflict in Darfur Darfur is a region in western Sudan. In 2003 Darfur rebel groups rose up against the Sudanese government. The rebels demanded that the government stop its unjust social and economic policies. The government reacted by raiding and burning villages. In the long conflict that followed, thousands were killed, and many were forced from their homes.

Refugees and IDPs Refugees are people who have left their country because they are in danger or have been victims of persecution. A major problem also exists with internally displaced persons (IDPs). An IDP is someone who is forced to flee his or her home because of danger, but who remains in his or her country.

World Refugee Day The United Nations (UN) World Refugee Day is observed every year on June 20. The events call attention to the problems refugees face.

People of South Sudan move to a new refugee camp to escape conflict and hunger. ▶

Paula Bronstein/Getty Images News/Getty Images

THERE'S MORE ONLINE

SEE a timeline of the crisis in Darfur • *WATCH* the Red Cross help refugees

These numbers and statistics can help you learn about the problems the Sudanese people face.

$1.25 a day

Although a peace agreement in 2005 brought some stability to the people of South Sudan, many terrible problems exist. More than 80 percent of the residents live on less than $1.25 a day. The country has the world's highest maternal mortality rate. About 50 percent of elementary school-age children do not attend school.

300,000

In 2003 rebellion broke out in Darfur, a region in western Sudan. Government militia attacked Darfur and the rebels. The United Nations estimates that as many as 300,000 people died in five years of conflict in Darfur. Violence still erupts at times, breaking the fragile peace.

More than 500,000

Refugees are people who have left their country because they are in danger or have been victims of persecution. In January 2011, 178,000 refugees were in Sudan, and 387,000 Sudanese who were living in other countries.

one point six million

In January 2011, more than 1.6 million Sudanese people were internally displaced.

two million

South Sudan seceded from Sudan in 2011 as a result of a peace treaty that ended decades of war that had killed 2 million people. The two countries have come close to war again. Disputes over control of territory led to armed conflict.

75%

Oil is a source of conflict between Sudan and South Sudan. About 75 percent of the oil is in South Sudan, but all the pipelines run north to Sudan. When disputes over oil erupted in 2012, Sudan bombed oil fields in South Sudan.

43.7 MILLION

By the end of 2010, about 43.7 million people of the world did not have a home. The number of refugees is the highest in 15 years.

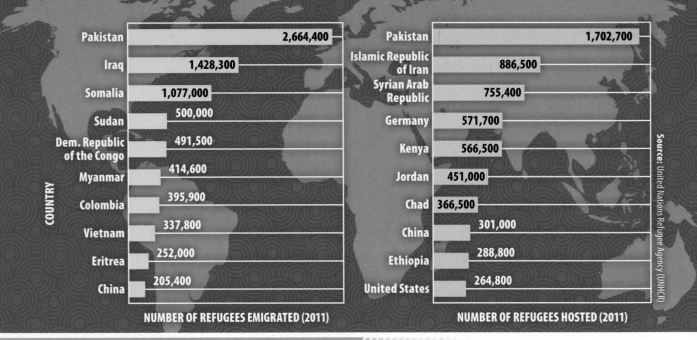

COUNTRY

Pakistan	2,664,400
Iraq	1,428,300
Somalia	1,077,000
Sudan	500,000
Dem. Republic of the Congo	491,500
Myanmar	414,600
Colombia	395,900
Vietnam	337,800
Eritrea	252,000
China	205,400

NUMBER OF REFUGEES EMIGRATED (2011)

Pakistan	1,702,700
Islamic Republic of Iran	886,500
Syrian Arab Republic	755,400
Germany	571,700
Kenya	566,500
Jordan	451,000
Chad	366,500
China	301,000
Ethiopia	288,800
United States	264,800

NUMBER OF REFUGEES HOSTED (2011)

GLOBAL IMPACT

MIGRATION OF REFUGEES Refugees are people who flee to another country because of wars, political unrest, food shortages, or other problems. The graph on the left lists the 10 major source countries of refugees and the number of refugees who emigrated from those countries in 2011.

The graph on the right lists the 10 major host countries. A host country is the country a refugee moves to. For example, more than 1.7 million refugees immigrated to Pakistan in 2011.

Sudan and South Sudan

The map shows the two countries, their national capitals, and disputed areas.

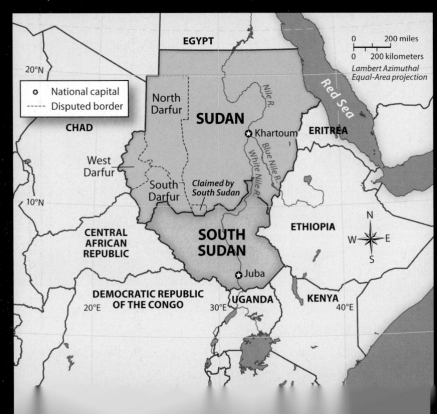

Thinking Like a
Geographer

1. **The Geographer's World** What are the major differences between Sudan and South Sudan?

2. **Human Geography** Why is oil a major factor in the conflict between Sudan and South Sudan?

3. **Human Geography** Imagine you are from Sudan and you come to live in the United States. Write a story about how you and your family learn to live in an American community.

Directions: Write your answers on a separate piece of paper.

1 Use your **FOLDABLES** to explore the Essential Question.

INFORMATIVE/EXPLANATORY WRITING Review the population map of East Africa at the beginning of the chapter. In two or more paragraphs, explain why people have settled in the locations indicated on the map.

2 **21st Century Skills**

INTEGRATING VISUAL INFORMATION Conduct research and write a paragraph about one of the national park systems in East Africa. Review a partner's paragraph using these questions to guide you: Did the paragraph include relevant details? Was there anything missing that you expected to find in the paragraph? Discuss the review of your paragraph with your partner. Revise your paragraph as needed.

3 **Thinking Like a Geographer**

IDENTIFYING Choose 1 of the 11 countries of East Africa. In a graphic organizer like the one shown, identify the capital city of the country and write two geographical facts about the country.

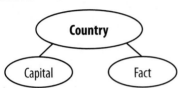

4 **GEOGRAPHY ACTIVITY**

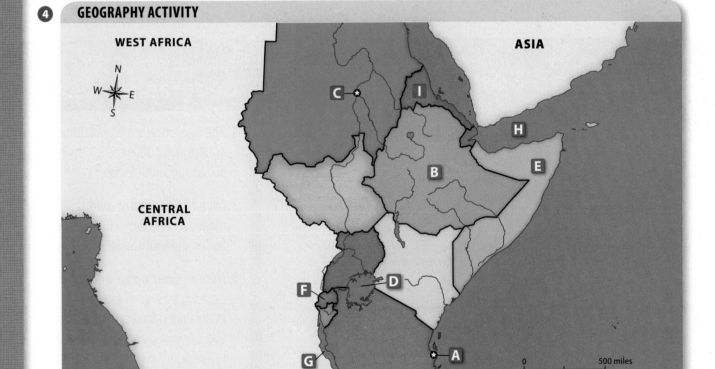

Locating Places
Match the letters on the map with the numbered places listed below.

1. Somalia
2. Eritrea
3. Lake Victoria
4. Khartoum
5. Rwanda
6. Gulf of Aden
7. Ethiopia
8. Lake Tanganyika
9. Dar es Salaam

REVIEW THE GUIDING QUESTIONS

Directions: Choose the best answer for each question.

1 The physical geography and landscape of East Africa are dominated by a series of geological faults collectively known as

A. the Ruwenzori Mountains.

B. the Great Rift Valley.

C. Kilimanjaro.

D. Jonglei.

2 Because of long periods of drought and overgrazing, agricultural land has turned into desert in a process called

F. irrigation.

G. desertification.

H. urbanization.

I. defoliation.

3 Throughout history, the countries of East Africa have been centers of

A. trade.

B. revolution.

C. oil exploration.

D. the slave trade.

4 Most countries in East Africa earned their independence from European colonial powers during which decade of the twentieth century?

F. the 1950s

G. the 1980s

H. the 1940s

I. the 1960s

5 What is the name of the nomadic people who herd cattle and build their mud-dung houses inside a kraal?

A. Samburu

B. Masai

C. Kamba

D. Meru

6 What is the main economic activity in East Africa?

F. manufacturing

G. tourism

H. agriculture

I. oil and gas production

DBQ ANALYZING DOCUMENTS

7 **ANALYZING** Read the following passage about economies in Africa:

"*Kenyan farmers, mostly small, are responsible for $1 billion in annual exports of fruits, vegetables, and flowers, a figure that dwarfs the country's traditional coffee and tea exports. . . . Rwanda, . . . long an importer of food, now grows enough to satisfy the needs of its people, and even exports cash crops such as coffee for the first time.*"

—from G. Paschal Zachary, "Africa's Amazing Rise and What It Can Teach the World" (2012)

What statement best explains the success of Kenya's farmers?

A. They produced a variety of crops that were in demand.

B. They produced more coffee and tea.

C. They exported cash crops for the first time.

D. They imported food from Rwanda.

8 **CITING TEXT EVIDENCE** Which sector of Rwanda's economy has seen success?

F. agriculture

G. industry

H. mining

I. service industries

SHORT RESPONSE

"*The world's biggest refugee camp, Dadaab, in northeastern Kenya marks its 20th anniversary this year. The camp, which was set up to host 90,000 people, now shelters nearly one-half million refugees. . . . The [United Nations] set up the first camps in Dadaab between October 1991 and June 1992, following a civil war[in Somalia] that continues to this day.*"

—from Lisa Schlein, "World's Biggest Refugee Camp in Kenya Marks 20th Anniversary" (2012)

9 **DETERMINING CENTRAL IDEAS** Why was a refugee camp needed?

10 **ANALYZING** What kinds of facilities would officials need to create to take care of tens of thousands of people?

EXTENDED RESPONSE

11 **INFORMATIVE/EXPLANATORY WRITING** In an essay, compare and contrast the countries of Somalia and Kenya. Use your text and Internet research to examine each country's physical geography, culture, average income, education levels, type of government, employment, and other factors that affect the way people live. Of the two, which country would you rather live in?

Need Extra Help?

If You've Missed Question	**1**	**2**	**3**	**4**	**5**	**6**	**7**	**8**	**9**	**10**	**11**
Review Lesson	1	1	2	2	3	3	3	3	2	2	3

From "Africa's Amazing Rise and What It Can Teach the World," by G. Paschal Zachary, Feb 25, 2012, http://www.theatlantic.com/international/archive/2012/02. Copyright © G. Paschal Zachary; "World's Biggest Refugee Camp in Kenya Marks 20th Anniversary," by Lisa Schlein, February 21, 2012. *Voice of America*, http://voanews.com

CENTRAL AFRICA

ESSENTIAL QUESTIONS • *How do people adapt to their environment?* • *How does technology change the way people live?* • *What makes a culture unique?*

A Congolese woman wears the traditional hairstyle that originated in the area long ago.

The Story Matters...

The region of Central Africa straddles the Equator, which creates tropical climates that support the growth of rain forests and savannas. The Congo River—Africa's second-longest river—has so many tributaries that it forms Africa's largest system of waterways. Abundant natural resources in Central Africa greatly influenced its history. Today, the resources are vital to helping nations in the region achieve and maintain stability.

FOLDABLES
Study Organizer

Go to the Foldables® library in the back of your book to make a Foldable® that will help you take notes while reading this chapter.

Central Africa's many rivers are a source of life for people of the region. The geography of the region is dominated by the rain forest basin of the Congo River.

Step Into the Place

MAP FOCUS Use the map to answer the following questions.

1 THE GEOGRAPHER'S WORLD
How many countries make up the region of Central Africa?

2 THE GEOGRAPHER'S WORLD
What island country is part of the region?

3 PLACES AND REGIONS What is unusual about the capitals of Congo and the Democratic Republic of the Congo?

4 CRITICAL THINKING
Analyzing Which cities in Gabon and Cameroon are located along the coasts? Think about their locations. What economic activities do you think are important in those cities?

ISLAND PARADISE A highway circles the scenic coast of São Tomé, a volcanic island, off the western coast of Central Africa. São Tomé forms part of São Tomé and Príncipe, a small island nation.

SULTAN'S COURT MUSICIANS Musicians perform traditional music at a palace of a sultan in Foumban, Cameroon. Music and dance are an important part of ceremonies and social gatherings in Central Africa.

Step Into the Time

TIME LINE Choose at least two events from the time line to describe the cause-and-effect relationship between natural resources and the slave trade in Central Africa.

1000 B.C. Iron Age spreads to Central Africa

1000 B.C.–A.D. 1100 Bantu people migrate to Congo rain forest

1700s Slave trade spreads to interior of continent

1470s São Tomé becomes the first port of the Atlantic slave trade

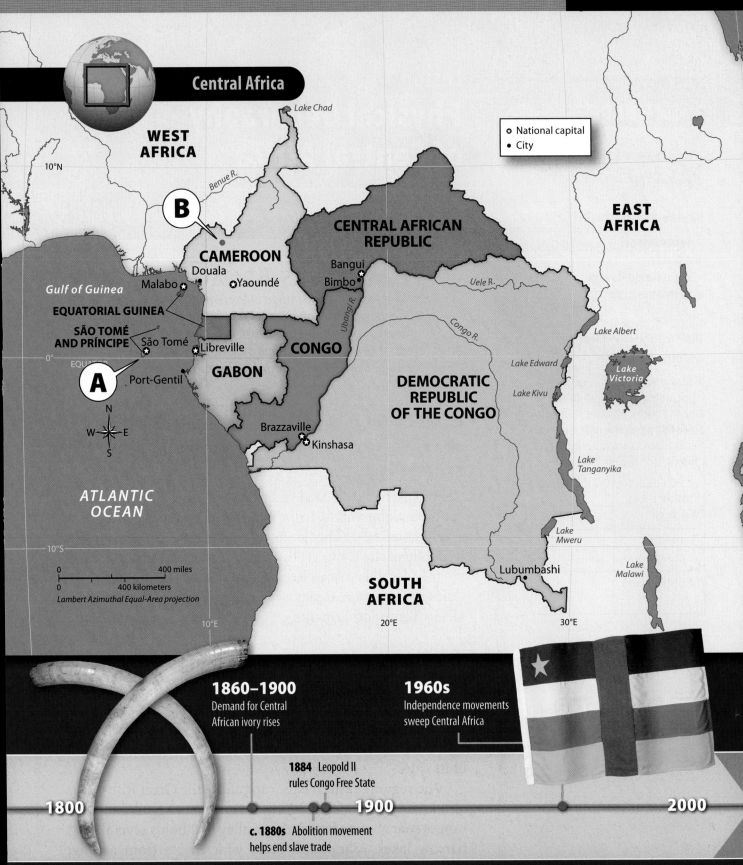

Central Africa

WEST
AFRICA

Lake Chad

⊕ National capital
● City

10°N

Benue R.

B

**CENTRAL AFRICAN
REPUBLIC**

EAST
AFRICA

CAMEROON

Douala

⊕Yaoundé

Bangui
Bimbo

Uele R.

Gulf of Guinea

Malabo

EQUATORIAL GUINEA

Ubangi R.

Congo R.

Lake Albert

**SÃO TOMÉ
AND PRÍNCIPE**

São Tomé

Libreville

CONGO

Lake Edward

*Lake
Victoria*

0°

EQUATOR

A

Port-Gentil

GABON

**DEMOCRATIC
REPUBLIC
OF THE CONGO**

Lake Kivu

N
W   E
S

Brazzaville
Kinshasa

*Lake
Tanganyika*

**ATLANTIC
OCEAN**

10°S

0   400 miles
0   400 kilometers
Lambert Azimuthal Equal-Area projection

*Lake
Mweru*

*Lake
Malawi*

Lubumbashi

**SOUTH
AFRICA**

10°E   20°E   30°E

1860–1900
Demand for Central
African ivory rises

1960s
Independence movements
sweep Central Africa

1884 Leopold II
rules Congo Free State

1800   **1900**   **2000**

c. 1880s Abolition movement
helps end slave trade

Reading **HELP**DESK (CCSS)

Academic Vocabulary

- potential

Content Vocabulary

- **watershed**
- **estuary**
- **slash-and-burn**
- **biodiversity**

TAKING NOTES: *Key Ideas and Details*

Find the Main Idea As you study the lesson, write a main idea about each topic on a graphic organizer. Then write details that support the main idea.

Topic	Main Idea/ Detail
Landforms and Waterways	
Climate and Vegetation	
Natural Resources	

Lesson 1
Physical Geography of Central Africa

ESSENTIAL QUESTION • *How do people adapt to their environment?*

IT MATTERS BECAUSE

Central Africa is smaller than many regions, but it holds a tremendous variety of geographic features. These features include a vast rain forest–covered basin, one of the world's greatest river systems, soaring mountains, and a deep rift valley marking the line along which Africa is splitting apart.

Landforms and Waterways

GUIDING QUESTION *What makes some landforms and waterways so important to the region?*

Central Africa is located in Earth's equatorial zone—that is, the area along and near the Equator. The region consists of seven countries. The largest of these is the Democratic Republic of the Congo (DRC). It dwarfs its neighbors, which are the Central African Republic, the Republic of the Congo, Cameroon, Gabon, Equatorial Guinea, and the island country of São Tomé and Príncipe.

Landforms

The dominant landform of Central Africa is the **watershed** of the Congo River. A watershed is the land drained by a river and its system of tributaries. At the center of the watershed is a depression called the Congo Basin. A rolling plain spreads across the center of the basin, and high plateaus rise on most of its sides.

The region's eastern edge runs along the Great Rift Valley, also known as the Great Rift System. Here, rugged mountain ranges soar above a broad, deep valley that holds several long, narrow lakes. Margherita Peak, which rises from a range

(l to r) ©Per-Anders Pettersson/Corbis; ©Ian Nichols/National Geographic Society/Corbis; ©Nigel Pavitt/JAI/Corbis; ©FINBARR O'REILLY/Reuters/Corbis

called the Ruwenzori (ROO-un-ZO-ree), reaches the lofty height of 16,763 feet (5,109 m). Margherita Peak is the highest summit in the region and the third highest on the entire continent, ranking after Kilimanjaro and Mount Kenya.

Along the Atlantic coast of Central Africa stretches a narrow lowland. Off the coast lie several important islands. Two of these islands form the country of São Tomé and Príncipe. Two other islands, called Bioko and Pagalu, belong to Equatorial Guinea. This country also includes several smaller islands, as well as a territory on the mainland known as Mbini.

Waterways

Six of the seven countries in Central Africa have coasts on the Atlantic Ocean. The Central African Republic is the region's only landlocked country.

The Congo River and its tributaries account for most of Central Africa's inland waterways. The source of the Congo lies in East Africa between Lake Tanganyika and Lake Malawi (also called Lake Nyasa). From there, the river flows about 2,900 miles (4,667 km) to its mouth on the Atlantic Ocean. Among African rivers, only the Nile is longer than the Congo. When measured by water flow, the Congo tops every river in the world except South America's Amazon.

Difficult Navigation

No other river system in Africa offers as many miles of navigable waterways as the Congo and its tributaries. A navigable river is one on which ships and boats can travel. The Congo River is not navigable for its entire course, however. Several series of cataracts and rapids interrupt the passage of ships on the river.

Perhaps the most significant of these interruptions occurs rather close to the mouth of the Congo River. Only about 100 miles (161 km) from the Atlantic Ocean lie cataracts that block seagoing ships from traveling farther inland. The seaport city of Matadi is found here. Downstream from Matadi, in the final part of its journey, the river widens into an **estuary**, a passage in which freshwater meets salt water.

Fishers in the Democratic Republic of the Congo use bamboo supports to lower their nets into the waters along the rapids of the Congo River.

▶ **CRITICAL THINKING**

Describing What makes navigation difficult on parts of the Congo River?

A scientist and animal trackers analyze camera trap video of gorillas in a Central African rain forest. A camera trap is a camera equipped with a special sensor that captures images of wildlife on film, when humans are not present.

▶ CRITICAL THINKING

Analyzing How might camera traps aid in protecting Central African wildlife?

The Congo River is important to Central Africa for several reasons. First, it provides a livelihood for people who live along its banks. They use the river's water for agriculture and depend on its fish for food. Second, the river is a vital transportation artery. Although the cataracts and rapids prevent ships from navigating the entire river, ship traffic connects people and places along various sections of the river. Finally, dams on the river generate hydroelectric power.

☑ **READING PROGRESS CHECK**

Identifying Give two reasons for the Congo River's importance to Central Africa.

Climate and Vegetation

GUIDING QUESTION *What are the prevailing climates in Central Africa?*

Because Central Africa is centered on the Equator, the climate in much of the region is tropical. Temperatures are warm to hot, and rainfall is plentiful. The amount of rainfall generally decreases as distance from the Equator increases. In the northern and southern parts of the region, dry seasons alternate with wet seasons.

Climate Zones

The belt of Central Africa that lies along the Equator has a warm, wet climate. Because of its location, this zone experiences little seasonal variation in weather and length of daylight. The midday sun is directly or almost directly overhead every day, and daytime temperatures are always high. Rainfall is abundant throughout the year, with totals greater than 80 inches (203 cm) in some areas.

To the north and the south of the region's equatorial zone lie tropical wet-and-dry climate zones. As the name suggests, these zones have both rainy and dry seasons. There is great climate variation within the zones. In the areas nearest the Equator, the dry season typically lasts about four months. In the areas farthest from the Equator, it might last as long as seven months.

Rain Forest

A tropical rain forest, the second largest in the world, covers more than half of Central Africa. In this rain-soaked realm, closely packed trees soar as high as 15-story buildings. The forest also holds a

tremendous variety of other plants, some of which are used in traditional medicines. Scientists think that only a fraction of the plant species in the forest have been identified. Most remain to be discovered.

The trees and other plants in the rain forest compete for survival. The interwoven crowns of the trees create a dense canopy, or roof, that blocks nearly all of the sunlight, leaving the lower levels in gloom. A tree seedling or sapling can only grow tall if a mature tree dies or if wind blows down some of its branches, creating an opening in the canopy. Because of the lack of sunlight, the forest floor has few of the small plants found in other types of forests. Instead, the forest floor is covered by dead leaves that decompose rapidly in the warm, wet conditions.

Savannas

In the northern and southern areas of Central Africa, rain forests give way to savannas, or areas with a mixture of trees, shrubs, and grasslands. The exact mix of vegetation depends on the length of the dry seasons. Human activity over thousands of years might be responsible for an increase in the size of the savanna areas. Experts believe that the change of forestland into savanna could be the result of clearing land through **slash-and-burn** agriculture.

This form of farming involves turning forestland into cropland. Trees and shrubs are cut down and burned in order to make the soil more fertile, at least temporarily.

Garamba National Park in the Democratic Republic of the Congo is home to the few surviving Northern white rhinoceroses.

Miners search for gold in an open-pit mine in the northeastern part of the Democratic Republic of the Congo. For more than a decade, rival groups in the DRC have fought each other for control of the country's rich natural resources, such as gold, diamonds, and timber.

Within a few years, the soil becomes depleted and farmers move on. The plots of land are left so that trees and shrubs can return. Then the farmers return and repeat the process of cutting, burning, and farming. Sometimes plots remain fallow—that is, not planted—for as long as 20 years.

Slash-and-burn agriculture provides a livelihood for many people. Others, however, point out that it destroys plant and animal habitat, creates air pollution, and is causing the world's tropical forests to shrink at an alarming rate.

Those who support protecting the environment worry that activities that harm rain forests threaten biodiversity. **Biodiversity** refers to the wide variety of life on Earth. These conservationists argue that if ecosystems are wiped out, numerous valuable species will be lost forever.

✔ READING PROGRESS CHECK

Describing Imagine that you live in northern Cameroon. You decide to move to a location in the Republic of the Congo very near the Equator. What changes in climate can you expect?

Natural Resources

GUIDING QUESTION *Which natural resources are important in Central Africa?*

Central Africa is rich in mineral, energy, and other resources. However, many of these resources have not yet been developed.

Mineral Resources

The greatest abundance of mineral resources in Central Africa is found in the Democratic Republic of the Congo. Within that country, the area richest in mineral resources is a province called Katanga. Katanga holds deposits of more than a dozen minerals, including cobalt, copper, gold, and uranium. Minerals mined in other areas of the country include diamonds, iron ore, and limestone.

Gabon produces more than a tenth of the world's supply of manganese. This hard, silvery metal is used in the manufacture of iron and steel. The country also produces uranium, diamonds, and gold, and it holds reserves of high-quality iron ore.

Bauxite and cobalt are among Cameroon's most significant mineral resources. Equatorial Guinea has deposits of uranium, gold, iron ore, and manganese. Most of these deposits have yet to be exploited. Diamond mining is an important industry in the Central African Republic. Rich deposits of uranium, gold, and other minerals could bring the country wealth in the future.

Why have the mineral resources in some parts of Central Africa remained underdeveloped? Political instability, civil conflict, and the high costs of investment have played a role. Perhaps the most important reason, however, is that the region lacks good transportation networks. In the Democratic Republic of the Congo, for example, the Congo River still serves as the major transportation artery. The landlocked Central African Republic also must use rivers for transport. It has few paved roads and no railways. Similarly, most roads in Equatorial Guinea are unpaved and there is no railway system.

Other Resources

The region is rich in resources other than minerals. In the Democratic Republic of the Congo, for example, rapids and waterfalls on the Congo River and its tributaries offer vast **potential** for hydroelectric power. The forest reserves of the DRC are rivaled by few countries in the world. Fish from the ocean and its fresh water systems provide another important resource.

Developing these resources, however, will affect the environment. Damming rivers for hydroelectric power results in large changes to river ecosystems. Deforestation and habitat change occur when forests are cut.

In the 1990s, large reserves of petroleum and natural gas were discovered under the seafloor off Equatorial Guinea's Atlantic Coast. The export of oil and gas products have boosted the country's economy.

Likewise, petroleum has been Cameroon's leading export since 1980. The country also has natural gas deposits, but the high cost of development has kept them untapped. Nearly all of Cameroon's energy comes from dams that generate hydroelectricity.

✔ **READING PROGRESS CHECK**

Identifying Central Issues What are two of the factors that have slowed development of Central Africa's rich natural resources?

Academic Vocabulary

potential possible; capable of becoming

Include this lesson's information in your Foldable®.

LESSON 1 REVIEW **CCSS**

Reviewing Vocabulary

1. Why might *slash-and-burn agriculture* be harmful to a country's land in the long term?

Answering the Guiding Questions

2. *Analyzing* What comparisons can you make between the Congo River in Central Africa and two of the other great rivers of the world: the Nile and the Amazon?

3. *Describing* What conclusion can you make about the prevailing climate conditions in areas near the Equator?

4. *Analyzing* How would you evaluate Central Africa's hydroelectric potential?

5. *Argument Writing* Write a letter to a friend to persuade him or her to invest in mineral production in Central Africa. In your letter, use some of the information you have learned in this lesson.

(l to r) Gary John Norman/Alamy; Mary Evans Picture Library/Alamy; UPPA/Photoshot; INTERFOTO/Personalities/Alamy

Reading **HELP**DESK ⓒⓒⓢⓢ

Academic Vocabulary

- **foundation**

Content Vocabulary

- **millet**
- **palm oil**
- **cassava**
- **colonialism**
- **missionary**
- **coup**

TAKING NOTES: *Key Ideas and Details*

Organize Use a diagram like this one to note important information about the history of Central Africa, adding one or more facts to each of the boxes.

Topic	Information
Products	Millet Sorghum
European contact	Slave trade

Lesson 2
History of Central Africa

ESSENTIAL QUESTION · *How does technology change the way people live?*

IT MATTERS BECAUSE

Central Africa's past is a fascinating and often tragic story involving migrations, slavery, exploitation by foreign powers, and the struggle for independence, stability, and prosperity.

Early Settlement

GUIDING QUESTION *How did agriculture and trade develop in Central Africa?*

Through most of the prehistoric period, the people of Central Africa were hunters and gatherers who survived on wild game and wild plants. Eventually, climate changes forced the people to develop new ways of life.

Development of Agriculture

Around 10,000 years ago, Earth's climate entered a dry phase. Vegetation patterns changed in response and led to the movement of people. The changes also intensified the struggle for survival. The region's inhabitants were forced to find ways to get more food from a smaller area of their environment.

Gradually, a transformation that historians call the agricultural revolution swept through the region, beginning in the north. People began to collect plants—especially roots and tubers—on a more regular basis. They developed and refined tools such as stone hoes that were specially designed for digging. They discovered that if they planted a piece of a root or tuber in fertile soil, a new plant would grow from it. Over time, the hunters and gatherers turned into farmers.

Cereal farming was the next agricultural development in Central Africa. In the savannas of the north, people began cultivating **millet** and sorghum, two wild grasses that produce edible seeds. Millet proved to be especially well-suited to the

area's climate. The crop thrives in high temperatures and is resistant to drought.

In addition to growing crops, Central Africa's early farmers cultivated trees and gathered their fruit. From the fruit of oil palms they made a cooking oil that was rich in proteins and vitamins. The nutrition boost provided by **palm oil** helped people to become healthier, and improved health brought about population growth.

The increase in the food supply from the practice of agriculture meant that people could live in larger, more settled communities. The agricultural revolution laid the **foundation** for village life and also for the development of items used in daily life.

Academic Vocabulary

foundation the basis of something

Using Mineral Resources

Since early times, people in Central Africa have made tools from stone. Around 3,000 years ago, they began using a new material that was far better in many ways: iron. Iron tools were expensive and could only be made by skilled artisans, but they were far more efficient and less brittle than stone tools.

In addition to iron, people in ancient Central Africa made use of other minerals, especially copper and salt. These resources played an important role in trade.

☑ **READING PROGRESS CHECK**

Determining Central Ideas Why was the agricultural revolution so important for the development of Central Africa as a region?

European Contact and Afterward

GUIDING QUESTION *How did colonization by foreign countries affect Central Africa?*

Regular contact with Europeans, which began in the 1400s, marked the start of a new era in Central Africa's history.

The Slave Trade

As European ships began reaching the Atlantic coast of Central Africa in the 1400s, the region began developing into one of the busiest hubs of the slave trade. The trade was driven because European colonizers demanded a large workforce for their huge plantations in the Americas. The slave trade would continue and grow for more than three centuries.

The first European country to become actively involved in the slave trade was Portugal. In the late 1400s, the Portuguese established a colony on the island of São Tomé in order to grow sugar.

A woman in Cameroon carries fruit from oil palm trees to market. The pulp from this fruit is used to make cooking oil. Some environmentalists fear that the creation of more palm oil plantations in the country will endanger the livelihood of small farmers and lead to the destruction of existing rain forests.

Gary John Norman/Alamy

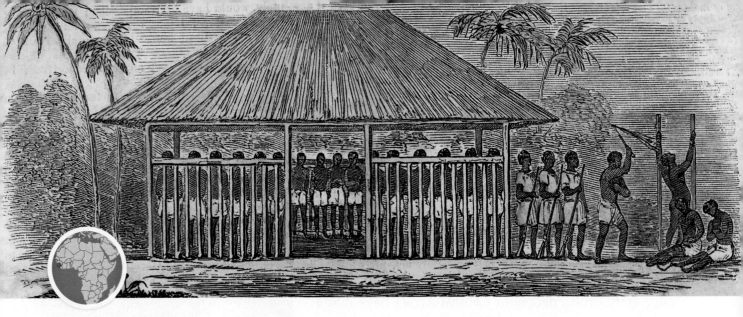

Slave traders held people who were captured for slavery in *barracoons,* or sheds. There, the captives stayed for several months before they were sold and shipped to the Americas. During imprisonment, the captives were chained by the neck and legs and often beaten.

▶ **CRITICAL THINKING**
Determining Word Meaning What was the Middle Passage?

Later, this island became a staging area for the transportation of African slaves to Portugal's main conquest in the Western Hemisphere: Brazil in South America.

Gabon served as one of the most important centers of the slave trade. Slaves were gathered in the country's interior and taken on boats to a coastal inlet called the Gabon Estuary. Some of these slaves were people who had been cast out of their own societies. Others had been captured in warfare. At settlements on the estuary, the slaves were held in enclosures known as barracoons until European ships arrived.

The slave trade was part of what is sometimes called the "triangular trade," named for the triangular pattern formed by the three stages of the trade. In the first stage, ships would sail to Africa with cargoes of manufactured goods such as cloth, beads, metal goods, guns, and liquor. These goods would be traded for slaves. In the second stage, known as the Middle Passage, the ships would carry their human cargo to the Americas. There, the slaves would be exchanged for goods such as rum, tobacco, molasses, and cotton that were produced on slave-labor plantations. In the third stage, the ships would return to Europe. Ships of the time were powered by wind, and each stage followed the direction of prevailing winds.

Adoption of New Crops

After European countries established colonies in the Americas, they brought some of the native plants back to Europe and to Africa. Two plants in particular had an important effect on farming and diet in Central Africa. These were cassava and maize.

The **cassava** plant has thick, edible roots known as tubers. Rich in nutritious starch, the tubers are used to make flour, breads, and tapioca. The plant thrives in hot, sunny climates and is able to survive droughts and locust attacks. Historians believe that Portuguese ships brought cassava to Africa from Brazil. Today, cassava is a staple food for many of Africa's people.

Maize, often called corn, is one of the most important staple grains of the Western Hemisphere. Today it is the most widely grown grain crop in the Americas. Maize was domesticated in prehistoric times, probably in Central America. It was carried around the world by Europeans after their discovery of the Americas.

Colonialism

Colonialism is the political and economic rule of one region or country by another country, usually for profit. European countries began to practice colonialism in the 1500s and 1600s. They first founded colonies in the Americas for economic gain.

Colonialism came much later to Central Africa. Exploration and settlement by Europeans was impeded by the difficulty of transportation, the presence of tropical diseases like malaria, and other challenges. In the second half of the 19th century, however, European presence in the region began to grow sharply.

In 1884–1885, Germany hosted a landmark conference of European countries in the city of Berlin. The countries attending the conference agreed on a plan for dividing Africa into colonies that could be exploited for European profit.

King Leopold II of Belgium was among the strongest supporters of the conference. He believed that a fortune could be made from rubber plants. Rubber was one of the most plentiful and valuable natural resources of Central Africa. After 1885, King Leopold took over a vast area that came to be known as the Congo Free State. The king held the area as a personal possession, and he did indeed make a fortune.

Belgium's King Leopold II (oval) turned his privately owned Congo territory into a large labor camp for the harvesting of rubber. African workers (below) were overseen by European officials. Nearly 10 million Africans died of overwork or cruelty under Leopold's direct rule.

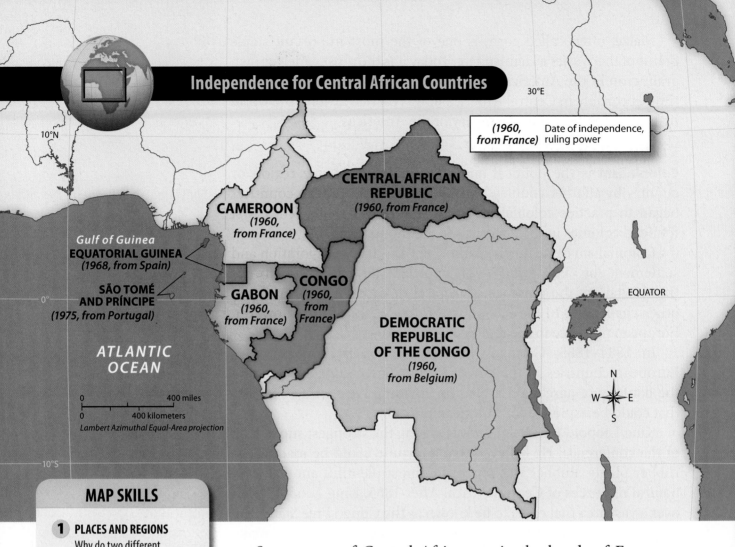

30°E

10°N

| (1960, | Date of independence, |
| from France) | ruling power |

CENTRAL AFRICAN REPUBLIC
(1960, from France)

CAMEROON
(1960, from France)

Gulf of Guinea
EQUATORIAL GUINEA
(1968, from Spain)

SÃO TOMÉ AND PRÍNCIPE
(1975, from Portugal)

GABON
(1960, from France)

CONGO
(1960, from France)

DEMOCRATIC REPUBLIC OF THE CONGO
(1960, from Belgium)

EQUATOR

ATLANTIC OCEAN

0 _____ 400 miles
0 _____ 400 kilometers
Lambert Azimuthal Equal-Area projection

N
W E
S

10°S

MAP SKILLS

1 PLACES AND REGIONS
Why do two different countries have Congo as part of their name?

2 THE GEOGRAPHER'S WORLD What country of Central Africa is landlocked?

Soon, most of Central Africa was in the hands of European colonizers. France gained control of what is now the Republic of the Congo, Gabon, and the Central African Republic. Spain colonized what is now Equatorial Guinea, while Germany ruled Cameroon. Portugal retained possession of São Tomé and Príncipe.

Europeans often justified their economic exploitation of Africa by claiming that their goal was to promote civilization and to spread Christianity. Europeans sent many **missionaries** to Africa in order to convert the native people.

At the same time, Europeans often treated the African workers under their control harshly. By the early 1900s, small revolts against French rule and the plantation-based economy were common. In the Congo Free State, the people suffered severe hardships and cruel treatment under King Leopold. Pressured by growing outrage from around the world, the Belgian parliament took over the vast area from King Leopold. It became an official colony of Belgium and was known as the Belgian Congo.

✓ **READING PROGRESS CHECK**

Determining Central Ideas How was Central Africa affected by a conference held in Germany in the mid-1880s?

Independent Countries

GUIDING QUESTION *What effects did gaining independence have on the countries of Central Africa?*

Near the middle of the 20th century, European countries became willing to grant independence to their African colonies. All seven of Central Africa's countries gained their independence in the period from 1960 to 1975.

A Wave of Independence

In 1960 France was the most important European colonial power in Central Africa. That year witnessed the independence of four French colonies: Gabon, the Republic of the Congo, the Central African Republic, and Cameroon, which France had gained from Germany during World War I. In the same year, the Democratic Republic of the Congo won independence from Belgium.

Many of these new countries experienced hard times after independence. Their people suffered through periods of ethnic conflict, harsh rule, and human rights abuses. In the Central African Republic, military officer Jean-Bédel Bokassa staged a **coup**. He ruled as a dictator and proclaimed himself the country's emperor. He brutally punished anyone who protested against his rule.

There have also been success stories, such as Gabon. Thanks in large part to its plentiful natural resources, Gabon has become one of the wealthiest and most stable countries in Africa.

Smaller Countries

Equatorial Guinea won independence from Spain in 1968. The country's first president, Francisco Macías Nguema, soon took over the government and became a ruthless dictator. In 1979, he was ousted by his nephew, who also ruled with an iron hand.

Portugal granted independence to São Tomé and Príncipe in 1975. With independence came great hope for the future. Like many of its neighbors, however, the country has been plagued by political instability and corruption.

☑ **READING PROGRESS CHECK**

Identifying Since the countries of the region gained independence, list at least two factors that have limited their political and economic progress.

FOLDABLES®
Study Organizer

Include this lesson's information in your Foldable®.

LESSON 2 REVIEW

Reviewing Vocabulary

1. Why was *millet* a suitable grain for planting in Central Africa?

Answering the Guiding Questions

2. ***Determining Central Ideas*** Why might tools made from iron be superior to tools made from stone?

3. ***Identifying*** What were two major motivations for the European colonization of Central Africa?

4. ***Analyzing*** In what way was the rule of Bokassa in the Central African Republic similar to the rule of Nguema in Equatorial Guinea?

5. ***Informative/Explanatory Writing*** Write a few paragraphs explaining the development of the slave trade in Central Africa.

networks

There's More Online!

 SLIDE SHOW Types of Housing in Central Africa

☑ **VIDEO**

Reading **HELP**DESK ⓒⒸⓈⓈ

Academic Vocabulary

- **characteristic**
- **depict**

Content Vocabulary

- **refugee**
- **trade language**

TAKING NOTES: *Key Ideas and Details*

Find the Main Idea As you study the lesson, create a chart like this one, and fill in at least two key facts about each topic.

Topic	Fact #1	Fact #2
People		
Daily Life and Culture		

Lesson 3
Life in Central Africa

ESSENTIAL QUESTION · *What makes a culture unique?*

IT MATTERS BECAUSE
Central Africa is characterized by tremendous diversity in terms of its people, population patterns, languages, arts, and daily life.

The People of Central Africa

GUIDING QUESTION *What are some of the differences found among the people of Central Africa?*

Many different ethnic groups live in Central Africa. Each group is united by a shared language and culture.

Makeup of the Population

Central Africa is home to around 105 million people, which is roughly one-third as many as the United States has. The Democratic Republic of the Congo is by far the most populous country: It holds more than two-thirds of the region's people. Compared to other parts of the world, Central Africa does not have a large population or a high population density. However, its countries have high population growth rates. In all seven countries, the median age is below 20, which means that children and teenagers make up about half of the population. By comparison, the median age in the United States is about 37.

Life expectancy at birth for the people of the region varies considerably. In the Central African Republic, for example, life expectancy is about 50 years, while in Equatorial Guinea and São Tomé and Príncipe, it is around 63.

Hundreds of ethnic groups live in the region. Cameroon and the DRC each are home to more than 200 different groups. Some groups spread into two or more countries. One

such group is called the Fang. Historians believe the Fang once dwelled on the savanna. In the late 18th century, they began a migration into the rain forests. Today they live in mainland Equatorial Guinea, northern Gabon, and southern Cameroon.

Another people found in the region is the Bambuti, sometimes called the Mbuti. The Bambuti live in densely forested areas of the Democratic Republic of the Congo. They are extremely short in stature: Adults average less than 4 feet 6 inches (137 cm) in height. They were probably the earliest inhabitants of an area known as the Ituri Forest. Historical records show that the Bambuti have lived in this area for at least 4,500 years as nomadic hunters and gatherers.

The population of Central Africa also includes many refugees. **Refugees** are displaced people who have been forced to leave their homes because of war or injustice. Between 1997 and 2003, for example, a brutal civil war devastated the Democratic Republic of the Congo. A huge number of refugees fled the conflict.

City and Country

Most of Central Africa's people live in the rural areas and make their living through subsistence farming. In the DRC, for example, almost two-thirds of the people live in the country.

Large urban areas dot the region, however. More importantly, the ratio of city-dwellers to rural residents is changing rapidly. For example, every year, the DRC's urban areas gain about 4.5 percent of the population. Many of the chief cities of the region are capitals of countries. Kinshasa, the capital of the DRC, has grown into a sprawling metropolis with around 9 million inhabitants. Just across the Congo River from Kinshasa sits Brazzaville, the capital of the Republic of the Congo. Brazzaville is home to about 1.6 million people.

Government buildings surround a busy central square in Brazzaville, capital of the Republic of the Congo. Brazzaville is located on the north bank of the Congo River, just across from Kinshasa, the capital of the neighboring Democratic Republic of the Congo. This is the only place in the world where two national capital cities are located on opposite sides of a river, within sight of each other.

MJ Photography/Alamy

Village women of the Hutu ethnic group prepare a meal in the eastern part of the Democratic Republic of the Congo. The Hutu are one of hundreds of ethnic groups in Central Africa.

▶ **CRITICAL THINKING**
Describing Why did trade languages emerge in Central Africa?

In a few countries in this region, more than half the people live in cities. In Gabon, for example, city-dwellers account for 86 percent of the total population. In the Central African Republic, the figure is 62 percent, and in Cameroon it is 58 percent. Cameroon's two major cities, Douala and Yaoundé, each have populations of around 2 million, roughly the population of Houston or Philadelphia.

Language and Religion

Because Central Africa's population is made up of hundreds of different ethnic groups, it is not surprising that hundreds of different languages are spoken across the region. Cameroon, which has been described as an ethnic crossroads, serves as an example of the region's linguistic diversity. Three main language families are used in Cameroon. In the north, where Islam has been a significant influence, the languages spoken are in the Sudanic family. The Sudanic-speaking people in this part of Cameroon include the Fulani, the Sao, and the Kanuri. The Fulani are Muslims. They began migrating into Cameroon from what is now Niger more than 1,000 years ago. In the southern part of Cameroon, people speak Bantu languages. In the west are found semi-Bantu speakers, such as the Bamileke and the Tikar.

During the colonial era, people searched for common linguistic ground, especially when they traded with one another. Certain **trade languages** emerged. Because France and Belgium had the most widespread interest in the region, French became the most common trade language of Central Africa.

Eye Ubiquitous/Glow Images

Central Africa is also diverse in terms of religion. In the Democratic Republic of the Congo, for example, roughly 50 percent of the people are Roman Catholic, 20 percent are Protestant, 10 percent are Muslim, and 10 percent belong to a local sect called the Kimbanguist Church. The remaining 10 percent follow traditional African religions, which are based on a core set of beliefs, including the existence of a supreme being, the presence of spirits in the natural world, and the power of ancestors and magic.

☑ **READING PROGRESS CHECK**

Describing Where do most of Central Africa's people live? How do they make their living?

How People Live

GUIDING QUESTION *What are some of the key aspects of daily life and culture in Central Africa?*

In Central Africa, daily life and culture reflect the variety of influences to which the region has been exposed.

Daily Life

Daily life in Central Africa is a blend of traditional and modern **characteristics**. This combination results mainly from the impact of colonialism. It also reflects the urban-rural division of the population.

In the countryside, most people practice subsistence farming. On small plots of land, they grow crops such as cassava, maize, and millet, and they raise livestock. They strive to grow enough crops and raise enough animals to feed themselves and their families. If any surplus remains, they may sell it for small amounts of cash. Most people in the countryside live in small houses that they have built themselves. Building materials include mud, sundried mud bricks, wood, bark, and cement. For roofs, people typically use palm fronds woven together or sheets of corrugated iron.

One of the most common methods of constructing houses is known as wattle-and-daub. Poles driven into the ground are woven with slender, flexible branches or reeds to create the wattle. The finished wattle is plastered with mud or clay, known as daub. Then, a palm-frond roof is placed on top of the house.

In much of rural Central Africa, women carry out the gathering, production, and preparation of food for the household. The men hunt, trap, and fish. In addition to growing food to eat, men and women may also grow crops for sale, such as coffee, cotton, and cocoa.

Okra, corn, and yams are staple vegetables in the diets of many Central Africans. Other important foods are cassava, sweet potatoes, rice, beans, and plantains. Game is popular, as are fish-based dishes. Peanuts, milk, and poultry add protein to many dishes.

Academic Vocabulary

characteristic a quality or an aspect

Many food-related customs of Central Africa may seem surprising. In some areas, for example, males eat in one room while females eat in another. Generosity and hospitality are so deeply ingrained in some cultures that hosts might offer guests an abundance of food while remaining hungry themselves.

In many Central African cities, traditional ways of life are yielding to contemporary lifestyles. Modernization, however, has not always come easily. Central Africa has little infrastructure, which includes the fundamental facilities and systems that serve a city, an area, or a country. In the Central African Republic, for example, only a few cities and towns have modern health care facilities. Other obstacles to modernization include poor transportation and educational systems.

Culture and Arts

The visual, verbal, and performing arts are important to the cultures of many ethnic groups in Central Africa. In the southwestern part of the DRC, for example, the people known as the Kongo produce wooden statues in which nails and other pieces of metal are embedded. The Yaka create highly decorative masks and figurines. The Luba people of the southeastern part of the country are known for their skillful carvings that **depict** women and motherhood. The Mangbetu people of the northeast are known internationally for their pottery and sculpture.

In the field of literature, several modern Congolese authors are internationally recognized as poets, playwrights, and novelists. They include Clémentine Madiya Faik-Nzuji, Kama Kamanda, Ntumb Diur, and Timothée Malembe.

The DRC's capital city of Kinshasa is known around the world for its thriving music scene. The most popular style is African jazz, known as OK jazz. This style originated in the nightclubs of Kinshasa in the 1950s.

In northern Cameroon, the Fulani people decorate leather items and gourds with elaborate geometric designs. In music, the country's southern forest region is known for its drumming. In the north, the focus is on flute music.

Equatorial Guinea was formerly a colony of Spain, and Spanish influence can be tasted in the country's cuisine. In Malabo, the capital city, Spanish styles are found in the architecture. The country has produced several writers whose Spanish-language works have become known around the world.

☑ **READING PROGRESS CHECK**

Determining Word Meanings What is the main goal of subsistence farmers?

Musicians of the Congolese Symphony Orchestra perform in Kinshasa, capital of the DRC.

▶ **CRITICAL THINKING**

Describing Why is Kinshasa widely regarded as one of the music centers of the world?

Junior D. Kannah/AFP/Getty Images

Regional Issues

GUIDING QUESTION *What are the greatest challenges confronting Central Africa?*

Central Africa's people and governments face many complex issues. Among the issues are the economy, the environment, political stability, and population growth.

Growth and the Environment

With great population growth comes the need for economic development. The economies of many countries in the region depend heavily on agriculture, logging, and mining. Central African countries such as Gabon and Equatorial Guinea export significant amounts of valuable hardwoods, including mahogany, ebony, and okoumé.

In general, economic development depends on economic growth. Economic growth usually means that more resources are used and more pollution and waste are produced. In Central Africa, economic activities such as mining and logging are taking place in areas of high biodiversity. Some conservationists fear that this biodiversity is being lost in the rush to exploit resources.

The result is tension between the forces of economic growth and the forces of environmental conservation. In some areas, these tensions have at times led to conflict between local people and outside groups.

Development in Central Africa also raises other questions. Should the profits from economic activities go to foreign corporations and investors, or should they remain with the national governments? How should the profits and economic benefits be shared by the people?

A nurse checks a patient's treatment chart at a clinic in Brazzaville, Republic of the Congo.

Include this lesson's information in your Foldable®.

☑ **READING PROGRESS CHECK**

Describing Why are some people critical of mining and logging activities in the region?

Pascal Deloche/Godong/Corbis

LESSON 3 REVIEW

Reviewing Vocabulary

1. Why was the use of a trade language helpful in Central Africa?

Answering the Guiding Questions

2. *Describing* What are two distinctive features of the Bambuti people, who live in the rain forests of the Democratic Republic of the Congo?

3. *Determining Central Ideas* What are some of the customs in Central Africa related to food?

4. *Analyzing* Why has the lack of political stability posed major problems for development in some Central African countries?

5. *Informative/Explanatory Writing* Write a paragraph or two in which you explain some of the economic issues confronting Central Africa.

What Do You Think?

Has the United Nations Been Effective at Reducing Conflict in Africa?

The 53 independent states of Africa share one continent, but the vast land holds different people and varied resources and economies. During the past six decades, armed political conflict, either within a nation or between nations, has been the norm rather than the exception for many Africans. Since 1945, the United Nations (UN) Security Council, in its resolve to maintain international peace and security, has mounted 29 peacekeeping missions in Africa. Have the efforts of the United Nations been effective?

No!

PRIMARY SOURCE

" Africa has thus been a giant laboratory for UN peacekeeping and has repeatedly tested the capacity and political resolve of an often self-absorbed Security Council. . . . Under the loose heading of peacekeeping, the UN launched an unprecedented number of missions in the post–Cold War era. But . . . hard times appeared after disasters in Angola in 1992, when warlord Jonas Savimbi brushed aside a weak UN peacekeeping mission to return to war after losing an election; in Somalia in 1993, when the UN withdrew its peacekeeping mission after the death of eighteen U.S. soldiers; and in Rwanda in 1994, when the UN shamefully failed to halt genocide against about eight hundred thousand people and instead withdrew its peacekeeping force from the country. These events scarred the organization and made its most powerful members wary of intervening in Africa: an area generally of low strategic interest to them. "

—Adekeye Adebajo, Director, Centre for Conflict Resolution, University of Cape Town, South Africa, in *UN Peacekeeping in Africa: From the Suez Crisis to the Sudan Conflicts*

A militiaman stands guard with a machine gun in Somalia. The East African country has had political upheaval since 1991. Islamic militia forces and UN-backed government forces compete for control of Somalia.

Troops from Thailand have contributed to UN–African Union peacekeeping efforts in the war-torn Darfur region of Sudan. In Darfur, the Thai troops conducted patrols and provided protection to civilians and humanitarian aid workers.

Yes !

PRIMARY SOURCE

❝ Bearing in mind the relatively young period of existence of the PBC [Peacebuilding Commission] (and the fact that inevitably it was experimenting in its first two years), Sierra Leone and Burundi—the first two countries placed on the PBC agenda—seem to have achieved some real success in consolidating peace. Burundi has made headway with its peace process, including the gains relating to its inclusive political dialogue. Sierra Leone has also chalked up significant milestones and emerged as a post conflict state on the tracks of peace consolidation, with clear reforms in socioeconomic and security sector aspects. Both countries have also had to deal with some potentially serious setbacks and in both cases it seems that the proactive involvement of the PBC has added value in overcoming those problems. ❞

—Security Council Report, Special Research Report No. 2, The Peacebuilding Commission, 17 November 2009

What Do You Think? DBQ

1 *Identifying Point of View* What evidence does the Yes answer contain to support its viewpoint? What evidence does the No answer contain to support its viewpoint?

2 *Distinguishing Fact From Opinion* The two sources state facts and opinions. Identify one fact and one opinion from each source.

Critical Thinking

3 *Analyzing* Weigh the evidence. Africa has had 29 peacekeeping missions in roughly 60 years. If the United Nations has to keep going back, have the missions been effective? Why or why not?

Directions: Write your answers on a separate piece of paper.

1 Use your FOLDABLES to explore the Essential Question.

INFORMATIVE/EXPLANATORY WRITING Some geographers believe the area of Central Africa that is now savanna was at one time forestland. What might have changed the physical composition of this area? Write a short essay discussing this hypothesis. How does this relate to the concept of people adapting to the environment?

2 **21st Century Skills**

IDENTIFYING Use the text and the Internet to find information about one of the countries of Central Africa. List important resources for that country.

3 **Thinking Like a Geographer**

ANALYZING Much of Central Africa depends on agriculture as a main economic activity. Why is a good transportation system important to an agricultural society?

4 **GEOGRAPHY ACTIVITY**

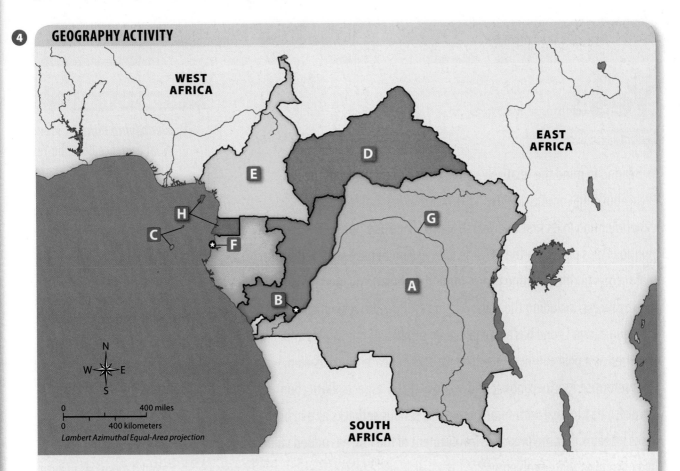

Locating Places

Match the letters on the map with the numbered places below.

1. Libreville
2. Congo River
3. Equatorial Guinea
4. Democratic Republic of the Congo
5. Brazzaville
6. Central African Republic
7. Cameroon
8. São Tomé and Príncipe

REVIEW THE GUIDING QUESTIONS

Directions: Choose the best answer for each question.

1 Choose Central Africa's primary landform.

A. islands

B. caves

C. Congo River watershed

D. steppes

2 Because Central Africa is centered on the Equator, its climate is

F. Mediterranean.

G. tropical.

H. arid.

I. mild.

3 Why were Europeans slow to establish colonies in Central Africa?

A. The area had no worthwhile natural resources to exploit.

B. They couldn't decide how to share the African countries.

C. Transportation was difficult, and they feared malaria and other diseases.

D. They didn't want to convert people.

4 The Republic of the Congo was a colony of which country?

F. Spain

G. France

H. Portugal

I. England

5 Most of the population of Central Africa makes a living by means of

A. subsistence farming.

B. factory labor.

C. crafting.

D. foreign trade.

6 Identify an obstacle to developing a modern economy in Central Africa.

F. diverse populations

G. commitment to traditional religions

H. primitive housing construction

I. insufficient infrastructure, poor transportation, and poor education

DBQ ANALYZING DOCUMENTS

7 **ANALYZING INFORMATION** Read the following passage about European colonial rule in Africa:

"*The manner in which colonial administrations governed virtually ensured the failure of Africa's transition into independence. Their practice of 'divide and rule'—favoring some tribes to the exclusion of others—served to [heighten] the ethnic divisiveness that had been pulling Africa in different directions for centuries.*"

—from David Lamb, *The Africans* (1984)

What group does Lamb blame for the problems African nations have had creating stable governments since independence?

A. the groups that led independence movements in Africa

B. the wealthy, more-developed nations of the world today

C. European colonial governments

D. African leaders since independence

8 **IDENTIFYING POINT OF VIEW** Why would favoring some tribes over others cause problems?

F. by draining the country of valuable resources

G. by making disfavored tribes dependent on the colonial government

H. by creating inequality that becomes entrenched

I. by forcing tribes to find allies outside their country

SHORT RESPONSE

"*The Congo's economy is based primarily on its petroleum sector, which is by far the country's greatest revenue earner. . . . A new potash mine . . . is expected to produce 1.2 million tons of potash . . . [used in fertilizer] per year by 2013. This will make Congo the largest producer of potash in Africa. . . . [One new iron ore mine] is thought to have ore reserves enough to be the world's third-largest iron ore mine.*"

—from "Republic of the Congo," State Department Background Notes

9 **DETERMINING CENTRAL IDEAS** What is the main idea of this passage?

10 **IDENTIFYING POINT OF VIEW** If oil is so valuable, why would the government of the Republic of the Congo bother developing potash and iron ore mines? Explain your answer.

EXTENDED RESPONSE

11 **ANALYZING** How was slavery in Central Africa and in the Americas similar?

Need Extra Help?

If You've Missed Question	**1**	**2**	**3**	**4**	**5**	**6**	**7**	**8**	**9**	**10**	**11**
Review Lesson	1	1	2	2	3	3	2	2	3	3	2

WEST AFRICA

ESSENTIAL QUESTIONS · *How does physical geography influence the way people live?* · *How do new ideas change the way people live?* · *What makes a culture unique?*

Young woman wears dress of traditional Kente cloth at a cultural festival in Ghana

Ariadne Van Zandbergen/Alamy

The Story Matters...

West Africa's diverse landscape includes desert, steppe, savanna, and tropical rain forests. Ancient rock paintings in Chad reveal details of the early herding societies who lived in this region. With rich deposits of gold and salt, trading kingdoms developed in Ghana and Mali. Trade also spread Islamic culture through much of the region. Gold attracted Europeans, whose colonial rule and slave trade impacted West Africa. Many of the independent nations of West Africa still struggle with civil war.

FOLDABLES®
Study Organizer

Go to the Foldables® library in the back of your book to make a Foldable® that will help you take notes while reading this chapter.

Natural Resources | Trade | Population Growth

WEST AFRICA ⓒⓒⓢⓢ

West Africa presents a rich variety of ethnic groups who speak many languages. Over the past 50 years, a number of the countries in the region have gained independence.

Step Into the Place

MAP FOCUS Use the map to answer the following questions.

1 **THE GEOGRAPHER'S WORLD** What are the two major lakes shown on the map?

2 **THE GEOGRAPHER'S WORLD** Which of the following countries is landlocked: Senegal, Burkina Faso, or Benin?

3 **PLACES AND REGIONS** What is the capital city of Liberia?

4 **CRITICAL THINKING** **Integrating Visual Information** If you are traveling from the capital city of Togo to the city of Bamako, in which direction are you traveling?

West Africa

TROPIC OF CANCER

MAURITANIA

★ Nouakchott

Senegal R.

CAPE VERDE

☉ ★ Praia

Dakar ★

SENEGAL

Banjul ★

GAMBIA

Bissau ★

Bamako ★

GUINEA-BISSAU

GUINEA

ATLANTIC OCEAN

Conakry ★

Freetown ★

SIERRA LEONE

LIBERIA

Monrovia ★

10°W

☉ National capital

Step Into the Time

TIME LINE View the time line to answer these questions: In what year did Liberia declare independence? What event occurred in 1787?

1300s Timbuktu is a center of Islamic culture

1000 B.C. Bantu people of West Africa begin migrations south and east

1000 — B.C. | A.D. — 1000 — 1300

250 B.C. Mali becomes center for trade

1324 Mansa Musa makes pilgrimage to Mecca

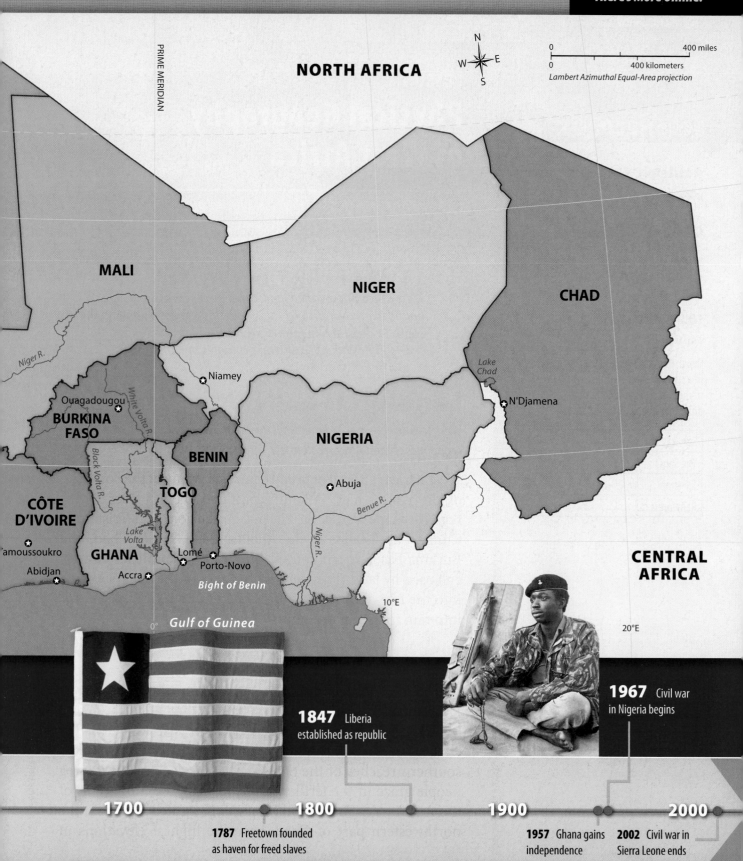

NORTH AFRICA

PRIME MERIDIAN

N
W + E
S

0   400 miles
0   400 kilometers
Lambert Azimuthal Equal-Area projection

MALI

NIGER

CHAD

Niger R.

● Niamey

Lake Chad

● N'Djamena

Ouagadougou ✪

BURKINA FASO

White Volta R.

BENIN

NIGERIA

TOGO

CÔTE D'IVOIRE

Black Volta R.

Lake Volta

GHANA

Lomé ✪

● Abuja

Benue R.

Niger R.

CENTRAL AFRICA

amoussoukro ✪

Abidjan ✪

Accra ✪

Porto-Novo

Bight of Benin

10°E

20°E

Gulf of Guinea

0°

1847 Liberia established as republic

1967 Civil war in Nigeria begins

1700   **1800**   **1900**   **2000**

1787 Freetown founded as haven for freed slaves

1957 Ghana gains independence

2002 Civil war in Sierra Leone ends

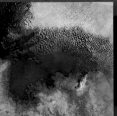

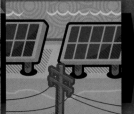

Reading **HELP**DESK (CCSS)

Academic Vocabulary

- **volume**

Content Vocabulary

- **basin**
- **harmattan**

TAKING NOTES: *Key Ideas and Details*

Organize As you read the lesson, fill in a chart like this one, discussing the dryness and vegetation in these different regions.

Sahara	
The Sahel	
Savanna	
Rain forest	

Lesson 1
Physical Geography of West Africa

ESSENTIAL QUESTION • *How does physical geography influence the way people live?*

IT MATTERS BECAUSE
West Africa has a varied landscape. Its vast grasslands are bounded on the north by the world's largest desert and on the south by coastal rain forests. West Africa provides an example of how cultures adapt to different climates and landforms.

Landforms and Bodies of Water

GUIDING QUESTION *In what ways do major rivers contribute to the economies of West African countries?*

West Africa is a land of broad contrasts. Many of the countries of the region have Atlantic coastlines. Four places in the region—Cape Verde, St. Helena, Ascension, and Tristan da Cunha—are islands in the Atlantic. Mali, Niger, Chad, and Burkina Faso are landlocked countries, or countries entirely enclosed by land. The southern reaches of the Sahara extend into the northern regions of the steppe, and savannas merge into rain forests in the south.

Landforms

Erosion has worn most of the land of West Africa into a rolling plateau that slopes to sea level on the coasts. West Africa has no major mountain ranges, but it has highland regions and isolated mountains. The Air, also known as the Air Massif, is a group of mountains in central Niger, along the southern reaches of the Sahara. The livestock of the Tuareg people graze in the fertile valleys between the mountains of the Air Massif. The Tibesti Mountains, located mostly in the northwestern part of Chad, have the highest elevations in

West Africa. The highest peak is Emi Koussi, an extinct volcano standing 11,204 feet (3,415 m) above sea level.

Southeast of the Tibesti Mountains, near Chad's eastern border, is an arid desert plateau region called the Ennedi. There are abundant wild game animals in the Ennedi. These animals attract a small population of seminomadic people who live here with their livestock during the rainy season. The Jos Plateau in central Nigeria is mostly open grassland and farmland.

In central Guinea is a highland region of savanna and deciduous forest known as the Fouta Djallon. It extends southeast to become the Guinea highlands, a humid, densely forested region.

Bodies of Water

The Niger River, West Africa's longest and most important river, originates at an elevation of 2,800 feet (853 m) in the Guinea highlands. It flows northeast toward the Sahara. Just beyond the town of Mopti in Mali, the Niger River enters the "inland delta." This is an area where the river spreads out across the relatively flat land into many creeks, marshes, and lakes that are connected to the river by channels. During the rainy season, the area floods completely. In the dry season, however, the waters recede, leaving fertile farmland.

As the Niger flows past Timbuktu, along the southern edges of the Sahara, the inland delta ends and the river returns to a single channel. As it flows through Niger and Nigeria, it joins its most important tributary, the Benue River, which doubles its **volume** of water. Where the Niger River reaches the Gulf of Guinea, it forms the Niger delta. This large area is a great **basin**, a lower area of land drained by a river and its tributaries. Along its 2,600-mile (4,184-km) course, the Niger provides water for irrigation and hydroelectric power. It is the main source of Mali's fishing industry, and it serves as an important route for transporting crops and goods.

George Holton/Photo Researchers

Academic Vocabulary

volume an amount

Tuareg people lead their camels and goats to an oasis. The nomadic group tend to their livestock in the valleys of mountains. At one time, the Tuareg controlled the caravan trade routes across the Sahara.

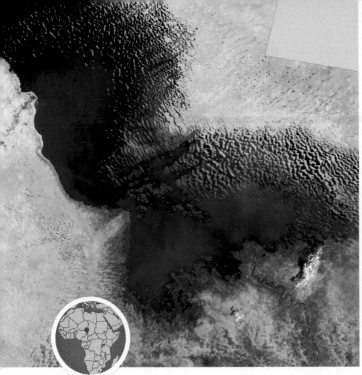
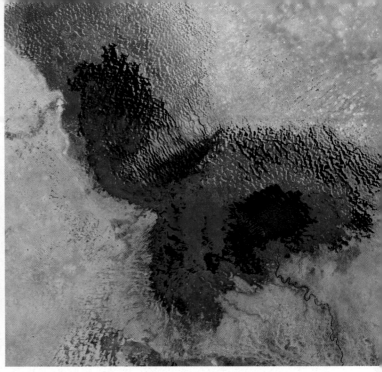

Satellite images show the Lake Chad area. Lake water is shown in blue, vegetation in red, and the surrounding desert in brown. The lake once was one of the largest bodies of water in Africa. The image on the left shows Lake Chad as it appeared in 1973. By 2007 (right), the lake had dramatically decreased in size.

▶ **CRITICAL THINKING**

Analyzing Taking water from the lake to use for irrigation is one reason that Lake Chad is smaller. What is another major cause of the lake shrinking in size?

The Senegal River rises in the Fouta Djallon in Guinea. Then it flows northwest to the Atlantic Ocean. For about 515 miles (829 km), the course of the Senegal River marks the border between the countries of Mauritania and Senegal.

The Black Volta River and the White Volta River originate in Burkina Faso and flow into Ghana. At a point near where the two rivers once met is the Akosombo Dam, which forms one of the world's largest artificial lakes—Lake Volta. The dam provides most of Ghana's electrical needs, and the lake provides water for irrigation.

Lake Chad covers portions of Niger, Nigeria, Chad, and Cameroon along the south part of the Sahara. The size of the lake varies from season to season, depending on the rainfall that feeds its tributaries. A series of droughts helped cause Lake Chad to shrink. Another cause was the amount of water taken from the lake and from the rivers that feed it to use for irrigating crops. People may be taking more water than water systems will be able to replace.

☑ **READING PROGRESS CHECK**

Describing Describe the landforms of West Africa.

Climate

GUIDING QUESTION *How do amounts of rainfall differ throughout West Africa?*

The climate of West Africa is diverse, from the harsh, arid Sahara in the north to the lush, coastal rain forests in the south. In between are vast stretches of grassland—the dry, semiarid Sahel and the lush, rainier savanna. The key characteristic of the climates throughout the region is its two distinct seasons: the wet season and the dry season.

Dry Zones

The Sahel is a semiarid region that runs between the arid Sahara and the savanna. Beginning in northern Senegal and stretching west into East Africa, the Sahel has a short rainy season. Annual rainfall ranges from only 8 inches to 20 inches (20 cm to 51 cm). This climate supports low grasses, thorny shrubs, and a few trees. The grasses are plentiful enough to support grazing livestock, such as cattle, sheep, camels, and pack oxen. It is important, however, that the herds do not grow too large. Too many animals leads to overgrazing and permanent damage to the grasslands. Overgrazing and too much farming result in desertification, the process in which semiarid lands become drier and more desert-like.

North of the Sahel is the Sahara. Daytime high temperatures are often more than 100°F (38°C) in the summer, though temperatures at night can drop by as much as 50°F. The few shrubs and other small plants that live in the Sahara must go for long periods without water. Some plants send long roots to water sources deep underground. After a rainfall, plants that have waited many months will suddenly flower, carpeting the desert with color.

From late November until mid-March, the dry, hot wind known as the **harmattan** blows through the Sahara with intensity. It carries thick clouds of dust that can extend hundreds of miles over the Atlantic Ocean, where it settles on the decks of ships. Harmattan winds contribute to the process of desertification.

Wet Zones

South of the Sahel, rains are more plentiful and feed the fuller, lusher plant life of the savanna. The savanna has two seasons—rainy and dry. The rainy season usually extends from April to September, and the rest of the year is dry. Annual rainfall reaches between 31 inches and 59 inches (79 cm and 150 cm). In some locations, though, annual rainfall can be as little as 20 inches (51 cm). This wide variability in rainfall makes human activity difficult.

The hot, dry wind that streams in from the northeast or east in the western Sahara is called the harmattan.

▶ **CRITICAL THINKING**

Describing During what months is the harmattan the strongest?

WEST AFRICAN ENERGY

West Africa is relatively rich in energy resources. These resources are not, however, equally developed. Also, each form of energy has benefits and drawbacks.

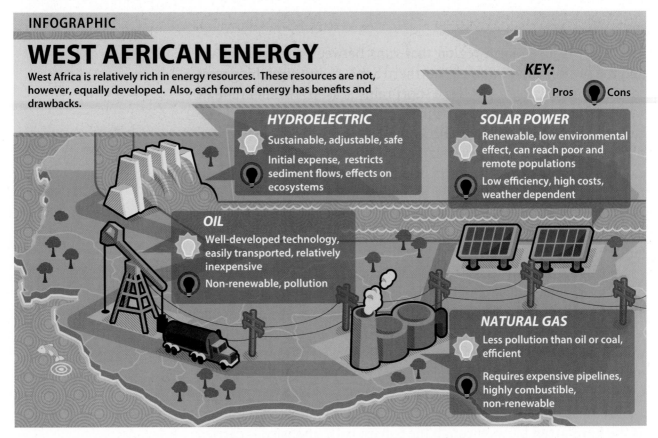

KEY: Pros Cons

HYDROELECTRIC
- Sustainable, adjustable, safe
- Initial expense, restricts sediment flows, effects on ecosystems

SOLAR POWER
- Renewable, low environmental effect, can reach poor and remote populations
- Low efficiency, high costs, weather dependent

OIL
- Well-developed technology, easily transported, relatively inexpensive
- Non-renewable, pollution

NATURAL GAS
- Less pollution than oil or coal, efficient
- Requires expensive pipelines, highly combustible, non-renewable

West Africa is rich in energy resources. Because of the increasing demand for energy, West Africa's importance as an energy supplier to world markets continues to grow.

▶ **CRITICAL THINKING**

Analyzing What form of energy has little or no effect on the environment?

Farther to the south, tropical rain forests cover the land. The rain forests, known for their broad-leaved evergreen trees and rich biodiversity, receive plenty of rain. Rain forests in Sierra Leone and Liberia can receive as much as 200 inches (508 cm) of rain per year.

☑ **READING PROGRESS CHECK**

Identifying What are the two major grassland areas of West Africa?

Resources

GUIDING QUESTION *How do West African countries provide for their energy needs?*

West Africa contains many resources, but some countries lack the money to develop their resources into industries.

Energy Resources

Nigeria is the region's biggest producer of petroleum. In 2010 Nigeria ranked among the top 10 in the world, producing oil at a rate of almost 2.5 million barrels per day. The petroleum industry has not been as quick to tap the country's vast natural gas reserves.

Chad has oil fields in the area north of Lake Chad. Benin has offshore oil fields, as does Ghana. The oil in Ghana was considered too expensive to extract until increases in oil prices made it profitable to drill.

Ghana's main sources of electricity are two dams on the Volta River: the Akosombo Dam and the dam at Kpong. The Organization for the Development of the Senegal River, which is made up of several countries of the region, manages the river's resources, including hydroelectric stations. The Senegal River provides about half of the energy used in Mauritania today. Hydroelectric power is vital to meeting the energy needs in Togo and Nigeria.

Other Resources

The gold trade attracted Portuguese explorers, who first visited the region in 1471. For centuries afterward, the country was simply called the Gold Coast and Europeans lived and worked there, building trading posts and forts. Gold mining remains important to the economy of Ghana, along with the mining of diamonds, manganese, and bauxite. Ghana also has unmined deposits of limestone and iron ore.

Gold is mined in Mali, Burkina Faso, and Nigeria, too. Togo mines phosphate and limestone, which are used as fertilizers and to make paper, glass, paint, and other everyday products. Togo also has promising gold deposits, but so far no gold-mining industry. Niger has a salt-mining industry. In addition to its important gold industry, Mali also mines salt and limestone, but many of Mali's mineral resources are untapped. These include iron ore and manganese, which are important in making steel.

Mauritania mines copper and iron ore, but many of the iron ore deposits have been depleted. Nigeria mines iron ore, tin, limestone, and small quantities of other minerals. Burkina Faso is one of the world's leading sources of manganese. Benin has deposits of iron ore, limestone, chromium ore, gold, and marble. Benin is a leader in the production of hardwoods, but most of the rain forests where this wood comes from have been cleared.

FOLDABLES
Study Organizer

Include this lesson's information in your Foldable®.

☑ **READING PROGRESS CHECK**

Analyzing Why is Benin's hardwood industry at risk? What is another industry in this region that has faced a similar problem?

LESSON 1 REVIEW

Reviewing Vocabulary

1. How can the *harmattan* winds influence climate and vegetation?

Answering the Guiding Questions

2. *Describing* Why has Lake Chad changed in size over time?

3. *Determining Central Ideas* Why is desertification an issue in West Africa?

4. *Identifying* Why did the government in Ghana change its mind about its offshore oil deposits?

5. *Informative/Explanatory Writing* Explain how desertification occurs and why it is such an important environmental issue in West Africa.

Reading **HELP**DESK (CCSS)

Academic Vocabulary

- **displace**
- **element**
- **revenue**

Content Vocabulary

- **imperialism**
- **secede**

TAKING NOTES: *Key Ideas and Details*

Describe On a chart like this one, write at least two different facts about the three ancient empires of Ghana, Mali, and Songhai.

Ghana	Mali	Songhai

Lesson 2

The History of West Africa

ESSENTIAL QUESTION • *How do new ideas change the way people live?*

IT MATTERS BECAUSE

West Africa was under the control of a series of wealthy trading kingdoms until the late 1800s, when Europeans seized control of their lands. Regaining their independence from Europe and establishing themselves in the modern world has been a challenge.

Ancient Times

GUIDING QUESTION What opportunities did Muslims from North Africa see in West Africa?

Early civilizations in West Africa learned to thrive in a variety of climates and landscapes, from the Sahara to the tropical rain forests. Throughout its history, the region was open to many migrations and invasions because of its resources.

Ancient Herders

We think of the Sahara as a hot, intensely dry place where few living things can survive. Ten thousand years ago, the Sahara was a much different place. Rock drawings from 8000 B.C. depict a world that looks more like a savanna than a desert. Drawings include lakes, forests, and large animals not seen in the Sahara in modern times: ostriches, giraffes, elephants, antelope, and rhinoceroses. Seminomadic people herded cattle and hunted wild animals. As the climate grew drier, fewer species of plants and animals survived the harsh environment. Many people moved south, following the retreat of the grasslands and the rain. During this period of desertification, people discovered that camels can survive without water for longer periods than cattle, sheep, or goats. Camels also can carry heavy loads for long distances. They were perfect domesticated animals for desert dwellers.

Movement of People

The Bantu people inhabited West Africa in ancient times. They had developed farming as early as 2000 B.C. A vast migration of Bantu people out of West Africa began around 1000 B.C. and continued for hundreds of years. Their farming practices allowed them to expand rather easily into areas occupied by hunter-gatherer groups. The Bantu became the dominant population in most of East and South Africa. The Bantu might have reached the East African coast as early as the A.D. 200s. They **displaced** many of the people who had lived on these lands before them.

Wherever the Bantu moved, they spread their culture, including three vital **elements**. First, the Bantu cultivated bananas, taro, and yams, which were originally from Malaysia. These crops thrived in the humidity of the tropical rain forests. Second, the Bantu spread their languages everywhere they settled. Today, more than 500 distinct Bantu languages are spoken by 85 million Africans. Third, the Bantu brought iron-smelting technology. They could create tools and weapons unlike any the native people had ever seen.

Trade Across the Sahara

For thousands of years, the Sahara was a barrier to contact between West Africa and North Africa. By the A.D. 700s, Arab Muslims had crossed the Sahara and conquered most of North Africa. They dominated the southern Mediterranean and controlled trade in that region, as well as Saharan trade routes into West Africa. Arab geographers slowly learned about Africa south of the Sahara. They realized that the region offered opportunities, not only in trade, but also in adding converts to Islam. One of the West African kingdoms they learned about was Ghana.

Ghana controlled the gold trade in the region. It traded gold for salt that Arab traders brought in from the Sahara. Salt was a very important trade good. It was the best preserver of food, and it was rare and difficult to acquire.

The Berber people had lived in North Africa long before the Arabs arrived. The Berbers resisted Islam for a while, but in time they converted. Several groups of Islamic Berbers joined together to become the Almoravids. They were fierce fighters, and they wanted to spread their new faith.

John Elk/Lonely Planet Images/Getty Images

Academic Vocabulary

displace to take over a place or position of others

element an important part or characteristic

Merchants sell slabs of salt in a market town in central Mali. Salt is plentiful in the Sahara. These conditions gave rise to the African salt trade of ancient times.

▶ CRITICAL THINKING

Describing Why was salt a valuable item for trade?

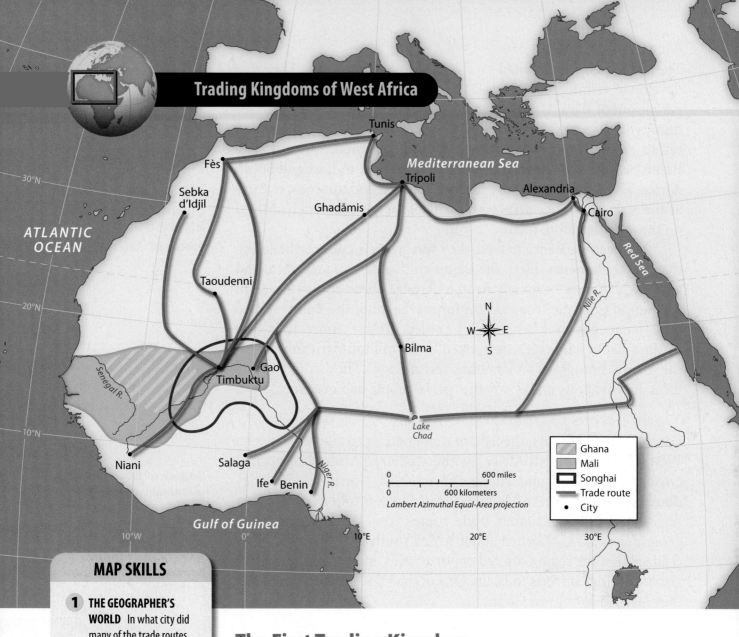

Mediterranean Sea

ATLANTIC OCEAN

Tunis

Fès

Sebka d'Idjil

Ghadāmis

Tripoli

Alexandria

Cairo

Red Sea

Taoudenni

Nile R.

Bilma

Senegal R.

Gao

Timbuktu

Lake Chad

Niani

Salaga

Niger R.

Ife

Benin

Gulf of Guinea

	Ghana
	Mali
	Songhai
	Trade route
•	City

0 — 600 miles
0 — 600 kilometers
Lambert Azimuthal Equal-Area projection

MAP SKILLS

1 THE GEOGRAPHER'S WORLD In what city did many of the trade routes merge?

2 PLACES AND REGIONS The early African kingdoms developed near gold-rich areas. What did a plentiful supply of gold allow them to do?

The First Trading Kingdom

The Ghana Empire was a powerful kingdom in West Africa during the Middle Ages. It had grown wealthy from its control of the gold trade. As the empire grew wealthy, it conquered many of its neighbors, including gold-rich lands to the south. Ghana built a strong trade with Muslim countries in North Africa.

In the A.D. 1000s, the Almoravids conquered Ghana. Almoravid rule over Ghana lasted only a few years, but that was long enough to damage the trade that kept the empire alive. With the arrival of so many Almoravids, there was now a larger population to feed. The dry climate could not support the sudden increase in agriculture, and this resulted in desertification: Land that at one time was fertile became desert. Ghana became weak, and its conquered neighbors began to break away.

☑ **READING PROGRESS CHECK**

Determining Central Ideas Why did the Ghana Empire collapse?

West African Kingdoms

GUIDING QUESTION *Why was the city of Timbuktu important to different trading kingdoms in West Africa?*

Several West African kingdoms sought to take control of the trade the fallen Ghana Empire had established in the region. Islam was taking hold in northern West Africa, and the most powerful of the new trading kingdoms were Muslim: Mali and Songhai.

Mali

The trading kingdom of Mali came after Ghana. Mali grew rich from the gold-for-salt trade. Its rulers conquered neighboring lands and built an empire.

The empire of Mali reached its height under the emperor Mansa Musa. The city of Timbuktu, which was already an important trading post, became a center for Islamic culture. Mansa Musa went on a historic pilgrimage to Mecca. By the time Mansa Musa returned home, the Islamic world knew there was a new and powerful Islamic kingdom south of the Sahara.

Songhai

After Mansa Musa's death, Songhai replaced Mali as the most powerful West African empire. A well-trained army and a navy that patrolled the Niger River made Songhai the largest of the three trading empires. Songhai now controlled the region's trade routes, had salt mines in the Sahara, and sought to turn their empire into the center for Islamic learning. The Songhai Empire eventually fell to the Moroccans, who seized the salt mines and destroyed the empire by the end of the 1500s.

Coastal Kingdoms

Slavery had been practiced in Africa for centuries. Muslims from North Africa and Asia had been buying enslaved people from south of the Sahara. As European colonists established colonies in the Western Hemisphere, they purchased enslaved people to do their labor. Small African kingdoms along the Atlantic coast became trading partners with Portugal, Spain, and Great Britain. Trade in enslaved persons became highly profitable in these kingdoms. When Europeans outlawed the slave trade, the kingdoms' economies began to fail.

✔ **READING PROGRESS CHECK**

Identifying How did the slave trade in West Africa change after the arrival of the Europeans?

Mansa Musa is shown sitting on his throne in this map of Africa from an atlas created in 1375.

Abraham Cresques/The Bridgeman Art Library/Getty Images

European Domination

GUIDING QUESTION *What were some of the factors that aroused European interest in exploring and colonizing Africa?*

In the late 1800s, several European countries were eager to expand into Africa with its many natural resources. After hundreds of years of self-rule, West Africans lost control of their lands to the Europeans.

Changing Trade

The British government outlawed the slave trade in 1807. Great Britain tried but could not keep other European countries from continuing to export slaves. The British founded the colony of Freetown in Sierra Leone in 1787 as a safe haven for runaway or freed enslaved persons. The Americans followed suit in 1822, founding Liberia as a home for freed American slaves.

The British encouraged development of the palm oil trade to make up for the loss of **revenue** from the slave trade. British traders and missionaries familiarized themselves with the trading culture of the Niger River. The British started to profit from their growing role in West African trade. Before long, France took notice.

Creating Colonies

Before the mid-1800s, the British and French posted military in North Africa, and the British and Dutch had settled in South Africa. In 1869 two events increased Europe's interest in Africa. First, the Suez Canal—an artificial waterway connecting the Mediterranean Sea to the Red Sea—opened. Second, diamonds were discovered in South Africa.

Academic Vocabulary

revenue income

Freetown, the largest city in Sierra Leone, is located in the western part of the country along the Atlantic coast.

▶ **CRITICAL THINKING**
Describing What is unique about how Freetown was founded?

Chris Jackson/Getty Images News/Getty Images

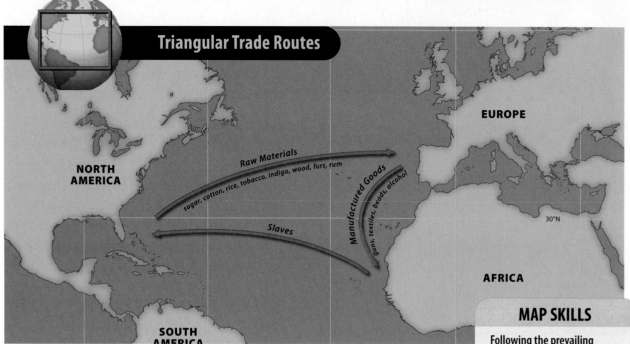

Triangular Trade Routes

EUROPE

NORTH
AMERICA

Raw Materials
sugar, cotton, rice, tobacco, indigo, wood, furs, rum

Manufactured Goods
guns, textiles, beads, alcohol

Slaves

30°N

AFRICA

SOUTH
AMERICA

MAP SKILLS

Following the prevailing winds, European ships sailed along the triangular trade route.

1 THE GEOGRAPHER'S WORLD Why were raw materials from the Americas shipped to Europe rather than to other parts of the world?

2 PLACES AND REGIONS Most European trade with Africa was conducted in West Africa. Why do you think this was so?

Before that time, Europeans had been in contact with the coastal kingdoms, but few had ventured into the interior of West Africa. This began to change as Great Britain, France, Portugal, Belgium, and the new, unified country of Germany began planning to carve out empires in Africa. Europeans followed a policy of **imperialism**, or seizing political control of other places to create an empire. At the Berlin Conference of 1884–1885, the European powers established rules for partitioning Africa.

The French and British claimed the most territory in West Africa. The modern countries of Benin, Burkina Faso, Chad, Côte d'Ivoire, Guinea, Mali, Mauritania, Niger, and Senegal were French colonies. Gambia, Ghana, Nigeria, Sierra Leone, and St. Helena were British colonies. Togo started out as a German possession, but Germany lost it in World War I.

Some colonies were ruled more harshly than others, but one thing was true about all of them: Europeans made the important decisions about how Africans lived. Settlers from Europe could set up farms on the most fertile land, even if Africans were forced off that land. Africans resisted, sometimes violently.

☑ **READING PROGRESS CHECK**

Describing What was the purpose of the Berlin Conference?

New Countries

GUIDING QUESTION *How does a ruler gain and hold political power?*

West African colonies began gaining independence in the late 1950s. The struggle for economic and political success was difficult because of the complex ethnic makeup of these countries.

The leader of Ghana, Kwame Nkrumah (center), meets with a citizen in 1959. Bediako Poku (standing) headed the Convention People's Party (CPP), a political party Nkrumah started. A referendum election in 1964 made the CPP the nation's only legal party, and Nkrumah was installed as president of Ghana for life. The military and police ousted Nkrumah from power in 1966.

Another factor working against political stability was the actions taken by the leaders. Often, they seized power and then used force to stay in power.

Ghana Leads the Way

The Gold Coast had been important to European trade since Portugal set up its first fort there in 1487. By the early 1800s, it was under British control, and in 1874 the Gold Coast became a British colony. The most important industries were gold, forest resources, and a newly introduced crop—the cocoa bean from which chocolate is made. By the 1920s, the Gold Coast supplied more than half the world's cocoa.

After the end of World War II, Dr. Kwame Nkrumah led protests calling for independence. Protests flared throughout the colony. In 1957 the Gold Coast gained its independence and became the Republic of Ghana. For many years, Ghana was troubled by conflict. In 1992 Ghana approved a new constitution that established a multiparty democracy. Since then, presidential elections have been peaceful, and Ghana has become a model of political reform in West Africa.

Former French Colonies

Throughout the 1950s, France made concessions to its West African colonies, many of which had instituted some form of self-rule. The independence of Ghana excited and inspired Africans, however. Cries for independence began erupting all over Africa. French Sudan and Senegal united to become the Republic of Mali, but Senegal broke away the following year, and Mali declared its own independence. In 1960 Mauritania, Niger, Côte d'Ivoire, Gambia, and Burkina Faso declared their independence. By 1961, all of France's colonies in West Africa were independent.

Nigeria

Nigeria had never been a country before the British combined two of their colonies to form the Nigerian Protectorate in 1914. Several different ethnic groups lived in this new territory. The Yoruba and Edo people had distinct territories in the south. The Hausa people were Muslims who lived in the north. The Igbo people were farmers who began to migrate east after British colonization.

When Nigeria gained its independence in 1960, tensions grew between ethnic groups. Because Britain had concentrated most of its development, including building schools, in southern Nigeria,

Mark Kauffman/Time & Life Pictures/Getty Images

most well-educated Nigerians lived in the south. They became leaders of the new government. Many of them were Igbo, who were mostly Christian. When they ventured north to govern the Hausa, they were met with hostility. Conflicts among the three major ethnic divisions in the country—the Hausa, the Yoruba, and the Igbo—grew violent. Thousands of Igbo living in northern Nigeria were massacred. As many as a million more fled to a region dominated by the Igbo. In 1967 the eastern region **seceded**, or withdrew formally, from Nigeria and announced that it was now the independent republic of Biafra. Nigerian forces invaded Biafra, and after more than two years of war, Biafra was in ruins. Starvation and disease may have killed more than 1 million people.

For long periods since then, the military has controlled the country's government. A new constitution was written in 1978. A year later, a democratically elected civilian government took office. That ended in a military coup in 1983. It was not until 1999 that another democratically elected president was able to rule Nigeria.

Civil War

Sometimes a military coup erupts into civil war. In late 1989, Charles Taylor led an invasion of Liberia. His aim was to depose the president, Samuel Doe. Ethnic conflict was at the heart of this struggle. Taylor and his rebels were of the Mano and Gio peoples. President Doe belonged to the Krahn people. After Doe's arrest and execution, Liberia endured seven years of civil war.

Conflict spilled over Liberia's borders into neighboring Sierra Leone. Thousands of civilians died, and many were forced to leave their homes. The civil war in Sierra Leone did not end until 2002. Estimates are that 50,000 died and another 2 million people lost their homes in the civil war. In 2012 Charles Taylor was brought to trial by a special court. He was found guilty of war crimes and crimes against humanity for his part in Sierra Leone's civil war.

☑ **READING PROGRESS CHECK**

Determining Central Ideas What reason might the French have had for letting their African colonies declare independence?

Thinking Like a Geographer

Religious Rivalry

In the A.D. 700s, Muslim people converted many West Africans to Islam. Over time, Islam became the dominant religion in northern parts of the region. Many West Africans along the coastal areas adopted Christianity after the arrival of the Europeans in the 1500s. Others continue to practice traditional African religions. Relations between these different religious groups have not always been good.

FOLDABLES
Study Organizer

Include this lesson's information in your Foldable®.

Natural Resources | Trade | Population Growth

LESSON 2 REVIEW

Reviewing Vocabulary

1. How did the Berlin Conference of 1884–1885 help achieve the goals of European *imperialism*?

Answering the Guiding Questions

2. ***Determining Central Ideas*** How did the Bantu culture influence the parts of Africa to which the Bantu migrated?

3. ***Identifying*** How did the Islamic world become aware of the wealth of the Mali Empire?

4. ***Describing*** Why did the British encourage development of the palm oil trade in African kingdoms on the Atlantic coast?

5. ***Analyzing*** Why did Nigeria face so many challenges when becoming an independent country?

6. ***Argument Writing*** Write a speech to the French government in the late 1950s about how important the independence of Ghana is and why France should release its colonies.

networks

There's More Online!

☑ **SLIDE SHOW** Daily Life in West Africa

☑ **VIDEO**

Reading **HELP**DESK **CCSS**

Academic Vocabulary

• **diverse**

Content Vocabulary

• **pidgin**
• **creole**
• **animist**
• **extended family**
• **nuclear family**
• **kente**
• **infrastructure**

TAKING NOTES: *Key Ideas and Details*

Compare and Contrast Use a graphic organizer like the one below to show some of the differences between life in a West African city and life in a rural area of West Africa.

City	Rural
•	•
•	•
•	•
•	

Lesson 3
Life in West Africa

ESSENTIAL QUESTION • *What makes a culture unique?*

IT MATTERS BECAUSE
West African countries have faced many challenges in the decades since they achieved independence. Even as they struggle to modernize and improve their economies, West Africans have been able to spread their artistic gifts to people around the world.

The People of the Region

GUIDING QUESTION Why are two or more languages spoken in some West African countries?

West Africa is the most populous region of Africa. Many different ethnic groups live here, and each group brings something unique to their country's culture.

Ethnic Groups

European countries that carved up Africa to add to their overseas empires had their own reasons for setting colonial borders where they did. They did not consider the borders of the different ethnic groups of the Africans who already lived on the land they colonized. As a result, the European colonies in Africa contained populations that were ethnically **diverse**. When African countries declared independence, the new national borders closely followed the old colonial borders, preserving that diversity.

Establishing a sense of national unity is difficult, however. Many people have a stronger identification with their ethnic group than with the country they live in. For example, the Hausa are citizens of Nigeria, but many feel closer ties to the people and culture of the Hausa than to the country of Nigeria.

Another factor working against national unity is that a West African ethnic group does not typically live in only one country. Some Yoruba people, for example, live in Nigeria. Other Yoruba live in Benin. No matter where they live, the Yoruba feel a closer connection to other Yoruba people than to Nigeria or Benin.

Languages

When West African countries achieved independence, the majority held on to the European language that had been used most in business and government during the colonial period. The European languages—English, Portuguese, and French—became the official languages of West African countries. In some cases, Arabic is also an official language.

Most ethnic groups retain their traditional languages, and many use that language more than they do the official European language. In some places, European and African languages have intermixed to form a **pidgin** language. A pidgin is a simplified language used by people who cannot speak each other's languages but need a way to communicate. Sometimes two or more languages blend so well that the mixture becomes the language of the region. This is called a **creole** language. Crioulo is the language spoken most often in Cape Verde. It is a creole language, part Portuguese and part African dialect.

Religion

Islam was introduced to North Africa in the A.D. 600s, and it spread southward with trade. It was well established in many parts of West Africa long before the arrival of Christian missionaries from Europe. Today, a sizable portion of West African populations are Muslim.

Glen Allison/Photodisc/Getty Images

Women talk near the Great Mosque in Djenne, Mali. The Great Mosque is the largest mud-brick building in the world. The first mosque on this site was built in the 1200s. The current mosque was built in 1907.
▶ CRITICAL THINKING
Describing In what region of Africa was Islam first introduced?

Lagos is Nigeria's largest city and ranks among the fastest-growing megacities in the world. During the 2000s, about 600,000 people have moved to Lagos every year.

▶ **CRITICAL THINKING**

Describing Are megacities the most common form of settlement in West Africa? Explain.

In some countries, however, Christianity is just as dominant as Islam, or more so. Even in countries where Christianity or Islam is dominant, many people still practice traditional African religions. These religions have their own rituals and celebrations that tie communities together. The beliefs are not written on paper but passed on from one generation to another. Often, the people believe in a supreme creator god and are **animists**, which means they believe in spirits—spirits of their ancestors, the air, the earth, and rivers.

Settlement Patterns

West Africa had few large towns until the colonial period. Even today, the most common settlement patterns in West Africa consist of scattered villages. Villages represent the homesteads of **extended families**, or families made up of parents, children, and other close relatives, often of more than two generations. The size of the villages and the population density depend on how much human activity the land can sustain. Water is an issue in the northern parts of West Africa, so populations are small and spread out.

Most of the largest cities in West Africa are capital cities, like Bamako, Mali, which more than tripled in population between 1960 and 1970, when droughts caused people to migrate from the countryside. The largest city in West Africa is Lagos in Nigeria, with an estimated 10.5 million people. Lagos was Nigeria's capital until 1991; Abuja is now Nigeria's capital city.

☑ **READING PROGRESS CHECK**

Determining Central Ideas What are two ways traditions have survived in West Africa?

©Carlos Cazalis/Corbis

Life and Culture in the Region

GUIDING QUESTION *Why are traditions more important in rural areas of West Africa than they are in West African cities?*

A variety of cultures thrive in West Africa, some of them traditional and some contemporary. Countries in the north—such as Mali, Mauritania, and Niger—are more influenced by the culture of North Africa than are their neighboring countries.

Daily Life

West Africa, with its complicated history and rich mixture of ethnicities, has a diverse culture. English or French may be the language of the cities, or of business and politics, but hundreds of other languages are still spoken. People in cities are more likely to wear Western-style clothing and to live and work in Western-style buildings. Far from the city, many people retain the traditions of their ancestors. These people are more likely to wear traditional clothing. Life in rural villages is built around the extended family. In cities, the **nuclear family**—parents with their children—is the more common family structure.

A worker weaves kente cloth in Ghana (above). The painted wood mask (below) comes from Burkina Faso.

City dwellers are far more likely to deal with a wide variety of people over the course of a day than are rural dwellers. Capital cities teem with people from different ethnic groups, races, and countries. They are more likely to speak the official language because that is the language they all have in common. In rural areas, ethnicity and tradition are still important. Devotion to traditional values keeps those languages and cultures alive even as the country is changing. The downside is that ethnic pride sometimes results in conflicts between neighboring ethnic groups.

The Arts

West Africans have created numerous unique and important works of art in many artistic fields. Traditional artwork, such as carved masks from Nigeria and Sierra Leone, are world famous. Another well-known traditional art form is **kente**, a colorful, handwoven cloth from Ghana. One of West Africa's most important artist-figures is the griot, a musical storyteller who is part historian and part spiritual advisor.

Dance is the most popular form of recreation in West Africa. Workers use dance to celebrate their skills and accomplishments; professional guilds have their own dances. Dance is used for its healing qualities, but people also dance to popular music in clubs.

Workers are on the job at a steel plant in Côte d'Ivoire. Like many other countries in the region, Côte d'Ivoire is developing its industries. Major manufacturing industries include bus and truck assembly and shipbuilding.

West African music blends a variety of many sounds, combining traditional and modern instruments. The Arabic influences of North African music mix with the music of sub-Saharan Africa, as well as with American and European rock and pop music.

West Africa has produced some of the world's finest writers. The most widely read of all African novels is *Things Fall Apart* by Chinua Achebe of Nigeria. Senegal's first president, Léopold Senghor, was a famous poet and lecturer on African history and culture.

☑ **READING PROGRESS CHECK**

Identifying Point of View Select an artistic field that you think best shows the culture of West Africa, and explain why you believe it does.

Challenges Facing the Region

GUIDING QUESTION *How did West African countries build up so much debt?*

Many West African countries face serious problems. To improve their situations, they must deal with these challenges and more: bad economies, corrupt governments, out-of-control population growth, disease, and poorly funded schools.

Government and Economics

The European powers that colonized Africa built the colonies' economies on a few key resources, such as petroleum, gold, peanuts, or copper. The **infrastructure**, or underlying framework of the colonies, was built around those key resources. Once the colonies achieved independence, it became important for them to develop a variety of different industries. Otherwise, any drop in the world price of a key resource would greatly affect a country's economy. Many foreign governments invested money in developing the

©nabil zorkot/dpa/Corbis

industrial base of the new African countries. In many cases, money from the loans was not used wisely. As a result, the invested money did not yield high returns. West African countries were not only dealing with struggling economies but also faced enormous debt.

The International Monetary Fund and the World Bank have declared that no poor or developing country should have to pay a debt it cannot possibly manage. Debt relief has been important for countries such as Ghana, which suffered from bad economic choices and political instability. Ghana is now a model for economic and political reform in West Africa. Nigeria benefited from debt relief as well, and was able to pay off the remainder of what it owed in 2006.

A major challenge for the region is that, as the population grows, the demand for food and jobs grows. An economy does not have a chance to grow if it cannot meet the needs of a growing population.

Health and Education

Thirty-four million people in the world are living with the HIV virus, and 22.9 million of them live in sub-Saharan Africa. Dealing with such a large number of HIV-infected people presents several challenges. The first challenge is to supply health care to the growing number of people who carry the virus. The second challenge is to reduce the number of new HIV infections, usually through education. The third challenge is to deal with the families and communities hurt by AIDS-related deaths.

West Africa has a high birthrate, and the population is growing quickly, but life expectancy is still short compared to other parts of the world. Health care is an issue in large cities, but it is an even greater issue in rural areas.

The population of West Africa is young. For West Africa to have a sound future, educational systems must effectively prepare young people for economic and social development. A lack of education funding means that some countries cannot afford to make updates or improvements to their schools.

Study Organizer

Include this lesson's information in your Foldable®.

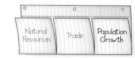

 READING PROGRESS CHECK

Analyzing Why is education such an important issue for the future of West Africa?

LESSON 3 REVIEW CCSS

Reviewing Vocabulary

1. How did the *infrastructure* of West African colonies lead to poor economies when they became independent countries?

Answering the Guiding Questions

2. *Analyzing* What are the advantages of a pidgin language?

3. *Describing* What is one potential downside to preserving traditional values in rural West Africa?

4. *Identifying* What are three major challenges to fighting AIDS in West Africa?

5. *Narrative Writing* Select a West African country and write a letter to the government of the country about the importance of investing in education. Explain how education will help solve the problems discussed in this lesson and how it will help preserve West African culture.

Directions: Write your answers on a separate piece of paper.

1 Use your **FOLDABLES** to explore the Essential Question.

INFORMATIVE/EXPLANATORY WRITING   Choose any three countries in this region and take a closer look at how the people who live there earn their living. Use the CIA Factbook or another Internet resource to list the top five occupations in each of the three countries and the average annual wage people earn for each. Arrange your data in a chart. Are any of those occupations directly related to the physical geography of the area? Explain in one or two paragraphs.

2 21st Century Skills

INTEGRATING VISUAL INFORMATION   In small groups, research the problems facing one of the countries of West Africa. Discuss possible solutions to one of the problems. Create a bulletin board display with pictures and captions to illustrate your solution to the problem.

3 Thinking Like a Geographer

DETERMINING CENTRAL IDEAS   After reviewing the chapter, choose five of the most important events in the history of West Africa. Place those events and their dates on a time line.

4 GEOGRAPHY ACTIVITY

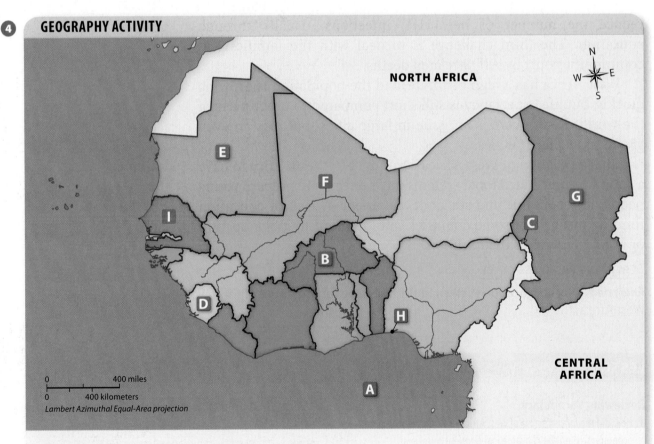

Locating Places

Match the letters on the map with the numbered places below.

1. Chad   **3.** Lagos   **5.** Mauritania   **7.** Burkina Faso   **9.** Sierra Leone

2. Niger River   **4.** Senegal   **6.** Gulf of Guinea   **8.** Lake Chad

REVIEW THE GUIDING QUESTIONS

Directions: Choose the best answer for each question.

1 The climate of northern West Africa can best be described as

A. desert.

B. savannah.

C. tropical rain forest.

D. marine west coast.

2 What attracted Portuguese explorers to Ghana in 1471?

F. salt

G. petroleum

H. lithium

I. gold

3 What did wealthy North African Muslim traders want from the people of West Africa?

A. gold and converts to Islam

B. salt and gold

C. diamonds and emeralds

D. ivory and camels

4 At the Berlin Conference of 1884–1885, France, Germany, Great Britain, Portugal, and Belgium decided to

F. fund secondary schools and universities in West Africa.

G. restrict the trade of enslaved people.

H. heavily tax coffee and cocoa grown in West Africa.

I. build empires by carving up and taking political control of African lands for themselves.

5 Traditional African religions are

A. the dominant religions in West Africa.

B. no longer practiced.

C. only practiced in Mali.

D. practiced, even in countries where Christianity or Islam is dominant.

6 When two or more languages blend and become the language of a region, the result is called a

F. pidgin language.

G. blended language.

H. creole language.

I. diverse language.

DBQ ANALYZING DOCUMENTS

7 DETERMINING WORD MEANINGS Read the following passage about the problem of desertification:

"Nomads are trying to escape the desert, but because of their land-use practices, they are bringing the desert with them. It is a misconception that droughts cause desertification. Droughts are common in arid and semiarid lands. Well-managed lands can recover from drought when the rains return. Continued land abuse during droughts, however, increases land degradation."

—from United States Geological Survey, "Desertification"

What does the passage mean in saying that nomads "are bringing the desert with them"?

A. Nomads bring their desert customs wherever they move.

B. Nomads create desert in new areas because of their land-use practices.

C. Nomads have the skills they need to survive in the desert.

D. Nomads can teach their way of life to other people.

8 IDENTIFYING What evidence does the passage give that droughts alone do not cause desertification?

F. the movement of nomads to new areas

G. the rate at which desertification takes place

H. droughts prevent desertification

I. the ability of lands to recover from drought

SHORT RESPONSE

"The people of Ghana have . . . put democracy on a firmer footing, with repeated peaceful transfers of power. . . . This progress . . . will ultimately be more significant [than the struggle for independence]. For just as it is important to emerge from the control of other nations, it is even more important to build one's own nation."

—from President Barack Obama, "Remarks to the Ghanian Parliament" (2009)

9 DETERMINING CENTRAL IDEAS How do "repeated peaceful transfers of power" show that democracy in Ghana is on a "firmer footing"?

10 IDENTIFYING POINT OF VIEW Why does President Obama think this achievement is more important than winning independence?

EXTENDED RESPONSE

11 ARGUMENT WRITING Ecotourism is often mentioned as a way that developing nations can use their natural resources to produce income while preserving those resources for future generations. Research the pros and cons of the topic, and decide whether or not ecotourism would be good for the countries of West Africa. Present your response in an essay.

Need Extra Help?

If You've Missed Question	❶	❷	❸	❹	❺	❻	❼	❽	❾	❿	⓫
Review Lesson	1	1	2	2	3	3	1	1	2	2	3

From "Desertification," http://pubs.usgs.gov/gip/deserts/desertification, Department of the Interior/United States Geological Survey. The USGS home page is http://www.usgs.gov.

SOUTHERN AFRICA

ESSENTIAL QUESTIONS · *How does geography influence the way people live?* · *How do new ideas change the way people live?*

The Story Matters...

From the steep slopes of the Great Escarpment to the plunging Victoria Falls, Southern Africa is filled with magnificent scenery and wildlife, which draw tourists from around the world. Southern Africa is also the continent's richest region in natural resources, including gold and diamonds. Many of Southern Africa's natural resources have become important to the global economy. Control over these vital resources has brought many great economic, political, and social changes to the region.

FOLDABLES
Study Organizer

Go to the Foldables® library in the back of your book to make a Foldable® that will help you take notes while reading this chapter.

Miner from Johannesburg, South Africa

Gallo Images - LKIS/Getty Images

SOUTHERN AFRICA

Most of inland Southern Africa is rich in resources and home to a wide variety of ethnic groups. The region's coastal and island countries are struggling to develop their economies.

Step Into the Place

MAP FOCUS Use the map to answer the following questions.

1 PLACES AND REGIONS
Luanda is the capital city of what country?

2 THE GEOGRAPHER'S WORLD
What country is located on the southern tip of the African continent?

3 THE GEOGRAPHER'S WORLD
What countries share a border with Zimbabwe?

4 CRITICAL THINKING
Integrating Visual Information Which of these places is the smallest in area: Lesotho, Gabarone, or Malawi?

PROVINCIAL CAPITAL The flat-topped Table Mountain overlooks the city of Cape Town, South Africa. Cape Town serves as the capital of the Western Cape province.

DESERT MAMMAL The meerkat is a member of the mongoose family. Only about 1 foot tall (30 cm), the meerkat stands by using its long tail for balance.

Step Into the Time

TIME LINE Based on events on the time line, predict the effects of Southern Africa's colonial past on the economy, government, and culture of the region.

A.D. 900
Kingdom of Great Zimbabwe established

1806 Britain gains control of Cape Colony

1800

1400s Mutapa Empire flourishes

1652 Dutch establish Cape Colony

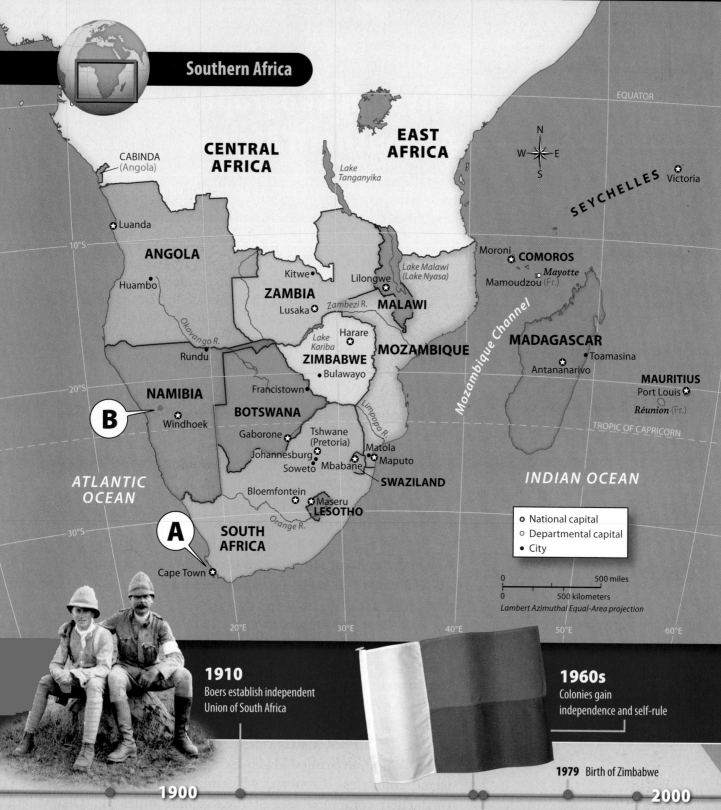

Southern Africa

EQUATOR

N
W · E
S

CENTRAL AFRICA

EAST AFRICA

CABINDA (Angola)

Lake Tanganyika

SEYCHELLES Victoria

10°S

Luanda

ANGOLA

Kitwe

Lake Malawi (Lake Nyasa)

Moroni **COMOROS**
Mayotte
Mamoudzou (Fr.)

Huambo

ZAMBIA

Lilongwe

MALAWI

Okavango R.

Lusaka

Zambezi R.

Harare

MADAGASCAR

Rundu

Lake Kariba

ZIMBABWE

MOZAMBIQUE

Toamasina
Antananarivo

20°S

B

Windhoek

NAMIBIA

Francistown

Bulawayo

Mozambique Channel

MAURITIUS
Port Louis
Réunion (Fr.)

TROPIC OF CAPRICORN

BOTSWANA

Gaborone

Tshwane (Pretoria)

Matola

INDIAN OCEAN

ATLANTIC OCEAN

Johannesburg
Soweto

Mbabane

Maputo

SWAZILAND

Bloemfontein

Maseru

LESOTHO

30°S

A

Orange R.

Limpopo R.

SOUTH AFRICA

Cape Town

National capital
Departmental capital
City

0   500 miles
0   500 kilometers
Lambert Azimuthal Equal-Area projection

20°E   30°E   40°E   50°E   60°E

1910
Boers establish independent Union of South Africa

1960s
Colonies gain independence and self-rule

1979 Birth of Zimbabwe

1900

2000

1886 World's largest gold deposits discovered

1962 Nelson Mandela sentenced to life in prison

1993 Interim constitution enumerates rights for all people

Reading **HELP**DESK

Academic Vocabulary

- **network**

Content Vocabulary

- **escarpment**
- **landlocked**
- **reservoir**
- **blood diamonds**
- **poaching**

TAKING NOTES: *Key Ideas and Details*

Summarize As you read, list important details about two countries of Southern Africa in a graphic organizer like the one below.

Country:		
Physical Geography		
Climate		
Natural Resources		

Lesson 1
Physical Geography of Southern Africa

ESSENTIAL QUESTION • *How does geography influence the way people live?*

IT MATTERS BECAUSE

Southern Africa is the world's leading producer of gold, platinum, chromium, and diamonds. Other minerals the region supplies, including uranium and copper, are also important in the global economy.

Landforms and Bodies of Water

GUIDING QUESTION *What are the dominant physical features of Southern Africa?*

The region of Southern Africa consists of the 10 southernmost countries on the African continent. It also includes four independent island countries and two French island territories in the Indian Ocean off Africa's east coast.

Southern Africa is bordered by the Indian Ocean on the east and the Atlantic Ocean on the west. The Cape of Good Hope at the southern tip of the continent is considered the place where the two oceans meet.

Several of the region's countries are fairly large. Angola and South Africa are each nearly the size of Western Europe and are the continent's seventh- and ninth-largest countries, respectively. Along with Namibia, Mozambique, Zambia, and Botswana, Angola and South Africa rank in the top 25 percent of the world's countries in land area.

The country of Madagascar occupies the world's fourth-largest island, also called Madagascar. The region's three other island countries—Comoros, Mauritius, and Seychelles—are tiny. Their combined area of 1,800 square miles (4,662 sq. km) makes them smaller than the state of Delaware.

Landforms

If Southern Africa's physical geography had to be described with one word, that word would be *high*. A series of plateaus that range in elevation from 3,000 feet to 6,000 feet (914 m to 1,829 m) cover most of the region. The northern plateaus extend from Malawi across Zambia and Angola. These plateaus are largely forested. Farther south, the plateaus are covered mainly by grasslands.

The plateau's outer edges form a steep slope called the Great Escarpment. In Angola, the **escarpment**, a steep cliff between a higher and a lower surface, runs parallel to the Atlantic Coast and continues through Namibia. Between the escarpment and the Atlantic Ocean lies a strip of desert called the Namib that is 80 miles to 100 miles (129 km to 161 km) wide. The Namib runs 1,200 miles (1,931 km) from southern Angola to western South Africa, where it merges with another desert, the Kalahari.

The Kalahari Desert is a vast, sand-covered plateau that sits some 3,000 feet (914 m) above sea level. It is bordered by even higher plateaus. The Kalahari covers much of eastern Namibia and most of Botswana. In some places, long chains of sand dunes rise as much as 200 feet (61 m) high. The sand in some areas is red because of minerals that coat the grains of sand.

South of the Kalahari Desert, much of the rest of Southern Africa is covered by a huge plateau that slopes from about 8,000 feet (2,438 m) in the east to 2,000 feet (610 m) in the west. At the southern tip of this plateau, the Great Escarpment breaks into several small, low mountain ranges. This group of ranges is known as the Cape Ranges. The ranges are separated from each other by dry basins called the Great Karoo and the Little Karoo.

As the Great Escarpment follows South Africa's coastline, it forms the Drakensberg Mountains. This is the most rugged part of the escarpment. Mountain peaks rise to more than 11,000 feet (3,353 m). A narrow coastal plain lies between the mountains and the Indian Ocean.

The Drakensberg is the highest mountain range in all of Southern Africa. The caves and rock shelters contain rock painting made by the San people. The San lived in the area for about 4,000 years.

©Martin Harvey/Corbis

As rivers spill from one plateau to the next, they create thundering waterfalls, such as the spectacular Victoria Falls.

▶ **CRITICAL THINKING**

Describing Where are the falls located?

The Drakensberg mountains parallel the Indian Ocean coastline for some 700 miles (1,127 km) through Lesotho and Swaziland, two **landlocked** countries in Southern Africa. Near Swaziland, the escarpment pulls back from the coastline to create a broad coastal plain that covers much of Mozambique. Northwestern Mozambique and the neighboring countries of Zimbabwe, Zambia, and Malawi lie at higher elevations west of the escarpment, on the plateau.

Bodies of Water

Three major river systems—the Zambezi, Limpopo, and Orange—drain most of Southern Africa. The Zambezi, which stretches for 2,200 miles (3,541 km), is the region's longest river. On the Zambia-Zimbabwe border, midway through its course, the Zambezi plunges over the spectacular Victoria Falls into a narrow gorge. Roughly a mile (1.6 km) wide and 350 feet (107 m) high, the falls are about twice the width and height of Niagara Falls in North America. Because of the heavy veil of mist that rises from the gorge, the area's indigenous people named the falls *Mosi-oa-Tunya* ("The Smoke That Thunders").

The Orange River is Southern Africa's second-longest river. It begins in the highlands of Lesotho and flows westward to reach the Atlantic Ocean. Its course marks the southern boundary of the Kalahari Desert. The region's third-longest river, the Limpopo, flows eastward in a large arc along South Africa's border with Botswana and Zimbabwe. The river then drops over the Great Escarpment to cross the plains of southern Mozambique to the Indian Ocean.

These three rivers, their tributaries, and Southern Africa's other rivers have carved a **network** of canyons and gorges across the plateaus. Dams have been built in the area to store water. Lake Kariba, Southern Africa's second-largest lake, is really a **reservoir**, or an artificial lake created by a dam.

The region's largest lake—and the third largest in all of Africa—is Lake Malawi (also known as Lake Nyasa), which forms Malawi's border with Mozambique and Tanzania. It is the southernmost lake of the Great Rift Valley which stretches for thousands of miles. Lake Malawi fills a depression, or hollow, that follows one of the rifts, or tears, in Earth's crust. Because of the great depth of the depression, Lake Malawi is one of the deepest lakes in the world.

A number of flat basins, called pans, can be found in Southern Africa. The salt deposits they contain provide nourishment for wild animals. Etosha Pan, in northern Namibia, is an enormous expanse of salt that covers 1,900 square miles (4,921 sq. km). It is the largest pan in Africa, and it is the center of Etosha National Park. The park is home to some of the greatest numbers of lions, elephants, rhinoceroses, and other large animals in the world.

✔ **READING PROGRESS CHECK**

Identifying Which type of landform is common in Southern Africa?

Climate

GUIDING QUESTION *What is the climate of Southern Africa?*

Southern Africa has a wide variety of climates, ranging from humid to arid to hot to cool. Nearly all of the region's climates have distinct seasons, with certain seasons receiving most of the rain.

Tropical Zone

The Tropic of Capricorn crosses the middle of Southern Africa. This places the northern half of the region in the Tropics. Northern Angola and northern Mozambique have a tropical wet-dry climate. Each area gets as much as 70 inches (178 cm) of rain per year. Most of it falls in the spring, summer, and fall—from October to May. The high elevation makes temperatures cool.

Thinking Like a Geographer

Pans

Pans are believed to be the beds of ancient lakes whose water evaporated over time. They are among the flattest known landforms. Small amounts of rain can flood large areas of their surface. It is this flooding that causes and maintains their flatness. Salt deposits form as rainwater pools slowly evaporate. *Why can a small amount of rain flood a large area of a pan?*

Academic Vocabulary

network a complex, interconnected chain or system of things such as roads, canals, or computers

---

Zebras are among the many animals that live in the national park that is part of the Etosha Pan.

▶ **CRITICAL THINKING**

Describing What is unique about the Etosha Pan?

People struggle to wade through the waters after a heavy rainfall in the coastal city of Maputo in southwestern Mozambique.

▶ **CRITICAL THINKING**

Describing How does the length of the rainy season in western Mozambique compare with the length in the northern part of the country?

Daily average temperatures range from the upper 60s°F (upper 10s°C) to the upper 70s°F (mid-20s°C). Along the coasts, temperatures are warmer.

Much of northern Mozambique's coastline is watered by rain-bearing winds called monsoons that sweep in from the Indian Ocean during the summer months. More than 70 inches (178 cm) of annual rainfall is common.

Parts of Angola and Mozambique have humid subtropical climates, as do Malawi, Zambia, and northeastern Zimbabwe. The rainy season here is shorter than in the tropical wet/dry zone, and also brings less rainfall. Most places average 24 inches to 40 inches (61cm to 102 cm) per year. Average temperatures are also slightly cooler. Nighttime frosts are not uncommon in July on the high plateaus of Zambia and Malawi. Temperatures on summer days in lowland areas, however, can exceed 100°F (38°C).

Temperate Zones

Much of South Africa, central Namibia, eastern Botswana, and southern Mozambique have temperate, or moderate, climates that are not marked by extremes of temperature. Most of these areas are semiarid. Summer days are warm—from 70°F to 90°F (21°C to 32°C), depending on elevation. Winters are cool, with frosts and sometimes freezing temperatures on the high plateaus.

Annual rainfall varies from 8 inches (20 cm) in some areas to 24 inches (61 cm) in others. Most of the rain falls during the summer, with very little the rest of the year. Droughts are common; in some places, they last for several years.

Lesotho, Swaziland, and eastern South Africa, including the Indian Ocean coastline, are much wetter. Temperatures are like those in the semiarid regions, but ocean currents and moist ocean air bring up to 55 inches (140 cm) of rain annually. Like elsewhere in the region, most of this rain falls in the summer.

Desert Regions

Western South Africa, western Namibia, and much of Botswana are arid. Along the coast, the Namib gets very little rain. In some years, no rain falls. But fog and dew provide small plants with the moisture they need to survive. Temperatures along the coast are mild, however, with daily averages ranging from 48°F to 68°F (9°C to 20°C). The

aridity, the fog, and the mild temperatures result from the cold Benguela Current that flows along the coast. This area is sometimes called the "Skeleton Coast" because many ships used to lose their way in the fog and run aground. Once ashore, the sailors rarely survived because of the lack of water in the sandy desert.

In inland areas of the Namib Desert, temperatures are hotter with summer highs from the upper 80s°F to more than 100°F (30°C to 38°C). In winter, freezing temperatures sometimes occur. During wet years, desert grasses and bushes appear. Much of the time, however, the Namib is home to vast areas of barren sand.

The Kalahari's location—farther inland than the Namib—and dry air make its temperatures more extreme than in the Namib. The Kalahari also gets a little more precipitation than the Namib.

☑ READING PROGRESS CHECK

Describing Why are temperatures in Southern Africa's tropical countries generally not hot?

Natural Resources

GUIDING QUESTION *What natural resources are found in Southern Africa, and why are they important?*

Southern Africa is the continent's richest region in natural resources. Mineral resources have helped the Republic of South Africa, in particular, to build a strong economy. In other countries, like Angola and Namibia, such resources provide the only source of wealth.

The landscape of the Skeleton Coast is made up of sand dunes, rocky canyons, and mountains. Dense fogs and cool sea breezes are characteristic of the area.

▶ **CRITICAL THINKING**

Describing How did the Skeleton Coast get its name?

©George Steinmetz/Corbis

More than one-half of the world's diamonds are harvested from mines, such as this one, in Southern Africa. Diamonds were formed deep in Earth thousands of years ago under extreme heat and pressure. Volcanic pressure brings them to Earth's surface.

South Africa's Resources

The Republic of South Africa has some of the largest mineral reserves in the world. It is the world's largest producer of platinum, chromium, and gold, and one of the largest producers of diamonds—both gems and industrial diamonds, or diamonds used to make cutting or grinding tools. These resources, along with important deposits of coal, iron ore, uranium, copper, and other minerals, have created a thriving mining industry. This industry has attracted workers and investments from other countries that have helped South Africa's industries grow.

Energy Resources

The Republic of South Africa, Zimbabwe, Botswana, and Mozambique mine and burn coal from their own deposits to produce most of their electric power. Mozambique has large deposits of natural gas as well, as does Angola. Angola is also one of Africa's leading oil producers. Namibia has oil and natural gas deposits, too, and they are slowly being developed. Oil and gas must be refined, or changed into other products, before they can be used.

The region's rivers are another resource for providing power. Zimbabwe and Zambia get electricity from the huge Kariba Gorge dam on the Zambezi River. Malawi's rivers and falls generate power for that country. Deforestation, however, allows more sediment to enter the rivers, which reduces the water flow and the electricity that the rivers produce. Mozambique, Zimbabwe, and Angola have not made full use of their rivers to provide power. Economic development and the standard of living in those countries have suffered as a result.

©Herve Collart/Sygma/Corbis

Minerals and Other Resources

Namibia is one of Africa's richest countries in mineral resources. It is an important producer of tin, zinc, copper, gold, silver, and uranium. It also ranks with South Africa and Botswana as a leading world supplier of diamonds. In the 1990s, rebels captured Angola's mines and sold the diamonds to continue a 20-year-old civil war against the government. In countries outside Southern Africa, groups have also mined diamonds to pay for rebellions and other violent conflicts. Diamonds used for this purpose are called **blood diamonds**.

Gold is a leading export for Zimbabwe. Mozambique has the world's largest supply of the rare metal tantalite, which is used to make electronic parts and camera lenses. Gold, platinum, and diamonds are mined there too, as are iron ore and copper. Much of Zambia's economy is based on copper and cobalt, although gold, silver, and iron ore are also mined. Zambia has some of the largest emerald deposits in the world. A small amount of rubies, sapphires, and a variety of semiprecious gems are mined in neighboring Malawi.

Malawi's most important natural resource is its fertile soil. The country's economy is based mainly on agriculture. Tobacco is its most important export. Exporting farm products is also a major economic activity in Zimbabwe. Lesotho and Swaziland have few natural resources. Most of their people practice subsistence farming, growing only enough to meet their needs.

Wildlife

Southern Africa is known for its variety of animal life. Wildebeests, lions, zebras, giraffes, and many other animals are found across the region. They live within and outside the many national parks and wildlife reserves that nearly every country has created to protect them. Tourists come from throughout the world to see these animals. **Poaching**, or illegally killing game, is a problem. Poachers shoot elephants for their valuable ivory tusks and rhinoceroses for their horns. Others kill animals to sell their skins and meat and to protect livestock and crops.

Include this lesson's information in your Foldable®.

 READING PROGRESS CHECK

Describing How does deforestation affect the energy supply in the region?

LESSON 1 REVIEW (CCSS)

Reviewing Vocabulary

1. Why is *poaching* against the law?

Answering the Guiding Questions

2. ***Describing*** How has damming Southern Africa's rivers benefited the people and countries of the region?

3. ***Identifying*** What are the rainfall and temperature differences between Southern Africa's tropical, temperate, and arid regions?

4. ***Describing*** For what resources is Southern Africa known throughout the world?

5. ***Narrative Writing*** Create a journal entry recording your observations and experiences during one day of a photo safari at Etosha National Park.

Reading **HELP**DESK CCSS

Academic Vocabulary

- **exploit**
- **grant**

Content Vocabulary

- **apartheid**
- **civil disobedience**
- **embargo**

TAKING NOTES: *Key Ideas and Details*

Sequence Create a time line like this one. Then list five key events and their dates in the history of the region.

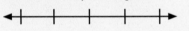

Lesson 2
History of Southern Africa

ESSENTIAL QUESTION · *How do new ideas change the way people live?*

IT MATTERS BECAUSE
Many of Southern Africa's resources have become important parts of the global economy. Political instability and unrest have sometimes disrupted the flow of products to world markets. Much of the instability and unrest is directly or indirectly the result of the region's colonial history.

Rise of Kingdoms

GUIDING QUESTION What major events mark the early history of Southern Africa?

Southern Africa's indigenous people have inhabited the region for thousands of years. Some lived as hunter-gatherers. Others farmed and herded cattle. Trade among the groups flourished. Ivory, gold, copper, and other goods moved from the interior to the east coast. There such goods were exchanged for tools, salt, and luxury items including beads, porcelain, and cloth from China, India, and Persia.

Great Zimbabwe

Around the year A.D. 900, the Shona people built a wealthy and powerful kingdom in what is now Zimbabwe and Mozambique. The capital was a city called Great Zimbabwe. (*Zimbabwe* is a Shona word meaning "stone houses.") As many as 20,000 people lived in the city and the surrounding valley.

Great Zimbabwe was the largest of many similar cities throughout the region. By the 1300s, it had become a great commercial center, collecting gold mined nearby and trading it to Arabs at ports on the Indian Ocean.

Great Zimbabwe was abandoned in the 1400s, possibly because its growing population exhausted its water and food

resources. The city's ruins show the Shona's skill as builders. Some structures were more than 30 feet (9 m) high. Their large stones were cut to fit and stay in place without mortar to hold them together.

The Mutapa Empire

In the late 1400s, the Shona conquered the region between the Zambezi and Limpopo rivers from Zimbabwe to the coast of Mozambique. Like Great Zimbabwe, the Mutapa Empire thrived on the gold it mined and traded for goods from China and India.

The Portuguese arrived and took over the coastal trade in the 1500s. They gradually gained control over the empire and forced its people to mine gold for them. In the late 1600s, Mutapa kings allied with the nearby Rozwi kingdom to drive out the Portuguese. Instead, the Rozwi conquered the Mutapa's territory and ruled it until the early 1800s, when it became part of the Zulu Empire.

Other Kingdoms

The Zulu leader Shaka united his people in the early 1800s to form the Zulu Empire in what is now South Africa. He built a powerful army and used it to expand the empire by conquering neighboring people. Shaka was killed in 1828, but his empire survived until the British destroyed it in the Zulu War of 1879.

A series of kingdoms rose and fell on the island of Madagascar from the 1600s to the 1800s. Some of the early kingdoms were influenced by Arab and Muslim culture. In the early 1800s, one king allied with the British on the nearby island of Mauritius to prevent the French from taking control of Madagascar. He eventually conquered most of the island and formed the Kingdom of Madagascar. French troops invaded the kingdom in 1895 and made it a French possession.

☑ **READING PROGRESS CHECK**

Identifying Which outsiders traded with Southern Africans before the Europeans arrived?

Shown are remnants of the walls of the Great Enclosure of the city of Great Zimbabwe. According to historians, houses of the royal family were located within the walls.

Identifying How did Great Zimbabwe become an important center of trade?

European Colonies

GUIDING QUESTION *How did Southern Africa come under European control?*

Around 1500, Portugal and other European countries began establishing settlements along the African coast. The first settlements were trading posts and supply stations at which ships could stop on their way to and from Asia. As time passed, the Europeans grew interested in **exploiting** Africa's natural resources and, as a source of labor, its people.

Clashes in South Africa

During the 1600s till about the 1800s, Europeans set up trading posts but did not establish colonies, which are large territories with settlers from the home country. One exception was Cape Colony, founded by the Dutch in 1652 at the Cape of Good Hope on the southern tip of what is now South Africa. The Dutch became known as Boers, the Dutch word for farmers. They grew wheat and raised sheep and cattle. Enslaved people from India, Southeast Asia, and other parts of Africa provided much of the labor.

The Africans did not like the Dutch pushing into their land, and soon they started fighting over it. By the late 1700s, the Africans had been defeated. Some fled north into the desert. Others became workers on the colonists' farms.

The Union of South Africa

Wars in Europe gave Britain control of the Cape Colony in the early 1800s. Thousands of British settlers soon arrived. The Boers resented British rule. Many decided to seek new land beyond the reach of British control. Beginning in the 1830s, thousands of Boers left the colony in a migration called the Great Trek and settled north of the Orange River.

In the 1860s, the Boers discovered diamonds in their territory. Then, in 1886, they found the world's largest gold deposits. British efforts to gain these resources led to the Boer War in 1899. The Boers were defeated and again came under British control. In 1910 Britain allowed the Boer colonies to join the Cape Colony in forming an independent country—the Union of South Africa. The small African kingdoms of Lesotho and Swaziland remained under British control.

Three Zulu leaders are shown holding shields and wearing traditional attire. The Zulu built a great empire, but during the 1800s, European settlers took control of their grazing and water resources. The Zulu population is about 9 million today, making them the largest ethnic group in the Republic of South Africa.

Academic Vocabulary

exploit to make use of something, sometimes in an unjust manner for one's own advantage or gain

Hulton Archive/Getty Images

Colonialism in Other Areas

While the British and the Boers competed for South Africa, other European countries were competing over the rest of Africa. In 1884 representatives of these countries met in Berlin, Germany, to divide the continent among themselves.

In Southern Africa, Britain gained control over what is now Malawi, Zambia, Zimbabwe, and Botswana. The Berlin Conference decided Portugal had rights to Angola and Mozambique. Germany received what is now Namibia, although South Africa seized the colony during World War I. Besides Madagascar, France controlled what is now Comoros. Mauritius and Seychelles were British colonies.

European control in Southern Africa continued for about the next 80 years. Not until the 1960s did the region's colonies begin to gain independence and self-rule.

☑ **READING PROGRESS CHECK**

Analyzing Which European country claimed the most territory in Southern Africa in the 1800s?

Academic Vocabulary

grant to permit as a right, a privilege, or a favor

Independence and Equal Rights

GUIDING QUESTION *What challenges did Southern Africans face in regaining freedom and self-rule?*

French rule in Madagascar ended in 1960, making it the first Southern African country to gain independence. Britain **granted** independence to Malawi and Zambia in 1964 and to Botswana and Lesotho in 1966. Swaziland and Mauritius gained their freedom in 1968, and Seychelles in 1976. Elsewhere, however, freedom was more difficult to achieve.

Boer soldiers fight from trenches at the siege of Mafeking in 1900. The siege, lasting more than 200 days, resulted in an important victory for British forces.

▶ **CRITICAL THINKING**

Describing Who were the Boers? Why were the Boer Wars fought?

Laws in South Africa limited the political rights of black Africans and set up separate parks, beaches, and other public places.

▶ **CRITICAL THINKING**

Describing Who controlled South Africa's government until World War II? Who controlled the government beginning in 1948?

The End of Portuguese Rule

While other European nations gave up their African colonies, Portugal refused to do so. Revolts for independence broke out in Angola in 1961 and in Mozambique in 1964. The thousands of troops Portugal sent to crush these revolts failed to do so.

By 1974, the Portuguese had grown tired of these bloody and expensive wars. Portuguese military leaders overthrew Portugal's government and pulled the troops out of Africa. Angola and Mozambique became independent countries in 1975 as a result. Fighting continued, however, as rebel groups in each country competed for control. Mozambique's long civil war ended when a peace agreement was reached in 1994. Peace was not finally achieved in Angola until 2002.

The Birth of Zimbabwe

After granting Malawi and Zambia independence, Britain prepared to free neighboring Zimbabwe, then called Southern Rhodesia. The colony's white leaders, who controlled the government, instead formed a country they called Rhodesia and continued to rule.

Rhodesia's African population demanded the right to vote. When the government resisted, a guerrilla war began. In 1979 the government finally agreed to hold elections in which all Rhodesians could take part. Rebel leader Robert Mugabe was elected president, and Rhodesia's name was changed to Zimbabwe.

Equal Rights in South Africa

After independence, the growth of South Africa's mining and other industries depended on the labor of black Africans, who greatly

©David Turnley/Corbis

outnumbered the country's whites. The white minority government stayed in power by limiting the black population's educational and economic opportunities and political rights.

English South Africans controlled the government until the end of World War II. Then a strike by more than 60,000 black mine workers frightened white voters into electing an Afrikaner government in 1948 that promised to take action. (Afrikaners are the descendants of the Boers. They speak a language called Afrikaans, which gives them their name.)

The new government leaders began enacting laws that created a system called **apartheid**—an Afrikaans word meaning "apartness." Apartheid limited the rights of blacks. For example, laws forced black South Africans to live in separate areas called "homelands." People of non-European background were not even allowed to vote. The African National Congress (ANC), an organization of black South Africans, began a campaign of **civil disobedience**, disobeying certain laws as a means of protest. The government's violent response to peaceful protests caused the ANC to turn to armed conflict. In 1962 ANC leader Nelson Mandela was arrested and sentenced to life in prison.

By the 1970s, apartheid-related events in South Africa had gained world attention. Countries began placing **embargos**, or bans on trade, on South Africa. Meanwhile, the struggle in South Africa grew more intense. In 1989 South Africa's president, P.W. Botha, was forced to resign. In 1990 the government, under Botha's successor, F.W. de Klerk, began repealing the apartheid laws. Mandela was released from prison in 1991. In 1993 a new constitution gave South Africans of all races the right to vote. The ANC easily won elections held in 1994, and Mandela became the country's president.

In 1995 the new government created a truth and reconciliation commission. Its task was to ease racial tensions and heal the country by uncovering the truth about the human rights violations that had occurred under apartheid.

✓ **READING PROGRESS CHECK**

Determining Central Ideas Why do you think South Africa's government created the apartheid system?

By 1994, South Africa's policy of apartheid was officially over. Nelson Mandela became the first black person to be elected president of South Africa. Mandela is shown voting for the first time in his life on April 27, 1994.

FOLDABLES
Study Organizer

Include this lesson's information in your Foldable®.

©Louise Gubb/Corbis SABA

LESSON 2 REVIEW (CCSS)

Reviewing Vocabulary

1. Why might some people disapprove of *civil disobedience* as a means of protest and of achieving change?

Answering the Guiding Questions

2. *Analyzing* How did some of Southern Africa's early people benefit from the region's natural resources?

3. *Identifying* Name five present-day countries in Southern Africa that were once controlled by Britain.

4. *Determining Central Ideas* Why was gaining independence especially difficult for Angola and Mozambique?

5. *Argument Writing* Write a paragraph explaining whether actions against the governments of Rhodesia and South Africa were justified.

Lesson 3
Life in Southern Africa

(l to r) ©STR/Reuters/Corbis; Ken Gerhardt/Gallo Images/Getty Images; Tom Cockrem/Photolibrary/Getty Images; ALEXANDER JOE/AFP/Getty Images

Reading **HELP**DESK CCSS

Academic Vocabulary

- **contact**
- **trend**

Content Vocabulary

- **utility**
- **thatch**
- **periodic market**

TAKING NOTES: *Key Ideas and Details*

Summarize Create a chart like this one. Then list information about Southern Africa on these three topics.

Population	
Culture Groups	
Health Issues	

ESSENTIAL QUESTION • *How does geography influence the way people live?*

IT MATTERS BECAUSE

Control over Southern Africa's vast and vital natural resources has been passed on to new leadership. Great economic, political, and social changes and challenges have accompanied this transfer.

The People of the Region

GUIDING QUESTION *Where do people live in Southern Africa?*

The population of Southern Africa is overwhelmingly black African. The largest white minority is in the country of South Africa, where whites represent 10 percent of the population. In almost every other country, whites and Asians make up less than 1 percent of the population. The region's black African population is made up of many different ethnic and culture groups.

Population Patterns

Southern Africa's countries vary widely in population. Fewer than 2 million people live in the small countries of Lesotho and Swaziland. South Africa, which surrounds both of them, has the region's largest population—about 49 million.

Population depends heavily on geography and economics. For example, Botswana and Namibia are much larger than Swaziland and Lesotho, but their populations are only slightly larger. Most Batswana, as the people of Botswana are called, live in the northeast, away from their country's desert areas. Similarly, most Namibians live in the northern part of their country, away from the arid south and west.

South Africa and Angola are about the same size. South Africa, the region's most industrialized nation, has three times as many people. In both countries, most people live in

cities. Angola's rural areas are thus much more thinly populated than rural areas in South Africa.

Mozambique, which is slightly smaller than Namibia and much smaller than Angola, has a population greater than those two countries combined. Most of Mozambique's 23 million people are engaged in farming, mainly along the fertile coastal plain.

Zambia is twice as big as Zimbabwe. Zimbabwe, with a population of about 12 million, has only 2 million fewer people. Both countries are largely rural, with only about one-third of their people living in cities. Large parts of Zambia are thinly populated.

Malawi is just one-third the size of Zimbabwe and one-sixth the size of Zambia, yet it exceeds both in population. With some 16 million people living in an area roughly the size of Pennsylvania, it is the region's most densely populated country. On average, every square mile holds more than 250 people.

Surprisingly, Malawi is also Southern Africa's most rural nation. Only 20 percent of its people live in cities. Its small size and large rural population mean that most of its farms are small. Most farm villages are not able to produce much more than what they need. As a result, Malawi is the region's poorest country. The average Malawian earns less than $350 per year.

MAP SKILLS

1 PLACES AND REGIONS What do the cities of Johannesburg, Durban, and Cape Town have in common?

2 THE GEOGRAPHER'S WORLD In general, which area of Southern Africa is more densely populated: eastern or western?

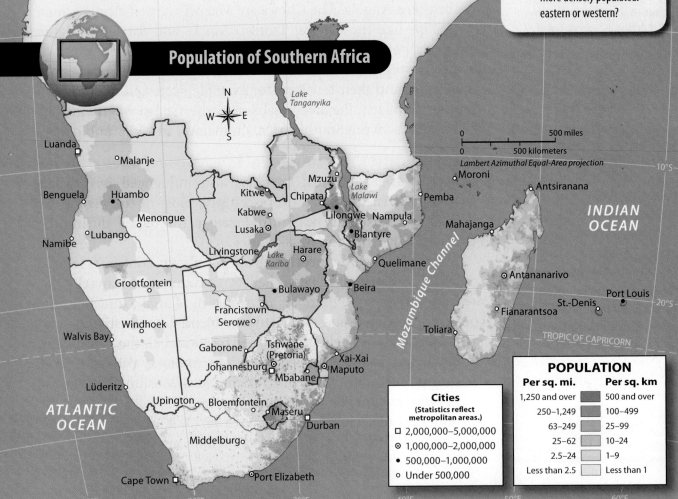

Population of Southern Africa

Cities
(Statistics reflect metropolitan areas.)
- □ 2,000,000–5,000,000
- ⊙ 1,000,000–2,000,000
- • 500,000–1,000,000
- ○ Under 500,000

POPULATION

Per sq. mi.	Per sq. km
1,250 and over	500 and over
250–1,249	100–499
63–249	25–99
25–62	10–24
2.5–24	1–9
Less than 2.5	Less than 1

500 miles
500 kilometers
Lambert Azimuthal Equal-Area projection

Ethnic and Culture Groups

Africans are not a single people. Southern Africa is home to many ethnic and cultural groups who speak several different languages. One group, the Shona, makes up more than 80 percent of the population of the country of Zimbabwe. South Africa's 9 million Zulu make up that country's largest ethnic group. More than 7 million Xhosa also live there, as do the Khoekhoe. Some 4.5 million Tsonga people are spread among the countries of South Africa, Zimbabwe, and Mozambique.

About 4 million Tswana form the major population group in Botswana. A similar number of Ovimbundu and 2.5 million Mbundu make up approximately two-thirds of Angola's population. A smaller group, the Ambo, live in Angola and Namibia. About half of Namibia's people belong to this ethnic group. The San, a nomadic people, live mainly in Namibia, Botswana, and southeastern Angola. The Chewa are Malawi's largest ethnic group.

Groups like the Chewa, Tsonga, Ambo, and San illustrate an important point about Southern Africa's history. When Europeans divided the region, they paid little attention to its indigenous people. The Chewa and their territory, for example, were split among four colonies. Similarly, the area inhabited by the Tsonga was divided by the borders between South Africa, Zimbabwe, and Mozambique.

Religion and Languages

Southern Africa's colonial past has also influenced its people's religious beliefs. In almost every country, most of the people are Christians. Christianity was introduced to the region during the colonial era by Christian missionaries.

In Angola, however, nearly half the population continues to hold traditional indigenous religious beliefs. Traditional African religions are followed by large numbers of people in Namibia and Lesotho, too. In Zimbabwe and Swaziland, a blend of Christianity and traditional religious beliefs is followed by about half the population.

Swaziland, Zambia, Malawi, and Mozambique also have large Muslim populations. Most of Mozambique's Muslims live on the coast, where **contact** with Arab traders led long ago to the introduction of Islam. Immigration from Asia explains Zambia's Muslim population, as well as its large Hindu minority.

Members of the Nazareth Baptist Church in South Africa take part in their annual pilgrimage to the mountain of Nhlangakazi. The church is also called the Shembe Church after its founder, Isaiah Shembe.

Think Again?

Southern Africa's large island country of Madagascar was settled by African people.

Not true. Most of Madagascar's people speak Malagasy, a language related to those spoken in Indonesia, the Philippines, and islands in the South Pacific. The language of Madagascar indicates that the island's early inhabitants probably came from that part of the world.

©STR/Reuters/Corbis

Portuguese remains the official language in Angola and Mozambique. English is an official language in most of the former British colonies. Its use, however, is mainly limited to official and business communications; nowhere is it widely spoken by the people. Instead, most speak indigenous languages. South Africa has 10 official languages besides English; Zambia has 7.

✅ **READING PROGRESS CHECK**

Determining Central Ideas What is the main religion practiced in Southern Africa?

Life in Southern Africa

GUIDING QUESTION *How do the various people of Southern Africa live?*

As in other regions of Africa, life differs from city to countryside. Many rural people continue to follow traditional ways of life. At the same time, urban and economic growth are challenging and changing many of the traditional ways.

Urban Life

Although most people in the region of Southern Africa live in the countryside, migration to cities grows because of job opportunities. Harare, Zimbabwe, has grown to more than 1.5 million, as have Lusaka, Zambia, and Maputo, Mozambique. Luanda, Angola's capital, is even larger: It holds some 4.5 million people. South Africa has four cities—Durban, Ekurhuleni, Cape Town, and Johannesburg—with populations of around 3 million or more.

Ken Gerhardt/Gallo Images/Getty Images

contact communication or interaction with someone

Shown here is a high-rise building under construction in the city of Luanda in Angola. Luanda is the country's main seaport and government center.

At an outdoor market in Lusaka, vendors come to sell handcrafted items. Food and entertainment are also available.

▶ **CRITICAL THINKING**

Describing What are periodic markets?

Urban Growth and Change

The rapid growth of some cities has strained public **utilities**—services such as trash collection, sewage treatment, and water distribution. Luanda, for example, has had many problems providing enough clean water for its many people. Outbreaks of cholera and other diseases have resulted from drinking polluted water.

The region's cities have a mix of many ethnic groups and cultures. An example is Johannesburg, where the wealth from nearby gold fields helped build one of the most impressive downtowns in all of Africa. Outside the central city are the white neighborhoods where about 20 percent of the city's population live. Some black South Africans have moved into these neighborhoods since the end of apartheid. Most, however, live in "townships" at the city's edge. These areas often have no electricity, clean water, or sewer facilities. Most of the region's large cities have shantytowns.

Johannesburg's role as a mining, manufacturing, and financial center has attracted people from around the world. Every black ethnic group in Southern Africa is present, as well. The white community is mainly English and Afrikaner. Large Portuguese, Greek, Italian, Russian, Polish, and Lebanese populations also live there. Indians, Filipinos, Malays, and Chinese live mainly in the townships. At least 12 languages are heard on city streets.

Tom Cockrem/Photolibrary/Getty Images

Family and Traditional Life

People who move to the cities must adjust to new experiences and a different way of life. In the countryside, traditional ways of life remain strong.

Rural villages are often small—consisting of perhaps 20 or 30 houses. Building materials, which vary by ethnic group, include rocks, mud bricks, woven sticks and twigs packed with clay, and **thatch**—straw or other plant material used to cover roofs.

In many cultures, all the people in a village are related by blood or marriage to the village's headman or chief. Men often have more than one wife. They provide a house for each wife and their children. Growing food crops is the main economic activity. Many families raise cattle as well, mainly for milk and as a symbol of wealth.

People in the countryside practice subsistence farming, growing the food they need to survive. Artwork sometimes provides a family with a source of cash. Wood and ivory carving are art forms that are generally practiced by men. Pottery-making is usually a woman's craft. In some cultures, both men and women make baskets. They sell the products in cities or at **periodic markets**—open-air trading markets held regularly at crossroads or in larger towns.

In recent times, more and more men have been leaving their villages to work at jobs in cities or mines. Although the money they send home helps support their families, this **trend** has greatly changed village life. Many villages now consist largely of women, children, and older men. Women have increasingly taken on traditional male roles in herding, family and community leadership, and other activities.

✔ **READING PROGRESS CHECK**

Citing Text Evidence Where in their countries do most Southern Africans live?

Southern Africa Today

GUIDING QUESTION *What challenges and prospects do the countries of Southern Africa face?*

Southern Africa's wealth of mineral, wildlife, and other resources may be the key to its future. Still, the region faces serious social, economic, and political challenges.

Health Issues

Life expectancy in Southern Africa is low. In the majority of countries, most people do not live beyond age 50 to 55. Lack of good rural health care is one reason, although many countries are trying to build or improve rural clinics.

Academic Vocabulary

trend a general tendency or preference

Members of South Africa's Ndebele tribe attend a gathering of traditional leaders from all over the country in November 2009 to honor former President Nelson Mandela.

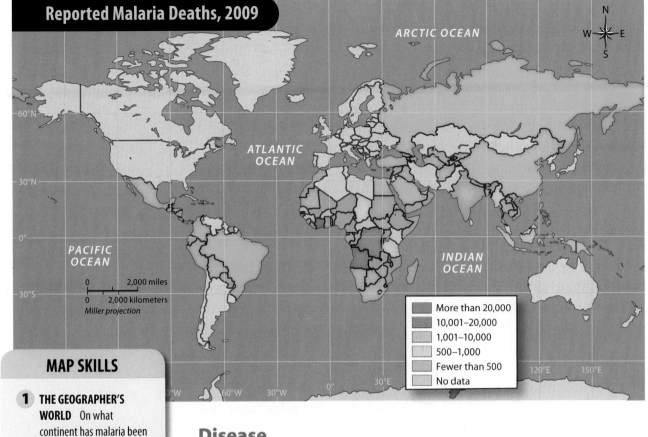

Reported Malaria Deaths, 2009

ARCTIC OCEAN

ATLANTIC OCEAN

PACIFIC OCEAN

INDIAN OCEAN

0   2,000 miles
0   2,000 kilometers
Miller projection

More than 20,000
10,001–20,000
1,001–10,000
500–1,000
Fewer than 500
No data

MAP SKILLS

1 **THE GEOGRAPHER'S WORLD** On what continent has malaria been responsible for the greatest number of deaths?

2 **THE GEOGRAPHER'S WORLD** In what regions of the world has malaria been responsible for the fewest deaths?

Disease

Malaria, a tropical disease carried by mosquitoes, is a problem in several countries. Dysentery and cholera, potentially fatal diseases caused by bacteria in water, are also widespread. So is tuberculosis. Malnutrition is a cause of death for many infants and young children.

Southern Africa has some of the highest rates of infant death in the world. In Angola, Malawi, and Mozambique, about 100 to 120 of every 1,000 children die in infancy. Elsewhere in the region, the figure is 40 to 60 per 1,000. (The infant death rate in the United States is 7 per 1,000.)

A major cause of death in children and adults is HIV/AIDS. Southern Africa has a higher HIV/AIDS rate than any other region in Africa. Swaziland, Botswana, Lesotho, and South Africa have the highest rates in the world. About one of every four adults (25 percent) in these countries is infected with this sexually transmitted disease, which women pass on to their children at birth. In the rest of the region, the adult HIV/AIDS rate averages between 11 and 14 percent. (In the United States, the rate is 0.6 percent.)

The high incidence of HIV/AIDS has disrupted the labor force by depriving countries of needed workers. It has also disrupted families through death, inability to work, or AIDS-related family issues. The disease has created millions of AIDS orphans, children whose mother and father have died from AIDS. The huge number of AIDS orphans is a major social problem.

Progress and Growth

Angola and Mozambique continue to rebuild the cities and towns, industries, railroads, and communications systems that have been damaged or destroyed by years of civil war. Oil exports in Angola and aluminum exports in Mozambique help finance this effort. So does the tourism that peace and stability have brought back to the beautiful beaches and resorts along Mozambique's coast.

Tourism at national parks has grown with the establishment of stable, democratic governments. Zambia and Malawi replaced one-party rule with more democratic forms of government in the 1990s. Botswana and Namibia have been strong democracies, respecting and protecting human rights, since independence. Only Zimbabwe and Swaziland continue to suffer economic decline and political unrest, largely due to repressive leaders.

Help From Other Countries

The United States has used economic aid to strengthen democracy in Southern Africa. Other U.S. programs have provided billions of dollars to pay for medications and care for AIDS sufferers and AIDS orphans.

Other countries and international organizations have also made huge investments in the region. Taiwan's development of a textile industry in Lesotho, for example, is giving some of that poor country's workers an alternative to employment in South Africa's mines.

Foreign investment, workers, and tourists have also returned to South Africa as it continues to recover from the effects of apartheid. South Africa remains the region's most industrial and wealthiest country. It also faces serious economic challenges. Many of its traditional African farming communities struggle in poverty, growing few if any cash crops. Its heavy reliance on the export of mineral and agricultural goods places it at risk if world demand or prices for the goods fall. These problems mirror the challenges that many other countries in Southern Africa also confront.

Include this lesson's information in your Foldable®.

☑ **READING PROGRESS CHECK**

Analyzing Why is life expectancy in Southern Africa so low?

LESSON 3 REVIEW

Reviewing Vocabulary

1. What did rural Southern Africans use clay and *thatch* for?

Answering the Guiding Questions

2. *Determining Central Ideas* How did colonialism and contact with traders influence religious beliefs in Southern Africa?

3. *Describing* What are rural and city life like for Southern Africa's black population?

4. *Analyzing* How and why has Southern Africa benefited from the growth of democracy in the region?

5. *Argument Writing* Write a letter to the editor of a Southern African newspaper explaining whether the region should continue to work for change.

Directions: Write your answers on a separate piece of paper.

① Use your **FOLDABLES** to explore the Essential Question.
INFORMATIVE/EXPLANATORY WRITING Write two paragraphs explaining how Southern Africa's resources place the region in a favorable position to develop trade with other countries.

② **21st Century Skills**
DESCRIBING Using information from the text and online, create a brief slide show of Southern Africa's energy resources and how the region uses the resources. Narrate the slide show, identifying the different countries' means of generating power.

③ **Thinking Like a Geographer**
DETERMINING CENTRAL IDEAS As a geographer, would you favor setting aside more or less land for game preserves in Southern Africa? Use a T-chart to list your pro and con arguments.

④ **GEOGRAPHY ACTIVITY**

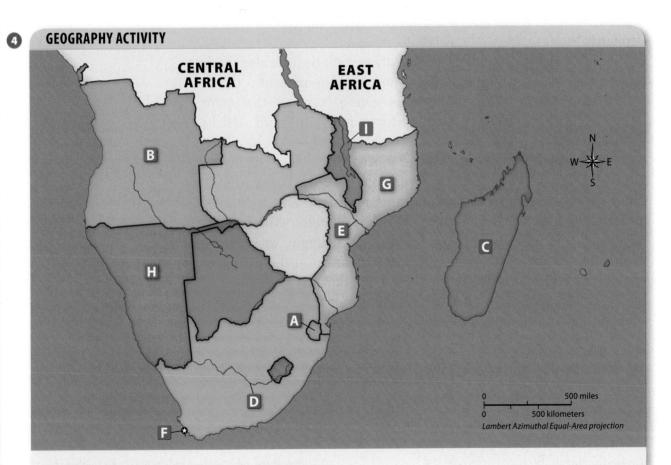

Locating Places
Match the letters on the map with the numbered places listed below.

1. Zambezi River
2. Madagascar
3. Angola
4. Cape Town
5. Orange River
6. Mozambique
7. Namibia
8. Swaziland
9. Lake Malawi (Lake Nyasa)

Chapter 23 ASSESSMENT CCSS

REVIEW THE GUIDING QUESTIONS

Directions: Choose the best answer for each question.

1. The country of Madagascar is
   A. a large plateau.
   B. Southern Africa's regional capital city.
   C. the world's fourth-largest island.
   D. the world's largest exporter of coconut milk.

2. Which is the longest river in Southern Africa?
   F. Kariba
   G. Congo
   H. Great Karoo
   I. Zambezi

3. Western South Africa, western Namibia, and Botswana have what climate zone in common?
   A. tropical
   B. desert
   C. Mediterranean
   D. steppe

4. The amount of hydroelectric power in this region has been reduced by
   F. deforestation.
   G. droughts.
   H. monsoons.
   I. civil disturbances.

5. South Africa's Afrikaners are descended from which population group?
   A. native Africans
   B. Boers
   C. Portuguese colonists
   D. Zambians

6. Which is the most densely populated country in Southern Africa?
   F. Zambia
   G. the Republic of South Africa
   H. Madagascar
   I. Malawi

DBQ ANALYZING DOCUMENTS

7 ANALYZING Read the following passage about the area around the Okavango River and the Kalahari Desert.

"*During dry periods [the Okavango Delta] is estimated to cover at least 6,000 square miles, but in wetter years, with a heavy annual flood, the Okavango's waters can spread over 8,500 square miles of the Kalahari's sands. Deep water occurs in only a few channels, while vast areas of reed beds are covered by only a few inches of water.*"

—from Cecil Keen, *Okavango*

As described in the reading, the Okavango is a

A. desert.
B. mountain.
C. river.
D. reed bed.

8 ANALYZING What can you infer about the land of the Kalahari from this passage?

F. It is sandy because it absorbs most of the water fairly quickly.
G. It is fairly flat because more of the water is shallow than deep.
H. It is wet most of the time because it lets the floodwaters stand.
I. It tilts to the west because that is where the deep channels form.

SHORT RESPONSE

"*Discouraged about the lack of results from their nonviolent campaign, Nelson Mandela and others called for an armed uprising . . . that paralleled the nonviolent resistance. That, too, failed to tear down the apartheid system, and in the end a concerted grassroots nonviolent civil resistance movement [together] with international support and sanctions [against the government] forced the white government to negotiate.*"

—from Lester R. Kurtz, "The Anti-Apartheid Struggle in South Africa"

9 DETERMINING CENTRAL IDEAS What were Mandela and others trying to achieve?

10 ANALYZING How did they eventually succeed?

EXTENDED RESPONSE

11 INFORMATIVE/EXPLANATORY WRITING Southern Africa has an abundance of wildlife, including animals, birds, fish, and exotic plant life. Tourists come from all over the world to see the animals, which live on animal preserves and in the wild. Do some research on travel in Southern Africa, then write an essay describing the experience of going on safari. Talk about which areas of the region you visited and what you saw, and what kind of accommodations you had on your safari.

Need Extra Help?

If You've Missed Question	1	2	3	4	5	6	7	8	9	10	11
Review Lesson	1	1	1	1	2	3	1	1	2	2	3

From THE ANTI-APARTHEID STRUGGLE IN SOUTH AFRICA (1912-1992), by Lester R. Kurtz, Ph.D., June 2010. © 2010 International Center on Nonviolent Conflict.

OCEANIA, AUSTRALIA, NEW ZEALAND, AND ANTARCTICA

McGraw-Hill
networks™

UNIT 7

EXPLORE the CONTINENT

THIS region lies almost entirely in the Southern Hemisphere, reaching from north of the Equator to the South Pole. The countries that make up this region contain an amazing variety of landforms and range in size from tiny islands to large continents.

1 **NATURAL RESOURCES** This region holds abundant natural resources. New Zealand's North Island has good farmland and pasture for grazing sheep and other animals. The location of New Zealand along the Ring of Fire provides geothermal energy. Australia is rich in precious metals, oil and natural gas, and fertile farmland. In Oceania, wind and solar energy are plentiful, as are fish and other seafood.

2 **ISLANDS AND REEFS** Oceania is made up of thousands of islands with different physical features. These islands were formed millions of years ago by underwater volcanoes. Coral islands, called atolls, are made up of reef islands surrounding lagoons. Off Australia's northeastern shore lies the spectacular Great Barrier Reef. By contrast, the water surrounding Antarctica's coasts freezes into thick plains of ice during winter. Other huge ice formations include glaciers and icebergs.

3 LANDFORMS Physical landscapes vary throughout the region. In New Zealand, the towering mountains of the Southern Alps rise on South Island. Australia is mostly flat. The Central Lowlands, however, contain a massive stone monolith known as Ayers Rock, or *Uluru* to the Aboriginal people of Australia. Antarctica is made up of one large, icy landmass and an archipelago of rocky islands.

FAST **FACT**

Australia is slightly smaller than the 48 U.S. states.

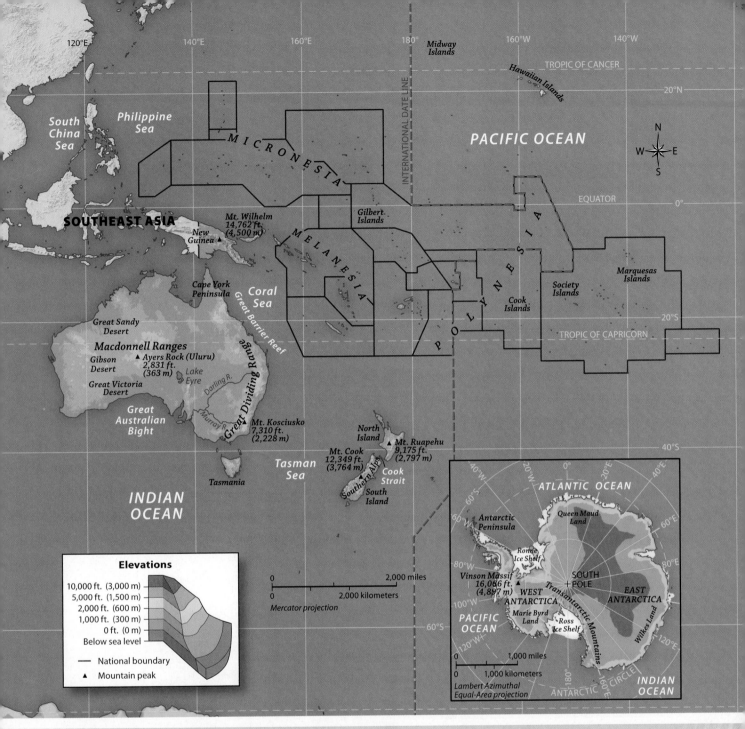

OCEANIA, AUSTRALIA, NEW ZEALAND, AND ANTARCTICA

PHYSICAL

MAP SKILLS

1 **THE GEOGRAPHER'S WORLD** Where are the region's major rivers located?

2 **PLACES AND REGIONS** How is the land elevation in Antarctica different from elevations elsewhere in the region?

3 **PHYSICAL GEOGRAPHY** Which deserts are found in Australia?

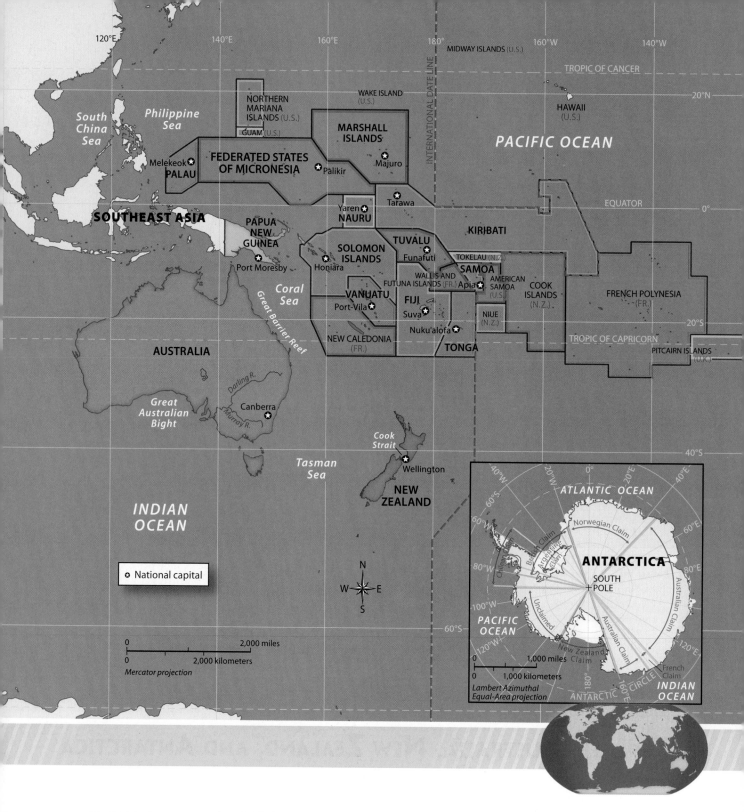

South China Sea

Philippine Sea

MIDWAY ISLANDS (U.S.)

TROPIC OF CANCER

WAKE ISLAND (U.S.)

NORTHERN MARIANA ISLANDS (U.S.)

GUAM (U.S.)

HAWAII (U.S.)

20°N

MARSHALL ISLANDS

FEDERATED STATES OF MICRONESIA

Majuro

PACIFIC OCEAN

Melekeok

PALAU

Palikir

INTERNATIONAL DATE LINE

SOUTHEAST ASIA

Yaren NAURU

Tarawa

EQUATOR

0°

PAPUA NEW GUINEA

KIRIBATI

SOLOMON ISLANDS

TUVALU

Honiara

Funafuti

TOKELAU (N.Z.)

Port Moresby

Coral Sea

VANUATU

Port-Vila

WALLIS AND FUTUNA ISLANDS (FR.)

SAMOA

Apia

AMERICAN SAMOA (U.S.)

COOK ISLANDS (N.Z.)

FRENCH POLYNESIA (FR.)

FIJI

Suva

Great Barrier Reef

NEW CALEDONIA (FR.)

Nuku'alofa

NIUE (N.Z.)

TONGA

20°S

TROPIC OF CAPRICORN

PITCAIRN ISLANDS (U.K.)

AUSTRALIA

Great Australian Bight

Darling R.

Canberra

Murray R.

Cook Strait

Wellington

40°S

Tasman Sea

NEW ZEALAND

INDIAN OCEAN

● National capital

N
W   E
S

0   2,000 miles
0   2,000 kilometers
Mercator projection

ANTARCTICA inset:
ATLANTIC OCEAN
Norwegian Claim
British Claim
Argentine Claim
Chilean Claim
Australian Claim
ANTARCTICA
SOUTH POLE
Unclaimed
Australian Claim
PACIFIC OCEAN
New Zealand Claim
French Claim
INDIAN OCEAN
ANTARCTIC CIRCLE
0   1,000 miles
0   1,000 kilometers
Lambert Azimuthal Equal-Area projection

POLITICAL

MAP SKILLS

1 THE GEOGRAPHER'S WORLD How far is Wellington, New Zealand, from Canberra, Australia?

2 PLACES AND REGIONS Which country controls Guam and the Northern Mariana Islands?

3 PLACES AND REGIONS Which country is located on half of a major island?

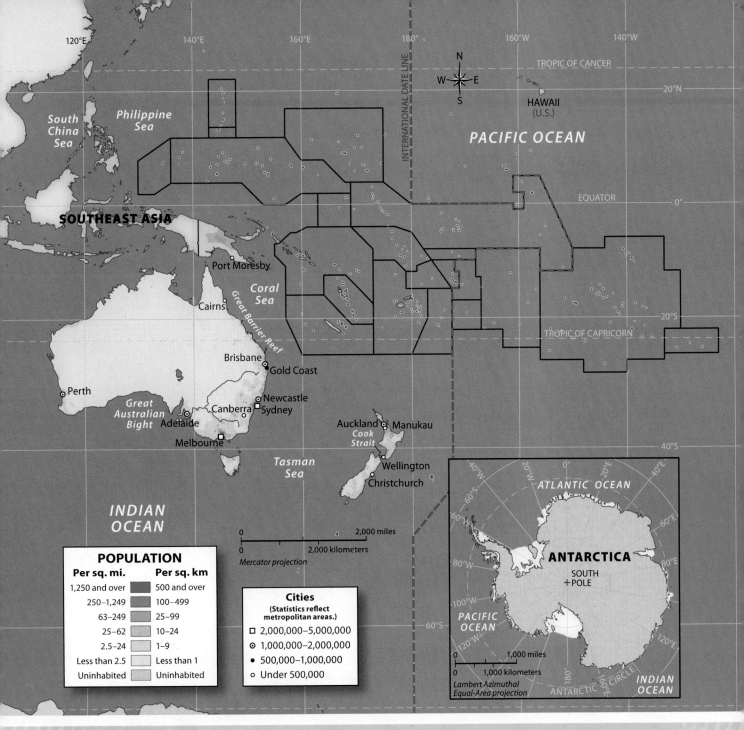

POPULATION DENSITY

POPULATION

Per sq. mi.	Per sq. km
1,250 and over	500 and over
250–1,249	100–499
63–249	25–99
25–62	10–24
2.5–24	1–9
Less than 2.5	Less than 1
Uninhabited	Uninhabited

Cities
(Statistics reflect metropolitan areas.)

□ 2,000,000–5,000,000
◉ 1,000,000–2,000,000
● 500,000–1,000,000
○ Under 500,000

0   2,000 miles
0   2,000 kilometers
Mercator projection

OCEANIA, AUSTRALIA, NEW ZEALAND, AND ANTARCTICA

MAP SKILLS

1 ENVIRONMENT AND SOCIETY Why do you think most people in Australia live along the country's east coast?

2 PLACES AND REGIONS How do population densities compare on New Zealand's South and North Islands?

3 ENVIRONMENT AND SOCIETY Why do you think Antarctica has no permanent human residents?

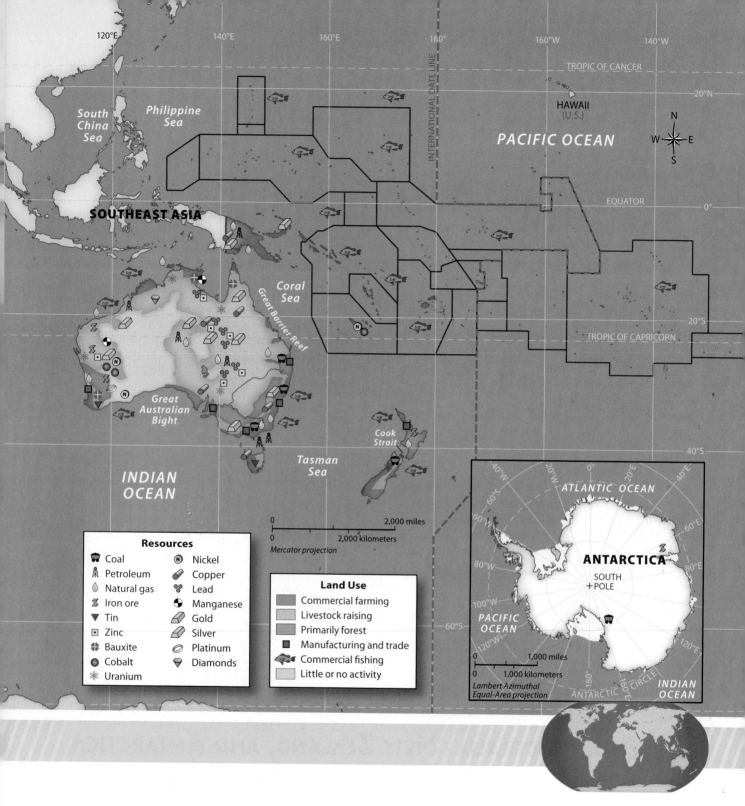

Resources

- 🛢 Coal
- ⚒ Petroleum
- 💧 Natural gas
- ⚒ Iron ore
- ▼ Tin
- ⊡ Zinc
- ⚜ Bauxite
- ● Cobalt
- ❄ Uranium
- Ⓝ Nickel
- ⬦ Copper
- ❦ Lead
- ◑ Manganese
- ◢ Gold
- ◿ Silver
- ⬭ Platinum
- ▽ Diamonds

Land Use

- Commercial farming
- Livestock raising
- Primarily forest
- Manufacturing and trade
- Commercial fishing
- Little or no activity

2,000 miles
2,000 kilometers
Mercator projection

1,000 miles
1,000 kilometers
Lambert Azimuthal Equal-Area projection

ECONOMIC RESOURCES

MAP SKILLS

1 ENVIRONMENT AND SOCIETY How is much of the land in Australia used?

2 PHYSICAL GEOGRAPHY What mineral resources are found in New Zealand?

3 HUMAN GEOGRAPHY What are the primary economic activities in Oceania north of New Zealand?

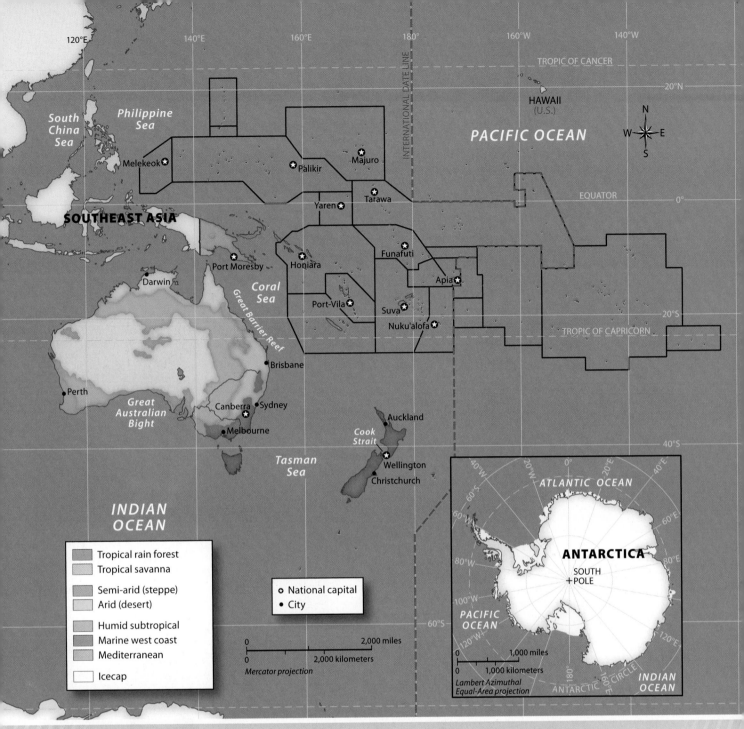

OCEANIA, AUSTRALIA, NEW ZEALAND, AND ANTARCTICA

CLIMATE

MAP SKILLS

1 **PLACES AND REGIONS** What climate zones are found in northern Australia?

2 **PLACES AND REGIONS** What climate zone is found throughout New Zealand?

3 **ENVIRONMENT AND SOCIETY** In what parts of the region could tropical crops be grown?

AUSTRALIA AND NEW ZEALAND

ESSENTIAL QUESTIONS · *How does geography influence the way people live?*
· *Why does conflict develop?* · *What makes a culture unique?*

The Story Matters...

Thousands of years ago, Asian and Pacific people began migrating to Australia and New Zealand. People from different European countries later migrated to the region because of its abundant natural resources. These diverse cultures are woven into the fabric of Australia and New Zealand.

Penny Tweedie/The Image Bank/Getty Images

A baby kangaroo nuzzles a young boy. The name "kangaroo" stems from aboriginal language.

FOLDABLES
Study Organizer

Go to the Foldables® library in the back of your book to make a Foldable® that will help you take notes while reading this chapter.

Australia

New Zealand

AUSTRALIA AND NEW ZEALAND

As the only place on Earth that is a continent and a country, Australia is unique. Located 1,200 miles (1,931 km) southeast of Australia, two large islands make up most of New Zealand's landmass.

Step Into the Place

MAP FOCUS Use the map to answer the following questions.

1 PLACES AND REGIONS
What city is Australia's national capital?

2 THE GEOGRAPHER'S WORLD
What sea separates Australia and New Zealand?

3 THE GEOGRAPHER'S WORLD
What reef lies off the northeastern coast of Australia?

4 CRITICAL THINKING Integrating Visual Information Look at the map. What can you infer from the location of Australia's major cities about where most Australians live?

A

HARBOR VIEW A view from the harbor's bridge provides a magnificent view of Sydney, Australia. Sydney is the country's largest city in population and a major world port and business center.

B

MOUNTAIN SPECTACLE A hiker looks toward Mount Cook, New Zealand's highest mountain. Mount Cook lies in the Southern Alps, a mountain range that runs the length of South Island.

Step Into the Time

TIME LINE Choose at least two events from the time line. For each event, write a paragraph describing some of the positive and negative effects that event had on the region.

c. A.D. 800–1300
Maori arrive in New Zealand from Polynesia

1770 Captain James Cook explores Australian coast

1800

B.C. 48,000 Aboriginal people begin migrating to Australia

1642 Explorer Abel Tasman lands on what is now Tasmania

Australia and New Zealand

SOUTHEAST ASIA

PACIFIC OCEAN

INTERNATIONAL DATE LINE

EQUATOR

Coral
Sea

OCEANIA

10°S

Coral Sea
Islands
(Australia)

20°S

TROPIC OF CAPRICORN

Great Barrier Reef

AUSTRALIA

Brisbane

Norfolk Island
(Australia)

Kermadec Islands
(N.Z.)

30°S

Lake
Eyre

• Perth

Darling R.

Adelaide

Canberra

Sydney

A

Melbourne

Murray R.

INDIAN
OCEAN

Tasmania

Tasman
Sea

North
Island

Auckland
Manukau

Cook
Strait

Lake
Taupo

40°S

B

Wellington

Christchurch

Chatham
Islands
(N.Z.)

South
Island

☼ National capital
• City

0   1,000 miles
0   1,000 kilometers
Mercator projection

NEW
ZEALAND

50°S

120°E

160°E

170°E

180°

1840
Maori grant Great Britain
possession of New Zealand

1900

1947 New Zealand
gains independence

2010 Earthquake in Christchurch,
New Zealand, causes severe damage

2000

1851 Discovery of gold draws thousands of
people from around the world to Australia

1872 Maori lose conflict,
land to European colonists

1951 ANZUS Pact signed

Reading **HELP**DESK

CCSS

Academic Vocabulary

- overall

Content Vocabulary

- **Outback**
- **monolith**
- **Aboriginal**
- **coral reef**
- **hot spring**
- **geyser**
- **drought**
- **marsupial**
- **eucalyptus**

TAKING NOTES: *Key Ideas and Details*

Summarize As you read about the physical geography of Australia and New Zealand, take notes on each section of the lesson using the graphic organizer below.

Heading	Main Idea
The Land	
Climate	
Plants/ Animals	

Lesson 1
Physical Geography

ESSENTIAL QUESTION · *How does geography influence the way people live?*

IT MATTERS BECAUSE
Australia and New Zealand have unique landscapes, plants, and wildlife not found in other parts of the world.

The Land of Australia and New Zealand

GUIDING QUESTION *What physical features make Australia and New Zealand unique?*

Australia is nicknamed "The Land Down Under" because it is located south of, or "under," the Equator. On a map, Australia looks like a large island, because it is not attached to any other landmasses. Australia, however, is a continent. It is the world's smallest continent, but it is the world's sixth-largest country. New Zealand is made up of two islands with a variety of landscapes, ecosystems, and climate zones. Australia and New Zealand also have some of the world's most unusual plants and animals.

Australia's Landforms

It might not seem that Australia is a flat continent based on images of its huge rock formations and mountain ranges. **Overall**, however, Australia has low elevation. This means that although the land is high in some places, most of its surface is low compared to the land on other continents. Australia generally has a dry climate. One-third of Australia is covered by deserts. Another one-third is semiarid.

Geographers divide Australia into three main geographic regions: the Western Plateau, the Central Lowlands, and the Eastern Highlands. In general, the Western Plateau is rocky

and dry, and the Central Lowlands are flat and rugged. The Eastern Highlands have a variety of high and low areas, forests, and fertile farmlands. In addition, coastal lowlands are found all around the continent, particularly in the north and east.

The Western Plateau makes up the western half of Australia. The plateau is rocky, with few water sources other than small salt lakes. Near the center of the continent are the interior highlands. Here, the Musgrave and Macdonnell mountain ranges rise above a huge expanse of flat plateau.

The Central Lowlands are a rough, dry region. Although a system of rivers runs through the rugged land, much of the region is desert. The most rural and isolated parts of the Central Lowlands are commonly called the **Outback**. Survival is a challenge for humans and animals in the Outback. Strong winds cause harsh dust storms across the desert plains. In many parts of the Outback, water is difficult to find. To get the water they need, ranchers must dig wells deep into the sun-baked earth.

One of Australia's most fascinating landforms is found in the Central Lowlands. A massive, solid stone called a **monolith**, measuring 1,100 feet (335 m) tall and 2.2 miles (3.5 km) long, stands alone on the desolate plain. This amazing monolith has two names. Many Australians know it as Ayers Rock. However, the first humans to live in Australia, the **Aboriginal** people, call it *Uluru*. Uluru is sacred to many Aboriginal people. Small caves along the base of the monolith contain ancient Aboriginal paintings and carvings. Today, Uluru is an official World Heritage site located within a protected national park.

The Eastern Highlands, called the Great Dividing Range, run parallel to Australia's east coast. The mountains and valleys of this range were formed by folding and uplifting movements that occurred in the past within Earth's lithosphere. Today, little movement occurs, and earthquakes are rare.

One of the most spectacular and complex ecosystems on Earth is located in the ocean waters just off Australia's northeastern shore. The Great Barrier Reef is a living **coral reef**, a giant community of marine animals called corals.

©John Baker/Corbis

Ayers Rock, or Uluru, in central Australia, is all that is left of a large mountain range that slowly eroded over millions of years.

▶ CRITICAL THINKING

Describing Why is Ayers Rock (Uluru) called a monolith?

Academic Vocabulary

overall as a whole; generally

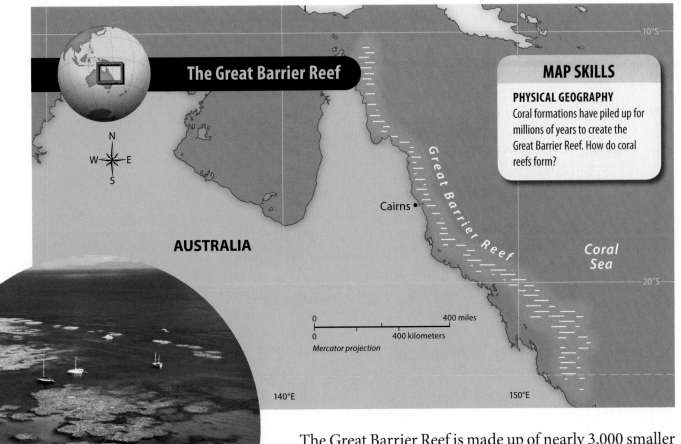

The Great Barrier Reef

MAP SKILLS

PHYSICAL GEOGRAPHY
Coral formations have piled up for millions of years to create the Great Barrier Reef. How do coral reefs form?

10°S

Cairns

Great Barrier Reef

AUSTRALIA

Coral Sea

20°S

0 ——— 400 miles
0 ——— 400 kilometers
Mercator projection

140°E   150°E

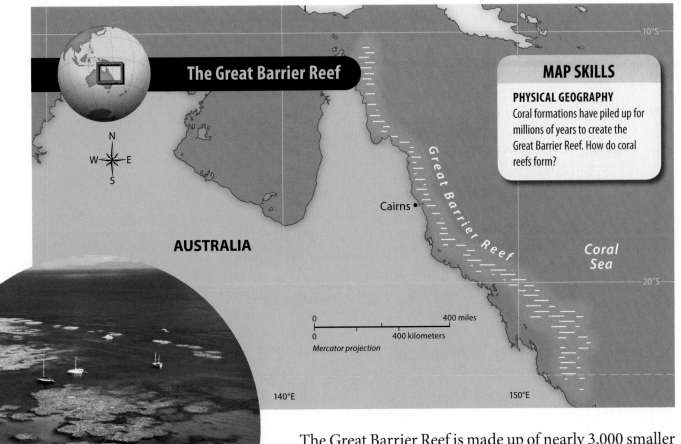
Bait Reef in northeastern Australia

The Great Barrier Reef is made up of nearly 3,000 smaller coral reefs extending for more than 1,250 miles (2,012 km). These reefs were formed over millions of years by natural actions of the biosphere. As corals lived, grew, and died, their hard skeletons built layer upon layer, forming large underwater formations called reefs. Coral reefs are held together by algae and tiny bits of plant and animal matter that become stuck in the reefs.

The Great Barrier Reef is teeming with marine life. Brightly colored fish and shellfish search for food and hide among the corals. Sponges, starfish, and anemones cling to the coral reefs, decorating them like living ornaments.

Just south of Australia's southeastern coast is a unique island called Tasmania. Mountains, valleys, and plateaus cover its surface. Some of the world's wildest, unexplored rain forests grow in Tasmania. These are cool temperate rain forests, which differ from the warm tropical rain forests in the northeastern part of Australia. Tasmania's broad Central Plateau is scattered with more than 4,000 shallow lakes.

Waterways of Australia

Most of Australia has a dry climate. Lack of rain means few large rivers flow across the land. Australia's largest rivers flow from the eastern mountains. The important Murray-Darling river system flows through a dry basin in southeastern Australia. The many rivers in the Murray-Darling system bring so much water to the area that the land is green and fertile.

Many of Australia's people, plants, and animals depend on underground aquifers for their water. Nearly one-third of all water used in Australia comes from underground sources. In a dry land such as Australia, water is a precious resource.

New Zealand's Landforms and Waterways

New Zealand is made up of two main islands called North Island and South Island plus many small islands. These islands are located in the southeastern Pacific Ocean, far from any other landmasses. Geographers believe that more than 60 million years ago, movements within the lithosphere pushed the land that would become New Zealand up out of the ocean. Over time, processes within the lithosphere and hydrosphere shaped the land. The movement of ice and liquid water has carved out basins and eroded rocks into unusual shapes. Volcanic eruptions and earthquakes created low hills, fertile valleys and plains, and rows of sharp mountain peaks.

Today, one-third of New Zealand's lands are protected as national parks and nature reserves. The country is known for its beautiful and unusual landscapes. New Zealand's main islands are similar in that both have many forests, mountains, and waterways. Local climates and other types of landforms, however, show the differences between the two islands.

Many of North Island's landforms were created by volcanic activity. A huge volcanic plateau makes up the center of North Island. Lake Taupo, the largest lake in New Zealand, formed millions of years ago in the massive crater left behind by a devastating volcanic eruption. Lake Taupo is surrounded by fertile plains and valleys. This land is enriched by volcanic soil, which makes good farmland and pasture for grazing. North Island's most productive farm and pasture lands are located in a region called Waikato.

Northeast of Lake Taupo is Rotorua, an area famous for its steaming hot springs, bubbling mud pools, and violent geysers. **Hot springs** are pools of hot water that occur naturally. Hot springs form in rocky areas when rainwater seeps into cracks in Earth's surface. Water is exposed to intense heat, and then bubbles back up to gather in surface pools. **Geysers** are hot springs that sometimes shoot hot water out of the ground. The water in hot springs and geysers is warmed by heat energy from deep within Earth.

Pohutu Geyser in New Zealand's North Island erupts about 20 times each day. The water it releases can reach as high as 100 feet (30 m).

▶ **CRITICAL THINKING**

Describing How has volcanic activity shaped New Zealand's North Island?

The stunning landscape of the Dark Cloud Range stretches across the southern end of New Zealand's South Island.

▶ **CRITICAL THINKING**

Analyzing Why does the southern end of South Island have many deep valleys, fjords, and lakes?

New Zealand's South Island is famous for its spectacular Southern Alps. These towering mountains are higher than the mountains on North Island. The Southern Alps cover hundreds of miles along the western side of South Island.

The southern end of South Island is filled with an incredible variety of landforms and environments. Millions of years ago, the movement of glaciers cut deep valleys into the rocky land. As the glaciers melted, clear, cold lakes formed. Some glacier valleys were so deep they were not completely filled until rising sea levels flooded them with ocean water more than 6,000 years ago. These water-filled landforms are called *fjords*. The fjords of South Island are such fascinating landforms that many are preserved within a huge national park.

The Ring of Fire

New Zealand's location in the southeastern Pacific Ocean puts it within the volcano-studded Ring of Fire. Active volcanoes and the frequent movement of tectonic plates along the Ring of Fire often result in earthquakes. In fact, geographers using special equipment detect more than 15,000 earthquakes in New Zealand every year. Most of these tremors are far too small for humans to feel. At least

100 of these yearly earthquakes, however, are large enough to be noticed by humans. Scientists predict that New Zealand will experience at least one severe earthquake each century.

☑ **READING PROGRESS CHECK**

Citing Text Evidence In what ways are the lands of Australia and New Zealand alike, and in what ways are they different?

Climates of the Region

GUIDING QUESTION *What types of climates and climate zones are found in Australia and New Zealand?*

Australia and New Zealand are located in the Southern Hemisphere, so their seasons are at opposite times of the year from seasons in the United States. For example, June, July, and August are winter months in the region. Summer months are December, January, and February.

Australia's Climates

The climate changes dramatically from one part of Australia to another. The northern third is in a tropical climate zone, and most of the other two-thirds are in a subtropical climate zone. Northern Australia, generally, has a warm, tropical climate. Winter months are dry, and summer months are rainy and hot. Seasonal monsoons can bring damaging winds and heavy rainfall.

Climate in most areas of the western, southern, and eastern regions changes with the seasons. Winters in Queensland, the Northern Territory, and Western Australia are warm and dry. Most rainfall in these regions happens during the long, hot spring and summer seasons. Coastal areas tend to be sunny and dry, with seasonal rains. The climate along the east coast is more humid, and the area receives more rainfall than the rest of the continent.

Much of central and western Australia, such as the Outback, has a desert climate with bands of semiarid steppes to the north, east, and south. These dry areas have extremely hot weather during much of the year. Weather in the Outback can change quickly from one extreme to another. Daytime temperatures in the Outback can reach 122°F (50°C), yet temperatures can fall below freezing at night.

Australia is the driest inhabited continent in the world. Only the icy continent of Antarctica receives less precipitation than Australia. A major problem in the dry areas is that long periods of little or no rain can result in **drought**.

Droughts are common in Australia and threaten the survival of wildlife, livestock, and farm crops. Low water reserves can lead to poor-quality drinking water for humans. Geographers can now predict some droughts by monitoring climate changes caused by El Niño. El Niño occurs every few years when global winds and ocean currents shift, affecting global rainfall patterns.

A cattle driver on a station, or ranch, in Australia's Northern Territory herds Brahman cattle, a breed that is well adapted to the area's climate extremes.

▶ **CRITICAL THINKING**

Analyzing How does climate in inland Australia compare with climate in coastal areas of the country?

New Zealand's Climates

New Zealand's climate is as varied as its land. Climates range from subantarctic in the south to subtropical in the north. Most areas on North Island and South Island have mild, temperate climates, however, with average daily temperatures between 86°F (30°C) and 50°F (10°C). For most of the year, temperatures on North Island are higher than temperatures on South Island. South Island experiences the hottest summers and the coldest winters in New Zealand. Both islands receive plenty of rain, keeping their forests and grasslands healthy.

Although average temperatures are generally mild and extremes of heat and cold are rare, snow does fall in New Zealand. The central region of South Island experiences the coldest winters of any part of the country. Highland zones, such as the Southern Alps, receive the most snowfall.

☑ **READING PROGRESS CHECK**

Describing What problems are caused by drought in Australia? Why are geographers monitoring climate changes there?

Plant and Animal Life

GUIDING QUESTION *What plants and animals are unique to Australia and New Zealand?*

If you have seen pictures of plants and animals from Australia and New Zealand, you already know that many of them are unusual. Why are plants and animals in this region so different from living things in other parts of the world?

The landmasses that would become Australia and New Zealand separated from other lands millions of years ago. The animals and plants living on these lands developed in isolation, separated from living things on other landmasses. Over millions of years, the plants and animals in Australia and New Zealand adapted to live in their own unique environments.

Plants and Animals of Australia

Bandicoots, kangaroos, and koalas are 3 of the 150 native species of marsupials in Australia. A **marsupial** is a type of mammal that raises its young in a pouch on the mother's body. Marsupials vary in size from tiny kangaroo mice to 6-foot-tall, gray kangaroos.

In the driest regions of Australia, water can be difficult to find. Over time, many native animals learned to get water from plants. The leaves, stems, and roots of desert plants contain water. Koalas adapted to eating only one type of plant—the leaves of the eucalyptus tree. **Eucalyptus** trees are native Australian evergreen trees with stiff, pleasant-smelling leaves.

Plants and Animals of New Zealand

Many of New Zealand's trees, ferns, and flowering plants are unique to New Zealand and cannot be found on other continents. Much of the land is covered in hardwood beech tree forests. Alpine plants such as sundew and edelweiss have adapted to the cold, windy, dry environments of South Island's mountain regions.

Lizards such as geckos are common sights in New Zealand. Other reptiles, such as the chevron skink, have become rare. Long ago, New Zealand was home to two types of flightless birds—the kiwi and the moa. The moa was hunted to extinction, but the kiwi was more fortunate. Today, the kiwi is a common sight on both islands and has become a symbol of New Zealand and its people.

☑ **READING PROGRESS CHECK**

Determining Central Ideas Explain why animals and plants native to Australia and New Zealand are different from living things in other parts of the world.

Road signs in rural Australia often warn motorists about kangaroo crossing areas.

▶ **CRITICAL THINKING**
Citing Text Evidence Why is the kangaroo found in Australia and nowhere else?

Include this lesson's information in your Foldable®.

Australia
New Zealand

©Paul A. Souders/Corbis

LESSON 1 REVIEW **CCSS**

Reviewing Vocabulary
1. What kinds of equipment might you need to explore a *coral reef*?

Answering the Guiding Questions
2. *Identifying* List three physical features that make Australia unique and three physical features that make New Zealand unique.

3. *Identifying* Identify one of the two locations in New Zealand that has a cold, snowy climate in winter.

4. *Citing Text Evidence* What native animal has become a well-known symbol of New Zealand and its people?

5. *Informative/Explanatory Writing* Write a paragraph explaining why either Australia or New Zealand would be an interesting place to visit. Use information from the lesson to describe landscapes, ecosystems, climates, and other features of either Australia or New Zealand.

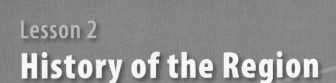

networks

There's More Online!

☑ **IMAGES** Aboriginal Rock Paintings

☑ **MAP** History of Australia

☑ **TIME LINE** Australian Gold Rush

☑ **VIDEO**

Reading **HELP**DESK

Academic Vocabulary

- **unify**

Content Vocabulary

- **boomerang**
- **dingo**
- *tikanga*
- *kapahaka*
- **station**
- **introduced species**
- **dominion**

TAKING NOTES: *Key Ideas and Details*

Determine Cause and Effect As you read about the histories of Australia and New Zealand, use a graphic organizer like this one to take notes about the causes and effects of human settlement in the region.

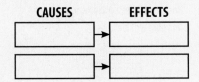

CAUSES → EFFECTS

Lesson 2
History of the Region

ESSENTIAL QUESTION · *Why does conflict develop?*

IT MATTERS BECAUSE
Australia and New Zealand have made many advances in democratic government and have close ties to the United States.

First People

GUIDING QUESTION *How and when did the first humans settle in Australia and New Zealand?*

Ancient tools, cave paintings, rock art, and fossilized human remains provide clues that humans traveled to this region from other places on Earth. There is still some debate, however, about where the early people came from and when they first arrived in the region. Most scientists agree that the lands of Australia and New Zealand existed for millions of years before the first human settlers arrived.

Aboriginals of Australia

Fossil evidence shows that humans began migrating to Australia at least 50,000 years ago. At that time, much of Earth's surface water was frozen, and sea levels were hundreds of feet lower than they are today. Areas that are underneath the ocean today were exposed above the surface of the water. This created land bridges and peninsulas over which groups of people left New Guinea and walked across to Australia. Other migrating groups from the Asian mainland may have traveled longer distances by boat. These early people journeyed far and wide across the land, even to the island of Tasmania. In time, humans settled the entire Australian continent.

The first people of Australia are known as Aboriginal people. They generally lived a hunter-gatherer lifestyle. These early people survived by gathering fruit, roots, and other

(l to r) ©Bill Bachman/Alamy; ©Stapleton Collection/Corbis; Frans Lemmens/The Image Bank/Getty Images; SuperStock/Getty Images; The Print Collector/Alamy

plant parts for food and by hunting animals. They developed a flat, bent, wooden weapon called a **boomerang**. Hunters threw the L-shaped boomerang to stun their prey. If the boomerang missed, it curved and sailed back to the hunter.

Early Settlements

Some hunter-gatherers were nomadic, moving from place to place, following animals or searching for water sources. Nomadic people did not practice farming. The only domesticated animals they owned were dingoes. **Dingoes** are a species of domestic dog first brought to Australia from Asia about 4,000 years ago. Today, wild dingoes roam free in the Outback and other parts of Australia. There is controversy regarding the animal's suitability as a pet.

Other Aboriginal people settled permanently in one location. Some settled in the rain forests of the northeast. Some made their homes in the mountains of the Great Dividing Range. Others traveled as far as the humid lands of the southeast and the island of Tasmania. Some even learned to live in the harsh, dry lands of the Outback. Eventually, Aboriginal settlers began farming the land.

Aboriginal Culture

Traditional Aboriginal culture takes many forms. Aboriginal peoples living in different parts of Australia developed their own languages, religions, traditions, and ways of life. It was difficult for separate tribes of native people to communicate with one another, because about 400 different languages were spoken across Australia. Yet, many of Australia's native people share a number of common beliefs and cultural traditions. Aboriginal culture is closely connected to the natural world. Australia's native cultures have traditional beliefs about the creation of Earth and of plants and animals. They use song, dance, poetry, drama, storytelling, and visual arts to retell the creation story known as "the Dreaming" or "the Dreamtime." The creation story tells how the Spirit Ancestors created the world around them, the universe, and the laws of life, death, and society.

The concept of Dreaming is central to many of Australia's native cultures' social structures and belief systems. Music, dance, storytelling, and other art forms remain an important part of Aboriginal culture.

An Anangu woman prepares wood for carving. In addition to wood carving, Australia's aboriginal artists work in media such as painting on leaves, rock carving, fabric printing, and sandpainting. The Anangu people live in an area extending from Uluru in Northern Territory to the Nullarbor Plain in southwestern Australia.

▶ **CRITICAL THINKING**
Determining Central Ideas What do the Aborigines see as the purpose of their art forms?

©Bill Bachman/Alamy

The Maori of New Zealand

The first humans to live on the islands of New Zealand were the Maori. The Maori people came to New Zealand much later than the Aboriginal people came to Australia. Historians believe that sometime between A.D. 800 and 1300, humans began traveling in canoes from Polynesian islands such as Tahiti to New Zealand. The Maori built villages and lived in tribal groups led by chiefs. Tribes traded and went to war with other Maori tribes. Nearly all early Maori settlements were located on North Island.

The Maori way of life had minimal impact on the environment. They fished, gathered plants, and hunted wild animals for food. Huge, flightless birds called moa were easy prey and became an important part of the Maori diet. Eventually, the Maori hunted all 10 species of moa to extinction. In time, the Maori introduced food crops such as taro and yams. Boating and diving were important parts of Maori life. Canoes carved from the trunks of massive trees were used for transportation and also as vehicles for warfare.

Maori Culture

The Maori have a spiritual belief system based on the concept that all life in the universe is connected. At the heart of Maori culture is *tikanga*. **Tikanga** are traditional Maori customs and traditions passed down through generations. Maori tradition says that tikanga come from *tika*, the "things that are true," which began with all creation at the dawn of time. According to tikanga, the past is always in front of an individual, there to teach and guide that person. The future is behind the individual, hidden and unknown. Tikanga is part of everyday Maori life, from building homes and preparing food and medicine to social customs and arts, such as *kapahaka*. **Kapahaka** is a traditional Maori art form combining music, dance, singing, and facial expressions.

Maori warriors sail in a war canoe below one of their lookout points on the North Island coast. To get to their battlefields, the Maori built large war canoes called *waka taua*. Each vessel held about 100 people and was up to 130 feet (40 m) in length. Viewed as sacred, the war canoes were elaborately carved with images of deities and ancestors.

▶ **CRITICAL THINKING**

Analyzing Why do you think the early Maori carved sacred images on their war canoes?

©Stapleton Collection/Corbis

One of the most important parts of any culture is language. Before Europeans arrived in New Zealand, the Maori people spoke different dialects of the same language. When New Zealand was colonized by the British, many Maori began speaking English. Over time, fewer and fewer people spoke the traditional Maori language, even in their homes. By the mid-1900s, the language had become so rare it was in danger of dying out. The efforts of Maori leaders and activists brought the Maori language back from the brink of extinction. During the 1970s and 1980s, the Maori language was reborn, as language recovery programs and schools across New Zealand taught younger generations to speak Maori. In 1987 Maori became an official language of New Zealand.

A unique part of Maori culture is the sacred practice of facial tattooing. Maori tattoos are permanent decorations on the body made by cutting designs into the skin, then rubbing black soot into the cuts. Traditionally, Maori boys were tattooed during puberty as a rite of passage into manhood.

Each individual's tattoo was unique, showing his ancestry, status in the tribe, military rank, profession, and family relationships. Maori tattooing is still practiced in New Zealand as an expression of cultural pride and identity. Other Maori art forms such as weaving, painting, and wood carving are also important parts of past and present Maori culture.

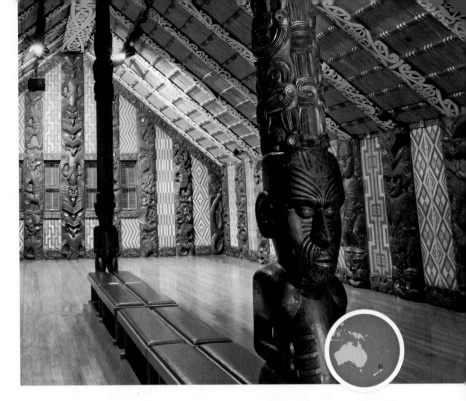

A Maori meeting house in New Zealand's Waitangi area displays detailed wooden carvings, all done by hand. Maori meeting houses have long served as centers for important community events and ceremonies.

▶ CRITICAL THINKING

Analyzing How do Maori meeting houses reflect Maori beliefs about their society?

☑ READING PROGRESS CHECK

Describing How did the early hunter-gatherers live?

Colonial Times

GUIDING QUESTION *What happened when Europeans came to Australia and New Zealand?*

Australia and New Zealand were colonized by the British. However, the British were not the first Europeans to visit the region. Explorers from other European countries had been sailing around Australia and landing on its shores for hundreds of years before the British arrived.

Europeans Come to the Region

During the 1600s, 1700s, and 1800s, Dutch, Spanish, French, and Portuguese explorers visited Oceania. Some of them mapped Australia's coastline. Some even went ashore to explore coastal areas and search for supplies. In 1642 the Dutch East India Company sent Abel Tasman on a mission to sail around the Australian continent. During this voyage of discovery, Tasman circled the island of Tasmania and then sighted the coast of New Zealand. More than a century would pass, however, before Europeans started colonizing the region.

British Australia

Perhaps the most well-known British explorer was the sailor Captain James Cook. He carried out three voyages in the 1760s and 1770s. Following Cook's explorations, the British government prepared to send settlers to the wild, unexplored lands of Australia. Most of the first colonists sent to Australia did not go by choice. In 1788 a group of 11 British ships, known as the First Fleet, landed on Australia's east coast. The crowded ships carried 778 convicted criminals from the British Isles. The First Fleet also included 250 soldiers and government officials. This was the first shipment of about 160,000 convicts sent to Australia during the next 80 years, due to a lack of space in England's prisons. Living conditions were terrible, and punishments were harsh for convicts held in Australia's cruel, filthy prisons. For many years, Australia was known to most of the world as a prison colony.

On April 29, 1770, Captain James Cook made his first landing in Australia at Botany Bay, near present-day Sydney. This imagined view of the landing was painted by E. Phillips Fox, an Australian artist of the early 1900s.

Beginning in the late 1700s, a settlement began to form in Sydney. Colonists were slow to come to the area, but in time, Sydney grew into a busy center of trade and industry. Settling the inland areas of Australia did not happen for many years. Only a few people other than escaping criminals dared to venture into the rough Outback. By the 1880s, however, most of the continent had been explored by Europeans.

In 1851 an English prospector found gold near Bathurst, New South Wales. Soon after, thousands of people from all over Australia were camped in the area, digging for gold. Gold was discovered in other parts of Australia, and word spread across the globe. People

from around the world flocked to Australia at a rate of 90,000 per year, all hoping to find a bounty of gold. Prospectors came from as far away as England, Ireland, China, and the United States.

Australia began to grow for other reasons, as well. When resources such as coal, tin, and copper were discovered, workers came for mining jobs. Business owners started and built shops and hotels wherever towns sprang up, and there were people to spend money. Small towns grew into cities. Farmers planted crops in Australia's most fertile areas. Ranchers brought sheep and cattle from overseas, and Australia's ranching industry was born. Vast ranches called **stations** covered millions of acres in the Outback and other areas. Today, millions of sheep and cattle live on ranches all across Australia.

A gold prospector sits outside a hut he built in the settlement of Gippsland, Australia. During the late 1800s, gold discoveries in southeastern Australia drew many prospectors to the area.

▶ CRITICAL THINKING

Describing How did the discovery of gold and other mineral resources contribute to Australia's development?

Challenges and Conflict

As humans from other parts of the world moved to Australia, they brought animals with them. Ranchers brought dogs for guarding and herding sheep. Wealthy landowners imported European rabbits to hunt for sport. In an effort to rid sugarcane fields of a destructive beetle, farmers brought in huge, poisonous cane toads. These and other nonnative animal species caused major problems. Rabbits multiplied quickly to a population of 1 billion. They ate so much grass and so many wildflowers that entire areas were left bare. The cane toads also multiplied, crowding out and killing native animal species. Animals that are not native to an area but are brought from other places are called **introduced species**. It is impossible to estimate how much damage has been done to Australia's environment by introduced species. Some introduced species are now under control, but others continue to cause problems for humans and native animals.

British settlers built their homes and farms all over Australia, often forcing native Aboriginal people off their land. Thousands of Aboriginal families and tribes were forced to leave lands where their ancestors had lived for generations.

In both world wars, Australia sent its soldiers to foreign battlefields in support of the British Empire. The heroism and sacrifices of Australian soldiers in these global conflicts attracted the attention of people in other parts of the world.

▶ **CRITICAL THINKING**

Analyzing How did involvement in both world wars change the way Australians viewed themselves and their country?

Academic Vocabulary

unify to unite; to join together; to make into a unit or a whole

European diseases and violence steadily reduced the Aboriginal population. The survivors had no choice but to live on rugged lands that European settlers did not want.

British New Zealand

Captain Cook explored the islands of New Zealand during the early 1770s. Cook reported to the British government that the fertile islands had many valuable natural resources and would be good places to colonize. Soon, British colonists and British, American, and French traders and whalers built settlements on North Island. At first, most relations between the Maori and foreign settlers were peaceful. In 1840 the British government, ruled by Queen Victoria, convinced Maori leaders to sign the Treaty of Waitangi. This treaty gave legal ownership and control of New Zealand to Great Britain, but it guaranteed protection and certain land rights to the Maori.

As Europeans continued to arrive in New Zealand, the Maori saw more and more of their land taken by foreign settlers. Maori society and ways of life weakened when British settlers brought new methods of farming and other features of European culture. Conflict between the Maori and British continued sporadically, until 1872 when many Maori were killed and they lost most of their land to the British.

As was happening in Australia, businesses, industries, farms, and sheep ranches were built across New Zealand. Sheep ranching changed the land, as native scrubland and forests were cleared to make pastures for livestock. Introduced species, such as rabbits, goats, pigs, deer, rodents, and feral cats, began destroying natural habitats. They killed many native animals and threatened the survival of entire species.

Independent Countries

In 1901 the six British colonies set up in Australia took action to **unify** as a federation. This action formed a political alliance between New South Wales, Queensland, Northern Territory, Western Australia, South Australia, and Tasmania. The former colonies set up the Commonwealth of Australia. The new country was a

dominion, a largely self-governing country within the British Empire. Like Canada, Australia had a form of government that blended a U.S.-style federal system with a British-style parliamentary democracy.

Throughout the 1800s, New Zealand residents pushed independence from Great Britain. The 1852 New Zealand Constitution Act recognized local governments in the six provinces, but New Zealand was still a long way from independence. In 1907 the British government named New Zealand an independent dominion with a British-style parliamentary democracy. Even before independence, New Zealand had made a number of political advances. In 1893 it became the first country in the world to legally recognize women's right to vote.

The Region in Contemporary Times

Australia and New Zealand were pulled into World War I and World War II through their ties to Great Britain. During World War II, Australian, New Zealand, and U.S. soldiers fought together in the Pacific region. This alliance created closer ties among the three countries.

After World War II, Australia became completely independent. Australia loosened ties with Britain and established closer ties to the United States and Asia. In 1951 Australia, New Zealand, and the United States signed a mutual security treaty called the ANZUS Pact. The treaty was meant to guarantee protection and cooperation among the three countries in case of military threats in the Pacific region.

In recent years, a huge increase in Asian immigration has led to more diversity in Australia and New Zealand. Today more people of Asian background live in New Zealand than native Maori. The governments of both countries have also continued to address Aboriginal and Maori rights and social concerns. The native people of Australia and New Zealand and their supporters continue to work for justice and equal rights under the law.

☑ **READING PROGRESS CHECK**

Determining Central Ideas How and when did Australia and New Zealand become independent nations?

Include this lesson's information in your Foldable®.

LESSON 2 REVIEW

Reviewing Vocabulary

1. Are *dingoes* native to Australia?

Answering the Guiding Questions

2. ***Identifying*** For whom was the island of Tasmania named?

3. ***Identifying*** Name one native species and one introduced species in Australia.

4. ***Determining Central Ideas*** In what ways were the colonization of Australia and the colonization of New Zealand alike?

5. ***Narrative Writing*** Imagine you are a young Australian Aboriginal or New Zealand Maori living during the time the first Europeans came to your homeland. Write a few paragraphs telling how you feel about the arrival of these foreign settlers. Whenever possible, include details from the lesson in your narrative.

Reading **HELP**DESK · CCSS

Academic Vocabulary

- **controversy**

Content Vocabulary

- **bush**
- **didgeridoo**
- **action song**
- **geothermal energy**
- **kiwifruit**
- **lawsuit**

TAKING NOTES: *Key Ideas and Details*

Summarize As you read the lesson, use a graphic organizer like the one below to write a short summary about each of the topics.

Topic	Australia	New Zealand
Natural Resources		
Economy		

Lesson 3
Life in Australia and New Zealand

ESSENTIAL QUESTION · *What makes a culture unique?*

IT MATTERS BECAUSE
The people of Australia and New Zealand are working to blend diverse populations successfully.

Life in the Region

GUIDING QUESTION *What is it like to live in Australia and New Zealand?*

European culture exercises the most influence in Australia and New Zealand, but indigenous cultures also play an important role. In recent years, Asian influences have increased in the region.

The People of the Region

Australia and New Zealand are multicultural lands. They have diverse human populations where different cultures, languages, and lifestyles are mixed together. New Zealand has had a diverse population for much of its history. Today, New Zealand's population is about 57 percent European, 12 percent Asian and Pacific Islander, and 8 percent Maori. Other groups account for the rest. Australia's population is much less diverse. About 92 percent of Australians are of European descent, 7 percent are Asian, and 1 percent are Aboriginal and other groups.

Religion and Culture

Christianity is the most common religion in the region. Also practiced in the region are Buddhism, Islam, Hinduism, and native religions. In addition, about 30 percent of the people in the region describe themselves as "nonreligious" or did not state a religious affiliation.

English is the official language in Australia. New Zealand has three official languages: English, Maori, and New Zealand sign language. The sign language, the main language used for communicating by members of the deaf community, became an official language in 2006.

The lifestyles in Australia and New Zealand are similar to modern American and British lifestyles. The residents drive cars and use public transportation. In their free time, Australians and New Zealanders shop, go out to eat, watch television, and go to movies. They keep pets such as cats and dogs. Outdoor activities, such as hiking, biking, running, boating, surfing, and swimming, are popular. Watching and playing sports such as football (soccer) and rugby are popular pastimes.

Urban Life

Human populations are unevenly distributed in the region. The vast majority of Australians and New Zealanders live in urban areas. A population map of Australia shows something interesting: The highest populations are concentrated in small land areas, while the smallest populations are scattered throughout the largest land areas. Approximately 89 percent of Australia's people live in cities and suburbs. Most New Zealanders, about 87 percent of the population, live in cities and suburbs.

Australia's largest cities are Sydney, Melbourne, Brisbane, and Perth, all with populations of more than 1 million. New Zealand's largest cities are Auckland, Christchurch, and Wellington. Life in these cities is busy. People who live in cities face everyday challenges—noise, traffic, urbanization and crowding, rising housing prices, pollution, and crime. Cities also offer an endless variety of culture, recreation, shops, restaurants, and entertainment. Urban residents must balance the challenges of city life with its many benefits.

A player from Australia's Queensland Reds charges forward during a rugby match between the Reds and New Zealand's Canterbury Crusaders at Suncorp Stadium in Brisbane, Australia. Sports like football (soccer) and rugby are popular in Australia and New Zealand.

Rural Life

Life in Australia's rural areas moves at a slower pace. Many individuals and families live alone on huge sheep or cattle stations. These people tend to be isolated, far from towns and other people. Farm life can be hard, with work from sunrise to nighttime. The term **bush** means any large, undeveloped area where few people live. The phrase "in the bush" can refer to any location that is wild, unsettled, and rough, such as the Australian Outback.

Bradley Kanaris/Getty Images Sport/Getty Images

A student on a remote sheep station in rural Australia takes part in a School of the Air lesson on her home computer. She can see and talk to her teacher by way of a live video cam hookup made possible by a satellite system.

▶ **CRITICAL THINKING**

Analyzing How has electronic technology improved the education of Australian students living in remote areas?

Rural areas in New Zealand are not as remote as those in Australia, because New Zealand has a much smaller land area. Lush pasture lands can feed herds of sheep and cattle on fewer acres than the dry Australian Outback. As a result, New Zealand farms are located closer together, and farm families have more contact with friends and neighbors. Many people in rural areas live near small towns, where they can shop and interact with others. In recent decades, the populations of New Zealand's small towns have been shrinking as more and more rural people move to cities.

Australian English

Australians are famous for their use of nicknames and slang. Australian English, called *Strine*, has a unique vocabulary made up of Aboriginal words, terms used by early settlers, and slang created by modern Australians. Common slang in Australia uses rhymes and word substitutions. For example, "frog and toad" is slang for "road," and "steak and kidney" is a slang nickname for the city of Sydney. The Australian people are nicknamed "Aussies." New Zealanders also have a common nickname. During World War I, New Zealand soldiers and Australian soldiers served together. The Australian soldiers nicknamed the New Zealanders "Kiwis." The nickname is still used today, even by New Zealanders.

Education

The people of Australia and New Zealand take pride in their educational systems. Both countries are home to well-respected universities that rank as some of the best schools in the world.

Australia is a huge continent with many isolated communities. Many students, especially in the Outback, use modern methods of communication to receive and turn in their lessons. Beginning in the 1950s, classes were conducted via shortwave radio, with students having direct contact with a teacher in town. Previously, students relied on mail service to deliver assignments. Today, the Internet provides quicker and more reliable delivery.

Aboriginal and Maori Culture

Australia and New Zealand have experienced revivals in Aboriginal and Maori cultures. In Australia, some Aboriginal storytellers still use oral tradition to pass down history and myths from one generation to the next. Traditional customs and tools such as the boomerang are still part of daily life. Boomerangs have been used for centuries as tools for hunting, as toys, and as weapons for hand-to-hand combat. Today, boomerangs are used for recreation and contests of skill. Another Aboriginal artifact still used today is a musical instrument called the didgeridoo. The **didgeridoo** is a long wood or bamboo tube that creates an unusual vibrating sound when the player breathes into one end. Aboriginal instruments such as the didgeridoo are part of modern Australian culture, helping to keep traditional music alive.

As part of the modern movement to restore Maori culture in New Zealand, performers created **action songs**. These performances combined body movement with music and singing, often with lyrics that celebrated Maori history and culture.

☑ **READING PROGRESS CHECK**

Identifying Points of View Based on information in the lesson, decide if the nicknames "Aussie" and "Kiwi" are offensive to Australians and New Zealanders, respectively.

Maori young people in the city of Christchurch, New Zealand, perform an action song. Both men and women perform action songs, using tight arm motions with straight, vibrating hands.

▶ **CRITICAL THINKING**
Determining Central Ideas Why do you think the performance of action songs is important to Maori people today?

Greg Balfour Evans/Alamy

Natural Resources and Economies

GUIDING QUESTION *What resources are important to the economies of Australia and New Zealand?*

Australia's Natural Resources

Australia is rich in valuable natural resources. Coal mining is a major industry in eastern and northwestern Australia and also in Tasmania. Iron ore is plentiful in the northwest. Gold discovered in Western Australia in the 1850s spurred a gold rush. Precious metals, including gold and silver, are still mined today. Large offshore oil and natural gas reserves are located in northern Australia and also in the Bass Strait between Australia and Tasmania. Australia exports some of its oil and natural gas.

The fertile farmland in southeastern Australia and other areas is one of the country's valuable natural resources. Wool, food crops, and other agricultural products are raised in many different parts of Australia. Timber and other products come from various species of trees growing across Australia. For example, eucalyptus trees are harvested for their wood, oil, resin, and leaves.

Australia's Economy

Australia's economy relies heavily on exports of its many natural resources. Coal, iron ore, and gold are Australia's three leading exports. Australia's economy also depends on its exports of meat, wool, wheat, and manufactured goods to countries all over the world. Australia's chief trading partners are China, Japan, South Korea, and the United States. Australia's manufacturing sector is not as strong as is the manufacturing sector in several East Asian nations, but Australia has enjoyed robust economic growth in recent years. Unemployment is relatively low. Australia ranks twelfth in world gas reserves and eleventh in world gas exports. Tourism continues to be a vital industry. More than 5 million people visit Australia each year, bringing revenue to local businesses and employing thousands of Australians. During the past few decades, Australia's film industry has grown to international status.

New Zealand's Natural Resources

New Zealand enjoys one important benefit of its location along the Ring of Fire: easy access to geothermal energy. **Geothermal energy** is naturally occurring heat energy produced by extremely hot liquid rock in Earth's upper mantle. As magma rises up through cracks or holes in Earth's crust, it heats the rock and water within the crust. Humans reach this heat and hot water by digging and drilling. The heat energy is used to warm homes, to provide hot water,

An Australian worker empties recently shorn wool into a bin for cleaning. Traditionally, Australian wool sold mostly in Europe and North America. Today, Australian wool suppliers rely increasingly on growing markets in China and other countries of Asia.

Identifying What countries are Australia's major trading partners?

Ian Waldie/Getty Images News/Getty Images

and to generate power. Another benefit of geothermal energy is that it is clean and does not pollute the environment. Hydroelectric power, derived from the energy of moving water, and windmills are other kinds of nonpolluting renewable resources.

Farmland is one of New Zealand's most valuable resources. Almost one-half of all land on the islands is used for farming and livestock grazing. The grass growing on steep hillsides tends to dry out as rainwater drains down to lower pastures. Low-lying lands can become too soaked with rain, which is not good for most crops. The most productive farmland in New Zealand is in places where the soil receives enough rain but is well drained. The dark soil of North Island's volcanic plateau also holds ribbons of mineral deposits. New Zealand's other valuable natural resources include coal, iron ore, natural gas, gold, timber, and limestone.

New Zealand's Economy

During the past two decades, New Zealand's economy has gone through a major transformation. What was once a farm-based economy has become an industrialized, free-market economy. New Zealand's government has plans to continue to increase production of wood and paper products, food products, machinery, and textiles. Agriculture is still a key part of the economy, though. Products such as beef, lamb, fish, wool, wheat, flowering plants, vegetables, and fruits are exported and shipped around the world. One notable food export is the **kiwifruit**, a type of gooseberry fruit that originated in East Asia but has become a symbol of New Zealand.

New Zealand is not as wealthy as Australia, but it has a strong economy: It is forty-eighth in the world in per capita gross domestic product (GDP). New Zealand's chief trading partners are Australia, China, and the United States. Because New Zealand depends on export income, low export demand can badly damage its economy. As in Australia, tourism is a major industry. After the release of the hugely popular *Lord of the Rings* films in the early 2000s, tourists from all over the world flocked to New Zealand. New Zealand's film industry continues to grow, producing top-grossing films that are viewed worldwide.

New Zealand produces specialized food products that are exported around the world. (Top) A fruit grower inspects his kiwifruit grown on supported vines. (Bottom) A factory worker prepares pieces of New Zealand's famous Egmont cheese for wrapping.

☑ **READING PROGRESS CHECK**

Distinguishing Fact From Opinion Is the following statement a fact or an opinion? *Tourism is an important industry in Australia and New Zealand.*

Activists protest the development of a gas project in a coastal area of Western Australia that is rich in petroleum and natural gas. The protestors claim the building of a gas plant and a port will damage offshore reefs and the fossil remains of prehistoric animals.

▶ **CRITICAL THINKING**
Citing Text Evidence What other environmental challenges do Australia and New Zealand face?

Academic Vocabulary

controversy a dispute; a discussion involving opposing views

Current Issues

GUIDING QUESTION *What challenges do the people of Australia and New Zealand face?*

Australia and New Zealand face challenges resulting from their locations, populations, climates, and physical geography. Just as the landforms and wildlife of the region are unique, so are the issues and problems facing Australia and New Zealand.

Indigenous Rights

The concern over Aboriginal people's rights has been an ongoing issue in Australia for well over a century. Aboriginal activist groups in Australia have filed several major lawsuits over land rights and environmental issues. A **lawsuit** is a legal case that is brought before a court of law. Many Aboriginal lawsuits have sparked **controversy** over the rights of the Aboriginal people versus the rights of big businesses to use the land and its resources. Some cases are still being decided. Similar human rights issues continue in New Zealand between New Zealanders of European descent and native Maori.

The Maori own only about 5 percent of the land in New Zealand. They petitioned the government to return their lost land. The government could not return land to the Maori without hurting the people who were living on it. The government agreed to pay the Maori for lost land and lost fishing rights. Payments continue, but they have been slow in coming.

Protecting Natural Resources

Protecting the environment and natural resources is important to many Australians and New Zealanders. Recent environmental issues in Australia are drought and limited water supplies, bushfires, and threats to the survival of the Great Barrier Reef. All of these are affected by global warming. For example, the Great Barrier Reef is affected as water temperatures in the oceans rise, resulting in the death of organisms that need a cooler climate. The corals are also sensitive to climate change and water pollution. Parts of the Great Barrier Reef have already died. Scientists are concerned that if global warming continues, this entire massive reef system will be lost.

In recent decades, many geothermal hot springs and geysers have disappeared in New Zealand, but many of the remaining hot springs and geysers are located in protected areas. Geographers believe human activities such as drilling for hot water and building power stations have destroyed many natural wonders. Such environmental issues are of great concern to Australians and New Zealanders.

Other Issues

Australia and New Zealand have low birthrates. They also have low death rates. Low birthrates result from families having fewer children. Low death rates result from Australia's and New Zealand's high life expectancies. While such rates are generally beneficial for a country, the combination produces a gradually aging population. It also creates a need for more workers to support the older population. People immigrating to Australia and New Zealand are filling some of these positions, which helps meet the need for care providers. Immigration is also changing the region's ethnic makeup.

Australia and New Zealand face many other issues and challenges. Both countries are affected by the overall health of the planet. Their economies depend upon trade with other countries. Their industries could not survive without strong economic relations with one another and without trade with other countries.

☑ **READING PROGRESS CHECK**

Describing In your own words, explain why survival of the Great Barrier Reef is threatened.

FOLDABLES
Study Organizer

Include this lesson's information in your Foldable®.

LESSON 3 REVIEW **CCSS**

Reviewing Vocabulary

1. What is the *bush* like?

Answering the Guiding Questions

2. *Identifying* What type of resource is *geothermal energy*?

3. *Describing* How is daily life in Australia and New Zealand similar to life in the United States?

4. *Identifying* Give three examples of resources found in Australia and three examples of resources found in New Zealand.

5. *Argument Writing* Think about the issues facing Australians and New Zealanders today. Write a paragraph in the form of an argument describing which issue you believe is the most important and why.

UNFRIENDLY Invaders

Invasive species are animals, plants, and infectious organisms that take over the natural environment of other species. Invasive species often harm the environment and cause the native species to decline in number. They can also harm the health of humans.

Cane Toads Cane toads were brought to Australia in 1935 to control beetles that attacked sugarcane crops. But cane toads contain toxins that poison many native animals that eat them. Cane toads also eat large numbers of honeybees that pollinate plants and crops. The toads carry diseases that can be passed on to other frogs and to fish.

> " **Wild rabbits are considered Australia's most widespread and destructive pest.** "

A Vast Number Why are cane toads so plentiful? Twice every year, female cane toads produce 8,000 to 35,000 eggs. That's many more eggs than the average frog lays. These eggs quickly hatch and form a school of tiny, black tadpoles. Many of the tadpoles do not survive. Those that survive can live for 10 to 40 years.

Animals That Prey Australia has many wild foxes and feral cats that prey on other animals, and rabbits that devastate vegetation. Their numbers are so great that the government says it cannot eliminate them. The goal is to reduce the damage they cause.

$4 Billion Per Year Invasive alien plants present problems for the economy. The cost of the damage and attempts to control the plants amounts to $4 billion per year.

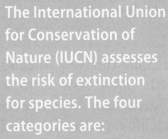

Did You Know ?

The International Union for Conservation of Nature (IUCN) assesses the risk of extinction for species. The four categories are:
- **Extinct**—Species has died out and no longer exists
- **Critically endangered**—Species faces a high risk of becoming extinct
- **Endangered**—Species faces a risk of extinction
- **Vulnerable**—Species is likely to become endangered unless circumstances improve

A worker weighs a cane toad at a collection point in Cairns , Australia. ▶

Brian Cassey/AP Images

THERE'S MORE ONLINE

HEAR cane toads • *SEE* invasive plant species • *WATCH* an animation on invasive species

These numbers and statistics can help you learn about the invasive species of Australia.

TWENTY-TWO
EXTINCT

There are 22 extinct mammals in Australia. *Extinct* means no more are left. Many other animals are in danger of dying out. Australia has more endangered species than any other continent.

The red fox, the feral cat, and the rabbit are probably responsible for the loss of 20 of the 22 extinct marsupials and rodents in Australia. Other animals that have become major agricultural and environmental problems are wild pigs, goats, and deer.

12 pounds

In many parts of Australia, all native mammals weighing up to 12 pounds (5.4 kilograms) are extinct. Nine species of mammals exist only on Australian islands that have no cat or fox population.

1974 Epidemic

Phytophthora root rot is an invasive disease that threatens many important crops and plant species in Australia. For example, a 1974 root rot epidemic destroyed more than one-half of all the avocado trees in eastern Australia.

3,300-MILE FENCE

The dingo looks like a dog and is the largest carnivorous animal in Australia. More than 100 years ago, Australians built a long fence to keep the dingo away from sheep flocks and other animals. At 3,300 miles (5,311 km) long, the dingo fence is the world's longest fence. Rabbit-proof fences were built to protect Western Australian crops and pasture lands.

No. 1

Wild rabbits are considered Australia's most destructive pest. European rabbits were introduced in Australia more than 150 years ago. The rabbit population grew huge because few animals prey on the rabbits. A virus in the 1950s killed many rabbits, but as rabbits built up an immunity to the virus, the rabbit population began to grow again. Today, millions of wild rabbits live in all parts of Australia.

27,000

In the 1800s, a weed called the prickly pear overran large areas, forcing many farmers off their land. Today, more than 27,000 invasive alien plants grow in Australia.

(t) ©Steve Parish Publishing/Steve Parish Publishing/Corbis; (cl) Nikolas Gregorkiewitz/Alamy; (cr) Greg Harold/The Image Works; (bl to br) Gerard Soury/Oxford Scientific/Getty Images; Tobias Helbig/Vetta/Getty Images; Michael Nichols/National Geographic/Getty Images; Louis Agassi Fuertes/Glow Images

VULNERABLE	ENDANGERED	CRITICALLY ENDANGERED	EXTINCT
Species is likely to become endangered unless circumstances improve	Species faces a risk of extinction in the wild; usually when fewer than 250 mature individuals exist	Species faces a high risk of becoming extinct	Species has died out and no longer exists
African Elephant, Polar Bear, **Great White Shark**	Blue Whale, Giant Panda, **Asian Elephant**	California Condor, Red Wolf, **Mountain Gorilla**	Zanzibar Leopard, Caspian Tiger, **Passenger Pigeon**

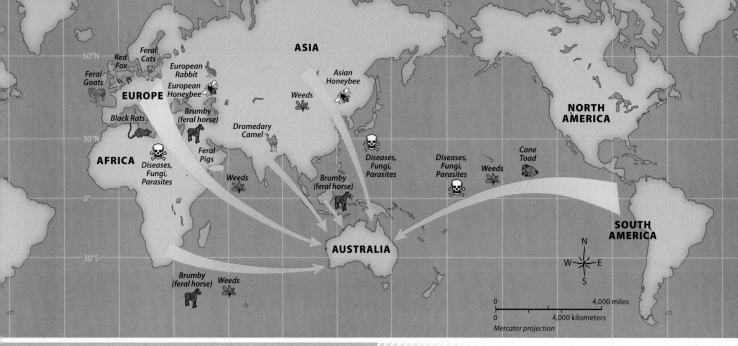

AUSTRALIA'S INVASIVE SPECIES

The map shows the location from where various species were brought to Australia. But why were they brought to Australia? To European settlers, Australia seemed strange. There were no familiar wild animals. In 1859 a rancher brought wild rabbits from England and set them free on his land. As the number of rabbits grew, businesses began to can rabbit meat to sell and used the skins and fur to make clothing and hats. When the rabbit population continued to grow, steps were taken to control them. In the 1950s, a virus was developed that killed most of the rabbits. However, rabbits became resistant to the virus, and the population grew again.

Global Species Extinction

Mass extinctions are time periods in the history of Earth when an extraordinarily large number of species go extinct. Today, many scientists believe the evidence shows a mass extinction is underway.

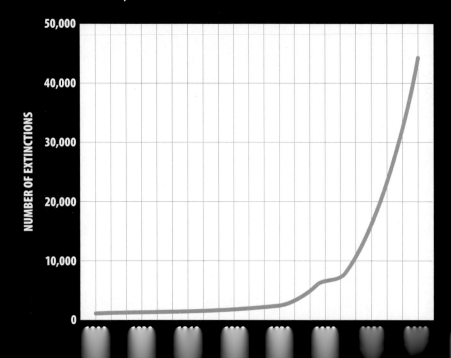

Thinking like a
Geographer

1. **Physical Geography** Why was the cane toad introduced into Australia? Why did the cane toad population grow so fast?

2. **The Uses of Geography** Find out what invasive species have been introduced into your area. Prepare a poster to show how to protect native plants and animals against invasive species.

3. **Environment and Society** Research to find information about an endangered species. Write about what is being done to protect these animals.

Directions: Write your answers on a separate piece of paper.

1 Use your **FOLDABLES** to explore the Essential Question.

INFORMATIVE/EXPLANATORY WRITING Review the physical map and the population map at the beginning of this unit. In two or more paragraphs, explain how Australia's physical geography has affected the country's settlement patterns.

2 21st Century Skills

INTEGRATING VISUAL INFORMATION Work in small groups to research the physical geography, people, and culture of each of Australia's three main geographic regions, New Zealand's North and South Islands, and Tasmania. Present the information as a slide show or a poster.

3 Thinking Like a Geographer

IDENTIFYING Create a two-column chart similar to the one shown here. Label one side Australia and the other side New Zealand. Use the chart to list the natural resources of each country.

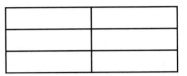

4 GEOGRAPHY ACTIVITY

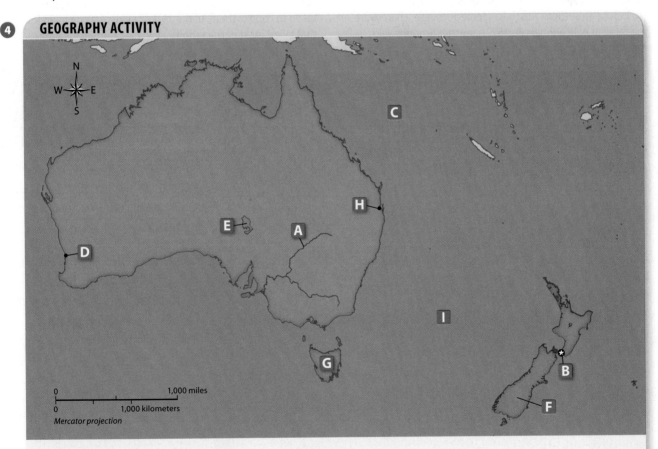

Locating Places

Match the letters on the map with the numbered places listed below.

1. Darling River
2. Wellington
3. Coral Sea
4. South Island
5. Perth
6. Lake Eyre
7. Tasmania
8. Tasman Sea
9. Brisbane

Chapter 24 ASSESSMENT CCSS

REVIEW THE GUIDING QUESTIONS

Directions: Choose the best answer for each question.

1. New Zealand can best be described as a land of
   - A. extremely diverse landscapes, ecosystems, and climate zones.
   - B. nomadic people.
   - C. harsh, dry deserts.
   - D. cold, barren landscapes.

2. How much of Australia is covered by desert?
   - F. one-half
   - G. one-fourth
   - H. two-thirds
   - I. one-third

3. The first people to live in Australia are called
   - A. Maori.
   - B. Indians.
   - C. Aboriginal people.
   - D. convicts.

4. The first people to inhabit New Zealand came from
   - F. Borneo across a land bridge.
   - G. Polynesia in canoes.
   - H. Australia in sailboats.
   - I. Britain in convict ships.

5. What is the basis for New Zealand's economy?
   - A. coal
   - B. geothermal energy
   - C. agriculture
   - D. fishing and hunting

6. What is the name of the bamboo instrument used by Aboriginal musicians to make traditional music?
   - F. boomerang
   - G. kiwi
   - H. marsupial
   - I. didgeridoo

DBQ ANALYZING DOCUMENTS

7 **ANALYZING** Read the following passage about new construction taking place in Australia:

"*Australia is set for an . . . $115 billion infrastructure boom as the nation adds ports and railways to feed China and India's appetite for coal and iron ore. . . . The demand, coupled with economic slowdowns in the U.S. and Europe, has helped make Australia the developed world's fastest-growing construction market.*"

—from David Fickling, "China Trade Spurs $115 Billion Australia Building Boom: Freight"

What factor spurred this upcoming building boom in Australia?

A. slumping economies in the United States and Europe

B. discovery of new sources of coal and iron in Australia

C. development of new uses for coal and iron ore

D. economic growth in China and India

8 **DETERMINING CENTRAL IDEAS** What economic trend does the Australian building boom demonstrate?

F. increasing productivity

G. global interdependence

H. growth of high-technology industries

I. growth of service industries

SHORT RESPONSE

"*In June 2010, the government signed a new agreement with the Maori over contentious [disputed] foreshore and seabed rights, replacing a 2006 deal that had ended Maori rights to claim customary titles in courts of law. Tribes can now claim customary title to areas proven to have been under continuous indigenous occupation since 1840. Maori tribes that secure a customary title will be granted title deeds, but cannot sell the property or bar public access to the area.*"

—from "New Zealand," FreedomHouse.org

9 **DETERMINING WORD MEANINGS** What does the term "continuous indigenous occupation" mean in the agreement?

10 **ANALYZING** How does the 2010 agreement protect the rights of Maori and non-Maori?

EXTENDED RESPONSE

11 **INFORMATIVE/EXPLANATORY WRITING** If you had the opportunity to relocate and live for a couple of years in Australia or New Zealand, which country would you choose? Explain your choice in a short essay. Be sure to consider such things as climate, landforms, recreation, cost of living, and employment opportunities in your writing.

Need Extra Help?

If You've Missed Question	**1**	**2**	**3**	**4**	**5**	**6**	**7**	**8**	**9**	**10**	**11**
Review Lesson	1	1	2	2	3	3	3	3	2	2	1

OCEANIA

©Charles & Josette Lenars/Corbis

A man of Papua New Guinea wears an elaborate ceremonial headdress.

ESSENTIAL QUESTIONS • *How does geography influence the way people live?*
• *What makes a culture unique?* • *Why do people make economic choices?*

The Story Matters...

Thousands of islands make up the three sections of Oceania in the Pacific Ocean—Micronesia, Melanesia, and Polynesia. Because of their location, the islands attracted Europeans. Oceania's colonization and occupation during World War II had a tremendous impact on the region. Many countries, including the United States, still have strategic military bases there.

FOLDABLES®
Study Organizer

Go to the Foldables® library in the back of your book to make a Foldable® that will help you take notes while reading this chapter.

763

OCEANIA (CCSS)

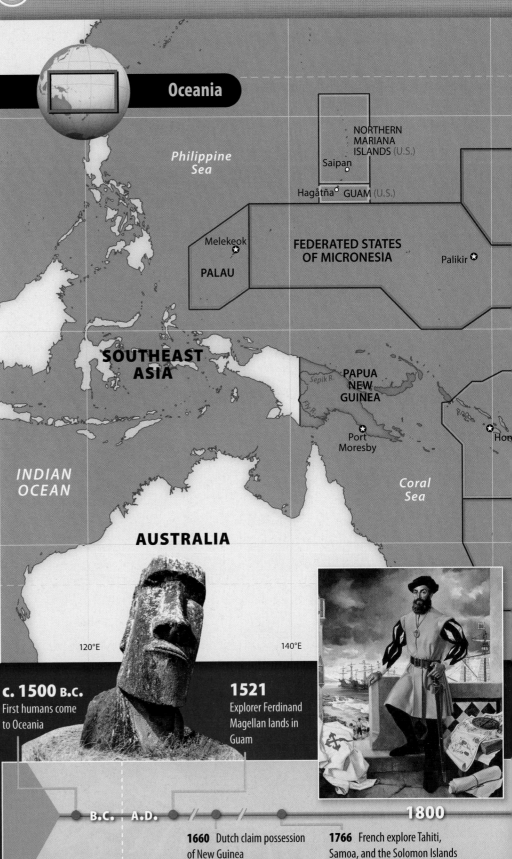

Thousands of islands differing in size and extending across millions of square miles of the Pacific Ocean are located in the region called Oceania.

Step Into the Place

MAP FOCUS Use the map to answer the following questions.

1 THE GEOGRAPHER'S WORLD What is Oceania's largest country in land area?

2 THE GEOGRAPHER'S WORLD In which direction would you travel to go from the Marshall Islands to the Cook Islands?

3 PLACES AND REGIONS What is the capital of Palau?

4 CRITICAL THINKING Describing Of the Solomon Islands, Tonga, and Samoa, which are located east of the International Date Line?

Oceania

Philippine Sea

NORTHERN MARIANA ISLANDS (U.S.)
Saipan
Hagåtña ○ GUAM (U.S.)

Melekeok ★
PALAU

FEDERATED STATES OF MICRONESIA
Palikir ✪

SOUTHEAST ASIA

Sepik R.
PAPUA NEW GUINEA
Port Moresby

Hon ★

INDIAN OCEAN

Coral Sea

AUSTRALIA

120°E   140°E

Step Into the Time

TIME LINE Choose two events from the time line to explain why many countries, including the United States, view the strategic location of Oceania's islands as an important resource.

c. 1500 B.C. First humans come to Oceania

1521 Explorer Ferdinand Magellan lands in Guam

B.C. A.D.

1660 Dutch claim possession of New Guinea

1766 French explore Tahiti, Samoa, and the Solomon Islands

1800

(l) Art Wolfe/Iconica/Getty Images; (r) DeAgostini/SuperStock

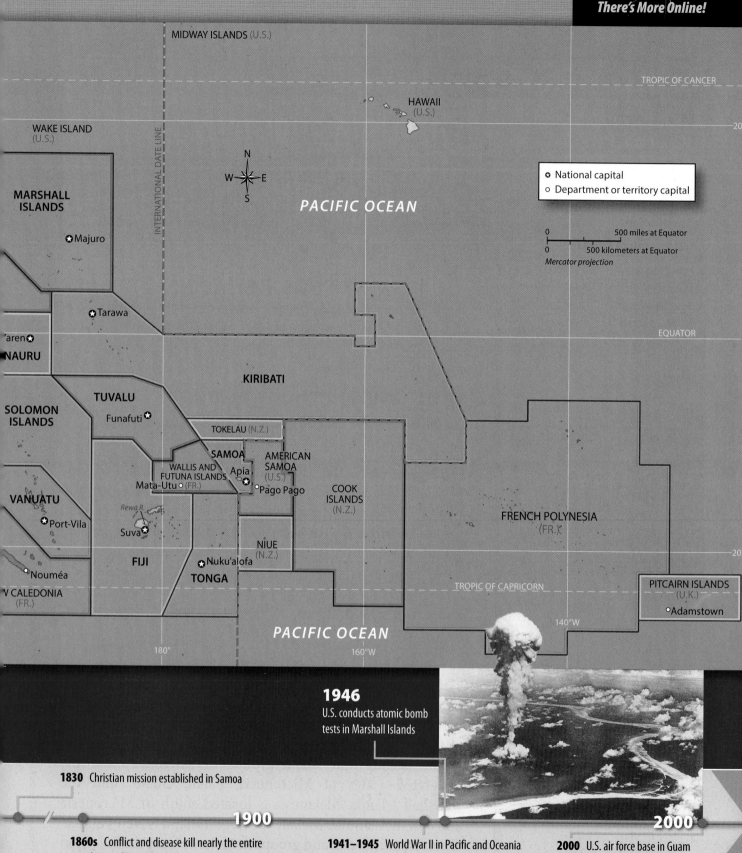

MIDWAY ISLANDS (U.S.)

TROPIC OF CANCER

HAWAII
(U.S.)

20°

WAKE ISLAND
(U.S.)

N
W—E
S

PACIFIC OCEAN

MARSHALL
ISLANDS

☉ National capital
○ Department or territory capital

Majuro

0   500 miles at Equator
0   500 kilometers at Equator
Mercator projection

Tarawa

EQUATOR

aren

NAURU

KIRIBATI

TUVALU

SOLOMON
ISLANDS

Funafuti

TOKELAU (N.Z.)

SAMOA

AMERICAN
SAMOA
(U.S.)

WALLIS AND
FUTUNA ISLANDS
Mata-Utu (FR.)

Apia

Pago Pago

COOK
ISLANDS
(N.Z.)

FRENCH POLYNESIA
(FR.)

VANUATU

Rewa R.

Port-Vila

Suva

20°

Nouméa

FIJI

NIUE
(N.Z.)

Nuku'alofa

TONGA

TROPIC OF CAPRICORN

PITCAIRN ISLANDS
(U.K.)

Adamstown

N CALEDONIA
(FR.)

140°W

PACIFIC OCEAN

180°

160°W

1946
U.S. conducts atomic bomb
tests in Marshall Islands

1830 Christian mission established in Samoa

1900

2000

1860s Conflict and disease kill nearly the entire
population of Easter Island

1941–1945 World War II in Pacific and Oceania

2000 U.S. air force base in Guam
stores conventional cruise missiles

 Lagoon Atoll

Reading **HELP**DESK · CCSS

Academic Vocabulary

- **capable**

Content Vocabulary

- **continental island**
- **archipelago**
- **high island**
- **low island**
- **lagoon**
- **atoll**

TAKING NOTES: *Key Ideas and Details*

Organize On a chart like the one below, fill in important facts about the physical geography for each of the areas.

Micronesia	
Melanesia	
Polynesia	

Lesson 1
Physical Geography of Oceania

ESSENTIAL QUESTION · *How does geography influence the way people live?*

IT MATTERS BECAUSE
The United States has many interests in Oceania, including marine resources and important shipping routes across the Pacific.

Landforms of Oceania

GUIDING QUESTION How did thousands of islands appear across the Pacific Ocean?

Oceania covers 3.3 million square miles (8.5 million sq. km) of the Pacific Ocean between Australia, Indonesia, and the Hawaiian Islands. An estimated 10,000 islands make up Oceania. Most of the islands are small, and many are uninhabited. The smallest inhabited island country is Nauru, which measures a mere 8 square miles (21 sq. km) of land area. Some islands in Oceania are located close together in clusters or island chains. Others stand alone, hundreds of miles from their nearest neighbors.

Islands, Nations, and Territories

Geographers divide Oceania into three sections, according to culture and physical location. The sections are called Micronesia, Melanesia, and Polynesia. Micronesia is located in the northwest section of Oceania, east of the Philippines. Major islands and island groups in Micronesia are the Federated States of Micronesia, Palau, Guam, and the Marshall Islands. Melanesia is located south of Micronesia and east of Australia. Melanesia's largest island is New Guinea. Other islands and island groups in Melanesia are the Solomon Islands, Vanuatu, Fiji, Tonga, and Samoa. Polynesia is a vast

section of Oceania located in the central Pacific Ocean. Major islands and island groups in Polynesia include French Polynesia, Kiribati, Niue, the Hawaiian Islands, and the Cook Islands.

The islands of Oceania range in size from New Guinea, which is 303,381 square miles (785,753 sq. km), to tiny rock outcroppings and patches of sand covering less than 1 square mile (2.6 sq. km). Oceania's islands vary in physical characteristics such as landforms and native plants and animals. The islands also differ in their forms of government. Some islands, such as Fiji and Palau, are independent countries. Other islands, such as American Samoa, Guam, and New Caledonia, are overseas territories that are under the jurisdiction of other countries. Australia, England, France, New Zealand, and the United States have island territories in Oceania. One island, New Guinea, is divided politically.

The Divided Island of New Guinea

New Guinea is the largest island in Oceania. New Guinea is a **continental island**, an island that lies on a continental shelf and was once connected to a larger continental landmass. New Guinea lies on the same continental shelf as Australia and was part of that continent during the past when sea levels were lower. New Guinea is now part of the Malay Archipelago. (An **archipelago** is a group of islands clustered together or closely scattered across an area.)

Traditional homes along the Sepik River in Papua New Guinea are thatched houses on stilts. In this swampy and isolated area, the dugout canoe is the only means of travel.

Children play in a highland village in the Owen Stanley Range of Papua New Guinea. The rugged mountain range is the southeastern part of a long mountain chain that stretches across New Guinea, the world's second-largest island after Greenland.

Physically, New Guinea is one island; however, the island is divided into two parts. The western part belongs to Indonesia. The eastern part is an independent country called Papua New Guinea. Both parts of the island have the same types of land features, plants, animals, climate, and resources. Only the eastern part of the island, Papua New Guinea, however, is considered part of Oceania. When describing the physical features of the island, we will use the name of the country, Papua New Guinea.

Papua New Guinea has a great many different landforms. A very large mountain range stretches across the island. This chain of rugged mountain peaks and glaciers dominates the inland areas of the island. Low mountains and fertile river valleys roll across northern Papua New Guinea. Papua New Guinea also has a northern coastal plain. In the south, swampy lowlands lead to the Owen Stanley Range, which is considered the "backbone" of Papua New Guinea.

The Smaller Islands

Oceania includes several physically different types of islands. **High islands** have steep slopes rising from the shore, higher landforms, and diverse plant and animal life. High islands are generally the largest and greenest of Oceania's small islands. The soil that covers high islands is fertile, and the climates are generally humid and rainy. Many have freshwater streams, rivers, and waterfalls. These conditions allow dense rain forests to grow. Tahiti and the Hawaiian islands are examples of high islands.

Low islands are smaller, flatter islands with sandy beaches. Low islands tend to have fewer forests and less diverse plant and animal life. Most of the islands scattered across Micronesia and Polynesia are low islands. The typical "desert island" described in books and seen in films could be classified as a low island. Some desert islands in Oceania are so low they just break the water's surface. The island country of Tuvalu is an example of a low island with an extremely low elevation. Tuvalu is gradually being eroded by ocean waters. With each passing decade, Tuvalu loses more surface area, as its land washes away with the tides. If sea levels continue to rise, Tuvalu and Oceania's other lowest islands will disappear below the water.

The islands of Oceania were formed by processes involving the lithosphere, the hydrosphere, and the biosphere. Most of Oceania's low islands were formed by a gradual process involving volcanic eruptions (lithosphere), erosion by water movement (hydrosphere), and the growth of coral (biosphere). This process began millions of years ago with the eruption of an undersea volcano. As lava from the volcano cooled, it formed a buildup of volcanic rock below the surface of the water. Corals started growing on the volcano, eventually forming large reefs that circled the volcano. Over time, the volcano crumbled and sank, while the coral reefs continued to grow higher and higher. The coral reefs built up layer upon layer until they grew above the water's surface. Waves crashing against the reefs eventually eroded channels, allowing ocean water to flood the center area of the island and form shallow pools called **lagoons**. As the reef aged, crumbled, and died, ocean waves deposited sediment, such as sand and tiny specks of plant and animal life, on the coral remains. The resulting landform is called an **atoll**, a coral island made up of a reef island surrounding a lagoon. If the center of an island does not fill with water, or if the atoll becomes completely covered with sediment over time, a desert island forms.

The majority of the high islands in Oceania were formed by underwater volcanoes. As lava from erupting volcanoes flowed into the ocean waters, it cooled and formed huge mounds of volcanic rock. Eventually the volcanic rock built up above the water's surface, forming high islands.

☑ **READING PROGRESS CHECK**

Identifying What are the names of Oceania's sections?

DIAGRAM SKILLS >

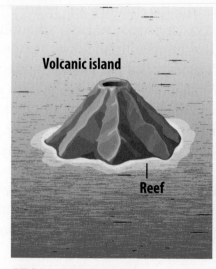
Volcanic island
Reef

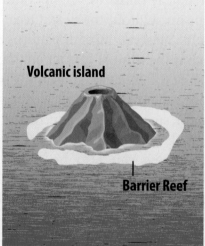
Volcanic island
Barrier Reef

Lagoon
Atoll

ATOLL FORMATION
South Pacific atolls are known for their beautiful coral reefs and marine life, which draw tourists from around the world. The typical atoll takes about 30 million years to form.

▶ CRITICAL THINKING
1. ***Describing*** What two processes are involved in atoll formation?
2. ***Analyzing*** How might tectonic plate movements affect the formation of an atoll?

French Polynesia is known for its black pearl industry. Oysters containing the black pearls are raised in underwater farms. Black pearls make up more than half of French Polynesia's exports.

▶ **CRITICAL THINKING**

Analyzing Why are other abundant ocean resources, such as fish and seafood, used by Oceania's people but not exported?

Academic Vocabulary

capable having the ability to cause or accomplish an action or an event

Climates of Oceania

GUIDING QUESTION *What factors affect climate in Oceania?*

Nearly all of Oceania's islands are located within the Tropics. Only a few, such as the Midway and Pitcairn Islands, lie north or south of this climate zone. Thus, most of Oceania's islands have warm, humid, tropical climates. The islands located outside the Tropics have mixed tropical and subtropical climates. Some islands experience local climate variations caused by elevation, winds, and ocean currents.

Papua New Guinea Climates

Papua New Guinea has two climate zones: tropical and highland. The lowland areas have warm to hot temperatures. The highland regions are much cooler. For example, the average daily temperature in the lowland coastal plains is 82°F (28°C). The average daily temperature in the highland mountains is 73°F (23°C). Papua New Guinea's climate is wet, with an average of 45 inches (114 cm) of rain falling on the island annually. February is the month with the most rainfall, and July has the least. Monsoon rains are common. Dense rain forests grow well in the warm, wet conditions, and much of the island is covered in trees and other forest plants. Papua New Guinea does, however, experience occasional droughts caused by the El Niño effect.

Climates of the Smaller Islands

The smaller islands scattered throughout Oceania have tropical and subtropical climates. Temperatures on the islands are generally warm throughout the year. Oceania's islands receive large amounts of rain. Seasonal rainfall patterns determine an island's wet season, which is the time of year when the heaviest rains fall. Islands located north of the Equator, such as the Marshall Islands and Palau, have a wet season from May to November. Islands located south of the Equator, such as Samoa and Tonga, have a wet season from December to April. The rainfall patterns occur under normal conditions. Heavy rains can also be caused by storms, such as typhoons. Typhoons cause intense winds and powerful waves **capable** of toppling trees and houses and eroding island shores. Only the island of Yap is affected by monsoon winds, which are brought by weather patterns of the western Pacific and Indian oceans.

✔ **READING PROGRESS CHECK**

Analyzing The Northern Mariana Islands experience a wet season from May to November. Based on this information, are the Northern Mariana Islands located north or south of the Equator?

Resources of Oceania

GUIDING QUESTION *What natural resources do the islands of Oceania possess?*

Most of Oceania's islands are small. Limited land area limits the amount of natural resources, such as minerals, to be found on land. Still, the islands have some valuable resources. Resources that the islanders trade or sell are essential to the islands' economies.

Papua New Guinea's Resources

The independent nation of Papua New Guinea has natural resources such as gold, copper, timber, fish, petroleum, and natural gas. Compared to the island's size, Papua New Guinea's natural gas reserves, discovered fairly recently, are large. They have the potential to benefit the nation's economy.

Resources of the Smaller Islands

Oceania's smaller islands have few valuable natural resources for trading on the international market. Limited land area is one reason for this; another reason is that low islands, based on coral, do not have rock foundations. It is in deep layers of rock that large deposits of metal ores, such as gold, are found. New Caledonia is also developing wind power by building large stands of wind turbines. This technology is beginning to spread to other islands. People living in rural areas of Kiribati and the Solomon Islands have begun using solar-powered lighting in their homes. Wind and solar energy are renewable, nonpolluting resources. Because of the sunny climates and ocean winds, these resources are plentiful throughout Oceania.

Some of Oceania's high islands have large trees that are used for timber, rubber, and other products. Soil quality varies from island to island. Some islands, such as those with rich volcanic soil, have excellent soil for growing farm crops. Other islands have poor-quality soil, making farming difficult. Fish and other seafood are important resources for the people in Oceania. Most islands use fish only for their own food, not for export.

☑ **READING PROGRESS CHECK**

Identifying What are two types of renewable resources in Oceania?

Include this lesson's information in your Foldable®.

LESSON 1 REVIEW (CCSS)

Reviewing Vocabulary
1. Why is New Guinea called a *continental island*?

Answering the Guiding Questions
2. *Identifying* Which countries have territories in Oceania?

3. *Describing* What is the difference between New Guinea and Papua New Guinea?

4. *Determining Central Ideas* In what ways are low islands and high islands alike and different?

5. *Analyzing* What factors affect the climates of Oceania?

6. *Analyzing* Why would people living in rural Kiribati and the Solomon Islands use solar-powered lights in their homes?

7. *Informative/Explanatory Writing* In your own words, write a paragraph describing how an atoll forms.

(l to r) Douglas Peebles/Alamy; Mark Harris/The Image Bank/Getty Images; PJF/Alamy; Patrice Coppee/StockImage/Getty Images

Reading **HELP**DESK (CCSS)

Academic Vocabulary

- **distinct**

Content Vocabulary

- **wayfinding**
- **trust territory**
- **possession**
- **pidgin**
- *fale*

TAKING NOTES: *Key Ideas and Details*

Summarize Using a chart like the one shown here, summarize important information from each section of the lesson.

History of Oceania	People of Oceania

Lesson 2

History and People of Oceania

ESSENTIAL QUESTION · *What makes a culture unique?*

IT MATTERS BECAUSE

Oceania has a unique culture but one that faces challenges.

History of Oceania

GUIDING QUESTION *How were the islands of Oceania populated?*

The first humans began settling in Oceania sometime around 1500 B.C. Historians believe the early settlers came from Southeast Asia, using available resources to build large sailing canoes. They filled their sturdy canoes with people, food, plants, and animals. Traveling from west to east and powered only by sails, the settlers crossed the waters of the Pacific Ocean.

The Polynesian Migrations

Their only way of navigating was to use the ancient practice of wayfinding. **Wayfinding** is a method of navigation that relies on careful observation of the natural world. For many thousands of years, humans have charted courses across the open ocean by watching the sun, the stars, and the movement of ocean currents and swells. Wayfinding was practiced long before the invention of navigation instruments, such as compasses and sextants. Even today, navigators practice wayfinding as a way to stay connected to Earth and to keep cultural traditions alive.

Over many centuries, people from areas such as the Philippines and Indonesia sailed from their homelands and settled the islands across Oceania. Many islands were uninhabited until settlers from other islands migrated farther into unexplored areas of Oceania.

The Coming of Europeans

European explorers began sailing through Oceania in the 1500s. They began colonizing the islands in the 1600s. Often, violent conflict broke out between colonists and native people. During the 1800s and 1900s, Christian missionaries came to thousands of islands in the region. Native people did not always welcome the missionaries. Many missionaries succeeded, however, in converting local populations to Christianity. Christian faiths are still widely practiced across Oceania.

Europeans had many reasons for wanting to colonize territories in Oceania. For practical reasons, the locations of many islands made them convenient stops for ships crossing the vast Pacific Ocean. Travelers wanted safe, reliable locations to restock their ships with food, drinking water, and other supplies. European powers were also interested in claiming resources.

Some Europeans mined gold and other precious metals from the islands. Some governments also saw the advantage of building military bases in Oceania. In fact, many islands in the region were occupied by Japanese, German, English, and American forces during World War II. Several battles were fought in the region. Unfortunately, some islands in the Pacific also became testing sites for nuclear weapons.

Contemporary Times

After World War II, many island colonies began to demand independence. Some independence movements involved conflict, but many islands were able to negotiate freedom with their former colonial powers. Some islands negotiated independence by "free association" with foreign powers.

A typical South Pacific canoe lies on a beach in Fiji. Early people, originally from the Asian mainland, sailed and settled in Oceania centuries before the arrival of Europeans in the 1500s.

▶ **CRITICAL THINKING**

Describing How were early people able to sail vast distances to settle Oceania?

Douglas Peebles/Alamy

The bicycles of worshippers are parked outside a picturesque Catholic church on the Fakarava atoll in the Tuamotu Islands. Built mostly of coral in 1874, the building is the oldest church in Polynesia. Christian faiths are widely practiced in Oceania today.

The island of Palau, for example, is an independent republic, but it has a voluntary free association with the United States. Palau has its own constitution and governs itself. Palau and the United States have an agreement that benefits both: Palau allows the United States to keep military facilities on one of its islands, and in return, the United States provides millions of dollars of aid money to Palau each year.

Other islands in Oceania have agreements of various kinds with foreign governments, including Australia, New Zealand, Great Britain, France, and the United States. The agreements generally involve use of land or other resources in exchange for military protection and economic aid. Relationships between island territories and foreign governments have different levels of political control and responsibility. A **trust territory** is one that has been placed under the governing authority of another country by the Trusteeship Council of the United Nations. The Marshall Islands were a trust territory until they gained independence in 1986. **Possession** is another name for a territory occupied or controlled by a foreign government and its people. French Polynesia can be classified as a possession because it is an overseas territory of France. Trust territories and possessions do not govern themselves but are run by foreign governments.

☑ **READING PROGRESS CHECK**

Describing Why did the first Europeans come to Oceania?

Mark Harris/The Image Bank/Getty Images

The People of Oceania

GUIDING QUESTION *What is life like in Oceania?*

Many people imagine the South Sea Islands as tropical paradises. There are many wonderful things about living in this beautiful region of the world. On the other hand, life in Oceania has many challenges.

The People of the Region

Oceania has one of the world's most diverse populations. So many different ethnic groups live on Oceania's thousands of islands that it is impossible to classify them all. Groups have their own **distinct** languages, cultures, and ways of life. Many islands are home to a wide range of ethnic groups. The most amazing diversity is found in Papua New Guinea.

Papua New Guinea has a total human population of more than 6 million. This is by far the largest population of any of Oceania's islands. Papua New Guinea's large population is condensed onto an island about the size of California.

The island country's population is made up of people from many different ethnic and native tribal groups. Natives of Papua New Guinea make up about 84 percent of the total population. The native population includes people from hundreds of tribal groups. The other 16 percent come from various backgrounds, including Polynesian, Chinese, and European.

The different groups have their own lifestyles, cultural traditions, beliefs, and languages. Geographers and language experts believe 860 different languages are spoken in Papua New Guinea. In other words, 10 percent of all languages known to exist are used in Papua New Guinea.

PJF/Alamy

Academic Vocabulary

distinct separate; easily recognized as separate or different

The *USS Bonhomme Richard* pulls into Apra Harbor in Guam. The *Bonhomme Richard* is an amphibious assault ship. These ships are able to land and aid forces on shore during armed conflict.

▶ **CRITICAL THINKING**
Describing What are the terms of agreement between foreign governments and the islands of Oceania?

This village is located on Viti Levu, the largest and most populous of Fiji's more than 300 islands. About 75 percent of Fiji's 600,000 people live on Viti Levu. The island measures about 65 miles (106 km) from north to south and 90 miles (146 km) east to west.

In spite of speaking many different languages, the people of Papua New Guinea are still able to speak to one another. Many of the people speak their own language, as well as a pidgin language. A **pidgin** is a simplified language that is used for communication between people who speak different languages.

Such a wide diversity has both positive and negative effects. For instance, Papua New Guinea's many different ethnic groups create a rich and varied culture on the island. An endless variety of music, foods, clothing, and artwork can be enjoyed. At the same time, serious problems, such as crime, ethnic discrimination, and violent conflicts among tribal groups, are common.

Papua New Guinea's population is denser in some parts of the main island than in others. In general, highland areas have higher population densities than do the lowlands and coastal plains. Only about 12 percent of the population live in urban areas. Isolated towns are hidden in rugged mountain areas. Thousands of distinct tribal groups live in small villages in the islands' many remote locations. A traditional folk saying in Papua New Guinea is, "For every village, a different culture."

The Culture of the Region

Life in the villages of Oceania is based on tradition. Many traditions involve fishing, diving, and celebrating battle victories. Another tradition is an important event in the lives of young people called coming-of-age ceremonies. In Polynesian cultures, ceremonies and celebrations include feasts and dancing. Like their ancestors, Polynesian people practice artistic wood carving and use the carvings to decorate their homes. Many traditional Polynesian and Micronesian cultures practice tatooing. Micronesian people use storytelling to retell history and to keep track of family heritage. The Melanesian people were the only traditional culture of Oceania known to use bows and arrows in hunting.

Patrice Coppee/StockImage/Getty Images

In the island nation of Samoa, traditional homes called **_fales_** are common. _Fales_ are open structures made of wood poles with thatched roofs. Local trees are used to make the poles, and palm leaves are used for roof thatch. The dwellings have no walls and are used mainly for shade and shelter from frequent rainfall.

Today, Oceania's many island cultures are mixtures of traditional and modern practices, beliefs, and lifestyles. Although they have adopted many Western attitudes, people see the value in continuing some of the traditional ways.

For example, Christianity is widely practiced in island communities, along with elements of traditional religions, such as songs, dances, and ceremonial costumes. Many island people wear Western-style clothing and hairstyles. Cell phones and laptop computers are common. Elements of local traditional cultures, including tribal tattoos, jewelry, and art forms such as wood carving, are common on many islands, as well.

Traditional celebrations are practiced throughout Oceania. Some traditional events, such as the Hawaiian luau, have become more modern in recent decades. Luaus were traditionally ritual ceremonies and feasts celebrating important events, such as victories in battle. Centuries ago, men and women ate in separate areas during luaus, and only chiefs ate certain foods. Some luaus were attended only by men. Today, luaus are banquets of traditional and modern foods eaten on a low table.

Maintaining elements of traditional cultures in their lives is important to the people of Oceania. Respecting and continuing certain traditions keeps cultures alive. Celebrating the traditional culture of their ancestors gives young people a sense of pride and identity. Making the past part of the present keeps people of all ages connected to their cultural heritage.

☑ **READING PROGRESS CHECK**

Analyzing Why might using a pidgin language be useful to the population of Papua New Guinea?

Include this lesson's information in your Foldable®.

LESSON 2 REVIEW

Reviewing Vocabulary
1. For what purpose did the early people of Oceania use *wayfinding*?

Answering the Guiding Questions
2. *Identifying* For approximately how long have humans been living on the islands of Oceania?

3. *Identifying* List three of the many groups of people who came to Oceania from other parts of the world.

4. *Determining Word Meanings* Would a *fale* be an appropriate home for the climate where you live? Why or why not?

5. *Distinguishing Fact From Opinion* Is the following statement about the culture of Oceania a fact or an opinion?

With more than 860 different spoken languages, Papua New Guinea has one of the most culturally diverse populations in the world.

6. *Narrative Writing* Write a short story from the perspective of a young person living a traditional lifestyle in a small village on one of the islands of Oceania. Include details about your daily life. Describe your home, your family and friends, the foods you eat, the work you do, and what you do for fun.

networks

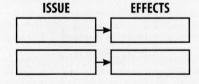

There's More Online!

☑ **IMAGES** Tropical Fish and Sharks in Tahiti

☑ **SLIDE SHOW** A Beach in Polynesia

☑ **VIDEO**

Reading **HELP**DESK CCSS

Academic Vocabulary

• **collapse**

Content Vocabulary

• **cash crop**
• **resort**
• **remittance**
• **MIRAB economy**

TAKING NOTES: *Key Ideas and Details*

Determine Cause and Effect As you read, use a graphic organizer like the one shown here to identify two important issues and describe the effects of the issue.

ISSUE		EFFECTS
☐	→	☐
☐	→	☐

Lesson 3

Life in Oceania

ESSENTIAL QUESTION • *Why do people make economic choices?*

IT MATTERS BECAUSE
The people of Oceania face economic challenges that affect the United States and other countries.

The Economies of Oceania

GUIDING QUESTION *How do the people of Oceania earn their living?*

The islanders of Oceania face difficult economic challenges. With small land areas, few valuable resources, and vast distances between islands, earning a living in Oceania can be difficult. The people of Oceania, however, have found ways to support their families. They also take great care in using the natural resources available to them. Despite challenges, Oceania's island communities have the potential for a bright economic future.

Papua New Guinea's Economy

Papua New Guinea has the most valuable natural resources in Oceania, other than Australia and New Zealand. The challenge is to locate, harvest, and transport the resources through Papua New Guinea's wild and rough terrain.

Gold and copper are Papua New Guinea's most profitable resources. Sales of gold and copper account for about 60 percent of the country's total export income. Other major exports are silver, timber, and agricultural products. **Cash crops** are crops grown or gathered to sell for profit. Papua New Guinea's cash crops include coffee, cacao, coconuts, rubber, and tea.

Mining and farming provide jobs for many people in Papua New Guinea. Industries such as timber processing, palm oil refining, and petroleum refining employ many

others. The tourist industry also provides many jobs. The vast majority of Papua New Guinea's people, however, live by subsistence farming. Most families raise their own food crops, such as yams, taro, bananas, and sweet potatoes. Some raise pigs or chickens for meat and eggs. Although the unemployment rate in Papua New Guinea is low, most people earn low incomes. The government of Papua New Guinea plans to increase exports of minerals and petroleum to strengthen the country's economy.

Economies of Small, Independent Countries

Smaller independent island countries throughout Oceania face many obstacles to economic development. With limited land, poor soil quality, and large populations, many islands must import much of their food, fuel, finished goods, and raw materials. Most islands import far more than they export—many import five or six times more goods and materials than they export. On the tiny island country of Tuvalu, for example, import values exceed export values by 200 to 1. People on the smaller islands raise what food they can by subsistence farming.

Some islands in Oceania raise limited cash crops, such as fruits, vegetables, sugar, nuts, coffee, tea, cocoa, and palm and coconut oils. The farms and plantations that produce cash crops employ some island residents. This type of agricultural work can be physically exhausting and usually pays very little.

Fish and other seafood are available across Oceania. Fishing operations on the smaller islands are usually small; fishers catch only enough to feed their families or to sell to local markets. Most island countries do not have the equipment or processing facilities for operating large fishing industries. Some island countries earn revenue by selling fishing rights to other countries. Japan, Taiwan, South Korea, and the United States are some of the foreign lands that pay for access to Oceania's marine resources.

On islands that have minerals and other marketable resources, many people work for industries such as mining, fishing, clothing, and farming. Some people make a living by using local materials, such as shells, wood, and fibers, to create art and to craft tools and artifacts.

Farmers in Oceania often grow taro, a root vegetable, brought from Southeast Asia centuries ago. The leaves and the root of the taro plant are widely used in South Pacific cooking.

▶ **CRITICAL THINKING**
Describing Why is subsistence farming widely practiced in Oceania?

Tourists go snorkeling at a beach resort in French Polynesia. Tourism is now the major industry in Oceania, creating jobs and bringing in money for many people of the region. Most of Oceania's tourists come from Australia, New Zealand, Japan, and other countries of the Pacific area.

Academic Vocabulary

collapse a sudden failure, breakdown, or ruin

Tourism is important to the economies of many small, independent countries in Oceania. Tourists come from all over the world to enjoy the sunshine, warm ocean waters, and panoramic views. **Resorts** provide comfortable lodging, food, recreation, and entertainment. Most resorts are located in beautiful natural areas, such as tropical beaches, mountains, and forests. Tourist businesses employ thousands of people. Without the revenue from tourism, many small island countries would be at risk of economic **collapse**.

Many people on the islands that have less-developed economies depend on income earned by family members living overseas. Thousands of young people have left Oceania in search of jobs in other countries. Most have settled in Australia, New Zealand, and the United States. Young workers support their families in Oceania by sending them the money they earn. Foreign-earned wages, called **remittances**, are vital sources of income for families. Workers employed overseas also send remittances to pay for community projects, such as schools. Workers contribute to their home countries' tourist industries when they return for visits.

Many of the islands of Oceania with less-developed economies are known as **MIRAB economies**. MIRAB is an acronym for *migration, remittances, aid,* and *bureaucracy.* Island countries with MIRAB economies are not able to fully support the needs of their people through their own resources and labor. These countries depend on outside aid and remittances from workers who have migrated to overseas areas. One example of a MIRAB economy is the Polynesian island nation of Tonga. Tonga raises a few export crops, such as squash, vanilla beans, and root vegetables. The tourist industry also brings in some revenue. Still, the most important source of income for most Tongans is remittances. Seventy-five percent of all Tongan families receive money from family members who live and work elsewhere.

Tonga and many other small island countries also depend on aid from foreign governments. The governments of Australia, Great Britain, Japan, New Zealand, and the United States have created international trust funds for many islands in Oceania. Foreign governments also give money directly to various island countries. Governments of MIRAB economies use some of the foreign aid for food, schools, water and sanitation, and other basic needs. Foreign aid also funds economic development by expanding local industries and exports.

Economies of America's Pacific Islands

American Samoa is a territory of the United States. Nearly all of American Samoa's economic activity involves the United States. The island group's chief industry is tuna fishing and processing, which employs 80 percent of its people. American Samoa produces a few cash crops, such as bananas, coconuts, taro, papayas, breadfruit, and yams.

The unemployment rate is high at nearly 30 percent. One cause of unemployment in American Samoa is the lasting impact of a 2009 earthquake and tsunami. The disasters caused terrible damage to the islands and their transportation systems, electrical systems, and businesses. Some industries were completely ruined, causing loss of jobs. Other industries, such as tourism, are starting to regain strength and show promise for future development.

Other U.S. territories in Oceania include Guam and Wake Island. Both islands are home to U.S. military bases. Local people are employed in transportation, housing, maintenance, food service, and other industries that serve the needs of military personnel. Guam also has a well-developed tourist industry.

☑ **READING PROGRESS CHECK**

Describing In your own words, briefly explain MIRAB economies.

Issues Facing the Region

GUIDING QUESTION *What challenges do the people of Oceania face?*

Despite signs of progress, the future is uncertain for the islands of Oceania. Three major areas of concern in the region are migration, economics, and the environment.

Human Migration

The movement of people from the islands of Oceania to other parts of the world is a major issue. So many young people have left Oceania to seek employment overseas that the populations of some islands have changed dramatically. The island of Tonga is an extreme example: Population estimates show that 50 percent of all Tongan people live abroad in foreign countries.

The major population shift has had negative effects. A large percentage of the people remaining in Tonga are children or older people who are unable to work. Migration has left many young and elderly Tongans without caregivers. The income that migrating workers send home helps.

Most Tongan workers who live abroad maintain strong ties to their homeland. With half of all Tongans now living overseas, however, more and more Tongan children are born in other countries. These children will grow up as natives of their adopted lands, not as native Tongans. Within the next few years, the majority of Tongan people will be born and raised overseas. This situation can weaken cultural ties to the homeland, as more foreign-born Tongans adapt to the cultures where they live.

Economics and Society

Another major issue in Oceania is the need for economic development. Many islands have slow economic development because of lack of resources. New industries cannot be planned and built without money to invest in growth and development. The need is urgent on islands where large numbers of people are migrating from farms and rural villages to cities to try to find work. With fewer people living in rural areas, fewer people are growing their own food. There is a continuing need across Oceania to buy food, energy resources, and raw materials from other countries. Many countries of Oceania import more goods and services than they export. When a less-developed country imports more than it exports, the entire economy is affected. Countries like to export more than they import because this creates jobs and demand for their goods and services.

Trash covers an otherwise-attractive beach on the island nation of Kiribati. Oceania, like other parts of the world, faces serious problems with waste pollution caused by an increasing population, rapid economic development, and the concentration of people in urban centers.

©George Steinmetz/Corbis

Tensions and Conflict

Although most island nations in Oceania are free of conflict, some unrest occurs. Crime and human rights abuses are problems on some of the islands. In Fiji, tensions between native Fijians and immigrants from India led to conflicts. The Solomon Islands have also seen conflict on issues. Disagreement on the issue of land rights led to conflict between the native people of Guadalcanal and people from the neighboring island of Malaita. Peace was restored with help from the United Nations.

Environmental Issues

The islands of Oceania face serious environmental issues, including climate change, deforestation, pollution, natural disasters, and declining fish populations. Many scientists and island residents view climate change as one of the most urgent and serious issues in the region. With continued global warming, sea levels are already rising. If this continues, many of Oceania's low islands could be completely covered by water. Some of the lowest islands are already experiencing surface flooding from rising oceans and eroding beaches.

Natural disasters such as earthquakes, tsunamis, typhoons, and resulting floods continue to threaten the islands. These events have the potential to destroy homes and claim lives, and they can also damage farmland and natural habitats.

Survival of Marine Animals

Commercial fishing companies have harvested so many fish from some areas that almost no fish remain for local people to catch and eat. When huge numbers of fish are caught at one time, the remaining fish are unable to reproduce quickly enough to restore populations. In time, entire populations of fish will be gone. Ocean pollution also threatens the survival of fish and other marine animals.

Include this lesson's information in your Foldable®.

☑ **READING PROGRESS CHECK**

Citing Text Evidence Why is climate change a serious concern for the islands of Oceania?

LESSON 3 REVIEW (CCSS)

Reviewing Vocabulary

1. What is the major difference between *cash crops* and subsistence crops?

Answering the Guiding Questions

2. *Identifying* List three ways the people of Oceania earn a living.

3. *Describing* How does aid from foreign countries benefit islands with MIRAB economies?

4. *Citing Text Evidence* The birthrate in Tonga has increased, but the island's population has decreased. Explain.

5. *Citing Text Evidence* What do you think is the most serious issue or challenge in Oceania today? Use information from the lesson in your explanation.

6. *Argument Writing* Using information from the lesson, write an argument either in favor of or against remittances. Explain why you believe remittances benefit or harm the cultures and economies of Oceania.

Directions: Write your answers on a separate piece of paper.

1 Use your **FOLDABLES** to explore the Essential Question.
INFORMATIVE/EXPLANATORY WRITING It is generally believed that inhabitants of Southeast Asia began to populate the islands of Oceania more than 2,500 years ago. Describe their method of navigation known as *wayfinding*.

2 **21st Century Skills**
DETERMINING CENTRAL IDEAS Work in a group to learn more about one of the island countries of Oceania. Put together a travel brochure to appeal to people who are interested in traveling there. Annotate your brochure with information from the text and from online sources.

3 **Thinking Like a Geographer**
ANALYZING How did the South Pacific's physical geography contribute to Oceania's cultural diversity?

4 **GEOGRAPHY ACTIVITY**

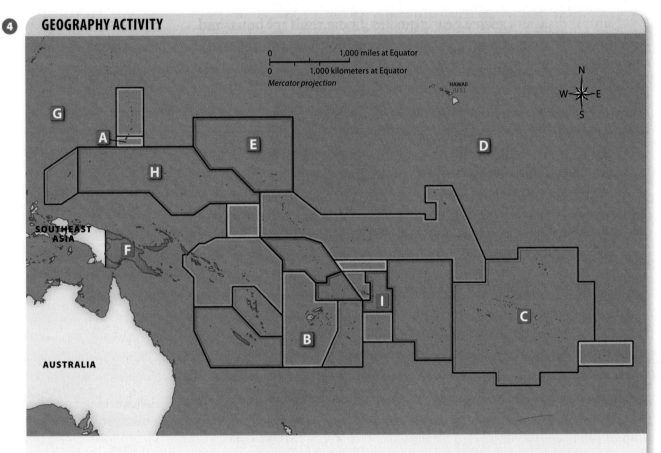

Locating Places
Match the letters on the map with the numbered places listed below.

1. Guam
2. French Polynesia
3. Papua New Guinea

4. Federated States of Micronesia
5. Philippine Sea
6. Marshall Islands

7. Fiji Islands
8. North Pacific Ocean
9. American Samoa

REVIEW THE GUIDING QUESTIONS

Directions: Choose the best answer for each question.

1 What is the smallest inhabited island in Oceania?

A. Micronesia

B. Nauru

C. Guam

D. Papua New Guinea

2 New Guinea is a continental island, which means that

F. like Australia, it is also a continent.

G. it is part of an archipelago.

H. it was at one time connected to a continent.

I. its people are highly educated.

3 From what region did the original settlers of the islands of Oceania come?

A. New Zealand

B. East Africa

C. Southeast Asia

D. the Bering Strait

4 Apart from Australia and New Zealand, which island of Oceania has the largest population?

F. American Samoa

G. Solomon Islands

H. Papua New Guinea

I. Tahiti

5 Why, despite abundant fish and seafood, do the islands of Oceania not export more of these resources?

A. Oceania does not have enough workers.

B. The nations do not have enough equipment or processing facilities.

C. Other countries steal the fish from Oceania's waters.

D. Mercury poisoning is a major threat.

6 Guam and Wake Island are territories of the United States that are maintained to support

F. movie filming.

G. forestry.

H. military bases.

I. the MIRAB program.

DBQ ANALYZING DOCUMENTS

7 **DETERMINING CENTRAL IDEAS** Read the following passage about Tonga's economy:

"*The remittances of cash and goods from migrants who live and work overseas . . . [keep] the Tongan economy afloat. . . . Remittances . . . accounted in [2002] for about 50 percent of [gross domestic product]. . . . Although individuals and families are the main benefactors, . . . overseas Tongans [also] regularly send back money to their villages and local institutions, . . . funding practical community projects.*"

—from Cathy A. Small and David L. Dixon, "Tonga: Migration and the Homeland"

What best explains why Tongans leave the island to find work?

A. Tongans traditionally have loved to travel.

B. Tonga's climate makes farming difficult.

C. Tonga's economy does not offer enough jobs.

D. Wages in Tonga are higher than elsewhere.

8 **IDENTIFYING** What is an example of a "practical community project"?

F. equipment for a village health clinic   H. the education of a cousin

G. better housing for a migrant's family   I. wedding presents to a sister

SHORT RESPONSE

"*The isolation created by the mountainous [landscape] is so great that some groups, until recently, were unaware of the existence of neighboring groups only a few kilometers away. The diversity, reflected in a folk saying, 'For each village, a different culture,' is perhaps best shown in the local languages. . . . Over 850 of these languages have been identified; of these, only 350–450 are related.*"

—from "Papua New Guinea," State Department Background Notes

9 **DETERMINING CENTRAL IDEAS** Explain the meaning of the folk saying quoted in the passage.

10 **ANALYZING** What other kinds of environments have a similar effect on groups of people?

EXTENDED RESPONSE

11 **INFORMATIVE/EXPLANATORY WRITING** The MIRAB system (migration, remittances, aid, bureaucracy) has helped the economies of some of the islands of Oceania. What are the disadvantages of the MIRAB system? In an essay, identify some of its flaws, and predict their future impact on the region.

Need Extra Help?

If You've Missed Question	**1**	**2**	**3**	**4**	**5**	**6**	**7**	**8**	**9**	**10**	**11**
Review Lesson	1	1	2	2	3	3	3	3	2	2	3

Originally published on the Migration Information Source, the online journal of the Migration Policy Institute, an independent, nonpartisan think tank in Washington D.C., dedicated to the study of the movement of people worldwide. (www.migrationinformation.org)

ANTARCTICA

ESSENTIAL QUESTIONS · *How does physical geography influence the way people live?* · *How do people adapt to their environment?*

South Korean research team endures icy conditions to gather biological material from sea on southeastern tip of Antarctica.

Yonhap/Newscom

netw⊕rks

There's More Online about Antarctica.

CHAPTER 26

Lesson 1
The Physical Geography of Antarctica

Lesson 2
Life in Antarctica

The Story Matters...

For centuries, explorers from many countries set out on dangerous expeditions, competing to be the first to reach the frozen continent of Antarctica. Antarctica, which means "opposite to the Arctic," is shared today by many nations whose scientists are researching its vital role in maintaining the world's climate balance.

FOLDABLES
Study Organizer

Go to the Foldables® library in the back of your book to make a Foldable® that will help you take notes while reading this chapter.

Extremes	The Antarctic Treaty	Global Importance

ANTARCTICA (CCSS)

The continent of Antarctica lies at the southern extreme of Earth, with its land lying beneath a massive ice cap. It is larger in size than either Europe or Australia.

Step Into the Place

MAP FOCUS Use the map to answer the following questions.

1 **THE GEOGRAPHER'S WORLD**
Which country's claim in Antarctica is larger: the Australian, the Argentine, or the French?

2 **THE GEOGRAPHER'S WORLD**
What sea is located between longitudes 100°W and 120°W?

3 **PLACES AND REGIONS** Which country's claim is adjacent to the Ross Sea?

4 **CRITICAL THINKING** **Analyzing** Think about the location of Antarctica in relation to other parts of the world. How might its location affect the development of natural resources?

A

ICY SUMMIT A climber approaches the summit of Mount Vaughn in the Queen Maud Mountain range. The range runs from Antarctica along the ocean floor and continues into South America as the Andes Mountains.

B

ANTARCTIC ISLAND Kayakers paddle near the rugged cliffs of Petermann Island off the west side of the Antarctic Peninsula. The island is home to large numbers of penguin colonies.

Step Into the Time

TIME LINE Based on events on the time line, write a paragraph explaining the international importance of Antarctica.

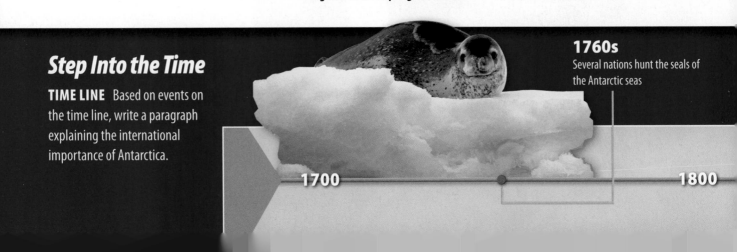

1760s
Several nations hunt the seals of the Antarctic seas

1700

1800

ANTARCTICA

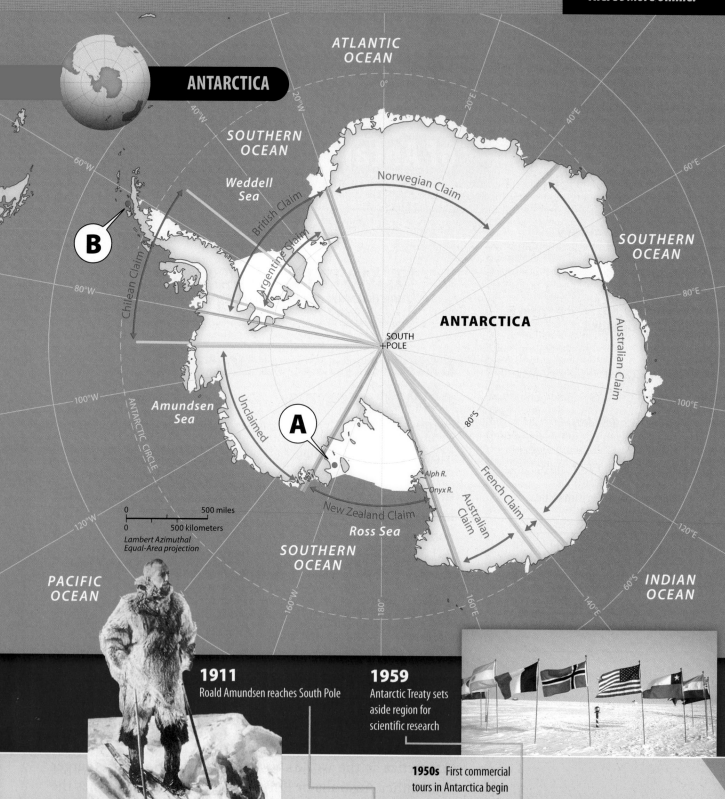

ATLANTIC
OCEAN

SOUTHERN
OCEAN

Weddell
Sea

Norwegian Claim

British Claim

Argentine Claim

Chilean Claim

B

SOUTHERN
OCEAN

ANTARCTICA

SOUTH
+POLE

Australian Claim

Amundsen
Sea

Unclaimed

A

Alph R.

Onyx R.

80°S

French Claim

Australian
Claim

ANTARCTIC CIRCLE

0   500 miles
0   500 kilometers
Lambert Azimuthal
Equal-Area projection

New Zealand Claim

Ross Sea

SOUTHERN
OCEAN

PACIFIC
OCEAN

INDIAN
OCEAN

1911
Roald Amundsen reaches South Pole

1959
Antarctic Treaty sets
aside region for
scientific research

1950s First commercial
tours in Antarctica begin

1900

2000

1820 Explorers from Russia, England, and
America claim first sighting of Antarctica

1929 Richard Byrd is first
to fly over the South Pole

1978 Antarctic Conservation Act
protects region's plant and animal life

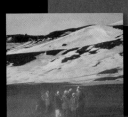

Reading **HELP**DESK

Academic Vocabulary

- **visible**

Content Vocabulary

- **ice sheet**
- **ice shelf**
- **calving**
- **iceberg**
- **katabatic wind**
- **lichen**
- **krill**
- **plankton**

TAKING NOTES: *Key Ideas and Details*

Summarize Using a chart like the one shown here, summarize the information presented in each section of the lesson.

Landforms and Water
Climate and Resources

Lesson 1
The Physical Geography of Antarctica

ESSENTIAL QUESTION · *How does physical geography influence the way people live?*

IT MATTERS BECAUSE
Antarctica plays an important role in maintaining world climate balance.

Landforms and Waters of Antarctica

GUIDING QUESTION *Is Antarctica the last unknown land region on Earth?*

More than 2,500 miles (4,023 km) south of the Tropic of Capricorn, far from the warm, sunny islands of Oceania, is the coldest, cruelest, darkest land on Earth. Nicknamed "the last continent" and "the bottom of the world," this frozen continent is Antarctica. Throughout most of history, humans did not know that this bitterly cold, windy, barren land even existed. Scholars and scientists had theories and ideas that a large "southern land" might be at the bottom of Earth. There was no proof that Antarctica was real, however, until explorers first sighted its icy shores in 1820. Exploration was, and still is, extremely difficult in this harsh land. It was not until 20 years after the first sighting that geographers confirmed that Antarctica is, in fact, a continent, not a mass of floating ice as some early explorers believed.

Size of Antarctica

Antarctica is the world's fifth-largest continent. It is larger than either Australia or Europe, and it is about 1.5 times the size of the United States. Antarctica is located in the Southern Hemisphere and is surrounded on all sides by the Southern Ocean. Together, Antarctica and the Southern Ocean cover

about 7 percent of Earth's surface. However, very few people reside, even for a short time, within these millions of square miles. The South Pole is located near the center of the continent.

Rock, Ice, and Water

Antarctica is called the highest continent because it has the highest overall elevation of any continent. Antarctica's surface is an average of 7,000 feet (2,134 m) above sea level. This figure measures the height of the **visible** surface of the continent, which is mostly solidly packed ice. While some of Antarctica's mountains rise above the ice, most of the land that makes up the foundation of the continent is much lower than the dense ice layer that covers it. In fact, the weight of so much ice has forced some of Antarctica's surface land far below sea level. Entire mountains are completely buried under layers of ice.

It might be difficult to believe, but the ice sheet covering most of Antarctica is 2 miles (3.2 km) thick in some places. An **ice sheet** is a thick layer of ice and compressed snow that forms a solid crust over an area of land. This vast ice sheet covers about 98 percent of Antarctica's surface. Scientists estimate that two-thirds of all freshwater on Earth is frozen in the Antarctic ice sheet. But it is important to remember that land lies beneath this ice sheet. By using high-resolution satellites and other specialized equipment, geographers have learned that Antarctica is one large landmass that also has an archipelago of rocky islands.

Many extremes of high and low elevation can be found across this frozen land. The highest point on the continent is Vinson Massif, which measures an incredible 16,066 feet (4,897 m) high. The Transantarctic Mountains stretch for 2,200 miles (3,541 km) across Antarctica. This enormous mountain range divides the continent into two regions: East Antarctica and West Antarctica.

Academic Vocabulary

visible able to be seen

Tourists bathe in a thermal pool, which is actually an inactive volcanic crater on Deception Island, Antarctica. Volcanic activity creates the thermal pools by heating underground water that comes to the surface.

The lowest point on the continent is deep within the Bentley Subglacial Trench. This low area is also the lowest place on the surface of Earth that is not under seawater.

Ice not only covers the land in ice sheets but also extends out into the ocean in what are called ice shelves. An **ice shelf** is a thick slab of ice that is attached to a coastline but floats on the ocean. It is a seaward extension of the ice sheet that forms a thick, hard plain of ice. During the dark winter months, temperatures fall to unbelievable lows in Antarctica. The air and water get so cold that the seawater next to the ice shelves freezes. This increases the size of the ice shelf. So much coastal water freezes into the ice shelves that Antarctica appears to double in size every winter. As the weather warms during the spring and summer months, parts of the ice shelves farthest away from the land break up into huge, floating islands of ice.

The process of ice breaking free from an ice shelf or glacier is called **calving**. The large chunk of ice that breaks off is called an **iceberg**. Icebergs are made of freshwater, not salt water. These massive bodies of ice slowly drift through frigid ocean waters with only their top portions showing. As much as 90 percent of an iceberg remains underwater. Because most of their mass is hidden below the water's surface, icebergs appear much smaller than they really are. This has led to accidents when ships have come close to what appeared to be small chunks of ice floating on the surface but then crashed into the icebergs' enormous, unseen bulk under the water. Icebergs are classified as chunks of ice larger than 16 feet (5 m) across. While many icebergs are the size of houses and mountains, some are the size of small islands. Satellite images show a huge crack in an Antarctic glacier that could produce an iceberg of 350 square miles (906 sq. km), the size of New York City.

In the middle of summer, a special-purpose ship called an icebreaker moves through an ice pack in the Ross Sea along the coast of Antarctica. The Ross Sea region has permanent year-round icy conditions, and a powerful icebreaker is the best vessel suited for its exploration.

▶ **CRITICAL THINKING**

Describing How do icebergs form?

©David Ball/Corbis

Antarctica is surrounded by stormy seas. These include the Weddell Sea, the Scotia Sea, the Amundsen Sea, the Ross Sea, and the Davis Sea. These icy waters were named by and for explorers and scientists who have mapped and studied the region. In recent years, hundreds of new species of marine animals have been discovered in the mysterious depths of these cold seas.

☑ **READING PROGRESS CHECK**

Determining Central Ideas Does the continent of Antarctica increase in size during the winter?

Researchers battle powerful winds as they carry out their research in the interior of Antarctica.

Climate and Resources

GUIDING QUESTION *Does any life exist on such a forbidding continent?*

How would you like to live in a place where summer brings a high temperature of only a few degrees above freezing?

A Dry, Frigid Climate

Because of its high latitude, Antarctica never receives direct rays from the sun, even when the Southern Hemisphere is tilted toward the sun. Thus, the sun's energy is spread out over a wider surface area than at lower-latitude places such as the Equator. The result is that very little heat energy reaches the surface of the land. In addition, being south of the Antarctic Circle, most of Antarctica receives no sunlight at all and is completely dark for at least three months of the year.

The high elevation across Antarctica affects its climate, as well. Strong, fast winds blow colder air down from high interior lands toward the coasts. These powerful gusts, called **katabatic winds**, are driven by the force of Earth's gravity.

©George Steinmetz/Corbis

A gentoo penguin leaps onto an iceberg to join its companion near Antarctica's Gerlache Strait. The gentoo is one of six penguin species that populate the Antarctic Peninsula and the many islands around the frozen continent. An orange beak, white head marking, and longer tail distinguish the gentoo from the other penguin species.

All of these factors combined produce an intensely cold climate. Temperatures are coldest in the interior highlands and warmest near the coasts.

Antarctica also has an extremely dry climate. The continent receives only 2 inches to 4 inches (5 cm to 10 cm) of precipitation each year. The extreme cold makes the climate even drier by limiting the amount of moisture the air can hold. This effect combined with lack of precipitation makes Antarctica the world's driest continent.

Antarctica plays a vital role in maintaining the global climate balance. The millions of square miles of ice covering Antarctica act like a giant reflector. The white, smooth surface of the ice shelf reflects sunlight. An estimated 80 percent of the solar radiation that reaches Antarctica is reflected back into the upper atmosphere. So, this continent-sized reflector reduces the amount of solar energy that is absorbed by Earth. On such a large scale, this effect lowers the overall temperature of the entire planet.

Plants and Wildlife

Antarctica's harsh climate limits the types of plants and animals that can live on its surface. Only 1 percent of Antarctica's land area is suitable for plant life. Plants found in Antarctica include mosses, algae, and **lichens**, which are organisms that usually grow on solid, rocky surfaces and are made up of algae and fungi. Only two species of flowering plants grow in the region: Antarctic hair grass and Antarctic pearlwort. Most plants are found on the Antarctic Peninsula and its surrounding islands. These are the warmest, wettest areas in the region. However, a few plants have been discovered in intensely cold inland areas. Scientists have found hardy species of mosses, lichens, and algae growing in the tiny cracks and pores of rocks in Victoria Land.

©Momatiuk - Eastcott/Corbis

Most of Antarctica's land animals are tiny insects and spiderlike mites. Most other land animals are only part-time visitors to the region. Weddell seals are water animals, but they only come to Antarctica's shores during summer months to raise their pups. Many species of seabirds, such as albatross, cormorants, and gulls, visit Antarctica in the summer to breed. Only the emperor penguin stays in the region year-round. These giant penguins are adapted to extremely cold temperatures and cannot survive outside of Antarctica's frozen climate.

Although the land of Antarctica is not home to many living things, the Antarctic seas are filled with life. Many kinds of seals, dolphins, fish, and other marine animals live in waters surrounding Antarctica. Whales come to the Southern Ocean to feed during the summer. Sea mammals and seabirds eat **krill**, tiny crustaceans similar to shrimp. Krill thrive in the waters of the Southern Ocean, feeding on even smaller life-forms called plankton.

Plankton are tiny organisms floating near the water's surface. Some types of plankton are single-celled bacteria; others are plants such as algae. Plankton, krill, fish, sea mammals, seabirds, and land animals are part of Antarctica's ecosystem. These creatures are essential parts of the food chain that supports life on the continent.

Resources of Antarctica

Several factors make it difficult or impossible to explore Antarctica for natural resources. The harsh climate creates dangerous conditions for human workers. The land that could hold mineral resources is buried by the thick ice sheet. Even reaching the interior of the continent with heavy mining equipment could prove impossible. As a result, mineral and energy resources that might exist have not been exploited. Some species of fish are plentiful in the Southern Ocean, and commercial fishing operations harvest fish from Antarctica's waters.

Include this lesson's information in your Foldable®.

☑ **READING PROGRESS CHECK**

Describing Where in Antarctica would you find the most living things? Explain why this is so.

LESSON 1 REVIEW **CCSS**

Reviewing Vocabulary

1. How are *krill* important to the survival of Antarctica's seabirds and sea mammals?

Answering the Guiding Questions

2. *Analyzing* Why do you think Antarctica was nicknamed "The Last Continent"?

3. *Determining Central Ideas* How are ice sheets and ice shelves alike and different?

4. *Identifying* What types of life-forms are found on the Antarctic Peninsula?

5. *Describing* Explain why Antarctica's mineral and energy resources are not commercially mined or harvested.

6. *Informative/Explanatory Writing* In your own words, explain why Antarctica's climate is so cold and dry. Use facts and details from the lesson in your writing.

Reading **HELP**DESK (CCSS)

Academic Vocabulary

- **acknowledge**

Content Vocabulary

- **treaty**
- **research station**
- **ozone**
- **remote sensing**

TAKING NOTES: *Key Ideas and Details*

Determining Central Ideas Using a chart like this one, take notes on the central idea from each section of the lesson.

Section	Central Idea
Sharing the Land	
The Scientific Continent	

Lesson 2
Life in Antarctica

ESSENTIAL QUESTION · *How do people adapt to their environment?*

IT MATTERS BECAUSE
Antarctica is a region shared by many nations. It is important to scientific research.

Sharing the Land

GUIDING QUESTION *How do people share common resources and protect unique lands?*

In 1911, two explorers, Roald Amundsen and Robert Scott, were in a race to be the first to reach the South Pole. Amundsen was a Norwegian explorer who had experience dealing with cold and snow. Scott was an English explorer who had commanded one previous expedition to Antarctica but had little experience traveling in the cold. Both men were driven to compete by a combination of curiosity and national pride. The two explorers and their teams made incredible journeys across the brutal land, losing men, ponies, and sled dogs to cold and starvation along the way. In the end, it was Amundsen's team that reached the South Pole first. Scott and his team arrived 34 days later.

The remarkable feats of Amundsen and Scott gave the world an insider's view of Antarctica for the first time. Their experiences and discoveries helped the scientists and explorers that came after them. Memorials to both men have been built in Antarctica.

The Antarctic Treaty

Antarctica is the only continent with no native population. Therefore, when the first humans visited the continent, there were no previous claims to the land. Several countries made territorial claims to sections of Antarctica's land, but these

claims were set aside by the Antarctic Treaty. A **treaty** is an agreement or a contract between two or more nations, governments, or other political groups. Signed by 12 countries in 1959, the Antarctic Treaty states that the continent of Antarctica should be used only for scientific research and other peaceful purposes. In later years, 33 additional countries signed the Antarctic Treaty. The treaty also sets rules for the management of the land and its resources. The Antarctic Treaty is an effort to protect Antarctica's environment and to prevent weapons testing and other military actions from being carried out on the continent. Other agreements such as the Antarctic Conservation Act relate to protecting the region's wildlife.

Regulating Relations

Nations including Argentina, Australia, Great Britain, New Zealand, and Norway still claim large sections of the continent as territories, but not all governments **acknowledge** these claims. Because Antarctica is not considered a nation or a country and it has no government, issues relating to the continent's use and protection are decided by the Antarctic Treaty System. This is a group of representatives from 48 countries that have interests in Antarctica. Important decisions affecting Antarctica's environment, wildlife, and resident scientists are made by annual meetings of the Antarctic Treaty System.

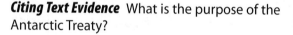 **READING PROGRESS CHECK**

Citing Text Evidence What is the purpose of the Antarctic Treaty?

©Deborah Zabarenko/Reuters/Corbis

Academic Vocabulary

acknowledge to recognize the rights, status, or authority of a person, a thing, or an event

The Amundsen-Scott station is a U.S. scientific research center at the South Pole. The original station was built in 1956 for the International Geophysical Year—a special period of international scientific investigations in Antarctica during 1957 and 1958. The rebuilt station today houses as many as 200 people in summer and about 50 in winter.

The Scientific Continent

GUIDING QUESTION *How and why do scientists study Antarctica?*

Because of its harsh climate, Antarctica is the only continent that has no permanent human settlement. Its largely unspoiled environment has made it a favorable place for scientific research.

Living in Antarctica

No one lives in Antarctica full time, but many people stay in this harsh world for months or even years at a time. During the summer, when the sun shines brightly and temperatures are cold but not deadly, about 4,400 people reside in Antarctica. When the temperatures fall and the constant darkness of winter sets in, this temporary population shrinks to about 1,100 people. Most of these people are scientists and support teams living in shelters on land, but some stay aboard ships near the coasts. People from all over the world come to the frozen continent to study climate change, oceanography, geology, and countless other sciences involving Antarctica's landforms, weather, plants, animals, and the ancient rocks that form the continental shelf below the land.

Why do most research scientists stay in Antarctica only during the summer season? Weather conditions during summer months are still cold and windy, but they are not nearly as severe as those in winter. Summer temperatures usually remain below freezing, but the sun shines 24 hours a day. In winter, powerful winds tear across the land from the mountains, creating blizzard-like conditions. Temperatures are so low that without protective clothing, humans can freeze to death in a matter of minutes. The ice shelves that form along the coasts prevent ships from coming anywhere near the land. For these reasons, summer is the best time for scientists to work in Antarctica. Only a small number of brave, patient researchers continue their work through the long winters, staying sheltered indoors as much as possible.

Governments with scientific interests in Antarctica have built research stations in different locations across the continent. A **research station** is a base where scientists live and work. These stations usually consist of a few

A scientist from South Korea collects meteorites on an ice field in the Ross Sea region of Antarctica. Meteorites are rocks from space that have fallen on earth. They provide clues about the history of our solar system. Antarctica is the best place in the world to collect meteorites. There, meteorites fall into ice rather than hard rock and so are damaged less. Also, Antarctica's icy environment makes them more visible and easier to collect.

▶ **CRITICAL THINKING**

Analyzing Why do scientists come to Antarctica despite its climatic conditions?

©YONHAP/epa/Corbis

September 1980   **September 1990**   **September 2000**   **September 2010**

Thinner Ozone ▬▬▬▬▬▬ Thicker Ozone

SOURCE: http://ozonewatch.gsfc.nasa.gov/

OZONE HOLE

Images show changes in the size of the ozone hole over Antarctica during a 30-year period.

▶ **CRITICAL THINKING**

Analyzing How has the size of the ozone hole changed over time? What do these changes reveal?

buildings set close together, each built with special insulation against the brutally cold and windy weather. Currently, Antarctica has about 50 research stations. Each station has living quarters for the scientists and other staff, such as medical doctors, cooks, and mechanics. Stations also have laboratories, other types of research facilities, and large storage areas filled with fuel, food, and equipment. Research stations need to have enough food stored away to feed teams of scientists for a year or more. Having extra food is important because violent storms can happen at any time, making it impossible for planes or ships to bring in fresh food and other supplies.

The Ozone Layer

One important area of study focuses on the ozone layer that is high in Earth's atmosphere. **Ozone** is a gas in the atmosphere that absorbs harmful ultraviolet radiation from the sun. Through decades of research and study, scientists discovered that the layer of ozone in our planet's atmosphere had thinned and decreased. This means that less ozone exists to protect Earth and life on Earth from damaging solar radiation.

The two locations on Earth where levels of ozone are the lowest are over the Arctic Circle and over Antarctica and the Southern Ocean. Loss of ozone affects all life on the planet. Scientists in Antarctica and other parts of the world are working to find a way to protect and preserve the ozone layer.

Studying Earth From Above

Have you ever wondered how scientists are able to figure out what lies below the surface of Earth? For example, how do scientists know what landforms lie under the thick Antarctic ice sheet? Scientists are able to study what they cannot see, such as layers of rock under the ice, by using an amazing kind of technology called remote sensing.

Remote sensing involves using scientific instruments placed onboard weather balloons, airplanes, and satellites to study the planet. Remote-sensing technology is used to collect data, allowing scientists to take images and measurements of objects and places that would be impossible to reach in person. Some remote-sensing equipment emits electromagnetic waves that create detailed images of landforms and bodies of water buried under miles of ice. This technology has allowed scientists to study Antarctica's rock foundation and the floor of the Southern Ocean. Remote sensing has helped geographers learn more about how Antarctica's land, water, and atmosphere interact and affect the global climate system.

Research Yields Clues

The research work done in Antarctica has resulted in many amazing discoveries. Scientists have learned about our planet's birth and evolution over time, and they have found clues about the origins of the universe. By comparing rocks and fossils found in Antarctica to those found on other continents, scientists have determined that the Antarctic continent was once part of a huge landmass called Gondwana. Antarctica, Asia, Africa, and other continents broke away from this supercontinent millions of years ago and drifted across Earth's surface as a result of plate tectonics.

Climate Change

Another major area of research in Antarctica is climate change. Through satellite imagery and local measurements, geographers have learned that the Antarctic ice sheet is shrinking. Many people are concerned that this shrinkage might be caused by global warming. This concern is twofold: First, scientists fear that the impact of climate change will permanently damage Antarctica's environment and ecosystems. Second, Antarctica is seen as a "global barometer," or an indicator of what is happening to the climate of the entire planet. Signs such as melting ice and shrinking icebergs tell geographers and scientists that global temperatures are rising quickly. As massive chunks of ice calve, or break away, from

Tourists enjoy close-up views of a female humpback whale and her calf. Humpback whales receive their name because they raise and bend their backs to begin diving. They gather in groups along the Antarctic coast, where they feed and breed. Because humpback whales are slow swimmers, tourist boats can easily approach them.

▶ **CRITICAL THINKING**

Describing How is climate change affecting the life of Antarctica's sea animals?

Hugh Rose/Danita Delimont/Alamy

Antarctica's glaciers, more icebergs clutter the surrounding waters. This makes traveling by ship in these areas more difficult and dangerous. Even more serious is the possibility that as these icebergs melt, global sea levels will rise. Rising sea levels would make survival impossible in many parts of the world.

Climate change is also affecting the food supplies of Antarctic land and sea animals. Scientists believe that global warming and changing water currents are affecting the amount of plankton and krill in the region. Recent studies have found low populations of these tiny water organisms in the waters around Antarctica. Without enough plankton and krill to eat, fish will die off or leave the area. This leaves seals and penguins without enough fish to eat. This disruption in the local food chain could have terrible consequences for Antarctica's animal life.

Taking a Tour

The bitterly cold, icy shores of Antarctica might not seem like a great place to spend a vacation. Still, about 6,000 tourists visit Antarctica each year. Most visitors are interested in seeing the incredible natural beauty of the area, taking photographs of wildlife, and visiting the research stations. Some are excited by the idea of visiting a harsh, dangerous place that few humans will ever see firsthand. Most tourists choose Antarctica because they want an adventure more than a vacation.

Thanks to the work of the scientists who have studied Antarctica, humans have come a long way in understanding this mysterious continent. However, there is still much to learn. Geographers and other scientists living and working in Antarctica are dedicated to their research. With each passing year, they discover more clues about our planet's past. They also use what they learn to make predictions about our planet's future. Antarctica's secrets are slowly being uncovered, but much remains unknown about this frozen land at the bottom of the world.

☑ **READING PROGRESS CHECK**

Determining Central Ideas Why do scientists think Antarctica was part of a larger landmass that included continents such as Africa and Asia?

Thinking Like a Geographer

Lake Vostok

After spending years drilling through more than two miles (3 km) of solid ice in Antarctica, Russian scientists reached the surface of a gigantic freshwater lake in February 2012. Named Lake Vostok, it is the largest of the hundreds of the continent's subglacial lakes—roughly the size of Lake Ontario in North America. Scientists are taking samples of the water hoping to find living organisms that could provide clues about the unusual environment.

Include this lesson's information in your Foldable®.

LESSON 2 REVIEW (CCSS)

Reviewing Vocabulary
1. Why is *ozone* high in the atmosphere important?

Answering the Guiding Questions
2. *Describing* Describe the 1911 race to the South Pole in your own words.

3. *Analyzing* What are some of the advantages of using remote sensing to study Earth?

4. *Identifying* What are the possible causes and effects of Antarctica's shrinking ice sheet?

5. *Narrative Writing* Imagine that you are a young scientist who has just arrived in Antarctica for the summer. Write a fictional journal entry describing your research and daily life at a research station. Whenever possible, use facts, details, and vocabulary terms from the lesson in your narrative.

What Do You Think?

CCSS

Is Global Warming a Result of Human Activity?

Scientists agree that Earth's climate is changing. They do not agree on what is causing the change. Is it just another natural warming cycle like so many cycles that have occurred in the past? Scientists who support this position cite thousands of years' worth of natural climatic change as evidence. Or is climate change anthropogenic—caused by human activity? Scientists who support this position cite the warming effect of rapidly increasing amounts of greenhouse gases in the atmosphere. Greenhouse gases occur naturally, but they also result from the burning of fossil fuels. Which side's evidence is more convincing?

No !

PRIMARY SOURCE

" The Intergovernmental Panel on Climate Change (IPCC), an agency of the United Nations, claims the warming that has occurred since the mid-twentieth century "is *very likely* due to the observed increase in anthropogenic greenhouse gas concentrations." Many climate scientists disagree with the IPCC on this key issue.

Scientists who study the issue say it is impossible to tell if the recent small warming trend is natural, a continuation of the planet's recovery from the more recent "Little Ice Age," or unnatural, the result of human greenhouse gas emissions. Thousands of peer-reviewed articles point to natural sources of climate variability that could explain some or even all of the warming in the second half of the twentieth century. S. Fred Singer and Dennis Avery documented natural climate cycles of approximately 1,500 years going back hundreds of thousands of years. "

—Joseph Bast and James M. Taylor, "Global Warming: Not a Crisis," The Heartland Institute

A scientist weighs an ice core sample on a glacier in Antarctica. Ice cores provide detailed information about changes in Earth's climate over many centuries.

TEXT: Climate Change 2007, The Physical Science Basis: Working Group I Contribution to the Fourth Assessment Report of the Intergovernmental Panel on Climate Change; FAQ 9.2, p. 702. Cambridge University Press.

Afternoon traffic flows along the streets of Beijing, China, on a smog-filled day. Despite efforts to improve air quality, the Chinese capital remains one of the world's most polluted cities.

Yes!

PRIMARY SOURCE

" It is very unlikely that the 20th-century warming can be explained by natural causes.... Palaeoclimatic reconstructions show that the second half of the 20th century was likely the warmest 50-year period in the Northern Hemisphere in the last 1300 years. This rapid warming is consistent with the scientific understanding of how the climate should respond to a rapid increase in greenhouse gases like that which has occurred over the past century, and the warming is inconsistent with the scientific understanding of how the climate should respond to natural external factors such as variability in solar output and volcanic activity. Climate models provide a suitable tool to study the various influences on the Earth's climate. When the effects of increasing levels of greenhouse gases are included in the models, as well as natural external factors, the models produce good simulations of the warming that has occurred over the past century. The models fail to reproduce the observed warming when run using only natural factors. "

—Contribution of Working Group I: The Physical Science Basis to the Fourth Assessment Report of the Intergovernmental Panel on Climate Change, 2007

What Do You Think?   DBQ

1 **Analyzing** What specific evidence is offered to support the position that warming is caused by human activity?

2 **Identifying Point of View** What specific evidence is offered to support the position that warming results from natural causes?

Critical Thinking

3 **Analyzing** What effect could belief in one viewpoint or the other have on people or governments?

Directions: Write your answers on a separate piece of paper.

1 **Exploring the Essential Question**

INFORMATIVE/EXPLANATORY WRITING In two or more paragraphs, explain why Antarctica can be described as a polar desert.

2 **21st Century Skills**

PEER REVIEW Work with a partner. Each partner will write a paragraph to answer the question: What do you think it's like to live in a cold climate like Antarctica? Check your partner's paper for meaning, grammar, and complete sentences. Discuss the review of your paragraph with your partner. Revise as needed.

3 **Thinking Like a Geographer**

LISTING Create a Venn diagram like the one shown here. Label one side West Antarctica and the other side East Antarctica. List some physical geography features that can be found in each. In the center of the diagram, list features found in both.

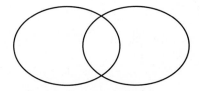

4 **GEOGRAPHY ACTIVITY**

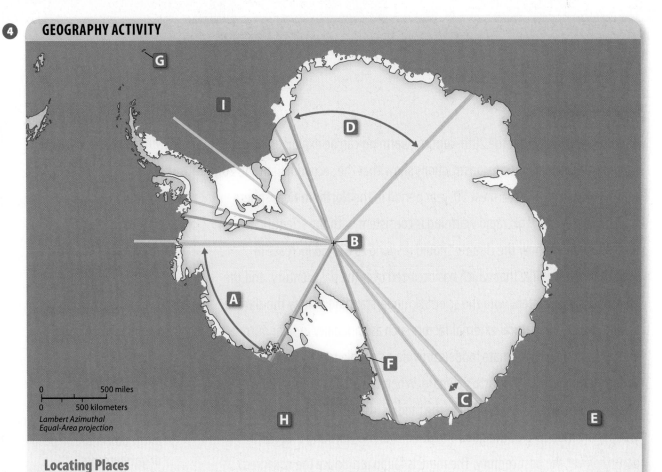

0   500 miles
0   500 kilometers
Lambert Azimuthal Equal-Area projection

Locating Places

Match the letters on the map with the numbered places or phrases listed below.

1. French Claim
2. Norwegian Claim
3. Southern Ocean
4. Unclaimed
5. Weddell Sea
6. Indian Ocean
7. South Orkney Islands
8. South Pole
9. Onyx River

REVIEW THE GUIDING QUESTIONS

Directions: Choose the best answer for each question.

1 How much of Earth's freshwater is frozen in Antarctic ice?

A. one-half

B. one-third

C. one-fourth

D. two-thirds

2 Who was the first explorer to reach the South Pole?

F. Ernest Shackleton

G. Roald Amundsen

H. Robert Scott

I. Richard Byrd

3 It gets so cold in Antarctica during the winter that salt water from the surrounding seas freezes into

A. huge ice sculptures.

B. ice shelves.

C. ice cliffs.

D. icebergs.

4 During the summer, birds, whales, and other sea mammals come to the Southern Ocean around Antarctica to feed on tiny crustaceans called

F. plankton.

G. shrimp.

H. krill.

I. lichens.

5 How does the Antarctic Treaty protect the continent's environment?

A. It prohibits drilling for oil in the Southern Ocean.

B. It restricts the number of people who can be there at any one time.

C. It prohibits tourism.

D. It reserves the continent for peaceful scientific research.

6 In which months of the year do researchers in Antarctica have to tolerate months of total darkness, frigid temperatures, and howling winds?

F. July, August, September

G. only December

H. all 12 months

I. April, May, June

DBQ ANALYZING DOCUMENTS

7 **DETERMINING CENTRAL IDEAS** Read the following passage about sea ice in the Arctic and Antarctic regions:

"*Sea ice differs between the Arctic and Antarctic, primarily because of their different geography. The Arctic is . . . almost completely surrounded by land. As a result, the sea ice that forms in the Arctic is not [very] mobile. . . . Antarctica is a land mass surrounded by an ocean. . . . Sea ice is free to float northward into warmer waters.*"

—from "Arctic vs. Antarctic," National Snow & Ice Data Center

What causes the differences between the two kinds of sea ice?

A. the different geographies of the two regions

B. the colder temperatures of Antarctica

C. different impacts of global warming

D. coldness of Arctic waters

8 **ANALYZING** Which of these events is more likely to occur to Antarctic sea ice than Arctic sea ice?

F. growing larger over time

G. piling up in high ice jams

H. slowly melting over time

I. being packed tightly with other ice

SHORT RESPONSE

"*Lars-Eric Lindblad led the first traveler's expedition to Antarctica in 1966. Lindblad once said, "You can't protect what you don't know." He believed that by providing a first-hand experience to tourists you would . . . promote a greater understanding of the earth's resources and the important role of Antarctica in the global environment.*"

—from "Tourism Overview," International Association of Antarctic Tour Operators

9 **CITING TEXT EVIDENCE** According to Lindblad, how is experience connected to environmental preservation?

10 **IDENTIFYING POINT OF VIEW** Do you agree with Lindblad? Why or why not?

EXTENDED RESPONSE

11 **INFORMATIVE/EXPLANATORY WRITING** Conduct research and then write a paper on what it is like to live and work on a research station in Antarctica. To find out what it was like to build these research stations under such harsh conditions and how the stations are equipped to handle the weather and protect their inhabitants from the coldest weather on Earth.

Need Extra Help?

If You've Missed Question	❶	❷	❸	❹	❺	❻	❼	❽	❾	❿	⓫
Review Lesson	1	2	1	1	2	2	1	1	2	2	2

Using **FOLDABLES** is a great way to organize notes, remember information, and prepare for tests. Follow these easy directions to create a Foldable® for the chapter you are studying.

CHAPTER 1: THE GEOGRAPHER'S WORLD

Describing Make this Foldable and label the top *Geographer's View* and the bottom *Geographer's Tools*. Under the top fold, describe three ways you experience geography every day. Under the bottom fold, list and describe the tools of geography and explain how a map is a tool. In your mind, form an image of a map of the world. Sketch and label what you visualize on the back of your shutter fold.

Step 1
Bend a sheet of paper in half to find the midpoint.

Step 2
Fold the outer edges of the paper to meet at the midpoint.

CHAPTER 2: PHYSICAL GEOGRAPHY

Identifying Make this Foldable and label the four tabs *Processes*, *Forces*, *Land*, and *Water*. Under *Processes*, identify and describe the processes that operate above and below Earth's surface. Include specific examples. Under *Forces*, give examples of how forces are changing Earth's surface where you live. Finally, under *Land* and *Water*, identify land and water features within 100 miles (161 km) of your community and explain how they influence your life.

Step 1
Fold the outer edges of the paper to meet at the midpoint. Crease well.

Step 2
Fold the paper in half from side to side.

Step 3
Open and cut along the inside fold lines to form four tabs.

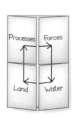

Step 4
Label the tabs as shown.

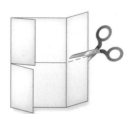

CHAPTER 3: HUMAN GEOGRAPHY

Analyzing Create this Foldable, and then label the tabs *Adaptations*, *Cultural Views*, and *Basic Needs*. Under *Adaptations*, describe how humans have adapted to life in two different geographic regions and describe population trends in each. Under *Cultural Views*, analyze what makes two different cultures unique. Finally, under *Basic Needs*, describe how your basic needs might be met in two different economic systems.

Step 1
Fold a sheet of paper in half, leaving a ½-inch tab along one edge.

Step 2
Then fold the paper into three equal sections.

Step 3
Cut along the folds on the top sheet of paper to create three tabs.

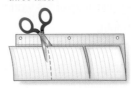

Step 4
Label your Foldable as shown.

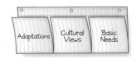

CHAPTER 4: THE UNITED STATES EAST OF THE MISSISSIPPI

Organizing After you create the Foldable below, write the chapter title on the cover tab and label the three small tabs *East and West*, *Geographic Barriers*, and *Diversity*. Under *East and West*, sketch an outline of the United States and draw the Mississippi River. Then list facts about the region. Under *Geographic Barriers*, give examples of physical features that were barriers to westward expansion. Under *Diversity*, explain how cultural diversity makes the United States East of the Mississippi a unique region.

Step 1
Stack two sheets of paper so that the back sheet is 1 inch higher than the front sheet.

Step 2
Fold the paper to form four equal tabs.

Step 3
When all tabs are an equal distance apart, fold the papers and crease well.

Step 4
Open the papers, and then glue or staple them along the fold.

Foldables® Library

CHAPTER 5: THE UNITED STATES WEST OF THE MISSISSIPPI

Identifying Make a three-tab book with three columns. Label the columns *Geography*, *History*, and *Economy*. Under each column heading, write: *Know* and *Learned*. Use the book to record what you know and what you learn about the western region of the United States.

Step 1
Fold a sheet of paper in half, leaving a ½-inch tab along one edge.

Step 2
Then fold the paper into three equal sections.

Step 3
Cut along the folds on the top sheet of paper to create three tabs.

Step 4
Label your Foldable as shown.

CHAPTER 6: CANADA

Identifying Follow the steps below, and then label the four tabs *North*, *South*, *Past and Present*, and *World Relations*. Describe and give examples of the geography of the northern and southern regions of Canada under either the *North* or the *South* tab. Under *Past and Present*, identify important people, places, and events from Canada's history. Explain Canada's relations with other countries under the *World Relations* tab.

Step 1
Fold the outer edges of the paper to meet at the midpoint. Crease well.

Step 2
Fold the paper in half from side to side.

Step 3
Open and cut along the inside fold lines to form four tabs.

Step 4
Label the tabs as shown.

Foldables® Library

CHAPTER 7: MEXICO, CENTRAL AMERICA, AND THE CARIBBEAN ISLANDS

Analyzing Make the Foldable below. Write the chapter title on the cover tab, and label the three small tabs *Gulf of Mexico*, *Civilizations*, and *Trade and Commerce*. Under *Gulf of Mexico*, explain how the gulf has affected life in the region. Include information on weather, tourism, and the economy. Under *Civilizations*, sequence and describe the major civilizations that developed in this region and their cultural influences. Finally, under *Trade and Commerce*, compare and contrast trade events that are important to the economy of the region.

Step 1
Stack two sheets of paper so that the back sheet is 1 inch higher than the front sheet.

Step 2
Fold the paper to form four equal tabs.

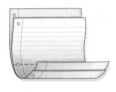

Step 3
When all tabs are an equal distance apart, fold the papers and crease well.

Step 4
Open the papers, and glue or staple them along the fold.

CHAPTER 8: BRAZIL

Organizing Create the Foldable below. On the back, write the chapter title and sketch a map of Brazil. Label the two front tabs *Valuable Natural Resources* and *Urban Population*. On your sketch, label Brazil's major geographic features. Under *Valuable Natural Resources*, outline when and where valuable natural resources were discovered and how the discoveries affected the native and colonial populations. Under *Urban Population*, discuss the impact of the population distribution.

Step 1
Bend a sheet of paper in half to find the midpoint.

Step 2
Fold the outer edges of the paper to meet at the midpoint.

CHAPTER 9: THE TROPICAL NORTH

Identifying Create the Foldable below. Label the cover *The Tropical North* and the layers *Geography*, *Foreign Influences and Resources*, and *Trade*. Under *Geography*, explain how geography and resources affect the countries and people of the region. Under *Foreign Influences and Resources*, explain how resources and foreign countries have impacted the Tropical North. Under *Trade*, explain how countries in the region are trying to expand trade and why.

Step 1
Stack two sheets of paper so that the back sheet is 1 inch higher than the front sheet.

Step 2
Fold the paper to form four equal tabs.

Step 3
When all tabs are an equal distance apart, fold the papers and crease well.

Step 4
Open the papers, and then glue or staple them along the fold.

CHAPTER 10: ANDES AND MIDLATITUDE COUNTRIES

Describing Make the Foldable below, and then label the top of the sections *Geography*, *Culture*, and *Economy*. Under *Geography*, explain how the Andes Mountains affect the lives of the people who live near or around them. Under *Culture*, describe the rise and fall of the Inca Empire and what it tells about the history of the region. Finally, under *Economy*, explain how the terrain and the resources available affect the way people live.

Step 1
Fold a sheet of paper into thirds to form three equal columns.

Step 2
Label your Foldable as shown.

Foldables® Library

CHAPTER 11: WESTERN EUROPE

Analyzing Follow the steps below to create a Foldable. Then sketch an outline of Western Europe on the back. Label important geographic features in the region. On the front, label the tabs as illustrated. Under *Waterways—Landforms*, explain how waterways and landforms have influenced the development of Western Europe. Under *Early Civilizations—Industrial Revolution*, sequence the cultural and technological changes that occurred. Finally, under *War—Post-War*, summarize the effects of war on the region and explain why the EU was formed.

Step 1
Fold the outer edges of the paper to meet at the midpoint. Crease well.

Step 2
Open and cut three equal tabs from the outer edge to the crease on each side.

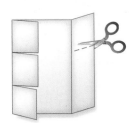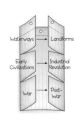

Step 3
Label the tabs as shown.

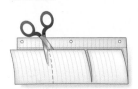

CHAPTER 12: NORTHERN AND SOUTHERN EUROPE

Describing Create the Foldable below, and label the anchor tab *Northern and Southern Europe*. Label the front of the tabs *Geography*, *History*, and *Culture*. Under *Geography*, describe how geography has influenced the lifestyles and economies in the region. Under *History*, explain the importance of the Silk Road and how trade with China influenced the people of the region. Finally, under *Culture*, identify different culture groups in each region.

Step 1
Fold a sheet of paper in half, leaving a ½-inch tab along one edge.

Step 2
Then fold the paper into three equal sections.

Step 3
Cut along the folds on the top sheet of paper to create three tabs.

Step 4
Label your Foldable as shown.

CHAPTER 13: EASTERN EUROPE AND WESTERN RUSSIA

Organizing Create this Foldable, and then sketch and label Eastern Europe, Western Russia, and the Ural Mountains on the back. On the front, label the top-left tab *Eastern Europe* and the top-right tab *Western Russia*. Under the tabs, describe landforms and natural resources found in each area. Label the two bottom tabs *Empires* and *Populations*. List the empires that once controlled this region and one important event from each. Finally, explain why populations are declining.

Step 1
Fold the outer edges of the paper to meet at the midpoint. Crease well.

Step 2
Fold the paper in half from side to side.

Step 3
Open and cut along the inside fold lines to form four tabs.

Step 4
Label the tabs as shown.

CHAPTER 14: EAST ASIA

Analyzing Create the Foldable below. Write the chapter title on the front and label the tabs *Mainland and Islands*, *Cultural Influences*, and *Economic Growth*. Under *Mainland and Islands*, describe the physical environment of the mainland and the islands. Under *Cultural Influences*, draw a three-circle Venn diagram and label the circles *China*, *Japan*, and *Korea*. Use the diagram to analyze similarities and differences in the history of these countries. Under *Economic Growth*, summarize the present economy of the region.

Step 1
Stack two sheets of paper so that the back sheet is 1 inch higher than the front sheet.

Step 2
Fold the paper to form four equal tabs.

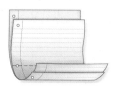

Step 3
When all tabs are an equal distance apart, fold the papers and crease well.

Step 4
Open the papers, and then glue or staple them along the fold.

| Economic Growth |
| Cultural Influences |
| Mainland and Islands |
| East Asia |

Foldables® Library

CHAPTER 15: SOUTHEAST ASIA

Identifying Sketch a map of Southeast Asia on the back of the Foldable and label the 11 countries discussed in the text. Label the top of each of the three columns: *Geography*, *History*, and *Cultural Diversity*. Use the sections to identify three geographic features that make Southeast Asia unique, make a time line of events that occurred during the history of the spice trade, and explain why Southeast Asia is such a culturally diverse region.

Step 1
Fold a sheet of paper into thirds to form three equal columns.

Step 2
Label your Foldable as shown.

CHAPTER 16: SOUTH ASIA

Organizing Make the Foldable below. Cut notebook paper into eighths to make small note cards that fit in the pockets. Label the pockets *Geography*, *History*, and *Economy*. Use small note cards to record information on major geographic features in this region, to record important historical events, and to document economic and political events that have affected the region over the last century.

Step 1
Fold the bottom edge of a piece of paper up 2 inches to create a flap.

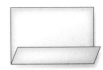

Step 2
Fold the paper into thirds.

Step 3
Glue or staple the flap on both edges and at both fold lines to form pockets. Label as shown.

CHAPTER 17: CENTRAL ASIA, THE CAUCASUS, AND SIBERIAN RUSSIA

Identifying Sketch an outline of Central Asia, the Caucasus, and Siberian Russia on the back of the Foldable and label the physical features. Label the top two tabs *Waterways—Landforms,* and list examples of each that form borders between countries within this region. Label the two middle tabs *Conquest—Independence,* and list examples of empires from the region and countries that are currently independent. Label the bottom two tabs *Rural—Urban*. Then describe and differentiate between rural and urban life in the region.

Step 1
Fold the outer edges of the paper to meet at the midpoint. Crease well.

Step 2
Open and cut three equal tabs from the outer edge to the crease on each side.

Step 3
Label the tabs as shown.

CHAPTER 18: SOUTHWEST ASIA

Describing On your Foldable, label the three tabs *Water, Civilization and Religion,* and *Oil and Water*. Under *Water*, describe the role of freshwater and salt water in the development of Southwest Asia. Under *Civilization and Religion*, explain why this region in called the "cradle of civilization." Under *Oil and Water*, describe and compare the importance of oil and water to the economy of the region.

Step 1
Fold a sheet of paper in half, leaving a ½-inch tab along one edge.

Step 2
Then fold the paper into three equal sections.

Step 3
Cut along the folds on the top sheet of paper to create three tabs.

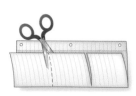

Step 4
Label your Foldable as shown.

CHAPTER 19: NORTH AFRICA

Organizing Label the rows *Geography*, *History*, and *Economy*. Label the columns *Know* and *Learned*. Use the table to record what you know and what you learn about the geography, history, and economy of North Africa.

Step 1
Fold the paper into three equal columns. Crease well.

Step 2
Open the paper and then fold it into four equal rows. Crease well. Unfold and label as shown.

CHAPTER 20: EAST AFRICA

Analyzing Write the chapter title on the cover and label the tabs *"Great" Things; Past Affects Present;* and *Daily Life, Literacy, and Health.* Under the first tab, describe the Great Rift Valley and the Great Migration and their impact on the economy of the region. Under the second tab, give examples of historical events that affect the region's current economy and politics. Under the third tab, compare two countries in East Africa using literacy rates and life expectancy.

Step 1
Stack two sheets of paper so that the back sheet is 1 inch higher than the front sheet.

Step 2
Fold the paper to form four equal tabs.

Step 3
When all tabs are an equal distance apart, fold the papers and crease well.

Step 4
Open the papers and then glue or staple them along the fold.

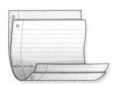

CHAPTER 21: CENTRAL AFRICA

Describing   Label the four tabs *Rain Forests*, *Savannas*, *Triangular Trade*, and *Rural vs. Urban*. Under the *Rain Forests* and *Savannas* tabs, differentiate between the climate and vegetation of rain forests and savannas. Under *Triangular Trade*, describe the three stages of the triangular trade route and explain why each was profitable for merchants. Under *Rural vs. Urban,* explain why you think the capital cities of many of the countries within this region are becoming huge metropolises.

Step 1
Fold the outer edges of the paper to meet at the midpoint. Crease well.

Step 2
Fold the paper in half from side to side.

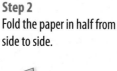

Step 3
Open and cut along the inside fold lines to form four tabs.

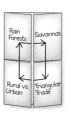

Step 4
Label the tabs as shown.

CHAPTER 22: WEST AFRICA

Analyzing   Label the first tab *Natural Resources,* and identify which resources you think should be protected and which should be developed. Label the second tab *Trade*. Summarize the importance of trade to the region and list three valuable trade goods. Finally, label the third tab *Population Growth*. Describe how and why rapid population growth is negatively affecting the economy of the region.

Step 1
Fold a sheet of paper in half, leaving a ½-inch tab along one edge.

Step 2
Then fold the paper into three equal sections.

Step 3
Cut along the folds on the top sheet of paper to create three tabs.

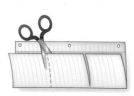

Step 4
Label your Foldable as shown.

CHAPTER 23: SOUTHERN AFRICA

Organizing Cut notebook paper into eighths to make small note cards that fit in the pockets. Sketch an outline of Southern Africa on the back of the Foldable and label geographic features of the region. On the front, label the pockets *Geography*, *History*, and *Economy*. On the note cards, record information about major geographic features, important historical events, and economic and political events that occurred in the region.

Step 1
Fold the bottom edge of a piece of paper up 2 inches to create a flap.

Step 2
Fold the paper into thirds.

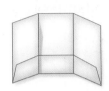

Step 3
Glue the flap on both edges and at both fold lines to form pockets. Label as shown.

CHAPTER 24: AUSTRALIA AND NEW ZEALAND

Describing Label the top section *Australia* and the bottom section *New Zealand*. Under the top-left tab, describe three geographic features of Australia. Under the bottom-left tab, describe the geography of New Zealand. Under the middle tabs, describe the ethnic groups in the regions. Under the top- and bottom-right tabs, describe issues facing Australia and New Zealand.

Step 1
Fold the outer edges of the paper to meet at the midpoint. Crease well.

Step 2
Label the tabs as shown.

Step 3
Open and cut three equal tabs from the outer edge to the crease on each side.

CHAPTER 25: OCEANIA

Identifying Write the chapter title on the cover tab, and label the three small tabs *Resources; Past Affects Present;* and *Daily Life, Literacy, and Health*. Under *Resources*, explain the impact of limited resources on Oceania's economy. Under *Past Affects Present*, explain how past colonization affects the economy and government of two countries in Oceania. Under *Daily Life, Literacy, and Health*, explain how the migration of young adults from Oceania to other countries is affecting the region.

Step 1
Stack two sheets of paper so that the back sheet is 1 inch higher than the front sheet.

Step 2
Fold the paper to form four equal tabs.

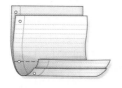

Step 3
When all tabs are an equal distance apart, fold the papers and crease well.

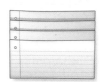

Step 4
Open the papers, and then glue or staple them along the fold.

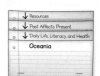

CHAPTER 26: ANTARCTICA

Describing Label the top three columns of your Foldable *Extremes, The Antarctic Treaty,* and *Global Importance*. Under *Extremes*, describe the geographic and climate extremes that make this continent unique. Under *The Antarctic Treaty* and *Global Importance*, summarize the history of Antarctica and the Antarctic Treaty, and explain the global importance of this protected continent.

Step 1
Fold a sheet of paper into thirds to form three equal columns.

Step 2
Label your Foldable as shown.

Gazetteer

A gazetteer (ga•zuh•TIHR) is a geographic index or dictionary. It shows latitude and longitude for cities and certain other places. Latitude and longitude are shown in this way: 48°N 2°E, or 48 degrees north latitude and two degrees east longitude. This Gazetteer lists many important geographic features and most of the world's largest independent countries and their capitals. The page numbers tell where each entry can be found on a map in this book. As an aid to pronunciation, most entries are spelled phonetically.

A

Abidjan [AH•BEE•JAHN] Capital of Côte d'Ivoire. 5°N 4°W (p. RA22)

Abu Dhabi [AH•BOO DAH•bee] Capital of the United Arab Emirates. 24°N 54°E (p. RA24)

Abuja [ah•BOO•jah] Capital of Nigeria. 8°N 9°E (p. RA22)

Accra [ah•KRUH] Capital of Ghana. 6°N 0° longitude (p. RA22)

Addis Ababa [AHD•dihs AH•bah•BAH] Capital of Ethiopia. 9°N 39°E (p. RA22)

Adriatic [AY•dree•A•tihk] **Sea** Arm of the Mediterranean Sea between the Balkan Peninsula and Italy. (p. RA20)

Afghanistan [af•GA•nuh•STAN] Central Asian country west of Pakistan. (p. RA25)

Albania [al•BAY•nee•uh] Country on the Adriatic Sea, south of Serbia. (p. RA18)

Algeria [al•JIHR•ee•uh] North African country east of Morocco. (p. RA22)

Algiers [al•JIHRZ] Capital of Algeria. 37°N 3°E (p. RA22)

Alps [ALPS] Mountain ranges extending through central Europe. (p. RA20)

Amazon [A•muh•ZAHN] **River** Largest river in the world by volume and second-largest in length. (p. RA17)

Amman [a•MAHN] Capital of Jordan. 32°N 36°E (p. RA24)

Amsterdam [AHM•stuhr•DAHM] Capital of the Netherlands. 52°N 5°E (p. RA18)

Andes [AN•DEEZ] Mountain system extending north and south along the western side of South America. (p. RA17)

Andorra [an•DAWR•uh] Small country in southern Europe between France and Spain. 43°N 2°E (p. RA18)

Angola [ang•GOH•luh] Southern African country north of Namibia. (p. RA22)

Ankara [AHNG•kuh•ruh] Capital of Turkey. 40°N 33°E (p. RA24)

Antananarivo [AHN•tah•NAH•nah•REE•voh] Capital of Madagascar. 19°S 48°E (p. RA22)

Arabian [uh•RAY•bee•uhn] **Peninsula** Large peninsula extending into the Arabian Sea. (p. RA25)

Argentina [AHR•juhn•TEE•nuh] South American country east of Chile. (p. RA16)

Armenia [ahr•MEE•nee•uh] European-Asian country between the Black and Caspian Seas. 40°N 45°E (p. RA26)

Ashkhabad [AHSH•gah•BAHD] Capital of Turkmenistan. 38°N 58°E (p. RA25)

Asmara [az•MAHR•uh] Capital of Eritrea. 16°N 39°E (p. RA22)

Astana Capital of Kazakhstan. 51°N 72°E (p. RA26)

Asunción [ah•SOON•see•OHN] Capital of Paraguay. 25°S 58°W (p. RA16)

Athens Capital of Greece. 38°N 24°E (p. RA19)

Atlas [AT•luhs] **Mountains** Mountain range on the northern edge of the Sahara. (p. RA23)

Australia [aw•STRAYL•yuh] Country and continent in Southern Hemisphere. (p. RA30)

Austria [AWS•tree•uh] Western European country east of Switzerland and south of Germany and the Czech Republic. (p. RA18)

Azerbaijan [A•zuhr•BY•JAHN] European-Asian country on the Caspian Sea. (p. RA25)

B

Baghdad Capital of Iraq. 33°N 44°E (p. RA25)

Bahamas [buh•HAH•muhz] Country made up of many islands between Cuba and the United States. (p. RA15)

Bahrain [bah•RAYN] Country located on the Persian Gulf. 26°N 51°E (p. RA25)

Baku [bah•KOO] Capital of Azerbaijan. 40°N 50°E (p. RA25)

Balkan [BAWL•kuhn] **Peninsula** Peninsula in southeastern Europe. (p. RA21)

Baltic [BAWL•tihk] **Sea** Sea in northern Europe that is connected to the North Sea. (p. RA20)

Bamako [BAH•mah•KOH] Capital of Mali. 13°N 8°W (p. RA22)

Bangkok [BANG•KAHK] Capital of Thailand. 14°N 100°E (p. RA27)

Bangladesh [BAHNG•gluh•DEHSH] South Asian country bordered by India and Myanmar. (p. RA27)

Bangui [BAHNG•GEE] Capital of the Central African Republic. 4°N 19°E (p. RA22)

Banjul [BAHN•JOOL] Capital of Gambia. 13°N 17°W (p. RA22)

Barbados [bahr•BAY•duhs] Island country between the Atlantic Ocean and the Caribbean Sea. 14°N 59°W (p. RA15)

Beijing [BAY•JIHNG] Capital of China. 40°N 116°E (p. RA27)

Beirut [bay•ROOT] Capital of Lebanon. 34°N 36°E (p. RA24)

Belarus [BEE•luh•ROOS] Eastern European country west of Russia. 54°N 28°E (p. RA19)

Belgium [BEHL•juhm] Western European country south of the Netherlands. (p. RA18)

Belgrade [BEHL•GRAYD] Capital of Serbia. 45°N 21°E (p. RA19)

Belize [buh•LEEZ] Central American country east of Guatemala. (p. RA14)

Belmopan [BEHL•moh•PAHN] Capital of Belize. 17°N 89°W (p. RA14)

Benin [buh•NEEN] West African country west of Nigeria. (p. RA22)

Berlin [behr•LEEN] Capital of Germany. 53°N 13°E (p. RA18)

Bern Capital of Switzerland. 47°N 7°E (p. RA18)

Bhutan [boo•TAHN] South Asian country northeast of India. (p. RA27)

Bishkek [bihsh•KEHK] Capital of Kyrgyzstan. 43°N 75°E (p. RA26)

Bissau [bihs•SOW] Capital of Guinea-Bissau. 12°N 16°W (p. RA22)

Black Sea Large sea between Europe and Asia. (p. RA21)

Bloemfontein [BLOOM•FAHN•TAYN] Judicial capital of South Africa. 26°E 29°S (p. RA22)

Bogotá [BOH•GOH•TAH] Capital of Colombia. 5°N 74°W (p. RA16)

Bolivia [buh•LIHV•ee•uh] Country in the central part of South America, north of Argentina. (p. RA16)

Bosnia and Herzegovina [BAHZ•nee•uh HEHRT•seh•GAW•vee•nuh] Southeastern European country bordered by Croatia, Serbia, and Montenegro. (p. RA18)

Botswana [bawt•SWAH•nah] Southern African country north of the Republic of South Africa. (p. RA22)

Brasília [brah•ZEEL•yuh] Capital of Brazil. 16°S 48°W (p. RA16)

Bratislava [BRAH•tih•SLAH•vuh] Capital of Slovakia. 48°N 17°E (p. RA18)

Brazil [bruh•ZIHL] Largest country in South America. (p. RA16)

Brazzaville [BRAH•zuh•VEEL] Capital of Congo. 4°S 15°E (p. RA22)

Brunei [bru•NY] Southeast Asian country on northern coast of the island of Borneo. (p. RA27)

Brussels [BRUH•suhlz] Capital of Belgium. 51°N 4°E (p. RA18)

Bucharest [BOO•kuh•REHST] Capital of Romania. 44°N 26°E (p. RA19)

Budapest [BOO•duh•PEHST] Capital of Hungary. 48°N 19°E (p. RA18)

Buenos Aires [BWAY•nuhs AR•eez] Capital of Argentina. 34°S 58°W (p. RA16)

Bujumbura [BOO•juhm•BUR•uh] Capital of Burundi. 3°S 29°E (p. RA22)

Bulgaria [BUHL•GAR•ee•uh] Southeastern European country south of Romania. (p. RA19)

Burkina Faso [bur•KEE•nuh FAH•soh] West African country south of Mali. (p. RA22)

Burundi [bu•ROON•dee] East African country at the northern end of Lake Tanganyika. 3°S 30°E (p. RA22)

C

Cairo [KY•roh] Capital of Egypt. 31°N 32°E (p. RA24)

Cambodia [kam•BOH•dee•uh] Southeast Asian country south of Thailand and Laos. (p. RA27)

Cameroon [KA•muh•ROON] Central African country on the northeast shore of the Gulf of Guinea. (p. RA22)

Canada [KA•nuh•duh] Northernmost country in North America. (p. RA6)

Canberra [KAN•BEHR•uh] Capital of Australia. 35°S 149°E (p. RA30)

Cape Town Legislative capital of the Republic of South Africa. 34°S 18°E (p. RA22)

Cape Verde [VUHRD] Island country off the coast of western Africa in the Atlantic Ocean. 15°N 24°W (p. RA22)

Caracas [kah•RAH•kahs] Capital of Venezuela. 11°N 67°W (p. RA16)

Caribbean [KAR•uh•BEE•uhn] **Islands** Islands in the Caribbean Sea between North America and South America, also known as West Indies. (p. RA15)

Caribbean Sea Part of the Atlantic Ocean bordered by the West Indies, South America, and Central America. (p. RA15)

Caspian [KAS•pee•uhn] **Sea** Salt lake between Europe and Asia that is the world's largest inland body of water. (p. RA21)

Caucasus [KAW•kuh•suhs] **Mountains** Mountain range between the Black and Caspian Seas. (p. RA21)

Central African Republic Central African country south of Chad. (p. RA22)

Chad [CHAD] Country west of Sudan in the African Sahel. (p. RA22)

Chang Jiang [CHAHNG jee•AHNG] Principal river of China that begins in Tibet and flows into the East China Sea near Shanghai; also known as the Yangtze River. (p. RA29)

Chile [CHEE•lay] South American country west of Argentina. (p. RA16)

China [CHY•nuh] Country in eastern and central Asia, known officially as the People's Republic of China. (p. RA27)

Chişinău [KEE•shee•NOW] Capital of Moldova. 47°N 29°E (p. RA19)

Colombia [kuh•LUHM•bee•uh] South American country west of Venezuela. (p. RA16)

Colombo [kuh•LUHM•boh] Capital of Sri Lanka. 7°N 80°E (p. RA26)

Comoros [KAH•muh•ROHZ] Small island country in Indian Ocean between the island of Madagascar and the southeast African mainland. 13°S 43°E (p. RA22)

Conakry [KAH•nuh•kree] Capital of Guinea. 10°N 14°W (p. RA22)

Congo [KAHNG•goh] Central African country east of the Democratic Republic of the Congo. 3°S 14°E (p. RA22)

Congo, Democratic Republic of the Central African country north of Zambia and Angola. 1°S 22°E (p. RA22)

Copenhagen [KOH•puhn•HAY•guhn] Capital of Denmark. 56°N 12°E (p. RA18)

Costa Rica [KAWS•tah REE•kah] Central American country south of Nicaragua. (p. RA15)

Côte d'Ivoire [KOHT dee•VWAHR] West African country south of Mali. (p. RA22)

Croatia [kroh•AY•shuh] Southeastern European country on the Adriatic Sea. (p. RA18)

Cuba [KYOO•buh] Island country in the Caribbean Sea. (p. RA15)

Cyprus [SY•pruhs] Island country in the eastern Mediterranean Sea, south of Turkey. (p. RA19)

Czech [CHEHK] **Republic** Eastern European country north of Austria. (p. RA18)

D

Dakar [dah•KAHR] Capital of Senegal. 15°N 17°W (p. RA22)

Damascus [duh•MAS•kuhs] Capital of Syria. 34°N 36°E (p. RA24)

Dar es Salaam [DAHR EHS sah•LAHM] Commercial capital of Tanzania. 7°S 39°E (p. RA22)

Denmark Northern European country between the Baltic and North Seas. (p. RA18)

Dhaka [DA•kuh] Capital of Bangladesh. 24°N 90°E (p. RA27)

Djibouti [jih•BOO•tee] East African country on the Gulf of Aden. 12°N 43°E (p. RA22)

Dodoma [doh•DOH•mah] Political capital of Tanzania. 6°S 36°E (p. RA22)

Doha [DOH•huh] Capital of Qatar. 25°N 51°E (p. RA25)

Dominican [duh•MIH•nih•kuhn] **Republic** Country in the Caribbean Sea on the eastern part of the island of Hispaniola. (p. RA15)

Dublin [DUH•blihn] Capital of Ireland. 53°N 6°W (p. RA18)

Dushanbe [doo•SHAM•buh] Capital of Tajikistan. 39°N 69°E (p. RA25)

E

East Timor [TEE•MOHR] Previous province of Indonesia, now under UN administration. 10°S 127°E (p. RA27)

Ecuador [EH•kwuh•dawr] South American country southwest of Colombia. (p. RA16)

Egypt [EE•jihpt] North African country on the Mediterranean Sea. (p. RA24)

El Salvador [ehl SAL•vuh•dawr] Central American country southwest of Honduras. (p. RA14)

Equatorial Guinea [EE•kwuh•TOHR•ee•uhl GIH•nee] Central African country south of Cameroon. (p. RA22)

Eritrea [EHR•uh•TREE•uh] East African country north of Ethiopia. (p. RA22)

Estonia [eh•STOH•nee•uh] Eastern European country on the Baltic Sea. (p. RA19)

Ethiopia [EE•thee•OH•pee•uh] East African country north of Somalia and Kenya. (p. RA22)

Euphrates [yu•FRAY•teez] **River** River in southwestern Asia that flows through Syria and Iraq and joins the Tigris River. (p. RA25)

F

Fiji [FEE•jee] **Islands** Country comprised of an island group in the southwest Pacific Ocean. 19°S 175°E (p. RA30)

Finland [FIHN•luhnd] Northern European country east of Sweden. (p. RA19)

France [FRANS] Western European country south of the United Kingdom. (p. RA18)

Freetown Capital of Sierra Leone. (p. RA22)

French Guiana [gee•A•nuh] French-owned territory in northern South America. (p. RA16)

G

Gabon [ga•BOHN] Central African country on the Atlantic Ocean. (p. RA22)

Gaborone [GAH•boh•ROH•nay] Capital of Botswana. (p. RA22)

Gambia [GAM•bee•uh] West African country along the Gambia River. (p. RA22)

Georgetown [JAWRJ•town] Capital of Guyana. 8°N 58°W (p. RA16)

Georgia [JAWR•juh] European-Asian country bordering the Black Sea south of Russia. (p. RA26)

Germany [JUHR•muh•nee] Western European country south of Denmark, officially called the Federal Republic of Germany. (p. RA18)

Ghana [GAH•nuh] West African country on the Gulf of Guinea. (p. RA22)

Great Plains The continental slope extending through the United States and Canada. (p. RA7)

Greece [GREES] Southern European country on the Balkan Peninsula. (p. RA19)

Greenland [GREEN•luhnd] Island in northwestern Atlantic Ocean and the largest island in the world. (p. RA6)

Guatemala [GWAH•tay•MAH•lah] Central American country south of Mexico. (p. RA14)

Guatemala Capital of Guatemala. 15°N 91°W (p. RA14)

Guinea [GIH•nee] West African country on the Atlantic coast. (p. RA22)

Guinea-Bissau [GIH•nee bih•SOW] West African country on the Atlantic coast. (p. RA22)

Gulf of Mexico Gulf on part of the southern coast of North America. (p. RA7)

Guyana [gy•AH•nuh] South American country between Venezuela and Suriname. (p. RA16)

H

Haiti [HAY•tee] Country in the Caribbean Sea on the western part of the island of Hispaniola. (p. RA15)

Hanoi [ha•NOY] Capital of Vietnam. 21°N 106°E (p. RA27)

Harare [hah•RAH•ray] Capital of Zimbabwe. 18°S 31°E (p. RA22)

Havana [huh•VA•nuh] Capital of Cuba. 23°N 82°W (p. RA15)

Helsinki [HEHL•sihng•kee] Capital of Finland. 60°N 24°E (p. RA19)

Himalaya [HI•muh•LAY•uh] Mountain ranges in southern Asia, bordering the Indian subcontinent on the north. (p. RA28)

Honduras [hahn•DUR•uhs] Central American country on the Caribbean Sea. (p. RA14)

Hong Kong Port and industrial center in southern China. 22°N 115°E (p. RA27)

Huang He [HWAHNG HUH] River in northern and eastern China, also known as the Yellow River. (p. RA29)

Hungary [HUHNG•guh•ree] Eastern European country south of Slovakia. (p. RA18)

I

Iberian [eye•BIHR•ee•uhn] **Peninsula** Peninsula in southwest Europe, occupied by Spain and Portugal. (p. RA20)

Iceland Island country between the North Atlantic and Arctic Oceans. (p. RA18)

India [IHN•dee•uh] South Asian country south of China and Nepal. (p. RA26)

Indonesia [IHN•duh•NEE•zhuh] Southeast Asian island country known as the Republic of Indonesia. (p. RA27)

Indus [IHN•duhs] **River** River in Asia that begins in Tibet and flows through Pakistan to the Arabian Sea. (p. RA28)

Iran [ih•RAN] Southwest Asian country that was formerly named Persia. (p. RA25)

Iraq [ih•RAHK] Southwest Asian country west of Iran. (p. RA25)

Ireland [EYER•luhnd] Island west of Great Britain occupied by the Republic of Ireland and Northern Ireland. (p. RA18)

Islamabad [ihs•LAH•muh•BAHD] Capital of Pakistan. 34°N 73°E (p. RA26)

Israel [IHZ•ree•uhl] Southwest Asian country south of Lebanon. (p. RA24)

Italy [IHT•uhl•ee] Southern European country south of Switzerland and east of France. (p. RA18)

J

Jakarta [juh•KAHR•tuh] Capital of Indonesia. 6°S 107°E (p. RA27)

Jamaica [juh•MAY•kuh] Island country in the Caribbean Sea. (p. RA15)

Japan [juh•PAN] East Asian country consisting of the four large islands of Hokkaido, Honshu, Shikoku, and Kyushu, plus thousands of small islands. (p. RA27)

Jerusalem [juh•ROO•suh•luhm] Capital of Israel and a holy city for Christians, Jews, and Muslims. 32°N 35°E (p. RA24)

Jordan [JAWRD•uhn] Southwest Asian country south of Syria. (p. RA24)

Juba [JU•buh] Capital of South Sudan. 5°N 31°E (p. RA22)

K

Kabul [KAH•buhl] Capital of Afghanistan. 35°N 69°E (p. RA25)

Kampala [kahm•PAH•lah] Capital of Uganda. 0° latitude 32°E (p. RA22)

Kathmandu [KAT•MAN•DOO] Capital of Nepal. 28°N 85°E (p. RA26)

Kazakhstan [kuh•ZAHK•STAHN] Large Asian country south of Russia and bordering the Caspian Sea. (p. RA26)

Kenya [KEHN•yuh] East African country south of Ethiopia. (p. RA22)

Khartoum [kahr•TOOM] Capital of Sudan. 16°N 33°E (p. RA22)

Kigali [kee•GAH•lee] Capital of Rwanda. 2°S 30°E (p. RA22)

Kingston [KIHNG•stuhn] Capital of Jamaica. 18°N 77°W (p. RA15)

Kinshasa [kihn•SHAH•suh] Capital of the Democratic Republic of the Congo. 4°S 15°E (p. RA22)

Kuala Lumpur [KWAH•luh LUM•PUR] Capital of Malaysia. 3°N 102°E (p. RA27)

Kuwait [ku•WAYT] Country on the Persian Gulf between Saudi Arabia and Iraq. (p. RA25)

Kyiv (Kiev) [KEE•ihf] Capital of Ukraine. 50°N 31°E (p. RA19)

Kyrgyzstan [s•gih•STAN] Central Asian country on China's western border. (p. RA26)

L

Laos [LOWS] Southeast Asian country south of China and west of Vietnam. (p. RA27)

La Paz [lah PAHS] Administrative capital of Bolivia, and the highest capital in the world. 17°S 68°W (p. RA16)

Latvia [LAT•vee•uh] Eastern European country west of Russia on the Baltic Sea. (p. RA19)

Lebanon [LEH•buh•nuhn] Country south of Syria on the Mediterranean Sea. (p. RA24)

Lesotho [luh•SOH•TOH] Southern African country within the borders of the Republic of South Africa. (p. RA22)

Liberia [ly•BIHR•ee•uh] West African country south of Guinea. (p. RA22)

Libreville [LEE•bruh•VIHL] Capital of Gabon. 1°N 9°E (p. RA22)

Libya [LIH•bee•uh] North African country west of Egypt on the Mediterranean Sea. (p. RA22)

Liechtenstein [LIHKT•uhn•SHTYN] Small country in central Europe between Switzerland and Austria. 47°N 10°E (p. RA18)

Lilongwe [lih•LAWNG•GWAY] Capital of Malawi. 14°S 34°E (p. RA22)

Lima [LEE•mah] Capital of Peru. 12°S 77°W (p. RA16)

Lisbon [LIHZ•buhn] Capital of Portugal. 39°N 9°W (p. RA18)

Lithuania [LIH•thuh•WAY•nee•uh] Eastern European country northwest of Belarus on the Baltic Sea. (p. RA21)

Ljubljana [lee•oo•blee•AH•nuh] Capital of Slovenia. 46°N 14°E (p. RA18)

Lomé [loh•MAY] Capital of Togo. 6°N 1°E (p. RA22)

London Capital of the United Kingdom, on the Thames River. 52°N 0° longitude (p. RA18)

Luanda [lu•AHN•duh] Capital of Angola. 9°S 13°E (p. RA22)

Lusaka [loo•SAH•kah] Capital of Zambia. 15°S 28°E (p. RA22)

Luxembourg [LUHK•suhm•BUHRG] Small European country bordered by France, Belgium, and Germany. 50°N 7°E (p. RA18)

M

Macao [muh•KOW] Port in southern China. 22°N 113°E (p. RA27)

Macedonia [ma•suh•DOH•nee•uh] Southeastern European country north of Greece. (p. RA19). Macedonia also refers to a geographic region covering northern Greece, the country Macedonia, and part of Bulgaria.

Madagascar [MA•duh•GAS•kuhr] Island in the Indian Ocean off the southeastern coast of Africa. (p. RA22)

Madrid Capital of Spain. 41°N 4°W (p. RA18)

Malabo [mah•LAH•boh] Capital of Equatorial Guinea. 4°N 9°E (p. RA22)

Malawi [mah•LAH•wee] Southern African country south of Tanzania and east of Zambia. (p. RA22)

Malaysia [muh•LAY•zhuh] Southeast Asian country with land on the Malay Peninsula and on the island of Borneo. (p. RA27)

Maldives [MAWL•DEEVZ] Island country southwest of India in the Indian Ocean. (p. RA26)

Mali [MAH•lee] West African country east of Mauritania. (p. RA22)

Managua [mah•NAH•gwah] Capital of Nicaragua. (p. RA15)

Manila [muh•NIH•luh] Capital of the Philippines. 15°N 121°E (p. RA27)

Maputo [mah•POO•toh] Capital of Mozambique. 26°S 33°E (p. RA22)

Maseru [MA•zuh•ROO] Capital of Lesotho. 29°S 27°E (p. RA22)

Masqat [MUHS•KAHT] Capital of Oman. 23°N 59°E (p. RA25)

Mauritania [MAWR•uh•TAY•nee•uh] West African country north of Senegal. (p. RA22)

Mauritius [maw•RIH•shuhs] Island country in the Indian Ocean east of Madagascar. 21°S 58°E (p. RA3)

Mbabane [uhm•bah•BAH•nay] Capital of Swaziland. 26°S 31°E (p. RA22)

Mediterranean [MEH•duh•tuh•RAY•nee•uhn] **Sea** Large inland sea surrounded by Europe, Asia, and Africa. (p. RA20)

Mekong [MAY•KAWNG] **River** River in southeastern Asia that begins in Tibet and empties into the South China Sea. (p. RA29)

Mexico [MEHK•sih•KOH] North American country south of the United States. (p. RA14)

Mexico City Capital of Mexico. 19°N 99°W (p. RA14)

Minsk [MIHNSK] Capital of Belarus. 54°N 28°E (p. RA19)

Mississippi [MIH•suh•SIH•pee] **River** Large river system in the central United States that flows southward into the Gulf of Mexico. (p. RA11)

Mogadishu [MOH•guh•DEE•shoo] Capital of Somalia. 2°N 45°E (p. RA22)

Moldova [mawl•DAW•vuh] Small European country between Ukraine and Romania. (p. RA19)

Monaco [MAH•nuh•KOH] Small country in southern Europe on the French Mediterranean coast. 44°N 8°E (p. RA18)

Mongolia [mahn•GOHL•yuh] Country in Asia between Russia and China. (p. RA23)

Monrovia [muhn•ROH•vee•uh] Capital of Liberia. 6°N 11°W (p. RA22)

Montenegro [MAHN•tuh•NEE•groh] Eastern European country. (p. RA18)

Montevideo [MAHN•tuh•vuh•DAY•oh] Capital of Uruguay. 35°S 56°W (p. RA16)

Morocco [muh•RAH•KOH] North African country on the Mediterranean Sea and the Atlantic Ocean. (p. RA22)

Moscow [MAHS•KOW] Capital of Russia. 56°N 38°E (p. RA19)

Mount Everest [EHV•ruhst] Highest mountain in the world, in the Himalaya between Nepal and Tibet. (p. RA28)

Mozambique [MOH•zahm•BEEK] Southern African country south of Tanzania. (p. RA22)

Myanmar [MYAHN•MAHR] Southeast Asian country south of China and India, formerly called Burma. (p. RA27)

N

Nairobi [ny•ROH•bee] Capital of Kenya. 1°S 37°E (p. RA22)

Namibia [nuh•MIH•bee•uh] Southern African country south of Angola on the Atlantic Ocean. 20°S 16°E (p. RA22)

Nassau [NA•SAW] Capital of the Bahamas. 25°N 77°W (p. RA15)

N'Djamena [uhn•jah•MAY•nah] Capital of Chad. 12°N 15°E (p. RA22)

Nepal [NAY•PAHL] Mountain country between India and China. (p. RA26)

Netherlands [NEH•thuhr•lundz] Western European country north of Belgium. (p. RA18)

New Delhi [NOO DEH•lee] Capital of India. 29°N 77°E (p. RA26)

New Zealand [NOO ZEE•luhnd] Major island country southeast of Australia in the South Pacific. (p. RA30)

Niamey [nee•AHM•ay] Capital of Niger. 14°N 2°E (p. RA22)

Nicaragua [NIH•kuh•RAH•gwuh] Central American country south of Honduras. (p. RA15)

Nicosia [NIH•kuh•SEE•uh] Capital of Cyprus. 35°N 33°E (p. RA19)

Niger [NY•juhr] West African country north of Nigeria. (p. RA22)

Nigeria [ny•JIHR•ee•uh] West African country along the Gulf of Guinea. (p. RA22)

Nile [NYL] **River** Longest river in the world, flowing north through eastern Africa. (p. RA23)

North Korea [kuh•REE•uh] East Asian country in the northernmost part of the Korean Peninsula. (p. RA27)

Norway [NAWR•way] Northern European country on the Scandinavian Peninsula. (p. RA18)

Nouakchott [nu•AHK•SHAHT] Capital of Mauritania. 18°N 16°W (p. RA22)

O

Oman [oh•MAHN] Country on the Arabian Sea and the Gulf of Oman. (p. RA25)

Oslo [AHZ•loh] Capital of Norway. 60°N 11°E (p. RA18)

Ottawa [AH•tuh•wuh] Capital of Canada. 45°N 76°W (p. RA13)

Ouagadougou [WAH•gah•DOO•goo] Capital of Burkina Faso. 12°N 2°W (p. RA22)

P

Pakistan [PA•kih•STAN] South Asian country northwest of India on the Arabian Sea. (p. RA26)

Palau [puh•LOW) Island country in the Pacific Ocean. 7°N 135°E (p. RA30)

Panama [PA•nuh•MAH] Central American country on the Isthmus of Panama. (p. RA15)

Panama Capital of Panama. 9°N 79°W (p. RA15)

Papua New Guinea [PA•pyu•wuh NOO GIH•nee] Island country in the Pacific Ocean north of Australia. 7°S 142°E (p. RA30)

Paraguay [PAR•uh•GWY] South American country northeast of Argentina. (p. RA16)

Paramaribo [PAH•rah•MAH•ree•boh] Capital of Suriname. 6°N 55°W (p. RA16)

Paris Capital of France. 49°N 2°E (p. RA18)

Persian [PUHR•zhuhn] **Gulf** Arm of the Arabian Sea between Iran and Saudi Arabia. (p. RA25)

Peru [puh•ROO] South American country south of Ecuador and Colombia. (p. RA16)

Philippines [FIH•luh•PEENZ] Island country in the Pacific Ocean southeast of China. (p. RA27)

Phnom Penh [puh•NAWM PEHN] Capital of Cambodia. 12°N 106°E (p. RA27)

Poland [POH•luhnd] Eastern European country on the Baltic Sea. (p. RA18)

Port-au-Prince [POHRT•oh•PRIHNS] Capital of Haiti. 19°N 72°W (p. RA15)

Port Moresby [MOHRZ•bee] Capital of Papua New Guinea. 10°S 147°E (p. RA30)

Port-of-Spain [SPAYN] Capital of Trinidad and Tobago. 11°N 62°W (p. RA15)

Porto-Novo [POHR•toh•NOH•voh] Capital of Benin. 7°N 3°E (p. RA22)

Portugal [POHR•chih•guhl] Country west of Spain on the Iberian Peninsula. (p. RA18)

Prague [PRAHG] Capital of the Czech Republic. 51°N 15°E (p. RA18)

Puerto Rico [PWEHR•toh REE•koh] Island in the Caribbean Sea; U.S. Commonwealth. (p. RA15)

P'yŏngyang [pee•AWNG•YAHNG] Capital of North Korea. 39°N 126°E (p. RA27)

Q

Qatar [KAH•tuhr] Country on the southwestern shore of the Persian Gulf. (p. RA25)

Quito [KEE•toh] Capital of Ecuador. 0° latitude 79°W (p. RA16)

R

Rabat [ruh•BAHT] Capital of Morocco. 34°N 7°W (p. RA22)

Reykjavík [RAY•kyah•VEEK] Capital of Iceland. 64°N 22°W (p. RA18)

Rhine [RYN] **River** River in western Europe that flows into the North Sea. (p. RA20)

Riga [REE•guh] Capital of Latvia. 57°N 24°E (p. RA19)

Rio Grande [REE•oh GRAND] River that forms part of the boundary between the United States and Mexico. (p. RA10)

Riyadh [ree•YAHD] Capital of Saudi Arabia. 25°N 47°E (p. RA25)

Rocky Mountains Mountain system in western North America. (p. RA7)

Romania [ru•MAY•nee•uh] Eastern European country east of Hungary. (p. RA19)

Rome Capital of Italy. 42°N 13°E (p. RA18)

Russia [RUH•shuh] Largest country in the world, covering parts of Europe and Asia. (pp. RA19, RA27)

Rwanda [ruh•WAHN•duh] East African country south of Uganda. 2°S 30°E (p. RA22)

S

Sahara [suh•HAR•uh] Desert region in northern Africa that is the largest hot desert in the world. (p. RA23)

Saint Lawrence [LAWR•uhns] River River that flows from Lake Ontario to the Atlantic Ocean and forms part of the boundary between the United States and Canada. (p. RA13)

Sanaa [sahn•AH] Capital of Yemen. 15°N 44°E (p. RA25)

San José [SAN hoh•ZAY] Capital of Costa Rica. 10°N 84°W (p. RA15)

San Marino [SAN muh•REE•noh] Small European country located on the Italian Peninsula. 44°N 13°E (p. RA18)

San Salvador [SAN SAL•vuh•DAWR] Capital of El Salvador. 14°N 89°W (p. RA14)

Santiago [SAN•tee•AH•goh] Capital of Chile. 33°S 71°W (p. RA16)

Santo Domingo [SAN•toh duh•MIHNG•goh] Capital of the Dominican Republic. 19°N 70°W (p. RA15)

São Tomé and Príncipe [sow too•MAY PREEN•see•pee] Small island country in the Gulf of Guinea off the coast of central Africa. 1°N 7°E (p. RA22)

Sarajevo [SAR•uh•YAY•voh] Capital of Bosnia and Herzegovina. 43°N 18°E (p. RA18)

Saudi Arabia [SOW•dee uh•RAY•bee•uh] Country on the Arabian Peninsula. (p. RA25)

Senegal [SEH•nih•GAWL] West African country on the Atlantic coast. (p. RA22)

Seoul [SOHL] Capital of South Korea. 38°N 127°E (p. RA27)

Serbia [SUHR•bee•uh] Eastern European country south of Hungary. (p. RA18)

Seychelles [say•SHEHL] Small island country in the Indian Ocean off eastern Africa. 6°S 56°E (p. RA22)

Sierra Leone [see•EHR•uh lee•OHN] West African country south of Guinea. (p. RA22)

Singapore [SIHNG•uh•POHR] Southeast Asian island country near tip of the Malay Peninsula. (p. RA27)

Skopje [SKAW•pyay] Capital of the country of Macedonia. 42°N 21°E (p. RA19)

Slovakia [sloh•VAH•kee•uh] Eastern European country south of Poland. (p. RA18)

Slovenia [sloh•VEE•nee•uh] Southeastern European country south of Austria on the Adriatic Sea. (p. RA18)

Sofia [SOH•fee•uh] Capital of Bulgaria. 43°N 23°E (p. RA19)

Solomon [SAH•luh•muhn] Islands Island country in the Pacific Ocean northeast of Australia. (p. RA30)

Somalia [soh•MAH•lee•uh] East African country on the Gulf of Aden and the Indian Ocean. (p. RA22)

South Africa [A•frih•kuh] Country at the southern tip of Africa, officially the Republic of South Africa. (p. RA22)

South Korea [kuh•REE•uh] East Asian country on the Korean Peninsula between the Yellow Sea and the Sea of Japan. (p. RA27)

South Sudan [soo•DAN] East African country south of Sudan. (p. RA22)

Spain [SPAYN] Southern European country on the Iberian Peninsula. (p. RA18)

Sri Lanka [SREE LAHNG•kuh] Country in the Indian Ocean south of India, formerly called Ceylon. (p. RA26)

Stockholm [STAHK•HOHLM] Capital of Sweden. 59°N 18°E (p. RA18)

Sucre [SOO•kray] Constitutional capital of Bolivia. 19°S 65°W (p. RA16)

Sudan [soo•DAN] East African country south of Egypt. (p. RA22)

Suriname [SUR•uh•NAH•muh] South American country between Guyana and French Guiana. (p. RA16)

Suva [SOO•vah] Capital of the Fiji Islands. 18°S 177°E (p. RA30)

Swaziland [SWAH•zee•land] Southern African country west of Mozambique, almost entirely within the Republic of South Africa. (p. RA22)

Sweden Northern European country on the eastern side of the Scandinavian Peninsula. (p. RA18)

Switzerland [SWIHT•suhr•luhnd] European country in the Alps south of Germany. (p. RA18)

Syria [SIHR•ee•uh] Southwest Asian country on the east side of the Mediterranean Sea. (p. RA24)

T

Taipei [TY•PAY] Capital of Taiwan. 25°N 122°E (p. RA27)

Taiwan [TY•WAHN] Island country off the southeast coast of China; the seat of the Chinese Nationalist government. (p. RA27)

Tajikistan [tah•JIH•kih•STAN] Central Asian country east of Turkmenistan. (p. RA26)

Tallinn [TA•luhn] Capital of Estonia. 59°N 25°E (p. RA19)

Tanzania [TAN•zuh•NEE•uh] East African country south of Kenya. (p. RA22)

Tashkent [tash•KEHNT] Capital of Uzbekistan. 41°N 69°E (p. RA26)

Tbilisi [tuh•bih•LEE•see] Capital of the Republic of Georgia. 42°N 45°E (p. RA26)

Tegucigalpa [tay•GOO•see•GAHL•pah] Capital of Honduras. 14°N 87°W (p. RA14)

Tehran [TAY•uh•RAN] Capital of Iran. 36°N 52°E (p. RA25)

Thailand [TY•LAND] Southeast Asian country east of Myanmar. 17°N 101°E (p. RA27)

Thimphu [thihm•POO] Capital of Bhutan. 28°N 90°E (p. RA27)

Tigris [TY•gruhs] **River** River in southeastern Turkey and Iraq that merges with the Euphrates River. (p. RA25)

Tiranë [tih•RAH•nuh] Capital of Albania. 42°N 20°E (p. RA18)

Togo [TOH•goh] West African country between Benin and Ghana on the Gulf of Guinea. (p. RA22)

Tokyo [TOH•kee•OH] Capital of Japan. 36°N 140°E (p. RA27)

Trinidad and Tobago [TRIH•nuh•DAD tuh•BAY•goh] Island country near Venezuela between the Atlantic Ocean and the Caribbean Sea. (p. RA15)

Tripoli [TRIH•puh•lee] Capital of Libya. 33°N 13°E (p. RA22)

Tshwane [ch•WAH•nay] Executive capital of South Africa. 26°S 28°E (p. RA22)

Tunis [TOO•nuhs] Capital of Tunisia. 37°N 10°E (p. RA22)

Tunisia [too•NEE•zhuh] North African country on the Mediterranean Sea between Libya and Algeria. (p. RA22)

Turkey [TUHR•kee] Country in southeastern Europe and western Asia. (p. RA24)

Turkmenistan [tuhrk•MEH•nuh•STAN] Central Asian country on the Caspian Sea. (p. RA25)

U

Uganda [yoo•GAHN•dah] East African country south of Sudan. (p. RA22)

Ukraine [yoo•KRAYN] Eastern European country west of Russia on the Black Sea. (p. RA25)

Ulaanbaatar [oo•LAHN•BAH•tawr] Capital of Mongolia. 48°N 107°E (p. RA27)

United Arab Emirates [EH•muh•ruhts] Country made up of seven states on the eastern side of the Arabian Peninsula. (p. RA25)

United Kingdom Western European island country made up of England, Scotland, Wales, and Northern Ireland. (p. RA18)

United States of America Country in North America made up of 50 states, mostly between Canada and Mexico. (p. RA8)

Uruguay [YUR•uh•GWAY] South American country south of Brazil on the Atlantic Ocean. (p. RA16)

Uzbekistan [uz•BEH•kih•STAN] Central Asian country south of Kazakhstan. (p. RA25)

V

Vanuatu [VAN•WAH•TOO] Country made up of islands in the Pacific Ocean east of Australia. (p. RA30)

Vatican [VA•tih•kuhn] **City** Headquarters of the Roman Catholic Church, located in the city of Rome in Italy. 42°N 13°E (p. RA18)

Venezuela [VEH•nuh•ZWAY•luh] South American country on the Caribbean Sea between Colombia and Guyana. (p. RA16)

Vienna [vee•EH•nuh] Capital of Austria. 48°N 16°E (p. RA18)

Vientiane [vyehn•TYAHN] Capital of Laos. 18°N 103°E (p. RA27)

Vietnam [vee•EHT•NAHM] Southeast Asian country east of Laos and Cambodia. (p. RA27)

Vilnius [VIL•nee•uhs] Capital of Lithuania. 55°N 25°E (p. RA19)

W

Warsaw Capital of Poland. 52°N 21°E (p. RA19)

Washington, D.C. Capital of the United States, in the District of Columbia. 39°N 77°W (p. RA8)

Wellington [WEH•lihng•tuhn] Capital of New Zealand. 41°S 175°E (p. RA30)

West Indies Caribbean islands between North America and South America. (p. RA15)

Windhoek [VIHNT•HUK] Capital of Namibia. 22°S 17°E (p. RA22)

Y

Yamoussoukro [YAH•MOO•SOO•kroh] Second capital of Côte d'Ivoire. 7°N 6°W (p. RA22)

Yangon [YAHNG•GOHN] City in Myanmar; formerly called Rangoon. 17°N 96°E (p. RA27)

Yaoundé [yown•DAY] Capital of Cameroon. 4°N 12°E (p. RA22)

Yemen [YEH•muhn] Country south of Saudi Arabia on the Arabian Peninsula. (p. RA25)

Yerevan [YEHR•uh•VAHN] Capital of Armenia. 40°N 44°E (p. RA25)

Z

Zagreb [ZAH•GREHB] Capital of Croatia. 46°N 16°E (p. RA18)

Zambia [ZAM•bee•uh] Southern African country north of Zimbabwe. (p. RA22)

Zimbabwe [zihm•BAH•bway] Southern African country northeast of Botswana. (p. RA22)

Gazetteer

GLOSSARY/GLOSARIO

- Content vocabulary words are words that relate to world geography content.
- Words that have an asterisk (*) are academic vocabulary. They help you understand your school subjects.
- All vocabulary words are **boldfaced** or highlighted in yellow in your textbook.

aboriginal • altitude

| ENGLISH | A | ESPAÑOL |

aboriginal a native people (p. 184); the first people to live in Australia (p. 733)

aborigen persona nativa (pág. 184); el primer pueblo que habitó en Australia (pág. 733)

absolute location the exact location of something (p. 21)

localización absoluta ubicación exacta de algo (pág. 21)

absolute monarchy a system of government in which the ruler has complete control (p. 477)

monarquía absoluta sistema de gobierno en el cual el gobernante detenta el control absoluto (pág. 477)

*****access** a way to reach a distant area (p. 183)

*****acceso** vía para llegar a un lugar distante (pág. 183)

*****accurate** without mistakes or errors (p. 44)

*****exacto** sin faltas o errores (pág. 44)

*****achievement** a great accomplishment due to hard work (p. 372)

*****logro** consecución importante que resulta de un trabajo arduo (pág. 372)

acid rain rain that contains harmful amounts of poisons due to pollution (p. 65)

lluvia ácida lluvia que contiene cantidades nocivas de venenos debido a la polución (pág. 65)

*****acknowledge** to recognize the rights, status, or authority of a person, thing, or event (p. 797)

*****admitir** reconocer los derechos, el estatus o la autoridad de una persona, cosa o suceso (pág. 797)

action song a song that combines singing and dancing to celebrate Maori history and culture in order to instill pride among Maori people (p. 751)

canción de acción canción que combina el canto y la danza para honrar la historia y cultura maoríes e inculcar orgullo entre el pueblo maorí (pág. 751)

*****adapt** to change a trait in order to survive (p. 336)

*****adaptar** cambiar un rasgo para sobrevivir (pág. 336)

aerospace the industry that makes vehicles that travel in the air and in outer space (p. 170)

aeroespacial industria que construye vehículos que viajan por el aire y el espacio exterior (pág. 170)

agribusiness an industry based on huge farms that rely on machines and mass-production methods (p. 170)

agronegocio industria basada en granjas extensas que dependen de máquinas y técnicas de producción masiva (pág. 170)

agriculture the practice of growing crops and raising livestock (p. 129)

agricultura actividad que consiste en cultivar la tierra y criar ganado (pág. 129)

alluvial plain an area built up by rich fertile soil left by river floods (p. 493)

llanura aluvial área formada por los sedimentos fértiles que dejan las inundaciones fluviales (pág. 493)

altiplano the high plains (p. 299)

altiplano meseta elevada (pág. 299)

altitude the height above sea level (p. 301)

altitud altura sobre el nivel del mar (pág. 301)

animist a person who believes in spirits that can exist apart from bodies (p. 686)

annex to declare ownership of an area (p. 159)

*annual yearly or each year (p. 165)

apartheid the system of laws in South Africa aimed at separating the races (p. 709)

aquifer an underground layer of rock through which water flows (p. 587)

archipelago a group of islands (pp. 181; 767)

*area a geographic location (p. 241)

atmosphere the layer of gases surrounding Earth (p. 44)

atoll a circular-shaped island made of coral (pp. 493; 769)

autonomy having independence from another country (p. 194)

axis an imaginary line that runs through Earth's center from the North Pole to the South Pole (p. 42)

animista persona que cree en espíritus que viven por fuera del cuerpo (pág. 686)

anexionar declarar la propiedad de un territorio (pág. 159)

*anual cada año (pág. 165)

apartheid sistema jurídico de Sudáfrica que establecía la segregación racial (pág. 709)

acuífero estrato rocoso subterráneo por donde corre el agua (pág. 587)

archipiélago grupo de islas (págs. 181; 767)

*área territorio geográfico (pág. 241)

atmósfera capa de gases que rodea la Tierra (pág. 44)

atolón isla coralina de forma anular (págs. 493; 769)

soberanía independencia respecto de otro país (pág. 194)

eje línea imaginaria que atraviesa el centro de la Tierra desde el Polo Norte hasta el Polo Sur (pág. 42)

B

balkanization to break a country up into smaller units that are often hostile to one another (p. 397)

basin an area of land that is drained by a river and its tributaries (p. 241)

bauxite the mineral ore that is used to make aluminum (p. 207)

*behalf in the interest of (p. 87)

*benefit an advantage (p. 203)

bilingual able to use two languages (p. 191)

biodiversity the wide variety of life on Earth (p. 648)

birthrate the number of babies born compared to the total number of people in a population at a given time (p. 72)

blood diamonds diamonds that are sold on the black market, with the proceeds going to provide guns and ammunition for violent conflicts (p. 703)

balcanizar fragmentar un país en partes más pequeñas, con frecuencia hostiles entre sí (pág. 397)

cuenca área de terreno drenada por un río y sus afluentes (pág. 241)

bauxita mineral metalífero que se utiliza para producir aluminio (pág. 207)

*a favor de en beneficio de (pág. 87)

*beneficio ventaja (pág. 203)

bilingüe que habla dos idiomas (pág. 191)

biodiversidad la amplia variedad de vida terrestre (pág. 648)

tasa de natalidad número de nacimientos comparado con el número total de habitantes de una población en un tiempo determinado (pág. 72)

diamantes sangrientos diamantes que se venden en el mercado negro y cuyas ganancias se utilizan para adquirir armas y municiones en conflictos violentos (pág. 703)

boomerang the flat, bent wooden tool of the Australian Aborigines that is thrown to stun prey when it strikes them and that sails back to the hunter if it misses its target (p. 741)

boycott to refuse to buy items from a particular country or company (p. 502)

brackish water that is somewhat salty (p. 398)

bush a rural area in Australia (p. 749)

búmeran utensilio de madera curvo y plano de los aborígenes australianos, que se lanza para aturdir a las presas cuando las golpea y regresa al cazador en caso de fallar el blanco (pág. 741)

boicotear rehusarse a comprar los artículos de un país o compañía en particular (pág. 502)

salobre agua algo salada (pág. 398)

brezal área rural de Australia (pág. 749)

C

caliph the successor to Muhammad (p. 592)

calving the process in which a section of ice breaks off the edge of a glacier (p. 792)

canopy the umbrella-like covering formed by the tops of trees in a rain forest (p. 241)

*capable having the ability to cause or accomplish an action or an event (p. 770)

cash crop a farm product grown for sale (p. 213)

cassava a tuberous plant that has edible roots (p. 652)

caste the social class a person is born into and cannot change (p. 499)

caudillo a person who often ruled a Latin American country as a dictator and was generally a high-ranking military officer or a rich man (p. 213)

central city the densely populated center of a metropolitan area (p. 257)

*channel a course for a river to flow through (p. 582)

characteristic a quality or an aspect (p. 659)

chinook a dry wind that sometimes blows over the Great Plains in winter (p. 154)

*circumstances conditions (p. 217)

city-state an independent political unit that includes a city and the surrounding area (p. 372)

civil disobedience the use of nonviolent protests to challenge a government or its laws (p. 503)

califa sucesor de Mahoma (pág. 592)

ablación proceso en el que un bloque de hielo se desprende del borde de un glaciar (pág. 792)

manto cubierta en forma de sombrilla formada por las copas de los árboles en una selva tropical (pág. 241)

*capaz que tiene habilidad para provocar o llevar a cabo una acción o un suceso (pág. 770)

cultivo comercial producto agrícola que se cultiva para la venta (pág. 213)

yuca planta tuberosa de raíces comestibles (pág. 652)

casta clase social en la que nace una persona y no puede cambiar (pág. 499)

caudillo persona que gobernaba un país latinoamericano como dictador; por lo general, era un oficial de alto rango o un hombre pudiente (pág. 213)

ciudad central centro densamente poblado de un área metropolitana (pág. 257)

*canal curso artificial por donde fluye un río (pág. 582)

característica cualidad o aspecto (pág. 659)

chinook viento seco que sopla a veces sobre las Grandes Llanuras en invierno (pág. 154)

*circunstancias condiciones (pág. 217)

ciudad-Estado unidad política independiente que incluye una ciudad y el área circundante (pág. 372)

desobediencia civil rebatir un gobierno o sus leyes mediante protestas no violentas (pág. 503)

Glossary/Glosario

civil rights the basic rights that belong to all citizens (p. 137)

derechos civiles los derechos fundamentales de todos los ciudadanos (pág. 137)

civil war a fight between opposing groups for control of a country's government (p. 595)

guerra civil lucha entre grupos opositores por el control del gobierno de un país (pág. 595)

clan a large group of people who have a common ancestor in the far past (p. 627)

clan agrupación extensa de personas que tienen un ancestro común en el pasado remoto (pág. 627)

climate the average weather in an area over a long period of time (p. 23)

clima tiempo atmosférico promedio en una zona durante un periodo largo (pág. 23)

coastal plain the flat, lowland area along a coast (p. 121)

llanura litoral planicie de baja altitud que bordea la costa (pág. 121)

*collapse a sudden failure, breakdown, or ruin (p. 780)

*colapso bancarrota, caída o ruina súbita (pág. 780)

*collapsing falling or breaking down (p. 551)

*colapsar caer o derrumbarse (pág. 551)

collective a farm that is owned by the government but run by a group of farmers who work together (p. 524)

colectiva granja de propiedad del gobierno que administra una cooperativa de granjeros (pág. 524)

collectivization a system in which small farms were combined into huge, state-run enterprises with work done by mechanized techniques in the hopes of making farming more efficient and reducing the need for farmworkers (p. 406)

colectivización sistema en el cual pequeños granjeros se integran a empresas gigantescas administradas por el Estado, en las que el trabajo se realiza mediante métodos técnicos con la esperanza de hacer más eficiente la agricultura y reducir la demanda de trabajadores agrícolas (pág. 406)

colonialism a policy based on control of one country by another (p. 212)

colonialismo política que se basa en el control o dominio de un país sobre otro (pág. 212)

colonist a person sent to live in a new place and claim land for his or her home country (p. 126)

colonizador persona enviada a establecerse en un nuevo lugar y reclamar territorios para su país de origen (pág. 126)

Columbian Exchange the transfer of plants, animals, and people between Europe, Asia, and Africa on one side and the Americas on the other (p. 214)

intercambio colombino traslado de plantas, animales y personas entre Europa, Asia y África, de un lado, y América, del otro (pág. 214)

command economy an economy in which the means of production are publicly owned (p. 96)

economía planificada sistema económico en el que los medios de producción son de propiedad pública (pág. 96)

*commodity a material, resource, or product that is bought and sold (p. 468)

*mercancía materia prima, bien o producto que se compra y se vende (pág. 468)

communism a system of government in which the government controls the ways of producing goods (p. 406)

comunismo forma de gobierno en la que el gobierno controla los modos de producción de los bienes (pág. 406)

compass rose the feature on a map that shows direction (p. 28)

rosa de los vientos convención de un mapa que señala la dirección (pág. 28)

*complex highly developed (p. 526)

*complejo muy desarrollado (pág. 526)

*component a part of something (p. 23)

*componente parte de algo (pág. 23)

*comprise to make up (p. 179)

*incluir integrar (pág. 179)

compulsory mandatory; enforced (p. 255)

condensation the result of water vapor changing to a liquid or a solid state (p. 64)

***conflict** a serious disagreement (p. 283)

coniferous describing evergreen trees that produce cones to hold seeds and that have needles instead of leaves (p. 179)

conquistador a Spanish explorer of the early Americas (p. 211)

***consist** to be made up of (p. 610)

constitution a document setting forth the structure and powers of a government and the rights of people in a country (p. 602)

constitutional monarchy a form of government in which a monarch is the head of state but elected officials run the government (p. 479)

***contact** communication or interaction with someone (p. 712)

***contemporary** of the present time; modern (p. 315)

contiguous joined together inside a common boundary (p. 149)

continent a large, unbroken mass of land (p. 52)

Continental Divide an imaginary line through the Rocky Mountains that separates rivers that flow west from rivers that flow east (p. 152)

continental island an island formed centuries ago by the rising and folding of the ocean floor due to tectonic activity (p. 767)

continental shelf the part of a continent that extends into the ocean in a plateau, then drops sharply to the ocean floor (p. 60)

***contribution** something that is given (p. 382)

***controversy** a dispute; a discussion involving opposing views (p. 754)

***convert** to change from one thing to another (p. 27); to change religions (p. 592)

***cooperate** to work together (p. 350)

compulsivo obligatorio; forzoso (pág. 255)

condensación cambio del vapor de agua a un estado líquido o sólido (pág. 64)

***conflicto** desacuerdo grave (pág. 283)

coníferas árboles perennes que producen conos para contener las semillas y tienen agujas en vez de hojas (pág. 179)

conquistador explorador español de América en sus inicios (pág. 211)

***consistir** estar hecho de (pág. 610)

constitución documento que establece la estructura y los poderes de un gobierno así como los derechos de las personas en un país (pág. 602)

monarquía constitucional sistema de gobierno en el que un monarca ostenta la jefatura del Estado pero el gobierno lo administran funcionarios elegidos (pág. 479)

***contacto** comunicación o interacción (pág. 712)

***contemporáneo** perteneciente al tiempo presente; moderno (pág. 315)

contiguo unido dentro de un límite común (pág. 149)

continente extensión de tierra grande e ininterrumpida (pág. 52)

divisoria continental línea imaginaria que atraviesa las montañas Rocosas para separar los ríos que fluyen hacia el oeste de los que fluyen hacia el este (pág. 152)

isla continental isla formada siglos atrás por el levantamiento y plegamiento del fondo oceánico resultantes de la actividad tectónica (pág. 767)

plataforma continental parte de un continente que se adentra en el océano en forma de meseta y luego desciende abruptamente hasta el fondo oceánico (pág. 60)

***contribución** algo que se entrega (pág. 382)

***controversia** disputa; discusión que involucra puntos de vista opuestos (pág. 754)

***convertir** cambiar de una cosa a otra (pág. 27); cambiar de religión (pág. 592)

***cooperar** trabajar en unión (pág. 350)

Glossary/Glosario

coral reef a long, undersea structure formed by the tiny skeletons of coral, a kind of sea life (p. 733)

arrecife coralino extensa estructura submarina formada por los diminutos esqueletos de los corales, una especie de vida marina (pág. 733)

cordillera a region of parallel mountain chains (p. 149)

cordillera región de cadenas montañosas paralelas (pág. 149)

cottage industry a home- or village-based industry in which people make simple goods using their own equipment (p. 508)

industria artesanal industria doméstica o aldeana en la cual las personas elaboran bienes sencillos utilizando sus propios equipos (pág. 508)

coup an action in which a group of individuals seize control of a government (p. 311)

golpe (de Estado) acción mediante la cual un grupo de individuos se apodera del control de un gobierno (pág. 311)

couscous a small, round grain used in North African and Southwest Asian cooking (p. 598)

cuscús cereal pequeño y redondo que se utiliza en la cocina de África del Norte y el Sudeste Asiático (pág. 598)

***create** to make (p. 151)

***crear** hacer (pág. 151)

creole a group of languages developed by enslaved people on colonial plantations that is a mixture of French, Spanish, and African (p. 289); two or more languages that blend and become the language of the region (p. 685)

criollo grupo de lenguas desarrollado por las personas esclavizadas en las plantaciones coloniales, que consiste en una mezcla de francés, español y africano (pág. 289); dos o más lenguas que se mezclan y convierten en la lengua de una región (pág. 685)

cultural region a geographic area in which people have certain traits in common (p. 86)

región cultural área geográfica donde las personas tienen ciertos rasgos comunes (pág. 86)

culture the set of beliefs, behaviors, and traits shared by a group of people (p. 82)

cultura conjunto de creencias, comportamientos y rasgos compartidos por un grupo de personas (pág. 82)

***currency** the paper money and coins in circulation (p. 101)

***moneda** dinero en billetes y monedas en circulación (pág. 101)

cyclone a storm with high winds and heavy rains (p. 494)

ciclón tormenta con vientos huracanados y lluvias torrenciales (pág. 494)

czar the title given to an emperor of Russia's past (p. 404)

zar título dado a un emperador de Rusia en el pasado (pág. 404)

D

dalit the lowest caste of Indian society; also called the "untouchables" (p. 509)

paria casta más baja de la sociedad india; también se le denomina "los intocables" (pág. 509)

***data** information (p. 158)

***dato** información (pág. 158)

de facto actually; in reality (p. 432)

de facto de hecho; en la realidad (pág. 432)

death rate the number of deaths compared to the total number of people in a population at a given time (p. 72)

tasa de mortalidad número de defunciones comparado con el número total de habitantes de una población en un tiempo determinado (pág. 72)

deciduous describing trees that shed their leaves in the autumn (p. 179)

caducifolios árboles que pierden sus hojas en el otoño (pág. 179)

***decline** to reduce in number (p. 170)

***declinar** reducirse en número (pág. 170)

define to describe the nature or extent of something (p. 517)

delta an area where sand, silt, clay, or gravel is dropped at the mouth of a river (pp. 62; 493)

democracy a type of government run by the people (p. 86)

demonstrate to show (p. 590)

dependence too much reliance (p. 218)

depict to describe or to show (p. 660)

desalinization a process that makes salt water safe to drink (p. 61)

desertification the process by which an area turns into a desert (p. 615)

despite in spite of (p. 276)

devolution the process by which a large, centralized government gives power away to smaller, local governments (p. 415)

dialect a regional variety of a language with unique features, such as vocabulary, grammar, or pronunciation (p. 83)

dictatorship a form of government in which one person has absolute power to rule and control the government, the people, and the economy (p. 87)

didgeridoo a large, bamboo musical instrument of the Australian aboriginal people (p. 751)

dike a large barrier built to keep out water (p. 336)

dingo a wild dog of Australia (p. 741)

displace to take over a place or position of others (p. 677)

distinct separate; easily recognized as separate or different (p. 775)

distort to change something so it is no longer accurate (p. 27)

diverse composed of many distinct and different parts (p. 256)

diversified increased variety to achieve a balance (p. 600)

dominate to control totally (p. 433)

definir describir la naturaleza o el alcance de algo (pág. 517)

delta área donde se deposita arena, sedimento, lodo o gravilla en la desembocadura de un río (págs. 62; 493)

democracia tipo de gobierno dirigido por el pueblo (pág. 86)

demostrar probar (pág. 590)

dependencia confianza excesiva (pág. 218)

representar describir o mostrar (pág. 660)

desalinización proceso que elimina la sal del agua para hacerla potable (pág. 61)

desertización proceso por el cual un área se transforma en un desierto (pág. 615)

a pesar de no obstante (pág. 276)

autonomía proceso mediante el cual un gran gobierno centralizado cede poder a gobiernos locales menores (pág. 415)

dialecto variedad regional de una lengua con características únicas, como vocabulario, gramática o pronunciación (pág. 83)

dictadura forma de gobierno en la que una persona detenta el poder absoluto para mandar y controlar al gobierno, el pueblo y la economía (pág. 87)

diyiridú instrumento musical de bambú, de gran tamaño, de los aborígenes australianos (pág. 751)

dique barrera grande construida para no dejar pasar el agua (pág. 336)

dingo perro salvaje de Australia (pág. 741)

desplazar tomar el lugar o la posición de otros (pág. 677)

distinto separado; fácilmente reconocible como separado o diferente (pág. 775)

distorsionar cambiar algo de modo que ya no es correcto (pág. 27)

diverso compuesto de muchas partes distintivas y diferentes (pág. 256)

diversificado variedad incrementada para lograr un equilibrio (pág. 600)

dominar controlar totalmente (pág. 433)

dominion a largely self-governing country within the British Empire (p. 747)

dormant still capable of erupting but showing no signs of activity (p. 208)

doubling time the number of years it takes a population to double in size based on its current growth rate (p. 73)

drought long period of time without rainfall (p. 737)

Dust Bowl the southern Great Plains during the severe drought of the 1930s (p. 167)

*dynamic always changing (p. 20)

dynasty a line of rulers from a single family that holds power for a long time (p. 440)

dominio país autónomo dentro del Imperio británica (pág. 747)

inactivo que aún es capaz de entrar en erupción pero no muestra señales de actividad (pág. 208)

tiempo de duplicación número de años que le toma a una población doblar su tamaño con base en la tasa de crecimiento actual (pág. 73)

sequía periodo largo sin lluvias (pág. 737)

Dust Bowl las Grandes Llanuras meridionales durante la fuerte sequía de la década de 1930 (pág. 167)

*dinámico en permanente cambio (pág. 20)

dinastía serie de gobernantes de una sola familia que detentan el poder por mucho tiempo (pág. 440)

E

earthquake an event in which the ground shakes or trembles, brought about by the collision of tectonic plates (p. 54)

economic system how a society decides on the ownership and distribution of its economic resources (p. 96)

ecotourism a type of tourism in which people visit a country to enjoy its natural wonders (p. 484)

*element an important part or characteristic (p. 677)

elevation the measurement of how much above or below sea level a place is (p. 29)

emancipate to make free (p. 254)

embargo a ban on trade with a particular country (p. 709)

emigrate to leave one's home to live in another place (p. 78)

*emphasis an expression that shows the importance of something (p. 599)

encomienda the Spanish system of enslaving Native Americans and making them practice Christianity (p. 281)

endemic specific to a particular place or people (p. 471)

terremoto suceso en el cual el suelo se agita o tiembla como consecuencia de la colisión de placas tectónicas (pág. 54)

sistema económico la forma en que una sociedad decide la propiedad y distribución de sus recursos económicos (pág. 96)

ecoturismo tipo de turismo en el cual las personas visitan un país para disfrutar de sus maravillas naturales (pág. 484)

*elemento parte o característica importante (pág. 677)

elevación medida de cuánto más alto o más bajo está un lugar respecto del nivel del mar (pág. 29)

emancipar liberar (pág. 254)

embargo prohibición de comerciar con un país específico (pág. 709)

emigrar abandonar el hogar propio para vivir en otro lugar (pág. 78)

*énfasis expresión que muestra la importancia de algo (pág. 599)

encomienda sistema español de esclavizar a los indígenas americanos y obligarlos a profesar el cristianismo (pág. 281)

endémico específico de un lugar o una persona en particular (pág. 471)

Glossary/Glosario **835**

environment the natural surroundings of a place (p. 23)

Equator a line of latitude that runs around the middle of Earth (p. 21)

equinox one of two days each year when the sun is directly overhead at the Equator (p. 46)

erg a large area of sand (p. 584)

erosion the process by which weathered bits of rock are moved elsewhere by water, wind, or ice (p. 55)

escarpment a steep cliff at the edge of a plateau with a lowland area below (p. 242)

*establish to start (p. 157)

estuary an area where river currents and the ocean tide meet (p. 301)

ethanol a liquid fuel made in part from plants (p. 155)

ethnic group a group of people with a common racial, national, tribal, religious, or cultural background (p. 83)

eucalyptus a tree found only in Australia and nearby islands that is well suited to dry conditions with leathery leaves, deep roots, and the ability to survive when rivers flood (p. 739)

evaporation the change of liquid water to water vapor (p. 63)

*eventually at a later time (p. 185)

*exceed to go beyond a limit (p. 274)

*expand to spread out; to grow larger (p. 550)

*exploit to use a person, resource, or situation unfairly and selfishly (p. 485)

export to send a product produced in one country to another country (p. 99)

extended family a unit of related people made up of several generations, including grandparents, parents, and children (p. 686)

extinct describing a particular kind of plant or animal that has disappeared completely from Earth (p. 160); describing a volcano that is no longer able to erupt (p. 208)

medioambiente entorno natural de un lugar (pág. 23)

ecuador línea de latitud que atraviesa la mitad de la Tierra (pág. 21)

equinoccio uno de dos días al año cuando el sol se halla situado directamente sobre el ecuador (pág. 46)

erg zona extensa de arena (pág. 584)

erosión proceso por el cual fragmentos desgastados de rocas son llevados a otra parte por acción del agua, el viento o el hielo (pág. 55)

escarpado acantilado pendiente, al borde de una meseta, que tiene debajo un área de tierras bajas (pág. 242)

*establecer comenzar (pág. 157)

estuario área donde convergen corrientes fluviales y la marea oceánica (pág. 301)

etanol combustible líquido que se fabrica a partir de vegetales (pág. 155)

grupo étnico grupo de personas con un antecedente racial, nacional, tribal, religioso o cultural común (pág. 83)

eucalipto árbol de hojas carnosas y raíces profundas que solo se encuentra en Australia e islas adyacentes. Se adapta bien a las condiciones de sequía y puede sobrevivir a las inundaciones fluviales (pág. 739)

evaporación cambio del agua en estado líquido a vapor (pág. 63)

*finalmente en un tiempo posterior (pág. 185)

*exceder traspasar un límite (pág. 274)

*expandir extender; agrandar (pág. 550)

*explotar utilizar a una persona, un recurso o una situación de manera injusta y egoísta (pág. 485)

exportar enviar un bien producido en un país a otro país (pág. 99)

familia extensa unidad de personas emparentadas conformada por varias generaciones, incluidos abuelos, padres e hijos (pág. 686)

extinto espécimen específico de una planta o un animal que ha desaparecido por completo de la Tierra (pág. 160); volcán que ya no puede entrar en erupción (pág. 208)

Glossary/Glosario

extract to remove or take out (p. 249)

extraer remover o sacar (pág. 249)

F

factor a cause (p. 416)

factor causa (pág. 416)

fale a traditional Samoan home that has no walls, leaving the inside open to cooling ocean breezes (p. 777)

fale casa tradicional de Samoa que carece de paredes, de manera que el interior queda abierto a las frescas brisas del océano (pág. 777)

fall line the area where waterfalls flow from higher to lower ground (p. 122)

línea de descenso área donde las cascadas fluyen de un terreno más alto a uno más bajo (pág. 122)

fault a place where two tectonic plates grind against each other (p. 54)

falla lugar donde dos placas tectónicas chocan entre sí (pág. 54)

fauna the animal life in a particular environment (p. 471)

fauna vida animal en un medioambiente específico (pág. 471)

favela an overcrowded city slum in Brazil (p. 257)

favela tugurio urbano superpoblado de Brasil (pág. 257)

feature a noteworthy characteristic (p. 210)

rasgo característica notable (pág. 210)

fellaheen the peasant farmers of Egypt who rent small plots of land (p. 597)

fellaheen campesinos de Egipto que arriendan pequeñas parcelas (pág. 597)

feudalism the political and social system in which kings gave land to nobles in exchange for the nobles' promise to serve them; those nobles provided military service as knights for the king (p. 343)

feudalismo sistema social y político en el cual los reyes entregaban tierras a los nobles, que a cambio prometían servirles; estos nobles proveían de servicio militar al rey como caballeros (pág. 343)

fishery an area where fish come to feed in huge numbers (p. 182)

pesquería zona donde los peces llegan a alimentarse en gran número (pág. 182)

fjord a narrow, U-shaped coastal valley with steep sides formed by the action of glaciers (p. 368)

fiordo valle costero estrecho, en forma de U con laderas escarpadas, formado por la acción de glaciares (pág. 368)

flora the plant life in a particular environment (p. 471)

flora vida vegetal en un medioambiente específico (pág. 471)

fossil water water that fell as rain thousands of years ago and is now trapped deep below ground (p. 563)

agua fósil agua que cayó en forma de lluvia hace miles de años y ahora se encuentra atrapada en las profundidades del subsuelo (pág. 563)

free trade arrangement whereby a group of countries decides to set little or no tariffs on quotas (p. 100)

libre comercio acuerdo por el cual un grupo de países decide imponer aranceles bajos a las cuotas o no fija ningún arancel (pág. 100)

free-trade zone an area where trade barriers between countries are relaxed or lowered (p. 219)

zona de libre comercio área donde las barreras comerciales entre los países se distienden o reducen (pág. 219)

frontier a region just beyond the edge of a settled area (p. 158)

frontera región inmediatamente posterior al borde de un área poblada (pág. 158)

Glossary/Glosario

fundamentalist a person who believes in the strict interpretation of religious laws (p. 595)

fundamentalista persona que cree en la interpretación estricta de las leyes religiosas (pág. 595)

G

genocide the mass murder of people from a particular ethnic group (p. 405)

genocidio asesinato masivo de personas de un grupo étnico específico (pág. 405)

geography the study of Earth and its peoples, places, and environments (p. 18)

geografía estudio de la Tierra y de sus gentes, lugares y entornos (pág. 18)

geothermal energy the electricity produced by natural, underground sources of steam (pp. 617; 752)

energía geotérmica electricidad producida por fuentes naturales de vapor subterráneas (págs. 617; 752)

geyser a spring of water heated by molten rock inside Earth that, from time to time, shoots hot water into the air (p. 735)

géiser fuente de agua calentada por rocas fundidas en el interior de la Tierra que, de vez en cuando, expulsa agua caliente al aire (pág. 735)

glaciation the process of becoming covered by glaciers (p. 367)

glaciación proceso en el cual los glaciares cubren zonas amplias del planeta (pág. 367)

glacier a large body of ice that moves slowly across land (p. 56)

glaciar masa de hielo enorme que se mueve lentamente sobre la tierra (pág. 56)

globalization the process by which nations, cultures, and economies become mixed (p. 89)

globalización proceso mediante el cual naciones, culturas y economías se integran (pág. 89)

granary a building used to store harvested grain (p. 188)

granero edificación en la cual se almacena el grano cosechado (pág. 188)

*grant to allow as a right, privilege, or favor (p. 707)

*conceder permitir como un derecho, privilegio o favor (pág. 707)

green revolution the effort to use modern techniques and science to increase food production in poorer countries (p. 508)

revolución verde esfuerzo por utilizar técnicas modernas y la ciencia para aumentar la producción de alimentos en los países más pobres (pág. 508)

gross domestic product (GDP) the total dollar value of all final goods and services produced in a country during a single year (p. 98)

producto interno bruto (PIB) valor total en dólares de todos los bienes y servicios finales producidos en un país durante un año (pág. 98)

groundwater the water contained inside Earth's crust (p. 61)

agua subterránea agua contenida en el interior de la corteza terrestre (pág. 61)

guerrilla a member of a small, defensive force of irregular soldiers (p. 309)

guerrillero miembro de una fuerza pequeña y defensiva de soldados irregulares (pág. 309)

H

hacienda a large estate (p. 281)

hacienda gran propiedad rural (pág. 281)

Glossary/Glosario

harmattan a wind off the Atlantic coast of Africa that blows from the northeast to the south, carrying large amounts of dust (p. 673)

hemisphere each half of Earth (p. 26)

*hierarchy a ruling body arranged by rank or class (p. 307)

hieroglyphics the system of writing that uses small pictures to represent sounds or words (p. 590)

high island a type of island in the Pacific Ocean formed many centuries ago by volcanoes and still having mountainous areas (p. 768)

hinterland an inland area that is remote from the urban areas of a country (p. 256)

Holocaust the mass killing of 6 million European Jews by Germany's Nazi leaders during World War II (p. 348)

homogeneous made up of many things that are the same (p. 380)

hot springs places where naturally heated water rises out of the ground (p. 735)

human rights the rights belonging to all individuals (p. 87)

hurricane a storm with strong winds and heavy rains (p. 122)

hydroelectric power the electricity that is created by flowing water (p. 616)

hydropolitics the politics surrounding water access and usage rights (p. 563)

harmattan viento de la costa atlántica de África que sopla del nordeste al sur, arrastrando consigo grandes cantidades de polvo (pág. 673)

hemisferio cada mitad de la Tierra (pág. 26)

*jerarquía cuerpo de gobierno organizado por rango o clase (pág. 307)

jeroglífico sistema de escritura que representa sonidos o palabras con dibujos pequeños (pág. 590)

isla alta tipo de isla del océano Pacífico formada hace muchos siglos por volcanes y que aún tiene zonas montañosas (pág. 768)

hinterland zona interior distante de las áreas urbanas de un país (pág. 256)

Holocausto exterminio masivo de 6 millones de judíos europeos ejecutado por los líderes nazis de Alemania durante la Segunda Guerra Mundial (pág. 348)

homogéneo compuesto por muchas cosas iguales (pág. 380)

fuentes termales lugares donde agua calentada por medios naturales brota del suelo (pág. 735)

derechos humanos los derechos que tienen todos los individuos (pág. 87)

huracán tormenta con vientos fuertes y lluvias copiosas (pág. 122)

energía hidroeléctrica electricidad producida por agua en movimiento (pág. 616)

hidropolítica política relativa al acceso al agua y los derechos de su uso (pág. 563)

I

ice sheet a large, thick area of ice that covers a region (p. 791)

ice shelf a thick layer of ice that extends above the water (p. 792)

iceberg a huge piece of floating ice that broke off from an ice shelf or glacier and fell into the sea (p. 792)

immigrate to enter and live in a new country (p. 78)

manto de hielo área extensa y gruesa de hielo que cubre una región (pág. 791)

plataforma de hielo capa gruesa de hielo que se extiende sobre la superficie del agua (pág. 792)

iceberg témpano gigante de hielo flotante que se desprendió de una plataforma de hielo o glaciar y cayó al mar (pág. 792)

inmigrar entrar a un nuevo país y vivir allí (pág. 78)

immunity the ability to resist infection by a particular disease (p. 281)

inmunidad capacidad de resistir la infección provocada por una enfermedad específica (pág. 281)

*impact an effect or influence (p. 312)

*impacto efecto o influencia (pág. 312)

imperialism a policy by which a country increases its power by gaining control over other areas of the world (p. 621)

imperialismo política mediante la cual un país aumenta su poder ejerciendo control sobre otras áreas del mundo (pág. 621)

import when a country brings in a product from another country (p. 99)

importación cuando un país ingresa un producto de otro país (pág. 99)

indigenous living or existing naturally in a particular place (p. 125)

nativo que vive o existe de modo natural en un lugar específico (pág. 125)

industrialized describing a country in which manufacturing is a primary economic activity (p. 346)

industrializado país en el cual la manufactura es la principal actividad económica (pág. 346)

industry the manufacturing and making of products to sell (p. 130)

industria manufactura y fabricación de bienes para la venta (pág. 130)

*inevitable sure to happen (p. 407)

*inevitable que sucederá con certeza (pág. 407)

inflation a sharp increase in the price of goods, sometimes caused by a shortage of goods (p. 410)

inflación incremento drástico en el precio de los bienes, a veces ocasionado por su escasez (pág. 410)

infrastructure a system of roads and railroads that allow the transport of materials (p. 688)

infraestructura sistema de carreteras y ferrocarriles que permite el transporte de materiales (pág. 688)

*inhibit to limit (p. 523)

*inhibir limitar (pág. 523)

*initiate to begin (p. 218)

*iniciar comenzar (pág. 218)

insular separate from other countries (p. 466)

insular separado de otros países (pág. 466)

*intense strong (p. 54)

*intenso poderoso (pág. 54)

*intertwine to become closely connected or involved (p. 443)

*entretejer unirse o envolverse estrechamente (pág. 443)

introduced species a nonnative species that is brought to a new environment (p. 745)

especie introducida especie foránea que se lleva a un medioambiente nuevo (pág. 745)

irrigate to supply land with water through ditches or pipes (p. 525)

irrigar suministrar agua a un terreno por medio de zanjas o ductos (pág. 525)

irrigation the process of collecting water and using it to water crops (p. 153)

irrigación proceso de recolección del agua para regar los cultivos (pág. 153)

*isolate to make separate from others (p. 125)

*aislar separar de otros (pág. 125)

isthmus a narrow strip of land that connects two larger land areas (p. 59)

istmo franja estrecha de tierra que conecta dos áreas de tierra más grandes (pág. 59)

Glossary/Glosario

K

kapahaka a traditional art form of the Maori people that combines music, dance, singing, and facial expressions (p. 742)

katabatic winds the strong, fast, cold winds that blow down from the interior of Antarctica (p. 793)

kente the colorful, handwoven cloth produced in Kenya (p. 687)

key the feature on a map that explains the symbols, colors, and lines used on the map (p. 28)

kiwifruit a small, fuzzy, brownish-colored fruit with bright green flesh (p. 753)

krill the tiny, shrimplike sea creatures that are eaten by whales and many other sea creatures (p. 795)

kapahaka manifestación artística tradicional del pueblo maorí que combina música, danza, canto y expresiones faciales (pág. 742)

vientos catabáticos vientos fuertes, rápidos y fríos que soplan desde el interior de la Antártida (pág. 793)

kente tela tejida de vivos colores que se fabrica en Kenia (pág.687)

clave elemento de un mapa que explica los símbolos, colores y líneas usados en este (pág. 28)

kiwi fruta pequeña y vellosa de tono marrón cuya carne es verde brillante (pág. 753)

krill crustáceo diminuto, similar al camarón, que comen las ballenas y otras criaturas marinas (pág. 795)

L

lagoon a shallow pond near a larger body of water (p. 769)

landform a natural feature found on land (p. 23)

landlocked having no border with an ocean or a sea (p. 698)

landscape the portions of Earth's surface that can be viewed at one time from a location (p. 19)

latitude the lines on a map that run east to west (p. 21)

lawsuit a legal action in which people ask for relief from some damage done to them by someone else (p. 754)

levee a raised riverbank used to control flooding (p. 120)

lichen tiny, sturdy plants that grow in rocky areas (p. 794)

lock a gated passageway used to raise or lower boats in a waterway (p. 119)

loess a fine-grained, fertile soil deposited by the wind (p. 435)

laguna pozo poco profundo cercano a una masa de agua mayor (pág. 769)

accidente geográfico formación natural que se encuentra sobre la tierra (pág. 23)

sin salida al mar que no limita con un océano o un mar (pág. 698)

paisaje partes de la superficie terrestre que se pueden observar a un mismo tiempo desde una ubicación (pág. 19)

latitud líneas sobre un mapa que van de este a oeste (pág. 21)

demanda acción legal en la que las personas exigen una indemnización por algún daño que les causó un tercero (pág. 754)

dique ribera elevada que sirve para controlar las inundaciones (pág. 120)

líquenes plantas pequeñas y resistentes que crecen en las zonas rocosas (pág. 794)

esclusa compartimento con puertas que se utiliza para subir o bajar los barcos en un canal (pág. 119)

loes suelo fértil y de granos finos depositado por el viento (pág. 435)

Glossary/Glosario

longitude the lines on a map that run north to south (p. 21)

longitud líneas sobre un mapa que van de norte a sur (pág. 21)

longship a ship with oars and a sail used by the Vikings (p. 374)

drakkar barco con remos y velas que utilizaban los vikingos (pág. 374)

low island a type of island in the Pacific Ocean formed by the buildup of coral (p. 768)

isla baja tipo de isla del océano Pacífico formada por acumulaciones de coral (pág. 768)

M

Manifest Destiny the idea that it was the right of Americans to expand westward to the Pacific Ocean (p. 158)

Destino Manifiesto ideología según la cual los estadounidenses tenían derecho a expandirse al oeste hacia el océano Pacífico (pág. 158)

map projection one of several systems used to represent the round Earth on a flat map (p. 28)

proyección cartográfica uno de los varios sistemas que se usan para representar la esfera terrestre en un mapa plano (pág. 28)

maquiladora a foreign-owned factory where workers assemble parts (p. 216)

maquiladora fábrica de propiedad extranjera donde los obreros ensamblan partes (pág. 216)

*margin an edge (p. 581)

*margen borde (pág. 581)

market economy an economy in which most of the means of production are privately owned (p. 96)

economía de mercado economía en la cual la mayoría de los medios de producción son de propiedad privada (pág. 96)

marsupial a type of mammal that carries its young in a pouch (p. 739)

marsupial tipo de mamífero que carga a su cría en una bolsa (pág. 739)

*mature fully grown and developed as an adult; also refers to older adults (p. 72)

*maduro adulto plenamente crecido y desarrollado; también se refiere a los adultos mayores (pág. 72)

megalopolis a huge city or cluster of cities with an extremely large population (p. 80)

megalópolis ciudad enorme o cúmulo de ciudades que tienen una población extremadamente grande (pág. 80)

Métis the child of a French person and a native person (p. 187)

métis hijo de un francés y una indígena (pág. 187)

metropolitan area having to do with a large city (p. 132); an area that includes a city and its surrounding suburbs (pp. 190; 257)

metropolitan área relativo a una ciudad grande (pág. 132); área que incluye una ciudad y los suburbios que la rodean (págs. 190; 257)

Middle Ages the period in European history from about A.D. 500 to about 1450 (p. 343)

Edad Media periodo de la historia europea que abarca aproximadamente del año 500 al 1450 (pág. 343)

millennium a period of a thousand years (p. 549)

milenio periodo de mil años (pág. 549)

millet a grass that produces edible seeds (p. 651)

mijo especie de pasto que produce semillas comestibles (pág. 651)

*migrate to move to an area to settle (p. 185)

*migrar trasladarse a un lugar para establecerse allí (pág. 185)

minority a group of people that is different from most of the population (p. 482)

minoría grupo de personas diferente a la mayoría de la población (pág. 482)

MIRAB economy a lesser-developed economy that depends on aid from foreign countries and remittances from former residents working elsewhere (p. 780)

economía MIRAB economía menos desarrollada que depende de la ayuda de países extranjeros y de las remesas de antiguos residentes que trabajan en otro lugar (pág. 780)

mission a Catholic-based community in the west (p. 157)

misión comunidad católica del Oeste (pág. 157)

missionary someone who tries to convert others to a certain religion (p. 654)

misionario persona que trata de convertir a otras a una religión específica (pág. 654)

mixed economy an economy in which parts of the economy are privately owned and parts are owned by the government (p. 96)

economía mixta economía en la cual unos sectores son de propiedad privada y otros son de propiedad del gobierno (pág. 96)

monarchy the system of government in which a country is ruled by a king or queen (p. 87)

monarquía sistema de gobierno en el que un rey o una reina gobiernan un país (pág. 87)

monolith a single standing stone (p. 733)

monolito piedra erguida de una sola pieza (pág. 733)

monotheism the belief in one god (p. 549)

monoteísmo creencia en un solo dios (pág. 549)

monsoon a seasonal wind that blows steadily from the same direction for several months at a time but changes directions at other times of the year (p. 494)

monzón viento estacional que sopla regularmente desde la misma dirección durante varios meses pero cambia de dirección en otras épocas del año (pág. 494)

Mormon a member of the Church of Jesus Christ of Latter Day Saints (p. 167)

mormón miembro de la Iglesia de Jesucristo de los Santos de los Últimos Días (pág. 167)

multinational a company that has locations in more than one country (p. 310)

multinacional compañía que tiene oficinas en más de un país (pág. 310)

mural a large painting on a wall (p. 217)

mural pintura de gran tamaño hecha sobre un muro (pág. 217)

myrrh a sweet perfume used as medicine in ancient times (p. 589)

mirra perfume dulce que antiguamente se empleaba como medicamento (pág. 589)

N

national park a park that has been set aside for the public to enjoy for its great natural beauty (p. 155)

parque nacional parque destinado al público para que disfrute sus grandes bellezas naturales (pág. 155)

*network a complex, interconnected chain or system of things such as roads, canals, or computers (p. 699)

*red cadena o sistema complejo e interconectado de carreteras, canales o computadoras, entre otros (pág. 699)

nomad a person who lives by moving from place to place to follow and hunt herds of migrating animals or to lead herds of grazing animals to fresh pastures (p. 585)

nómada persona que vive trasladándose de un lugar a otro para seguir y cazar manadas de animales migratorios o para conducir rebaños de animales de pastoreo hacia pastos frescos (pág. 585)

nomadic describes a way of life in which a person or group lives by moving from place to place (p. 156)

nómada forma de vida en la que una persona o grupo vive trasladándose de un lugar a otro (pág. 156)

nonrenewable resources the resources that cannot be totally replaced (p. 95)

recursos no renovables recursos que no se pueden reponer por completo (pág. 95)

nuclear family the family group that includes only parents and their children (p. 687)

familia nuclear grupo familiar que solo incluye a padres e hijos (pág. 687)

nuclear proliferation the spread of control of nuclear power, particularly the knowledge of how to construct nuclear weapons (p. 503)

proliferación nuclear expansión del dominio de la energía nuclear, en particular el conocimiento para construir armas nucleares (pág. 503)

O

oasis a fertile area that rises in a desert wherever water is regularly available (p. 529)

oasis área fértil que se desarrolla en un desierto cuando hay una fuente regular de agua (pág. 529)

*occupy to settle in a place (p. 184)

*ocupar establecerse en un lugar (pág. 184)

*occur to happen or take place (p. 244)

*ocurrir suceder o acontecer (pág. 244)

oligarch a member of a small ruling group that holds great power (p. 410)

oligarca miembro de un pequeño grupo gobernante que detenta gran poder (pág. 410)

oral tradition the process of passing stories by word of mouth from generation to generation (p. 630)

tradición oral forma de transmitir historias de generación en generación, mediante la palabra hablada (pág. 630)

orbit to circle around something (p. 42)

orbitar moverse en círculo alrededor de algo (pág. 42)

Outback the inland areas of Australia west of the Great Dividing Range (p. 733)

Outback zonas del interior de Australia ubicadas al oeste de la Gran Cordillera Divisoria (pág. 733)

outsourcing hiring workers in other countries to do a set of jobs (p. 508)

subcontratar contratar trabajadores en otros países para que hagan una serie de trabajos (pág. 508)

*overall as a whole; generally (p. 732)

*global como un todo; generalizado (pág. 732)

ozone the certain kind of oxygen that forms a layer around Earth in the atmosphere; it blocks out many of the most harmful rays from the sun (p. 799)

ozono tipo de oxígeno que forma una capa alrededor de la Tierra en la atmósfera; bloquea el paso de los rayos más dañinos del sol (pág. 799)

P

Pacific Rim the countries bordering the Pacific Ocean, particularly Asian countries (p. 483)

Cuenca del Pacífico países que bordean el océano Pacífico, en particular los países asiáticos (pág. 483)

pagan someone who believes in more than one god or someone who has little or no religious belief (p. 374)

pagano persona que cree en más de un dios o cuya creencia religiosa es escasa o nula (pág. 374)

palm oil an oil that is used in cooking (p. 651)

aceite de palma un aceite que se usa para cocinar (pág. 651)

pampas the treeless grassland of Argentina and Uruguay (p. 242)

pampas praderas sin árboles de Argentina y Uruguay (pág. 242)

*parallel running side by side with something; following the same general course and direction (p. 121)

*paralelo que corre lado a lado con algo; que sigue el mismo curso y dirección generales (pág. 121)

Glossary/Glosario

Parliament the national legislature of England (now the United Kingdom), consisting of the House of Lords and the House of Commons (p. 345)

Parlamento asamblea legislativa nacional de Inglaterra (hoy Reino Unido) integrada por la Cámara de los Lores y la Cámara de los Comunes (pág. 345)

pastoral describing a society based on herding animals (p. 522)

pastoril sociedad que vive del pastoreo de animales (pág. 522)

peacekeeping sending trained members of the military to crisis spots to maintain peace and order (p. 193)

pacificación envío de miembros entrenados de las fuerzas armadas a sitios críticos para mantener la paz y el orden (pág. 193)

periodic market an open-air trading market that springs up at a crossroads or in larger towns (p. 715)

mercado ambulante mercado al aire libre que se instala en una aldea o en pueblos más grandes (pág. 715)

permafrost the permanently frozen, lower layers of soil found in the tundra and subarctic climate zones (p. 518)

permacongelamiento capas bajas del suelo, permanentemente congeladas, que se encuentran en la tundra y las zonas de clima subártico (pág. 518)

pharaoh the name for a powerful ruler in ancient Egypt (p. 588)

faraón nombre dado a un poderoso gobernante en el Antiguo Egipto (pág. 588)

phosphate a chemical salt used to make fertilizer (p. 587)

fosfato sal química utilizada para producir fertilizantes (pág. 587)

pidgin a language formed by combining parts of several different languages (pp. 685; 776)

pidgin lengua formada por la combinación de partes de varias lenguas distintas (págs. 685; 776)

pilgrimage a journey to a sacred place (p. 343)

peregrinación viaje a un lugar sagrado (pág. 343)

plain a large expanse of land that can be flat or have a gentle roll (p. 58)

llanura gran extensión de tierra plana o con ligeras ondulaciones (pág. 58)

plankton plants or animals that ride along with water currents (p. 795)

plancton plantas o animales que se desplazan con las corrientes de agua (pág. 795)

plantation a large farm (p. 213)

plantación granja grande (pág. 213)

plateau a flat area that rises above the surrounding land (p. 58)

meseta área plana que se eleva por encima del terreno circundante (pág. 58)

poaching illegal fishing or hunting (pp. 633, 703)

caza furtiva pesca o caza ilegal (págs. 603, 733)

polder the land reclaimed from building dikes and then draining the water from the land (p. 336)

pólder terreno ganado al mar a partir de la construcción de diques y el posterior desecado de la tierra (pág. 336)

*policy a plan or course of action (p. 503)

*política plan o curso de acción (pág. 503)

polytheism the belief in more than one god (p. 549)

politeísmo creencia en más de un dios (pág. 549)

population density the average number of people living within a square mile or a square kilometer (p. 76)

densidad de población número promedio de personas que habitan en una milla cuadrada o un kilómetro cuadrado (pág. 76)

population distribution the geographic pattern of where people live (p. 76)

distribución de la población patrón geográfico que muestra dónde habita la gente (pág. 76)

possession an area or a region that is controlled by another country (p. 774)

posesión zona o región controlada por otro país (pág. 774)

Glossary/Glosario

postindustrial describing an economy that is based on providing services rather than manufacturing (p. 357)

posindustrial economía que se basa en la prestación de servicios, no en la manufacturación (pág. 357)

potential the possibility (p. 649)

potencial posibilidad (pág. 649)

precipitation the water that falls on the ground as rain, snow, sleet, hail, or mist (p. 48)

precipitación agua que cae al suelo en forma de lluvia, nieve, aguanieve, granizo o rocío (pág. 48)

primate city a country's main city that is so large and influential that it dominates the rest of the country (p. 481)

ciudad principal la ciudad más importante de un país, tan grande e influyente que domina el resto del país (pág. 481)

Prime Meridian the starting point for measuring longitude (p. 21)

primer meridiano punto de partida para medir la longitud (pág. 21)

productivity the measurement of what is produced and what is required to produce it (p. 98)

productividad medición de lo que se produce y lo que se requiere para producirlo (pág. 98)

*project a planned activity (p. 588)

*proyecto actividad planificada (pág. 588)

province an administrative unit similar to a state (p. 178)

provincia unidad administrativa similar a un estado (pág. 178)

pueblo a town built by the Pueblo people in the American Southwest (p. 156)

pueblo poblado construido por las tribus pueblo del sudeste estadounidense (pág. 156)

pueblo joven shantytown with poor housing and little or no infrastructure built outside a large metropolitan area (p. 313)

pueblo joven barrio marginal con viviendas precarias y poca o ninguna infraestructura, construido en las afueras de una gran área metropolitana (pág. 313)

R

rain forest a dense stand of trees and other vegetation that receives a great deal of precipitation each year (p. 241)

selva tropical formación densa de árboles y otra vegetación que recibe una gran cantidad de precipitación todos los años (pág. 241)

rain shadow an area that receives reduced rainfall because it is on the side of a mountain facing away from the ocean (p. 49)

sombra pluviométrica zona que recibe pocas precipitaciones porque se halla en la ladera de una montaña que está en el lado contrario al océano (pág. 49)

Raj the period of time in which Great Britain controlled India as a part of the British Empire (p. 502)

Raj periodo durante el cual Gran Bretaña controló India como parte del Imperio británico (pág. 502)

*ratio the relationship in amount or size between two or more things (p. 286)

*ratio relación en cantidad o tamaño entre dos o más cosas (pág. 286)

*rational reasonable (p. 375)

*racional razonable (pág. 375)

recession a time when many businesses close and people lose their jobs (p. 385)

recesión época en que muchos negocios cierran y las personas pierden sus empleos (pág. 385)

refugee a person who flees a country because of violence, war, persecution, or disaster (p. 78)

refugiado persona que huye de un país por la violencia, una guerra, una persecución o un desastre (pág. 78)

reggae a traditional Jamaican style of music that uses complex drum rhythms (p. 221)

regime a style of government (p. 595)

region a group of places that are close to one another and that share some characteristics (p. 22)

*regulate to control something (p. 351)

reincarnation the belief in Hinduism that after a person dies, his or her soul is reborn into another body (p. 500)

relative location the location of one place compared to another place (p. 20)

relief the difference between the elevation of one feature and the elevation of another feature near it (p. 29)

remittance the money sent back to the homeland by people who have gone somewhere else to work (pp. 221; 780)

remote sensing the method of getting information from far away, such as deep below the ground (pp. 32; 800)

Renaissance the period in Europe that began in Italy in the 1300s and lasted into the 1600s, during which art and learning flourished (p. 375)

renewable resources a resource that can be totally replaced or is always available naturally (p. 95)

representative democracy a form of democracy in which citizens elect government leaders to represent the people (p. 87)

research station a base for scientific research and observation, often in a remote location (p. 798)

reservation an area of land that has been set aside for Native Americans (p. 161)

reserves a large amount of a resource that has not yet been tapped (p. 401)

reservoir an artificial lake created by a dam (p. 699)

resort a vacation place where people go to relax (p. 780)

resource a material that can be used to produce crops or other products (p. 23)

reggae género musical tradicional de Jamaica que utiliza complejos ritmos de tambor (pág. 221)

régimen estilo de gobierno (pág. 595)

región agrupación de lugares cercanos que comparten algunas características (pág. 22)

*regular controlar algo (pág. 351)

reencarnación creencia del hinduismo según la cual el espíritu de una persona muerta renace en otro cuerpo (pág. 500)

localización relativa la ubicación de un lugar comparada con la de otro (pág. 20)

relieve diferencia entre la elevación de una formación y la de otra formación cercana (pág. 29)

remesa dinero enviado al país de origen por personas que se han ido a trabajar a otro lugar (págs. 221; 780)

detección remota método para obtener información muy lejana, como de las profundidades del subsuelo (págs. 32; 800)

Renacimiento periodo de Europa que comenzó en Italia en el siglo XII y finalizó en el siglo XV, durante el cual florecieron el arte y la cultura (pág. 375)

recursos renovables recursos que pueden reponerse totalmente o siempre se encuentran disponibles en la naturaleza (pág. 95)

democracia representativa forma de democracia en la que los ciudadanos eligen líderes de gobierno para que representen al pueblo (pág. 87)

estación de investigación base para la investigación y observación científicas , por lo general ubicada en un sitio remoto (pág. 798)

reservación territorio que ha sido destinado a los indígenas americanos (pág. 161)

reservas gran cantidad de un recurso que aún no ha sido explotada (pág. 401)

embalse lago artificial creado por una presa (pág. 699)

centro vacacional lugar de vacaciones donde las personas van a descansar (pág. 780)

recurso materia prima que se puede utilizar para obtener cultivos u otros productos (pág. 23)

Glossary/Glosario

*revenue the income generated by a business (p. 133)

revolution a complete trip of Earth around the sun (p. 42); a period of violent and sweeping change (p. 212)

rift to separate two pieces from one another (p. 610)

Ring of Fire a long, narrow band of volcanoes surrounding the Pacific Ocean (p. 54)

rural describes an area that is lightly populated (p. 77)

Rust Belt the area of the Midwest, Mid-Atlantic, and New England where many factories closed during the 1980s (p. 139)

*renta ingresos generados por un negocio (pág. 133)

revolución recorrido completo de la Tierra alrededor del Sol (pág. 42); periodo de cambio violento y radical (pág. 212)

escindir separar dos partes entre sí (pág. 610)

Cinturón de Fuego banda larga y estrecha de volcanes que rodean el océano Pacífico (pág. 54)

rural área poco poblada (pág. 77)

Rust Belt zona del Medio Oeste, Atlántico Medio y Nueva Inglaterra donde muchas fábricas se cerraron durante la década de 1980 (pág. 139)

S

samurai a powerful, land-owning warrior in Japan (p. 444)

scale the relationship between distances on the map and on Earth (p. 29)

scale bar the feature on a map that tells how a measured space on the map relates to the actual distance on Earth (p. 28)

scrubland land that is dry and hot in the summer and cool and wet in the winter (p. 370)

secede to withdraw from a group or a country (p. 683)

semiarid having lower temperatures and cooler nights than hot, dry deserts (p. 546)

separatist a group that wants to break away from control by a dominant group (p. 194)

serf a farm laborer who could be bought and sold along with the land (p. 404)

service industry a type of business that provides services rather than products (p. 139)

shield a large area of relatively flat land made up of ancient, hard rock (p. 179)

shogun a military leader who ruled Japan in early times (p. 444)

*significant important (p. 151)

silt small particles of rich soil (p. 582)

samurái poderoso guerrero japónes dueño de tierras (pág. 444)

escala relación entre distancias en un mapa y en la Tierra (pág. 29)

escala numérica elemento cartográfico que muestra la relación entre un espacio medido sobre el mapa y la distancia real sobre la Tierra (pág. 28)

chaparral territorio seco y caliente en el verano, y frío y húmedo en el invierno (pág. 370)

separarse retirarse de un grupo o un país (pág. 683)

semiárido que tiene temperaturas más bajas y noches más frías que los desiertos cálidos y secos (pág. 546)

separatista grupo que quiere sustraerse del control de un grupo dominante (pág. 194)

siervo trabajador agrícola que podía comprarse y venderse junto con la tierra (pág. 404)

industria de servicios tipo de negocio que provee servicios en vez de productos (pág. 139)

escudo extensa área de terreno relativamente plano formado por rocas duras y antiguas (pág. 179)

sogún líder militar que gobernaba Japón antiguamente (pág. 444)

*significativo importante (pág. 151)

limo pequeñas partículas de suelo fértil (pág. 582)

Glossary/Glosario

*similar having qualities in common (p. 203)

sitar a long-necked instrument with 7 strings on the outside and 10 inside the neck that provides Indian music with a distinctive sound (p. 507)

slash-and-burn agriculture a method of farming that involves cutting down trees and underbrush and burning the area to create a field for crops (p. 249)

smallpox an often-fatal disease that causes a rash and leaves marks on the skin (p. 308)

smelting the process of refining ore to create metal (p. 342)

solstice one of two days of the year when the sun reaches its northernmost or southernmost point (p. 45)

souk a large, open-air market in North African and Southwest Asian countries (p. 597)

spatial Earth's features in terms of their places, shapes, and relationships to one another (p. 18)

*sphere a round shape like a ball (p. 26)

sphere of influence an area of a country where a single foreign power has been granted exclusive trading rights (p. 444)

*stable staying in the same condition; not likely to change or fail (p. 285)

standard of living the level at which a person, group, or nation lives as measured by the extent to which it meets its needs (p. 98)

staple a food that is eaten regularly (p. 210)

station a cattle or sheep ranch in rural Australia (p. 745)

steppe a partly dry grassland often found on the edge of a desert (p. 397)

*strategy a plan to solve a problem (p. 407)

*structure an arrangement of parts (p. 451)

subcontinent a large landmass that is part of a continent (p. 492)

subregion a smaller part of a region (p. 116)

*similar que tiene cualidades en común (pág. 203)

cítara instrumento de cuello largo con 7 cuerdas en la parte exterior y 10 dentro del cuello, que da a la música india un sonido distintivo (pág. 507)

agricultura de tala y quema método agrícola que consiste en talar árboles y rastrojos y quemar el área despejada para crear un campo de cultivo (pág. 249)

viruela enfermedad, por lo general mortal, que causa sarpullido y deja marcas en la piel (pág. 308)

fundición proceso mediante el cual se refinan minerales para producir metales (pág. 342)

solsticio uno de dos días al año cuando el sol alcanza su máxima declinación norte o sur (pág. 45)

zoco mercado grande al aire libre propio de África del Norte y los países del Sudoeste Asiático (pág. 597)

espaciales características de la Tierra en cuanto a sus lugares, formas y relaciones entre sí (pág. 18)

*esfera figura redonda como una pelota (pág. 26)

esfera de influencia área de un país donde se le ha concedido a una sola potencia extranjera derechos comerciales exclusivos (pág. 444)

*estable que permanece en la misma condición; algo que es improbable que cambie o decaiga (pág. 285)

estándar de vida nivel en que vive una persona, grupo o nación, medido según la capacidad de satisfacer sus necesidades (pág. 98)

alimento básico alimento que se consume habitualmente (pág. 210)

estación rancho de ganado vacuno o lanar del área rural de Australia (pág. 745)

estepa pradera parcialmente seca que se encuentra con frecuencia al borde de un desierto (pág. 397)

*estrategia plan para resolver un problema (pág. 407)

*estructura organización de las partes (pág. 451)

subcontinente gran masa de tierra que forma parte de un continente (pág. 492)

subregión parte más pequeña de una región (pág. 116)

subsistence farming/agriculture a type of farming in which the farmer produces only enough to feed his or her family (pp. 484; 629)

agricultura de subsistencia tipo de agricultura en el que los granjeros producen apenas lo suficiente para alimentar a su familia (págs. 484; 629)

sultan the ruler of a Muslim country (p. 475)

sultán gobernante de un país musulmán (pág. 475)

surplus extra; more than needed (p. 210)

excedente sobrante; más de lo que se necesita (pág. 210)

sustainability the economic principle by which a country works to create conditions where all the natural resources for meeting the needs of society are available (p. 101)

sostenibilidad principio económico según el cual un país crea condiciones para que estén disponibles todos los recursos naturales que satisfacen las necesidades de la sociedad (pág. 101)

T

taiga a large coniferous forest (p. 518)

taiga gran bosque de coníferas (pág. 518)

tariff a tax added to the price of goods that are imported (p. 290)

arancel impuesto añadido al precio de los productos importados (pág. 290)

technology any way that scientific discoveries are applied to practical use (p. 30)

tecnología cualquier forma en que los descubrimientos científicos se aplican para un uso práctico (pág. 30)

tectonic plate one of the 16 pieces of Earth's crust (p. 53)

placa tectónica uno de las 16 partes de la corteza terrestre (pág. 53)

temperate zone a region with a climate that is neither too hot nor too cold (p. 245)

zona templad región cuyo clima no es ni muy frío ni muy caliente (pág. 245)

territory the land administered by the national government (p. 178)

territorio tierra administrada por el gobierno nacional (pág. 178)

thatch a bundle of twigs, grass, and bark (p. 715)

techo de paja armazón de ramas, pasto y corteza (pág. 715)

thematic map a map that shows specialized information (p. 30)

mapa temático mapa que muestra información especializada (pág. 30)

*theory an explanation of why or how something happens (p. 345)

*teoría explicación de por qué o cómo ocurre algo (pág. 345)

tierra caliente the warmest climate zone, located at lower elevations (p. 205)

tierra caliente la zona climática más cálida, ubicada en elevaciones bajas (pág. 205)

tierra fría a colder climate zone, located at higher elevations (p. 205)

tierra fría zona climática fría, ubicada entre elevaciones altas (pág. 205)

tierra templada a temperate climate zone, located at mid-level elevations (p. 205)

tierra templada zona de clima templado, ubicada entre elevaciones medias (pág. 205)

tikanga Maori customs and traditions passed down through generations (p. 742)

tikanga costumbres y tradiciones maoríes transmitidas de generación en generación (pág. 742)

timberline the elevation above which it is too cold for trees to grow (p. 149)

límite forestal elevación por encima de la cual hace demasiado frío para que los árboles prosperen (pág. 149)

topsoil the fertile soil that crops depend on to grow (p. 168)

mantillo suelo fértil del cual dependen las plantas para crecer (pág. 168)

tourism the industry that provides services to people who are traveling for enjoyment (p. 133)

turismo industria que presta servicios a las personas que viajan por placer (pág. 133)

trade deficit a situation that occurs when the value of a country's imports is higher than the value of its exports (p. 454)

déficit comercial situación que ocurre cuando el valor de las importaciones de un país es superior al valor de sus exportaciones (pág. 454)

trade language a common language that emerges when countries trade with each other (p. 658)

lenguaje comercial lenguaje común que surge cuando los países comercian entre sí (pág. 658)

trade surplus a situation that occurs when the value of a country's exports is higher than the value of its imports (p. 454)

superávit comercial situación que ocurre cuando el valor de las exportaciones de un país es superior al valor de sus importaciones (pág. 454)

trade winds the winds that blow regularly in the Tropics (p. 277)

vientos alisios vientos que soplan regularmente en los trópicos (pág. 277)

traditional economy an economy where resources are distributed mainly through families (p. 96)

economía tradicional economía en la que los recursos se distribuyen principalmente entre las familias (pág. 96)

transcontinental describing something that crosses a continent (p. 187)

transcontinental que atraviesa un continente (pág. 187)

*transform to change something completely (p. 64)

*transformar cambiar algo por completo (pág. 64)

trawler a large fishing boat (p. 371)

trainera barco pesquero grande (pág. 371)

treaty an official agreement, negotiated and signed by each party (p. 797)

tratado acuerdo oficial, negociado y firmado por las partes (pág. 797)

trench a long, narrow, steep-sided cut on the ocean floor (p. 60)

fosa depresión larga, estrecha y profunda del fondo oceánico (pág. 60)

*trend a general tendency or preference (p. 715)

*tendencia inclinación o preferencia general (pág. 715)

tributary a small river that flows into a larger river (p. 119)

tributario río pequeño que desemboca en uno más grande (pág. 119)

tribute money paid by one country to another in surrender or for protection (p. 621)

tributo dinero que un país paga a otro por sometimiento o para obtener protección (pág. 621)

Tropics an area between the Tropic of Cancer and the Tropic of Capricorn that has generally warm temperatures because it receives the direct rays of the sun for much of the year (p. 243)

trópicos zona entre el trópico de Cáncer y el trópico de Capricornio que generalmente tiene temperaturas cálidas porque recibe los rayos directos del sol la mayor parte del año (pág. 243)

trust territory an area temporarily placed under control of another country (p. 774)

territorio en fideicomiso área puesta transitoriamente bajo el control de otro país (pág. 774)

tsunami a giant ocean wave caused by volcanic eruptions or movement of the earth under the ocean floor (p. 54)

tsunami gigantesca ola oceánica provocada por erupciones volcánicas o movimientos de la tierra bajo el lecho oceánico (pág. 54)

Glossary/Glosario

Glossary/Glosario

tundra a flat, treeless plain with permanently frozen ground (p. 181)

tundra llanura plana y sin vegetación cuyo suelo permanece helado (pág. 181)

U

*****ultimate** most extreme or greatest (p. 477)

*****supremo** extremo o mayor (pág. 477)

*****uniform** not varying across several parts (p. 369)

*****uniforme** que no cambia en varias partes (pág. 369)

*****unify** to unite; to join together; to make into a unit or a whole (p. 746)

*****unificar** unir; juntar; integrar en una unidad o totalidad (pág. 746)

*****unique** unusual (p. 259)

*****único** inusual (pág. 259)

upland the high land away from the coast of a country (p. 396)

tierra alta tierra elevada de un país, alejada de la costa (pág. 396)

urban describes an area that is densely populated (p. 77)

urbana área densamente poblada (pág. 77)

urbanization when a city grows larger and spreads into nearby areas (p. 80)

urbanización cuando una ciudad crece y se expande hacia las áreas adyacentes (pág. 80)

utility the infrastructure provided by companies or governments such as electricity, water, and trash removal (p. 714)

servicios públicos infraestructura que proveen algunas compañías o gobiernos, como electricidad, agua y recolección de basuras (pág. 714)

V

*****vary** to show differences between things (p. 191)

*****variar** mostrar diferencias entre cosas (pág. 191)

*****via** on the way through (p. 190)

*****vía** en el camino hacia (pág. 190)

*****visible** able to be seen (p. 791)

*****visible** que se puede ver (pág. 791)

*****volume** an amount (p. 671)

*****volumen** cantidad (pág. 671)

W

wadi a dry riverbed that fills with water when rare rains fall in a desert (p. 546)

vado lecho seco de un río que se llena de agua cuando ocasionalmente llueve en un desierto (pág. 546)

water cycle the process in which water is used and reused on Earth, including precipitation, collection, evaporation, and condensation (p. 63)

ciclo del agua proceso en el cual el agua se usa y reutiliza en la Tierra; incluye la precipitación, recolección, evaporación y condensación (pág. 63)

wayfinding the system of navigating a foreign place through observation of natural phenomena, such as the movement of the sun and stars (p. 772)

orientación sistema que consiste en navegar por un lugar desconocido con la ayuda de la observación de fenómenos naturales, como el movimiento del sol y las estrellas (pág. 772)

weathering the process by which Earth's surface is worn away by natural forces (p. 55)

meteorización proceso mediante el cual la superficie terrestre se deteriora por la acción de fuerzas naturales (pág. 55)

welfare capitalism a system in which the government is the main provider of support for the sick, the needy, and the retired (p. 384)

capitalismo de bienestar sistema en el cual el gobierno es el principal proveedor de ayuda para los enfermos, necesitados y jubilados (pág. 384)

westerlies strong winds that blow from west to east (p. 338)

vientos del oeste vientos fuertes que soplan de oeste a este (pág. 338)

*widespread commonly occurring (p. 558)

*extendido que ocurre con frecuencia (pág. 558)

Y

yurt a large, circular structure made of animal skins that can be packed up and moved from place to place; used as a home in Mongolia (p. 532)

yurta estructura grande y circular, hecha de pieles de animales, que se puede empacar y trasladar de un lugar a otro; se usa como vivienda en Mongolia (pág. 532)

Glossary/Glosario

Index

Index

Index

Index

Index

Index

Index

Index

Index

Index

Index

Index

Index

Index

McGraw-Hill Networks™ meets you anywhere—takes you everywhere.
Go online at connected.mcgraw-hill.com.

Circle the globe, travel across time. How do you access networks?

1. Log on to the internet and go to connected.mcgraw-hill.com
2. Get your User Name and Password from your teacher and enter them.
3. Click on your networks book.
4. Select your chapter and lesson. Start networking.

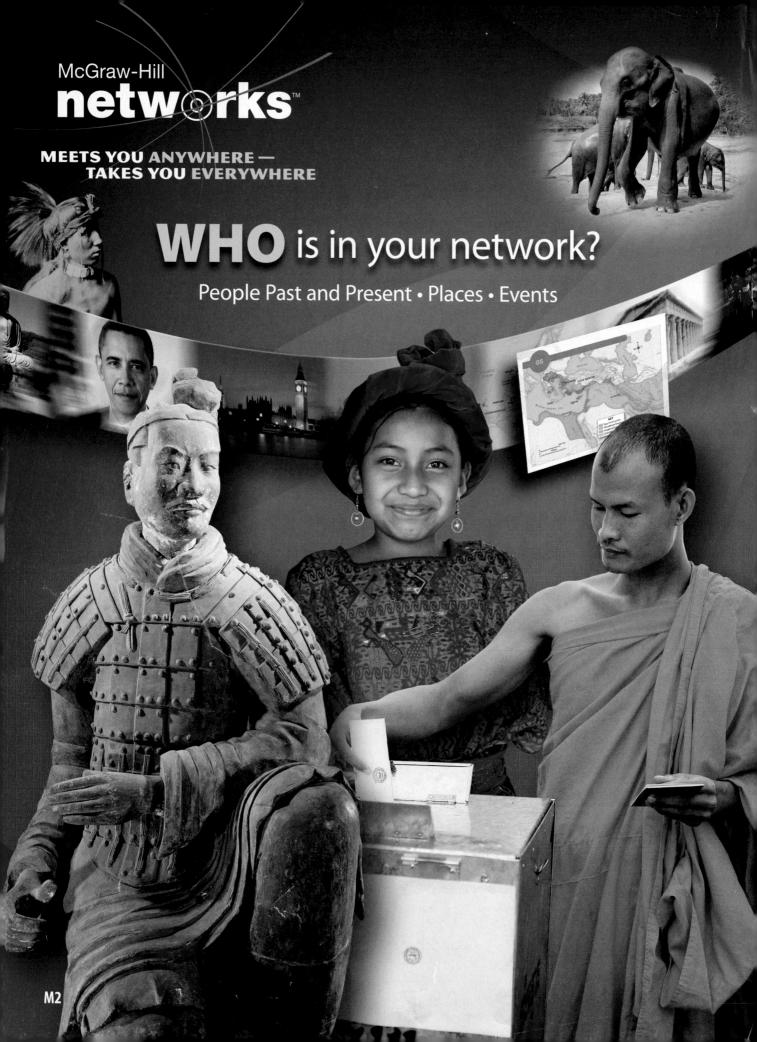

HOW do you learn?

Read • Reflect • Watch • Listen • Connect • Discover • Interact

start **network**ing

McGraw-Hill

netw⊙rks™

A Social Studies Learning System

DISCOVERING **WORLD GEOGRAPHY**

Discovering World Geography **READING ESSENTIALS & STUDY GUIDE**

**MEETS YOU ANYWHERE —
TAKES YOU EVERYWHERE**